They Made America

Given
In memory
Of
Tom Dague

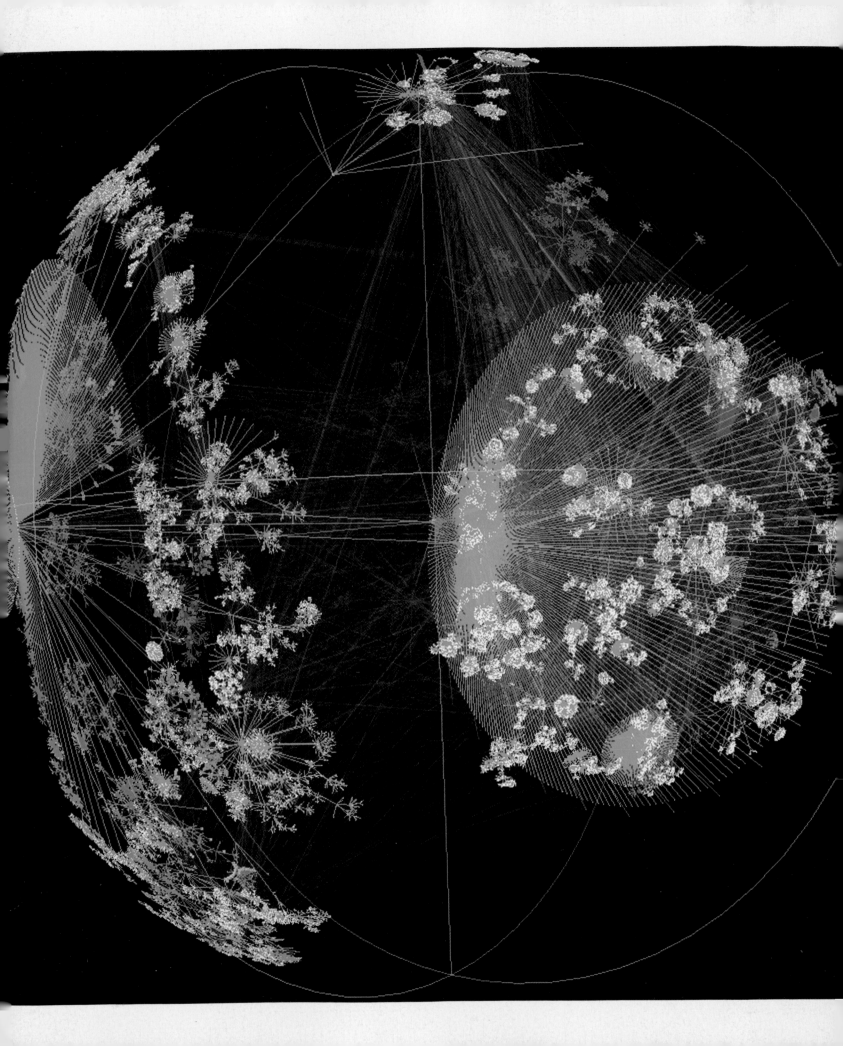

From the Steam Engine to the Search Engine:
Two Centuries of Innovators

They Made America

Harold Evans

with Gail Buckland and David Lefer

Little, Brown and Company

New York   Boston

Copyright © 2004 by Harold Evans Associates LLC

Library of Congress Cataloging-in-Publication Data
Evans, Harold.
        They made America / Harold Evans, with Gail
Buckland and David Lefer.—1st ed.
                p.   cm.
        ISBN 0-316-27766-5
        1. Inventors—United States—History. 2. Inven-
tions—United States—History. I. Buckland, Gail.
II. Lefer, David. III. Title.
T39.E83 2004
609.2'273—dc22   2003065954

Little, Brown and Company
Time Warner Book Group
1271 Avenue of the Americas, New York, NY 10020
Visit our Web site at www.twbookmark.com

First Edition

10 9 8 7 6 5 4 3 2 1

Q-Ver

Design concept by Rina Migliaccio
Designed by Wendy Byrne

Printed in the U.S.A.

ENDPAPERS: SEEDS OF THE FUTURE: Patent
drawings submitted by Sam Colt, Samuel Morse,
Isaac Singer, Elisha Otis, Cyrus McCormick,
George Eastman, Edwin Armstrong, Henry Ford,
Orville and Wilbur Wright, Philo Farnsworth,
Robert Noyce and Edwin Land. All patents issued
since 1790 by the United States Patent Office can
be found in a searchable Web database at
http://www.uspto.gov/.

PAGE 2: The Internet jellyfish: Can you imagine
what the World Wide Web looks like? This is a map
from Young Hyun, a researcher at the Cooperative
Association for Internet Data Analysis, using three-
dimensional noneuclidean spheres. It shows
600,000 links between 535,000 Internet modes—
a fraction of the millions of computers tethered
together through cyberspace.

PAGES 6–7: A clean room at Intel, Albuquerque,
New Mexico, 2003, by Greg Miller.

ONE OF THESE MEN IS AN INNOVATIVE GENIUS.

CAN YOU PICK HIM OUT?
(See page 236)

To the Memory
of My Brother
The Reverend
Frederick Albert Evans

THE AUTHOR GRATEFULLY
ACKNOWLEDGES THE
GENEROUS SUPPORT OF THIS
PUBLICATION BY
THE SLOAN FOUNDATION

ADVISERS TO
THEY MADE AMERICA

DANIEL KEVLES, Stanley Woodward
Professor of History at Yale University

VICTOR MCELHENY,
Massachusetts Institute of Technology

MERRITT ROE SMITH,
The Leverett and William Cutten
Professor of the History of
Technology at Massachusetts Institute
of Technology

THEY MADE AMERICA

Acknowledgments

There were times in the writing of *They Made America* that felt like traversing the continent in a Conestoga wagon before the innovations of steamboat and railway. I was lucky that the entire terrain had been so well scouted by David Lefer, my gifted chief researcher, who went ahead fearlessly. He identified the peaks and the swamps; and if I am caught in one of the latter it is entirely my fault. Lewis and Clark did it in three years in 1804–6; it took us five, but we worked in as close a harmony. I thank him. Of course, my family—my wife, Tina, and my children, George and Isabel—deserve a special citation, not just for allowing me to undertake another long journey after *The American Century,* but for tolerating my bringing to the dinner table whatever strange and wonderful character was my current obsession.

Gail Buckland, the leading photographic historian, had been with me on that earlier, and longer, adventure writing *The American Century*. She brought to this book what she brought to that: diligence, scholarship, authority and initiative. As well as identifying revealing photographs, she took a full part in all the discussions on the contributions of those I considered for inclusion. Her administrative skills were much in evidence throughout. Rina Migliaccio began the design of this book with verve before her appointment to *People* magazine took her away. We were fortunate in securing the services of Wendy Byrne, who came gallantly to the rescue with the fine eye and scrupulous attention to detail she brought to *The American Century.* Nothing I could do (rather a lot!) seemed able to shake her calm. All this effort was correlated by Cindy Quillinan, my editorial assistant, who combined a cool capacity for organization with acute perceptions. Peter Wohlsen was cheerfully tireless in special editorial research at many levels and innovative in his tracking of a multiplicity of images.

I owe a special debt to the resource and enterprise of Sarah Frank and Jonathan Marder, who introduced me to the Sloan Foundation and WGBH Television, Boston. The Sloan Foundation is renowned for the care and imagination it has brought to remedying Alfred Sloan's complaint that "too often we fail to recognize and pay tribute to the creative spirit." I am grateful to the program director Doron Weber and the Sloan trustees—in several regards. In the first place, their grant enabled me to deepen my researches. In the second, their early support got the WGBH television series off the ground. And in the third, Sloan introduced me to three peerless academic advisers who rode shotgun on the entire project: Daniel Kevles, the Stanley Woodward Professor of History at Yale; Victor McElheny at the Massachusetts Institute of Technology; and Merritt Roe Smith, the Leverett and William Cutten Professor of the History of Technology at MIT. They lent freely of their scholarship to guide me in the selection of innovators, invaluably suggested sources, pointed out omissions and obfuscation in my essays and altogether sustained me in trying to portray the innovators and their innovations in the broad context of American history. I am forever in their debt. Once again, though, I lay exclusive claim to any writing that may be judged in error.

The collaboration with WGBH to produce a four-hour documentary companion series for public television was of immense value to me in writing the book. Margaret Drain, vice president of national programming at WGBH, saw the point of the enterprise from the first moment. Her support and discerning contributions were crucial. Mark Samels, executive producer of the WGBH History Unit, brought a discerning eye to the project, as did Carl Charlson, the series producer of *They Made America,* and supervising producer Sharon Grimberg. The discussions with them in Boston, facilitated by the wit and authority of Professor Daniel Kevles, who joined the group, gave me a number of new perspectives. The program producers and their associates conducted some interviews of value, and I am grateful for their diligent questioning as well as for the animating imaginations brought to the lives we chose to dramatize. I refer to the documentary production teams of Carl Charlson and Cathleen O'Connell, Daniel McCabe and Megan Gelstein, Patricia Garcia-Rios and Morgan Faust, and Linda Garmon and Caroline Toth.

In addition to Sloan, the television series was supported by the Ewing Marion Kauffman Foundation and the Arthur Vining Davis Foundations. It was a powerful stimulus to our researches and filming when Olympus agreed to provide exclusive corporate funding. I must especially thank Mark Gumz, chief executive, for his pungent observations from the trenches of business leadership, and all his colleagues at Olympus whose enthusiasm was heartening. It was particularly fascinating to attend their annual managerial meeting and see firsthand how rubber met the road in a company that lives on innovation.

The origin of *They Made America* lay in conversations about American business I had in 1998 with Michael Lynton, formerly of the Penguin Group and AOL and now with Sony;

with my good friend Ken Lerer; with Pat Mitchell, the visionary head of the Corporation for Public Broadcasting; and then with the legendary Laurence Kirshbaum, chairman and CEO of Time Warner Book Group. Larry was both a competitor (in hardback publishing) and an ally (in paperback publishing) during my seven years as the president of the Random House Trade Group. When I became a full-time author, his enthusiasm for celebrating the innovators in American history attracted me to Time Warner Book Group. I am grateful for his inspiration—and his shrewdness in placing me at Little, Brown, first with Sarah Crichton, and then the thoughtful publisher, Michael Pietsch. My editor, Geoff Shandler, was an editorial colleague when I was at Random House, and I learned firsthand what many authors there and at Little, Brown know full well, that Geoffrey, now Little, Brown's editor in chief, has the mildest of manners but the sharpest of minds. Liz Nagle, his assistant, was invaluable in the closing stages. Marilyn Doof gave generously of her prodigious experience in production. Betsy Uhrig marshaled the copyediting, and I am grateful to all of those whose blue pencils saved me from sins too numerous to mention.

The literary lawyer Bob Barnett facilitated all the publishing arrangements with his customary deftness. I benefited greatly at several stages from the wisdom of my own lawyer, James Goodale.

Parts of the manuscript were kindly read by a number of people and I thank them all for their comments. They were R. E. G. Davies at the Smithsonian, the wizard of aviation history; editors Jon Karp and Bob Loomis, my old friends at Random House; the writer Robert Lenzner at *Forbes* magazine; Jonathan Mahler; Jon Nagel; Peter Petre at *Fortune;* Professor Allen Weinstein; and Mort Zuckerman, editor in chief at *U.S. News & World Report*. I benefited from early observations by Jack Welch and by Bruce Greenwald at Columbia University. John Huey, the former editor of *Fortune* and now the editorial director of Time Inc., is a brilliant editor—and provocateur—and I thank him in both roles, and for his assistance with the wondrous Time Inc. archives, along with Pamela Wilson in the archives department. Randall Rothenberg at *Strategy and Business,* for whom I wrote a business column, and his onetime colleague Michael J. Wolf, then at Booz Allen Hamilton, now at McKinsey, accelerated my determination to explore innovation in American history, as did Richard Snow, the editor of *American Heritage*. Suzy Wetlaufer made some penetrating observations at an early stage, drawing on her experience as the editor of the *Harvard Business Review*.

David Lefer and I have very many people to thank for the generous way they contributed information, ideas and criticism. I am especially grateful to the family of the late Gary Kildall—his first wife, Dorothy McEwen, and their children Kristin and Scott—for their agreement to let us draw from Gary's unpublished manuscript of his life. They deserve the appreciation of everyone interested in the much-obscured origins of the personal computer. Tom Rolander, Gary's closest friend, shared generously of his experiences working many innovative years with Gary. John Wharton, the former chip designer and computer industry analyst, has written with authority and insight on Kildall and served his memory well. I am grateful for the hours he spent and for his introductions.

Shortly before his death, Marvin Bower, the founder of McKinsey, gave David a last interview from which we learned much. Thanks to McKinsey historian Bill Price for the introduction.

The following people also gave invaluable assistance in the making of this book, either by reading text or imparting information to me or to my researchers, and I thank them all: Michele Aldrich, Pam Alexander, Ken Auletta, Dane Baird, Julie Berkin, James Birkenstock, David Boies, Sergey Brin, David Brown, John Seely Brown, William Brown, Bob Buderi, A'Lelia Bundles, Frederick Busch, Nikko Canner and Katzenbach Partners, Lee Chen, Joan Ganz Cooney, Brenda Coughlin, Raymond Damadian, Gerry Davis, Lynn Downey, Esther Dyson, Michael Eisner, Susan Ferraro, Alex Fukunaga, Richard Grennan, J. C. Herz, Dr. Robert Holtzman, Eric Hsia, Clive Irving, Rick Kaplan, Joel Klein, Neeraj Kochhar, Leonard Lauder, Dr. Yvonne Lui, Kevin Maney, Roger Martin, Tom Michalak, Jacqui Morby, Lloyd Morrisett, Beth Noveck, Michael Ovitz, Larry Page, Ginny Papaionnou, Pete Peterson, Nicole Phillips, Harold Platt, Jane Plitt, Sherrie Rollins, Wheeler Ruml, Amy Salzhauer, Reese Schonfeld, Tony Schwartz, Charles Seife, Tamar Shay, Jim Sherwood, Craig Silverstein, Henry Tannas, Alan Tse, Ted Turner, James Utterback, Alberto and Gioa Vitale, Duncan Watts and Hannah Yang.

There are a number of references in *They Made America* to libraries and in particular the New York Public Library. Who can count how many innovations have been inspired and informed by the treasures in the library, and I for my part readily acknowledge that this book could not have been written without the NYPL and the dedicated guidance of other librarians at Pace University, Columbia University, Yale University, New York University, Harvard Business School, the New-York Historical Society, the New York Society Library, the Quogue Village Library and Westhampton Beach Library, the French Library in Boston and the Bancroft Library at the University of California (Berkeley), among others.

Nothing can substitute for the knowledge and experience of a good librarian. I thank Paul LeClerc, president and CEO of the NYPL, for access to the Allen Room and for nourishing that great institution in the face of budget restraints. I am proud to think *They Made America* will find a place there.

Harold Evans,
New York, 2004

America's Genius

"YOU SEE THINGS; AND YOU SAY, 'WHY?' BUT I DREAM THINGS THAT NEVER WERE; AND I SAY, 'WHY NOT?'"

—GEORGE BERNARD SHAW, *BACK TO METHUSELAH* (1921)

NEW AMERICANS have been coming to the North American continent for four centuries in the hope of building new lives, free of the restrictions of the Old World. The newness and vastness of the surroundings, the shock of unfamiliar environments and the shortage of ready hands impelled an almost frantic drive by the early settlers for practical innovations that would make life less tenuous and more agreeable. Understanding just what innovation is and how it comes about is a vital subject for the 21st century, when intensifying competition from around the world requires Americans to innovate as briskly as did those first adventurers.

I have been interested in the adaptive genius of Americans since the misty morning 50 years ago when I walked the shoreline of the James River to retrace the steps of the first English settlers in Virginia. Around the original site of the Jamestown palisaded fort, I watched archaeologists sift through the soil where they had then isolated 50,000 fragments of the first English colony of 350 years before: an earthenware oven; a swept hilt rapier; pieces of ivory chessmen; a small caltrop with four wicked spikes to throw in the path of Spanish cavalry; scissors, needles and thimbles; a branding iron; hundreds of candle snuffers and bottles of gin; and an ice pit for storing food. Captain John Smith brought 104 settlers in three sailing ships, anchoring on May 14, 1607. The first colony of 117 men, women and children, settled in the Roanoke wilderness 10 years earlier by Sir Walter Raleigh, had vanished, but the Jamestown settlers survived Indian hostility and the "starving time."

They did it by innovating. In Jamestown today, visitors can see how they made good use of the chain mail and breastplates they brought to fight a Spanish army that never came. It was too cumbersome for Indian warfare, so they cut up the armor and recycled the parts as cooking pans. When their exports of silk, glass, sassafras and soap ashes were not enough to pay for their essential supplies from the motherland, they concentrated on John Rolfe's innovation, the crossing of an indigenous plant with seed from the West Indies to produce the first and long-sustaining export: tobacco well suited to the soil of Virginia and appealing to tastes in London. After 1776, the political innovations of these newly independent Americans gradually brought reality to the promise of individual liberty. The story of that evolution has been amply told in a number of classic histories and biographies; I added to the extensive literature with my own account of the flowering of freedoms in the second hundred years from 1889–1989 (*The American Century*). On the other hand, much less attention has been paid to the story of the practical innovations by which the Americans over two centuries used their freedom to provide comfort and security, and so came to advance the well-being of all mankind. The purpose of this book is to depict some of the principal creators of those innovations.

There were not many of them in the early days when the destiny of the new republic was in balance, but there were new generations of innovators among the thousands, then millions, of newcomers. It is commonly said that these later immigrants brought their dreams. In fact, they brought ours. They brought to fruition the hidden promise of America. It is obvious enough that the energies of the new masses cleared the wilderness/planted the corn/laid the railways/reaped the wheat/spun the cotton/erected the cities/dug the canals/forged the steel/built the bridges/set the factories humming. But they brought more than muscle. They had fled class-ridden conformity or outright tyranny, so they tended to be of an awkward, questioning disposition. And unnoticed among the

for INNOVATION

millions of these ambitious self-selected risk-takers from all parts of the world were individuals exceptionally willing to dare. Their gifts for innovation accelerated America's progress over two centuries. For the most part, they did not come with any special secret, any patented invention, any great wealth or connections. When they disembarked, blinking in the bright light of the New World, they had no idea what their destinies would be. The magic was in the way they found fulfillment for themselves—and others—in the freedom and raw competitive excitements of the republic.

Innovation, the concept and activity that made Dr. Johnson shudder, has turned out to be a distinguishing characteristic of the United States. It is not simply invention; it is inventiveness put to use. Herbert Boyer was not content with splicing a gene in a university laboratory; he risked academic odium by going into business to mass-produce man-made hormones. Cyrus McCormick was not the only farmer to invent a reaper, but he was the one who initiated the financing mechanisms that made it possible for hundreds of thousands of farmers to afford the invention. The sorely neglected genius of radio, Edwin Armstrong, went into the marketplace himself rather than see his invention of FM radio shelved by RCA's desire to maintain its income stream from the manufacture of AM radios. Ida Rosenthal did not invent the brassiere, nor even the famous Maidenform "I dreamed" campaign, but she put all the pieces together in production and marketing so that her husband's invention reached millions of women. Theodore Maiman, having invented the first working laser on May 16, 1960, described it as "a solution looking for a problem" because so few appreciated its manifold possibilities; he ended up founding his own companies. He was first an inventor and then an innovator.

This crucial difference between invention and innovation was borne in on me on my return to England in 1957. As a young science reporter, I visited the government-funded National Physical Laboratory at Teddington, and they showed me where their senior researcher Robert Watson Watt had in 1935 invented the radar system that was to help the Royal Air Force win the Battle of Britain. His former colleagues remarked with chagrin on how swiftly this British invention had been taken up and exploited in the United States after 1939, laying the foundation for the great electronics industry. It was the same story with antibiotics, following Alexander Fleming's 1928 discovery of penicillin; with Maurice Wilkes's pioneering efforts in developing the first commercial application of the computer at the offices of J. Lyons and Company in 1951; and with the jet engine. All these British inventions were superseded by the innovative energies of America. Frank Whittle designed and patented his gas turbine to produce jet propulsion in 1930, when he was only 24; the first test run was made at Rugby on April 12, 1937, and the Pioneer jet first flew on May 15, 1941. The inertia of the British Air Ministry and the skepticism of the National Academy of Sciences delayed production of the Whittle Meteor jet fighter plane until 1943. The secret blueprints were given to the United States as part of the Allied war efforts—and the United States came to dominate jet engine manufacture. Whittle, impressed by American openness and enthusiasm for innovation, ended up as yet another enriching immigrant to the United States, finally working as a research professor at the U.S. Naval Academy.

Practical innovation more than anything else is the reason America achieved preeminence while other well endowed landmasses lagged or failed. America's emergence from a rural backwater to a position of dominance is not to be explained by the access to physical resources or population, since Russia, China, Australia, Canada, Brazil, Argentina and South Africa were also richly endowed but failed to develop anywhere near as rapidly. The Americans laid rails across their continent long before the Russians and Canadians. The speed of American adoption of new ideas is manifest in the vignette of Asa Whitney, who arrived in England in 1830 to buy items for his fancy-goods business. Fifty years after American independence, England, hearth place of the industrial revolution, was so far ahead with the railway as to inspire a lifelong awe in this proud Yankee. He took a ride on the new Liverpool and Manchester Railway in 1830, traveling at a velocity previously unthinkable; he guessed it was 46 miles an hour. But within Whitney's lifetime, America caught up and surpassed Britain and everyone else in its railway engineering. Ten years after Whitney returned home to extol the wonders of George Stephenson's Rocket, there were 3,312 miles of track in America, more than in all of Europe, and Whitney's passionate campaign to build a transcontinental railway to the Pacific was no longer seen as an unfortunate consequence of traveler's delirium.

One of the purposes of this book is to note transforming connections such as this, to see the innovator in the context of his times as both a legatee and an explorer. There are many eureka moments, but antecedents always matter. Jack Kilby at Texas Instruments and Robert Noyce at Intel no more plucked the idea of the integrated circuit out of thin air than Robert Fulton "invented" the steamboat on a sunny day in Paris. Thomas Edison introduced electric power into cities, but it was his immigrant clerk Samuel Insull who found a way to make power cheap enough for everyone. Insull, in turn, depended on the innovations of George Westinghouse in alternating current—but Westinghouse had no concept of the nexus of technology and marketing exploited by Insull.

All these men were innovators: They were entrepreneurs in action. It has been said that a scientist seeks understanding and an inventor a solution, to which we might add that an innovator seeks a universal application of the solution by whatever means. The case of Alexander Graham Bell is illustrative. He was not an innovator. He was the discoverer of how sound waves could be converted into undulating electric current. It was certainly a marvelous moment on the evening of March 10, 1876, when his young assistant Thomas Watson heard Bell's voice down the wire, "Mr. Watson, come here, I want you!" but as Watson later remarked, the Bell Company phone was calculated more to develop the American voice and lungs than to encourage conversation. Bell did not solve the problem. He did not make any further contribution to the technology of the telephone or to the manifold microinventions necessary to make it an effective instrument by means of automatic switchboards, loading coils, carrier currents, marketing and the like. The problem of indistinct and muffled sound was solved by Thomas Edison (with Charles Batchelor). They produced an effective carbon button transmitter for the rival Western Union so that when Western Union pooled the patents of Edison and Bell's rival Elisha Gray it had a real working telephone. Then Theodore Vail presided over the amalgamation of Western Union and the Bell Company to create the American Telephone and Telegraph Company. It was Vail who foresaw the potential of a national long-distance system and worked to overcome myriad technical, political and bureaucratic obstacles with such effect that on January 25, 1915, he was able to sit in his convalescent retreat on Jekyll Island, Georgia, and listen to Bell in New York repeat his original request of 1876 to Thomas Watson in San Francisco (to which

Watson replied that it would take him a week to get there now). Vail was also the initiator of a research facility that in 1925 became the Bell Labs, the genesis of decades of inventiveness including the transistor in 1947 and the Telstar I communications satellite in 1962.

Vail was an innovator. So was Samuel Morse, though he was not the first to invent a practical electromagnetic telegraph. The scientist Joseph Henry preceded him, but the gentle Henry had no interest in developing it for commercial application. Morse did. He was the innovator of the telegraph. Chester Carlson cooked up chemicals on his kitchen stove in Queens in 1938 so as to transfer a dry mark from one piece of paper to another. No commercial organization was interested. The nonprofit Battelle Memorial Institute in Columbus, Ohio, took the invention a stage further from 1944. Joseph C. Wilson, new to the presidency of his father's Haloid Corporation, a maker of photography products, sent his friend Sol M. Linowitz, a public-spirited lawyer just out of the navy, to look at it. "We went to Columbus to see a piece of metal rubbed with a cat's tail," said Linowitz. From 1947 to 1960 Wilson took his company to the brink of extinction, investing $75 million in the strange device, which became the Xerox machine, one of the most successful products ever made. Wilson was the innovator.

Thomas Edison is thought of as America's foremost inventor, with 1,093 patents in his name, but his most important work was translating the insight of invention into the practical reality of innovation through the long process of development and commercial introduction. He exhorted his associates: "We've got to come up with something. We can't be like those German professors who spend their whole lives studying the fuzz on a bee." A score of experimenters before Edison had worked on heating a filament to incandescence, and any day one of them might well have succeeded, but Edison's transcendent innovation was to understand that the lightbulb he invented would be a mere novelty unless he could find a way to integrate it into an economical and safe electrical system. The simple act of flicking a light switch in offices and homes depended on a complex of dynamos, cables and numerous connections that all had to be devised, costed and manufactured. Edison had also to fulfill the entrepreneurial role of raising the money, arranging the legal rights-of-way and cultivating the market. Edison was a supreme innovator.

Invention without innovation is a pastime. Patents are important, very much so in some industries, like pharmaceuticals, and hardly at all in others, like machine making, but their role, like that of the inventor, has been overplayed. A patented invention is only a beginning. Less than 10 percent of patents turn out to have commercial importance, according to a study for the Lemelson-MIT Program, and less than 1 percent have the seminal importance of, say, John Vaught's ink jet for Hewlett-Packard in 1975, or for that matter Eli Whitney's humble cotton gin nearly two centuries before.

Some of the innovators in this book were inventors who carried their inventions into patents and to fulfillment in society; some of them invented nothing. A handful made scientific discoveries, but few of them were versed in any branch of pure science. Their distinctive quality is not that they filed a patent or elaborated a formula. It is that somehow they got their hands on the most important ideas and turned them into commercial realities with enormous impact. Originality is not the prime factor; effectiveness is. Samuel Slater and Francis Cabot Lowell did not invent the machinery that made Massachusetts a center for cotton manufacturing at the turn of the 18th century. They stole it from the British.

The innovators featured in these pages are rich in their diversity. It gives a flavor of where lightning strikes in egalitarian America to mention that they include a trucker, a portrait painter, a cobbler, a Harvard professor, a deck boy, an immigrant seller of fruit and vegetables, a drug dealer, a frontiersman fleeing Indians, a hairdresser, a street peddler, a billboard salesman, a flour miller, an illiterate daughter of slaves, a '60s rebel on the streets of San Francisco, a beach taxi pilot, a seamstress, a piano salesman, a foreman in a power plant, a U.S. Navy seaman with nothing to do on a warship at the end of World War II, a playboy, a radio ham, a hardware store keeper, a clerk and of course a couple of bicycle mechanics.

A surprising number of these innovators can be described as democratizers making it possible for the whole population to enjoy goods and services previously available only to the elite. Amadeo Giannini opened banking to the common man. Before George Eastman, photographers practically needed a Ph.D. in chemistry to develop and print their pictures. The digerati long scorned those with AOL e-mail accounts, but Steve Case brought millions to e-mail and the Internet. Georges Doriot and then Michael Milken liberated entrepreneurs with a good business plan from needing personal connections with the wealthy. Gary Kildall and Ken Olsen extended access to the computer beyond a select priesthood of engineers. Pierre Omidyar created a democracy of supply and demand with eBay. Raymond Smith transformed casinos from dark smoky rooms peopled by men to public places of entertainment. Juan Trippe and Donald Burr ran airlines that opened up the world to everyman.

Some may say that this is a romantic illusion, that these creative democratizers were merely catering to the masses for the sake of higher profits. None of them, to be sure, sought penury in the service of the public, but immersion in the lives of innovators over several years leaves the impression that money making was not a sustaining motivation. It appealed to some—no doubt to Robert Fulton, who had spent so much of his time sponging off others. But Henry Ford would have made more money in his early days if he had done what his partners wanted and made cars for the rich. Giannini went to great lengths to avoid a personal fortune, buoyed by a deep populism born of his family's early struggles. A desire to be

God's agent in the service of mankind was uppermost in the actions of Morse, Vail, Lewis Tappan (credit rating), Theodore Judah (the transcontinental railway), Olsen and Martha Matilda Harper (beauty parlors). And John Wanamaker was mindful of his Christian ethics when he replaced the customary haggling of the pre–Civil War period with price tags to render equality of opportunity to all shoppers in his famous Philadelphia store.

In all the innovators I call democratizers, altruism was no doubt diluted by vanity, the desire to be acclaimed as a benefactor, to be acknowledged by one's professional peers—and there is nothing wrong with that. With Edison and Edwin Armstrong, the satisfaction of intellectual curiosity was paramount, with John Fitch (the steamboat) it was social recognition, with Madam C. J. Walker (beauty treatments) an affirmation of racial pride. Whatever conclusion one reaches about motivation, the democratizing instinct is evident in many innovators' successes.

These innovators are heroes and benefactors, but they are not saints. I thought it important to portray them as they were, with the inhibitions and prejudices of their times—and some vices of their own invention. It is a disservice to our understanding of them to bring the eraser of political correctness to their portraits—though the companies they founded do not always share this view. Many companies helped me as best they could within the limits of what they had in their archives and memories. Few companies cherish their history as well as United Technologies, which has taken care to preserve the origins of Otis elevators. It is revealing to see the Otis company ledgers from the 1860s with the pencil marks of engineers recalculating each time how much wire rope must be used, the number of ratchets, engine specifications, the size, shape and weight of platforms and the number of wrought-iron binders. Elevator making was clearly a craft then, not a result of mass production. On the other hand, the Bank of America was openly obstructive to my inclusion of its founder, Amadeo P. Giannini, because in his profile I mention an anti-Semitic epithet he once used. Giannini was an important innovator with a profound influence for the good. There was and is no suggestion that the Bank of America today, the largest in the country, has any prejudice. Giannini himself was a supporter of the state of Israel, and a splendid character, but it is necessary to see all these innovators in the context of their times, not as alabaster figures.

I have concentrated on the significant innovations in a considerable variety of disciplines over two centuries, from John Fitch's steamboat service on the Delaware to Larry Page and Sergey Brin's electronic service on Google, including a number of innovators who are known but ill served, such as Oliver Evans, developer of the high-pressure steam engine and the first automatic production line. Clearly, no one study can encompass every innovator in meaningful depth. There are thousands of ingenious Yankees who fiddled with bits and pieces of machinery whose incremental practical improvements were critical to American progress but whose names are lost to history. And we must never forget that the gifted few innovators drew deeply on the resources of millions of everyday workers. In the wake of the Great Exhibition at the Crystal Palace in London in 1851, where American innovations were star attractions, the English Parliament sent the noted manufacturer Joseph Whitworth and the educator George Wallis to learn what they could about "the American system." They concluded that the key to American progress was "the widespread intelligence which prevails among the factory operatives." Wallis attributed the "inventive disposition" to "the attention paid to the education of the whole people by the public school system." Sam Colt longed with much vituperation for Yankee mechanics when he opened (and later closed) his gun factory in London.

The selection of the 70 innovators in the following pages, plus identification of 100 more in a gallery toward the exit, was made after examining the achievements of many hundreds of innovators, with the research assistance of David Lefer and guidance from three academic advisers nominated by the Sloan Foundation; they are not to be held in any way responsible for my final choices. It will be apparent that I have not confined myself to technology, except in the broad sense of the term defined by the social scientist Daniel Bell that technological progress consists of all the better methods and organization that improve the efficiency of both old capital and new. This can be many things, the development of a scientific discovery, the combination of elements from several inventions, but also a reorganization of labor, a new concept of banking, trading or marketing. The Wright brothers are here, of course, but so also is the small boy who was among the cheering crowds watching Wilbur fly along the Hudson on that magnificent day in 1907: Juan Trippe, who went on to inaugurate mass air travel at Pan American. I do not pretend to be able to fathom all the complexities of all the sciences, still less translate every nuance into popular terms, but I have attempted to describe the technicalities as far as seemed necessary for a work of social history. Collectively, I believe the profiles shed new light on the processes of innovation. Of all the main subjects, the questions I asked were: Why this person and why at this time in America? What were the antecedents and the context of the innovation? What is common among these individuals and what is unique to each? What was the role of government? Were they moved by money or by any ideals beyond the urge to translate thought into a service or a product everyone would come to want?

I readily acknowledge that the book is light on the contributions of women, African Americans and other minorities. It is because the history of innovation is light on them. Raising the capital for large-scale innovation was impossible for black men like Frederick McKinley Jones, even though he did win 60 patents and invent refrigerator trucks. Tremendous odds were overcome by other black innovators like Madam Walker and Garrett Morgan. Women had to overcome prejudices that for the most part kept them in the kitchen, the fitting room and the beauty parlor. Russell Simmons, An Wang, Berry Gordy and Oprah Winfrey are in the forefront of a more hopeful trend. Donna Dubinksy, pioneer of the PalmPilot, and Carleton S. (Carly) Fiorina, reinventor of Hewlett-Packard, suggest how much was lost by relegating women to innovating for other women.

There are a number of rewarding biographies of inventors, but few focus on them as innovators, and science and technology have altogether suffered neglect in standard history texts (a deficiency finally addressed in 2003 by the two-volume textbook *Inventing America* cited in the bibliography). Mitchell Wilson's illustrated volume *American Science and Invention* entertainingly reviewed scientific achievements—rather than innovation—but that was 50 years ago. In more recent years, the literature on innovation has been growing. The financial historian Robert Sobel has explored the careers of 9 entrepreneurs, Richard Tedlow of 7, the writer David Brown a modern 35, but no popular book so far as I know has attempted, as does *They Made America,* to explore the histories of innovators over two centuries, to delve into the personal and the technical, to see how one influences the other and at the same time to set these individuals in the context of their times with contemporary illustration.

The book is divided into three parts. Part 1, concerned with the period of mechanical technology, identifies the key innovators from the early days of the republic to the Civil War. Three legal innovations provided a sympathetic framework: the Supreme Court ruling liberating interstate commerce, the federal patent law and the invention of the corporation to limit liability and facilitate the raising of capital for projects. Alexander Hamilton, the secretary of the treasury, produced his prescient report in December 1791 arguing that America's destiny lay in encouraging domestic manufacturers with high tariffs, immigration and "new inventions particularly those which relate to machinery." As his biographer Ron Chernow writes, Hamilton, the prophet of the capitalist revolution in America, was "the messenger from a future that we now inhabit." But Congress, dominated by agricultural interests, failed to act. The innovators who first broke through the inertia were self-taught self-starters ready to try their hand at anything. It is emblematic of the times that America's leading maker of pencils was Henry David Thoreau, better remembered as a transcendentalist philosopher, who chose to describe himself 10 years after graduating from Harvard as a carpenter, a mason, a glass-pipe maker, a house painter, a farmer, and a surveyor, as well as a writer and pencil maker. The workshop revolutionaries have never had the attention afforded the political revolutionaries, but they set America on a new course.

Part 2 begins with the end of the Civil War and the start of the second industrial revolution in which systems that were mechanical become electromechanical. The country moved from a "folk culture," as Daniel Boorstin put it, to a "mass culture." Immigrants arrived in the millions bearing a gene for change while American enterprise grew international business empires. The period is commonly noted as one in which the main thrust of research came more and more to be carried out by professionals in research and development (R&D) labs attached to large corporations, government and university departments, but individual innovators remained prolific. They worked for the most part independently of the developments of science and theoretical knowledge, and they made signal strides in the organization of complexity (Ford and IBM to mention two).

Part 3 is the digital age from the 1960s to the present, when intellectual technology becomes paramount. Disdaining the hippies in California in the '60s, Gordon Moore, the Silicon Valley innovator, remarked, "We were the real revolutionaries," and he was right. America has become what Peter Drucker calls an emerging "entrepreneurial society," with an information/service economy fathered by Moore and others. Business structures have become more networked than hierarchical. Mass production is evolving into mass customization. "Lifestyle" marketing is replacing a class-based economy. America is coming full circle. The vertical disintegration of industry and the Internet are facilitating more innovation from hundreds of smaller concerns and individuals.

One powerful force for innovation is missing in my survey: government. The image of the entrepreneur taking on the world, celebrated in the novels of Ayn Rand, still has a grip on the American business imagination. There is truth in that ideal, but it tends to obscure how much creativity has been stimulated by government, directly and indirectly. Land grants and loans were essential for the railroads and the interstate highways. Massive U.S. government support of research led to America's advances with semiconductors and hence the transistor. The Defense Department originated the Internet.

It was tempting to include in this book creative figures in government who accomplished all this and more, but that is another book; the men and women profiled here all ventured in the world of commercial risk. Nonetheless, all of us and the millions of Americans who innovated and sustained innovation were the beneficiaries of particularly enlightened initiatives in the public arena. Consider Justin Smith Morrill (1810–1898), the Vermont senator who had to leave school at 15 He introduced acts in 1862 and 1880 by which the federal government granted millions of acres to states for the establishment of agricultural and engineering colleges (long before European universities acknowledged engineering as a profession). Consider Harry Colmery of the American Legion, Congressman John Rankin of Mississippi and Congresswoman Edith Nourse Rodgers of Massachusetts, who initiated the G.I. Bill signed by President Roosevelt in 1944. The results in individual fulfillment and American prosperity are well cele-

brated in Tom Brokaw's book *The Greatest Generation*. Ken Olsen was a G.I. graduate, as was Douglas Engelbart, the pioneer of the computer mouse and graphic interface software. While two million American G.I.s metamorphosed into engineers, scientists and managers, higher education was still unduly restricted in Britain with the emphasis on the arts. Only a very select few—2 percent of the population—were then admitted to universities, and the proposal to open the doors wider provoked outcry from the establishment: "More Means Worse" was the rallying cry of its leading organ then, the *Times* newspaper. It was an aristocratic cast of mind that did a great disservice to Britain's innovative potential.

Democracy in the form of equal opportunity works. It is, of course, the American people who are the ultimate innovators, a faith reflected in the political, cultural and business institutions they have created and sustained. The innovators in my analytical biographies are players on the stage of a perpetual revolution. Where it may go next is the subject of speculation about nanotechnology, biotech, artificial intelligence and cheap renewable energy, all of which sounds exciting, but if the history of innovation teaches us anything it is that the greatest innovations are unpredicted. Caryl P. Haskins, the celebrated president of the Carnegie Foundation, reminded the country in a report to the president in 1965 that as late as 1929 it was still widely believed that the Milky Way constituted our entire universe: "Only within the last decade have we become fully aware that this galaxy of ours is in fact but one among millions or perhaps billions of such galaxies, stretching to distances of which the world of 1920 or even 1950 could have had little conception." We are, in that respect, in the same position as the men and women who first set foot on the beach at Jamestown in 1607.

For the immediate future, Americans must be concerned that their long worldwide supremacy through innovation is challenged as never before. In May 2004, the *New York Times* raised a front-page alarm that the United States was losing its dominance in the sciences. William J. Broad reported that the U.S. share of its own patents has fallen over decades to only 52 percent; its share of Nobel Prizes has similarly fallen to 51 percent, and American scientific papers are no longer in the majority in learned journals. Shirley Ann Jackson, the president of the American Association of the Advancement of Science, deplored a declining interest of young Americans in science careers and asked, "Who will do the science of this millennium?" Edison and Armstrong, and many other inventors and innovators, said their imaginations were excited in the first place by reading popular biographies of Faraday and Marconi and others. It would be gratifying if the exploits of the innovators who made America described in this history did something to spark the ambitions of the next generation to make a new America.

Harold Evans
New York, 2004

PATHFINDERS TO A NEW CIVILIZATION

The steamboat opened a great era of innovation by American originals—frontiersman, miller, soldier, artist, peddler, gunsmith. Dreamers and doers all

WHAT DOES IT TAKE to get a nation moving? The new Americans, concentrated on the fringe of the East Coast, were shut off from much of the riches of their vast continent. The distances were preposterous, the mountain ranges prodigious. The source of motive power on the land was the muscle of man and horse, and on the waterways it was wind and current. There were few roads, no railways, no telegraph; millions of acres of cotton went unpicked, tons of wheat went rotten for want of harvesting. All that was to change with dramatic speed. In the first section, we begin with the innovators of the steamboat services that opened the West and end with the visionaries of the railway that bound a continent. In the years between these transforming innovations, we acquire the sewing machine and the revolver; the reaper and the elevator; rubber and oil; the bicycle and credit rating; made-in-America cotton dresses and blue jeans; and the world's first automatic production line.

The seminal innovation of a commercial steamboat service, transcending the limitations of nature and opening the West, was the culmination of the work of four very different men: John Fitch, an eccentric frontiersman; the artist Robert Fulton partnered with a landed aristocrat, Robert Livingston; and an inventive miller-engineer, Oliver Evans, whose high-pressure engine took the steamboat to a whole new level.

NEW ORLEANS, 1883: The North sent its manufactures down the Mississippi-Ohio river system and the South sent back its harvests of rice, molasses, sugar, lumber and above all cotton. In the romantic painting by William Walker, the side-wheeler *Natchez* (left background) taking on cargo may be the same *Natchez* that made it upstream in 1870 from New Orleans to St. Louis in a record of 3 days, 18 hours, 14 minutes.

The HEROES Who

The American Declaration of Independence

Twas only one of three landmarks in 1776. In Glasgow that year, on March 8, James Watt unveiled the first commercial model of his condensing steam engine, the fulcrum of the industrial revolution, and from the same Scottish city a few days later Adam Smith published his *Wealth of Nations,* the foundation of a new era of economic thought on both sides of the Atlantic. He analyzed and extolled the virtues of manufacturing, with its division of labor, of free trade and the benefits to society from reasonable men pursuing self-interest without much restriction by government.

When the 13 states became the United States with the peace of 1783, America was an empty land, an agrarian nation with only half as many people (four million) as the mother country. No city was a tenth the size of London. The new Americans had endured a long war and dissension; they had barely begun to realize how great were the natural resources they could now exploit or even to decide whether they wanted to do so. The thoughts that made pulses beat faster were pastoral; the heroes of popular culture were generals and statesmen, clergymen and landed gentry. Adam Smith concluded that no manufactures "for distant sale" had ever been established in America because of the lure of uncultivated land. He noted that as soon as a producer of goods—Smith called him an "artificer"—had acquired more stock than he needed, he did not extend his own business. He was not tempted by large wages and the easy subsistence this might bring. "He feels that an artificer is the servant of his customers from whom he derives his subsistence; but that a planter who cultivates his own land, and derives his necessary subsistence from the labor of his own family, is really a master and independent of all the world." Colonialism had also fostered a habit of mind inimical to manufacturing and industry. The British imperial practice, known as mercantilism, had been to regard all colonies as sources of raw materials, not places for manufacturing.

The fomenters of the American Revolution were more or less of the same mind. They were men of property, imbued with the notion that society was best sustained by farming, fishing and trading; manufacturing was envisaged as women at home making cloth, rugs, soap and garments, men fashioning furniture, shovels and chains, and itinerant tinkers, smiths and carpenters filling the gaps left by the cottage workshops. Capitalism was not in their vocabulary, and if it had been it would have been as a dirty word. Benjamin Franklin constantly inveighed against the individual accumulation of wealth. In the 27 specific complaints in the Declaration of Independence, the founding fathers said nothing about the injustice of England's unpopular curtailment of American manufacturing or methods of financing it. The principal writer had clear ideas on what kind of society America should become: "While we have land to labour then," wrote the Virginian Thomas Jefferson in a letter in 1781, "let us never wish to see our citizens occupied at a workbench, or twirling a distaff. Carpenters, masons, smiths are wanting in husbandry; but for the general operations of manufacture, let our workshops remain in Europe." Gouverneur Morris foresaw a time when America "will abound with mechanics and manufacturers," but he and Alexander Hamilton were relatively isolated in seeing the potency of the industrial revolution gathering force in England. John Adams of Massachusetts clung to land as the only true wealth, turning aside, to his loss, Abigail's wifely advice to invest in securities. Even Franklin, businessman, scientist and inventor, exalted agriculture and looked down on trade.

Everything turned on individual enterprise. The national government was weak, and the laissez-faire ideas of Adam Smith had taken root. George Washington, in his first message to Congress in 1790, recommended "giving effectual encouragement to the introduction of new and useful inventions from abroad," but he could not get Congress to fund a national university. Alexander Hamilton, and especially his assistant secretary at the Treasury, Tench Coxe (1755–1824), pleaded in vain for the allocation of public money to encourage invention and manufacturing. Several states advertised bounties for the introduction of machinery, or the production of such known items as wool cards, sulfur, wire and fabrics, but those pockets did not have deep linings. The number of state charters granted to enable business concerns to raise money did double from 1786 to 89 by comparison with 1781 to 85, but capital was meager, skill scarce and the general atmosphere depressing.

How was it then that this backward, dozy America led the world in developing the steamboat? It is true that the seminal steamboat service, developed by Robert Fulton in 1807,

Got America Going

employed a British low-pressure Watt-Boulton engine, but by 1830 the flourishing Mississippi Basin steamers were powered by high-pressure engines of original American invention. It is true also that geography was a midwife. America's vast river systems and lakes, with forests yielding fuel on the run, offered more scope for the steamboat than Britain's relatively constricted internal waterways, which were flanked not so much by forests as faster roads for stagecoaches. Still, Britain was the leading maritime nation, with plenty of opportunities for steamboat entrepreneurs in intracoastal and cross-Channel trade. The vicissitudes of its weather were far less violent. And it had engineering and financial muscle to spare. William Symington (1763–1831) had a steamboat up and running on a Scottish ornamental lake as early as 1788 with an engine of his own design. It says something again about the significance of individuals that Symington lost interest when his financial backers withdrew in 1803 and nobody followed up.

A negative factor in England lay in the positive achievement of James Watt and Matthew Boulton in manufacturing steam engines of Watt's design. The 25-year monopoly the partners held, an extension of the original patent, was a major discouragement to other potential experimenters. The partners were eager to defend their rights, resistant to joint ventures. Who can blame them? It took the full 25 years to recoup the initial investment and finance the long battles in court. Additionally, Watt himself, so crucial a figure in the industrial revolution, lent his prestige to sustained skepticism about the potential of steam for navigation.

In the end, the character of America's steamboat pioneers lies at the heart of the country's early ascendancy in steam navigation. John Fitch (1743–1798), who launched the first practical steamboat in 1787, was too ignorant to know of Watt's misgivings and too headstrong to care if he had known. And where Symington faltered without a patron, Fitch persevered against all the odds. The very different characters of the magnetic Robert Fulton, his calculating partner Robert Livingston and the rebellious Henry Shreve were critical to the development of the steamboat—and the steamboat was the entering wedge of the industrial revolution in the Ohio valley and the Midwest. The machine shops and foundries that made steam engines and iron for the new steamboats attracted a fruitful concentration of skilled mechanics to Pittsburgh and Cincinnati, to Wheeling, Louisville and later St. Louis. One set of numbers gives an idea of the accelerating pace. In the ten years from 1809-19, the gross tonnage of steamboats built rose from

1,000 to 17,000, but in 1830 the figure was 64,000—and 202,000 in 1840.

By 1830, with the exhilarating success of the steamboat, Americans were eager to follow England in the epic innovation of the railway. The new spirit was optimistic and even bumptious, imbued with the idea that America could surpass the world in its inventions. Jefferson was not immune to the nascent doctrine of American perfectibility. In 1785 he was writing to Abigail Adams in Paris, beseeching her to send him two sets of fine linen tablecloths and napkins from England, "better and cheaper than here." By 1812 and the war with England, he was rhapsodizing about the textile machines he had installed at his estate by which two 12-year-old girls and two women were making all his family's linen, cotton and woolens: "Our manufacturers," he brags, "are very nearly on a footing with those of England."

The movement from defeatism to exuberance, from confinement to expansion, took 50 years, but there were two events that can be marked as red-letter days. The first was on Monday, March 3, 1824, for which the steamboat was the catalyst. In one of the most profound legal rulings in American history, a great judicial innovator, John Marshall, chief justice since 1801, ended a steamboat monopoly imposed by New York State, but his judgment affected more than navigation rights. It altogether liberated the conduct of business across the United States (pages 46–47).

The second emancipating event was in 1838, when the inventors were finally accorded protection under a national patent law. Before 1790 they had to win exclusive licenses— state by state—for varying terms. The first federal patent law in 1790 simplified matters, but it merely set up a registry of claims without examination. Patents could be registered without any proof of originality so that several people might hold a patent for the same idea. Inventors still had to spend time and money in defense of their property. Cyrus McCormick (1809–1884) was 25 when he won a patent in 1834 for the grain reaper he had invented at the age of 22, but after the expiration of the basic patent in 1848 he was engaged for the rest of his life in trying to protect his improvements. The prolific inventor and innovator Oliver Evans was so dismayed by a judge's ruling that patents were against the public interest that he went home and destroyed his papers.

The early innovators featured in this first section endured much in changing the atmosphere of America and setting the nation on a new course.

JOHN FITCH He was a frontiersman whose life was often at risk from Indian war parties, but he escaped with an idea that became the Delaware River's first steamboat

1743–1798

HUNTER AND HUNTED: Delaware Indian chief and John Fitch. The only known likeness of Fitch is this posthumous woodcut.

Fitch made the first commercial steamboat out of his nightmare of being chased by an Indian war party (above). Seventeen years before Fulton's steamboat, John Fitch's aptly named *Perseverance* (below) steamed thousands of miles on the Delaware.

THE STORY OF the steamboat properly begins with John Fitch, not yet 40, striding the dense wilderness forests of the Ohio Valley in 1781–82. He is a tall, dark-visaged frontiersman in a beaver hat who plants his feet like an Indian straight on the line of path and daily covers 40 miles faster than a man on horseback. He climbs a bluff, sits admiring the milewide Ohio River below and has an epiphany:

"I contemplated that beautiful river rolling full tide to toward the ocean, and reflecting on its immense length from its head to the ocean, I thought it impossible that God had in his wisdom created a river of such length and irresistible current, without giving man some power of overcoming the force of the water and being able to navigate up as well as down."

He had no idea steam might be such a power. Fitch had done almost everything a young man might do growing up in colonial and revolutionary America. He had been put to work on a Connecticut farm at ten, sailed miserably before the mast, apprenticed himself to a couple of exploitative clockmakers, set up a brass foundry, learned to be a good silversmith, started a potash business, abandoned his high-tempered wife and two children, served as a lieutenant in the Continental Army, sold tobacco and beer to soldiers, run a gun factory, speculated in land warrants in Kentucky, and in the Ohio Valley he was

ALL HIS OWN WORK: Fitch had walked and sailed the wilderness of what was then called the North-West Territory—the rivers and forests of Ohio and Pennsylvania. He adapted and corrected earlier maps, engraved his own version, printed it on a borrowed cider press, had it hand-colored by a young woman and in 1785 went out and sold it himself, copy by copy.

THE SHAPE OF THINGS TO COME: Fitch showed off his steamboat to the delegates to the Constitutional Convention charting the course for the new United States in Philadelphia in August 1787.

surveying land and trading with settlers. All this, but he had never heard of James Watt's invention or the installation of his first steam engine in a British factory in 1776, still less seen a steam engine. For the immediate moment he was too busy staying alive.

The Ohio Valley seethed with hostile Indians of the Delaware tribe. Soon after his contemplation of the river he was drifting down it on a flotilla of rafts with a handful of other traders, when a party of 30 Delawares spotted them and rushed into their war canoes. The white men rowed away for all their life. They escaped—a ball intended for Fitch lodged in a cask of flaxseed—but in March 1782 Fitch was ambushed again, this time trying to run a flatboat upriver with a cargo of flour he hoped to sell to settlers. Two of the party of war-painted Delawares, led by "Captains" Buffaloe and Crow, went

on board to plunder Fitch's boat and scalp two of his companions shot dead in the skirmish. Fitch wrapped himself in a camlet cloak and coolly asked permission to take a nap. He later recalled:

"Unfortunately Captain Buffaloe had made himself too free with our whisky and I had not lain many minutes, than I heard an Indian Speakeing in Broken Language, 'Teak! Teak!' I opened my eyes and rose up on end and shook my head and said, 'No!' He said again, 'Teak!'" and drew his tomahawk a fair blow to sink it into my head. I looked him full in the face, and felt the greatest composure to receive it than I ever felt to meet death, unless it is since I began the steamboat."

The steamboat? This was Fitch reflecting in his autobiography many years later. Not until three years after his capture did he first begin to think of trying to make a boat move by the power of steam, but when he did, his experiences with wilderness Indians were a seed come to germination. He escaped the tomahawk because Captain Crow stopped the descent of the drunken warrior's arm, but Fitch could not escape his own nature. He was conflicted, supremely confident that he was set apart for a glorious destiny but all too well aware that he was regarded as uncouth, a misfit; in a bout of introspection once, he described himself as "in wretchedness, haughty, imperious, insolent to my superiors, tending to petulance." When the prisoners reached the Delawares' village, four of Fitch's companions were put to death. Yet Fitch did not know how to submit. The warriors insisted he take part in a frenzied dance. He declined. Captain Buffaloe demanded his pants in exchange for a breechcloth. He refused. A chief offered his wife to Fitch as solace for a doomed man's last night on earth. Fitch spurned her. The Delawares moved from anger to bewilderment mixed with fear. Who was this strange, cantankerous prisoner with the awkward face and crazy spirits in his head? They handed him over to the British outpost at Detroit: The American victory at Yorktown had ended the war six months before, but the British retained a number of forts pending full U.S. compliance with the 1783 treaty to end the war. Fitch was a captive for nine months.

On his release by the British, he organized a company to survey and acquire lands in the North-West Territory, north of the Ohio, drew a fine *Map of the Northwest*, engraved it on copper, printed it and wandered east offering copies for sale. His exertions left him with an arthritic knee. One spring Sunday morning in 1785, having rented out his horse, he hobbled home from a religious meeting in the village of Neshaminy and was resentful when he was passed by a carriage. Out of the resentment an inspiration surfaced: "What a noble thing it would be if I could have such a carriage without the expense of keeping a horse." And suddenly the solution was clear to him: steam! He had heard now of a steam engine pumping water in a mine in New Jersey. "Nothing but able Mechaniks is required to make the prize sure," he wrote euphorically.

He had nerve. He had a talent for visualization but only the vaguest ideas how steam could be put to use. He had still never heard of the inventors Thomas Newcomen or James Watt until the village vicar showed him an encyclopedia plate of a Newcomen engine. He soon enough realized the incompatibility of a steam engine of that bulk and the rough, rutted roads, but the smooth river, now . . . Those Delaware braves would never have caught him! A steam engine could propel a boat against wind and current and outrun the fastest war canoes! He realized that the bulky

Newcomen engine would sink the kind of boat he could build, but he was not a man to let the wildest of dreams die. Since England had banned the export to America of any technology or skill, he set out to design and build his own steam engine from scratch, one light enough not to sink a small boat and somehow adapted to achieve traction against water resistance.

Fitch worked with manic intensity on research and fund-raising. He rode to the handful of state capitals in the East, petitioning the legislators for a monopoly license to steam the state's waters on the grounds that he was the first to come up with the idea.

He would have got nowhere without the happy circumstance that he had a drinking partner, a hearty German-born clockmaker by the name of Henry Voight, who, like Fitch, was a religious radical, a Christian Deist dismissive of the divinity of Jesus. Voight was eager to let God guide his hand in inventing an engine and offered to extend the drinking partnership into professional waters. Fitch went around Philadelphia offering shares in The Steamboat Company at $20 each. They were taken by a hatter, a grocer, a physician, an ironmonger, an antiquarian, the geographer general of the United States, a Quaker farmer, a manufacturer and half a dozen tavern keepers and merchants. At the time, America was in a sharp postwar recession and all that Fitch could raise from these adventurous souls was $300. With this he and Voight had to design and make a boat and engine. They had a 45-foot skiff ready by the spring of 1786. Astonishingly, by August the two amateur engineers had succeeded in making a tiny working model of an engine with a three-inch cylinder as the boiler—and they had contrived to have the steam work on both sides of the piston, something James Watt had managed to do only after 15 years of hard work.

But how was a full-scale engine to push the skiff through the water? Fitch's first thought had been that the reciprocating motion of the piston should be converted by a ratchet to the rotary motion of a paddle wheel. In his second draft, influenced by "Gentlemen of Learning and Ingenuity," notably Benjamin Franklin, he unwisely abandoned paddle wheels. Fitch and Voight,

testing their skiff without an engine, then tried various rowing contraptions on the river, one of them a chain carrying perpendicular boards round and round. They sweated away at the cranks, with minimum movement but maximum enjoyment among the watching professional boatmen. Fitch took to his bed with "West Indian produce," fretting how he could explain away to the directors the expenditure of $60. Maybe the hemp or rum had an effect: When the watchman in the silent street called "one o'clock," Fitch leapt out of bed to pick up a quill and transcribe his fevered imagination. He was seeing those Delaware war canoes again, but this time the downward movement of a crank drove the paddles through the water to the stern and the upward movement of the crank returned the paddles through the air to the bow.

The partners first successfully tried out the mechanism with muscle power. The next step was to power the paddles with a bigger boiler, which meant finding the money for a 12-inch cylinder. Wheedling the few hundred pounds needed for that from the townsmen was harder than rowing against the Delaware current. Fitch got on his horse again to beseech a round of state legislators for cash to finish the engine and for monopoly licenses to reduce the risk. He didn't get the state cash, but he got the licenses from Delaware (February 3, 1787), New York (March 9) and Pennsylvania (March 28), and then he raised just enough money from friends to put the engine together.

On August 22, 1787, Fitch went to the Constitutional Convention in Philadelphia, where the leading citizens of all the states gathered in their satin breeches, colored coats and lace ruffles. Fitch—somewhat out of place in his black frock coat—buzzed around inviting delegates to inspect his strange craft at the Front Street wharf: six paddles both sides of a chimney in the middle blowing smoke. "There was very few of the convention but called to see it," he wrote, but a few delegates risked going aboard and affected to be vastly entertained when the engine sputtered to life, the paddles cleaved the water and the boat moved off. It bucked the current of the Delaware at the rate of two and a half miles an hour. The next day Fitch had a letter from Dr. Johnson, a leading Connecticut delegate:

"Dr. Johnson presents his compliments to Mr. Fitch and assures him that the exhibition yesterday gave the gentlemen present much satisfaction. He himself and no doubt other gentlemen will always be happy to give him every countenance and encouragement in their power which his industry and ingenuity entitles him to."

Fine words, but not a cent of money came from any of the states he had begged. Fitch was bitter. "There is such a strange infatuation in mankind that it seems they would rather lay out money in balloons and fireworks, and be a pest to society, than to lay it out in something that would enrich America at least three times as much as all that vast country north west of the Ohio."

Fitch was reduced to traveling the countryside cleaning clocks, his clothes in tatters, but the proceeds he and Voight raised were enough to build a 60-foot boat with an 8-foot beam. It was to be propelled now not by Indian oars, but by paddleboards at the stern powered by a lighter, more compact engine devised by Voight. The new engine raised steam by heating a grillwork system of tubes in an 18-inch cylinder, an invention that saved three and a half tons of brickwork on the conventional boiler. (There are various claimants to the invention of the tubular boiler, among them James Rumsey, John Stevens and Nathan Read, but Fitch and Voight were probably the first to put one to work.)

Everything came together on April 16, 1790. The wood fire blazed, the steam rose, the boiler held, the pistons churned, the paddleboards dug into the water at 76 strokes a minute, the boat shook and shuddered and Fitch and Voight moved euphorically upstream. They outpaced several large sailboats and strongly manned rowboats. Fitch exulted: "We reigned Lord High Admirals of the Delaware; and no other boat in the River could hold its way with us. . . . Thus has been effected by little Johnny Fitch and Harry Voight one of the greatest and most useful arts that has ever been introduced into the world; and although the world and my country does not thank me for it, yet it gives me heartfelt satisfaction." On the second Sunday in May of 1790, they risked inviting passengers. *The Gazette of the United States* reported from Burlington, New Jersey, upriver from Philadelphia on the way to Trenton, that "the ingenious Mr. Fitch accompanied by several gentlemen of taste and knowledge in mechanics came from Philadelphia in three hours and a quarter with a head wind, the tide in their favor." On their return, they proceeded down the river by accurate observation at the rate of "upwards of seven miles an hour." Fitch next showed off in front of Water Street, Philadelphia, for the benefit of the governor and council of the state, and on the fine day of June 16 gave them all a ride. In a measured trial, a speed of eight miles an hour was recorded.

That summer Fitch's company ran services between Philadelphia and Bordentown, traveling 2,000 to 3,000 miles altogether at from six to eight miles an hour. They did the 38-mile run to Trenton in an hour and a half. This was faster than by sail, though still somewhat slower than a fast stagecoach on the good roads alongside the river, so they competed by charging half the price and served beer, sausages and rum in a pretty little cabin.

Alas for our heroes, the Delaware traffic was too little to sustain the undercapitalized company. Tradespeople were conservative, not yet ready to risk their persons or their goods to madcap modernity. It cost Fitch 30 shillings to carry people from Philadelphia to Trenton, Bordentown, Bristol and Burlington, but there might only be seven passengers paying a total of 20 shillings. There was little room for freight. Fitch had made what turned out to be a very good estimate of the profits that would accrue carrying 120 tons of cargo steaming the Mississippi "from New Orleans to the Illinois," but he was confined to Philadelphia, where he had raised his capital. Instead of giving up on the Delaware, however, he worked on a bigger, faster boat, aptly named the *Perseverance*. A storm wrecked it on its moorings in October 1791.

His little company bickered endlessly what to do next, but two important observers had watched Fitch. One was a rich landowner aged 35, John Stevens (1749–1838), of whom more in a moment. The other was Aaron Vail, an American diplomat home from service in France. Vail won Fitch a French patent, and in the spring of 1793 Fitch sailed joyously for France—straight into the adventures and misadventures of Voltaire's Candide. Instead of building his boat, he was caught in the turmoil of the French Revolution, staring in horror at Dr. Guillotine's daily work. When he reached the furnace in Nantes, where the castings for the engine were to be made, he ran into counterrevolutionary peasants, priests and scythe-bearing burghers marching into the city. The Loire, where the boat was supposed to be launched, was soon enough filled with thousands of their bodies. Vail sent him to England to see if he could buy an engine from the Birmingham factory of Boulton and Watt. He was refused an export license and the English blockade of France cut him off from his sponsor.

Fitch came home to further frustration. He expected to claim 1,600 acres of land he had at much risk secured by government warrant and recorded in his own name. The Indians who had held him captive 12 years before had finally been subdued, but his land along the Ohio was occupied by squatters. Fitch had no money to argue his ownership in court. He worked on a little model steamboat, three feet long, with paddle wheels and brass machinery "polished in a neat, workmanlike manner," but his dreams crowded mockingly in. Lonely and embittered, he drank copiously. He saved up opium pills prescribed for insomnia, and one summer night at the age of 55, he swallowed them all, precisely right in his dying prediction: "The day will come when some more powerful man will get fame and riches from my invention, but no one will believe that poor John Fitch can do anything worthy of attention."

Rich John Stevens was such a man, as different from Fitch as imaginable. He was an arrogant and rigid patrician; he insisted all his life that his wife call him Mr. Stevens. His Hudson River estate at Castle Point, most of what is now Hoboken, New Jersey, was staffed with slaves and furnished with elegant pieces from Europe. He was an epicurean and a dandy, but he had a law degree and he was an avid reader of scientific papers. After he had seen Fitch's boat on the Delaware near Burlington, he had hurried to Philadelphia to inspect it. He could easily have backed Fitch, but he was a snob. He fancied himself an inventor (with some justification in both steamboats and railways) and he never had the slightest scruples about stealing other men's work without giving them any credit. While Fitch was running his ferry in 1789, Stevens, with no vessel to his name, had tried to suborn the New York legislature to transfer Fitch's rights to him. He had failed and Fitch's rights ended up, on his death, with a man even more powerful and cunning than Stevens: Stevens's brother-in-law,

Robert Livingston (1746–1813). Livingston was a technical nincompoop whose vanity as an inventor outran his competence, but as a fixer he was in an Olympian class. He was one of the key men in revolutionary America, variously a judge, chancellor of New York, a congressman and a diplomat; it was Livingston as chancellor who administered to oath George Washington on Wall Street, New York, in 1789 and set the United States on its course. He had no difficulty persuading the New York legislators to give him Fitch's 20-year license for the Hudson in 1798, winning a monopoly on the promise that by 1802 he would run a steamboat "on new and advantageous principles" with a minimum speed of four miles an hour against the ordinary current of the Hudson. Despite Fitch's example, the promise seemed so fanciful to the legislators that they indulged Livingston in a gale of amusement.

Livingston formed a triple alliance: with Stevens, who understood mechanical matters but did not have the skill to make anything with his own hands, and with Nicholas J. Roosevelt, the son of a New York shopkeeper, who did have the craftsmanship and mechanics from England in his foundry at Belleville, New Jersey. It would have been a perfect coalition if Livingston had not persisted in his genius for throwing a monkey wrench in the works. He treated engineers, as historian James Flexner observed, like servants commanded to lay a table. He overrode Roosevelt's scheme for paddle wheels vertically at the side in favor of horizontal wheels. The result was a vessel called the *Polacca*, which in March 1799 wheezed out three miles an hour in still water before its boiler sprang a leak. It had to be abandoned—nine full years after Fitch's Delaware triumph. If he was to keep his license, Livingston would have to find a technical genius—and someone not cowed by his arrogance.

THE SOUTHERN GENTLEMAN

FITCH'S CLAIM to be the inventor of the steamboat is often challenged by Rumseians—supporters of James Rumsey (1743–92)—among them George Washington, Benjamin Franklin and Thomas Jefferson. In contrast to the unkempt Fitch, Rumsey was a well-dressed, courteous southern gentleman who liked to flourish a lace handkerchief from his sleeve. He was more technically gifted than Fitch. He kept a boardinghouse in the resort town of Bath, West Virginia, but won patents for steam engines for grist- and sawmills, and spent hours "dreaming of impossible things." His boat relied on hydraulics: Steam would pump water in at the bow and flow it out at the stern, where its reaction against the current would force the boat forward. It was a form of jet propulsion advocated by Franklin and hence acclaimed, but in this case the emperor had no clothes—it was impractical, flawed among other things by the amount of energy required to pass water through the boat. Rumsey did not show his steamboat on the Potomac until December 3, 1783, three months after Fitch's moment of glory. After a second sailing on December 11, Rumsey's boat never moved again, his inventiveness betrayed by the materials and engineering of the day. Sent to England by his supporters in 1788, he impressed Matthew Boulton, the partner of James Watt and manager of the Boulton and Watt works in Birmingham, enough to be offered an engine, but he overplayed his hand in the negotiation. He struggled to build an engine with English mechanics, but before he could prove his *Columbian Maid* on the Thames he died of a stroke at 49, poignantly while delivering a lecture on his inventions.

ROBERT FULTON His passion was to blow up warships, but his enduring triumph was in the peaceful art of commerce, the creation of the world's first successful steamboat services

1765–1815

WHEN FITCH was experimenting with his paddleboat on the Delaware, 20-year-old Robert Fulton was only a block away from the river, painstakingly arranging wisps of someone's hair in a decorative locket. He had learned the craft of painting miniatures from an immigrant English jeweler at the corner of Second and Walnut Streets, and now had his own little studio on Front Street. He was dexterous, focused and meticulous; it helped his little business that he was also graceful and charming, in appearance a sexually ambiguous Adonis with dark tumbling curls. It is more than likely he saw Fitch's boat in 1786. He certainly could not have avoided hearing about it. Whatever, his youthful aspirations were artistic, not mechanical or commercial, and when he did take an interest in steam navigation he took his time. He is the tortoise of the steamboat race.

Fulton was only eight when his father died just before the outbreak of the Revolutionary War. He was a tailor in Lancaster, Pennsylvania, who tried farming, lost everything and came disconsolately back to his first trade. Fulton was brought up in genteel poverty. His mother managed to have him taught by a Quaker builder-cum-schoolmaster. He frequented William Henry's famed Julian Library in Lancaster, where he browsed on publications like Ward Young's *Mathematical Guide,* Mott's *Treatise on Mechanical Powers* and *The Gentleman's Magazine*. He also hung around Lancaster's gun shops as they hammered out rifles for Washington's army, developing a fascination with the manufacture of instruments of death that was his life's abiding obsession.

It was another pattern in Fulton's life that he was desperate to avoid the poverty that had been visited on his mother, and that money came rather mysteriously to him at difficult times. From youth to middle age, he attracted benefactors, usually older men captivated by his looks and by his intelligence, especially as it manifested itself in his unabashed appreciation of their own virtues, but their identities were not always clear. Nobody quite knows who gave him the money to put down on an £80 sterling farm for his mother in 1786, nor how he found the wherewithal, when he began spitting blood, for his sojourn the same year in the spa at Warm Springs, Virginia (now West Virginia), nor the source of the 40 guineas (the equivalent of $210) he took with him on his paid passage to London in the summer of 1786. Nothing like this kind of money was to be found in miniatures.

Before he sailed, someone in Lancaster had given him a generous introduction to the Lancaster expatriate Benjamin West, who had been a court painter to King George III. The kindly West and his wife, Elizabeth, took him into the bosom of their family. West found him cheap lodgings, offered appraisals of his art and introduced him to men of rank and learning. Fulton charmed them all. Their enjoyment of him did not extend to paying him to paint their portraits, so he had to eke out his money, borrowing and begging as he went along. Only four years later, in 1790, did he feel able to tell his mother what it was like: "Many, many a silent, solitary hour have I spent in the most profound study, anxiously pondering how to make funds to support me. . . . Thus I went on near four years, happily beloved by all who knew me, or I had long e'er now been crushed by Povertie's cold wind and freezing rain." In his fifth year, he told his mother he had eight works accepted by the Royal Academy "with every posable mark of Approbation." In fact, the academy accepted two, and the melancholy theme of these paintings reflects a mood of morbid expectation—Mary, Queen of Scots under confinement and Lady Jane Gray the night before her execution. The truth was he was living off his friendships.

At 25, Fulton was presented with a morally tantalizing opportunity, an invitation to accept the patronage of a pariah of society, the scandalous young Viscount William Courtenay of Powderham Castle in Devonshire. Courtenay, later the ninth earl of Devon, was a racy figure, a transvestite referred to as "Kitty" in a notorious sodomy case; in 1811 he was to flee to New York, where, a mid-19th-century biographer says, "every door was closed against him except that of Fulton." The handsome Fulton risked the gossip to go and live at Powderham. He stayed three and a half years. He was soon bored with preserving Courtenay's features for posterity, copying famous works of art and hunting the countryside for more titled foxes. Sometime in Devon he confronted the reality he could never admit to his beloved mother: He would never make his way as an artist. He was proud, but he could confront his limitations because he had the audacious certainty that he was destined for greatness if only he could find the key to the door. He watched workers cut marble at a quarry on Courtenay's estate and straightaway invented a mechanical saw; it won a gold medal. His host was a promoter of a canal to link the Bristol and the English Channels. Very well, Fulton would come up with a superior solution to the principal problem of a canal, the change in gradients.

Handsome Robert Fulton was the James Bond of innovators, an artist and adventurer of complex character. This portrait by Charles Willson Peale was made in 1807, on the eve of his steamboat triumph, when he was 42. He regarded steamboats as "useful and honorable amusements," but blowing up warships as the "favorite offspring of my scientific pursuits."

The chairman of Courtenay's company was the inventive third earl of Stanhope, Charles Mahon, an irascibly brilliant aristocrat. His plan called for a series of locks, flooded and gated chambers to raise or lower vessels when water levels changed. In a letter of November 1793, Fulton sketched an alternative scheme, not original but vividly expressed, for dispensing with the expense of constructing locks. Small canal boats with wheels would be pulled up inclined planes by a falling counterweight. "Should your Lordship be so kind as to favor me," entreated Fulton, "One hundred Pounds would put me in Motion."

Stanhope said no. It is a mark of Fulton's adventurous American spirit, conceit if you like, that he was not in the least put out when the earl added that the young man did not know what he was talking about: "I doubt whether you will do well to pursue Mechaniks at present as a Profession." Their encounter was the beginning of a long, stimulating relationship in which Fulton's surf beat against the sand dunes of Stanhope's scientific skepticism. Fulton spent the next two years proving Stanhope wrong on canals. To refine his ideas, he went to Manchester in northern England, the focal point of the canal mania gripping the country and the center also of political and social reform in aristocratic Britain. He soaked up the ferment of humanitarian ideas bubbling from new friends younger than himself: Robert Owen dreaming of the industrial utopia that would make him famous, and the poet Samuel Coleridge. Fulton adopted an elevated vision of technology as the salvation of mankind. The 158 pages and 17 engravings of his *A Treatise on the Improvement of Canal Navigation* (1796) vaulted over Stanhope's niggles with nothing less than a plan for the transportation needs of the whole world. The essence of it was to conceive of canals as capillaries rather than arteries. He envisaged boats of low tonnage on networks of small canals connecting scores of communities rather than a few. They would be cheap to build with a ditch-digging machine of his invention, and less restricted by terrain because he had devised prefabricated aqueducts to cross valleys and the machinery to pull small boats up hills: The

GALLERY: Fulton's 1813 drawing of himself (bottom) exudes the wit and bravura of someone confident enough to mock himself. The top portrait, by James Sharples Sr., shows a bewigged Fulton. The bust done in 1803–4 by Jean-Antoine Houdon is referred to in the text. In the background of the splendid painting by Benjamin West (1806) is a ship exploding after being hit by one of Fulton's "torpedoes."

counterweight on his inclined plane would be huge buckets of water dropped down a vertical shaft and emptied at the bottom so they could easily be pulled up again. The treatise won a royal patent on June 3, 1794. It marked the metamorphosis of Fulton, artist, to Fulton, civil engineer. The talents he had perfectly suited the new career he had found for himself. He was skilled enough to make a drawing in scale. He had an intuition for mathematics so he could calculate the feasibility of an idea. His ability to translate his spatial visions into sketches and precise specifications meant that others could fabricate them: Several of his aqueducts from prefabricated parts were built. The sketches were promotional gold; they had such an air of reality they made it look as if a scheme were already up and running.

Fulton veered between advocating his canals as an altruistic public service the state should subsidize and a project for private enrichment. He sent his treatise to George Washington, elaborating the public benefits, and tried to interest Stanhope. Alas, the earl had the disdain of the already rich for those who want to eat while coming to the rescue of humanity. He recoiled also when Fulton tried to excite him with the commercial potential of a small-canal enterprise in America linking New York, Philadelphia and Baltimore. Stanhope, however, confided his hopes of building a steam warship of 200 tons for the British navy and the eager Fulton said he, too, had "some ideas" for steamboats. The American inventor James Rumsey (see panel) had arrived in London about the same time as Fulton and the two had become friends. A few months after Rumsey's death, Fulton had fooled around with a model boat in which a paddle at the stern mimicked "the spring in the tail of a Salmon." He sent Stanhope a crude drawing and on November 4, 1794, he wrote to the English engine makers Boulton and Watt asking the price of a rotary engine of three or four horsepower "which is designed to be placed in a boat."

Boulton and Watt never replied, and Fulton's imagination was—as always—too fertile to be put on hold. He convinced himself that revolutionary France would be a better arena for "the adventurer armed with fortitude," as he now described himself, and ripe for a national network of small canals. Fulton arrived in Paris in the springtime of 1797, halfway through the corrupt, futile post-terror regime called the Directory, crossing the channel during a temporary break in the wars with England. He stayed at a pension on the Left Bank and for the first time a woman other than his mother entered his life. He fell in love with another guest, a vivacious and daring older woman in her 40s. This was Ruth Barlow, the wife of Joel Barlow, the wealthy American entrepreneur, poet, aesthete, voluptuary, intellectual and diplomatic troubleshooter who was away in Algeria negotiating the release of Americans captured by Barbary pirates (charismatic to the end, he was to die retreating with Napoleon's army from Moscow). There was no need for subterfuge about the intense relationship that developed between Fulton and Ruth. On the contrary, Joel and Ruth were not unused to sharing lovers. When Barlow returned, they both took the dashing 31-year-old Fulton to their bosoms, and then to live with them in a grand mansion by the Jardin du Luxembourg. It was a ménage à trois shocking to New England sensibilities but not seen as much of an innovation in Paris; after all, the French had long ago fashioned the phrase.

The French were not in the mood for Fulton's canal schemes. War was the raging obsession. Fulton, ever the opportunist, came up with another grand vision in which his moral and his mercenary impulses cohered. The moral vision was the creation of a prosperous new world order based on free trade. In 1793 in Devon he had witnessed the massing of British warships at Torbay for war with Napoleonic France and he genuinely hated the war's subsequent toll on commerce, industry and civil liberties. The new world order he called for demanded liberty of the seas, to which end he offered to annihilate the most powerful navy in the world. ("Make no small plans" might have been a phrase specially minted for Fulton.) With money from a Dutch patron, he invented Nautilus, a submarine disguised as a boat, very much based on David Bushnell's American Turtle, and a mine he called a torpedo that could place against warships. On November 9, 1799, a

coup d'état overthrew the Directory and installed Napoleon as first consul. Fulton offered his submarine and mine to Citizen General Bonaparte's ministers, asking £160 for each gun on any big British ship he destroyed. To show it off, he wriggled into his 20-by-5-foot tin tube on June 13, 1800, and, with a companion, plunged it under the Seine in two dives of around 20 minutes each. On September 12, this James Bond of innovators risked his life again, making an abortive attack on two English brigs anchored off Cherbourg. In 1801 the French treasury put up a token sum ($2,500), but Fulton found it too hard to navigate the Nautilus underwater and in August he resorted to three longboats carrying mines he intended to place against British ships. The British were too alert, and once again he had to abort the mission.

Fulton was still dreaming of sea warfare when he finally caught up with his destiny, or rather his destiny caught up with him, in 1802. The Barlows entertained on a splendid scale in their mansion. It was a place where the top people mingled— assorted members of the Directory; the Marquis de Lafayette; Napoleon's foreign minister, Prince Charles de Talleyrand; the Montgolfiers of ballooning fame; Count Constantin de Volney; and the most prominent expatriates, including the revolutionary hero Tom Paine. At one of the Barlow dinners, Fulton had alarmed Talleyrand by rhapsodizing about boats moved by steam. "I was overwhelmed with sadness," recalled Talleyrand, "for I could not but feel he was mad." At a dinner party thrown in February or March 1802—nobody was ever sure of the date—one of the Barlows' guests was the newly arrived U.S. minister plenipotentiary, one Robert Livingston. He had been sent to Paris by President Jefferson to negotiate with Napoleon for the right of American vessels to sail the lower Mississippi, territory then owned by France. Livingston was 55 now, hard of hearing and out of sorts, and frustrated by the intrigues at Napoleon's court: He spoke no French. It was a relief to fall into conversation—in English—with the courteous Fulton, who bubbled with enthusiasm. It was one of history's most critical one-on-one meetings. Livingston had got nowhere with his earlier dreams of a steamboat on the Hudson

and he did not enjoy being regarded as an eccentric. (The French Academy of Science called the steamboat "an insane idea, a gross blunder, an absurdity." The church declared it a heresy: "In the beginning fire and water were separated by especial ordinance from the Creator, and man has no right to join together what He has put asunder.") Fulton and Livingston happened to meet just when both were at a loose end, both dependent on Napoleon's pleasure. By the time the carriages were summoned, the two "madmen" had a handshake to build a steamboat together.

Barlow was suspicious. He saw Livingston as a great man but an inveterate promoter who could not be trusted to keep Fulton's secrets and Barlow talked of funding the project himself. But Fulton saw clearly how Livingston's political clout might further any one of his schemes. Fulton was dizzy with excitement, but he dealt coolly enough with the imperious Livingston. He talked a great invention, blinding the older man with his engineering know-how, his vigor and his dexterity with mathematics. The role he loftily consigned to Livingston was to find a way around the British export laws to get a Boulton and Watt engine to Fulton's specifications. Fulton had no intention of spending his intellect inventing a steam engine as Fitch, Stevens and Livingston had done. Why bother when Watt and Boulton had a whole factory doing that? Fulton was not concerned with being original. Innovators don't have to be original. They have to be effective. Fulton had no compunction about taking bits and pieces of ideas from everyone. He had the same attitude in borrowing from David Bushnell for his Nautilus. He wrote a perceptive passage that applies to so many innovations, then and now: "As the component parts of all new machines may be said to be old . . . the mechanic should sit down among the levers, screws, wedges, wheels, etc. Like a poet among the letters of the alphabet considering them as the exhibition of his thoughts; in which a new arrangement transmits a new idea to the world."

Fulton sat down among the plans and specifications of everyone who had tried to make a steamboat: Fitch first of all, Rumsey, Stevens, New Hampshire's Samuel

Morey, the Scotsman William Symington and early French failures. He attempted to sift every bit of research with the aim of identifying engineering principles. He pored over recently published experiments on water resistance. His calculations had as much to do with economics as energy. From the first, he aimed to design a fairly large steamboat that would carry enough people and cargo to pay its way on the Hudson and be still more profitable on the Mississippi. He even commissioned a model from the celebrated instrument maker Etienne Calla.

So much for calculation. For inspiration, Fulton borrowed Ruth from Joel. On a lovely spring day early in April 1802, he sat Ruth by his side in a phaeton drawn by white ponies and jogged off with her to the fashionable spa of Plombières in the Vosges Mountains. The two of them were there for the whole season, abandoning both Joel and Livingston in Paris.

It was a strange thing to do. Biographer Cynthia Phillips thinks the self-portrait he painted at this time is suggestive of anguish. "The two halves of his pale face do not match. The right side is more open, the expression at once benign, aloof and frightened. The left side is compressed, its expression controlled and crafty. The distortions are deliberate. It is a portrait of the inner man. He wanted to end his dependence on the Barlows and on Livingston but he could not solve the tensions created by his divided loyalties." Ruth and Robert found relief in dancing and riding; he read to her in her bath. Joel, left to oversee the model, was happy enough indulging in passion by proxy. He amused himself writing the absent couple long erotic letters salivating in childish language about Ruth "foolin' with toot."

Livingston, fretting at Napoleon's court, was not amused. The putative partners, who had no formal agreement yet, had strong reasons to combine, but relations were edgy. Livingston was a natural controller of others, Fulton a proud free spirit. He gambled on testing Livingston's patience because at this stage he had still not given up hope of selling naval warfare to the French (or any other taker), and he reckoned Livingston needed him more than he needed Livingston. Calla's model arrived at Plombières at the end of May. It

RUTH BARLOW: Fulton's daring soul mate

was three feet long, eight inches wide. Fulton stopped up a small stream to make a 66-foot pond—a forerunner of the modern testing tanks—and systematically timed various arrangements of paddleboards, screw propellers, sculls and chains. Josephine Bonaparte and her little coterie were amused to see a grown man playing around with a clockwork boat for hours and hours. *Quel enfant!*

The best method, Fulton concluded, was to have flat, boardlike oars on an "endless chain" run over a pulley at bow and stern, the power communicated to crank wheels on each side of the cylinder. It was very like Fitch's first effort. But when Barlow wrote to say that a French inventor, Desblancs, was exhibiting a similar arrangement in Paris, Fulton settled on side wheels like the waterwheels in a flour mill. Again, he was not much concerned with originality. "Although the wheels are not a new application," he told Livingston, "yet if I combine them in such a way that a large proportion of the power of the engine acts to propel the boat in the same way as if the purchase was upon the ground, the combination will be better than anything that has been done up to the present and it is in fact a new discovery."

On his return to Paris, Fulton had a difficult negotiation with Livingston before the two finally signed a partnership on October 10, 1802. This memorable agreement committed Fulton to building a 120-by-8-foot boat in New York to carry at least 60 passengers to Albany. Livingston was to provide the cash for a prototype and the political clout to protect the New York monopoly, Fulton the design and management and they would share the profits. If the project failed, Fulton had to return to Livingston half his investment.

The partners agreed to build a prototype in France for a trial on the Seine in 1803. Livingston's efforts to get an engine from Britain failed, so Fulton risked leasing an engine and boiler in Paris. The boiler blew up. They leased another and by mid-May 1803 had installed it in a 75-foot boat at dock in the Seine. On the stormy eve of

His *North River Steamboat of Clermont* was already at work when Fulton drew this perspective of machinery in 1809 for a patent application.

the attempt, there was a loud banging on Fulton's door: "A messenger awakened me from the docks exclaiming, 'Oh, sir, the boat has broken in pieces and gone to the bottom.' I felt a greater despondency than I had ever known before."

Perhaps, as some papers reported, it was sabotage by jealous boatmen. Maybe the cumbersome makeshift low-pressure engine was too heavy for the structure. The distraught Fulton rushed down to the river through the rain without bothering to put on a coat. He worked all night and all day, without pausing for food or rest, exhausting himself to recover the engine from the river bottom.

They were ready again by August 9, 1803. A French newspaper reported: "At six o'clock in the evening helped by only three persons he put the boat in motion with two other boats in tow behind it, and for an hour and a half he afforded the strange spectacle of a boat moved by wheels like a cart, these wheels being provided with paddles or flat plate, and being moved by a fire engine." Fulton's boat puffed along the Seine at about three miles an hour. It was, said the *Journal des debats, "un succès complet et brilliant."* The speed was a fraction of the 16 miles an hour Fulton had hoped for, but it was a promising debut. It refuted the conventional wisdom, most recently proclaimed by the celebrated architect and designer Benjamin Latrobe in Philadelphia, that steam navigation was essentially impracticable. Yet now, with success foreseeable, a curious thing happened.

Nothing.

More specifically, the partners lost their concentration on the steamboat. Livingston pulled off a coup in the talks with Napoleon. He seized the chance for the United States to buy 565 million acres of what is now America at less than three cents an acre. The Louisiana Purchase he negotiated (along with James Monroe) doubled the size of the United States, and Livingston went exultantly back to his New York estate to breed merino sheep. For his part, Fulton took a roundabout route to London,

**Partner and fixer
Robert Livingston**

NAPOLEON AND FULTON, 1804: Napoleon got tired of Fulton's foxiness. "This man Fulton has an avaricious disposition. We will have no more to do with his plunging machine."

arriving there in May 1804. He would stay there two and a half years.

Fulton's return to England came about after he had been approached in Paris by a certain Mr. Smith, who was a secret service agent for William Pitt's government. In England, Fulton himself assumed the alias Robert Francis to thwart Napoleon's spies. Disappointed by the French, he was now dedicated to blowing up their navy on behalf of their English enemies, for a fee. (So much for blowing up the British navy on behalf of free trade.) The fee he looked for was a small fortune for every French ship he blew up—and permission for Boulton and Watt to make a steam engine to his design. (Pitt agreed to that, though not to an export license.) Fulton's marine warfare was tried in raids on the French fleet at Boulogne and Calais, and for show he blew up a brig in Dover. But the adventure ended in bad blood. Admiral Horatio Nelson was the nemesis. His English fleet defeated Napoleon at Trafalgar in October 1805, assuring England of supremacy at sea, and Pitt abandoned "Robert Francis's" scheme for subma

rine warfare. Fulton was furious. His pent-up frustration betrayed the cynicism that was always suborning his idealism. He tried to blackmail the British with the threat that he would sell his explosive devices to an enemy. He finally settled on a payment of £12,000, or about $60,000. He did so with much petulance, but it was ample to ensure a comfortable living—and as a bonus the British government also sanctioned the export of the precious engine.

Still, Fulton was simply in no hurry to get on with the steamboat. The stereotype of the single-minded innovator succeeding by perseverance just does not fit Fulton. He was easily distracted. The steamboat was within reach, but in a letter to Barlow he said it was "half as important as the torpedo system." And then there was love. He enjoyed again a ménage à trois with the Barlows, who had moved to the elegant No. 9 Bedford Square. Not until December 1806 did Fulton follow his engine to New York, a return to his native country after an absence of more than 15 years. The engine was waiting for him in the customhouse. The monopoly was due to expire in four months, but Fulton went off to Philadelphia for a month with the Barlows and kept Livingston in suspense.

The recent biographer Kirkpatrick Sale has an interesting comment on the complexities of Fulton in midlife. He compares a self-portrait in oils with a sculpted bust in marble made at the same time by Jean-Antoine Houdon, known for his psychological insights in stone (page 28). The painting, he suggests, captures Fulton's character in his 42nd year—confident but with only a hint of inner strengths and bearing "unmistakable traces of sorrow and fright, an aura of what might be despondency." The Houdon bust, on the other hand, he suggests, shows Fulton as he would become, a powerful figure "fixing a steely gaze on the world, the shadow of a smile that might almost be sardonic on his lips, a face not untroubled but with a willful force behind it." It makes sense.

Livingston won an extension of two years on the Hudson license, but Fulton became excited by an even more glittering prospect. Livingston's acquisition of Louisiana and the Floridas, and with them navigation rights on the Mississippi, threw open the West for settlement and trade. If they could only succeed on the Hudson, they could dominate trade with the West. A reenergized Fulton rallied the shipwrights at Charles Browne's shipyard on the East River. His tall, elegantly dressed figure could be seen every day down among the beams with the craftsmen, cheering them on with his zest and good humor, his readiness to listen to them and redesign details on the run. The language of the onlookers, on the other hand, Fulton wrote, was "uniformly that of scorn, or sneer or ridicule." By early summer he had installed the engine and two 15-foot circular wooden paddle wheels in a long narrow boat, 146 feet long by 12 feet wide with a 15-foot smokestack, altogether two and a half times the size of Fitch's boat. It made the river men laugh even more—everyone knew the Hudson needed an oversize gaff rig and broad beam. They called his boat Fulton's Folly. "Never," wrote Fulton, "did a single encouraging remark, a bright hope or a warm good wish, cross my path." He cheered himself by blowing up a derelict brig at anchor in the waters off the Battery. Even as the climax of the steamboat neared, he could not shake his obsession with marine warfare. He beseeched President

Harriet, heiress and wife, painted by Fulton on ivory. It was a cool marriage. Fulton was still in love with Ruth Barlow.

Jefferson to be ready to do to the British navy what Fulton had hoped the British would do to the French (or the French to the British, depending on the season). Jefferson sent a naval delegation to watch.

Finally, on August 9, four years to the day since his Seine test, Fulton tried out his incomplete boat on a short trip. He achieved three miles an hour. Noting the strength of the axles at full power, he had more than doubled the size of the original three-feet-by-eight-inch paddleboards. He told Livingston, "Whatever may be the fate of steamboats for the Mississippi, my thing is completely proved for the Mississippi and the object is immense—please forward me 1000 or 1,500 dollars as soon as possible." Kirkpatrick Sale's calculation is that both men put about $10,000 into the boat. They could not raise interest in anyone else they approached.

The plan for the debut voyage to Albany was to steam first to Livingston's vast estate 110 miles upriver, rest the night and complete the final 40 miles the next day. The big day was Monday, August 17, 1807. No paying passengers were invited, just a small group of friends and Livingston relatives. The city's newspapers were above this nonsense; only one carried a small announcement of the attempt. Fulton's Folly attracted a gaggle of spectators to the North River wharf in little Greenwich Village to watch the adventurers. The men

boarding were elegant in spotless ruffles, the ladies in fine bonnets. The historian James Flexner says they gave off an air of sophisticated disdain, representing as they did the most powerful political and social force in New York State, but Fulton himself writes: "There was anxiety mixed with fear among them. They were silent, sad and weary. I read in their looks nothing but disaster, and almost repented of my efforts."

Around 1 p.m., Fulton shouted his orders to his English engineer and his captain. The paddle wheels splashed and *Clermont* moved away from the wharf—and then it stopped. "I elevated myself upon a platform," Fulton wrote. "I stated I did not know what the matter was, but if they would be quiet and indulge me for half an hour, I would either go on or abandon the voyage for that time. I went below and examined the machinery, and discovered that the cause was a slight maladjustment of some of the work. In a short time it was obviated." The boat moved off again up the Hudson—and kept going. Black smoke from the engine, then fired by coal rather than wood, trailed like a triumphant banner all the way upriver to nightfall. Candles were lit in the cabins; the ladies tried to sleep on improvised cots, the men on the vibrating deck. Flexner paints a pleasing picture of Fulton standing alone on the bow gazing at the stars far from his earthly strivings for "money and fame and social position and love, all the prizes for which he had struggled so painfully and long." Next morning, the boat was still steaming steadily north, a plangent volcano coughing smoke and sparks. One rustic reportedly rushed home, locked the doors and shouted that the devil was going up the river in a blazing sawmill. Fulton's first biographer, his friend Cadwallader Colden, says crews on other boats on the river "prostrated themselves and besought Providence to protect them from the approaches of the horrible monster." The passengers aboard the monster were merry as they steamed through the Highlands; the ladies and gentlemen of the Fulton and Livingston families, both of Scottish heritage, gathered in the stern just a few feet above the rushing water and regaled the woodlands with "Ye banks and braes o'bonny Doon."

The *Clermont*, as it came to be called, makes a landing at Cornwall on the Hudson in 1810. By this time Fulton had established regular service and was building an empire.

Fulton dropped anchor at Clermont at 1 p.m. on August 18, 110 miles in 24 hours, a little faster than the license required. He was emotionally spent; Livingston stood on deck and gracefully congratulated his partner as an inventor whose name would descend to posterity as a benefactor of the world. Livingston averred that before the end of the century it was "not impossible" that vessels might make the voyage to Europe solely on steam. Then he is supposed to have added extra bubbles to the champagne with the announcement that Fulton was to join the Livingston family: Livingston's cousin Harriet had accepted his proposal of marriage. (It is more likely that he met Harriet for the first time on arrival at Clermont and the betrothal was months later.) Harriet was 25, a harpist and painter, and rich, and Fulton was 42, heading for riches. He was no doubt still in love with Ruth Barlow, but his marriage to Harriet in 1808, duly consummated, was a neat emotional knot in a partnership of mutual commercial interest. Livingston's license would have been forfeit without Fulton, but without Livingston's license Fulton would have been unable to launch any steam vessel on the Hudson. Fulton could not even prevent others from copying his steamboat; he did not possess a patent until 1809, and then shakily.

The *Clermont* steamed off again the next morning, arriving at Albany at 5 p.m. The trip had taken 32 hours at an average speed of nearly five miles an hour. By comparison, trips from New York City to Albany by sail could take three to nine days depending on the vagaries of the wind. For the return trip on August 21, Fulton attracted five paying passengers willing to risk the widespread expectation that the boiler would blow up. They arrived back in New York City in two hours' faster time, overtaking many sloops and schooners, and were cheered along the riverbanks. At West Point, the corps of cadets of the newly established U.S. Military Academy turned out to watch.

Two weeks after the maiden voyage, Fulton advertised the first trip north for paying passengers, hired a cook and waiter, and stocked up on beef and chicken, eggs, watermelon, sugar, rum and brandy. He ran a schedule of two round trips a week. Each trip attracted more passengers; by October 1 there were 60, and on a November trip the vessel was overcrowded, with more than 100 people onboard. That month, with ice impeding travel, Fulton set about rebuilding and enlarging his boat with "three excellent cabins or rather rooms, containing 54 berths with kitchen, larder, pantry, Bar, and steward's room." Fulton's imagination was complemented by a pragmatic temperament. As a painter under instruction he had got used to painting over and reworking canvases. So, too, in his career as an innovator. For all his proud attempts to be scientific about the steamboat, the calculations for the *Clermont* had misled him: The flat-bottomed original boat was so narrow as to be unstable. The rebuilt version was five feet wider. When jealous sloop captains made a point of running alongside in attempts to break off a paddle wheel, he built guardrails. There were frequent breakdowns, but because he had made meticulous notes and always kept his eye open when out on the water, he was able to trace and repair his errors. His steamboats were always a work in progress.

The speed Fulton had attained with the *Clermont* was fast enough to retain the New York license, fast enough to compete with the river sloops, but slower than Fitch had run his service on the Delaware. Weather and water conditions are hard to compare, but had the Lords of the Delaware been competing on the Hudson on that maiden voyage, Fulton would have been 52 miles behind when Fitch and Voight reached Albany. Of course, their homemade engine might have failed, but Fitch and Voight traveled 2,000 miles in their first commercial season. It compares well with Fulton's 1,200 to 1,400 miles before he laid up his boat in November after six weeks in operation.

So why did Fulton succeed where Fitch failed? For one thing, although the populations of New York and Philadelphia were comparable, the geography of New York favored the steamboat. The hilly, winding roads of the Hudson Valley were less amenable to competitive coaches than the roads along the Delaware. Technology was also particularly important. Not only did Fulton's own design improve the Boulton and Watt engine, but he had the benefit of the nascent industrial revolution—nearly 20 years of incessant Yankee ingenuity in mechanics and advances in the quality of materials. Third, capital. Fitch was forever short of money, whereas Fulton and Livingston were able to invest for success. Fourth, the pragmatic Fulton's personal gifts were superior, as a designer and organizer. And, fifth and very important, Fulton unashamedly ascended the steps laid by earlier generations. An innovator's essential contribution may be to realize the promise of the known.

The successes that distinguish Fulton describe an exponential arc. Fitch's boat lost money, Fulton's made it. Livingston reckoned that in a year they would both earn a profit to $3,500, given three sailings a week. One of the new boats would produce from $8,000 to $10,000 over a season. "Another boat which will cost us $15,000," Fulton told his partner, "will also produce us $10,000 a year . . . the only method I know of gaining 50 or 75 per cent." They plowed the profits back into bigger and better boats to beat off com-

petitors with the velvet glove of fancier service and the iron fist of lawsuits.

In this manner, Fulton built and operated 21 successful boats and created a steamboat empire. Five years after the maiden voyage, the Fulton-Livingston interests operated eastern services on the Hudson, Delaware, Potomac and James rivers and Chesapeake Bay, and on the western rivers, the Mississippi and the Ohio. Fulton built workshops in New Jersey and

Facsimile of a letter written by Robert Fulton: "Estimate for the expence of a steam ferry boat for one year," January 22, 1810

a factory in Pittsburgh, at the head of the Ohio and a hotbed of steam engine manufacture. He and his associates set up Nicholas Roosevelt to build a steamboat to serve the entire length of the rivers from Pittsburgh to New Orleans with the benefit of a Fulton-Livingston monopoly license for the lower Mississippi. (Fulton ought to have attempted the Mississippi earlier, but in these last years of his life he was once again preoccupied with naval warfare. New York City and Congress, in the panic of the 1812 war, built a massive floating fortress to Fulton's design, the precursor of steam

warships. It worried the British enough for them to try to kidnap or kill Fulton; they staged an unsuccessful commando raid on a house where he had intended to sleep.)

Roosevelt took the helm of the *New Orleans* (319 tons) in October 1811, as it churned out of Pittsburgh in an attempt to reach New Orleans, a maiden voyage of 2,000 miles on which nobody in Pittsburgh dared book a passage. Low water on the Ohio delayed him for days at Louisville (where his wife, Lydia, gave birth to a son). There was only five inches' clearance when Roosevelt risked the plunge into the rapids below the city, then, as they neared the Mississippi, the first of the New Madrid earthquakes struck. The Mississippi floodwaters forced the Ohio to drain backward. Still they continued. They steamed on a vast, featureless muddy lake surrounded by a flooded forest out of which shot a large Chickasaw Indian war party paddling furiously toward them. Roosevelt released the safety valve on the steam and the *New Orleans* gradually outran the screaming Indians in a manner that would have given joy to John Fitch. It was with much relief that Roosevelt docked in New Orleans on January 12, 1812. His actual steaming time was 259 hours at just over eight miles an hour. It was a triumph, but he realized his boat did not have the power to ascend the river with all its risks; she operated between the Gulf and Natchez until hitting a snag and sinking above Baton Rouge in 1814.

Fulton never did manage to send a steamboat all the way from New Orleans to the mouth of the Ohio or to Pittsburgh, but his ambitions were limitless. He contemplated nothing less than his control of steam navigation throughout the civilized world. He sought from the Russian tsar the exclusive privilege of running steamboats between St. Petersburg and the nearby Baltic naval station at Kronshtadt, and made an agreement with an Englishman in India to introduce steamboats on the Ganges. He wrote to Thomas Jefferson in

1813 that in a few years he would have a line of steamboats from Quebec to Mexico and St. Marys.

The Fulton-Livingston steamboat empire was altogether much more than an achievement of technology; in fact, technology was the least of it. The all-important initial engine, after all, was an import. The singular accomplishment was one of modern management. Single-handedly, Fulton ran what would now be departments of engineering, personnel (he quickly fired his first two captains), finance, law, public relations, promotion, marketing, advertising, training, research and development—and customer service. Among other things, Fulton began the tradition of glamour and hedonism we always associate with steamboats (and later transatlantic liners). He described his third boat, the *Paragon*, 170 feet long by 28 feet wide, as a floating palace with mahogany staircases, ornamental paintings and a sumptuous dining room for 150. His attention to detail and comfort was impressive: "The cabins," he wrote in a magazine article, "are lighted by large skylights so as to be perfectly airy, and are elegantly furnished with carpets, looking glasses, etc. The meals are served on china. Every upper berth, except for a few near the wheels, has a large window, and each has a shelf for the reception of a hat and clothes. The curtains which are of fringed muslin with silk drapery are so contrived that the cornice to which they are fixed draws out, and thus forms a little closet in which a person may dress without being seen from the cabin." He promulgated detailed rules: "As the comfort of all persons must be considered, neatness and order are necessary; it is therefore not permitted that any person shall smoke in the ladies' cabin or the great cabin, under a penalty, first of one dollar and a half, and a half for each half hour they offend against this rule; the money to be spent in wine for the company."

Fulton's last boat, *Chancellor Livingston,* was a fitting tribute to his partner, who had died in 1813. It was Fulton's biggest and best, 526 tons, with huge cargo space, against the *Clermont*'s 100. He did not live to see the launching in 1816. Fulton badly missed Livingston, increasingly exposed as he was to the squabbling among Livingston's

PHOENIX GOES TO SEA

JOHN STEVENS, who had been abandoned by his brother-in-law Livingston in favor of Fulton, still dreamed of creating a steamboat empire. He was not a mechanic but he had good ideas. Two years before the *Clermont* made its maiden voyage, he made a number of crossings of the Hudson with *Little Juliana*. It was not much bigger than a rowboat, but it was distinguished by two screw propellers. On Fulton's return to New York, Stevens was offered a chance to join the partnership because they were looking for extra investors. He turned down a fifth share on the grounds that Fulton's excessively narrow boat was inferior to one he proposed to build; he recoiled also from the condition that he would have to acknowledge Fulton's primacy. He valued the accolade of "inventor" too much to agree to bend his proud knee, and

resolved to defy the law that accorded the partnership a monopoly. Five months after Fulton's maiden voyage, he began building his rival 100-foot steamboat, with a low-pressure engine and paddle wheels, and in 1808 ran services between New York and Brunswick. Livingston moved to seize the boat by order of the New York courts. Stevens took fright. To escape, he risked sending his boat 240 miles to Philadelphia, 150 miles of the journey in the open ocean. Now named the *Phoenix*, it steamed out on June 10, 1809. Eleven days were in heavy seas and storms, but on the 13th day the *Phoenix* reached the shelter of the Delaware. It was the world's first seagoing voyage by a steamboat. Just ten years later, the *Phoenix* captain, Moses Rogers, took the 320-ton *Savannah* from Georgia to Liverpool in 29 days. Much of the voyage was under sail, but it was suggestive of the steamship crossings to come.

heirs at a time when rivals were everywhere and the efforts were intensifying to challenge Fulton's patent and break the Fulton-Livingston monopoly on the Hudson and the lower Mississippi. More important, perhaps the sagacious Livingston would have saved Fulton from humiliation in January 1815, when, as proof of his precedence in the use of side wheels, he produced in court in Trenton, New Jersey, a letter he said he had written to Lord Stanhope in 1793. It was a stunning moment when the opposing counsel held the paper to the light and revealed a watermark of 1796. Fulton survived the confrontation, with a brilliant peroration on his service to mankind from his émigré Irish lawyer, Thomas Addis Emmet, but he did not

survive the sequel. At the time, the Hudson was partly frozen, the ferry to New York unable to reach the slip. A shivering Fulton filled in a dank three hours showing his companions round his workshop on the Jersey shore. Eventually, they reached the edge of the ice mass in a small boat. The burly Emmet's weight was too much for the last few yards to the ferry on foot across the ice. He plunged into the freezing water. Fulton grabbed him by the arm. The lawyer was rescued, but the price for Fulton was pneumonia, from which he died on February 23, 1815. New York bestowed his name on a street and a market, and the world paid homage to "the inventor of the steamboat." He was less than that—and much more.

OLIVER EVANS

Few have heard of him, but it was this farm boy who first got America moving with his high-pressure steam engine

1755–1819

Trick questions:

- Who created and installed the world's first automatic production line?
- Who built the first wheeled vehicle to move under its own power on American roads?
- Who built the first amphibious vehicle?
- Who designed and manufactured America's first effective high-pressure steam engine for factories and steamboats?

TRICK QUESTIONS because only one name is required for all four answers. One man introduced all these innovations at the dawn of the American republic: Oliver Evans. The country boy who grew up to look like a plump Lord Byron was a poet of mechanical principle. His automatic production line was in operation long before Henry Ford was born. He foresaw the age of the railway and the steamboat, and his high-pressure steam engine advanced America toward it. Evans is not a figure in the pantheon inhabited in the popular mind by Ford, Edison, Fulton and Bell, yet he was an original, the new nation's first notable innovator.

"A man's useful inventions," wrote Benjamin Franklin, "subject him to insult, robbery and abuse." Evans collected all three. As a self-taught innovator who shared generously of his insights, he endured much from snobs, fools and thieves. The leading men in Philadelphia and Wilmington who could have done much to advance his ideas regarded him as a social inferior. John Stevens, the wealthy New Jersey landowner, realized that Evans could assist his own ambitions to build a steamboat, but when

EVANS: Inspiration from the shot of a pistol

Evans let him have the details of his engine, Stevens went public with spiteful criticisms, partly in an attempt to claim priority. Evans was not saintly enough to take all this with a sweet smile. He had black moods of furious resentment and at times his tongue was provocatively sharp. His natural disposition, though, was kindly. He was obsessive but he would pause in his ceaseless experimenting to make toys for the seven children of his marriage to Sarah Tomlinson, a farmer's daughter. He had a passion to bequeath the new generation the knowledge he had been denied. He educated a generation of mechanics with books on steam engineering and milling.

The origin of Evans's genius is a puzzle. He was imbued with a mechanical imagination but also an analytical mind that could translate ideas into practical machines. There was not a trace of this anywhere in his bloodline or upbringing. His father, originally a shoemaker by trade, was a descendant of a literary man, the spellbinding Welsh preacher Evan Evans (1671–1721), a pauper scholar at Oxford University who came to America to be the rector of Christ Church, Philadelphia; his mother's side had Dutch and Swedish ancestry. None of Oliver's seven brothers or four sisters showed any interest in scientific mystery. There were no machines on the 200-acre Delaware farm where he grew up during the Revolutionary War; the only books at home seemed to have been the Bible and a spelling guide. There were no free or public schools, but the clarity of Evans's writings suggest he had a teacher of English and arithmetic,

and by the time he was 14 he knew himself well enough to direct his speculative mind to practical matters. Until the U.S. Corps of Engineers (and later Military Academy) was formed at West Point in 1802, there was no school in America where an aspiring civil engineer—the term was unknown—might learn the elements of engineering or mechanical drawing or even hear of the latest inventions.

Wheelwrights and millwrights were the engineers of the day. Evans apprenticed himself to a wheelwright and wagon maker in Newport, but every chance he had he studied whatever he could of mechanics and mathematics. His master begrudged the expense of a candle at night, so the apprentice did his reading by the flickering light of the day's wood shavings. He kept his eyes and ears open, too. At Christmas in 1772, there occurred an incident that seems straight out of the mythology of great men, but it is in all Evans's accounts of his work. One of his brothers told him about a blacksmith's boy who had fun in the forge. The boy put water in a gun barrel, rammed down a tight wadding and then put the butt end of the gun in the smith's fire. The compressed steam in the cylinder ejected the wadding "with as loud a crack as if it had been gunpowder." To Evans it wasn't a joke. It was evidence, as he supposed, of a previously undiscovered source of energy. "It immediately occurred to me that there was a power capable of propelling any wagon, provided that I could apply it; and I set myself to work to find the means of doing so." He was all of 17.

Perhaps if he had known more, he would have achieved less. He would have reconciled his excitement to the received wisdom about steam engines. He did not know that a young man destined for fame, James Watt, had considered and rejected high-

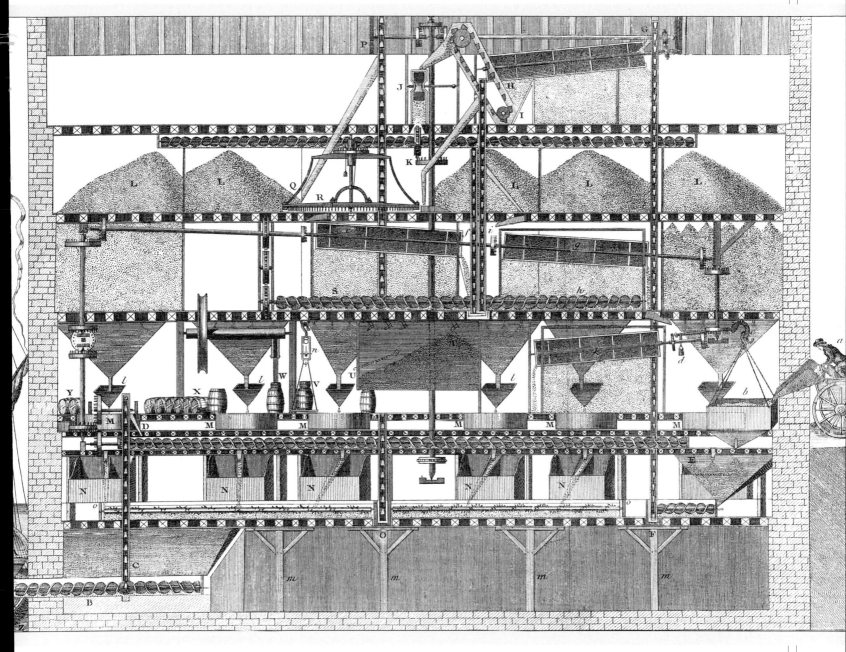

THE SHAPE OF THINGS TO COME: In a mill at Red Creek, Newport, Evans invented and installed history's first automatic integrated production line. Water-powered conveyor belts and bucket elevators carried wheat through the processes from boat or wagon through to flour, untouched by human hand. This diagram is of a mill built at Occoquan, Virginia, in 1795, for his friend Thomas Ellicott and described in Evans's *Young Mill-Wright and Miller's Guide.*

pressure steam in favor of condensed low-pressure steam. He did not know that Cornish tin mines had been pumped free of water for 50 years by a low-pressure steam engine, the critical strategic invention of an English country blacksmith, Thomas Newcomen. Nor did he know that two years before he was born, an engine like Newcomen's had been set to work pumping a copper mine in New Jersey by an English immigrant. When someone gave Evans a book describing Newcomen's engine, he was astonished—not by Newcomen's brilliance but by the sense of missed opportunity—and he was unabashed in his reaction: "He's doing it the wrong way!" He felt the same when he eventually heard about Watt's 1769 English patent for an engine that would be a major improvement on Newcomen's.

The Newcomen and Watt engines were both low pressure (see panel). What amazed Evans was that the low-pressure engine failed to exploit all the energy of steam when it was heated and confined at the kind of high pressure that blew the wadding out of the gun in the smithy. The young Watt had been excited like Evans about the potential of superheated high-pressure steam; he had mentioned the potential of steam for driving a steam carriage on the common roads in both his patents of 1769 and 1784, but he had concluded that steam expanded by heat beyond the boiling point of water would burst any conceivable container. Given the craftsmanship and materials of his day it was a sensible conclusion, but Evans, feeling his way in the dark, was neither discouraged by doubt nor stimulated by precedent. (Two other things he did not know were that the German engineer Jacob Leupold had incorporated a design for a high-pressure engine in his *Theatrum Machinarium Generale* [1723–39] and that in 1770, when Evans was 15, a French artillery officer, Nicolas-Joseph Cugnot, had used high-pressure steam to run a three-wheeled gun carriage, the world's first steam tractor, albeit discarded as a failure.)

Evans told anyone who would listen that a high-pressure engine could be half the size and weight of a low-pressure engine and deliver more power for half the fuel. His figures were a little optimistic, but the principle was sound. As early as 1778, he envisaged a high-pressure engine driving a paddle wheel boat as well as a wagon, predicting that a man then living would see the western waters covered with steamboats and that by steam power a child would be able to travel by land from Philadelphia to Boston in one day.

The youthful Evans experimented with models, but to attempt the powerful engine of his dreams required tools and ironwork of a quality he could not obtain or afford to ensure safety, not least in the middle of the Revolutionary War so hotly fought nearby around the then American capital of Philadelphia. Nor did he soon solve the problem of converting the up-and-down motion of a piston in a cylinder to a rotary motion. Watt did that in 1781, and in 1782–84 he used steam to drive a piston down as well as up—but it was still steam at low pressure. The difference between Evans and Watt was that for all his disappointments Evans never relinquished his faith in the feasibility of high-pressure steam. Of course, Watt was deflected by success. The partnership with Matthew Boulton of the Soho Works in Birmingham that made them both rich was for the production of low-pressure engines, and Watt remained convinced to the end of his long life that any man who tried to mobilize "strong steam" deserved a public hanging.

At 22, Evans got a job in a textile factory, drawing wire from a bar of iron and bending it into one of the teeth in the textile cards by which snags were combed out of wool and cotton. Happily, the tedium of the work proved subversive. Bored by the slowness of making teeth one by one,

he contrived a machine he was sure would turn out thousands by the minute. His father's family was sure he was off his head. So was the blacksmith whose help he beseeched. A justice of the peace in Newport, one George Latimer, heard of the nonsense young Evans was proposing and intervened—on his side. The blacksmith, cowed by Latimer's authority and cheered with a bottle of rum, made Evans's wireformer. Everything worked. The machine turned out teeth at the rate of 1,000 a minute, then 3,000. The buoyed-up Evans applied to the legislature for $500 to start a wire mill and drily recorded, "The committee, like the parson in the fable, lavished eulogiums, but would lend no money." He accepted $200 from the Wilmington card makers for rights to make more of his machines on condition he was given a share of the profits. The Wilmington men sold the basic idea of Evans's machine to other card makers without a cent more coming to him.

Around the time the British withdrew from Philadelphia, Evans opened a village store in 1782 on the Eastern Shore of Maryland at Tuckahoe. He and his younger brother Joseph were storekeepers for only a few months, but it was there, serving the farmers and millers, that he experienced an epiphany as significant as hearing about the corked gun barrel. He saw the methods by which millers turned wheat into flour and he was disgusted by the crudity, waste and dirtiness. A mill customarily employed four men and a "hopper boy" to clean, grind, cool, sift and pack. One strong man carried three-bushel sacks up ladder-like stairways and dumped the grain into a cylindrical "rolling screen" that freed the grain of its chaff and dirt. A boy raked the cleaned grain into a funnel or hopper leading to the second-floor millstones, driven by waterpower. The warm, moist ground meal was packed into buckets and hoisted to the third floor. It was spread out on the floor to cool and dry. A third man pushed the dried meal into a vertical chute. Gravity carried it into a cylinder covered in bolting cloth. Once sieved, the flour fell into a chest and a fourth man shoveled it out into barrels. In all this labor, dust was raised, grain was trampled underfoot and the flour quality was very mixed.

AMERICAN POWER: This drawing dated 1812 is the only known representation of Evans's definitive steam engine, the Columbian. The piston rod, extending up from the steam cylinder, is guided in a straight vertical line by the top beam C, which is pivoted at its left end and near its center. This arrangement is still called the Evans straight-line linkage.

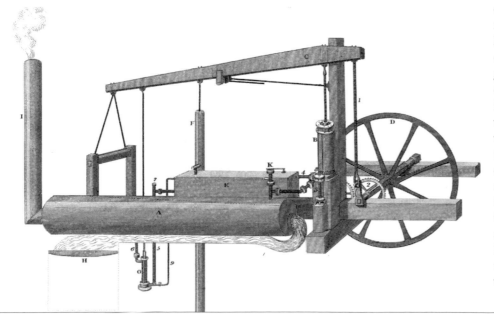

Evans was excited by the notion that he could do all this automatically in one continuous process, untouched by hand—or foot. With two brothers, he took over an old mill in a pleasant vale on the bank of the fast-flowing Red Clay Creek, Newport. They worked in secret for two years. His ideas proved excruciatingly difficult to execute, and they had to open in 1785 as a conventional flour mill. The newly married Evans lay in his bed at night running over and again in his mind how devices might be contrived, how the whole process could be kept going. He recalled seeing a picture of a chain pump for raising water out of a ship's hold. He could imagine lifting grain by the same principle in a series of small buckets fastened to a leather belt stretched over a lower and upper pulley. But how to lift and then disgorge grain without any human intervention was a problem of a different order. When finally a model functioned as he intended, he found it hard to get carpenters to translate sketch and model accurately.

Once the endless bucket elevator worked, emptying its contents on the top floor, he set about devising a "hopper boy" to take the warm, moist grain from the buckets, spread it on the floor and gradually rake it into the bolting hopper. But how could a machine both spread and gather at the same time? It was like being challenged to rub one's head and stomach in contrary motions. Evans wrote that it seemed absurd, "and the discovery caused months of the most intense thinking for the absurdity always presented itself to baffle and deter me." Months was a masterly understatement; it was at least seven years before he completed his five devices, including a horizontal conveyer using an Archimedes screw for the first time to move solids and a "drill" to push the grain along a trough into gravity-driven descenders.

The five devices were ingenious, but it was the manner of putting them together that was revolutionary. They constituted history's first automatic integrated production line. This was a conceptual leap from a single automatic machine. The textile, nail and card wire machines coming into use did not represent continuous production; the processes were interrupted to supply materials. In Evans's mill what went

RIVAL STEAM ENGINES

THE ENGINES put to work in America by Oliver Evans and in England by Richard Trevithick were *high pressure*, as distinct from the *low-pressure* engines patented by James Watt and manufactured in England by Matthew Boulton at his Soho, Birmingham, factory.

Watt's engine was a radical improvement on an engine made by Thomas Newcomen to pump water out of mines. In the Newcomen-type engine, steam was released into a cylinder to push up a piston. A jet of cold water condensed the steam, creating a vacuum under the piston. The pressure of the atmosphere on top of the piston, some 15 pounds to the square inch (psi), forced it back down; put it another way, it was sucked into the vacuum.

The inefficiency lay in the fact that the cold condensing water had to be heated up again to make steam. While that was going on, the engine was making no strokes; five or six cycles a minute was about the maximum. As a result, Newcomen's engine converted less than 1 percent of thermal energy into mechanical energy.

James Watt more than doubled Newcomen's efficiency by creating a vacuum in a condenser separate from the cylinder. Steam, having been driven up the piston, was sucked into this adjoining vacuum. In the cylinder, atmospheric pressure forced the piston down again.

Such low-pressure engines yielded only modest power. The only way to increase the power while maintaining the steam at 212 degrees was to increase the amount of steam for condensation, hence the size of the cylinder, the size of the boiler and the whole apparatus of heavy timber beams, pots and pipes. The idea that such a Brobdingnagian monster might power a boat or carriage justified the skeptics.

The high-pressure engine applied very hot steam directly to the piston, exploiting the principle that the power of steam increases geometrically with an increase in heat. Every extra 30 degrees of heat beyond the point at which water boils doubles the power of confined steam; doubling the heat increases the power of steam about 100 times. Watt believed such high pressure was dangerous because boilers could not be built strong enough.

The high-pressure engine working at ten times atmospheric pressure, 150 psi, could give ten times the power for half the size of engine. Oliver Evans wrote, "Steam presents us with a faithful servant, at command in all places, in all seasons; whose power is unlimited; for whom no task is too great nor yet too small; quick as lightning in operation; docile as the elephant led by a silken thread."

in as raw grain came out as smooth flour untouched by hand, and it was uncontaminated by hopper dust, footmarks or insects. A mill was no longer constrained by the number of three-bushel sacks of grain a man could carry on his back up a ladder. Evans's elevator belt of circulating wooden buckets dealt with 300 bushels an hour. One man could easily do the work of five.

It was a stupendous achievement. The economic advantages for all kinds of production became apparent only gradually, but for a miller the economic effects were immediately visible if he was prepared to

open his eyes—something too much of an effort for the self-satisfied Quaker millers of Wilmington whose large mills lined Brandywine Creek. A committee of them condescended to come to Red Clay Creek. Evans took care to be working in a neighboring hay field so they saw a "ghost" mill working without human intervention. He thought it would convince them to adopt his machinery, but they reported the whole contrivance "a set of rattle-traps unworthy of men of common sense." Evans explained the mechanisms in his book *The Young Mill-Wright and Miller's Guide,* and his originality was recognized by the grant of state licenses from the legislatures of Delaware, Maryland, Pennsylvania and New Hampshire, the only real protection an inventor could expect before the federal patent act of 1790. Still, the system was hard to sell. Evans had no success when his brother traveled through Pennsylvania, Delaware, Maryland and Virginia offering a license free to the first miller in a county who would adopt the Evans way.

The turning point was in May 1789, when Evans himself traveled to see the four Ellicott brothers who had mills on the Patapsco River above Baltimore. They were progressive Quakers, a cut above the diehards on the Brandywine. Evans made a sale and a useful discovery: Jonathan Ellicott had invented a screw conveyer that Evans was to adopt himself. When the Ellicotts used Evans's methods to make 325 barrels of flour a day at the then huge total saving of $32,500 a year, the Brandywine millers and others followed in lucratively large numbers—100 by 1792. George Washington bought a $40 license, valid for one pair of millstones, to automate the Dogue Creek mill on his Mount Vernon estate.

In 1791, all of 36 years old, afire with ideas and supporting a growing family, Evans moved to Philadelphia, one of the wisest things he ever did. It was the most culturally advanced city in America, full of skilled artisans. Evans opened a store with a blacksmith's shop in an annex and let it to "Messrs. Clark and Coffee, workmen of the first rate ability and integrity" with the idea that they would fabricate ironwork for millers and help with his own steam ambitions. The Pennsylvania legislature had liked what it heard about milling in 1787 but blanched when Evans, buoyed up by the committee's cordiality, began describing how he could propel carriages by steam on the public highway. "They conceived me as deranged, because I spoke of what they thought impossible." Maryland, however, tolerated him as a harmless eccentric. They gave him a steam carriage license for 14 years. They never thought he would exploit it, and nobody was willing to finance his attempts, but Maryland's sanction encouraged him not to give up altogether. When he went to see the Ellicotts, he showed them his drawings for a steam wagon. They declined to invest, but they

in turn endorsed his enthusiasm with a little trick they had worked out. It was water in a heated gun barrel again, but this time the steam ejected through a small hole caused a small wheel to revolve rapidly for about five minutes.

Despite the lack of support, Evans now had a stimulating environment and enough of his own money—$3,700—to try and turn his imaginings of 17 years into a versatile working steam engine. His machine tools were crude—a foot lathe to bore his cylinder, a drill and some blacksmith and brass-foundry equipment—but by 1802 his blacksmiths had turned the idea of his youth into a stationary engine with a 6-inch cylinder, 18-inch stroke. In February 1803

MANY FIRSTS: Evans astonished Philadelphia with his dual-purpose Orukter Amphibolos in 1805. It was a steam engine that moved along the road and into the river, where it became a steamboat. The two artist's conceptions were made at different times, but both show the Columbian engine.

readers of the Philadelphia newspaper *Aurora* were invited to visit Evans's store to witness "a new Era in the History of the STEAM ENGINE," which was Evans's new engine driving a screw mill to grind and break 12 tons of plaster of paris in 24 hours. He also excited "much attention," he tells us, by having his engine drive 12 saws in heavy frames to saw 100 feet of marble in 12 hours.

All of this was achieved in the face of scathing denunciations by "experts." Benjamin Henry Latrobe, the notable architect and professor of mathematics at the University of Pennsylvania, reported to the American Philosophical Society in Philadelphia that Evans's ideas about high-pressure steam were absurd; the society rejected the attack, though it accepted Latrobe's conclusion that steamboats would prove impractical. In 1803 a group of entrepreneurs with more faith contracted for Evans to build an 80-foot steamboat for the Mississippi; Evans was sure

the power of a high-pressure engine was best suited to overcome the strong currents of the Mississippi and the Ohio. In 1803 the engine was delivered to New Orleans for installation in the hull, but before the boat was ready a spring flood tide carried it half a mile inland. It was judged impractical to move engine and boat back, so the engine was taken out and put to good use in a sawmill.

In 1804 Evans won a federal patent for his engine. He longed to show off the elephant's tricks, and the Lancaster Turnpike Company offered a good target. It took the company three days to carry cargo the 80 miles from Lancaster to Philadelphia in six-horse Conestoga wagons. Evans told them

he could build a steam wagon that would carry 100 barrels of flour, as much as five Conestogas, in only two days. He would thus triple their profits.

The idea of steam on a turnpike startled the directors. The steam engines they knew about were low-pressure mammoths, heavy with brick ovens, iron pipes and pots. And something like that—if it were at all possible to put it on wheels—would simply frighten the horses. They rejected the proposal.

Evans got his chance when the Philadelphia Board of Health wanted the docks dredged. In March 1804 he proposed using a steam engine he would make and mount on a scow 12 feet wide by 30 long with a chain of buckets to bring up the mud. His engine was compact, delivering five horsepower from a cylinder of only 5 inches and a piston stroke of 19 inches, but the vessel he built weighed 17 tons. How could he get it from his shed to the water a mile and a half away? By steam power, of course. Evans put wheels on his scow and on a July day in 1805, the doors of his workshop opened and the extraordinary machine trundled out, circled the waterworks building and drove straight into the Schuylkill River. The steam, at a pressure of 30 pounds a square inch (psi), that had driven the road wheels now drove the paddles of the scow, which steamed downriver, leaving all the sailing craft behind. Thousands of astonished spectators gathered in Philadelphia's Center Square to see Evans and his machine rise up from the river and drive back to his shed.

His Orukter Amphibolos, as he chose to call it, meaning amphibious digger, was the first amphibian, the first steam dredger and America's first automotive vehicle. It was something of a stunt, in the judgment of technology historian Eugene Ferguson, because Evans did no more work on the vehicle, but it did dredge the harbor for several years, and Evans had demonstrated the principle that a steam engine of the necessary power could be made small enough to drive wheels on water and land. He drafted the prospectus for a limited company to build steam carriages, predicting engines "will propel boats against the current of the Mississippi and wagons on turnpike roads with great profit." He called his venture The Experiment Company, but he was under too much pressure to be able to take it further: Congress, succumbing to the machinations of various millers, failed to renew the 14-year patent for the automatic mill he had been granted in 1790. It was a stunning turn of events and he was understandably lugubrious about his prospects: "Having now made perhaps the greatest improvement, and most useful invention on steam engines ever produced by any one man, I expect to be attacked from all quarters: in every state in the union will, no doubt, be found one or more inventors who have made the same invention, as was the case with my improvements on merchant flourmills after they were published." He was right. Mechanics in Pittsburgh and elsewhere copied and adapted his engine without acknowledgment.

He was 50. His eyesight was failing. Legal bills were piling up in defense of his engine and his automatic mill. He had spent

his last dollar building his engine and working on a second book, his *Young Steam Engineer's Guide*. In frustration, he abandoned it and published part of it under the title *The Abortion of the Young Steam Engineer's Guide*. This remarkable book included as an afterthought a description of a vapor-compression refrigeration system, including an expander, cooling coil, compressor and condenser. (This was almost 30 years ahead of Jacob Perkins, who took out a patent on just such a device in 1834. It may just be a coincidence that he knew Evans in Philadelphia, but it was all there in the book Evans put out in 1805.)

It took Evans three years of campaigning to have his legitimate rights restored. In 1808 Congress passed an act for the Relief of Oliver Evans, signed into law by President Jefferson. Evans could now demand recompense from all the millers who had profited by his innovation. His bitterness expressed itself in raising his license fees. The Baltimore millers, in particular, continued to defame him: "Few if any of these [Baltimore] millers," wrote a merchant, "are inclined to give the pompous blockhead, Oliver Evans, the credit for inventing any of the useful contrivances in milling for which he now enjoys a patent." Evans had to go back to court. He prevailed in an affirmation of his originality, though Congress, in extending his patent for a further 14 years, specified he must not raise license fees any more.

Evans, now at last growing richer every year, spent his new money and his energies building the first steam engineering factory, the Mars Works, for the production of steam engines, typically small cylinders and a long stroke. In 1811–12, with his son George as partner, he opened the first steam engine manufactory in the

West in Pittsburgh. Altogether Oliver and George profitably built 100 stationary high-pressure steam engines for American workshops throughout the Atlantic states. Ranging in size from 8 to 70 horsepower, they rolled iron, milled flour, made paper; one was installed in the Navy Yard in Washington, D.C., and another (operating at 200 pounds of pressure) in the Fairmont Waterworks in Philadelphia. After the New Orleans fiasco, Evans did not involve himself in steamboat navigation again, but his type of engine came into general use on western waters. He saw his engines with 150 psi drive big steamboats running at nine miles an hour. He would have won a beaver hat if he had taken up Robert Fulton on his friendly challenge in 1812 to run a boat in still water at nine miles an hour; Fulton maintained that eight pounds of pressure and a speed of six miles an hour were about the upper limit.

Evans remained prolific to the end of his life. He devised a solar boiler, a flour press, a gas lighting system, a multiple-effect evaporator and a machine gun. He invented and installed a central hot-air heating system in a Middletown, Connecticut, textile mill, feeding the exhaust heat from a steam engine into a network of small radiators to keep the building warm. In 1814, during the war with England, he outlined for President Madison an invincible steam-powered ironclad warship that was superior to Robert Fulton's floating fortress or anything built until John Ericsson's *Monitor* in 1862. His impulse was always to follow the ideas in his teeming brain wherever they might lead rather than to follow the money. He became convinced that to move forward faster America needed government sponsorship of research.

Evans died in his 64th year. He was buried at Zion Episcopal Church near the Bowery in lower Manhattan; when the church was sold, he was interred in a crypt with others at Murray Hill, and when that was sold, his body was moved to an unmarked common grave in Trinity Cemetery, Broadway, at 157–158th Street. Here, somewhere, lies Oliver Evans, vindicated by history but insufficiently remembered by historians and the fellow Americans he longed to benefit.

HENRY MILLER SHREVE
Legal and physical barriers to navigation impeded the growth of the West until the "master of the Mississippi" freed the waterways

1785–1851

Captain Henry M. Shreve invented the snag boat to batter and saw through the tangled "rafts" impeding and endangering navigation on the great river systems. Lloyd Hawthorne's painting portrays him clearing the Great Raft on the Red River, 1833–38.

CAPTAIN HENRY SHREVE spent much of his life with his hand on the helm of one boat or another, navigating the frightening wilderness reaches of the Mississippi. Like the river he mastered, he was a man of long silences and sudden energies. In the eyes of Robert Fulton, he was an outlaw. To the river towns of the West, he was a hero. He defied the Fulton-Livingston monopoly on the lower Mississippi, his boats were seized and his spare, brooding figure, five foot eleven inches topped off with a black Quaker hat, became familiar striding along the levee to the courthouse in New Orleans. Fulton and Nicholas Roosevelt were the first pioneers on the lower Mississippi, but as a captain, litigator and innovator, Shreve made his own singular contribution to steam navigation and to the opening of the whole Ohio-Missouri-Mississippi system so essential to the rapid growth of the West.

Rivers and adventure were his life from the age of three, when in 1788 his father, having fought as a colonel in the Revolutionary War, led a party of 29 across the Delaware to settle with his wife and six children on the western Pennsylvania frontier. The farm was close to the confluence of the Allegheny and the Monongahela flowing into the Ohio. Every day the growing Shreve saw the keelboats, crude rafts, barges and flatboats drifting west, packed with emigrant families, cattle and horses, and frequently he watched also, rifle ready, for Indian war parties. (Flatboats were downstream flood-tide vessels, sold for firewood at end of a run. The keelboats and barges could be towed upriver by men on the bank, or poled in shallower water or could catch a favorable wind with a sail.) His father died when he was 13, and after a year or two helping out on the farm, the boy took a job loading river cargo. By the time he was 21, he had saved enough to build a boat of his own, a 35-ton keelboat.

What Shreve did with his keelboat was prescient. He had read a newspaper account of the rich prospects for fur trapping on the upper Missouri reported by the Lewis and Clark expedition, and realized that beaver pelts, in particular, offered the prospect of a cargo of high value with little bulk. Instead of taking his boat on the conventional trading route from Pittsburgh down the Mississippi to New Orleans, he determined to go upriver to St. Louis.

He hired ten tough French boatmen, "half-horse, half-alligator" fellows with muscles hardened by years of dragging and poling boats against the currents. He set off down the Ohio in October 1807, winning the respect of the crew with his knowledge of the channels and shoals. Some three weeks and 1,132 river miles later, he reached the Mississippi, "the great Mississippi, the majestic, the magnificent Mississippi, rolling its mile-wide tide along, shining in the sun." Mark Twain's majesty was for Shreve and his men a homicidal maniac. It took a frantic effort for them to resist being swept downstream by the headlong current. As a result, the remaining 170 wilderness miles to St. Louis took them as long to navigate as the Ohio, and every stretch threatened danger from pirates and hostile Indians and sandbanks and submerged tree stumps that could sink a keelboat in a flash.

Shreve arrived early in December in St. Louis, then no more than a village seven blocks by nine. The bales of beaver and otter pelts he took back on the long strenuous pull up the Ohio to Pittsburgh, and then overland to Philadelphia, were the first furs to reach the city from St. Louis and very profitable. Others followed his path. In 1809 he took his boat even farther north up the Mississippi to trade Pittsburgh manufactures for tons of lead from the Sac Indians, a business until then monopolized by the British trading whiskey and rum. His profit on that single run was a staggering $11,000. Two years later he launched a 95-ton keelboat for his now flourishing business between Pittsburgh and New Orleans, but late in 1811, as he sweated slowly up the Ohio with a crew of 40, he saw a cloud on his future: smoke rising from the *New Orleans* on its maiden voyage, waiting in Louisville for a crest of water to run the falls.

Shreve had taken a look at the *New Orleans* when it was being built in Pittsburgh and had irritated its builder and captain, Nicholas Roosevelt, by doubting the capacity of the low-pressure 24-horsepower engine for the tough currents of the Mississippi. But he knew he had to switch to steam. He joined a group of men at Browns

ville, led by the mechanic-inventor Daniel French, to build three small low-powered steamboats, and he captained the biggest of them, a 75-ton stern-wheeler called the *Enterprise,* on an urgent mission to New Orleans. General Andrew Jackson was improvising a defense of the city and the Mississippi Valley against a surprise attack by British forces in the War of 1812. Three keelboats with supplies had left Pittsburgh in October. Shreve, loaded up with ordnance, got the *Enterprise* to the besieged city ahead of them and was immediately pressed into war service by Jackson, thereby escaping arrest for infringing the Fulton-Livingston monopoly. Shreve went back upriver to find the keelboats, then evacuated women and children 50 miles to an upstream plantation. On return, he accepted Jackson's dangerous charge to ferry supplies to Fort St. Philip in the delta below New Orleans. He armored the sides of the *Enterprise* with cotton bales and succeeded in creeping through both ways under cover of darkness and fog. When the British attacked the city in close ranks on January 5, 1815, Shreve was on the breastworks with the Tennessee and Kentucky marksmen who cut the enemy down for a decisive victory in the last major engagement of the war.

On the other hand, the battle with the Fulton-Livingston interests, represented in New Orleans by Robert Livingston's lawyer brother, Edward, had just begun. On May 6, 1815, as Shreve prepared to cast off, he was served legal notice to stay in port. Another steamboat of the Daniel French group, the *Despatch,* was seized loading a cargo of sugar and molasses for the Ohio. The captain had no bond to offer and was compelled to yield the ship. Shreve was shrewder. Quietly, in 1814 he had hired Abner L. Duncan, the leading lawyer in New Orleans. When the marshal served the papers on Shreve, Duncan had already posted a bond, and Shreve was free to adventure the *Enterprise* northward against the booming spring flood tides to Louisville. He reached Louisville in 25 days and then Pittsburgh in another 9. He was a pathfinder again. This was the first time a steamboat had ascended the rivers.

Not yet 30, Shreve made a decisive move for the future of the western steamboat. When French would not join him, he went ahead on his own and built the first steamboat with a high-pressure noncondensing engine. It was rated at 100 horsepower, four times the power of the *New Orleans*. He set the engine horizontally on the main deck and connected it directly to the paddle wheel instead of through a beam or flywheel, establishing a pattern for a generation of western steamboats both in the style and the placing of the engine. It was the largest steamboat yet built, a 400-ton sidewheeler, 150 feet long, and he called it the *Washington* in tribute to his father's friend. He put the passenger cabins on the second deck and topped it all off with twin smokestacks and a pilothouse, a double-deck pattern that developed into a distinctive profile on the Mississippi. Shreve risked a high-pressure engine because the risks of navigating the river with a low-powered engine were in his experience so much greater. He knew that only with a powerful engine could he hope to ascend safely against rapid currents, churn through sud-

den shoals and bars, dodge lethal surfing debris and evade other vessels, particularly in the narrow and short channels he called "chutes" where boats sometimes collided. He had been right about the *New Orleans*. The weakness of its low-pressure engine confined the vessel to the New Orleans–Natchez trade, albeit profitably.

Shreve was inspired by Oliver Evans (page 36), who had established the first steam engine manufactory in the West in Pittsburgh in 1811–12 in partnership with his son George. Evans had demonstrated that the high-pressure engine cost half as much to build as a low-pressure condensing engine, took less space and weighed a fraction of the low-pressure engine, a more important consideration for navigators of the shallow waters of western rivers than the extra fuel burned. Thus Shreve's *Washington* was only 5 tons compared with around 100 tons for the engine on the Fulton group's successful *Aetna* (360 tons, 1815). Shreve and Evans seemed to have

been proved wrong on the trial voyage in July 1816. One of the boilers exploded, killing 14 men and hurtling Shreve overboard. He recovered to the jeers of the naysayers, found the problem in his design of one safety valve, remedied it and had the *Washington* ready to embark for New Orleans on September 24, 1816.

Unsurprisingly, no passengers bought tickets. Unsurprisingly also, Edward Livingston had a marshal on the levee to arrest Shreve when the *Washington* docked in October. Livingston had already compelled the *Constitution* to leave New Orleans without her cargo. On Duncan's advice, Shreve refused bail and was led off amid angry cries from townspeople. A jury found for him, and then Livingston made him a proposal. If he would stop his defiance and place his boat in the Fulton-Livingston line, they would reward him with a half interest in the whole company. Such an offer promised Shreve a comfortable life. He declined. Another court battle in New Orleans in March 1817 was preceded by mass meetings in the river towns denouncing the eastern "swindlers" attempting to limit the road to market on "the common highway of the West." Judge Dominick A. Hall, of the U.S. District Court for the Louisiana District, no doubt took heed of the clamor. In April, he dismissed the case against Shreve on grounds that his court had no jurisdiction.

Shreve was a sensation on the return upriver to Louisville. He completed his memorable voyage in only 25 days—three months faster than keelboats and ten days faster than the more heavily laden *Aetna*. "This was the trip," a writer said in 1819,

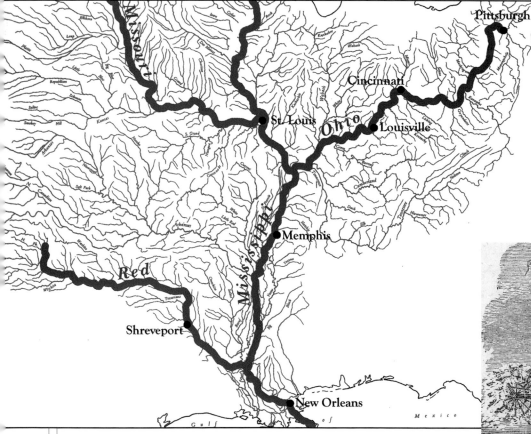

**THE RIVER ARTERIES OF AMERICA: Going upstream was the problem.
Fourteen died when Shreve's revolutionary *Washington* (right)
blew up on a trial voyage in July 1816.**

"that convinced the despairing public that steamboat navigation would succeed on the western waters." Merchants and migrants found cargo costs halved and halved again, fully vindicating John Fitch's unbelieved predictions of 1787. For Shreve it was an astonishing business. Two round trips from New Orleans to Louisville enabled him to pay the capital cost of the boat and its running costs, and still give a surplus of $1,700. He built a line of steamboats; in 1824 one of his vessels, the *President,* fulfilled his ambition to reach Louisville in only ten days out of New Orleans.

Profits running at 40 percent, and feebler enforcement of the monopoly, led to a boom in steamboat building with high pressure engines. In 1817 there were 17 vessels on the "high seas of the western country"; in 1820 there were 69 — and hundreds more in the wake of the Supreme Court ruling in 1824 ending all legal restrictions on freedom of movement.

What the highest court rulings could not do was clear the rivers of physical obstructions. When from time to time currents undermined the banks, thousands of trees plunged into the river. Some (called "planters") became anchored in the mud of the riverbed, concealed beneath the surface like huge lances. Others (called "sawyers") bobbed up and down in the channels. "Snags," as the obstructions were called, accounted for three-fifths of all steamboat accidents. No fewer than 58,000 snags were identified in the lower Ohio, the Mississippi and the Missouri and Arkansas rivers. For stretches, great accumulations of lumber and driftwood blocked passage altogether and diverted major rivers into unnavigable channels.

It was Shreve who came to the rescue. Appointed superintendent of western river improvements in 1826, when he was 41, he went at the dangers and blockages as he went after everything else. He thought hard and then hit hard: He invented the snag boat. His first and the model for a generation was the *Heliopolis,* which he unleashed in 1829, built in just one year after Congress voted the War Department funds for river clearance. It was a steam-powered monster with giant claws, cranes and a heavy battering ram of an iron-sheathed snag beam. Huge trunks some-

The meaner western steamboats merited the definition of "an engine on a raft with $11,000 worth of jig-saw work." But the grander vessels were floating palaces. This is the *Island Queen* excursion steamer leaving Louisville, Kentucky.

times six feet in diameter were prodded up and fed into a powered sawmill on deck. Shreve successfully directed hundreds of men for several years. By 1832 not a single boat was lost to snagging on the Ohio-Mississippi rivers. Passage to Louisville from New Orleans was cut by days; countless lives were saved.

The "Great Raft" of Louisiana's Red River was a more stupendous challenge. As historian Edith McCall writes, Shreve stood "almost alone" in thinking it could ever be cleared. Hundreds of years in the making, it was a tangled mass of decaying wood and brush stretching bank to bank for 160 miles with few breaks from the present-day Alexandria, Louisiana. No canoe, white or Indian, had penetrated it for 50 years. It was so solid in parts that hunters and horsemen crossed it unaware of the stream beneath them. Shreve built a compact twin-hulled snag boat, the *Archimedes,* recruited 160 men, three steamers and a dozen flatboats, and flung his armada at the chaos.

Mosquitoes and alligators plagued them in the swamps. Four men died of cholera.

In the first season, Shreve cleared the lower Red of hundreds of obstructions for 70 miles so it could run freely to the Mississippi; settlers followed the snag boat. By the last of his five seasons Shreve had cleared 300 miles of Red River, opening up some of the richest corn and cotton land in the nation. Congress maddened him by intermittently closing off funds for the work. In April 1838, when he arrived to find work stopped and seven steamboats stuck in the final stages, he rode his horse into the town of Washington, Arkansas, persuaded the bank to let him have $7,147.50 for the work and had the river flowing freely by May 4. It was a satisfying moment when steamboats passing upriver gave a celebratory whistle as they passed his work camp on Bennett's and Cane's Bluff. Soon, the camp would be called Shreveport, the Confederate capital at the end of the Civil War and in modern times a thriving city, the second largest in the state.

THE STEAMBOAT WAR
that opened the country to competition

HENRY SHREVE'S REVOLT IN the West against the Fulton-Livingston monopoly had the support of the river towns. In the East, states bordering New York State were up in arms at its presumption in deciding who could or could not operate steamboats on contiguous waterways. Several states retaliated by banning entry into their ports of any "ships operated by fire or steam" that operated under a license from New York. Enmity between New York, New Jersey, Pennsylvania, Connecticut and Ohio over the monopoly was so bad, said one lawyer, that they were "almost on the eve of civil war."

Four characters figured in the denouement, which had profound implications for the future of business in the United States. They were Thomas Gibbons, a rascally, quick-tempered Georgian lawyer notorious for being a Loyalist in the Revolutionary War; Aaron Ogden, a Revolutionary War hero and one-year governor of New Jersey; an unknown coarse young steamboat captain by the name of Cornelius Vanderbilt; and a brilliantly clear-sighted innovator, Chief Justice John Marshall. Fire and ice.

Gibbons (1757–1826) never bothered to correct the impression that he had killed a man in a duel; in fact, both duelists missed their shots, but Gibbons always had murder in his heart. In 1817, while living in New Jersey, he acquired a little steam ferry, *Mouse of the Mountain,* which he was able to run freely on the Raritan River since its waters did not cross state lines. His ferry from New Brunswick to Elizabethtown Point near the northwest corner of Staten Island connected with a service operated by Ogden (1756–1839). Ogden paid a license fee to the Fulton-Livingston monopoly so that his steamboat, the *Atlanta,* could take passengers on New York waters to and from Elizabethtown Point and New York. Gib-

bons and Ogden formed a partnership, but it was short-lived. Gibbons, whose wife had walked out accusing him of keeping prostitutes in the family home, had turned angrily on his son-in-law, saying he had seduced his daughter before marriage. Ogden tried to step in as a friend of the family. Gibbons exploded at this interference. He went round to Ogden's home and posted a menacing note. In turn, Ogden had Gibbons arrested for trespass. Naturally, Gibbons challenged Ogden to a duel. Ogden declined, so Gibbons came up with another reprisal: He would run his *Mouse of the Mountain* across New York Bay along the same route plied by Ogden's *Atlanta.*

Ogden went to court in New York, asking for an injunction, and Gibbons argued that he had every right to sail across the bay. He had a United States coastal license, granted under the Federal Coasting Act of 1793. This, he claimed, allowed him to "navigate the waters of any particular state by steamboats." Predictably, the New York courts ruled against Gibbons. Enraged once again, he soon found someone who relished a fight as much as he did. Cornelius Vanderbilt (1794–1877), on his way to becoming a legend, was only 24. With $100 advanced by his parents, when he was barely 16, he had bought a small sailing vessel; at 23 he had three schooners. To everyone's astonishment he sold up to work for the volatile Gibbons. Vanderbilt drove the crew of Gibbons's little *Mouse of the Mountain* relentlessly and once it made a profit persuaded Gibbons to invest in a bigger and grander vessel, named *Bellona* after the Roman goddess of war. That, too, was soon the subject of a court injunction.

Vanderbilt had been given the honorary title "Commodore" for operating a ferry between Staten Island and the Battery in lower Manhattan during the War of 1812. If running the British blockade had not scared him, a New York court order was a

breeze. He and Gibbons kept on sailing. New York officials were hesitant to call their bluff by seizing the *Bellona,* lest New Jersey officials seize a New York ship. Gibbons remained safely outside New York jurisdiction in his New Jersey home, so New York went after Vanderbilt. State officials spent months trying to arrest him as he docked the *Bellona.* Robert Albion writes, "With impudence and resourcefulness, he managed to land his passengers at one place or another along the New York waterfront, despite all the efforts of sheriff and process-servers." Vanderbilt constructed a secret cabin in the ship where he could hide should New York officials decide to board ship. One Sunday officials pounced on Vanderbilt when they caught him lounging on a New York pier. After hauling him to court, they were forced to let him go; it turned out Vanderbilt had not been piloting the *Bellona* that day.

The confrontation escalated. Vanderbilt kidnapped one New York deputy and threatened to send him to New Jersey. Meanwhile, Gibbons convinced the New Jersey legislature to authorize the arrest of anyone from New York who arrested a citizen of New Jersey for violating the monopoly. He next persuaded New Jersey legislators to invoke their right to regulate travel on the lower Hudson. In 1824 the legal suits ended up in the U.S. Supreme Court before Chief Justice John Marshall (1755–1835) amid excitement and puzzlement about how even America's most celebrated judge could untie the knots. Could Gibbons's U.S. federal license override New York State's license? What about Ogden's right to protection of his contract with the monopoly? By what right anyway could the Supreme Court decide between two states, each with its sovereign powers? The sharpest legal minds in America were engaged in the battle. Gibbons hired Daniel Webster, then 42, and William Wirt, 52, a former United States attorney general. Ogden was represented by two other famous lawyers, Thomas J. Oakley, an attorney general of New York, and Thomas A. Emmet, the burly, eloquent Irishman who had been saved from drowning by Robert Fulton.

Marshall was a figure out of American myth, born one of 15 children in a log cabin in the foothills of the Blue Mountains, a soldier in the Revolutionary War who rose by sheer merit to be secretary of state to President John Adams. Adams had appointed him chief justice just before relinquishing the presidency to Thomas Jefferson, who distrusted Marshall's cleverness in the cause of the federalism Jefferson detested. "When conversing with Marshall, I never admit anything," wrote Jefferson to James Madison. "So great is his sophistry, you must never give him an affirmative answer or you will be forced to grant his conclusion. Why if he were to ask me if it were daylight or not, I'd reply, Sir, I don't know, I can't tell." Marshall's contemporary William Wirt described the chief justice as tall, meager, emaciated, loose-jointed, inelegant in dress, his attitudes pervaded with great good humor and hilarity, "while his black eyes possess an irradiating presence which proclaims the imperial powers of the mind that sits enthroned therein." In 23 years as chief justice, Marshall had led a united court in expansion of federal power at the expense of the states, but the court had also invalidated laws that might encourage wildcat competition and it had treated contracts as sacrosanct.

Arguments in the case, which began on February 4 before a packed court, centered on Article I, Section 8 of the Constitution, known as the commerce clause, which gives Congress the "power . . . to regulate commerce . . . among the several states." It was far from settled in 1824 whether Congress had an exclusive power and whether it was confined to the regulation of tariffs, customs and other measures imposed at interstate boundaries. Emmet, for Ogden, argued that granting Congress exclusive power would "make a wreck of state legislation." He observed that states had always exercised control over aspects of commerce and cited several examples, including the

traffic of slaves, quarantine laws, inspection laws and the regulation of trade with Indians. Since states had concurrent power to regulate what the federal government failed to regulate, Emmet claimed, the New York monopoly was fully valid. He concluded with a quote from the *Aeneid,* Virgil's famous epic of the foundation of Rome, showing that New York State had, as M. Baxter puts it, "wisely fostered Fulton's genius and thereby promoted the interests of the nation."

Webster stuck to one simple course throughout: Congress and only Congress had the right to regulate interstate commerce. He declared, "Henceforth, the com-

JOHN MARSHALL, INNOVATIVE CHIEF JUSTICE: His steamboat ruling was seminal for federal initiatives over a vast range, from railroad regulation to minimum wages and the outlawing of racial segregation.

merce of the States was to be a unit." Its commerce "was to be described in the flag which waved over it E PLURIBUS UNUM." Webster hammered repeatedly at this idea: "All useful regulation does not consist in restraint; and that which Congress sees fit to leave free, is a part of its regulation, as much as the rest." In other words, just because Congress did not mention it did not mean it could not regulate it. Webster argued for what was called exclusive power theory over the concurrent power theory used by Ogden's lawyers. The case for Gibbons was closed by Wirt, who had first come to fame as a prosecutor in the treason trial of Aaron

Burr. His main thrust was that Gibbons's coasting license gave him the right to sail anywhere in the United States, including inland waters. Although there was little in the statute to support this interpretation, it impressed observers. Chancellor John J. Crittenden wrote, "I heard from Wirt the greatest display that I have ever heard at the Bar since the days of Patrick Henry."

Marshall was ruminating on the case at his boardinghouse ten days after arguments ended, on February 19, 1824, when in the darkness he slipped on ice over a cellar door. He was "deprived of his senses for a quarter of an hour," concussed and also suffered a dislocated shoulder. He was in his 69th year and when he finally appeared on March 5, his arm was in a sling. It took him 45 minutes to read the unanimous opinion in a voice that was so low and feeble the courtroom had to crane to hear what he said, but it resounded across the land and down the years. Marshall's judgment was carefully defined. He decided the case not by abolishing the notion of concurrent power but by accepting the validity of Wirt's broad interpretation of Gibbons's coasting license. It was, however, a judgment typical of the way Marshall carefully planted acorns he hoped would grow into great oaks. The judgment came to be read as meaning that none of the several states was entitled to impose burdens on any business or in any way to restrict interstate commerce. It was more than a triumph for Gibbons. It was the emancipation of American business, a potent factor in the fusion of the American people into a single nation—and a single marketplace. The immediate effect was to open the waterways of America to all kinds of steamboat entrepreneurs at a time when the waterways were the main arteries for trade—by November, 46 steamers were plying New York waters by comparison with 8 in March.

Ogden found his license was worthless and competition escalated. He lost his home and in 1830 was imprisoned for debt for a time. Gibbons died rich and friendless. And Vanderbilt, of course, went his own ferociously innovating way for the next half-century in steamships and railways, to be richer in the 1870s than everyone else in America combined.

ELI WHITNEY

In ten days on a plantation he made a model of a machine that changed the South forever—the cotton gin. Then he became the godfather of the machine age

1765–1825

Whitney's gin exploded the production of cotton—and the import of slaves. In 1791, the entire U.S. cotton output was 4,000 bales. It more than quadrupled in four years, and in 1801 was an astounding 100,000 bales. The gin increased labor productivity 50-fold, but more hands were needed for the millions of acres planted with the new cash crop. By 1810 more than 200,000 Africans had been forced into slavery in the South. In New England, the Whitney effect was heard in the ever-busier whirr of the carding and spinning frames in the new textile mills.

THE FARM BOY Eli Whitney did not learn to read quickly or easily, but he understood figures very well. He was just a boy at the start of the Revolutionary War, and it gave him an early schooling in the opportunities of commerce. England's ban on trade with the American rebels led to all sorts of shortages, and nails, normally imported, became hard to find and expensive. Eleven-year-old Eli put his big hands to good use in his father's forge on the family farm in Westboro, Massachusetts, making all the nails they needed, and then he

CATHARINE GREENE (1755-1814): The heroine of the Revolutionary War who was Whitney's patron. He idolized her for years; only after her death did he marry.

ventured into the countryside to sell them at a profit.

By 15, he was an embryonic capitalist. Having seen what the market would pay, and what he could produce on his own, he calculated that his profit margins justified taking on labor. He did not tell his father what he was up to when he borrowed the family horse and rode off to find some able person he could afford. He was away three days; the following week he had a man making nails with him at the farm, the cost of his room and wages carefully included under "overheads." It was a good little business until the end of the war, when England dumped nails by the ton. Whitney switched to making hatpins. He had noted the new vogue for ladies to hold down their bonnets with three pins instead of silk bows tied under the chin. His cut-price pins secured

him a near monopoly in his section of the state, but, acutely aware how vulnerable he was to the vagaries of fashion, he opened a line in walking canes as well.

Such attention to potential vulnerabilities was natural to Whitney, motherless from the age of 12; she had died giving birth to his brother. At 19 he resolved to qualify himself for better things with a college education. His father thought it was too late for that and the family budget could ill afford the fees, but Eli took jobs teaching in several local schools, earned enough to pay for tuition at Leicester Academy and won admittance to Yale in May 1789. He was 23, geriatric for a Yale student. He helped to pay his way by repairing equipment around the college and graduated in law when he was 27. He was a late entrant for a legal career, without a glittering enough record or money to buy into a partnership, but his decision to go to Yale proved the wisest of his life. It admitted him to the magic circle of Yale alumni.

Another Yale graduate, Phineas Miller, one year older, put him in touch with a plantation owner in South Carolina who was looking for a family tutor. Whitney set off from New Haven in 1792 full of apprehension. He was seasick on the packet to New York, the boat ran on the rocks at Hell Gate and then, within an hour of arriving exhausted in the city, he shook hands with a friend who had smallpox. He had to take the precaution of undergoing "variolation," infection with a mild version of the disease, the precursor of Edward Jenner's introduction of vaccination with cowpox (1798), which left him with "only a dozen pock." For a world-class worrier, it took courage to get on a boat again for the long voyage from New York to Savannah, which he had heard was rampant with all diseases. He was very seasick again—"eat nothing but what I puked up immediately"—but cheered by the companionship onboard of Miller and Miller's employer, Catharine Greene.

Mrs. Greene is more important in the Whitney story than commonly acknowledged. She was a vivid heroine of the Revolutionary War. She was married to General Nathanael Greene and stuck it out with the remnants of General Washington's army during the bitter winter at Valley Forge in 1777–78; she won Washington's

devotion and his gratitude for the way she kept up spirits among the starving and the sick. General Greene had pledged his own land and fortune to supply and clothe his soldiers. After the war, he was awarded confiscated land north of Savannah, which became known as Mulberry Grove Plantation, though the government did not otherwise liquidate its debt to him. When he died, cash poor, at the age of 44, Catharine, then only 31, fought a dilatory Congress six years for the money owed to him; she had five children to bring up. This grant of $47,000 had just been agreed upon when Miller introduced her to Whitney, and he was immediately entranced. Miller, formerly a tutor to the three Greene daughters, managed Mulberry Grove for her; unbeknownst to Whitney, Miller and Widow Greene had just become secretly betrothed—secretly because they did not want to risk complicating the grant from Congress.

Greene invited Whitney to stay on her plantation until he crossed the river to his teaching job. Whitney was overjoyed by life at Mulberry Grove. He had exchanged the chores of a frosty New England farm and the spartan chills of a room in New Haven for a lazy lotusland of servants, sunshine and exotic flowers and fruits. He marveled at the oranges, pomegranates, olives, figs, pecans, nectarines and peaches, but most of all the company, which for a time moderated his misanthropic view that it was "a damn'd kind of world." He con-

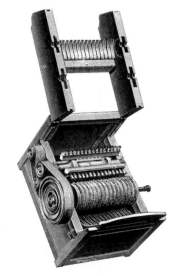

WHITNEY'S COTTON GIN: He spent years of his creative genius fighting pirates.

trived an excuse about his pay not to take up the South Carolina tutoring position, and the Miller-Greenes were glad to have him around, fixing things. Among other things, he made Greene an embroidery frame.

Out of this relationship came Whitney's first great invention, as profound in its effect as it was simple in its conception: the cotton gin (short for "engine") that is part of every schoolchild's catechism. Cotton grown on the coast was profitable, since seed and long-staple lint were amenable to separation when put through two rollers like a clothes wringer, a roller gin. But these varieties of cotton would not grow inland. Instead, a different variety grew profusely, a short-staple cotton with sticky green seeds that defeated the roller gins. Green-seed cotton had to be picked by hand, one man needing a whole day to separate a single pound of lint from three pounds of cotton. Whitney heard from Greene and Miller and the gentry of the state who dropped in for leisurely visits how impossible it was to make a living from green seed. What a pity it was, everyone sighed over the Madeira, that there was no easier way of extracting the cotton. Greene is supposed to have exclaimed, "Gentlemen, apply to my young friend, Mr. Whitney. He can make anything."

It took Whitney all of ten days in the barn to make a little model of the machine that was to revolutionize the South and galvanize the textile factories in the North. It was a simple hand-cranked revolving drum cylinder with hooks to pluck the fiber from the seeds. The restraining wires of a sieve held the seeds back. A faster rotating brush cleaned the lint. One slave could turn it by hand and produce ten times as much cotton as another picking by hand, and it was cleaner, too. One man with a gin driven by a horse or waterpower could produce more than could 50 hand pickers.

There are many stories about how Whitney devised his model gin. The most romantic he retailed was that he watched a cat shoot out its claws through a fence to catch a chicken and retrieve only a pawful of feathers. Another version has him watching a slave for hours and then replicating the movements as the man held the seed with one hand while teasing out the cot-

At his back Whitney always heard disaster hurrying near: "If I should have good success, I shall make something handsome, but I may have bad luck and lose all. The world is a Lottery in which many draw Blanks, one of which may fall to my share."

ton with the other. He described how he came to make the teeth out of wire: "One of the Miss Greenes had brong out a coile of iron wire to make a bird cage and being embarrassed for want of sheet iron and seeing this wire hung in the parlor, it struck me I could make teeth with that." Greene herself is said to have suggested the brush, but her indisputable and essential contribution would be to finance the firm Miller and Whitney set up to exploit his invention.

When Whitney showed off the model to a few of Greene's friends, one of them offered him 100 guineas to buy it outright. Everybody tut-tutted about the need for secrecy. It was pious nonsense—it was like inventing the wheel and expecting nobody to talk about it. Word spread like fire and at one point Whitney's little barn workshop

was raided. He was still deluding himself that his secret was safe when a month after proving his first full-scale machine, in May 1793, he went to Philadelphia to petition Secretary of State Thomas Jefferson for a patent, and then to New Haven to set up a factory. In a letter to his father explaining an uncharacteristic silence of five months, he wrote: "I wish you, Sir, not to show this letter nor communicate anything of its contents to any body except My Brothers & Sister; enjoining it on them to keep the whole a profound secret."

Green-seed cotton was quicker than patent clerks. By the time President Washington signed the patent authorization on March 14, 1794, backdated to November 6, 1793, hundreds more acres had been planted in anticipation of being harvested by the amazing gin. Instead of selling the

MULBERRY GROVE: Whitney made a revolution in the little barn. The drawing was made in 1911 from someone's recollection; the river had no steamboats when Whitney was there.

gins or licensing their manufacture, the partners asked planters to bring their crops to ginning stations they set up. Two things went wrong, first bad luck, then bad judgment. Whitney performed wonders creating tools and machinery to make the gins in New Haven and bag the cotton, but disasters prevented him from manufacturing gins fast enough for the astounding avalanche of seed. Scarlet and yellow fever epidemics swept New Haven in 1794, forcing him to close the plant; then a fire in March 1795 destroyed 20 finished machines and his original tools. The efficiency of his manufactory, when it was operating normally, is manifest in that only seven months after the fire he had retooled the factory and shipped 26 machines.

The misjudgment was Miller's idea that planters would pay one-third of the profit to have their cotton ginned for them. The charge was too high and resentment sanctified piracy. Damned Yankees! What was this newfangled patent law anyway? It was natural that the frantic planters, fearing to miss a license to plant money, should do anything they could to bootleg or copy a Whitney gin to deal with the thousands of tons of seed that would ruin them if unprocessed. Those who realized they were infringing Whitney's rights took a chance that the southern courts would be sympathetic to them. They won their bet, for decades. In 1803, when the cotton-growing states repudiated all their agreements with Whitney, cotton earned the planters $10 million and the man whose machine enriched them was penniless. It all confirmed Whitney's view of the deprav-

ity of human nature. He seethed: "It is a solemn truth that many of the Citizens of Georgia are amassing fortunes, living voluptuously and rolling in splendor by the surreptitious use of the gin while I am chained down to this spot, struggling under a heavy load of debt."

Whitney exhausted his and much of Widow Greene's fortune in ten years of patent battles involving 60 separate suits. Georgia was the most unjust. He wrote to the inventor Robert Fulton: "I had great difficulty to prove that the machine had been used in Georgia and at the same moment there were three separate sets of this machinery in motion within fifty yards of the building in which the court sat and all so near that the rattling was distinctly heard on the steps of the courthouse." When he went to petition the South Carolina legislature, the planters conspired to have him arrested and put in prison, a shameful act that rebounded against them.

The southern states finally paid some money in 1807 following a crushing judgment for Whitney by Justice William Johnson in the United States court in Georgia. By then Miller was dead of blood poisoning from a prick by a thorn, and Whitney was caught up in an even more original adventure.

THE GUNMAKER'S LITTLE TRICK AND BIG IDEA

THE NEW YEAR of 1801 was not very festive for Eli Whitney. His nerves were too on edge to celebrate anything. His 35th birthday present in December had been a summons to the new capital of Washing-

ton to justify his expenditure of thousands of U.S. dollars for thousands of muskets he had not produced. He left his New Haven workplace early in January, rehearsing in his mind the arguments he would make to convince the lame-duck president, John Adams, President-elect Thomas Jefferson, the secretaries of war and treasury, leading Congressmen and officials that he was not a charlatan.

Whitney had won fame when he was only 27 as the inventor of the cotton gin—but not the fortune that ought to have gone with it. By 1797 bankruptcy and ignominy loomed for the golden boy. He brooded by himself in a rented room near his forlorn workshop. The magic triangle of Mulberry Grove had been broken. His patron, the vivacious Catharine Greene, and fellow Yalie Phineas Miller refused to subsidize him anymore. Whitney, who may have been in love with Catharine, sank into self-pity. At Thanksgiving, he wrote: "This day has been spent thro' Connecticut in giving Thanks, talking and preaching Politics, sleighing, dancing, laughing, eating Pumkin-Pie, Grinding Salt, Kissing the Girls, etc. etc. As for myself [this day] has been spent in that kind of anxious solitude with which I have been for some time been encumbered." It is a testament to his largeness of spirit and the depth of his creative impulses that he began to use his isolation to concentrate on ideas. He was determined not to let his creations again fall afoul of pirates and the patent litigation in which he was still ensnared, so he looked for opportunities to do business with the U.S. government. To begin with, he made a printing press with dies and offered the drawings to the Treasury as the ideal instrument for executions of legal documents. The supervisor had just given a contract to someone else, but he passed the drawings to the secretary of the treasury, Oliver Wolcott. This was a happy circumstance. Wolcott was the official who in 1794 had blessed Whitney's proof of his cotton gin—and he was another Yale alumnus. On the supervisor's letter of rejection to Whitney, Wolcott scribbled a note recording his high opinion of "the ingenuity and talents of Mr. Whitney."

The single phrase was potent. It encouraged Whitney to take a huge gamble on his mechanical gifts. Congress had voted to make ready for war with France, appropriating the huge sum of $800,000 for arms. Whitney had never made a gun in his life but in a vibrant letter of May 1, 1798, written directly to Wolcott, he offered to use machinery to make 10 or 15,000 "stand" of arms (a stand is musket, bayonet, ramrod, wiper and screwdriver) and to do so in double-quick time, much more quickly and efficiently than the gunsmiths of the Connecticut Valley, who traditionally crafted each weapon by hand. Even the national armory at Springfield, Massachusetts, and

all means in my power and I trust effectually against the mischiefs of too great a competition."

On June 21, 1798, Whitney was offered a contract for 10,000 French-type Charleville muskets. Wolcott signed contracts with 27 gunmakers altogether, for a total of 32,000 muskets, but Whitney's part in this was the single largest financial transaction in the business of the country. It was an amazing deal, achieved only because the war scare was the catalyst for fusing the manic desperation of a wronged inventor and the uncharacteristic adventurism

He had parried inquiries from Wolcott with excuses—"I must not only tell the workmen but must show them how every part is be Done"—but more effectively with a vision. In 1799 he wrote, "One of my primary objects, is to form tools so the tools themselves shall fashion the work and give to every part its just proportion—which when once accomplished will give expedition, uniformity and exactness to the whole." He closed with a seductive metaphor: "In short, the tools which I contemplate are similar to an engraving on a copper plate from which may be taken a great

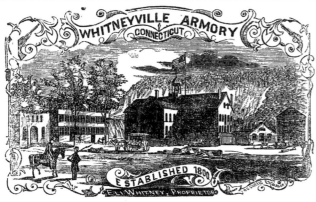

Whitney's little deception in 1801 was the foundation of a renowned arms business carried on by his son.

the new one at Harpers Ferry, Virginia, had never produced 1,500 firearms in a year. Whitney wrote: "I am persuaded that Machinery moved by water adapted to this Business would greatly diminish the labor and facilitate the Manufacture of this Article. Machines for forging, rolling, floating, boreing, Grinding, Polishing, etc. may all be made use of to advantage." Wolcott was influenced by his personal regard for Whitney but also by philosophy. He was a loyal protégé of Alexander Hamilton's, as committed as Hamilton was to government's essential role in promoting manufacturing enterprise, and he had visions of Whitney establishing an advanced armory for continuous large-scale production. In any event, there is a letter from Wolcott to Whitney saying a greater number of proposals had come from New England than he expected, "but I shall however guard by

of a bureaucrat. Whitney was promised $13.40 for each musket, making a total of $134,000, worth many millions now, and thousands of dollars along the way to keep him in business. In signing the contract, Whitney bound himself to deliver all 10,000 by September 30, 1800, and the first 4,000 in only 15 months.

When he made his offer to Wolcott, Whitney had identified a source of power for his machinery in a log-dammed waterfall in New Haven, but in 1798 he did not own the site and he had no money to buy it, no workmen skilled in ordnance, no designs for guns or buildings, no machines and no assured supply of the high-quality iron he would need. Nearly three years later, on that bleak January day, the eighth, four months after the end of the contract, he had no muskets, either. His factory had delivered not a single one.

number of impressions imperceptibly alike."

Wolcott had been persuaded, and besides, almost all the other contractors were behind or bankrupt; only 1,000 of the 32,000 muskets ordered had been delivered. But Wolcott was no longer in charge at the Treasury on the morning of January 1801 when Whitney made his entry into a room of dignitaries in blue coats, knee breeches and silk hose, assembled most likely in the newly occupied president's mansion. He took a large box with him and laid out its contents on a table. It was not a musket but all sorts of anticlimactic bits and pieces—or so it seemed for a few moments. Then he surprised the observers, including Wolcott's more skeptical successor, by quickly assembling the bits into fine new muskets. He picked apparently at random among ten different firelocks and with a screwdriver fitted them to ten muskets. On the testimony of Thomas Jefferson, he also assembled the actual firelock mechanism from a random selection of the internal pieces (tumbler, sear, hammer, lock plate, etc.), a far more impressive accomplishment, since it was the most delicately calibrated part of the weapon. In a letter introducing "Mr. Whitney of Connecticut, a mechanic of the first order of ingenuity," Jefferson told Virginia's governor, James Monroe: "He has invented molds and machines for making all the pieces of his locks so exactly equal that take 100 locks to pieces and mingle their parts and the hundred locks may be put together as well by taking the first piece that comes to hand. This is of importance in repairing, because out of 10 locks e.g. disabled for want of different pieces, 9 good locks may be put together without employing a smith."

Elizur Goodrich, a Congressman and onetime Yale law tutor, reported from Washington that his friend Whitney had met "universal approbation." All judges and inspectors were united, he said, in a declaration that his machine-made muskets were superior to any imported or made at home. Whitney got the government's blessing and more money. Jefferson and others, including the inspector of small arms, another Yale man by the name of Colonel Decius Wadsworth, believed they had seen the augury of the machine age, the production of machine-made parts so uniform they could be rapidly assembled by hands as unskilled as those operating the machinery. Whitney's demonstration entered folklore as the birth of the American system, the apotheosis of what historian Daniel J. Boorstin called the Know-How Revolution.

How much does it matter, then, that the show for Adams and Jefferson and company was a fake?

Whitney had secretly marked the parts that were meant to go together. In 1966 the Smithsonian's Edwin A. Battison identified the Roman numeral VI on the inside lock of a Whitney musket in the gun museum in New Haven, "unnecessary if the parts had been interchangeable. They were not uniform and they were not freely interchangeable." (Marks like this were made when the parts were fitted soft to facilitate reassembly after hardening.) Robert S. Woodbury, a scholar of technology, reported on a test of a number of Whitney arms that found "they were not interchangeable in their parts . . . in some respects; they were not even approximately interchangeable." These criticisms have to be put in context. They do not mean that those who testified to Whitney's achievement of "uniformity" of parts at the beginning of the 19th century were liars. Timothy Dwight, for instance, independently reported on a visit to Whitneyville, where he saw the machinery at work "producing locks so similar, that they may be transferred from one lock and adjusted to another. . . . By the application of the same principle, a much greater uniformity has also been given to every part of the muskets." But uniformity then meant within a thirtieth of an inch (compared with a thousandth of an inch by the end of the century). Whitney's technical drawings at Yale show metal parts, molds, hammers and gears and some parts of his musket measured to within one thirtieth of an inch. Battison also doubts, however, that Whitney invented a true milling machine to cut metal precisely to a template. Whitney used a hollow mill to make screws and a circular saw, says Battison, but he argues that the "Whitney" milling machine, the oldest still surviving, was not a wholly original work. He portrays Whitney as part of a broader-based, more gradual development of machine methods in a workable system of interchangeable parts.

Qualifications about Whitney's originality and his machinery are the fine print necessary to appreciate, once again, the social and technical context of innovation. Fifteen years before the famous demonstration, Jefferson, then a diplomat in France, took himself along to the workshop of the inventive Parisian gun maker Honoré LeBlanc. He reported to Washington in August 1785: "He presented me with the parts of fifty locks taken to pieces, and arranged in compartments. I put several together myself, taking pieces at hazard as they came to hand, and they fitted in a most perfect manner. He affects it by tools of his own contrivance." A committee of the French Academy of Science verified LeBlanc's claims, and the year Whitney entered Yale, 1789, Jefferson brought back a box of his muskets. Nobody was much interested. Whitney may well have learned of LeBlanc's work from Jefferson, who had become a friend after his purchase of a cotton gin for his own plantation. But while LeBlanc preceded Whitney, the interchangeable system did not take wing with LeBlanc any more than steamboat services took off with John Fitch. LeBlanc, too, had an order for 10,000 muskets—from Napoleon's ministers—but they quite suddenly stopped the process on the philosophical and political grounds that harmonious products could not emerge from many hands: LeBlanc's use of unskilled labor challenged both the craft labor monopolies and the government's control over them. The English naval engineer Samuel Bentham (1757–1831) also pioneered the concept of precise uniform parts for wooden pulleys for sailing ships, but it was in America, not England or France, that the ideal of interchangeability dramatized by Whitney was most notably accepted and pushed by the political and industrial leadership. Indeed, Whitney's initiative may have so commended itself to Jefferson he might well have remained just as enthusiastic if he had tumbled to Whitney's artful dodge of marking parts. They met for dinner the night before. Who knows what passed between them?

Whitney thus did not quite achieve what folklore says he did on January 8, but it remains a landmark moment. The demonstration, limited by the exigencies of the present, enabled him to map the future. The time and money he was allowed, in part thanks to his good connections, proved well spent. If he was eight years late delivering the final muskets of the 10,000, he did it for the price agreed and to a standard Colonel Wadsworth described as superior to any he had ever seen, and Whitney justified the endorsement by promptly fulfilling a contract in 1812 for 15,000 firearms.

If Whitney was perhaps more limited in his technical achievements than presumed, he was unlimited in his vision of mechanization and interchangeability, and untiring in advocacy. His persistence with his network of friends was central to persuading federal government ministers to press for interchangeable parts in the

CONVENTIONAL INNOVATOR: Whitney enforced as strict a discipline on others as he did on himself. He ran his Rock Mill model village on strict moral principles. Only in viewing his special world of mechanics and technology, wrote biographers Jean Mirsky and Alan Nevins, was Whitney willing to allow any deviation from the orthodox.

manufacture of firearms, a leading industry after the war of 1812. In asking for bids for 20,000 pistols at $7 each, the War Department in 1813 wrote into the contract with Connecticut's Simeon North (1765–1852) that "the component parts are to correspond so exactly that any limb or part of any one Pistol may be fitted to any other Pistol of the twenty thousand." It was at Whitney's place in New Haven in 1815 that directors of the national armories and Colonel Wadsworth, by then chief of the U.S. Ordnance Department, agreed on a formal strategy for standardizing the manufacture of muskets. They decided that the national armory at Springfield should make special inspection gauges to enable even an unskilled worker to see if a piece fell within the tolerance of allowable error. Of crucial importance was the government's backing from 1819 for the ebullient innovator from Maine, John H. Hall (1781–1841). He was allowed independent use of the facilities at the fractious Harpers Ferry Armory and in 1824 produced a breech loading rifle with functionally interchangeable parts so well machined and die-forged that the metal parts could

be mixed and remounted on 100 new stocks.

Whitney was both a visionary and an incomparable manager. He reduced the complexities to a succession of simple processes so as to make a really effective division of labor among his largely unskilled workmen. MIT's Merritt Roe Smith, who has reservations about Whitney's technical originality, suggests that his business innovations have been too much overlooked. The national armories and other large private firms in textiles rarely calculated their costs properly. Manufacturers typically took their profit rule-of-thumb by adding a dollar or two to the cost of raw materials and labor. Whitney was as meticulous in keeping a record of every cent spent as he was in dress and deportment. He insisted that the price of a musket cover the cost of insurance, wear and tear, and interest paid on borrowed money, in addition to materials and labor. The result was that he made good profits while others hovered at the margin or vanished.

Whitney was ceaseless in his pursuit of a mechanical synthesis. Not until he was 52, eight years before his death, did the Old

Bachelor find time for courtship, marriage to a woman 20 years his junior, and children. His only son, Eli Whitney Jr., when grown up carried on the flourishing business.

Samuel Colt gave the Whitney factory his famous order for 1,000 Walker pistols in 1847, and in 1858 Oliver Winchester began to use the facility to make Winchester rifles. By then a stream of experts was coming from England to examine "the American system" of which Whitney was a key progenitor.

It was to be another 100 years before machines as complex as steam turbines and aircraft engines could be made entirely by other machines. But the idea of manufacturing nearly identical parts, propagated and imperfectly practiced by Whitney and so effectively carried out by John Hall in 1824 and Simeon North in 1828, was the foundation of the machine tool industry and of mass production, which could put a sewing machine and a pocket watch in every home, a harvester on every farm, a typewriter in every office—and hundreds of thousands of rifles in the hands of Civil War soldiers.

SAMUEL SLATER
Dressing America

SAMUEL SLATER'S mind was set on espionage soon after he read an advertisement by the Pennsylvania legislature in his hometown newspaper in Derbyshire, England. The advertisement offered a bounty to anyone who could instruct America in the latest arts of textile manufacture. Slater, then 21, worked in a factory at Belper designed by a onetime barber and wig maker, Richard

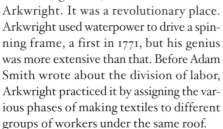

He opened America's first true factory.

Arkwright. It was a revolutionary place. Arkwright used waterpower to drive a spinning frame, a first in 1771, but his genius was more extensive than that. Before Adam Smith wrote about the division of labor, Arkwright practiced it by assigning the various phases of making textiles to different groups of workers under the same roof.

Slater would have been prosecuted if anyone had known what he was doing while he worked as an overseer. With the hope of winning a reward in America, he was committing to memory the inventions of Arkwright, and also the spinning machines of James Hargreaves and Samuel Crompton. It was a crime punishable by prison to export machinery or knowledge, and textile workers were not allowed to emigrate. Slater got around that by taking himself to London in September 1789, where he boarded a boat as a "farm boy." Not long after arriving in New York, he wrote to Moses Brown (1738–1836), an engagingly eccentric and generous Quaker merchant in Providence, Rhode Island, who had been looking for a manager able to run a waterpowered mill at Pawtucket with the bits and pieces of Arkwright machinery collected by Brown and his partner, William Almy. Slater was made a remarkable offer. If he could build working machines similar to Arkwright's, he could have one half of the net profits and one-half ownership of

the machinery; as science writer Mitchell Wilson put it, there is no other record of a man applying for a job by letter and getting pretty well the entire plant by return mail.

From April 1790 Slater worked secretly with craftsmen in the Pawtucket mill, building anew and salvaging what he could from the Arkwright replicas. By December, he had three carding and roving machines up and running, and two spinning frames with 72 spindles. The initial power was from workers on a treadmill, but early in 1791 he had the machinery linked to a waterwheel, making Pawtucket the first mill in America powered by water. It was the foundation of an extensive business—the American textile industry—since competitors rushed to imitate.

Slater was also the first to install the factory system in America. The subdivision of work into simple sections was essential to cope with the shortage of artisans. Slater was only copying a concept learned in England, but he was imaginative and progressive in his management. He divided the work into such simple sections that children aged four to ten could do it—and did. Such child labor sounds outrageous today, but it was the tradition for children in America to be put to work around the farm as soon as they could walk, and Slater's "family" system was popular. The overseers did not beat and degrade the children as they often did in English mills, and mothers and fathers were glad to have their children around them while they earned good wages and were well fed.

Slater and his six sons and his in-laws went on to open more mills; he was one of the first to use steam power in a mill he built in Providence, Rhode Island. As for his plagiarism, Slater was unabashed. He was asked by President Andrew Jackson, visiting Old Slater Mill, if it was true that "you taught us how to spin so as to rival Great Britain in her manufactures." Slater is said to have replied, "Yes, sir. I suppose that I gave out the psalm, and they have been singing the tune ever since."

FRANCIS CABOT LOWELL
The Lowell Girls

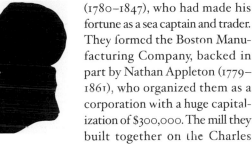

FRANCIS CABOT LOWELL was another innovative smuggler, as brilliant a pathfinder as the bonfire he lit in Harvard Yard that got him rusticated. Lowell was only 14 then and he was dead at 42, but he made two considerable innovations in his short life. Stricken with sickness when he was 36, he took a long trip to England and was stunned by the number of textile mills and their machinery. Like Slater, he memorized the machinery, but he had an ambition to go further than replication.

On his return to Boston in 1813, he communicated his ambition to the clever mechanic Paul Moody (1779–1831) and his brother-in-law Patrick Tracy Jackson (1780–1847), who had made his fortune as a sea captain and trader. They formed the Boston Manufacturing Company, backed in part by Nathan Appleton (1779–1861), who organized them as a corporation with a huge capitalization of $300,000. The mill they built together on the Charles River at Waltham was the first to convert raw cotton into cloth by power machinery within the walls of one building.

Lowell was an innovator in another direction, his importation and adaptation of the humanitarian social ideas of English factory reformer Robert Owen. Lowell insisted on good working and living conditions for the mainly female workers, ideas that survived his death. His partners, riding a wave of successes, carried the tradition with them when they secretly bought water rights to set up operations by the 30-foot Pawtucket Falls on the Merrimack River. They called the city they built around the mills by the name of Lowell. They built two other manufacturing towns in Manchester, New Hampshire, and Lawrence, Massachusetts, but it was Lowell whose name became synonymous for a time with decency in industry—and the high productivity that was one of its results.

The hundreds of women in the mills, mostly girls from good farm families, worked a 12-hour day, but such was the attention paid to their housing, health, welfare and education that they regarded it as prestige employment. To be a "Lowell Girl" was to be a subject of envy and admiration. Charles Dickens, a fierce critic of the conditions of factory workers in England, and not inclined to find much he liked on his visit to America in 1840, was impressed when he visited Lowell. "There was as much fresh air, cleanliness and comfort, as the nature of the occupation would possibly admit of."

The photographs of the neat mill girls and women from the American Textile History Museum, in Lowell, are from 1850–75.

The Comanche warriors were spectacular horsemen who finally met their match in Colt's repeating revolver. Arthur F. Tait, *The Pursuit*, 1856.

SAM COLT A reckless spendthrift who created his own legend—and a mass market

1814–1862

FIFTEEN TEXAS RANGERS rode into mortal peril on a hot June day in 1844. Scouting the hill country by the Pedernales River, they came across a war party of 80 Comanches, the fiercest of the Plains Indians and the finest horsemen in the West. A Comanche could hang by one heel over the side of a horse running at full speed, shielding himself from view while he shot arrows with deadly effect.

The Rangers, outnumbered five to one, did what they normally did in such circumstances: They dismounted so they might fire their single-shot muzzle-loading rifles more accurately. The Comanches whooped forward, sure of overwhelming the small band of white men, since it would take 20 seconds for the Rangers to reload—by which time either on foot or on horseback the mass of Indians would be upon them. In the time it took to ram a powder charge and lead ball down a barrel, a Comanche could shoot six arrows or run 150 yards with spear and tomahawk.

But the Rangers did not frantically reload and dig in. To the consternation of the Comanches, they remounted and charged back at the war party, firing with every finger on their hands—or so it seemed. The Rangers had equipped themselves with a new weapon, a repeating pistol made by the Patent Arms Manufacturing Company in Paterson, New Jersey. It could fire five shots without reloading. Thirty warriors fell. One Ranger, who still had charges left in his revolver, chased the chief to the top of a hill and shot him from 30 yards. The rest of the war party fled. Stone Age weaponry had finally met its match. It was an encounter that offered hope to settlers and spelled doom for the Plains Indians.

The Patent Arms Company was no longer making guns when Captain Jack (John Coffee) Hays and his men fought the Comanches. The factory had shut down in 1842, its principal founder broke and living hand to mouth in New York; "I hardly knew," he said, "where the dinner of tomorrow would come from."

This was Samuel Colt—or rather Sam Colt: There is much in a name and he rejoiced in the euphonic impact of "Sam Colt," the single broad syllable of his first name, with its promise of an outdoor man

ready for action, marinated with a surname of equine virility. Sam Colt was 26 when the creditors shuttered Patent Arms. He was only 47 when he died, but in the 21 years remaining to him, in which the Pedernales encounter was a critical turning point, he became one of the richest men in America and a legend—a legend in which it was hard to tell fact from fable. His loving widow, Elizabeth, who outlived him by 43 years, perpetuated the image of Sam Colt as a heroic individual who valiantly and selflessly conquered every adversity with God's

THE BEGINNINGS: Parts of Colt's revolver model of 1831, whittled out of pine on his voyage to India. Many models followed. Colt's Model 1860 Army, or Holster, .44-caliber revolver, weighing only 2 pounds 11 ounces, was the army's prized weapon. The famous "Peacemaker" was a Colt .45 Single Action Army (SAA) revolver, which used metallic cartridges for the first time instead of powder and ball. Colt did not actually create the company's most famous weapon, released in 1873, but its internal mechanism was based on his design.

help. She elaborated it in stained glass and marble, in museum, monument and literature, in a church of remembrance and ceremonials, beginning with the extravaganza of a military funeral that "Colonel" Sam—never more than a weekend militia soldier—had designed in every detail for his triumphal exit: his own band in brilliant Prussian-style uniforms marching with muffled drums, colors reversed; a procession of 1,500 workmen wearing black armbands; his body, adorned with fresh camellias, carried over the snow in a silver-mounted steel casket to a lakeside tomb.

The first headline in the Colt myth is "Early Genius Surmounts Emotional Trau-

mas." At 25, his father, Christopher, had narrowly escaped being booted out of Hartford as a ne'er-do-well by saving a young lady on a runaway horse who happened to be the daughter of the richest man in town, Major John Caldwell. Two years later he married Sarah Caldwell and had an up-and-down business life as a West Indies trader and sales agent. Sarah bore him seven children, of whom Sam Colt was the second youngest. She was an indulgent mother, but when Sam was only seven she died of consumption and his father married an unfeeling disciplinarian who soon evicted all but the youngest. Sam was packed off to a farm at 11. He must have been hurt by the breakup of the happy family, but it is hard to accept Elizabeth's later picture of Sam meeting all his disappointments with his hand on his Bible, a model of Congregationalist modesty, unselfish self-reliant fortitude, industry, weary toil, perseverance and self-denial. A cautionary letter from his father to Sam at 18 tells him to "depend on your own exertions" and "use the most rigid economy." Whatever the cause, Sam Colt's subsequent behavior was that of a spoiled child who thought the world owed him a living.

Colt's early technical aptitude and interest in lethal weapons is not to be doubted. It was romanticized by Elizabeth, who has him at the age of seven "sitting under a tree in the field, with a pistol taken entirely to pieces, the different parts carefully arranged around him." (This pistol was said to have been given to him by his dying mother, who, as the daughter of a Revolutionary War hero, had no qualms about arming her children.) It is hearsay, of course; Elizabeth herself was not yet born. But the teenage Colt did learn enough from a sort of encyclopedia of science called the *Compendium of Knowledge* to build a galvanic battery and make gunpowder. In late June 1829, while he was working in a textile factory in Ware, Massachusetts, he put up a notice: "Sam'l Colt will blow a raft sky-high on Ware Pond, July 4." A festive Independence Day crowd gathered. While he busied himself behind some bushes with his wired connections, the raft floated away from the underwater mine. A sudden explosion drenched spectators in holiday finery with plumes of pond water and mud. Legend has it that they were ready to tar and feather the perpetrator until he was rescued by a young mechanic, one Elisha Root, who would eventually play a key role in Colt's fortunes. Root wanted to know how the current got through the water without shorting. The 15-year-old Sam reportedly told Root, "Simple, I wrapped the wire in tarred cloth." Next Independence Day, Sam Colt was a student at Amherst Academy organizing a pyrotechnic display that set a school building on fire.

Escaping Amherst's punishment is the most likely explanation behind the second headline in the Colt myth: "Runaway Sailor Boy Invents the First Revolver." He did not run away from home. His father fixed him up with a voyage to India as an apprentice sailor and spent $90 equipping him with checked sailing clothes and seagoing gear including a dollar jackknife. Wielding this jackknife on the deck of the *Corvo,* so the story goes, he whittled a model of the first gun with a revolving chamber, following a day's frustration firing at porpoises and whales off the Cape of Good Hope. Actually, a rotated chambered breech was a feature of the flintlock designed by Elisha Collier of Boston in 1813 and patented in England in 1818. Young Colt certainly saw Collier's gun in India or London; British armed forces in India had Collier and Co. repeating firearms. But though Colt's enemies maintained that he simply copied Collier, Collier's chamber had to be rotated by hand. In Colt's six-chamber revolver, cocking the hammer caused the cylinder to rotate and line up a fresh charge with the barrel. Practical self-contained metallic cartridges had not yet been invented. Each of the six chambers in the cylinder had to be stuffed with gunpowder and a lead ball; the hammer struck a percussion cap, a relatively new invention that would ignite the powder and fire the projectile. What inspired him to this genuine invention of a cylinder rotating automatically remains an important mystery. Some versions say it was from watching the turning of the ship's wheel or the ship's capstan for raising and lowering the anchor. Both were secured in place with a pawl and ratchet, a device Colt used for turning and locking his rotating cylinder.

SIX-GUN SAM: The exuberant Sam—never Samuel—Colt, who never fired a shot in anger. He was only 47 when he died. He was worth $15 million, about $300 million today, and enjoyed his wealth, but he was propelled by other motives: "Money," he said, "is trash I have always looked down on." He wanted to bequeath a great school of technology and mechanical skills to Hartford but became disenchanted with the way the city was run.

Colt returned to Hartford and showed his parents the fortune he had found at sea. They were not impressed by his flimsy pieces of wood and urged him to sign on for a 30-month whaling voyage. Sam insisted he was onto something big. He went back to work in the textile factory but persuaded his father to advance a local gunsmith $15 for working models of a pistol and rifle. In 1832 Sam took these prototypes to Washington to show his father's friend Henry Ellsworth, who also happened to be the U.S. commissioner of patents. Ellsworth wrote to Christopher Colt on February 20, 1832: "Samuel is now here getting along very well with his new invention. Scientific men and the great folks speak highly of the thing. I hope he will be well rewarded for his labors. I shall be happy to aid him. He obtained $300 at the bank with my endorsement." Ellsworth's valuable advice to Sam was not to apply for a patent until he had improved the experimental models, but to establish priority by taking out a legal caveat specifying his intention to do so while leaving the prototypes in the Patent Office.

For a really good gunsmith, 17-year-old Sam needed much more than the $300 he had talked Ellsworth into sanctioning. He bolstered his finances by inviting people to make fools of themselves. Working with the chemist in the Ware textile factory, he had learned of the mood-altering qualities of nitrous oxide, or laughing gas. He ingeniously made a portable lab and reinvented himself on the lyceum and fairground circuit as the "celebrated Dr. Coult of New York, London and Calcutta," offering sniffs of the stuff at 50 cents a time. An advertisement he wrote for a newspaper in Portland, Maine, in October 1832, informed the audience that laughing gas "produced the most astonishing effects upon the nervous system; that some individuals were disposed to laugh, sing, and dance; others to recitations and declamations, and that the great number had an irresistible propensity to muscular exertion, such as wrestling, boxing and with innumerable fantastic feats." Clearly, Sam Colt had discovered the gift for showmanship that was to be such an important part of his innovative life. He made $10 a day—and spent $11 a day, another perennial feature of his mode of operating until he became rich.

In 1834 Colt moved to Baltimore to be amid a large population of mechanics. There are letters from his father repeatedly offering to help with loans and contacts, and Christopher went with him to inspect machinery in New England's armories, as documented in the inventor's own diary. He was a helpful father, though the myth, with its emphasis on hardship, asks us to accept Sam's version that his time in Baltimore was a difficult period when "his father with other friends opposed him." Colt's alliances in Baltimore were a perfect expression of the symbiosis of his business career. Joseph Walker was a fly-by-night musical impresario who staged performances by Dr. Coult. John Pearson was a capable gunsmith Colt invited to make prototypes of repeating firearms. When Dr. Coult went on the road to keep the enterprise afloat, he left Walker in charge and nagged him from afar with detailed queries and orders. Dr. Coult's laughing gas was a riot. He performed to sell-out crowds from Montreal to New Orleans, but he spent all the good money he earned. He was constantly on the run from creditors and he kept Pearson in misery for two years. "The money you sent me

WIDOW ELIZABETH COLT:
keeper of the legend

won't pay all your bills," Pearson wrote in one of many letters of protest. Colt was about as good at spelling as meeting his debts. His response to Pearson was: "Make your expenses as lite as possible. . . . Don't be alarmed about your wages, nothing shal be rong on my part, but doo wel for me & you shal fare wel." Biographer William Edwards is not unduly harsh in saying 22-year-old Sam was immature and selfish. "Sam had this habit of using people as stepping stones to success, and it was to get him hated by some, and disliked or feared by many." As to the erratic spelling—Colt would later maintain that anyone who spelled a word the same way more than once had no imagination.

On February 25, 1836, Colt tore open a big brown manila envelope to find his American patent, No. 138. The embittered Pearson's work had been so good—he later claimed he was the true inventor—that Colt was ready to manufacture revolvers and rifles in quantity. He borrowed $1,000 from his father. His cousin Dudley Selden, a New York lawyer, invested and so did 33 others, mainly family members and friends, to a total of $230,000 ($17.7 million today). On March 5, Selden, the designated treasurer and general manager, oversaw a bill in the New Jersey legislature incorporating the Patent Arms Manufacturing Company. The very next day, at San Antonio, the 187 Texans and Tennesseans fighting for an

independent republic were all dead at the Alamo, and General Santa Anna's triumphant Mexican army was marching north to subdue the American colonies in East Texas. It looked a propitious moment to launch a gun manufactory.

Nevertheless, the cousins soon fell out. Colt had been unable to get Pearson to work with them, so he had hired Pliny Lawton, a manager in the woolen-textile factory in Ware where his father worked. Lawton, unfamiliar with special-purpose metalworking machine tools, found it difficult to adapt Pearson's craftsmanship to volume production, and, over Selden's objections, Colt kept hindering that by tinkering with the design of what was now a five-shot revolver as well as with the first product to be made, an eight-shot ring lever–operated revolving rifle. Colt was also ready to spend much more of the investors' money in pursuit of the only potential for mass-market sales at that time, the U.S. military. At 22, he was an impressive, handsomely bearded man, powerfully built like his father, and he bustled through every door in Washington, lobbying politicians and bureaucrats in the cause of equipping the army with his guns. He wangled an audience with President Andrew Jackson to show off his repeating rifle, his main focus then, but Jackson was too much enamored of the single-shot flintlocks of his battle days. Colt threw himself lavishly into social life in Washington, drinking mightily with any politicians he thought could help him. Selden was shocked and exasperated and said that only Colt could made the prototypes work. "You use money," he blasted, "as if it were drawn from an inexhaustible mine. I have no belief in undertaking to raise the character of your gun by old Madeira." Selden was particularly enraged when he learned that Colt had pawned some of the display weapons to subvent his drinking. "I know not what you may think of the morals of this business, but it seems to me not much better than putting your hand in a man's pocket." He threatened to have his cousin arrested. Colt talked his way out of trouble.

When the army held competitive gun trials at West Point in the summer of 1837, Colt's wild public relations exercises in Washington had worked well enough to secure a place. A few of the Paterson guns,

Captain Samuel H. Walker went down with guns blazing in the war with Mexico—but not before he had saved Sam Colt's whole enterprise. His Colt 1009 revolver (above) made a great impression on President Polk.

however, were less successful than his lobbying: One exploded on June 21 and the Army Ordnance Board concluded that although Colt's repeaters "may be usefully applied in special cases," it would be imprudent for them to abandon the standard single-shot breech-loading flintlock musket and pistol. Colt seethed about "the fat head of the ordnance bureau," and biographies tend to caricature the officer in charge, Colonel George Bomford, as "facing boldly backwards," but the Ordnance Board was unanimous and Colt's Paterson weapons did have serious production flaws. Multiple discharges, caused when the ignition flame or powder discharge spread to other chambers, were apt to be injurious or possibly lethal; the revolvers were hard to dismantle, with too many parts, and often clogged; and Colt's long arms had even more production defects. A marine in Florida had been killed testing a rifle; his commanding officer wrote to Colt and the War Department to say Colt's concept was admirable, but the manner of manufacture was "infamous." Colt was not blind to these flaws; his efforts to resolve them were one of the reasons production was interrupted for changes. He was at the same time utterly unscrupulous. He asked Selden for money to bribe officials, which appalled

his cousin. "The suggestion with respect to Colonel Bomford," Selden said, "is dishonorable in every way."

The Paterson directors were right to be suspicious of the extravagances and ethical flexibility of the company's young founder. But they were wrong to assume that their guns would sell themselves. In Florida, soldiers of the Second Dragoons and settlers were fighting a losing guerilla war against Indians resisting deportation from their ancestral lands. Lt. Col. W. S. Harney wrote the Patent Arms Company with an urgent request to try out 100 repeating rifles. Colt's immediate impulse was to board a steamer for Florida and take the guns with him. The directors were deeply concerned that he might just pawn the company property to pay his bills and said he could not go. As usual, Colt was deaf to the word. The rifles went with him when he boarded a steamer to Camp Jupiter, near present-day Palm Beach.

The Second Dragoons rejoiced in a weapon that could fire 16 penetrating balls accurately in 31 seconds, and a triumphal Colt looked forward to returning with an army draft for $6,250 to show Selden and the doubting directors. He also sold and presented some pocket pistols to officers. The directors were not amused, upon his return, when he told them a fancy story about how a small boat taking him to shore had capsized, he had been near drowning for four hours and, alas, his trunk containing the draft had sunk beneath the waves. It happened to be true. The army was convinced and issued a duplicate draft. The Dragoons went on to devastate the Seminole and Spanish Indians of the Everglades, feats duly celebrated in press reports written or paid for by Colt. The Texas navy and

the Texas Rangers followed up, buying some holster model, or "Texas Paterson," revolvers, but the military bureaucrats in Washington remained adamantly opposed to general purchases for the army.

Colt was sure the despised desk soldiers were taking bribes from other arms makers. He campaigned strenuously when a second military test in 1840 went against him. It was downhill from there. The turmoil in the economy that had heralded their launch made it impossible to attract more financing. Colt was arrested by a New Jersey creditor and had to be bailed out by Selden. On June 19, 1840, the directors stopped production. The company had made some 3,000 pistols and 1,500 rifles, shotguns and carbines, but the average unit cost was $50, around $3,000 in modern dollars. Colt and Selden broke up.

The biographical judgments on Colt's Paterson period tend to be rough. The critical biographer and curator William Hosley calls him "an abrasive opportunist" who "betrayed the trust of family members and friends who invested." Certainly Colt had been a chaotic manager and reckless, but some redeeming qualities in his character began to emerge from the wreckage. Humility is not associated with thrusting extroverts, but ambition is a corrective. Colt longed to know what he didn't know. In 1842 he settled in New York, without any assured income, and dedicated his prodigious energies to self-improvement. He read texts on chemistry and mechanical engineering. He took a studio apartment at New York University and associated with other inventors, scientists and artists there and in the New-York Historical Society. He made a particular friend of another inventor just then emerging from similar setbacks: Samuel Morse, inventor of the electric telegraph (page 71). The two men shared Democratic sympathies, which is to say that in midcentury they were philosophically libertarian (but hungry for government subvention), egalitarian (for white men) and expansionist. In politics they were pro-Union but opposed to Lincoln. They were against interfering in the southern "way of life," i.e., slavery. (Colt's view was that it was so inefficient the South would come to abandon it.) Their attitudes are unappealing to modern generations, but they were widely shared in those perplexing times.

Colt did experiments in New York with Samuel Morse. He entered in the exhibitions of the American Institute for the promotion of mechanical skill and won several awards of elegant gold medals. He also plotted feverishly to save his older brother John from a hanging for murder (John, or a planted corpse of someone else, was found stabbed to death in his cell, suicide, it was said, after a mysterious explosion quite disrupted the public execution).

In his six years in New York, Colt never missed an opportunity to hustle the desk soldiers in Washington as tensions waxed and waned between the United States and Mexico over the future of the now independent republic of Texas. He would do anything to revive interest in his guns, including paying for favorable references in the press. He received a generous U.S. government grant of $50,000 to experiment with submarine mines and staged a dramatic demonstration before the president, members of Congress, the media and the public. It was skillfully choreographed and captured in oils by an eminent painter he commissioned. But even though he staged five successful explosions, he was unable to sell the system to the navy. He had a modest success selling the army a design for waterproof cartridges, and he floated a telegraph communications company, but all this left him, he wrote, still "poor as a church mouse."

It was during this generally low period in Sam Colt's career that word trickled through of the Rangers' triumph against the Comanches. He tried to get a letter through to Captain (now Colonel) Hays without earning a reply. In May 1846 the

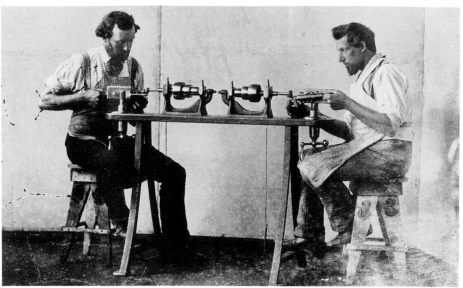

COLT'S MACHINES: Three photographs from glass-plate negatives, taken between 1856 and February 4, 1864, when fire destroyed the East Armory, Coltsville: screw head slotters (top); single spindle drill (middle); and lock frame jigging machines (right).

COLT'S MORALS: Colt was glad to have government contracts, but his credo was in a letter to his half brother, who asked help in getting a government job. "To be a clerk or office holder under the pay and patronage of the government is to stagnate ambition and hope.... I have never forgotten a saying, almost the first that I remember in life—it is better to be the head of a louse than the tail of a lion. Its sentiment took deep root in my heart and has controlled and shaped my destiny. If I can't be first, I won't be second in anything."

long-running tension with Mexico over Texas exploded in Congress's approving a declaration of war by President Polk. Colt every day expected an inquiry about his guns. None came.

In November the near-desperate Colt heard that a certain Rangers captain, Samuel Walker, had been appointed a captain of Mounted Rifles in the U.S. Army and was in New York City raising money to buy arms and equipment for his men. Captain Walker! Colt remembered he had been with Hays in the battle with the Comanches and wrote to him at once: "I have so often herd you spoaken off by gentlemen from Texas that [I] feel sufficiently aquainted to trouble you with a few inquires regarding your expereance in the use of my repeating Fire Arms. . . . I hope you will favor me with a minute detail of all occasions where you have used and seen my arms used with a success which could not have been realized with arms of ordinary construction." New York University had not taught him spelling, but life had taught him to pursue the smallest opportunity.

Captain Walker more than obliged. "Sir, I take great pleasure in giving you my opinion of your revolving patent arms," wrote Walker on November 30. "The pistols which you made for the Texas Navy have been in use by the Rangers for three years, and I can say with confidence that it is the only good improvement I have seen. The Texans who have learned their value by practical experience, their confidence in them is unbounded, so much so that they are willing to engage four times their number. Without your pistols we would not have had the confidence to have undertaken such daring adventures. They can be rendered the most perfect weapon in the World for light mounted troops which is the only efficient troops that can be placed upon our extensive Frontier to keep the various warlike tribes of Indians and marauding Mexicans in subjection."

Even Colt, in his fantasies, could not have dreamed of a better testimonial from a better source. Walker was a hero. He had led Rangers south of the Rio Grande in fights with Mexicans, survived Santa Anna's suicide lottery and a brutal seven months as a prisoner of war, returned to chase Comanches, suffered a near-fatal wound from

a Comanche lance thrust, recovered and returned to action yet again. He was a slim, mild man who did nothing for show, the opposite of the jovial Colt, but he knew he owed his life to the showman's revolver. The two men were nearly the same age (Colt 32, Walker 31) and when they met they became instant friends. Walker was ready to lend his expert knowledge to make the gun even better and Colt was ready to listen. Walker wanted a weapon that would knock a Comanche off his horse at 100 yards and stop a Mexican cavalryman in headlong charge. Colt persuaded the quiet captain that before going to the Mexican War, he should redesign the gun with him and then campaign in Washington for Colt weapons for his fellow fighting men. It was a masterstroke. The two men redesigned the five-shot gun of .36 caliber into a .44-caliber six-shooter with a nine-inch barrel that would fire both lead balls and the new oval-shaped bullets and be easier to load. On December 7, 1846, Walker carried this massive "hand cannon" into the office of President Polk, along with firsthand testimony as to how superior Mexican forces armed with antiquated flintlock rifles had been shredded by Colt-armed Texas Rangers. Polk was so impressed he took Walker and the .44 along to his new secretary of war. The Texan leader Sam Houston volunteered his endorsement, for which fulsome thanks came from the "poor devil of an inventor." Late that same month, the poor devil got an official order from the chief of ordnance for the U.S. Army—$25,000 for "one thousand revolving pistols," delivery in three months. The order Colt had sought for half his life came when he had no guns to sell and no factory or machinery to make them.

In a frenzy of action, Colt called on gunsmiths in three states, pleading and cajoling the likes of Edwin Wesson and Eliphalet Remington to make parts for him as a priority over whatever else they were doing, and then begging Eli Whitney Jr. to assemble them at the Whitneyville arms factory. Twenty-six-year-old Whitney had been in control of the armory founded by his father (page 48) only since 1842. Reluctant to get involved with a man of such erratic reputation, he declined, but Colt wore him down. When the assembly proved trickier

than imagined, Colt doubled the payment to the machinists to make them work through to midnight and raided other factories. "I shall not save one dollar out of the contract," he told Walker, and it was probably true. The guns were not ready when Walker returned to the war with his company of men in May 1847, but at last, on June 26, Colt was able to rush him a pair of the Whitneyville-Walker Colts. *Rush* in 1847 was a relative term; not until October did the weapons reach Walker in Mexico. He was overjoyed. "They are as effective as a common rifle at one hundred yards," Walker wrote, "and superior to a musket at even two hundred yards. All the cavalry officers are determined to get them." Four days later, on October 11, 1847, his Colts blazing away, Walker led 250 men against 1,600 Mexicans entrenched in the town of Huamantla. They routed the Mexican lancers, but at the end of the battle Walker was dead.

America had lost a hero but won a war and a land settlement of incalculable worth. With peace on February 2, 1848, Mexico granted the United States all of Texas up to the Rio Grande as well as New Mexico, California and parts of Utah, Nevada, Arizona and Colorado. Colt's guns had played a not inconsiderable role in helping to assemble the modern coast-to-coast continental America—a big building block in the West created by salesmanship on top of the one in the South Robert Livingston had achieved by diplomacy.

For himself, Colt was at last in sight of his dream, a mass market for his guns. The army ordered another 1,000 repeating pistols and Colt abandoned Whitneyville. Amid much bad blood over the proprietary rights in gauges, patterns and tools, and his poaching of two German gunsmiths, he took the equipment to his hometown of Hartford to set up his own interim workshop with $5,000 borrowed from a banker uncle and credit from a few other Hartford businessmen. He exulted: "I am working on my own hook and have sole control and management of my business and intend to keep it as long as I live without being subject to the whims of a pack of dam fools and knaves styling themselves a board of directors." Colt went into overdrive. He extracted pleasing pith from any testimonial he could get, something he had learned from perusing advertisements for patent medicine. He wooed powerful politicians with beautifully engraved sets of pistols. He assiduously cultivated the newspapers. It was just as well. A gun he sent to Senator—and former General—Thomas Rusk burst during a demonstration before military officials. Colt blamed the military testers. Rusk calmly asked for another, and the press was benign. The Colt Walker revolver did in fact prove to have deficiencies and the Army Ordnance Board was on his case. Colt redesigned it in early 1848, shortening the cylinders by half an inch so they would not take so much powder and cutting the long barrel to seven and a half inches (on the advice of the detested Ordnance Board). He also adjusted the rifling inside the barrel so that the rotation of the bullets was increased just before they left the muzzle. This version, dubbed by modern-day collectors as the First Model Dragoon, became one of his bestsellers.

Colt's genuine ambition to make a perfect gun, which he put above immediate profit, was one of several keys to his eventual success, though on the way he well-nigh drowned in red ink. A second was his infatuation with machinery, a seed planted in the visits with his father to inspect the Connecticut Valley armories. By 1849 the region had become the Silicon Valley of the 19th century. The new religion was machined uniformity of interchangeable parts for rapid assembly, a quest stimulated by the federal government's insistence on the free exchange of information among its contractors. The American system, as it came to be known, was crucial to America's emergence as an industrial power and gun making was in the vanguard of the movement. Colt's stroke of brilliance was in realizing the importance of adopting and improving this system in the steam-powered factory in Hartford he moved into in January 1849, the first ever year he made a profit ($75,000, or about $4.4 million today). The third element was his eye for a good man. He tempted Elisha King Root to join him, promising Root he could fix his own compensation. Root, supposedly the kindly rescuer mechanic at the Ware Pond explosion, was by now famous as an inventive and inspiring machinist, with specialized knowledge of drop-forging technology

he had developed at the nearby Collinsville Axe Factory. It was a coup to recruit him. The two men worked as well as Colt and Walker had on the design and redesign of revolvers and repeating long arms, but also on the system of manufacture. They are credited with designing and patenting machines for forging, boring cylinders, rifling and cutting metal, but the real contribution of Root, according to MIT historian of science and technology Merritt Roe Smith, was helping Colt introduce machinery that had been developed elsewhere and organizing its integration in a sophisticated production facility. Total interchangeability remained elusive, but together Root and Colt got to the point where 80 percent of the work could be done by machine, with hand filing for final fitting by skilled artificers with precision gauge systems.

The machines were expensive and the whole operation would have fallen apart if Colt had not been able to find thousands and then hundreds of thousands of orders. He was operating just about the first factory in America to achieve volume production with precision machinery and gauges. Competitors and enemies who had never heard of the concept of economies of scale thought Colt must be cheating or heading for bankruptcy when he lowered unit prices on big orders. Of course, he was not. This Hartford Sam Colt, blessed by luck and Mars, was no longer the reckless spendthrift of Paterson. Like Henry Ford, he continually pushed to lower the costs of his revolvers and repeating long arms through innovations in production and, like Eli Whitney, through obsessive monitoring of expenditures in raw materials and labor. In 1859 a Colt revolver cost $19 wholesale ($1,250 today), about a third of the Paterson price. He left the daily operation of the factory to Root while he brewed elixirs from his genius for marketing. He found a vast new market on top of the military—the homesteaders moving west into and through Indian territory, the forty-niners rushing to the California goldfields, the Mormons chased from state to state to what is now Utah, the Texas cowboys guarding the herds against rustlers, the lawmen in the booming frontier towns. Initially, Colt may not have understood the tactical significance of his revolver, but now

he had the inspiration to associate his six-gun indissolubly with a romantic, adventurous image of this expanding American West. He promulgated legends of reckless men gambling their lives on their Colts. He propagated popular slogans: "God created men equal, Col Colt made them equal. . . . There is more law in a Colt six-gun than in all the law books." He distributed tens of thousands of pamphlets and illustrated broadsheets. He recycled every story of derring-do. He created a network of retail commission representatives and sold guns at discounts through military officers. Giving full rein to his aesthetic sense, he engaged engravers and carvers to decorate grips and metal surfaces. He roll-engraved shoot-out dramas on the cylinders of his revolvers, including the Pedernales encounter. He commissioned George Catlin to produce 12 paintings and 6 mass-market lithographic prints of his frontier adventures using Colt's repeating rifle and the Dragoon wielded by the artist on horseback hunting buffalo—perhaps the earliest use, William Hosley observes, of celebrity endorsements. He even trademarked his scrawling signature. In his manifold exuberance, he basically invented modern branding, employing the leading practitioner of the day, one Edward N. Dickerson of New York City. He fought his competitors with price and lawsuit, but his worldwide branding was more important than his belligerence. His myriad brochures often stated "Beware of counterfeits & patent infringements." It is no accident that the French for revolver remains le Colt.

Colt's idea, it was said at the time, was making the world aware he was in it. He became one of the most traveled Americans of his day, sailing the Atlantic seasonally in pursuit of sales and fame. In Constantinople he palmed his way into the presence of the Ottoman Empire's Sultan Abdul Mejid I, gave him a magnificently gold-inlaid and cased Dragoon revolver and casually mentioned that the Russians were

SAM COLT ON THE WARPATH: He was brilliant in associating his six-guns with the romantic, adventurous image of the West. This is one page from his advertising broadside around 1854: "Colt's Patent Repeating Pistols. Manufactured at Hartford, Conn. Beware of Counterfeits & Patent Infringements."

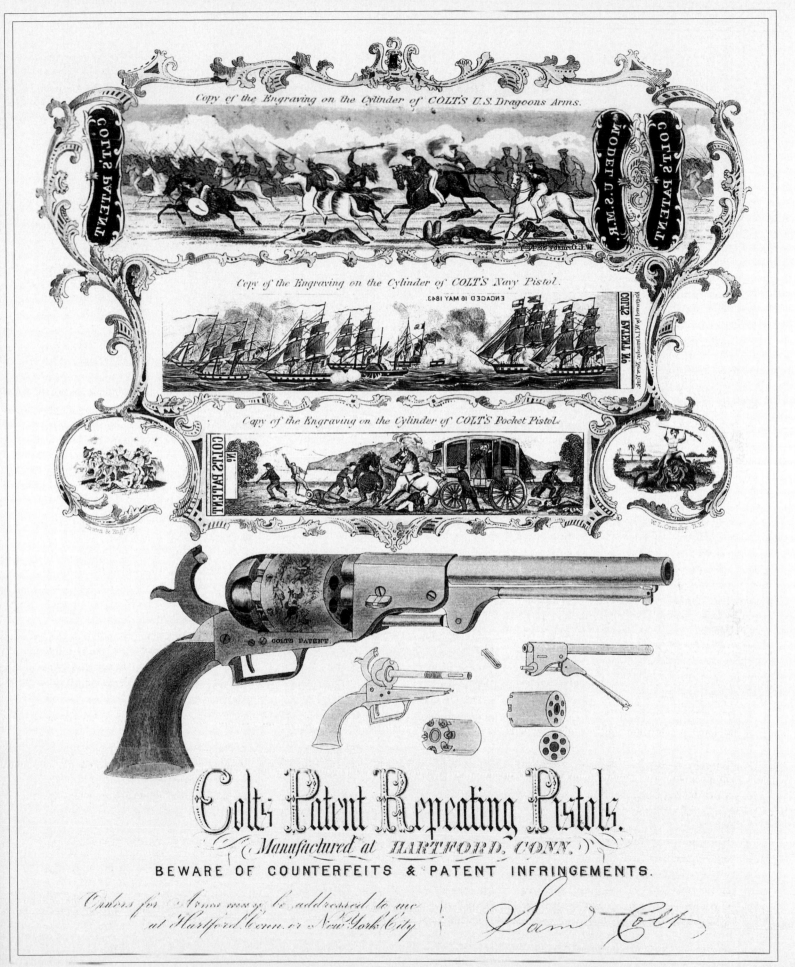

Copy of the Engraving on the Cylinder of COLT'S U.S. Dragoons Arms.

Copy of the Engraving on the Cylinder of COLT'S Navy Pistol.

Copy of the Engraving on the Cylinder of COLT'S Pocket Pistol.

Colts Patent Repeating Pistols.
Manufactured at HARTFORD, CONN.
BEWARE OF COUNTERFEITS & PATENT INFRINGEMENTS.

arming themselves with his revolvers (omitting to mention that he had told the Russians the Turks were buying). He returned to Hartford with a Turkish order for 5,000 revolvers. He obtained patents for new models in London, Paris, Brussels, Berlin and Vienna. He organized a dazzling presentation at the Great Exhibition of the World's Industry at the Crystal Palace in London in 1851. An English reporter noted, "None were more astonished than the English to find themselves so far surpassed in an art in which they had practiced and studied for centuries." The following year he became the first American industrialist to manufacture overseas, opening a factory in London. He closed it after the end of the Crimean War in 1856, but while it operated it awed Charles Dickens: "This little pistol which is just put into my hand will pick into more than two hundred parts, every one of which parts is made by machine. To see the same thing in Birmingham and in other places where firearms are made almost entirely by hand labor, we should have to walk a whole day visiting many shops carrying on distinct branches of the manu-

facture." Colt easily underbid British arms makers and added insult to injury by caparisoning commerce in the raiment of learning. He gave lofty lectures on the ineffable superiority of the American system and graciously acknowledged the wisdom of the institutions that conferred honor on an innovative Yankee at the court of Queen Victoria. On the strength of his high profit margins, Colt opened a new factory in Hartford in the winter of 1855–56, employing 1,000. He treated them fairly by the standards of the day, and in the dangerous business of explosives his safety record was altogether creditable. It was the world's largest private armory, but he was also secretly buying up 200 acres of meadowland on the Connecticut River floodplain for his most audacious project: Coltsville. His dream was an industrial utopia, a community built around a vast armory that would turn out 150 to 200 guns a day. The riverside complex, paying homage to the opulent mansion called Armsmear that Colt built on the hilltop beyond, had workshops and parkland, orchards and houses and classrooms. It had a farm to grow its

own food, a reservoir and waterworks to supply its water, a river port dock and harbor where the Sam Colt schooner could tie up alongside his steamboat ferry. It had a railroad depot, its own gasworks, a tobacco warehouse and a whole colony of German immigrants living in replica Swiss cottages with their own beer garden: Potsdam Village. It was an example of Colt's imaginative opportunism. When he built a two-mile dike to protect Coltsville from the floods the townspeople were sure would destroy him, he planted his dike with French willows and then, on an Arcadian whim, recruited willow workers from the basket-weaving districts of Prussia, built a copy of a willow weavers' village near Potsdam and started a willow manufactory that transformed the whim into a profitable business making 100 varieties of wicker furniture.

When Coltsville was finished in 1856, Colt topped it with a double flourish: a blue onion dome, a gesture of appreciation to his Turkish customers, and on top of that a rampant colt sculpted in blue gilt zinc, a less polite gesture to the old money of the Hartford establishment, the financial elite and merchants and Whig politicians, the kind of people who had snubbed his father. Colt often threw his weight about. He tried to hold the town hostage for tax relief. He scoffed at its careful ways as plodding provincialism. His steam hammers pounded its tranquility; his importation of foreign workers mocked its comfortable but smug Anglo-Saxon homogeneity. He valued Germans most of all. He spent $300,000 outfitting them with instruments for an oompah band and with uniforms for Colt's Armory Guard, the latter quite troubling to the good citizens when the guard marched smartly through town to a dress parade in the square, Colt revolving rifled muskets at the ready.

Throughout it all, Colt's lifelong obsession with the military remained an abstraction. He never fired a gun at anyone or served in combat. The "Colonel" was an honorary title bestowed as a political favor, and he was essentially amoral when it came to the cataclysm of the Civil War, which broke out in the last year of his life. He cynically referred to the weapons he kept selling to southerners right up to the last minute, and maybe beyond, as "my latest work of Moral Reform." In 1861 the *New*

For their photograph, around 1875, the legendary Jesse (left) and Frank James put on their best suits, but would have felt underdressed without their long-barreled Colts.

Hollywood, a century after Colt, was still finding big audiences for the American version of the morality play he propagated. This is William S. Hart (center) in *The Gun Fighter* (1917). Nowadays this kind of thing is called product placement.

York Times basically accused him of treason. Perhaps as a penance, as soon as the war started, he had 500 men in Colt's Rifle Regiment drilling on the South Meadows to join the Union Army, only to have his commission revoked by the governor and his regiment disbanded, due to differences between Colt and Governor W. A. Buckingham and the troops as to electing their own officers, and whether or not they would be a national or a state unit.

On his premature death in 1862, Colt had enjoyed only five years of married life to Elizabeth Jarvis. Eleven years his junior, she was a shy and high-minded daughter of an Episcopalian minister; to marry her, Colt converted from Congregationalist to Episcopalian, a step up in the Hartford social hierarchy. She saw in him "my ideal of noble manhood, a princely nature, an honest, true and warm-hearted man," and he tried to live up to her ideal in his patronage of civic amenities and cultivation of the arts. He took her to Europe on a six-month honeymoon

and he finagled an invitation to the czar's coronation at the Bolshoi Theater. Only one of their five children, Caldwell, survived to adulthood. Colt was so distraught by the successive deaths that he communed with his horse, according to his coachman. "Master," he told Mrs. Colt, "was talking to the horse about the little master in the graveyard beyond."

Of course, Colt's legacy went far beyond family. Some see his influence negatively and blame him for exploiting people's fears to create a gun culture. How many hundreds, even thousands, died in western shoot-outs can only be guessed. Colt made 600,000 guns in his lifetime. If one in ten of them killed somebody that would be 60,000 deaths. But the other perspective is relevant. How many lives did Colt save in violent times? He was stout in his defense of his guns as instruments of peace. "The good people of this world are very far from being satisfied with each other and my arms are the best peacemakers." His

weapons were used by scores of killers but prized by vulnerable thousands in violent times. Chinese miners and railroad workers of small stature carried Colts against intimidation and murder. Isolated settlers exposed to bad men, Indians and wolves were voted $50,000 by Congress so they might buy guns to defend themselves.

Whether Colt's revolver defended or retarded civilization is endlessly arguable, but there can be no doubt his advances in precision manufacturing and his iconic marketing methods advanced American industrialization and marked a coming of age of the ideal of American individualism. The year he opened his great armory on the floodplain, Walt Whitman published his "Song of Myself," which perfectly expressed the restless spirit of the times Sam Colt helped to fashion:

What is known I strip away,
I launch all men and women forward
    with me into the Unknown.

SAMUEL FINLEY BREESE MORSE

He had a grand vision as an artist, but it was as a scientist that he transformed the world, annihilating time and distance with the telegraph

1791–1872

IN 1815, at the age of 34, the New York artist Samuel Morse was mixing high in Washington. He was there to begin a splendid assignment that could only add to his growing reputation. The Marquis de Lafayette was making his immense journey around the United States, 50 years after the American Revolution in which he had fought so gallantly, and he had agreed to sit for a portrait by Morse commissioned by the City of New York. Morse, a tall, handsome and gregarious man with strong political convictions, enjoyed himself at a grand levee for the new president, John Quincy Adams, on the eve of his first day at the easel. He had left his dear wife, Lucretia, in New York, where she was recovering from the birth of their third child, and the next day he rejoiced to have a letter from his father saying "your dear wife is convalescent."

She was not. She was already dead. By the time the news reached Morse and he arrived home, after six days and nights of constant travel over wretched roads, she was in her grave.

Such random cruelties of time and distance left an enduring mark on his imagination. His father died the following year, and then his mother. Morse was so depressed he sailed for Europe, where he was perpetually anxious for news of

Samuel Morse's invention of the electric telegraph was grounded in grief on the death of his first wife. This portrait, taken around the time of his success, nonetheless captures his brooding melancholy. It is fitting that it is a daguerreotype. Morse met the inventor Louis Daguerre in Paris and was inspired to make his own daguerreotypes in New York. He dismissed the protests that technology would destroy painting.

the three young motherless children he had left in the care of his brothers. He immersed himself in the Louvre, imagining a gallery containing his representations of paintings like the *Mona Lisa* and *The Last Supper* and other masterpieces untraveled Americans would never have seen in original or copy. He gave lodgings

AN EARLY TELEGRAPH: In this 1847 version made by Morse's assistant Alfred Vail a stylus embossed Morse code on paper tape pulled through the machine by weights. The big electromagnet (the coil at center right) received the signal and attracted the lever arm, which had a stylus at one end—the vertical stylus can be seen just to the right of the gearing.

to a young art student from Georgia, Richard Habersham. He spent time in Italy with James Fenimore Cooper, who was becoming the first famous American novelist. One night in Paris he was at Cooper's home near the Louvre with Mrs. Cooper, their daughter, Susan, and Habersham, all sitting by their reading lamps, when Morse fretted again about the slowness of the mails. "The French semaphore is better than the mail system but it is not fast enough," he said. "The lightning would serve us better." Then he speculated that an electric spark could somehow be used for sending a message. The Coopers thought he was light-headed.

The deep-seated emotional impetus to his exasperated musings took concrete form in 1832, when he was aboard the French packet *Sully,* his completed *Gallery of the Louvre* with him. The "flash of genius," as he put it, came to him soon after the *Sully* had left Le Havre on its long voyage across the Atlantic. He was at dinner with a few people and the conversation turned to experiments in electromagnetism. It had been established that an electric current could magnetize a bar of iron shaped as a horseshoe. Turning the current off then demagnetized the bar, so here was a means of converting electrical energy into controllable mechanical work. Some years before, the French physicist André Ampère had suggested that an electromagnet could cause a needle to move to a letter or number. His idea, which he did not pursue, was to assign a separate circuit to each letter or number so that the on-off movements of the needle would formulate a message. Charles Johnson, a young Boston chemist at the dinner table on board the *Sully,* was asked whether the length of a wire retarded the speed of electricity. He said it did not, whereupon Morse exclaimed, "If the presence of electricity can be made visible in any part of the circuit, I see no reason why intelligence may not be transmitted instantaneously by electricity." (Johnson seemed to have been unaware of the negative report of Peter Barlow, an English researcher. Barlow had concluded in 1824 that the needle telegraph was impractical because the deflections inevitably faded sharply when the current had to travel more than 200 feet to the magnet.)

Morse spent the rest of the monthlong voyage happily pondering how an electrical signal might yield a mechanism for the transmission and reception of words. He had an inventive mind; in 1817, when he was 26, he and his brother had patented a flexible piston pump for fire engines, and he had built a marble-cutting machine. His excitement about a telegraph is a remarkable commentary on the value of ignorance. He imagined himself the first to think of the idea, but a number of German and English scientists had preceded him by several years in practical experiment, and stunning breakthroughs in the basic science had been made known just a few months before by a true American scientific genius, the onetime boy actor and country schoolmaster Joseph Henry in Albany, New York. Morse's exclamation aboard the *Sully* was itself a curious echo of something already said by Gustav Theodor Fechner of Leipzig. In a textbook in 1829, Fechner wrote how by the use of multiple insulated wires between Leipzig and Dresden, "intelligence could be instantaneously transmitted from one city to another." But Morse certainly knew nothing of Joseph Henry's work. As a student at Yale from 1805 to 1810, Morse had observed some experiments in electricity, but electromagnetism had not been discovered and elaborated then. In 1827 he had heard some lectures on the subject by James Freeman Dana of Columbia College, but his sketches on the *Sully* show he did not understand how a battery made electricity. He could have saved himself a great deal of labor if he had researched the science, but, again, perhaps it was as well that he did not. "Had I supposed at the time that the thought had ever occurred to any other person," he wrote years later, "I would never have pursued it." He was so confident that he would soon string the world with telegraph wires that when he disembarked from the *Sully* in 1832, he said to the captain, Samuel Pell, "Well, Captain, should you hear of the electric telegraph one of these days, as the wonder of the world, remember the discovery was made on board the good ship *Sully*."

As soon as he landed in New York, Morse began trying to turn his shipboard sketches into reality. He might still have been seduced by art. His talent justified his expectation that he would be selected to paint one of the four remaining blank panels in the totunda of the Capitol. He was blocked; it seems that

John Quincy Adams, by now an influential member of the House of Representatives, bore him a grudge for his political opinions and worked to block the appointment. Morse was a Nativist, later a stalwart of the Know-Nothing party, which exploited popular emotions against the wave of Catholic immigration from Europe. He wanted to deny citizenship to the hundreds of thousands of immigrants from Ireland. "They are pouring the scum of their populations into America like pouring dirty water in clean," was the way he expressed it to Habersham. Morse was dismayed when the rebuff over the rotunda was accompanied by the failure of the *Gallery of the Louvre* to make waves. He had little money for his telegraph experiments. His only income was a trickle from painting portraits. His appointment in 1835 as a professor of literature of the arts and design at the nascent New York University carried no salary. Student fees earned him less than he needed to pay for the rooms he took at the university, overlooking Washington Square. Ashamed of his penury, he waited until darkness to bring food to his workroom and cook it there among his paints and the building blocks of his telegraph: a lathe, a clutter of iron bars, coils of wire, rolls of paper, parts of galvanic batteries and bottles of chemicals.

His prototype telegraph did not include a manual key or dot-dashing code for letters of the alphabet. It was a cumbersome marriage of a painter's canvas stretcher and a printer's composing stick with current from a battery at the transmitting end and an electromagnet at the receiving end. He designed a hand-cranked lever to move over the coded teeth; it had an electrical contact at the tip that made and broke the circuit for short and long periods. These signals operated an electromagnet he had fashioned at the receiving end of the wire. Here he expected the impulses to activate a pencil so that it marked paper with horizontal lines, long and short, to correspond to the length of the spaces in the originating printer's stick. The short and long lines were initially to represent numbers and the numbers to signify words; Morse had numbered the words in an entire dictionary (not yet Morse code as we know it).

When he finally made all the connections and turned his hand crank, nothing

happened. No signal went through. He tinkered a lot more over several days. Eventually, the mechanics worked perfectly, but he could not send a signal more than 20 feet. It was a happy circumstance that he was in a university. He dropped into a lecture by the professor of geology and mineralogy Leonard Gale and invited the professor to his workroom. Gale took a look at Morse's apparatus and saw at once that was it made without benefit of the discoveries of Joseph Henry. In the vocabulary of the day, Morse was using the battery of one large cell to produce a large quantity of electricity (current) but not a great intensity of current (voltage), and intensity is needed for transmitting at a distance. In addition, Morse's receiver was feeble. He had made his electromagnet simply by coiling a few wisps of wire around a horseshoe-shaped bar of soft iron, whereas Henry, teaching mathematics and science at the Albany Academy, up the Hudson from New York, had been progressively demonstrating uniquely powerful magnets in which he wrapped the bar in multiple coils of silk-insulated copper wire. Henry was so far ahead of Morse—and everyone else—that in 1831 he had literally rung the bell on the electric telegraph, four years before Morse's experiments. Henry had connected several cells in a series to get a higher voltage and connected the wire to a magnet looped with more than a hundred coils. The signal ran through a mile of wire strung around the walls of the academy and caused an armature to strike a bell. It was the first operational magnetic telegraph, meaning the transmission of intelligence at a distance. Henry knew the implications of what he had achieved, but he had not the slightest interest in applying his science to the mechanics of making a machine carry commercial messages. He saw his role as a pure scientist, an explorer of the unknown, an enlightener of men; it was for others to reduce scientific discoveries to articles of commerce.

Gale so improved Morse's circuit, working to Henry's principles, that they sent signals over increasing distances. Gale told Morse they could not hope to reach much more than 20 miles, but Morse was imbued with the spirit of the times. Americans could do anything they put their mind to. "If I can succeed in working a magnet ten

PATHFINDER: Joseph Henry (1797-1878), actor, country schoolteacher, scientific adviser to Lincoln. Henry made the first operational magnetic telegraph but had not the slightest interest in designing a commercial machine.

miles," he kept saying, "I can go round the globe." He told Gale they could rig up a relay system in which a hairbreadth of electromagnetic movement at ten miles would trigger another circuit, and then another, through indefinitely great distances. It was a central insight that had been anticipated by Henry and the English experimenters, but Morse worked out all the practical applications.

His was a lonely life as a widower, his children scattered among relatives. His temperament was volatile, his health uncertain. One day of euphoria was followed by another in deep gloom. He kept hearing news from Europe suggesting he might be beaten. He had had enough of the lonely-inventor-in-the-garret syndrome. What matter if he had to share credit or profit, he must have still more help! He found it that recession-plagued fall of 1837 in a restless and somewhat morose 29-year-old graduate of New York University, Alfred Vail, who went to the same church as Morse

and shared his Nativist inclinations. On September 2 Vail had happened to visit the lecture hall where Morse was showing some professors how he could send a signal 1,700 feet. Vail watched in amazement as the pencil moved, went home to draw telegraph lines across a map of America and came back with his father, Stephen Vail, the owner of an iron and brass works in Speedwell, New Jersey. The arrangement was that Morse would take on Alfred as his assistant in return for an investment of $2,000. Morse made both Gale and Vail partners in his enterprise, Vail with a one-quarter stake, and Gale with an eighth, on the understanding that they would not claim any intellectual rights in what they achieved. Morse was in luck. Not only did he get money, he got a rare talent in Vail. At Speedwell, Vail made a working model for public demonstration and for Morse's patent application.

As the trio simplified the machine, Morse completed the code that bears his

name. Instead of the dots and dashes transmitting numbers coded for words, he evolved dot-dash coding for each letter of the alphabet. He took account of the incidence of the letters in ordinary usage so that the commonest took the least time to transmit. On January 24, 1838, at the university, Morse successfully transmitted the code over two five-mile lengths of wire. In February he packed up his equipment and took it to Philadelphia to show the scientific committee of Philadelphia's Franklin Institute, which long played a key role in evaluating inventions. They were impressed. The next stop was Washington. He took the capital by storm with his "thunder and lightning jim crack," according to a local newspaper. Everyone crowded to see this new wonder, including President Martin Van Buren and his cabinet, congressmen, foreign ministers and academics.

Given the warm reception, the economic utility of the telegraph and an 1836 vote in Congress endorsing the principle of some national system, the sequel to all the excitement was a cruel blow to Morse. Francis O. J. Smith, the chairman of the House Committee of Commerce, enthusiastically turned on the current, proposing a grant of $30,000 to build 50 miles of telegraph. It did not get very far because the economy was too depressed and some congressmen impeded the legislation on the grounds that Morse was surely insane.

Smith's patronage turned out to be a mixed blessing. Francis O. J. was a suave young lawyer of 32, known to history as "Fog" Smith for his double-dealing. He secretly took up a fourth share in the partnership (now divided Morse 9, Smith 4, Vail 2 and Gale 1) and did not declare his pecuniary interest. In the absence of a vote on the investment, Smith paid for Morse to go with him seeking financing in Europe and patents that would protect Morse while his American application was pending (it was granted in 1840). On this trip, Morse met another artist-inventor struggling for recognition, Louis Daguerre, whose technical breakthrough was the daguerreotype, but apart from a French patent it was a fruitless trip for advancing the telegraph. And Smith proved a devious and vituperative partner.

Morse had the same attitude to photography as he had to pure science—rejoice in discovery and take what you can from it. He was not interested in messing around in the mysteries of emulsions and paper. He did not advance the mysteries of photographic chemistry—but he did advance photography in America. On his return, he had a camera built for him, took daguerreotype portraits, opened a studio with Professor John Draper and trained some 20 students in the new system, including the young Mathew Brady. He had no time for the protests that technology would destroy painting. "Art is to be wonderfully enriched by this discovery," he wrote. Nature's "pencil," with its ability to record the minutest detail, would shame the slovenly painter's daubs and enrich an artist's store of images. Similarly, his essential contribution to the electric telegraph was his ability to borrow from science what he imperfectly understood.

Morse found some consolation in photography on his return to New York, but he had entered a grueling period. Nobody was putting up money to develop his invention. Morse would not borrow against its prospects; he would rather go hungry than assume debt. He was once again alone in his workroom studio. He could no longer afford Vail, whose family business had run into difficulties. Gale had taken a new job. Perhaps even the redoubtable Morse would not have persisted if he had known then that he would have five long years of penury and disappointment, but he was sustained day to day by his faith in God. Though sick with anxiety, he nonetheless kept steadily improving his transmitter. He also demonstrated that electric signals could travel underwater.

Morse was not, as he thought, the first in the world to prove this, but it was quite an achievement for an undernourished man of 51 to prove it the way he did. On an October day, he and an assistant rowed across the mile or so from New York's downtown Battery to Governors Island, trailing a two-mile wire he had coated with pitch, tar and rubber. It took them all day and into the evening, when Morse sent a clear signal back to the city. Even so, misfortune dogged him. In the public demonstration the next day, he sent and received a few characters, and then a ship in the harbor hooked the wire with its anchor. He had to cancel the show and suffer the jeers.

The one bright spark in the five years was the encouragement of Joseph Henry. Morse asked Henry if he might call on him "as a learner." He spent an afternoon and evening with the great man at Princeton in May 1839, and took back Henry's ideas for improving his circuits and relays. When Congress resumed consideration of a telegraph bill in 1842, Morse went to Washington armed with his wires and batteries, but also with a letter from Henry testifying that Morse's system was superior to two needle telegraphs patented in England by Charles Wheatstone and William Fothergill Cooke. Morse's machine, Henry told everyone, was "the most beautiful and ingenious instrument I have ever seen."

Day after day the gaunt Morse haunted the Capitol galleries with nearly every cent he had in the world in his pocket, listening to a mix of wonderment and jokes about mesmerism and mad inventors. The bill was agreed in the House by 89 to 83; without the enterprising attitude of the Whig administration, and the Vail family's efforts to win six New Jersey votes, the bill would have failed. Furthermore, as the session wound down it seemed that the press of business would keep the Senate from endorsing the bill. On the last day of business, March 3, 1843, there were 140 bills ahead of the telegraph. Only in the last hour did it squeak through.

The sum of $30,000, Smith's original figure, was now voted for a trial line of 40 miles from Baltimore to Washington. Morse was to be paid $2,500 a year as superintendent, Vail $3 a day. The sinister Smith awarded himself the contract and spent nearly $20,000 laying the first few miles of wire underground. He did not stipulate any insulation for the wires. The 36-year-old contractor, a sometime carpenter called Ezra Cornell, who had designed a trenching machine, found the line uselessly riddled with shorts. With only $10,000 left and limited time, Cornell urged Morse to let him string bare wire above ground with glass drawer knobs to insulate the wire from the wooden poles. Henry endorsed the idea, and Cornell, starting in Washington on April 11, 1843, began following the Baltimore and Ohio (B&O) railway line with 24-foot chestnut poles 200 feet apart all the way to Baltimore. Fabulously, he fin-

ished ahead of schedule on May 23. ("The Telegraph will go on," he predicted, "and when it does I shall go with it." His successes enabled him to found Cornell University.)

On May 24, 1843, Morse in the Supreme Court chamber in the Capitol telegraphed to Vail at the B&O depot in Baltimore, and Vail telegraphed back, the famous message suggested by the daughter of his friend the commissioner of patents: "What hath God wrought!" Still, man needed more convincing. The means were at hand. Morse had a flair for publicity. Both the Whig and Democratic parties were holding their conventions in Baltimore to nominate presidential candidates for 1844. Morse promoted the idea of a race between his telegraph and the railroad. More than two hours before the delegates arrived by rail in the capital with their news from the convention that Henry Clay and Ted Frelinghuysen had been nominated, Morse transcribed a ribbon of code unwinding before him and announced: "The train for Washington has just left Baltimore. The ticket is Clay and Frelinghuysen." Only a small group heard this demonstration of lightning news, but the reverberations never ceased.

A few weeks later, it was the turn of the National Democrats. Vail and Cornell had their instruments in a warehouse of the railroad depot in Baltimore, Morse in a room below the Senate chamber. He had said he would have the news before anyone, post it in the rotunda and announce it from his window. Congressmen crowded around to hear Morse say that Van Buren had won most votes in the first round but not the two-thirds required that year to protect the interest of the South. He reported every vote instantly as Lewis Cass of Michigan rose and former President Van Buren fell, then announced that on the ninth ballot a new man, James K. Polk of Tennessee, had 44 votes. Morse's bulletins were read aloud on the floor of the Senate. The excitement could not be contained. The Senators adjourned and rushed to Morse's window. He read from his ribbon of telegraph paper: "Polk is unanimously nom. 3 cheers were given in convention for restoring harmony." And three cheers, yelled the Democrats, for Samuel F. B. Morse.

Pencil in Holder

Cap Squarely on Head

Hair Trimmed

Black Four-in-Hand Tie

Working Kit in Pocket

Coat Buttoned Top to Bottom

Clean Hands and Face

Sleeves Correct Length

Uniform Pressed and Spotless

Puttees Shined

No Worn Heels

High-Top Shoes Polished

CORRECTLY UNIFORMED
WESTERN UNION MESSENGER

THE VITAL LINK: Repairing the telegraph was among the most dangerous jobs in the Civil War, frequently carried out under fire. The military strung 15,000 miles of wire to supplement the commercial systems. More than 300 operators were casualties.

Western Union came to dominate the telegraph business. It recognized that its messenger boys were the public face of the company so it insisted on every detail of appearance down to "clean hands and face."

The telegraph would not have started when it did without that initial government subvention. Time and again, the federal (and sometimes state) government was the only agency willing to support a new technology when it was still at a risky, very uncertain stage: canals, guns and railroads in the 19th century, later air transport and computers. Morse always believed the telegraph, having been funded by government, should also be run by government as a national system like the mails to prevent wasteful duplication and the dissemination of falsehood. With a fine public spirit he told the postmaster general, Mr. Cave Johnson, he would yield all his patent rights for

$100,000—but Mr. Johnson could not see beyond the end of one telegraph pole. Disappointed, Morse and his fellow owners fell back to organizing their own private stock company, the Magnetic Telegraph Company, to build and operate a line from New York to Washington—and defend the patents. The usual suspects tried to steal Morse's invention and rob him of revenues. He won every case and an extension of the patent.

Throughout, Morse recognized his limitations in business as he had recognized them in science; and again he delegated wisely. He appointed Amos Kendall, former postmaster general and a confidant of President Andrew Jackson's, to promote private development of his invention. Kendall's benign efforts to make Morse rich, a biographer noted, were constantly thwarted by Morse's open-ended benefactions and the ease with which he could be duped by swindlers. But the Morse telegraphs spread as fast as men could string wire. Newspapers signed up, and six New York newspapers formed the Associated Press to pool the expense of telegraphing foreign news. The railroads used the telegraph for signaling and the safe dispatch of trains. There was the chaos Morse had foreseen. Within five years there were 12,000 miles of telegraph lines in America run by 20

At 80, Morse was like an Old Testament prophet and had fire-and-brimstone passions to match. He was also a major philanthropist. He cofounded Vassar College in 1861.

different companies. With no traffic system, a dozen operators would try to send messages all at once on the same wire. Hiram Sibley (1807–1888) brought some order to it all in 1856 by merging the New York and Mississippi Printing Telegraph with a number of smaller companies to form Western Union Telegraph Company. An undersea telegraph cable linking America and Europe, after a false start in 1857, was achieved in 1868.

THE ELECTRIC TELEGRAPH system Samuel Morse developed transformed the world, and the development of it changed him, too. Who would have predicted that the artist who was arguably the finest portrait painter of his day in America would in midlife abandon art for technology? Or that the family bad boy who had never been any good at school would turn out to have a spectacular talent for getting businesses started? Or that the young man so eager for God's good graces that he nearly became an Episcopalian priest could find it in himself to be as devious and unscrupulous as a robber baron?

In his long life, Samuel Finley Breese Morse was all of these things. In his youth, the son of a celebrated Congregationalist minister, he was a promising but penniless artist who elevated friendship and his sense of moral duty above pecuniary gain. He was a fluent talker and a natural leader. He inspired New York artists to rebel against their domination by overweening patrons in the American Academy. He founded the National Academy of Design in New York, dedicated to promoting and training American artists. He threw himself, without reward, into a campaign to shut down New York theaters that allowed immoral French dancing and other "licentiousness," writing to his brother, "I feel satisfied that while engaged for God he will not suffer me to want."

His telegraph, it was aptly said at the time, annihilated distance and time, but for most of his life he stayed philosophically in the same place. At 80, a commanding six-foot figure with flowing beard and flashing blue eyes, he looked like an Old Testament prophet and had fire-and-brimstone passions to match, but they were not much different from those of the handsome painter and sculptor in his 20s gossiping with artists, actors and poets over Madeira wine and music in bohemian London, or 20 years later when he painted in Paris and supervised a Sunday school. His tenant in Paris, Richard Habersham, was enraptured to find that Morse endorsed the contention that slavery was an American institution to be preserved at all costs. Back in New York, Morse took his friend to an abolition meeting so they might be excited by mutual disgust at what the "fools and fanatics" were saying. He ran unsuccessfully for mayor of New York in 1836 and 1841 as a Nativist. As he advanced into his 60s, Morse mellowed just a little in the embrace of a young family. A widower for 23 years after the tragic death at the age of 25 of his first wife, he married Sarah Griswold, 26 years his junior, who bore him four children. He moderated his lifelong suspicions enough to accept a Catholic honor in Spain, Knight Commander of the Royal Order of Isabella. Still, a journalist who visited him and admired "his twinkling eye, his sly humor, his vivacious talk, his steady hand, his elastic step," felt compelled to add: "It must be honestly confessed that his manners, his spectacles, his red silk handkerchief and his dreadfully bad politics are peculiar signs of a gentleman of the old rather than the new school."

By modern attitudes, Morse may have had a narrow view of America, but a grand vision had inspired his enormous historical canvases and his second life as an inventor and innovator. The scholar Brooke Hindle has argued that the new technologies of the telegraph and the steamboat were notably advanced because artists with a conspicuous talent for spatial thinking were in the forefront. They were practiced at imagining and depicting multiple elements in varying three-dimensional complexities, Robert Fulton as a painter and designer, John Fitch as a topographic-map maker and Morse with his paintings—*The Dying Hercules* (1812), *The House of Representatives* (1822–23), 86 individual portraits and his virtuoso *Gallery of the Louvre* (1832). Morse, like Fulton, could also see beyond the immediate instrument to its vast potential for the world. He also had the energy, intelligence, organizational skill and above all perseverance to realize his ambition. These qualities, much more than scientific insight, explain his success. It is a curious coincidence that he was the first to report to America the technical breakthrough by Louis Daguerre, who was also deficient in specialized scientific knowledge, of physics and chemistry, but persevered long enough to fix photographic images on metal plates.

At 46 and beyond, Morse the artist was in hard-driving, corner-cutting mode as an innovator. He was willing as ever to acknowledge his celestial debt but not the terrestrials whose scientific and technical achievements made his telegraph possible. In a patent system, which put a premium on originality, it is understandable that he would not risk compromising his claim, but even when he was secure from challenge his pride of achievement was so intense he could not bring himself to set the record straight.

In the climax of Morse's long and valiant struggle, the one sour note was his desire to keep secret the origin of the receiving magnet and circuit. Pending his patent (granted in 1846), Morse in 1845 allowed Vail to publish a book that slighted the contribution of Henry, well on his way to a central position in American science. When Henry, in turn, was called as a witness in a suit where Morse's patents had certainly been infringed, he felt obliged to defend his own contribution in science. What he said was severe but fair: "Morse did not make a single original discovery in electricity, magnetism or electromagnetism, applicable to the invention of the telegraph. I have always

considered his merit to consist in combining and applying the discoveries of others in the invention of a particular instrument and process for telegraphic purposes." The distinction between discovery and invention infuriated Morse. The emphasis on originality encouraged by the patent system discouraged Morse from acknowledg-

MODEST VAIL: Alfred Vail (1807-1859), Morse's collaborator. Vail's grandchildren had a tombstone inscribed claiming he was the inventor of the telegraphic dot and dash alphabet. Vail himself never claimed that, but he contributed much. In 1844 he changed the registry pencil to a stylus that perforated paper. And it was Vail who suggested replacing the composing stick with a key. By the mid-1850s operators discovered that rather than wait for the message to be registered on paper, they could transcribe signals six times quicker by listening to the clicking of the armature against the register's electromagnet.

ing the DNA of innovation—that even the most contriving mind always has a double helix of debt to earlier discoverers and unknown patient translators in the mechanical arts. In January 1855 Morse wrote that he was not indebted to Henry "for any discovery in science bearing on the telegraph." That was just plain false. Nor was Morse ever willing to specify what Vail and Gale had contributed. Vail did not make claims. He always said he preferred to maintain the "peaceful unity of invention."

In his aggressive spirit, Morse was no more than representative of the spirit of the times. And when he became wealthy in

his 60s, he did as John Rockefeller and Andrew Carnegie were to do. He gave much of it away. Poor artists enjoyed his benevolence, as did Yale University and Vassar College, which he cofounded in 1861 with a Poughkeepsie brewer named Matthew Vassar; Morse built himself a mansion, Locust Grove, on the Hudson River nearby.

Toward the end of Morse's life, at one of the many functions in his honor, his conscience, giving wings to a subliminal sense of guilt, found a comforting refuge. He told the banqueters in London that the honors for a great invention or discovery would often fall on one person, and then he addressed the equity of this: "How significant is it that time and more research bring out other minds and other names to divide and share with him the hitherto exclusive honors? And who shall say that is not eminently just? Did Columbus first discover America, or does Cabot, or some more ancient adventurous Northman dispute the honor with him? Is Gutenberg or Faustus or Caxton, the undisputed discoverer of the art of printing? Does Watt alone connect his name with the invention of the steam engine, or Fulton with steam navigation?"

The "consoling" lesson Morse commended from this "voice of history" was that man was ever but an instrument and the chief honor was always due to God. Before he died, Morse knew he had conferred a nervous system on America and the world. Telegraph wires were approaching 250,000 miles in the United States, and the continents were linked by 100,000 miles of undersea cables. Of the telegraph, he said at yet another banquet: "If not a sparrow falls to the ground without a definite purpose in the plans of infinite wisdom, can the creation of an instrumentality, so vitally affecting the interests of the whole human race, have an origin less humble than the Father of every good and perfect gift?" Or, to adapt a famous phrase:

```
•--  ••••  •-  -
W    H     A   T
--•  ---   -•
G    O     D
••••  •-  -  ••••
H     A   T   H
•--  •-•  ---  -  --•  ••••  -
W    R    O    U  G    H     T
```

CYRUS McCORMICK The marriage of his mechanical reaper and his easy credit system fed America and freed labor for the factories. He was a skinflint and a fighter, and an architect of American big business

1809–1884

He gambled on the potential of the West for machine harvesting and won. The advertisement " 'Westward the Course of Empire Takes Its Way' with McCormick reapers in the Van" was a fair summation of history. He was a passionate advertiser: "Trying to do business without advertising is like winking at a pretty girl through a pair of green goggles. You may know what you are doing, but no one else does."

THE RISE OF Cyrus McCormick parallels the rolling settlement and enrichment of the West in the second half of the 19th century, of which he was a primary facilitator. At the age of 22, in an isolated little hollow in the Virginian mountains, he invented the first practical mechanical reaper, and then by adventurous manufacturing, innovative financing and imaginative marketing, he enabled many thousands of farmers to harvest the great plains of America and end for good the recurrent fear of famine. His reaper was the greatest single step toward the American mechanization of agriculture. But it was more. As well as enabling America to feed itself and then millions around the world, the reaper freed labor for the industrial revolution, and for the preservation of the Union. Without the reaper, only at the risk of starving the population could hundreds of thousands of young men have been spared from the land to serve in the Northern armies. In 1830 three million of the total labor force of just over four million worked in agriculture; the reaper began the process by which today less than 5 percent of the working population does that.

The episode that best illustrates the character of the man who brought all this about concerns the sum of $8.70.

In March 1862, when he was 53 and a millionaire, McCormick was returning to Chicago from Philadelphia with his 28-year-old wife, Nettie, two infant children, two nurses and nine trunks. A few minutes before the train was due to leave Pennsylvania Central Station the conductor confronted him with a request for $8.70 for "excess baggage." McCormick asked the conductor to explain why the same railway had carried the same nine trunks from Washington to Philadelphia without demanding a cent. The conductor would not budge. McCormick herded his family retinue back onto the platform, whereupon the conductor waved his flag and the train steamed off west with the McCormick luggage still on board.

The president of the railroad telegraphed to have the trunks held at Pittsburgh for the next train, but the wire got through too late and the baggage went to Chicago, where it was promptly incinerated—stored in a station room burned to

Competition was Cyrus McCormick's middle name. Nothing mattered more to him than beating the other man, whatever the cost. He left $10 million. Half was disbursed in philanthropy.

the ground by a shaft of lightning. Over the next 18 years the Pennsylvania Central Railroad Company came to see the lightning as a symbol for its experience in dealing with McCormick. He asked for $7,193, mainly in compensation for jewelry he had given Nettie before their marriage, and she submitted a careful list of their other losses. The railroad denied any responsibility and then, when it lost a court hearing, it signified its determination to appeal all the way to the U.S. Supreme Court.

McCormick fought through five trials at legal costs to both sides far beyond the $7,193 at issue. As an investment of time

and money, it was a dead loss, but to McCormick it was a matter of principle. He felt his personal integrity was impugned by the railroad's refusal to accept his account of the incident, and personal integrity was something he cherished more than all the reapers he ever made. The long delay, to be related, between his inventing the reaper and selling it, was in part a reflection of his reluctance to put his name to something that could not be guaranteed to do what it was supposed to do.

Whenever McCormick thought a principle was involved he was boneheadedly inflexible whatever the odds against him.

In fact, the greater the odds, the more he fought. His early biographer Herbert Casson characterized this as a reflex: McCormick always followed "the line of *most* resistance." The disbursement of petty cash was a neuralgic point. McCormick was comfortable spending large sums—on speculative ventures, on buying patents, on salaries, on helping out Civil War victims, on Presbyterian seminaries and secular schools—but he had a lifetime hatred of parting with any money for little things, and when he became rich his frugality was inflamed by the fear that he would be imposed on because of his wealth. Every minor bill was to him the start of a negotiation. He bargained for cheaper meals at his regular Chicago restaurant. He rebuked employees who used a word in a telegram he thought not absolutely essential. He roasted a senior man for writing to him on such good paper it risked adding to the cost of the letter's postage.

This tight fist did not reflect an obsession with wringing every last cent from the business. McCormick's biographer William T. Hutchinson writes: "Success to him did not mean the accumulation of money. In the many hundreds of his letters, profits are given a subordinate place among the objectives of season's business." His focus was always on beating someone, whatever the cost. He spent $90,000 fighting objectors to an extension of his original patent, in return for $40,000 in royalties.

But while McCormick was utterly pugnacious, he was bullheaded. The baggage incident occurred during the Civil War, in which he was suspected of having southern sympathies as a slave-owning Virginian and a prominent stand-pat Democrat. The war was so painful for him he went to Europe for nearly two years, so he did not press his suit until 1867, two years after the end of the war. His habit was to think things through slowly and deliberately, writing down all the points pro and con until he could forge the conclusions in solid iron.

He insisted on every understanding being written down and honored to the letter. Such forthrightness and rigidities disqualified him for politics, as he found when he ran for Congress in 1864. He could not bring himself to utter "weasel words." Later in life, he seldom decided anything important without consulting Nettie. He had first become enchanted by her voice in a Presbyterian choir; she learned how to bend his will by apparent compliance, and he more and more listened to her sweet counsels of moderation and generosity.

McCormick was at his best when his associates were certain defeat was imminent. He was indomitable even in death. When the U.S. Supreme Court awarded him the compensation he claimed and the board of Pennsylvania Central finally concluded it should get back to running a railway, Cyrus McCormick had been dead a year. Nettie collected the compensation, $18,060.79, their original claim of $7,193 plus 23 years' interest. She wrote to their son, Cyrus Jr., "You will see (by reading the opinion) that your dear father's course from first to last is approved and vindicated."

He enjoyed a good deal of posthumous vindication. Throughout the 1850s, '60s and '70s he had pushed to make more reapers, and more reapers, and more reapers, against the resistance of both his younger brother-partners—William, who died in 1865, and the youngest, Leander, who ran their Chicago factory. His brothers thought the 5,000 reapers a year achieved in 1859 an "enormous business." Cyrus looked to produce 10, 20, even 40,000. Though he appreciated the economies of scale that gradually reduced the cost of a $120 reaper from $55 then to $18 by 1877, his real motive was dominance. McCormick did not want to stay in the business unless his company could clearly be the premier world's harvest machine manufacturer; on the eve of the Civil War there were in the United States alone 100 such companies. That, he concluded, could not stand.

Leander had a point that his brother's incessant design changes for market appeal added to costs and delayed production, but emotion soured relations, too. Leander and his sisters resented the hero worship Cyrus had come to receive—and invite; they started suggesting that their father was the true inventor. Between the consolidator and the expansionist, the unsung superintendent and the icon, it became very heated. In an 1873 letter to Leander, Cyrus wrote: "I can not permit you to come to *my room* to continue and perpetuate your calumnious and bullying abuse of an 'old scoundrel' and 'old scamp,' as you have at different times termed me, while I told you that nothing your mouth could utter or fists gesticulate could induce me (at my age) to foist myself by a personal encounter with you."

The Great Fire of Chicago in October 1871 destroyed McCormick's first factory, but it gave him the opportunity at the age of 62 to express his faith. Under the darkening clouds of recession, he built a gigantic steam-heated, gaslit complex that could be justified only by sales of many more thousands of machines: four huge five-story buildings, a 300-horsepower engine, a boardinghouse and 40 workmen's cottages; he watched it all erected day after day astride his favorite horse amid the mud and clamor. In 1880, when the openly rebellious Leander wanted to limit production to 15,000 machines, Cyrus fired him.

It was a just action and decisive for the transition of the enterprise from a family firm to a great corporation. Leander was able but limited. In the judgment of technology historian David Hounsell, he "operated the reaper works as though it were a large country blacksmith shop," uninterested in the expertise in large-scale production accumulated in the New England armories. Cyrus was not an expert in machine tools, but his instinct that his factory could produce far more was right, and his selection of a replacement for Leander demonstrated an acute awareness of what was happening to American manufacturing. The mechanic he recruited, Lewis Wilkinson, was a graduate of the Colt "American system" geared to high-volume production with precision machine tools. He rapidly expanded output. In the year of his death, the old man could rejoice in making more than 50,000 machine implements (including binders), and within five years his son, Cyrus Jr. (1859–1936), had doubled that. In 1902 the McCormick Harvesting Machinery Co. had so much of the business it was the dominant company in a merger to found the International Harvester Company.

NONE OF the 100 or so spectators in July 1832 in farmer John Ruff's wheat field near the small town of Lexington, the county seat of Rockbridge, Virginia, could have had the faintest idea of what would come from the exertions of the young man with two horses pulling a weird contraption of wood and metal with a black man by his side. Ruff had agreed to let Cyrus McCormick attempt to cut his wheat out of a gesture of respect for the father, Robert, a battling Scotch-Irish farmer-inventor with 1,700 acres of land, nine slaves and a blacksmith's shop.

Robert was known to have failed with a reaper of his own design that left his field a tangled mass of cut and uncut grain. Cyrus, after a private trial run in 1831, had contrived a reaper that dealt with his father's failure by introducing a roughly triangular wedge to keep the standing growth clear of the grain cut when swept into the blade. It was a very simple idea that nobody else happened to have hit upon. It worked to separate the wheat on Ruff's field, but Ruff was quickly alarmed by the vibration of the machine on his rough ground. "Here! This won't do," he yelled. "Stop your horses. You're rattling the heads off my wheat."

One of the 75 or so amused spectators was William Taylor, later a candidate for the governorship of Virginia. "I'll give you a fair chance, young man," he told the dejected Cyrus. "That field on the other side of the fence belongs to me. Pull down the fence and cross over." Taylor's field was more level than Ruff's. By sundown Cyrus McCormick's mechanical reaper had cut six acres of wheat in the time it would have taken six strong men with scythes. It was not a perfect operation, but he had succeeded where scores of experimenters had failed. Six mechanical elements came to be the essential features of all reapers and they were all embodied in McCormick's 1832 design: the wedge; a single main power wheel; a row of projecting wire fingers to hold the grain to be cut by the blade; a large revolving flail of thin parallel slats to push the heads of the grain gently toward the reciprocating blade and then catch the grain and lift it onto a platform, where it could be raked off by a man walking alongside. In these first demonstrations, the man raking the sheared stalks by McCormick's side was a boyhood companion of his age by name of Jo Anderson—one of the family's nine slaves. (McCormick never forgot him and gave him his freedom before the Civil War. He was not an abolitionist. He favored gradual but not immediate emancipation, believing the Bible did not discountenance human bondage; it was an attitude common enough among those Democrats anxious to preserve the Union.)

The wedge apart, the elements in McCormick's reaper were not original. McCormick's distinction was to put them together in an effective manner, just as Robert Fulton had combined the work of the earlier steamboat pioneers. There was a long way to go to make the first crude reaper effective in all circumstances. It was awkward on lumpy ground like Ruff's, its knife was vulnerable to stones and wet grain clogged the slit in which the knife slid back and forth. Still, that first public demonstration was a critical augury and acclaimed as such in the locality. It was an invention that offered a cornucopia of benefits. The limitation on the food supply, and the size of farms, was the short window of the grain harvest. What could not be harvested in two weeks of backbreaking work, sunrise to sunset, went to rot. All around the world, millions of families labored to harvest grain with hand tools that had changed little in a thousand years, the sickle and the scythe. In a full day, five scythemen and ten helpers might harvest ten acres of grain. Even with McCormick's first crude reaper, that could be done by one man driving a horse with eight helpers, a dramatic saving in manpower or time. Cyrus would later say he realized at once he had invented a machine that was worth a million dollars, but, as the writer Harold Livesay observes, he was simply indulging in "the omniscience of hindsight." At the time, he was bothered by the imperfections and determined not to advertise until he had removed them. He did not patent his reaper until three years later, in 1834, and only then because he read about a mower-reaper patented by Ohio's Obed Hussey, a sailor turned inventor who sported an intimidating eye patch. Hussey went ahead, manufacturing his reaper for sale. It was to be six years before McCormick felt his reaper sufficiently improved to merit another public demonstration.

During these years, young McCormick by no means abandoned the reaper, but his priority in the mid-1830s was the farming of 500 acres his father gave him, and then an ambitious venture to make iron from a local deposit of the ore. The family hoped for quick profits from iron at $50 a ton; since iron was also the costliest material in a reaper, it was a two-edged gamble. But the foundry they built eventually failed in the long fallout from the 1837 financial panic. The recession ruined the land-rich, cash-poor McCormicks, as it ruined the Good-

A McCormick's reaper, first tried in Virginia, began the revolution that led to today's wheat being threshed, cleaned, stacked and weighed without being touched by hand. This reaper, built for the harvest of 1848, went through many improvements: a seat for the driver, a sheet of zinc on the platform to reduce the labor of raking, stronger wheels and more effective cutting blades.

McCORMICK'S PATENT VIRGINIA REAPER.

The above cut represents one of M'CORMICK'S PATENT VIRGINIA REAPERS, as built for the harvest of 1848. It has been greatly improved since that time, by the addition of a seat for the driver; by a change in the position of the crank, so as to effect a direct connection between it and the sickle, (thereby very much lessening the friction and wear of the machinery,) by dispensing altogether with the lever and its fixtures;) by board ribs on the reel, (which operates more gently on the grain than the round ones;) by a sheet of zinc on the platform, (which very much lessens the labor of raking;) by an increase of the size, weight and strength of the wheels of the machine, and by improvement made on the cutting apparatus.

D. W. BROWN,
OF ASHLAND, OHIO,

Having been duly appointed Agent for the sale of the above valuable labor-saving machine (manufactured by C. H. McCormick & Co., in Chicago, Ill.,) for the Counties of Seneca, Sandusky, Erie, Huron, Richland, Ashland and Wayne, would respectfully inform the farmers of those counties, that he is prepared to furnish them with the above Reapers on very liberal terms.

The Wheat portions of the above territory will be visited, and the Agent will be ready to give any information relative to said Reaper, by addressing him at Ashland, Ashland County, Ohio.

Ashland, March, 1850.

CREDIT AND CONFIDENCE: McCormick pioneered credit sales and after-sales service. Agents like D. W. Brown had to be a farmer's best friend.

years, but the one consolation was that it cut the price of iron just when the family's predicament forced Cyrus to turn his full attention to improving and then making reapers for sale. Hussey was ahead in what came to be called the War of the Reapers from 1839 to 1847, but his machine was less good at cutting grain than grass and he had inflicted all his experiments on his early purchasers.

McCormick's father and two brothers, and Anderson, began handcraft production in the family blacksmith shop at Walnut Grove in 1840 while Cyrus saddled up and went selling. He sold only three locally that year at $100 each, and none the next. Instead of selling a faulty machine, as Hussey continued to do, he experimented with improvements, notably a serrated edge to the cutting blade that enabled his reaper to cut wet grain almost as well as dry. The sacrifice of selling time was well judged. When the rivals met before festive crowds for the first reaper battle in history in 1844, in the lower James River region of Virginia, a shower drenched the grain and Hussey could not cut at all. In a return match the following harvest, McCormick cut 17 acres and Hussey 2.

It was only in 1842, ten years after his first machine, that McCormick sold a

reaper outside his neighborhood: A total of seven were ordered by farmers in Ohio, Illinois, Wisconsin, Iowa and Missouri. To deliver machines to the western states, however, was costly and laborious. Walnut Grove was 100 miles from the nearest railway, 60 from a canal. Each bulky machine had to be put on a horse-drawn cart to Scottsville, Virginia, transferred to a canal boat to Richmond, loaded onto a freighter to New Orleans, lifted onto a steamboat ascending the Mississippi and unloaded for final horse-drawn delivery. He started subcontracting as the orders flowed—50 in 1844, by which time the reaper had become of real practical value, then 190 in 1846.

McCormick's trademark mode of operating, prolonged deliberation and determined follow-through, is manifest in the most important decision he made in his early 30s, to follow the wheat and locate production somewhere closer to his markets. But where? The two largest wheat states in the Union in 1840 were New York and Virginia. Hussey had located one factory at Baltimore and another at Auburn, New York. One morning, with $300 in his belt, McCormick started not north but west, on a 3,000-mile scouting expedition by stage, canal barge and steamboat. Anyone traveling west at this propitious time in the American expansion could not have failed to remark the potential for machine harvesting in the flat prairie vistas of Illinois, Indiana, Ohio and Wisconsin and the empty lands that were to become Kansas, Minnesota and Nebraska. A geographer might have noted that a swampy Illinois settlement of 18,000 people, prone to epidemics of cholera and smallpox, was nonetheless ideally placed on Lake Michigan, with a canal connection to the Illinois River and the Mississippi, and an inviting landscape for the tendrils of railway lines feeling the way west. McCormick made the same observation and chose Chicago for his factory. His decision was not entirely based on the city's potential. Chicago had a devoutly active Presbyterian community, and that was an influence. "I see a great deal of profanity and infidelity in this country," he wrote, "enough to make the heart sick." Like Goodyear and Morse, he knew God was on his side. Perhaps more important, of all the McCormick reapers made by subcontractors, the best had come from Chicago's Charles M. Gray and Seth Warner. The accident of the subcontractors' competence thus attracted him to a location that before long would fulfill its promise as one of the nation's greatest manufacturing and distribution hubs. None of his rivals saw the potential as he did, few understood how the center of grain production was moving inexorably west, none appreciated as early as he did the magnetic vitality of the American dream. He had acted swiftly to place a bet on the future and it was about to be paid off. The ink was not dry on his real estate contracts when immigration soared from fewer than 100,000 a year to 430,000 by 1854.

His first Chicago factory, near the mouth of the Chicago River, was a two-story 40 by-100-foot building with a ten-horsepower steam engine. It turned out 500 reapers for the 1848 harvest. This was the year the earliest of his three patents (1834, 1845 and 1847) expired, opening the market to still more competition, but in 1851 he made 1,000 reapers and 23,000 in 1857 (at a profit of $1.25 million). Into the mid-1850s, McCormick made changes in the basic machine every year. It was as much an act of marketing as inventing. He sold the improvements as a "new model" reaper, as automobile manufacturers came to do. His wisest move, however, was to relinquish the role of inventor. In the 20 years after the Civil War there was acceleration in agricultural innovation that outpaced the creativity of any single inventor. Hundreds of small-town mechanics and rival companies struggled to put two men on a machine that would automatically deliver cut grain in a form where it could be bound in wire or twine. In 1868 there were no fewer than 80 patents for automatic binders. The McCormick Company stayed competitive with the mower, the self-rake reaper, then the harvester and the harvester binder principally by buying patents and companies and organizing a little espionage. Meanwhile, McCormick manufactured a revolution in marketing.

First, he put his good name on the line by offering a personal guarantee. Farmers could have their money back if that year's model did not, say, cut at least 1.5 acres an hour. "One Price to All and Satisfaction Guaranteed" was the message hammered home all over the West in the proliferating newspapers of rising circulation. Advertising on this scale and of this insistence was something new. McCormick's second innovation was buy now, pay later. Few farmers, limited in acreage, had $120 to $130 in cash before a harvest, and local banks would not give credit. His offer was $30 down and the balance within six months on full satisfaction (occasionally he allowed two years). He made a man-to-man appeal: "It is better that I should wait for the money than that you should wait for the machine that you need." It was a risk lending money to thousands of strangers across the country. McCormick was reluctant to hazard goodwill by taking a farmer to court, but he was not soft. Agents were told their own incomes were at risk if they misjudged creditworthiness, and they were exhorted to have "spunk," "grit" and "sand" when they chased delinquents. "Spit on your fists," they were told, "and make it hot for them." In 1860 he was owed $1 million. He was charging 6 percent then, the same rate of interest he paid on his borrowings, but he was shrewd not to squeeze profit out of the financing. Huge profits, at least 35 percent, lay in volume sales, and volume was what he got. This is the kind of strategic thinking that distinguishes the great entrepreneur from the mere moneymaker.

McCormick backed the flairful sales effort with an enormous service organization he invented as he went along. He regarded his personal guarantee as sacrosanct, and woe betide any sales agent who left a harvester struggling to introduce a skeptical horse to a rattletrap. Agents had to be the farmers' best friend. The loose arrangements of the early years developed into a system of closely regulated franchised dealerships, the forerunner of automobile dealerships today.

Almost to the end, McCormick drove himself 14 hours a day. He would be at his desk at daybreak, cogitating, preparing for the endless patent litigation, pondering the problems of exporting to Russia and Canada, Argentina and Australia. Around 7 a.m. Mrs. McCormick would read Scripture to him, and then he would breakfast on corn mush and milk and walk to work briskly and beautifully tailored. Often he had to catch a train for a legal hearing in some city where he would listen to the patent experts as he shaved in his hotel room. Every minute counted. Both he and Nettie instructed their children to write down how long they spent getting dressed and cleaning teeth. Time was a tangible to him, like the dimes he begrudged on every hotel bill. But in his last 25 years the always-careful calculator came more and more to share Nettie's belief that money should be put in the service of others during a lifetime rather than saved for a legacy. Among other philanthropies, he bought a cabin and a lot "lying well to the sun" for Anderson.

Hutchinson writes that after 1855, McCormick most frequently saw a farmer from the window of a railroad car, but that was a mark of his good sense. He could have stayed in the harvest field, he could have stayed in the machine shop, but if he had done so, he would not have become the emperor of agricultural machinery when both the technology and the marketplace were exploding and the patent wars were escalating from red-hot to white-hot, embattling lawyers as luminous as Abraham Lincoln, William Seward, Edwin Stanton and Roscoe Conkling. McCormick's most important innovation, though it may not have been consciously realized, was to isolate himself, to step out of the trenches and away from field headquarters, to survey not just a single battlefield but the whole campaign in America and the world. His competitors initially won more orders overseas, but when he turned his full attention to foreign markets, his lifelong habit of scrupulous observance of his agreements proved a superior asset. His signature was honored in the leading banking houses; his integrity (more than his inventiveness) was fertile ground for the growth of the International Harvester Company 20 years after his death. The young man guiding a team of four through the Virginian wheat field with a prayer in his heart became a pathfinder for the great American corporation.

ISAAC MERRITT SINGER His name is synonymous with the sewing machine. He led a scandalous life, but his mechanical insights and the acumen of his partner, Edward Clark, created the first successful American multinational

1811–1875

SUPPOSE YOU ARE a businessman and you hear in the club one lunchtime that there is a golden opportunity to invest with an innovative manufacturer, call him Mathews. He is a genial, well-dressed giant with a mellifluous voice, his enterprise is expanding, he is full of brilliant ideas and the numbers he shows you are magic. Would it make any difference at all to know that Mathews is a polygamist of violent temper and no scruples who has abandoned his wife, lived a decade under false names as the head of three other households, fathered at least 18 bastard children, choked the mother of 10 of them into unconsciousness, brutally beaten a daughter, threatened to murder a son and tricked his founding partner out of the business?

Of course it would. Sensible people would run a mile, out of prudence if not moral revulsion. But it would mean running away from a fortune—and from the future. Isaac Mathews, or Isaac Singer, to give him his real name, or Isaac Merritt, to note another name he adopted to conceal his betrayals, was a rascal, or a scoundrel, if you like, but an inventor-innovator with a golden touch. At 40 he was worth nothing. At 44 he was a millionaire, and at 64, after 20 years of wild spending, he was still able to leave an estate worth $150 million at today's values. He liked to say, "I don't give a damn for the invention, the dimes are what I am after," but although his little sewing machine made him rich, it also achieved a lasting renown surpassing all the sensations associated with its long-

The iconic Singer, one of millions. By 1877, the Singer Company controlled three quarters of the world sales of sewing machines.

forgotten inventor, and a universal esteem he never knew. It was an engine of industry, clothing a nation and its armies; and it was the first mass-market consumer durable, as indispensable as the personal computer today. Its domestic ubiquity was very American; it came to be as much a feature of the affluent middle-class parlor as the tenement kitchen, a symbol of achievement and self-reliance and classless in a way that the carriage and the automobile have never been. It was a democratizing influence. James Parton was exaggerating only a little when he wrote in the *Atlantic Monthly* in 1867 that it was "one of the means by which the industrious laborer is as well clad as any millionaire."

The reprobate inventor-innovator improved the lives of millions of people in the cities and small farming communities of America, in the mud-hut villages of Africa,

in the teeming townships of Latin America, in the bazaars of Asia. Wherever you went in the world in the second half of the 19th century and into the 20th, the Singer was the star. There is a telling photograph from Morocco in which a tribeswoman carries a brand-new Singer on her head, as reverently as if it were a crown. In settlements of the American West, when families put on their Sunday best and stood in the sunshine to be photographed amid a display of all their worldly wealth, the Singer had pride of place.

The sewing machine—the Singer most of all but also its competitors—was a manifold creator of wealth and work. The cornucopia of raw cotton from Eli Whitney's invention of the gin fed straight into the hungry power looms of Samuel Slater and Francis Lowell and then as cloth into Isaac Singer's chattering shuttles. The machinery vastly increased productivity, but also employment. In the women's clothing industry alone, for example, 112,000 were at work in 1905, by comparison with only 5,729 in 1850. Of course, the Singer was a paradox, one of the two great labor-saving inventions of the 19th century (the other being the typewriter) and at the same time the instrument for the exploitation of labor. Too often seamstresses at their machines were crowded in hundreds of unsanitary and unsafe factories. The most notorious

The imposing, impossible Isaac Merritt Singer. He could not control his lusts and rages, but his invention improved the lives of millions.

consequence was the fire at the Triangle Shirtwaist Factory in New York City in 1912, when 46 women trapped on the ninth floor jumped to their death and another 100 charred bodies were found inside. But that outrage in turn strengthened the International Ladies' Garment Workers Union and gave muscle to social reformers; in New York State alone it resulted in 56 landmark legislative reforms of hours and conditions affecting all industries.

There was paradox, too, in Singer the man and the Singer Manufacturing Company. The factory built at Mott Street, New York City, in 1858, was fireproof. Singer's company, in fact, was a model of the Victorian virtues of straight dealing and fidelity he did not practice in his personal life. He was for the most part, too, a generous father and provider; he did not drink, he did not disown his illegitimate children, his wives loved him and many people out of the eye of the hurricane found him a lot of fun.

It was just that he could not control his lusts and his rages. Singer's secret life was actually more extravagant than the fanciful sketch of Mathews. It was almost as if he were playing a role as a libertine. When he left home in Oswego, upstate New York, at the age of 12, unloved by his German immigrant father's second wife, he had odd jobs, including one as an apprentice machinist, but from the age of 18 and the rest of his life he was in love with drama and one woman or another. "Singer loved women for their bodies," writes biographer Ruth Brandon, "and not for anything as abstract and uninteresting as their minds." He had at least six common-law and legal wives—in the days before photography, still less paparazzi, represented a threat to privacy. In 1830, when he was 19, he took as his first wife Catharine Maria Haley, who was only 15. He walked out on her and two children in Rochester seven years later and set up in Baltimore with an 18-year-old, Mary Ann Sponsler, whom he persuaded to live with him unmarried. She was the counterfeit Mrs. Singer, the dutiful consort faithfully living Singer's lie for more than 20 years of prolific childbearing in various small towns and cities, and for a time acting in his traveling repertory troupe, but in New York, for almost all of their 11 years

from 1849, he secretly and simultaneously had a third and a fourth family in the same city. In the third family he was Mr. Merritt and the father of one child by Mary Walters; in the fourth he was Mr. Mathews and the father of five children by another Mary, a Ms. McGonigal. In any bed, "Mary!" was a conveniently safe utterance.

The very inconvenient event that led to Mary Sponsler's near strangulation was the screaming scene she created one beautiful August day in 1860. Taking the air down fashionable Fifth Avenue, ensconced in one of the six Singer family carriages, she hap-

Five women at least thought Isaac Singer was their husband. This one, Isabella, was his last (and legal) wife. Does she resemble Lady Liberty?

pened to glance out at a happy couple in an open carriage going the opposite way. They were "Mr. and Mrs. Mathews"—that is to say Isaac Singer and Mary McGonigal—stunned to find the afternoon air suddenly a whirlwind of vituperation directed at them. When Mary Sponsler returned home (a palace at 14 Fifth Avenue), an apoplectic Singer was waiting to beat her senseless; their daughter Voulettie, who tried to intervene, was pounded into unconsciousness. To escape the obloquy of his conviction—he was bound over six months for assault—Singer took refuge in Europe, bearing with him Ms. McGonigal's 19-year-old sister, Kate. He left a typically illiterate note with

an untypically humble message: "My private affairs (though justly merited) hangs heavily upon me and my soul sicends [sickens] as the prospects before me and for the well fare of all concerned to try and [make] my load of grief as light as possible . . ."

The consolation he found in Europe was a ravishing half-French, half-English beauty in her 20s, Isabella Boyer. Millions have been as overcome as Singer by contemplation of the exquisite profile of Isabella; the dates do not quite fit, but his biographers Ruth Brandon and Don Bissell are convinced that in Paris, as a much-sought-after widow, she posed for the sculptor Frédéric-Auguste Bartholdi as he fashioned the fair face of the Statue of Liberty.

Singer proposed marriage. It required a divorce from Catharine, the first Mrs. Singer. To appear as the innocent party, he tried to bribe their son, William, to lie about all that had happened. When William balked, his father exploded, threatening that he would kill him even if he "hung as high as Haman the next hour"; he never forgave William for his loyalty to his mother.

Nevertheless, the divorce went through and Singer married Isabella on June 13, 1863, in an Episcopalian ceremony in the Church of St. John the Evangelist, Waverly Place, in New York. She bore him the first of their six children 12 days later.

THE FOUNT of Singer's midlife carnival was the sewing machine he invented in a few days in Boston in November 1851. Hand-cranked mechanical devices for sewing together pieces of cloth had already been imperfectly invented in Germany (1755), England (1795), France (1830) and America (1834–46); the French one was good enough to provoke tailors to riot, breaking the heart of the inventor as well as his machine. The illusion that plagued early inventors of the sewing machine, as with the reaper, was that success lay in contriving a mechanical replication of the movement of the human hand. It was an exercise in futility. One has only to think of the difference between the hand movements in writing and typing to appreciate the point. For sewing to be mechanized it was necessary to break away from trying to replicate the nimble fingers of the seamstress, to remove all memory traces of hand-sewing and think conceptually of the desired end. Singer had this imaginative faculty for selective "seeing" and so did two of the American pathfinders who preceded him in the evolution of a reliable machine for stitching fabric. Why should a machine be restricted to using one thread as the seamstress is? Why should it be necessary to pass the full length of thread through the cloth at every stitch? And, again, although the eye of a hand-sewing needle is necessarily at the far end from the sharp point, why must that separation be replicated in a machine?

In 1834 Walter Hunt of New York City made such a leap in lateral thinking. In his little machine shop down a narrow alley in Abingdon Square, he devised a machine for stitching cloth with two threads from two separate sources, one a needle on a vibrating arm and the other a transverse shuttle fed by an unwinding bobbin. His needle had the eye at the sharp point so that when the operator cranked the gear the needle's thread was stabbed into fabric just enough to create a slack loop of thread on the underside. The shuttle on the underside pulled another thread through the slack loop. The shuttle's return to its starting position then tightened the two threads, forming a lockstitch.

Hunt, an altruistic Quaker, never pursued his invention because his 15-year-old daughter, Caroline, recoiled from the thought that it would put seamstresses out of work. Hunt's mind teemed with so many ideas, he moved cheerfully on to the next; he was the epitome of the absentminded professor. (This is the man credited with conjuring up the first safety pin in three hours from bits of wire, in a pause from inventing a breech-loading rifle, a variety of velocipedes, metallic cartridges, conical bullets, paper collars and a variety of machines to sweep streets and sharpen knives.) Ten years after Hunt abandoned his machine, a lame Boston machinist, Elias Howe Jr. (1819–67), unaware of Hunt's work, devised and later patented a machine with the same three elements of a lockstitch, an eye-pointed needle and a shuttle. The legend is that it was fast enough for him to beat five of the swiftest hand sewers in a race in Quincy Hall in Boston, but

SONG OF THE SHIRT: Thousands of new jobs were created in an expanded clothing trade—but exploitation was commonplace, dramatized by an artist for Frank Leslie's *Illustrated Newspaper* under the headline "The Female Slaves of New York—'Sweaters' and Their Victims."

Howe had trouble feeding the fabric, and the cost of his machine, at $300, enough to hire 30 seamstresses for a year, was too expensive to attract financing. In 1847 he sailed to try his luck in England.

No fewer than ten letters of patent for sewing machines were granted to other inventors between 1849 and 1850. The leading seller was a two-thread lockstitch machine designed by a Boston tailor, Sherburne Blodgett, in which a shuttle revolving in a continuous circle delivered thread as the needle began its upward revolution. A Boston machinist, Orson P. Phelps, turned out 120 of the Blodgett machines in his steam-powered workshop at 19 Harvard Place, but he was beside himself trying to keep them in working order. The rotary action took a twist out of the thread, thereby weakening it, and thread breaks were maddeningly frequent.

Into this workshop in June 1850 strode 38-year-old Singer. He was not interested in the sewing machine. He was renting space in Phelps's workshop because Boston

was the center of the book manufacturing trade and he needed somewhere to demonstrate to printers and publishers a machine he had invented for carving printer's type out of metal and wood. He was down on his luck. Through no fault of his, an exploding steam boiler in rented space in New York had wrecked his first machine (and killed 60 people). He was penniless, his expanding family with Mary Sponsler crammed into two rooms at 130 East 27th Street. He was wholly dependent on the last few dollars of his companion in Boston, George Zieber, a former publisher of very modest means, naively eager to become a capitalist. Zieber had seen Singer's type machine before its destruction and had raised the money to build a second one. He had also taken pity on the Singer family, sending them ten dollars a week for food while he and Singer were in Boston, "fighting the battle of life" together, as Zieber put it.

Nobody rushed to buy the type carver. The hot summer months went by. In

September Phelps approached the inventor as he sat dejectedly on a pile of boards by his idle machine. There was more future, he told him, in solving the problems of the sewing machine. The response, according to Phelps, was bitter: "Good God!" said he. "Phelps! Do you think I would leave this ponderous machine and go to work upon a little contemptible sewing machine?" And again: "What a devilish machine! You want to do away with the only thing that keeps women quiet, their sewing!" He grumpily agreed to take a look at Phelps's contemptible machine only because the Zieber-Singer partnership was running out of money. He saw at once what was wrong with Blodgett's rotary shuttle. He sketched a redesign for a table-mounted machine with a vertically reciprocating needle bar from an overhanging arm. He suggested a novel mechanism for feeding the fabric, called a yielding presser foot. While the fabric was held down by a flat metal plate, a wheel with short pins would rotate vertically under the plate, its pins engaging the cloth and feeding it past the needle. Phelps thought he could make a model of this machine for $40. Now Singer's blood was up. Zieber's was not. He had sunk $3,000 to $4,000 into the unfortunate carving machine. What was he to tell his backers? But the amiable Zieber was no match for Singer in full flow. Somehow, the embryonic capitalist came up with another $40.

On September 18, 1850, the three men signed an agreement "to become co-partners in an Improved Sewing Machine, to be called the Jenny Lind Sewing Machine." Phelps was to contribute "his best mechanical skill" in making the model for a patent and have exclusive manufacturing rights, Zieber to "attend to the business," and Singer to contribute "his inventive genius." All improvements made to the machine by Singer "shall belong to and be the property of the said parties thereto." They called themselves I. M. Singer & Company.

Singer drove Phelps and his workers hard behind locked doors. "I worked at it day and night," the histrionic Singer told

SEW AT HOME: The first machines were made for clothing factories. Isaac Singer was early to woo housewives.

an interviewer years later, "sleeping but three or four hours a day out of the 24 and eating generally but once a day, as I knew I must make it for forty dollars or not get it at all." It must have been a madcap scene, the machinists trying to make something of the impulsive inventor's sketches chalked on bits of board, Singer flaring up when something went awry, then seeking to cheer the winter of their discontent by solo performances from his Shakespearean repertory. The charm that never failed with the ladies won the machinists. On the 11th day around nine o'clock in the evening, as Singer told it, the parts were all ready to be put together. "It did not sew; the workmen exhausted with almost unremitting work, pronounced it a failure and left me one by one. Zieber held the lamp, and I continued to try to work the machine, but anxiety and incessant work had made me nervous and I could not get tight stitches. Sick at heart, about midnight, we started for our hotel."

Singer's and Zieber's accounts of the melodrama vary only a little. "On the way," says Singer, "we sat down on a pile of boards and Zieber mentioned that the loose loops of thread were on the upper side of the cloth. It flashed upon me that we had forgot to adjust the tension on the needle thread." Zieber says they had reached the hotel and gone to bed before realizing what

was wrong. In any event, they went back to the deserted workshop around 1 a.m. Zieber: "I held my finger against the thread upon the face plate of the machine to keep it up from the point of the needle, which caused a tight seam to be produced. We then arranged what was called the 'spring pad.'" Singer says: "We sewed five stitches perfectly and the thread snapped but that was enough. At three o'clock the next day the machine was finished." He had found a way to stop the thread breaking.

Singer's model was the first practical sewing machine, altogether superior to anything that had been produced before. He freed the hands of the sewer by introducing a treadle. By slight foot pressure, an operator could maintain the machine in motion; gears translated the pressure to produce the different movement of the horizontal shuttle and the vertical needle. He kept making improvements as they began manufacturing late in 1850. Hand sewing from an accomplished seamstress could be as fast as 40 stitches a minute on simple work. But a good shirt required more than 20,000 stitches, nearly a full day's work. At Howe's demonstration of the machine, the top speed was 250 stitches a minute, an operator had to stop after only 18 minutes to move the material and there were frequent thread breaks. Operating Singer's machine at 900 stitches a minute, an average seamstress could make four shirts a day. Andrew B. Jack of the Massachusetts Institute of Technology, in a groundbreaking 1959 study for Harvard's Research Center in Entrepreneurial History, identified ten essential mechanical features of a viable and versatile sewing machine, not one of which could be forfeited without seriously limiting its utility. Only Singer perceived the importance of all ten and incorporated them in a machine, and two of the features were innovations: the presser foot and the means of sewing a curved as well as a straight line. "Credit for the invention of the sewing machine," Jack concluded, "must go to Isaac Singer."

Singer was of the same mind. Over the objections of his partners, he took the model and applied for the patent in his name alone—and brooded on how he could get full control. He sold 30 machines at $100 each to a New Haven shirtmaker. He put the money in his pocket and used it to bully Phelps into selling out for $4,000 — "By God, he shall remain in the business no longer!" Then he drove out another investor once he had served his purpose of plugging a hole in their cash flow. He was, surprisingly, reticent about calling the machine a "Singer" in their advertising, no doubt because he relished the theatrical aroma of "Jenny Lind," the fabled singer. But Zieber argued that "Singer" suggested the machine sang, and so it was that a famous brand name was given life.

The world did not beat a path to I. M. Singer and Company's door. Too many people had been burned by failed sewing machines. As a result, Singer had to beat a path to the world's door, which he did with a gusto of benefit to all the emerging competitors. The energy and charisma that fired on too many cylinders in his domestic life found release in barnstorming the marketplace. While Zieber raised money dollar by dollar to keep them afloat, the inventor as impresario sang Thomas Hood's lachrymose "The Song of the Shirt" (1843) at country fairs and circuses and engaged pretty young women to show everyone how easy it was to sew the Singer way. He segued from buying clothes for his children at Smith and Conant's store on Broadway into convincing the management they were in urgent need of two Singer machines at $125 each. They gave up a shop-front window to demonstrate their forward thinking. At one machine, Singer installed his teenage son Gus, his first child with Mary Sponsler, at another a lady tailor.

It was a clever stunt, but among the passersby who paused to see this new wonder was Elias Howe, a disconsolate widower not long home with little in his pocket after being cheated by an English manufacturer of corsets and umbrellas. He was resolved not to be cheated again. He held the patent for the lockstitch, which was the stitch in Singer's machine and in many others. Howe, then working as a machinist, went to see Singer to demand that Singer pay him $25,000 for the rights or quit the business on pain of being sued. The Singer Company, early in 1851, did not have

ELIAS HOWE, INVENTOR: Singer threatened to kick him downstairs. That was a mistake. Howe's machine was not as good as Singer's, but he held a crucial patent.

that kind of money. Singer, towering angrily over the tiny Howe, threatened to kick him down the steps of the machine shop. It was a stupid error. Howe's machine was not a patch on Singer's, but his patent on the lockstitch required a timely negotiation. Soon afterward, Howe found a financial backer to support his claims in the courts.

Howe was the catalyst for the momentous partnership of Isaac Singer and Edwin Clark in June 1851. Two years before, when he arrived in New York City, Singer had gone to the New York law offices of Jordan, Clark and Company for title advice on his type-carving machine. The senior

partner in Clark's law practice, Ambrose L. Jordan, recently attorney general for the state of New York, would have nothing to do with Singer personally. All he had seen was a poorly educated loudmouth.

Jordan's partner, Edwin Clark (1811–1882), however, had seen a mechanical genius—and an opportunity to make money for himself. The two men were of the same age, 39, when they met but polar opposites. Clark was a cold fish. Society's respect counted much for him. The son of a pottery manufacturer, he had been educated at Lenox Academy, Massachusetts, where he was said to have read every one of the school library's 500 books, and then graduated from Williams College, which normally turned out clergymen. He had a churchly air and taught Sunday school. His wife, the boss's daughter, Caroline Jordan, would not have that "nasty brute" Singer in their house and they never did. The Clarks and Jordans were typical of the New York Edith Wharton was later to characterize in her novels, worthy Anglo-Saxon families drawing up the chaise longues against the new entrepreneurial class and its vulgar ostentation.

Singer, for his part, resented the disdainful gentility of the Clarks as he made them exceedingly rich. "Did you ever see Clark with his wig off?" he roared. "He is the most contemptible looking object I ever saw with his wig off." When he became wealthy, Singer built a monster canary yellow coach for family excursions in Central Park. You can just hear the reaction of straitlaced old-money matrons: "Well, there's a thing! Nine horses, three abreast! A liveried footman and guards! And a band on top! And all those children waving! And room, they say, for twenty people inside and water closets and a nursery! Disgusting!"

Yet Singer faced a predicament in 1851. He had no money to defend the suit Howe brought against him, and other vexatious suits were in the wind. Even if he could prove his originality in a trial, which he was sure he could do, the prospect was ruin in legal costs. So both men held their noses and did a deal. Clark agreed to give free legal and financial services, and Singer yielded him a third share in the business. The partnership between these two men was to be one of the most creative of the 19th century. Innovation is often advanced in a relationship between an originating and an enabling intelligence: Sam Colt and Elisha Root, Richard Sears and Alvah Roebuck, Henry Ford and James Couzens, Steve Jobs and Steve Wozniak. The history of the Singer-Clark relationship suggests that the relationship need not be based on affection; partners do not have to like so much as complement each other.

Zieber learned about the arrangement only when he went to consult Clark after a row with Singer. It was an unwise choice of counsel given that he knew Clark had previously advised Singer, but poor good-hearted Zieber was in a state of nervous collapse. He had had a perfectly reasonable request to make of Singer, which was to spell out the verbally agreed upon portion of the Phelps interest that would accrue to him. He was "utterly dumbfounded," he said, by Singer's eruption. "What do you mean?" Singer had shouted. "By God, you've got enough! You shan't have any more!" Clark coolly told Zieber he had no rights to the Phelps share; he, Clark, would assume it shortly on becoming the third partner. Zieber was dumbfounded all over again.

As soon as U.S. Patent 8,294 was approved on August 12, 1851, Clark and Singer split the rights 50:50, excluding Zieber but promising him a third of the profits. He never saw them. On December 15, 1851, when he was confined to bed with a raging fever, he was visited first by Clark, then a day later by Singer. Compassion was not the impulse. According to Zieber, Singer said, "The doctor thinks you won't get over this. Don't you want to give up your interest in the business altogether?" It was a lie. The doctor had said no such thing to Singer. Zieber wrote: "I was very much startled as to what he told me and not in a fit state of mind to think much about business. I believed he was telling me the truth, and did not then suspect that it was a trick probably concocted between him and his other partner." The fastidious Clark was assuredly part of the plot. The next day Zieber sat up in bed to sign a legal document drawn up by Clark by which Zieber gave up all his rights for $6,000.

Zieber recovered and repented. His apology to himself thereafter was that he wanted to go his Maker knowing he had paid his debts on earth. He may also have nursed doubts about the viability of the business he had helped to start. Later he accepted crumbs from the rich man's table, editing a Singer newsletter and opening an office in Brazil. The $6,000 was a cheap buyout, considering that within a year the company assets were worth half a million dollars, but Zieber's judgment and resolve suggest he would not have been a fit manager of a company struggling to find customers and protect itself from competitors and patent infringers.

The biggest cloud on the landscape was the persistent Howe, who had beaten all other lockstitch manufacturers into paying him a license fee and found money to make a few demonstration models. Zieber said Singer "raved to put his foot on the neck of Howe"; Clark dismissed Howe as "a humbug who never invented anything of value." Both Clark and Singer underrated him.

Howe gave as good as he got in what the sensational new newspapers liked to call the Sewing Machine War, one of mutual insults in advertisements and news stories, and endless lawsuits. Singer tracked down the aging Hunt to rebuild his machine from bits and pieces found in a blacksmith's shop to prove that Hunt, not Howe, had invented the lockstitch. The court ruled the submission invalid on the ground that Hunt had let his rights lapse. In July 1854 the Singer Company had to agree to pay Howe a $25 royalty on every machine it sold. In turn, Singer sued Howe for breaching Singer patents in the way he had modified his original concept for his exhibition models. Everybody piled on with suit and countersuit: Hundreds of patents for sewing machines or elements had been issued (900 by 1867), claiming improvements, so the prospect before all the parties was one of infinite litigation similar in its burden to the reaper wars that had taken so much of Cyrus McCormick's time and money. The Boston lawyer-president of another sewing machine company, Orlando B. Potter, came up with the innovative solution in 1856. The seven major manufacturers, with Howe in train, swiftly agreed to pool their patents so they could make best use of all the detailed inventions and expand without fear of litigation. Members paid the Great Sewing Machine Combination $15 on each machine sold, but there was no pooling of any other interest. They were all free to set their own prices and compete as hard as they could on merit. It made Howe a rich man—he received $5 on every machine anyone sold—but the margins were high enough for Singer to grin and bear it: his $125 machine cost him only $23. It also cleared the way for the fullest play of Singer's technical and promotional talents and Clark's vision.

The breakthrough was selling into the home. Singer, unlike Howe, had always envisaged a big domestic market. His initial machines were too heavy and expensive for families, but from early on in his profuse advertising, he featured a housewife sewing clothes in her sitting room. The $125 Singer could charge a clothing manufacturer was about a quarter of a whole year's income for the average family, and nobody was going to buy a machine weighing 130 pounds that might have to be moved every suppertime. Singer worked at reducing weight and had a $50 machine for the home in 1856, much lighter (though still heavier than Wilson's six-and-a-half-pound competition). Somehow, between all his romantic duties, Singer spent long hours in a new factory in Mott Street, New York City, designing successively lighter, faster, less expensive and more versatile home models for seams, hems, tucks, bends, quilts, gathers, braids and embroidery. One of the mechanics he worked with later enthused that he was "companionable, a good story teller . . . his genius for acting came into good play. The world was made brighter for his presence."

Despite a common impression, neither Singer nor the company as it developed was as much in the forefront of the machine-based "American system" of manufacturing as were Colt and the armorers and indeed Wilson-Wheeler. Until the 1860s Singers were manufactured one by one without special machine tools and jigs. Singer had read all the engineering literature, even if he was barely literate, but preferred the quality he obtained from hand filing.

Clark matched Singer's technical virtuosity with two dazzling commercial developments. McCormick of reaper fame had initiated short-term preharvest financing for farmers. In 1856 Clark pushed the concept to its logical and revolutionary next step of longer-term installment buying. A customer could have a machine for as little as five dollars down and three dollars a month for 16 months. Clark's second coup was to invite consumers to trade in their older sewing machines, any make, for a new, improved Singer. The trade-ins were destroyed to forestall a competitive second-hand market and to deny competitors spare parts. As sales soared from the hundreds into the thousands, Clark supported elaborate showrooms staffed with solicitous young women, widespread branch offices with repair mechanics and instructors as well as salespeople, and swift distribution of parts, a model for the burgeoning home appliance industry and later the automobile industry. The Supreme Court was a liberator, as it had been in the *Ogden v. Gibbons* steamboat war: In 1876, in *Welton v. Missouri,* it struck down as unconstitutional a 30-year-old Missouri antipeddling law that restricted sales of out-of-state products. By 1879 Clark had dispensed with all independent merchants and was selling through 530 retail stores.

Clark sold succeeding creations of Singer's on an ascending curve. Wives were told, with justification in good times, that they could earn $1,000 a year making clothes on the tabletop Singer with the foot treadle. By the time Ebenezer Butterick, a Yankee tailor, started selling dress patterns in 1863, Singer had become the most popular brand of machine. When the patent pool expired in 1877, Singer Company accounted for more than half the 500,000 machines made in a single year and had a monopoly of three-fourths of the entire world market.

This was Clark's achievement. He surpassed Colt—and everyone else—in establishing foreign-based manufacturing on a large scale and sound footing. In *The Emergence of Multinational Enterprise,* Mira Wilkins writes, "Singer was the first American international business, anticipating the Standard Oil companies, General Electric, National Cash Register, and International Harvester." When the company celebrated its centennial, it had sold 100 million machines in 67 countries and had 90 percent of the world market. The credit for this international bonanza lay in Clark's careful long-range planning and single-minded determination. His paradigm was do your own financing, invest in quality, go where the people are of whatever nationality and sell, sell, sell—sell not just a machine but a way of life.

The always fractious kinship of Clark and Singer did not last long after the scandal of 1860, when, with Singer hiding in Europe, Clark was left trying to pretend that nothing untoward had happened. In June 1863 he and Singer acrimoniously dissolved the partnership. A corporation was formed, the Singer Manufacturing Company, with Clark in control of its $550,000 of assets and 22 patents. Singer, with assured income from 40 percent of the stock, was 51 and eager for a new life with Isabella. Snubbed on his return to New York with Isabella, he lived out his life in opulence in Paris and finally in Paignton, Devon, near the socially conscious English seaside resort of Torquay. Here he built a mansion in the style of a florid French villa, which he called the Wigwam, presided immaculately as the Santa Claus grandfather of an eddying brood and threw epic parties. The last, which he planned ahead in detail, took place on July 23, 1875. Two thousand people, hundreds of men, women and children, lined the roads, bowing their heads as his funeral cortege, nearly three-quarters of a mile long, moved toward town. Church bells tolled and flags flew at half-mast, their machine stitches intact in the breeze.

EDWARD CLARK: A coldly calculating temperament and a perfect complement for the hot-blooded Singer.

CHARLES GOODYEAR

He was in and out of jail for debt, but only his dedication to solving one of nature's great mysteries gave us the indispensable vulcanized rubber

1800–1860

CHARLES GOODYEAR was always moved to tears by news of drownings at sea. He had once read an item that 20 people drowned every hour, and throughout his life, impelled by heartfelt indignation, he was never to give up designing and testing all kinds of ways to save mariners: "Must men be drowned because their fathers were?" he asked. In 1834, walking in Manhattan, he spotted a life preserver in the window of the Roxbury India Rubber Company in New York. He thought the air valve defective, so he bought the life preserver belt and was back weeks later to offer the company an improved valve.

The agent's response was to lead him to the warehouse behind the showroom where strange misshaped objects filled the shelves. The deformities were life belts, wagon covers, coats and shoes, melted and stuck together. So bad was the smell from decomposed rubber, the company had had to bury hundreds of pairs of shoes in a pit at the factory.

Rubber had come late—and disastrously—to American commerce in the 1830s. It was only belatedly discovered to be ruined by changes in temperature and solvents. In the winter rains, rubber products "stiffened to armor" and in summer they were clammily adhesive. Exposed to grease, oil or acid, anything made of rubber decomposed and gave off a disgusting smell. Biographer Charles Slack compares what happened to the dot-com frenzy of the 1990s: "Just as speculators drove the value of untested, unprofitable dotcom start-up companies to surreal levels during the 1990s, thousands of nineteenth-century New Englanders poured their savings into the miracle substance called rubber without knowing anything much about it. In both

cases, the education was hard-earned and painful." Thousands of angry customers returning sticky rubber shoes and collapsing rubber boats wiped $2 million off the value of rubber companies. The Roxbury Company was one of the few clinging to life. In England, a decade before Goodyear

Goodyear married Clarissa Beecher on August 24, 1824. Her father kept a hostelry that was for a time the headquarters of General George Washington. She was described as "of a peculiarly amiable temper, endowed with great fortitude, and sustained by a sincere religious faith," qualities she needed in years of privation. She was the mother of nine children, five surviving.

walked into the Roxbury store, Charles MacIntosh and Thomas Hancock had established a modest business in raincoats (Macs) by rolling a thin sheet of treated rubber between two pieces of cloth. Hancock had also sold thin strips of rubber elas-

tic to replace laces in gloves, coats and boots, and for use as erasers (to this day called rubbers in England). But that was it. Hancock had seen endless other commercial possibilities in this strong material that could be molded to any shape or size, did not conduct electricity and was impervious to water, but try as he might, and he was an ingenious and systematic experimenter, by 1830 he had given up thinking it was possible to overcome the critical limiting susceptibility of rubber.

It took all of five minutes for Charles Goodyear to satisfy himself he could solve the problems that had defeated everyone else on both sides of the Atlantic. He had no training in chemistry. He just had a presentiment, he writes, that he was "the instrument in the hands of his Maker." God, in wondrous ways, would lead him where reason could not.

Goodyear is the most humbling and the most exasperating of the early inventor-innovators. He is a Dickensian hero going nobly into a world of cynics and thieves, never straying from his own true purpose to enrich mankind. He is also a fanatic, so improvident that he is time and again confined in the squalor of a debtors' prison while his devoted wife and family starve and his infant children fall sick and die. He dedicates his life to the pursuit of a great invention and then neglects to secure the rights in the world's most important market.

To this strange man, we owe rubber—that is to say we owe much of our modern

The striking daguerreotype portrait of Goodyear was made in 1852, the year of his most decisive court victory, in which Daniel Webster routed the piratical Horace Day.

civilization to myriad smelly little experiments in attic rooms and kitchens carried out by a man who had not the faintest idea of the organic chemistry he was meddling with to convert raw rubber to practical use. The dissonance between the obscure origins of what we now call vulcanized rubber and its ubiquity is monumental. Imagine the rubber in your daily life. Perhaps you are accustomed to rising on an alarm, turning on the light, taking a shower, stepping on the scales, pulling socks from a washer-dryer and shoes from the closet. Dressed, you cross the kitchen floor to the refrigerator, take a look at the morning television, slot a bagel in the toaster, grab a briefcase and go out to your car. The car radio tells you to beware of a traffic jam on the freeway. You park at the railway station and take the train. At the workplace, you take the elevator to the eleventh floor, boot up the computer for the day's e-mail and resolve to go to the gym during the lunch break.

None of these daily activities, and a thousand more, would be possible without vulcanized rubber. Without it, the planet would shudder to a dead stop. Maybe we could get by without the obvious rubber in tires, sneakers and overshoes, floor tiles, mattresses, waterproofed roofs and raincoats, condoms, tennis and golf balls, dinghies and life jackets. But we could not even begin the day without the rubber we do not see: the rubber in electronic instruments, in the commutator bearings of electric motors, in transmission belts and assembly lines, in shock absorbers and insulating seals and gaskets. It would be a day without electrical power and lighting, automobiles, printed circuit boards, bicycles, radios and televisions, phones, railway trains, baby carriages, washing machines, toasters, vacuum cleaners, airplanes. To say nothing of elastic bands.

Goodyear himself, ignorant as he was in scientific understanding, was inspired in his visions of how rubber could permeate our daily lives. He gave the book he wrote in 1853 the mind-numbing title *Gum-Elastic and Its Varieties,* but behind the dull brick facade is an Ali Baba's cave, his illustrations of hundreds of pleasures posterity would find in utilitarian rubber. He was an obsessive but he was also a romantic.

CHARLES GOODYEAR was blessed with inventive genes. He was the seventh inventor in four generations of Goodyears. He was raised in a clapboard saltbox house close to a little mill on a tributary of the Naugatuck River (near present-day Waterbury, Connecticut), where his father, Amasas, made scythes, spoons, clocks and buttons of metal and pearl—the first pearl buttons manufactured in the colony. Everyone in the Connecticut Valley "made things," but Amasas, a fifth-generation descendant of one of the founders of the Colony of New Haven, was exceptionally creative. He invented the first closed lamp for burning oil and patented a "spring" steel fork that lightened the backbreaking labor of lifting hay and manure with a heavy wrought-iron fork. Charles proudly called it "one of the greatest improvements ever made in farming implements." He did not have his father's mechanical gifts. His contribution as he advanced into his teens was to bubble with ideas for implements his father might make.

Charles was little over five feet, a small pale boy, physically frail but self-confident and precocious in scholarship. He went away for a year to a school run by a clever Congregational minister, and for an extra year of schooling at home his father engaged a private tutor, William C. De-Forest, a big bluff young man with a taste for commerce. The tutor, only 20 himself, was as impressed by the boy's brilliance as he was irritated by his boastfulness. When Charles was 17, Amasas sent him to Philadelphia as an apprentice to an importer of English hardware and tableware. By some accounts, Amasas was anxious to divert his son from becoming a minister in the Congregational Church. Amasas was a faithful attender at the Congregational church and the family read the Bible together, but he was less of a puritan than most of the congregation, whereas Charles acquired piety and nurtured a tendency to self-abnegation. Biographer Richard Korman says he was drawn to the Book of Job, "a story of torment and faith that would fascinate him later on as his own life became a chronicle of tribulation." Whatever, when he came home in 1821, gaunt from digestive ailments diagnosed as "dyspepsia," he was exalted more by

commerce than church. He joined his father's business, selling what Amasas made, and married Clarissa Beecher, an innkeeper's daughter.

Five years later, A. Goodyear and Sons was doing so well, Charles convinced the more prudent Amasas that Charles and his younger brother Robert should open a store in Philadelphia—and not just any old store. This was to be America's first retail domestic hardware store and it would have a second distinction. The four years dealing with English imports had given Charles a marketing idea: Made in the U.S.A. Everything in the store would be from Goodyear or other Americans.

The Goodyears prospered for four years, and then hammer blows fell. Political maneuvers over the 1828 presidential nomination inflicted ruinously high duties in the "Tariff of Abominations," particularly damaging to New England and southern interests. Charles had granted liberal credit to southern farmers. As the country went down the roller coaster of the business cycle, debtors defaulted. He gave them more credit. He borrowed from De-Forest. Instead of limiting his liabilities by declaring bankruptcy, he struggled on, valiantly but foolishly putting all the family's assets in jeopardy. In the end, everything had to go: the store, the inventory, Amasas's precious workshop and their patent rights.

Charles Goodyear, at the age of 33, was sure he had more to offer than selling steel forks and kettles. He decided to be an inventor, and despite periodic depressions it was a propitious era for selling ideas. The steam engine in boats, trains and factories, and the energies of thousands of immigrants—600,000 in the 1830s—were changing America from a rural society of agriculture and handcrafts to an urban one of enterprise, machinery and manufacture. He won four patents for minor improvements (in spoons, faucets, air pumps and buttons), but in the four years 1830–34, the Goodyears lived hand to mouth on what the pawnshop would advance on their furniture and silverware and finally Clarissa's trinkets. One morning his first daughter, Ellen, woke to find her father missing, and her mother's explanation was to become routine: He had been taken to prison for

Goodyear's obsession with saving lives at sea was reflected in many designs for rubber boats and flotation devices.

owing money. Imprisonment for debt was arbitrary; however trivial the sum, a debtor was put away at the behest of a creditor until redemption. For six years thereafter, Goodyear was to be seldom a whole year out of jail in one town or another and never out of danger of arrest anywhere in Pennsylvania, New York, Connecticut or Massachusetts. Debts were sold to collectors, so Goodyear never knew who could have him arrested; often, anyway, he had no good record of the original obligation. Perhaps he picked up the virus of reckless indebtedness from the merchants who had defaulted on him, but his mind was too restless, his ambition too great, to be hemmed in by "the want of pecuniary facil-

ities," the euphemism threaded through his memoir.

In 1834, on return to Philadelphia from his epiphany in the rubber store, he was sure he would soon find his fortune. Rubber, a white milky latex from several hundred varieties of trees in the jungles of Latin America, was shipped into the United States as bricklike blocks, or "biscuits"; the liquid latex did not travel well. Both Hancock in England and, coincidentally, the Roxbury Rubber Company ground the raw rubber before thinning it in turpentine to get malleable dough. The first thing Goodyear did was take a few lumps of raw rubber into Clarissa's kitchen, knead it for hours, then mix it with turpentine and spread it

out with a rolling pin. However many hours he tried, he could not remove the stickiness. He would have to find some chemical that reacted with the rubber better than turpentine. Over weeks, he tried adding powders, then as many chemicals as he could lay hands on. To keep going with these quite random trials, he borrowed money again and again. "He could not have been a more prolific borrower," writes Korman, "if he had picked his lenders up by their legs, turned them upside down and shaken all the money out of their pockets and purses." Arrested once again, he carried the rolling pin and additives into prison with him until his father and brothers could redeem the debt. He did not pause in the family's third move in five years, this time back to familiar Connecticut but the uncongenial environment of a cottage in New Haven's run-down Sodom Hill.

Of all the materials, powdered magnesium yielded the most promising result, a smooth white texture on the surface. With money borrowed from a New Haven stove maker, Charles recruited his wife and daughters Ellen, ten, and Cynthia, eight, and four young women to stitch cambric soaked in his mixture into several hundred pairs of shoes. They were ready by the spring of 1835. He added decoration but prudently set them aside to await the test of summer heat.

In the meantime, he was overjoyed to get his hands on some 50 casks of pure liquid latex left on an importer's shelf. The day he was due to begin experimenting with the fluid, a newly arrived boy from Ireland, by the name of Jerry, announced that he had preempted his master by dipping a pair of trousers in the latex overnight; they had dried beautifully and he now stood before everyone in splendid array to prove, Goodyear wrote, "that a Yankee was not so quick at inventing as an Irishman." Jerry got on with his work, mixing sap by the kitchen fire. Before long he was crying for help. He could not get up: his legs were glued to his pants and his pants were glued to the chair.

"Sticky" Jerry had painfully anticipated the disappointing results Goodyear got with latex in subsequent tests. It was the same sequel with the treasured shoes. When he came to inspect them, they were one mass of melted gum. That was the end of money from his backer and the beginning of another cruel period. Their toddler son, William, died (the third lost child), and an eviction order forced the family of four into an even humbler home. But nothing would divert Goodyear. Retaining his faith in "the Great Creator," he left his grieving family penniless in Connecticut's Salem (shortly after renamed Naugatuck) while he pursued the phantom formula in New York. His onetime tutor, William DeForest, thought it an act of madness. He was shocked to climb the three flights of stairs to Goodyear's tiny hole of a room on Gold Street in Manhattan and find a shadow of a man fidgeting unkempt and gaunt amid his kettles, gum shellac and odiferous chemicals. DeForest tried to persuade him to go home and fend for his family by reentering the hardware business.

It was a hopeless plea. Goodyear was borne along by a manic certainty that he could crack the code. He moved feverishly from one mixture to another. Lime seemed promising, so he carried jugs of slaked lime on his shoulder three miles to a mill in Greenwich Village, where he boiled it with turpentine, magnesia and rubber. It was a triumph for five minutes. Any drop of fruit juice made it as adhesive as ever. Then he had a rare lucky break. Stripping bronze paint from a rubber drapery with nitric acid, he was annoyed that the rubber became discolored. He threw it away. A few days later he had a second thought. He hunted for the piece in the garbage and found that the discolored surface had lost the troublesome stickiness. At last! He was sure he had the secret. He closeted himself in his little room, brewing solutions of nitric acid. The searing fumes knocked him unconscious and he spent six weeks recovering. The consolation was that this acid gas process was effective in removing surface adhesiveness so he could contemplate making thin rubber sheets (his nitric acid, unknown to him, actually had an important residual trace of sulfuric acid).

Goodyear, the business evangelist, having personally secured U.S. Patent 240 on June 17, 1837, by filing at the Patent Office in Washington, convinced yet another backer that this apparent success, while limited, was Eldorado. William Ballard, a merchant he had met at the New York Mechanics' Institute, installed him in an abandoned rubber factory on Staten Island and opened a showroom near Broadway at Eleventh Street. Goodyear used some of the investment money to move his whole clan to a cottage by the factory—Clarissa, his six-year-old son, Charles Jr., and two daughters, his father and mother, and his brother Robert with his four children, too. They all pitched in to make life preservers, aprons, caps, piano covers; Clarissa deco-

Just look what rubber can do . . . Goodyear embellished his 1855 book, *Gum-Elastic,* with scores of examples, among them a life preserver—the origin of his lifelong passion—and a ball, hat, boat, umbrella, boot, rope, faucet, bucket and gymnastic ring.

rated schoolroom globes and gloves. All of the items would eventually become ruinously sticky in hot weather.

The financial crisis across the United States in 1837, in which 618 banks failed, wrecked Ballard. Goodyear clung to the wreckage, and the family did its best to keep manufacturing as the money ran out. They were soon back to subsistence levels, fishing in the river for their supper. Anything they had of value they pawned. His revered father searched the streets for lumps of coal that might have fallen off delivery wagons. Goodyear took to walking around Manhattan as Rubber Man. Strangers, he tells us, were advised how to recognize him: "If you see a man who has an India rubber cap, stock, coat, vest and shoes, with an India rubber purse without a cent of money in it that is he."

Any knowing person walked rapidly in the other direction. With so many bankrupt companies, rubber in the East had the taint of fool's gold in the West. But once Goodyear had someone by the lapels he was beguiling. His manifest sincerity surpassed his eccentricity.

An eager young man rescued him from the misery of Staten Island. William F. Ely had met Goodyear at the Mechanics' Institute when the inventor was honored with a silver medal for draperies, exhibited before they became viscous. Fresh out of the army, Ely wanted to go into business with a $10,000 legacy and he was bowled over by Goodyear's enthusiasm. They set up Goodyear and Ely's Fancy Rubber Establishment—but not in New York. Goodyear took the steamship to Boston, leaving his family with only 50 cents, because Boston was the home of the one of the few surviving rubber companies, Roxbury, whose life preserver had first infatuated Goodyear. On one of his trips to New York, he had called at the Roxbury shop and shown the company's friendly 35-year-old leader, John Haskins, his nitric acid samples. Haskins had agreed that if "Charly" got himself to Boston he could use the skeleton company's plant and credit.

Haskins's partner and foreman, Edwin Chaffee, had invented a wondrous machine, a 30-ton calender they called the Monster, whose steam-heated rollers masticated raw rubber into thin sheets. Goodyear applied his acid gas process and then decorated table covers, draperies and ladies' capes for Ely to sell. He sent samples to the renowned Yale chemist Professor Benjamin Silliman (1779–1864), who gave him a testimonial, and then started selling licenses to the process in the sincere belief that he had solved the problem, at least for thin rubber and rubberized fabrics of slight thickness. He had not. With the arrival of warm weather, the beautiful items became gooey. He had made enough money to bring his family to Boston, but he had outlived his welcome at Roxbury. In the summer of 1838 he took the reins of a chaise and, with his father at his side, drove 15 miles through the countryside to the Eagle mill by the canal at East Woburn to see what the prospects were at the Eagle India Rubber Company housed there.

Nathaniel Hayward, the former foreman and now the owner of the struggling mill, opened the door. They were a study in contrasts: the shrewd but illiterate Hayward, a bluff, square-built young man of open countenance, and Goodyear, the smooth talker, only 6 years older but looking 60. Hayward, with a wife and two young children, was tired of chasing rainbows and ready to sell the mill and his know-how in exchange for an annual salary of $800. The crucial thing in the deal was the odor of rotten eggs. Hayward was producing smooth-surfaced rubber by adding sulfur, and Goodyear, smelling it around the mill, wanted to know more. The upshot, in November 1838, was that he paid Hayward to be allowed to patent the sulfur additive in the name of Goodyear. Neither man knew it, but the rubber macromolecule, a stupendously long chain of carbon and hydrogen atoms, was uniquely susceptible to sulfur in certain conditions. Finding those conditions was the heartbreak. Goodyear persuaded the post office in Boston to give him an order for 150 mailbags and announced to the world that this was the big moment when all his prophecies would come true. The bags, made from a compound of rubber, sulfur and lead,

looked beautiful. In two weeks of hot weather, the insides were a glutinous mess.

It was impossible not to wonder if rubber was inherently unworkable. Was it not long past time to go back to hardware? Clarissa, his brothers and even his loyal father, Amasas, worried that all the vicissitudes—three dead children, poverty and jail—were God's punishment for making an idol of vanity. Call it determination, call it arrogance, call it divine inspiration, Goodyear could no more admit defeat in 1839 than stop breathing; his mind, he wrote, was "buoyant with new hopes and expectations." But so it had been for 14 years. What happened next with his rubber has to be told in his own words. Speaking of himself in the third person, he wrote:

"At the dwelling where he stopped whenever he visited the manufactory at Woburn, the inventor made some experiments to ascertain the effect of heat upon the same compound that had decomposed in the mailbags and other articles. He was surprised to find that the specimen, being carelessly brought in contact with a hot stove, charred like leather. He endeavored to call the attention of his brother, as well as some other individuals who were present, and who were acquainted with the manufacture of gum elastic, to this effect, as remarkable, and unlike any known before. The occurrence did not at the time appear to them to be worthy of notice; it was considered as one of the frequent appeals that he was in the habit of making, in behalf of some new experiment. He, however, directly inferred that if the process of charring could be stopped at the right point, it might divest the gum of its native adhesiveness throughout."

Exactly what happened remains a puzzle. There are many versions of this event of January 1839. Richard Korman credits Ely with first noticing the change on a patch left on top of a stove at the Boston shop. Did Goodyear get the idea from him? He never mentions Ely in his

THE HONORED PRISONER: Goodyear's pride, the Vulcanite Court at the Crystal Palace. In Paris in December 1855, Emperor Napoleon III made Goodyear a Chevalier de la Légion d'Honneur. The ironic bookend to his career is that he received the Légion medal in a prison cell—from the hands of his oldest son, Charles Jr., while languishing four days in Clichy Debtors' Prison. Royalty receipts due had been too late arriving in France to pay creditors on time, and gendarmes had arrested the chevalier in his hotel. Testimony at an 1864 appeal for an extension of the patent was that Goodyear left $200,000 in debts, but that may well have been a hard-luck story to soften the heart of the patent commissioner. Goodyear's second son, William, maintained that his father was worth hundreds of thousands of dollars.

memoir. Others have it happening at a country store. Yet another has Goodyear guiltily hiding a patch in Clarissa's oven. The flaw in most of the stories is that vulcanization—for such it was—is a more or less slow process. Whatever happened and to whom, it was a decisive episode. Impregnated with sulfur, the heated rubber did not melt, it was stronger, it did not smell and it was more elastic. The quest Hancock and others had for decades pursued and abandoned was now a reality in the paradox that heat, always considered the enemy of rubber, was its salvation.

Goodyear rejected the notion that this was a lucky accident and what he wrote is germane to many of the "accidents" in the history of invention and innovation: "Like the falling of an apple, it was suggestive of an important fact to one whose mind was previously prepared to draw an inference from any occurrence which might favor the object of his research."

The distance between scientific realization and business fulfillment, between invention and innovation, is often long and cruel. Exultation is followed by exhaustion and depression. Nobody believed poor mad Charly had made a breakthrough. Why should they? He had proclaimed so many;

and there was much more to do before his invention was ready for industrial manufacture and marketing. He brought the whole family to Woburn from Roxbury but ran out of money. He had to sell the mill. Still, he absolutely refused to abandon his quest. Charles Slack, a not uncritical biographer, justifiably comments, "During this period Goodyear's greatness fully emerged." He worked incessantly over the next two years to perfect his process, trying to fathom the permutations of sulfur and heat that made all the difference. Sheet rubber immersed in melted sulfur and maintained at a temperature of 120 degrees did not vulcanize; immersed for an hour at 140 degrees, it might. He begged furnace time wherever he could in local factories and shops. The man whose invention was worth multiple millions was tolerated as one would a slightly touched but harmless mendicant. At home, when Clarissa withdrew loaves from the oven, he would put in his samples and watch to see the effect of one hour's, two hours', three hours' and six hours' baking. He boiled mixtures of lead, rubber and soap in her saucepans, roasted specimens in the open fire, held them by the spout of a boiling teakettle. In short, he drove everyone crazy. There were happy evenings when he read to the family from the Bible, but yet again they were trapped in the cycle they hoped they had left behind: Clarissa and the children foraging for fuel, digging up half-grown potatoes, fearing the knock at the door could any day be their father's arrest for debt. He pawned the children's books for five dollars. Kindly neighbors gave flour and castoffs to the pitiful family with the unfortunate father. Goodyear himself waded into the waterways in search of turtles and bullfrogs for stew; during the snowstorms of 1839–40 he was observed digging in the drifts for firewood.

In October 1839 Silliman at Yale certified that in experiments he had found Goodyear's treated rubber did not melt in heat or stiffen in cold, but nobody was interested. In April 1840, on a money-foraging trip to Boston with Hayward, Goodyear was once again a miscreant, jailed for petty debt; paper he signed had been sold off at a discount and he had no idea whom the creditors were. "I have fallen into

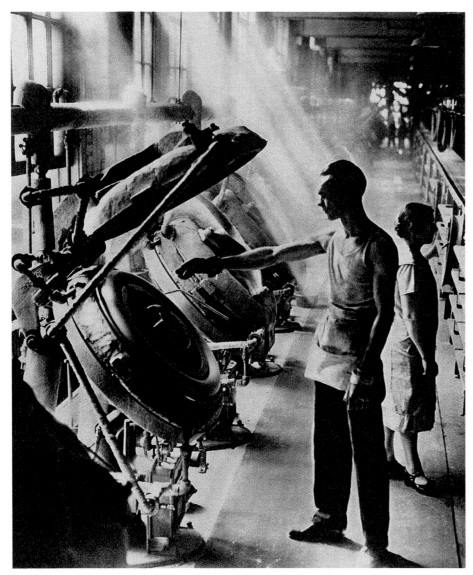

Vulcanizing a tire in an iron mold subjected to high-pressure steam. Thousands of scientists have worked to improve on the lonely inventor in the kitchen, but Goodyear's discovery of vulcanization, combining sulfur with heat and an accelerator element, remains the essence of rubber manufacture. The rubber is nowadays mechanically impregnated with sulfur about one-tenth of its weight.

THE GOOD NAME GOODYEAR: Goodyear's first son, Charles Jr., sold the Goodyear name in 1865, the year the patent expired, to a company in San Francisco that then called itself the Goodyear Rubber Company. It made almost everything except tires. Thirty eight years after Goodyear's death, Frank Sieberling, a farmer's son in Akron, Ohio, borrowed $3,500 from a brother-in-law to start making tires (in violation of various tire patents) and called his company Goodyear Tire and Rubber in honor of Charles Goodyear. The conflict between the two "Goodyear" companies was resolved in 1927. In Akron today, a statue of Charles Goodyear graces the headquarters of Goodyear Tire and Rubber, a multibillion-dollar company—with competitors like Firestone.

the hands of harpies," he wrote beseechingly to Ely, "and I do not suppose they will now leave me alone." No sooner had Ely sent the $75 for his release than he was jailed for several weeks in worse conditions in Cambridge, among lunatics and criminals. Walking the last ten miles of his return home to Woburn, he was met at the door with the news that William Henry the 2nd, the two-year-old son he had left happily playing, was dying. There was no money for a coffin or a carriage. The grief stricken father wrapped his son in a sheet and carried him to a wagon for the mile-long journey to the cemetery; the rest of the family trailed alongside. Shortly afterward, a despondent Amasas and his youngest son

and daughter-in-law adventured to Florida, where they all died of yellow fever.

It was the lowest point in Goodyear's odyssey. So much suffering, so little to show for it. In desperation, he wrote yet another begging letter to William DeForest. His old tutor had blossomed as a woolen manufacturer in Naugatuck and was now "family," having married Goodyear's widowed younger sister, Harriet. Fifty dollars came back in the fall. Goodyear never used other people's money to better effect. He spent the $50 going to New York to show his latest samples to the merchants William Rider and his brother Emory. Impressed, they set him up in a factory in Springfield, Massachusetts. Here he devised a six-by-eight-

foot cast-iron oven where objects placed on a rotating spindle like a rotisserie could receive uniform heat throughout. There were many setbacks. A shoemaker, Horace Cutler, was furious that overshoes were burned and blistered. Hayward, who was helping, was despondent. Gradually, however, Goodyear's perennial optimism was rewarded. He managed to have more and more complex objects emerge evenly vulcanized. He invented shirred rubber, a combination of leather and elastic thread. On December 6, 1841, he felt confident enough to file notice of intent to apply for a patent. He did not follow up until July 5, 1843, because he wanted to be close to perfection. The formula he elaborated was 25 parts rubber, 5 parts sulfur, 7 parts white lead, and spirits of turpentine rather than raw turpentine, heated at 270 degrees Fahrenheit, though he refrained from specifying for how long. He neglected to apply for a patent in England but entrusted a sample to an English émigré on the way to London, Stephen Moulton, to see if he could raise £50,000 for the manufacturing rights there.

Money began to flow to Goodyear from vulcanized manufactures and licensing as rubber's reputation was rehabilitated. He paid off $35,000 in debts. His patent, USP 3633, was granted on June 15, 1844, and should have made him one of the wealthiest men in America. He could have earned a fortune making and licensing shoes and waterproof clothing. Instead, he decided he had a sacred duty to explore what more vulcanization could achieve in the "philanthropic and humane department." By that he meant originating new articles for saving lives. Page after page of *Gum-Elastic* is devoted to different kinds of boats, life preservers, buoyant travel bags and even frogman suits, but also to fire hose, escape rope and medical instruments. On July 18, 1844, Goodyear accepted an offer from DeForest and others of $50,000 in cash for unlimited, though not exclusive, manufacturing rights on condition he had assured access to the factory for further research. DeForest and Emory Rider opened a factory in Naugatuck, where Amasas had started his button factory. The Naugatuck India Rubber Company, superintended by Goodyear's brother Henry, turned out suspenders, shoes, elastics and clothing. It grew rapidly and in time it became the United States Rubber Company and later Uniroyal.

It was pride and his sense of equity more than the familiar "pecuniary considerations" that induced Goodyear to join actions against the inevitable infringers of his patent. The samples he sent to England with Stephen Moulton had been shown to Thomas Hancock, who betrayed the trust. Hancock painstakingly analyzed the material, discovered Goodyear's secret and applied for an English patent himself. Hancock was upheld in the English court on the simple grounds that he had filed eight weeks before Goodyear got around to it, but the presiding judge found Hancock's actions "not handsome." A droll English comment was that Hancock's claim "may be seen as a new discovery of a fact already known — a novel solution of a problem which was known to be soluble, since it had already been solved." (Hancock behaved dishonorably, but at least he conducted some experiments, and one of his associates gave us the word *vulcanization,* after Vulcan, the Roman god of fire.) The only contribution made by Horace Day was to the annals of villainy. Day was a failing rubber manufacturer in Brunswick, New Jersey, a bully, a liar, a brazen thief and an unscrupulous litigator. He bribed the disaffected shoemaker, Cutler, to steal Goodyear's secrets and then claimed to be the original inventor of the whole process of vulcanization. The legendary Daniel Webster vanquished the rascal in the Great India Rubber Case, but Day still grew wealthy from rubber; he was the manufacturer of the rubber boat that collapsed on John Charles Frémont in the middle of the Great Salt Lake in Utah in 1843.

Goodyear's apogee came in the 1850s, when he was feted in London and in Paris; lionized in Europe, he stayed seven years. In London in 1851 he rode with Clarissa and four of their five surviving children in a splendid carriage with driver and footman in red livery to arrive at the Great Exhibition at the Crystal Palace, where they walked through his spectacular Vulcanite Court exhibit. He had spent $30,000 to stage his show of draperies, wall coverings, furniture and giant balloons, portraits painted on rubber sheets, medical instruments, inkstands, fans, boats, shoes, book bindings and dolls. His court was visited by millions and won six awards.

The acclaim buoyed his fading health. Contemporary press reports described him as "small, thin, sallow, and nervous." He hobbled along with the help of a cane with a sculpted rubber head. It was said he was suffering from gout; in fact, he had been poisoned by all the lead used in his experiments. Clarissa nursed him devotedly through another long illness, but it was Clarissa who succumbed. She was not quite 49 when she suffered a stroke in London on May 30, 1853. On the first anniversary of her death, Goodyear married a 20-year-old Englishwoman, Fanny Wardell. Supported for so long and heroically by a wife and family, he could not bear to be alone.

Goodyear lived a happy six years with three children by Fanny but still did not stop experimenting. He saw licensed rubber companies sprout in six eastern states with an output valued in 1860 at around $6 million, and that was just the smallest of beginnings: Before the end of the century rubber was indispensable for tires, first solid, then pneumatic. On June 15, 1858, the man who made it all possible enjoyed a valedictory vindication, proclaimed in heroic terms by the U.S. Patent commissioner, Joseph Holt. Holt was a rising man — he was later to be war secretary in the Buchanan administration and head the investigation into the assassination of Lincoln. Asked to grant an extension of Goodyear's 14-year patent, he did not merely say yes to an extension to 1865 but took the occasion to condemn the objectors Day and Hancock as sordid pirates who had conducted guerilla warfare against a public-spirited inventor. As for Goodyear's ledgers, said the grandiloquent commissioner, inventors had ever been distinguished by a total want of what was called business habits. "They fling from them the petty cares of the mere man of commerce as the lion shakes a stinging insect from his mane." Goodyear's account books might not have had the precision and symmetry of a merchant clerk's, but he was "a brilliant and impulsive genius, his diligence without parallel in the annals of invention."

It took years to sort out those accounts, but the wealth Charles Goodyear created for the world is beyond calculation.

DEMOCRACY ON WHEELS

ALBERT AUGUSTUS POPE (1843–1909)

THE BICYCLE has done more for the emancipation of women than anything else, said Susan B. Anthony. If we accept that judgment by the 19th century's most redoubtable campaigner for women's rights, the gallant liberator must surely be Albert Pope, who introduced the bicycle to America. As an individual, he would not likely have met with the approval of Susan Anthony, since his diary of his time as a Civil War soldier is frank in its descriptions of the nights he spent making love to young women, rebels and Unionists alike. He was a handsome volunteer, brevetted a lieutenant colonel, who was commended for leading a charge into the teeth of artillery fire.

Pope was from a Boston family of merchants. When he came out of the war, only 22 but ever after insisting on using the honorific "Colonel," he put his $900 savings into little businesses (shoe materials and air pistols) and was a millionaire by the time he was 30. He had never seen a bicycle of any kind, still less ridden one, before he visited the 1876 Centennial Exposition in Philadelphia. The English had taken over where the French bicycle inventors left off, and Pope was intrigued by his first sight of an English bone shaker: the "penny farthing," where the rider perched himself over a high front wheel rimmed in solid rubber and pedaled the axle directly without a chain to the small rear wheel ("penny farthing" after the largest and smallest English coins of the day). Pope had two questions: How on earth could anyone ride such a thing? And why should the English have all the fun?

He caught an early steamship to England, sneaked into factories to spy on production methods and adventured into the English countryside on a high-wheeler. He found exhilaration in the experiences that were later to provoke Mark Twain to say, "Get a bicycle. You will not regret it if you live." In 1877, with fearless male riders in mind, Pope tested the market with a variety of imports, then decided on the style he would manufacture himself, a Duplex Excelsior. On a spring day in 1878, he packed his Excelsior onto a railcar to Hartford, where, biographer Stephen B. Goddard tells us, he "jumped astride the high-wheeled contraption and pedaled uphill along Hartford's bumpy earthen streets." His destination was the Weed Sewing Machine factory, a repository of the skill and machines developed in New England's armories. Pope, having seen England's handcrafted production, gave Weed an order for 50 bicycles to be made with machine-produced interchangeable parts, cutting costs and time. By 1881 Weed was turning out 1,200 cycles a month.

Pope sold cycling. He could afford to popularize the pastime because he had bought the basic high-wheel patents and earned ten dollars on every cycle sold by licensed competitors. In 1883 he invented the trade show, an idea picked up later by the automobile industry. He organized cycling clubs; he sponsored poster competitions; and he recognized the appeal of the new monthly magazines—he bankrolled a number of publications, including *The Wheelman,* edited by young Sam McClure, the future founder of muckraking journalism. When New York City banned cycling in Central Park in 1880, for fear of upsetting the horses and impeding carriages, Pope fought in the courts and legislature (and won equal access after seven years).

Pope was doing so well with his "ordinary" Columbia, he nearly missed the big opportunity when the English, between 1884 and

Pope had fine artists competing for posters popularizing the bicycle.

1888, developed the "safety" bicycle as we know it today, with two equal-sized wheels, chain drive and inflatable pneumatic tires. Albert Overman was the first to make an American safety in 1887 (without pneumatic tires), but Pope powered to the front of the pack as a bicycle boom really took off. Some 20,000 high-wheelers had been sold; sales of the safety were 1.2 million by the mid 1890s. Pope was in the vanguard as a manufacturing innovator. Taking over the Weed factory for full-time cycle production, he carried out the industry's first "scientific testing," measuring just how much stress a loaded wheel could take. When Overman cut prices—and quality—Pope maintained the quality of the Columbias but sold cut-price bicycles under the label of the Hartford Cycle Company, concealing the connection to Pope Manufacturing.

The safety cycle was what Susan Anthony had in mind. Pope made a Ladies' Safety Cycle that was easier to mount and ride, and thousands of women took to the road, shedding corsets and long voluminous petticoats in favor of skirts split at midcalf and daring bloomers. A woman on a cycle was politics in motion, an expression of independence and the right to vote.

In the *Atlantic Monthly* in 1898, the anthropologist William McGee rhapsodized on the bicycle as an expression of American values. It was "a practical machine for the multitude"; it had "broken the pernicious differentiation of the sexes and rent the bonds of fashion"; and it had developed "individuality, judgment and prompt decision on the part of its users." Those latter qualities were much needed on the appalling roads. Pope campaigned for highway investment, financed engineering studies at MIT and pushed the measures that led eventually to the U.S. Bureau of Public Roads.

When the bicycle boom abruptly ended around 1896, Pope had already established a motor vehicle division to make electric, steam and gasoline automobiles and motorcycles. The 2,092 automobiles he manufactured in 1899 represented nearly half the cars made that year; he won races with his gasoline Pope-Toledo but bet most heavily on the electric Pope-Waverly, with women in mind again. His car business proved to be 30 miles of bad road. The Pope companies went into receivership in 1913, but by then he had been dead for four years.

EDWIN DRAKE His backers pulled the plug, but he found the golden key to the world's energy—drilling for oil

1819–1880

THE TINY VILLAGE of Titusville, tucked away in a beautiful valley in northwest Pennsylvania, did not get many visitors. The 125 residents were especially intrigued by the stranger who alighted from the stagecoach one day in December 1857. He looked like a *Harper's Weekly* caricature of Abraham Lincoln, his face gaunt under a silk stovepipe hat, his black frock coat flecked with mud after a long journey by rail and road from Connecticut. He was actually five foot ten, six inches shorter than Lincoln, but he carried himself with a similar stiff dignity and, importantly, he brought to Titusville one of Lincoln's qualities of character: determination. He signed in at The American Hotel, where the room clerk greeted him as "Colonel Drake," and a fair slice of Titusville ensconced itself in

THE STRANGER IN TOWN: Edwin Drake's deep-seated black eyes, framed by his black beard, gave him a melancholy air, but the locals found he was a good storyteller.

the lobby that evening as Edwin C. Drake puffed on his skinny long Pittsburgh stogie, took another nip of whiskey and—like Lincoln again—told funny stories.

But what produced the most laughter was the fact that this distinguished military man was wasting his time looking over a little oil seep leased by the Brewer and Watson sawmill to the upstart Pennsylvania Rock Oil Company. In Titusville everybody knew that the few gallons of smelly crude oil that could be soaked up from the surface of Brewer's spring with blankets— a practice begun by the Seneca Indians— were hardly worth the effort. The local

sawmill skimmed enough to lubricate its large circular saw, and occasionally provide a smoky source of light, but the only person in the state who seemed to have found a regular use for small amounts of oil was Samuel Kier in midstate Tarentum. He crudely refined whatever quantities turned up in his salt wells and peddled it as a cure for rheumatism, cough, ague, toothache, corns, neuralgia, indigestion and liver problems. Eight-ounce bottles of Kier's Natural Remedy sold all over the United States.

Old Ebenezer Brewer, who owned part of the Titusville sawmill and also ran the Titusville bank, recommended the medicinal use of oil to his doctor son, Francis, newly graduated from Dartmouth, who began using it for his lumbago. Not long after, Francis alarmed his father by helping form the Pennsylvania Rock Oil Company, which paid $5,000 to lease 100 acres of Brewer farmland and acquired oil rights on another 12,000 acres. "You are associated with a set of sharpers," Ebenezer told his son, "and if they have not already ruined you, they will soon do so if you are foolish enough to let them do it."

Ebenezer knew it was impossible to extract much oil, and he didn't trust a couple of lawyers from New York, George Bissell and his partner Jonathan G. Eveleth, the prime movers in the new company. Bissell looked like a villain, with long black hair and scowling eyebrows, but he was in

Everybody wanted a well. This is Bull Run Oil Creek, Titusville, Pennsylvania, in 1864, after one of the commonplace fires.

fact a brilliant Dartmouth College scholar and linguist and a farsighted businessman. He and Eveleth rescued oil from quack medicine by sending a sample for analysis to Benjamin Silliman, professor of chemistry at Yale. Silliman's report was enthusiastic. As good an illuminant as any the world knew might be made out of the rock oil, said Silliman, and it would also yield gas, paraffin and lubricating oil: "In short your company have in their possession a raw material from which, by simple and not very expensive process, they may manufacture very valuable products." Even with

Drake with his good friend Peter Wilson, a Titusville druggist who signed a note so that Drake could borrow $500 to continue work on the well.

Silliman's epochal report, Bissell could not find buyers for the stock; nobody was convinced oil could be found in quantities enough to justify the expenditures. As a result, Bissell was forced into the arms of a genuine sharper, James Townsend, president of City Savings Bank in New Haven, who raised the money and soon enough found a lever to oust all the founding partners from management.

Drake came into the picture after the discovery that Mrs. Ebenezer Brewer and Mrs. James Rynd, whose names were also on the title deeds, had not signed the original conveyance allowing Bissell and Eveleth to exploit the land. Townsend urgently wanted someone presentable to go down to Titusville to secure the missing signatures. In the old Tontine hotel in New Haven where Townsend lived, he met

another guest who seemed just right for the mission, Edwin Drake. Drake was not then, and never was, a colonel. He was a 38-year-old widower who had never risen higher than railway conductor, forced to retire because neuralgia of the spine made it too painful for him to stand and collect tickets. But he had intelligence beyond his common school education, a courtly manner, and he could travel most of the way to Titusville on his free rail pass. Townsend solved the issue of social status simply by addressing him as Colonel on letters sent to Titusville in advance of Drake's arrival, and the title stuck. For his own good reasons, Townsend convinced his colonel to invest his life's savings of $200 in the company.

When Drake returned with the signatures, Townsend used Drake's position as a stockholder to carry out a complicated coup that ended Bissell and Eveleth's management role, gave Townsend control via a new entity called Seneca Oil, and made Drake general manager at an annual salary of $1,000. This was barely more than he had earned on the railroad, but the property was leased to him and another stockholder for 15 years in return for royalty of 12 cents on each gallon of oil Drake could produce; as a bank president, Townsend thought it prudent not to be prominently associated with something as speculative as oil. Without an effective way to extract it from the earth in quantity, it was a scientific curiosity but commercially of little interest.

Drake, to this point, had been Townsend's puppet, but from the moment he returned to the muddy streets of Titusville in April 1858, he was very much his own man. He knew nothing of oil, nothing of engineering, nothing of geology; he was barely conversant with business procedures, but he knew how to make friends and his intelligence was served by an amazing tenacity. He rented room and board at $6.50 a week for his daughter and new wife, Laura, and mined for local knowledge, attending socials, yarning round the coal stove in Reuel Fletcher's general store, playing pinochle on an overturned barrel. His first move was to ride a borrowed horse into the next town to buy picks and shovels and then hire workmen at a dollar a day

to dig trenches. The common view was that oil dripped from coal in the contiguous hills and digging trenches would lead to a vat. Drake's first great contribution to the future was his early conviction that this was wrong, that oil lay trapped in rocky reservoirs beneath the ground and the only way to get at it was to drill a hole, exactly the way people mined for salt. Indeed, on his first trip to Titusville he had stopped off in Syracuse to look over the salt wells there, and the seed may have been planted then. In a memoir, he wrote, "Within ten minutes after my arrival upon the ground with Dr. Brewer, I had made up my mind that it [petroleum] could be obtained in large quantities by Boreing as for Salt Water. I also determined that I should be the one to do it."

Of course, everybody claims to have first thought of drilling. Some writers give the credit to Bissell, inspired it is said by the picture of an artesian well on a bottle of Kier's Natural Remedy, but nowhere in his papers does Bissell mention the story, and when he was running the company the emphasis was on mining and digging. It was certainly Drake's own decision, after a few weeks, to phase out skimming and drill down to bedrock. His intuitive understanding of where oil collected gave him confidence to press on in the face of widespread skepticism and a shortage of both cash and technical knowledge. He got on his horse again to see if he could find a driller in the town of Tarentum, where Kier had his salt wells. He had heard that most drillers were drunks and was pleased to find perhaps the only teetotaler in the business who was also willing to try and drill to 1,000 feet: No salt well had ever been dug anywhere near that deep.

The driller agreed to come in July, then never showed up. He knew Drake was off his head. Drake went off again in search of an expert driller and a steam engine. For $500 he acquired a six-horsepower steam engine of the kind used on Ohio and Allegheny steamships. Seneca Oil, at Townsend's behest, voted him another $1,000 for expenses, a total of $2,000 to date, and made him president. Drake designed and built an engine house, hoping to be ready to try his gamble at the beginning of September. But a second borer who

The prosperous oil trade, Pittsburgh, 1869, the year a destitute Drake was tramping the streets of New York City.

promised to come got cold feet, and by mid-November Drake did not know where to turn. The cold and the sticky yellow mud everywhere made it a miserable winter. "I set myself down very uneasy, to wait for Spring," he wrote. "I never saw such winter as they have in that part of Penn." His neuralgia flared up. Money from New Haven dried up. There was little food for his family. One hand at the sawmill gave Drake a credit slip for the gristmill to get a 100-pound bag of flour while storekeeper Fletcher let him run up a $300 bill and joined with Peter Wilson, a young druggist, to cosign a $500 personal loan. In February, racked with pain, Drake went back to Tarentum, a 100-mile shaking by buggy, to hire a third borer. He never showed up either. Throughout western Pennsylvania, "the Colonel" was now known as Crazy Drake.

April 1, 1859, was technically the end of his contract. He wrote to Townsend that he was determined to find a borer. From the comforts of New Haven, it looked a forlorn business. Seneca Oil had invested $2,000 and did not want to put up any more. Among the directors, only Townsend kept faith. He sent $500 from his own pocket. That same month, Drake got a letter from a salt well operator in Tarentum who said the Colonel might be able to hire a blacksmith, one William A. Smith, "Uncle Billy," of Saline, a small town outside Tarentum. Again Drake made the 100-mile trip. He found a short, broad man of few words, but this time they turned out to be true. Uncle Billy promised to come in four weeks. In mid-May, his fingers crossed, Drake sent a large wagon, and Smith arrived with his self-forged tools

weighing 100 pounds, his 15-year-old son, Samuel, and his daughter, Margaret.

When they began work that June, it was intense six days a week. Drake designed and built a 30-foot-high derrick, 3 square feet wide at the top, 12 square feet at the bottom, and it went up with out-of-work local lumberjacks among two dozen people heaving to raise the pine structure. Dr. Brewer derisively gave out cigars, saying, "Have one on me. They didn't cost a cent. I traded oil stock for them." But Drake believed in what he was doing; he used his own salary to meet expenses, though he was still deeply in debt to people in town. Much as they drilled their hole, however, Drake's team could not get to the bedrock. Water surged in faster than they could pump it out. It seemed as though the usual salt-boring techniques would not work.

Even before the internal combustion engine multiplied the demand a billion times, oil was prized as a fuel for lamps and a lubricant.

Smith was offered a job for four dollars a day in another town but had grown protective of his employer. He told his son, "I can't quit Drake now."

Once again, Drake's intuition came to the rescue. His revolutionary idea was that they should stop digging beyond the 16 feet already done, but try to drive an iron pipe through quicksand and clay to the underlying rock. He rushed off to Erie, bought a cast-iron pipe three inches in diameter, half an inch thick, and assembled 50 feet of it on site. It broke at 10 feet. He switched to soft iron, one and a half inches thick, and hammered it down with a white-oak battering ram. This pipe hit rock at 32 feet. He put the drill into the pipe, itself a novel way of drilling at the time, harnessed it to his steam engine and drilled three feet that day—and the next and the next, into the following week, and the week after. Locals who had been interested stopped coming by to watch "Drake

fooling away his time and money." Townsend, too, lost heart. In late August, he wrote a letter to Drake telling him to shut down at once, enclosing a final $500 for outstanding bills.

Our age of instant communication would have ended the adventure then and there, but receipt of Townsend's mail was dependent on the stagecoach from Erie. It ran only twice a week, so Drake remained ignorant that the last of his supporters had deserted him. He just kept drilling. The story generally told is that at the end of the day on Saturday, August 28, work stopped when the drill had reached 69 feet. Drake strictly observed the Sabbath, but Uncle Billy is said to have come back on the Sunday, glanced down the pipe and noticed something glistening on the surface below. He took the leftover end of the pipe, plugged one end to seal it like a cup and sent it down on a stick. The dipper came back filled with oil. Billy's

son, Sammy, ran barefoot to tell Drake. He would not move. The Sabbath was the Sabbath.

Uncle Billy remembered things differently, and he is more plausible. According to Smith, Drake was in the derrick on the Saturday when at 69 feet the drill slipped six inches into a crevice and the jars stopped working. "I noticed fluid rising in the drive pipe and called Drake's attention to it. He said, 'What does it mean?' I said, 'That's your fortune coming.' I lifted about half a gallon of oil."

Monday, August 30, was certainly the day Drake and Smith realized their bonanza. Drake attached 20 feet of his pipe to a common hand pump and brought up eight barrels of oil. He took Margaret Smith's bathtub from the engine house shanty and filled that, too, then built a wooden tank to hold about 25 barrels. Soon he was drawing 400 gallons of pure oil every 24 hours.

It was at this moment that Townsend's letter arrived. It was all such a close-run thing. Had the letter arrived the week it was supposed to, the well would have been abandoned, just a hole in the ground. Had Drake drilled only a few yards away in any direction, he would have failed to strike oil; to find oil, he would have had to drill an improbable hundred more feet, which would have taken at least another month. Had he been unwilling to invest his own credit and his own good name in town, the enterprise would have coughed to an end in the winter of 1858–59.

But now, in the halcyon summer days in the valley, the Great Oil Rush was on. Silliman's endorsement of petroleum as an illuminant and lubricator and substitute for coal was well-known in the region. Within days, there was a full-scale stampede for land in the valley, leases, machinery. A magazine writer of the time observed, "Merchants abandoned their stores, farmers dropped their plows, lawyers deserted their offices and preachers their pulpits." Those who had scoffed at Drake were frantic to drill. The population of the valley exploded. Oil derricks sprouted everywhere on the pastoral landscape, suddenly denuded of trees cut down to make timber for oil barrels. Bissell got advance word over the new-fangled telegraph, bought up all the Seneca Oil stock he could find and hastened to Titusville to buy leases. One sawyer with 200 acres of poor land made $2 million from leases and another $1 million in royalties. Newspapers proclaimed the Age of Illumination had dawned.

Where was Drake in all the excitement? After the strike, he went fishing, bought a pair of loud pantaloons and a horse, then went back to his well. Local friends who immediately started buying up leases tried to get him to join the game. He declined. He did not even consider developing the 25 acres he owned. The obstinacy, the single-minded dedication that had brought success, was still driving him to concentrate on finishing what he had begun. He bought some tools and, ignoring the hubbub all around him, worked quietly with Uncle Billy to make the well more efficient. On October 7, Uncle Billy took a lamp too near their great new tank, and in an instant everything was ablaze. Drake refused to be

discouraged. "The oil is still there," he told the disconsolate Smith. He put up a new derrick and got started again on November 7 producing 32 barrels a day.

Drake's resilient spirit was betrayed by the moneymen in New Haven. Townsend's fellow directors, the very same men who had prematurely pulled the plug, were impatient to profit. In the early days, before refineries were built and a proper market organized for oil as an illuminant and lubricant, there was something of a glut, and they thought Drake was not nimble enough to exploit his own discovery. They may well have been right in that; he was not a skilled salesman. Nevertheless, their treatment of him remains shameful. In the spring of 1860, they demoted him from president to agent, and then in 1863, when oil was $7 a barrel, they fired him altogether, claiming the company had never made a cent. The people of Titusville were more honorable. They recognized that Drake had brought untold prosperity to the region and made him a justice of the peace, at a salary of $3,000, which ironically meant he was called on to notarize leases on some of the biggest wells.

It was a bit too much to bear. Drake left the area with an estimated $15,000 to $20,000 in savings, but his lack of business acumen was manifest in the sale of his 25 acres for $12,000; a few months later they fetched $90,000 and in two years closer to $300,000. He failed to patent his invention of the driving pipe. Within a year he had lost all his money in a Wall Street oil-stock brokerage firm. Bissell, by contrast, became a very rich man but begrudged Drake his fame.

The trials of those grim winters in Titusville caught up with Drake. Living in New Jersey with Laura and a family of four children, he fell sick. Laura took in sewing to keep them on a diet of potatoes and salt. In 1869 an old Titusville friend and oil buyer, Zeb Martin, ran into Drake looking for work in New York City, limping along the wharves near the Customs House, wearing the same black coat he had worn ten years before. He spread word of "Colonel Drake's" condition among Drake's old friends, now rich. The men assembled in Corinthian Hall and raised an immediate $3,000 in pledges to "the man who laid the

foundation of so many splendid fortunes and pioneered the way to the grand spectacle of industrial activity witnessed in all the region." They gave more the following year, and in 1873 the state of Pennsylvania voted an annuity of $1,500 "to the said E. L. Drake or to his widow in the event of the death of the said Drake." He died in 1880 in Bethlehem. Seventy-two years later, his body was moved to Woodlawn Cemetery, Titusville, where he lies with Laura. A bronze sculpture of Drake donated by a Standard Oil baron is embraced by panels of stone incised with the story of how this one man "enriched the State, Benefited mankind, Stimulated the mechanic Arts, Enlarged the pharmacopeoia . . ." The tribute says, in part: "He triumphantly vindicated American skill. And near this spot, Laid the foundation of an industry . . . He sought for himself neither wealth nor social distinction, content to let others follow where he led . . . bequeathing to posterity the fruits of his labor and industry . . ."

Would oil have been discovered without Drake? Some believe it was only a matter of time, since others were thinking on the same lines. But the easy assumption overlooks the originality of his intuition about underground reservoirs of oil, the adamantine nature of his will, the depth of the skepticism he faced, the early capitulation of every one of the men-of-the-world investors, the brilliance of his simple invention of a driving pipe—and his luck in striking oil at such a shallow level. Uncle Billy maintained for years that no one would have attempted drilling again if Drake had failed after so much effort. At the end of his life, Drake himself wrote: "I do not say it egotistically, but only as a matter of truth, that if I had not done what I did in favor of developing Petroleum, it would not have been developed at that time. The suffering and anxiety I experienced, I would not repeat for a fortune. If I had not done it, it would not have been done to this day."

There is no need to spin the gossamer threads of might-have-been. The immutable fact is that it was Edwin Drake and Edwin Drake alone who released what historian Herbert Asbury calls "the golden flood of petroleum" and it was Edwin Drake who thereby transformed the world.

LEVI STRAUSS: He caught the American bug. He went west hoping to sell his bundles of blankets, spades and boots.

LEVI STRAUSS What kind of innovation lasts 150 years? The kind made by an adventurous peddler, because his product came to typify the democratic spirit of the West—blue jeans

1829–1902

FRESH OFF THE BOAT from Bremen in Germany in 1847, 18-year-old Levi Strauss was soon walking the streets of New York laboring under the 100-pound weight of two packs, one on his chest and one on his back. He carried bolts of cloth, yarns, needles, scissors, buttons, combs, books, shoes, blankets and kettles, and he carried them as far out as Pelham in Westchester County, knocking on doors and hoping to make a sale. He picked up a few words of English as he went along. On the road during the week, he slept in barns and stables and ditches. He was always back in lower Manhattan on Friday for the Jewish Sabbath.

Levi had crossed the Atlantic with his mother and three sisters, packed for a month in the cramped and dank quarters of steerage, the latter part on a diet of dried lentils and salt pork when their supply of kosher food ran out. "Levi" was a changed name. His original name of Loeb was all right in German but awkward in English, and he meant to get on in this new world where Jews were treated like everyone else. He had grown up in a village in the forested countryside of Buttenheim in upper Franconia north of Nuremberg, the son of Hirsch, a peddler, and Hirsch's second wife, Rebecca Haas. It was a beautiful place, but Jewish families were forbidden to own land, limiting them to a few trades like peddling, and the vague threat of pogroms haunted

their every day. On the death of their father in 1845, two of Levi's half brothers, Jonas and Louis, immigrated to America and they were busy street peddlers by the time Levi arrived. Jewish peddlers were known and often welcomed across America in the first part of the 19th century. Many Jewish peddlers did well enough to find more lucrative occupations. Some became doctors or other professionals. A few former peddlers, like Benjamin Altman, Adam Gimbel and Meyer Guggenheim, became tycoons.

After three years on the street, Jonas and Louis in 1848 were able to open a small wholesale business at 203½ Division Street in an area called Hebrew Market, but Levi had caught the American bug: He was restless. Everyone in America, it seemed, was forever on the move. Levi's sister Fanny married another Jewish immigrant peddler, David Stern, and was off to St. Louis, Missouri. Levi went to live in Louisville and tried his luck humping his brothers' supplies in the Kentucky hills.

He had not been at it long when the whole country was convulsed by the California gold rush. In 1849 alone, 80,000 prospectors flooded into the state. Soon afterward, Fanny and David Stern were drawn to San Francisco, and Levi decided to join them there in a trading partnership. Before heading out to San Francisco, he went all the way back to New York to stock up from his brothers' business. It is indica-

Stars anointed the workingmen's pants with glamour. Portrait of Marilyn Monroe by Andre de Dienes, 1950.

tive of the adventurous character of the boy from a Bavarian village, and the questing spirit of the age, that he should undertake the trip loaded up with everything he imagined California miners might need. There was no railway across the country. Some went by land, with all its dangers of thirst and Indian raids; many went by the malarial Isthmus of Panama; others sailed round Cape Horn. All these routes took tedious and dangerous months. Levi chose to put his bundles of blankets and spades and boots on a sailing ship to Panama in the first week of February and then on the Pacific Mail Steamer *Isthmus* to San Francisco. He did not disembark until March 14, 1853, just after his 24th birthday and his grant of American citizenship.

San Francisco was a riotous place, 399 saloons and 28 breweries, and 1,200 known murders that year in a population of 70,000, plus another 2,400 newcomers who vanished never to be seen again alive. More than 1,000 ghost ships lay in the harbor, abandoned by crews bitten by the gold bug. There were already 117 dry-goods stores in the gold rich but merchandise poor city, where a blanket worth $5 in New York could fetch $40. Levi and Stern set up a "dry goods and clothing" wholesale business at 90 Sacramento Street on the pilings close by the wharf. They had an edge on the competition, with supplies assured from Levi's brothers, who now had a large store at 165 Houston Street in New York, but Levi and Stern also posted a boy with a telescope in the San Francisco hills and, on his alert of approaching ships, they bedded down in their warehouse. At dawn, they would rush to the first ship docking to bid for cargo up for auction. Being first onboard was often the difference between a good and a bad week.

It is not entirely clear how Levi came up with the idea for what became his principal product—too many records were lost in the San Francisco earthquake and fire of 1906—but he could hardly have failed to notice that the miners, lumberjacks, teamsters and ranchers on the streets were raggedly clad in trousers with threadbare seats, holed knees and pockets ripped from jamming in gold nuggets, gloves, tools. Legend has it a miner complained: "Should'a brought pants. Pants don't wear worth a

hoot in the diggin's. Can't get a pair strong enough to last." The story goes that Levi then tailored a pair of pants from canvas intended for a Conestoga wagon. It is not likely, given that he was a wholesaler supplying retailers, and he would not have had tailors on the premises. Who it was who cut the first hard-wearing denim pants and in what style is unknown to history, but when the pants arrived (the material probably from the New York warehouse), we do know that Levi sold them as "waist high overalls." The miner in the myth is said to have paid in gold dust worth six dollars. Certainly his affection for them became widely known. More men went into stores asking for the same waist-high overalls. His brothers sent Levi blue denim, twill weave durable cotton, which he passed on to outside tailors to cut and sew.

The work pants soon became the center of a flourishing wholesale store, and Levi began his lifelong tradition of philanthropy: The Orphan Asylum society of San Francisco records a donation of five dollars in 1854. When a major banking crisis in 1855 closed businesses, Levi kept in the black by adventuring into the mining camps in the hills above Sacramento. (Leland Stanford, one of four merchant-progenitors of the transcontinental railway, had plied the same routes.) The partners hired traveling salesmen as trade revived and moved into successively larger premises; in 1857 Levi's half-brother, Louis, came out from New York to help. In February 1861 it was recorded that the company sent no less than $59,732.24 worth of gold to New York to pay for supplies. In 1866 Levi spent $25,000 to adorn new headquarters of Levi Strauss and Co. at 14-16 Battery Street with gaslight chandeliers, a cast-iron front and one of the first freight elevators. He dealt in fancier goods now, Irish linen and Belgian lace and Italian shawls along with blankets and women's clothing and the rugged work clothes. The store would no doubt have continued that way, successful but unexceptional, but for a remarkable congruence seized and exploited by Levi.

Another Jewish immigrant, who had come to America some seven years after Levi, changed his name and come west, was struggling to make a living in Reno, Nevada.

Jacob Youphes, born in Riga in 1831, was now Jacob Davis. In 16 years of roving North America, he had failed as a tailor, a miner, a brewer and a tobacconist, and was trying his luck again as a tailor. He lived in a shack by the recently finished transcontinental railroad track making a few horse blankets, wagon covers and tents. In December 1870 the wife of a woodcutter asked him to make pants that would not rip. The strongest material Jacob had was ten-ounce duck twill, which he usually used for making tents. When the wife came with three dollars and the woodcutter's measurements, Jacob sat cross-legged on his bench and sewed the heavy, hard-to-cut cloth while she waited. He fussed with the stitches because he knew that many workmen's clothes tore at the pockets. Some rivets lay on the table. He routinely used them to hold horse blankets together. The thought just struck him, he testified later in a patent trial, "to fasten the pockets with those rivets." He hammered them into the corners of the back and front pockets, and later watched the riveted woodcutter trudge into the hills with his ax over his shoulder. "I did not make a big thing of it," he told the court during one of many later patent suits. "I sold those pants and never thought of it for a time."

The next month four men came to Jacob's shack for riveted pants. In February 1871 he sold ten pairs and in March a surveying party bought a dozen. Men who wore them were walking advertisements. After 18 months he had sold 200 pairs in white duck cloth or blue denim. The hand-to-mouth Jacob had been able to buy the initial material only because his supplier was a trusting soul: Levi Strauss. He habitually let hardworking customers run up a line of credit. Jacob recognized the value of his little invention, but it cost money he did not have to apply for a patent; his wife, Annie, told him not to waste any more of their money on such madcap schemes, since he already had two patents and the couple were still poor. So it was that on July 2, 1872, the inventor wrote to Levi with a proposition. His German was still better than his English, and his letter, which he dictated to the town druggist, reflected the inflections of his Eastern European Jewish accent:

You could tell a gold miner in California in the 1850s—the man whose pants had holes in the knees, ripped pockets and a threadbare seat. The daguerreotype by George H. Johnson, *Mining on the American River, near Sacramento,* suggests the conditions that wreaked the havoc and led to the demand for strong pants—for Levi's.

The secret of them Pents is the Rivits that I put in those Pockets and I found the demand so large that I cannot make them up fast enough. I charge for the Duck $3.00 and the Blue $2.50 a pear. My nabors are getting yealouse of these success and unless I secure it by Patent Papers it will soon become a general thing. Everybody will make them up and there will be no money in it.

Tharefore Gentlemen, I wish to make you a Proposition that you should take out the Latters Patent in my name as I am the inventor of it, the expense of it will be about $68, all complit and for these $68 I will give you half the right to sell all such clothing Revited according to the Patent, for all the Pacific States and Teroterious, the balince of the United States and half of the Pacific Coast I resarve for myself. The invesment for you is but a trifle compaired with the improvement in all Coarst Clothing. I use it in all Blankit Clothing such as Coats, Vests and Pents, and you will find it a very salable article at a much advents rate. . . .

These looks like a trifle hardley worth speakeing off but nevertheless I knew you can make a very large amount of money on it. If you make pents the way I do you can sell Duck Pents such as the Sample at $30 per doz. And they will readily retail for $3 a pair.

Pants that could fetch $30 a dozen wholesale were gold dust. Levi's company pants sold for only $10 a dozen. A baser person than Levi Strauss would not have hesitated to steal Jacob's idea—the patent suits are populated by such scoundrels—but Levi accepted Jacob's offer. He paid for a patent application on behalf of Davis and Levi Strauss and Co. for "improvement in fastening seams . . . in order to prevent the seam from starting or giving away from the frequent strain or pressure." The Patent Office rejected the application. It was considered too similar to other patents using rivets to hold together clothing. Rivets had been used on soldiers' boots during the Civil War. Levi paid for ten months of haggling over wording, and three amendments later, Jacob and Levi had their patent—139,121—approved on May 20, 1873.

Jacob, back in Reno, had meanwhile been trying to sell pants on his own, with only a handful of orders. He realized he needed the superior business skills, literacy and capital of Levi; and the two men of roughly the same age—Levi was 44, Jacob 42—got on well. On April 26, 1873, a month before the patent was finally approved, Jacob sold his half share of it to Levi Strauss and Company, closed his railroad-side shop and moved with Annie and their six children to a good house in a fashionable district of San Francisco to begin a new salaried life as the head tailor and foreman of production. A devastating depression started in 1873, but the first blue jeans started running wild after the first sale in June. By the end of the year, the company had earned $43,510. No fewer than 20,000 men were out on the streets and in the hills sporting their Levi Strauss pants with a dis-

tinctive orange seam thread that Jacob had introduced to match the color of the rivets. Davis was running himself ragged, supervising the cutting of the blue denim and its delivery to individual seamstresses, so Levi decided demand was not best met by dispersed workers. Risking a fair amount of capital, he set up a factory in Fremont Street with 60 seamstresses on the spot, each one sewing a complete pair of pants from 15 pieces of cloth. The best could make five pairs a day, earning $3, the same level of earnings as a bricklayer or mechanic. It represented an increase in productivity, though not as much as would have been achieved by the American system, popular in the East, of dividing the work process into very small, specialized tasks. Levi could have cut his costs by employing cheaper Chinese labor (as Charles Crocker had done building the Central Pacific). Whether Levi shared the general xenophobia or just thought it good business is unclear, but he made a point of advertising his discrimination: "Our riveted goods . . . are made up in our Factory, under our direct supervision, and by WHITE LABOR only." Levi's campaign lasted until the 20th century. (He did employ one Chinese man to perform the exacting job of cutting denim and duck canvas with a long knife. It took strength to cut through the layers of cloth and endurance to do nothing but that all day. Every white worker hired for the job had quit.)

After David Stern died in January 1874 at the age of 51, Levi ran the business with his half-brother Louis; William Sahlein and Jonas Strauss were on the board. Levi was 45 then, and pictures show a man of five feet six inches with trimmed beard and heavy-lidded eyes. He was a vigorous bachelor who just happened never to have found Mrs. Right. He shared a home at 317 Powell Street with his widowed sister, Fanny, and her children. Fanny, a shrewd and ambitious woman, wanted her children to inherit the business Levi had built with the help of her husband, so she hoped he would not marry; she certainly provided a comfortable home, and it is said that liaisons were arranged with married women on the understanding between all the parties there would be no offspring.

Levi was generous to his synagogue. He qualified his religious observance by working on the Jewish Sabbath, Saturday, because the state of California banned Sunday work and he did not want to lose two days. But he was not like the tyrannical Gradgrind in Dickens's *Bleak House.* He would leave home at 9 a.m., wearing the standard businessman's outfit of the day, a black broadcloth suit with split tailcoat and a top hat of Japanese silk. He stopped off to talk to friends and customers and arrived at work at 10 a.m. He would check the previous day's sale and then visit the noisy factory floor where women—Italian, Spanish and Irish—worked at rows and rows of sewing machines in one big room. He made it a habit to talk with people at all levels of the company—clerks, bookkeepers and seamstresses. He urged everyone to call him "Levi," not "Mr. Strauss." Those who caught his eye were promoted. He would spend the rest of the day with his bookkeeper and at meetings for other business interests he had developed in real estate, gas and railroads, but he would invariably walk home at 5 p.m. He often took dinner with friends in a luxurious dining room at the St. Francis hotel. His nephews Nathan and Sigmund Stern, known to enjoy having fun, occasionally joined him. They would drink champagne and puff on Havana cigars afterward. Sigmund in the early 1890s introduced Levi to the flamboyant William Randolph Hearst, freshly expelled from Harvard for making an obscene gesture at a professor. Hearst, who enjoyed attending Levi's parties, brought along the dangerously witty Ambrose Bierce, a provocative writer on Hearst's first newspaper, the *San Francisco Examiner.*

By 1876 sales had climbed to $200,000 a year. Sales representatives traveled to Mexico, Hawaii, Tahiti and New Zealand. By 1880 the company had 250 workers and sales of $2.4 million. A decade later, the total was 450 workers, with an additional 85 employees in the offices. The pants had become *the* western dress code. Levi relished the way his clothing served the needs of the workingman. He emphasized it in his advertising: "These goods are specially adapted for the use of FARMERS, MECHANICS, MINERS and WORKING MEN in general." He had a guarantee sewn on to the back of the pants promising "a new pair FREE" if the current pair ripped. The picture of two horses trying to pull apart a pair of pants became a trademark. Infringers of the original patent were thwarted in successful court actions. The Levi's (as they became known early in the 20th century) were valued for their strength, but they were also on the way to becoming a symbol of democratic equality.

Levi was a millionaire several times over by then, but he had no time for mere accumulation. In his only known interview, in 1895, at the age of 66, he told the *San Francisco Bulletin:* "My happiness lies in my routine work. . . . I do not think large fortunes cause happiness to their owners, for immediately those who possess them become slaves to their wealth. They must devote their lives to caring for their possessions. I don't think money brings friends to its owner. In fact, often the result is quite contrary."

Louis Strauss died in San Francisco in 1881 and Jonas in New York in 1886. In that year, only 12 years after David Stern's death, Levi stepped back from his day to day role, handing the burden to Fanny's four sons, Jacob, Sigmund, Louis, and Abraham. He and his nephews officially incorporated the company in 1890, the year that the lot number "501R" was first used to designate the denim waist overalls—and the year the patent ran out, freeing others to imitate the design. Levi remained active in philanthropy as a trustee of the Pacific Hebrew Orphan Asylum and Home and the Eureka Benevolent Society. He established 28 perpetual scholarships to the University of California, four from each congressional district in the state. On his death at the age of 73 in 1902, he left much to Hebrew and Roman Catholic orphanages and the Emanu-El Sisterhood, but most of his $6 million went to his four nephews, as Fanny had hoped. The Stern brothers ran the business for years after their uncle's death, and then it passed to Sigmund Stern's son-in-law, Walter Haas, and his family. Robert Haas and his uncle Peter, leaders of the firm in 2003, are the fifth generation of Levi Strauss's extended family to own and run the company.

Levi achieved immortality with his jeans—though jeans is a word he never

Levi called his working pants "waist high overalls." They were popular with both sexes. These women railway workers in Wyoming are wearing his full-body overalls.

used. It was not until the 1960s, 60 years after his death, that the company came to use the term. They were still essentially western wear, with only 10 percent of sales in the East until the 1950s. But as a universal popular culture emanated from Hollywood, something happened to transform Levi's from clothing into a ubiquitous cultural statement all around the world, one of independence, rebellion, equality, freedom. It began in the '30s, when the West truly captured America's and the world's imagination through the movies. Levi's came to be associated less with rough labor and more with the romance of the free-roaming cowboy. Each generation since, from World War II to the rebellious '60s, and the leisure era, has vested Levi's with mythic qualities. *Life* magazine noted the effect in 1974, when U.S. production of 450 million yards of denim a year could not

keep up with demand: "The jeans that once encased the scrawny rumps of cowboys and gold miners of the American West have become the standard garb of the world's youth. They're the favorite off-duty clothes of fashion models from L.A. to St. Tropez. By buying up bales of fashionably tattered used ones, Britain and the Continent have made millionaire exporters of U.S. ragmen."

Seventy-five million pairs of genuine Levi's were sold that year; they were copied more than any other piece of clothing. How curious that a long-dead poor immigrant should be hailed as a world leader of fashion. It is a commentary on the social attitudes of the time that in England the men whose names came to be commemorated in clothing were aristocrats—Cardigan, Wellington and Raglan—but in America it was a street peddler of workaday clothes.

ELISHA OTIS
He was a country craftsman dogged most of his life by bad fortune, but he changed cities and the way we live and work when he invented the safety elevator

1811–1861

IT IS A STRIKING FACT that over the course of every 72 hours the Otis Elevator Company shuttles the equivalent of the world's population into space and brings them down again. There are 1.4 million Otis elevators on five continents, in skyscrapers and silos, in malls and apartment houses, in the White House and in the Kremlin, in Buckingham Palace and in the Vatican. It is just as striking that country craftsman Elisha Graves Otis, who began it all in 1852, had to pick himself up off the floor as many times as he did, had no head for business and never quite appreciated the significance of his invention. Like Sam Colt, Elisha Otis was into middle age before he was successful, like Colt he died young; unlike Colt, he had little interest in amassing a fortune. He just wanted to fix things.

Elisha Otis was the archetypal Yankee tinkerer. His name endures, but he is representative of thousands unknown to history who advanced the mechanical arts by fingertip instinct. They mended and adjusted, altered and created: They invented on the run and never bothered to claim ownership by patent. It took seven years for Otis to get around to patenting his foolproof way of stopping a runaway elevator.

He was the youngest of six children, raised on a farm in the hamlet of Halifax, Vermont, by a devout mother, Phoebe, and a father, Stephen, respected for his ingenuity with machines and his integrity: He was a four-term state legislator and a justice of the peace. Elisha "had no taste for a farmer's life," his son Charles wrote in 1911. Instead, he liked hanging about the village blacksmith's. He had not graduated from high school when his father agreed he could leave home at the age of 19 to help his

brother, Chandler, a master builder in Troy. In 1834 he married Susan A. Houghton. His first bad break, working in all weathers, was a bout of pneumonia that nearly killed him. Once recovered, Otis took a job driving wagons, and the earnings allowed him to take his wife and three-year-old son back to the Vermont hills. By the side of the Green River he built a house and a dam for waterpower for a gristmill of his design. The young couple had another son. It was idyllic, but the mill made no money. He turned his hand to making wagons and carriages, at which he did fairly well, but his conversion of the gristmill into a sawmill failed to attract enough customers. The

Otis cut the supporting rope and the hoist fell—but only for a few inches. His demonstration of his safe elevator was the crowd-puller in 1854 at the second New York World's Fair on a site that is now Bryant Park on 42nd Street. One hundred fifty years after Elisha's first hoist, the stunt was reenacted in New York with Otis elevators as a thriving part of United Technologies.

fates seemed to conspire against Elisha. His wife died, leaving him with two boys, aged seven and two. Then his health failed again.

In 1845 he made a fresh start. He married Betsy A. Boyd and moved his family to Albany, New York, for a job as a mechanic with Otis Tingley, a manufacturer of bedsteads. He was 34 and in his element among the lathes and saws. The Puritanism instilled in him by his mother gave him "a furious sense of imperfection," in the words of his biographer Jason Goodwin. "Almost everything he saw could be done better—faster, cheaper, and more accurately—and he went at it with the energy of a man possessed." A skilled woodcutter working all day at a lathe could make the pine board rails for joining up the sides and ends of 12 beds. Otis invented a machine that enabled an unskilled laborer to turn out the rails four times faster—enough for 50 beds a day. Tingley paid him a $500 bonus, and with the money Elisha went into business on his own again. He leased a building nearby on the banks of Patroon's Creek, designed and built a turbine to maximize the waterpower, and made and fixed machines, including his rail turner. In odd moments, he worked on making an automatic bread-baking oven and a new type of brake that would allow engineers to stop trains instantly. Just when his machine shop was succeeding, he had his fortunes plucked from him: The City of Albany cut off his power, diverting the stream for the growing population's fresh water. He attempted to manufacture carts, but his malign rhythm was unbroken and he failed.

Elisha Otis, up to his eyes in debt but with faith in God's will, faced the hazards of pioneering in the emergent industrial East as resiliently as the homesteaders then

No French company would bid for the curving elevator between the first and second tiers of the Eiffel Tower in 1887. Otis's two sons reckoned that it was worth attempting even at a loss.

heading west into dried-out plains and Indian ambush. Once again, he uprooted his family for an opportunity, first as a mechanic in Bergen, New Jersey, and then as a manager in Yonkers, New York, where a trio of investors commissioned him to convert a sawmill into yet another bedstead factory. It was here that Elisha almost casually changed the world. We can imagine him arriving by river steamboat at the tin-roofed, three-story brick factory built with its 40-foot chimney on a jetty by the side of the Hudson. In 1856 he is a full-bearded man of 40, his six feet or so given vertical emphasis by a black stovepipe hat. He stands at the stairs on the ground floor of the building amid heavy lumber and bricks and wood shavings, looking up and pondering how he will best raise his new machinery and furniture to the upper levels. There is nothing at all original in his decision to install a hoist, a lifting contrivance based on the principle of countervailing

force; the Romans built "elevator" shafts to hand-haul gladiators and Christians to their fates in the arena. By 1850 several American factories had caught up with the concept of the powered hoist first introduced in a British cotton mill in 1830. It lifted bales on a platform by running ropes from the platform to an overhead drum and a belt from the drum to the mill's steam engine. Elisha would have seen such a hoist in a furniture factory at 275 Hudson Street in New York owned by Benjamin New-house, one of the investors in Yonkers; Goodwin tells us the Boston firm of George H. Fox was shipping similar freight hoists to various parts of the country.

Elisha knows the hoist he envisages will save time and money. He can already see it carrying his precious machinery to the third floor, but his imagination works overtime. He can also see it crashing to the ground—he has heard of hoists failing because the ropes frayed, or the belt to the

engine broke, or the carrying platform was overloaded—and his life had been marred by enough ill luck not to take chances.

The automatic safety system he devised was very simple, though its execution owed much to his experience building carriages. He cut a series of notches in the hardwood ascending guide rails on either side of the platform cabin. Then he took the flat-leaf spring from the underside of a carriage and screwed it to the roof of the platform cabin. He ran the hoisting ropes through the spring. He reasoned that tension in the ropes pulling the platform up would keep the spring open. If the platform went into free fall for any reason, he hoped the release of the tension in the ropes would allow the spring to push outward into the notched guide rails. He set the hoist in motion and cut the ropes. The platform dropped—but only a few inches. The springs bit into the notches and the cabin came to a jarring stop. The ratchet and the spring served as the basis for all of Elisha Otis's later safety elevators, as they came to be called.

What Elisha Otis did was similar to what Robert Fulton or Henry Ford did: He combined existing elements in a new concept. Everything he used in his hoist was familiar. He thought so little of it he neither patented it at once nor pressed the factory owners for a big bonus, and was happy enough to be rewarded with a small old cannon left on the premises. Nor did he try to sell an automatic hoist to someone else. The customers came to him. First, there was Benjamin Newhouse. The freight elevator at his furniture factory had recently fallen several stories and killed one workman and severely hurt another. That elevator had a common safety device, a rotating pinion; in an emergency, the operator was supposed to pull a lever to activate the pinion. But Newhouse's operator had not acted quickly enough. Newhouse ordered two Otis automatics at $300 each. A picture frame factory ordered another. Elisha installed all three by November 1853.

It would be untypical of the Otis story if we could now say the elevator business took off. It did not. When the Yonkers bedstead factory faltered and failed a few months later, Newhouse allowed Elisha to start an elevator business on the premises as the Union Elevator and General Machine Works Company. His capital was a second-hand lathe, a three-horsepower steam engine, a drill press, a forge, a few vises—and his two sons, destined for a critical role. Weeks went by in 1853, then months into 1854, without a single new order. The journals of his older son, Charles, now 18, show a continuing frustration with his father's yen for experimentation rather than concentration. Waiting for orders that did not come, Elisha did more work on his ideas for bread ovens, train brakes, steam plows and railroad bridges

The good luck—at last!—was the opening of the second New York World's Fair held by the New York Chamber of Commerce on a site that is now Bryant Park on 42nd Street. The celebrated showman P. T. Barnum had helped to draw 50,000 in 1853 with an imposing statue of George Washington on horseback. In 1854 he was looking for a stunt to draw the crowds again.

Thus it was that hundreds of people at the opening of the fair in May 1854 looked up in alarm as 42-year-old Elisha Otis, borne aloft in his hoist, slashed with a saber at the retaining ropes. Barnum had admonished "those prone to fainting" to "take out your salts." Crowds were known to scream when the elevator fell; none of the thousands at the fair had ever seen a safety elevator. When it fell for only a second or two, they cheered the imperturbable Elisha as he swept off his stovepipe hat and with a bow announced, "All safe, ladies and gentlemen. All safe." He did it every hour of every day of the fair, a strict and sober Yankee in a bur-

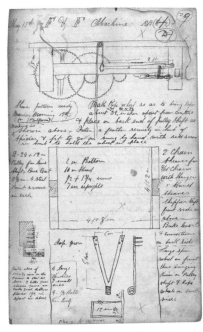

Every elevator Otis, and then his sons, made was designed by hand for the site.

gundy-colored topcoat with velvet lapels who had found a taste for show business.

Barnum paid him $100 for the stunt. It was a brilliant way to publicize the invention. Alfred Wilde of the Harmony Cotton Mills at Cohoes, New York, was in the audience and took the steam ferry to the Yonkers factory, where Charles wrote out a $300 order for a "No. 2 Hoist Machine." Elisha installed his elevator for four more companies. In 1855 he sold 15 elevators, for $5,605, and sales doubled again the following year and doubled again the year after that, with orders from New York, South Carolina and Massachusetts. By 1856 Elisha had installed 53 freight elevators. The shape of things to come was an initiative from a smart department store and the ingenious

response of Elisha. Manhattan's Haughwoot and Co., which sold French china and glassware from a five-story building at the corner of Broome Street and Broadway, wanted a passenger elevator to whisk its customers to its upper floors, but unlike the factories installing hoists, it did not have a central steam engine. Elisha developed a small steam engine specifically for an elevator; Haughwoot's customers were on the fifth floor in less than a minute.

This first independent elevator power source had a double significance. It meant that elevators could be installed in hotels, apartment buildings and offices without central power systems: There was no electricity as yet, and in 1854 the first electric elevator, in a Baltimore office building, was 27 years away. Second, Elisha devised a three-way steam valve so the engine could be put swiftly into forward, neutral or reverse, allowing the elevator to move up or down or stand still. A change of direction with a large central engine involved complicated maneuvers with belts and pulleys. The Otis elevator car could have its direction changed swiftly from inside the car.

Throughout all this, Elisha Otis continued his compulsive inventing, to the irritation of Charles, who was 23 in 1858. He was not impressed by his father's award of a patent for a steam plow (1857) and a rotary oven (1858). He called them "hundred thousand dollar Air Castles." He had to concede that the oscillating steam engine was a triumph—father and son worked on improving it for a patent in 1860—but, inflicted with frequent changes as a child, he was obsessed with stability. Charles evidently made his sentiments public, for Elisha had his son sign a gag order, stating, "It is understood that I am not to volunteer advice or opinions concerning that part of the business not placed in my charge." The gag order did not cover his journals, where he vented his aggravation: "No sooner will daylight appear than father will break loose again and kick me heels over head if he can, forgetting all the past, go crazy over some wild fancy for the future and get into debt worse than ever."

Elisha lived on credit most of his life, though honorably. A credit report of the time, possibly from Lewis Tappan (page

118), described Elisha as "a very conservative and conscientious manufacturer" who "for a considerable time made a practice of not entering into one contract until the last one taken was completed to everyone's satisfaction." His good name won wide recognition over emerging competitors. Some of the top companies in New York chose Otis: Steinway and Sons, A. T. Steward, Sharps Rifle Company, American Express and Lord and Taylor. Yet when Otis died of diphtheria on April 8, 1861, his business was not in great shape. He owed $8,200 and his estate was valued at only $5,000.

Posthumously, however, he was very lucky. His two sons, Charles and Norton, did not have his inventive genes but they had the keen business sense and the managerial skills to erect many stories on his foundation as the coincidence of elevator and steel-framed high-rises that sprouted in Chicago and elsewhere allowed cities to expand upward as well as outward. Charles took care of the office and the engineering, poring over tables and graphs into the small hours, while the garrulous Norton went out selling and oversaw production. They stayed in the fore-front of technology by investing in machinery and hiring professional engineers schooled in the laws of electricity and hydraulics. They bought out advanced companies that might be a threat: 30 companies by 1914. They decorated passenger elevators lavishly with sofas, tables, drapes, upholstery and mirrors. They advertised heavily and aggressively. Any time a rival elevator company suffered an accident, Charles and Norton headlined advertisements "Criminal Recklessness" and "Wholesale Manslaughter" and played to the popular-press taste for gore. The Barnum moment vividly in their memories, they knew the value of publicity. When Gustave Eiffel set out to build his tower in Paris in 1887, no French company would bid for the job of installing a curving elevator between the first and second tiers of the tower; it was as tricky as M. Eiffel himself. The Otis brothers diverted their best engineers. It was an unprofitable contract worth only $22,500, but the acclaim for the Otis elevators was worth a hundred times as much. Otis, like Colt, who first won fame at the 1851 Crystal Palace exposition in London, became a brand. It seemed natural that his name would be on the elevators

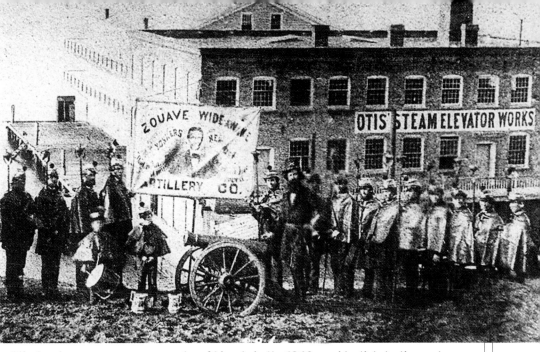

Otis, hand on canon, was a supporter of Lincoln in the 1860 presidential election and raised a small militia. The gun and carriage, valued at $58.65, were part of his payment for making the first safety elevator in 1853 for Benjamin Newhouse.

in the Chrysler and Empire State buildings.

Elisha Otis, unlike Colt, was not an initiator of machined volume production. What stands out reading his diaries and company ledgers in the Otis archives is how much each elevator had to be tailored to fit a particular architectural space—and how, amazingly, Elisha did much of it without a blueprint. He made a few sketches of elevator platforms and listed needed parts, but mostly his inventions, Mozart-like, sprang fully formed out of his head. Charles observed: "He could invent, design and construct a perfect working machine or improve anything to which he gave his mind, without recourse to any of the modern drafting methods. He needed no assistance, asked no advice, consulted with no one and never made much use of pen or pencil." He was able to foresee where the stress points would be and built his elevators so that they would not wear out quickly, a selling point in days when customers were generally required to make their own repairs. When it came to installing an elevator, he would instruct the workers orally or work alongside them. The company logs after Elisha show a noticeable change, being filled with detailed sketches, diagrams and measurements, but even into the 1860s, 1870s and 1880s, the company ledgers, brittle old pages covered with cigarette burns, show engineers recalculating each time how much wire rope must be used, the number of ratchets,

engine specifications, the size, shape and weight of the platforms, the number of wrought-iron binders. Gears, shafts and struts were all made in different sizes. The tradition of craftsmanship did not end until the 1970s, according to archivist Michelle Aldrich, when architectural and engineering methods became more standardized.

Elisha's diaries are more revealing of his character than of his business calculations. He sought moral truth. He wrote maxims to himself: "Give a beggar a horse and he will ride to the devil." And next to such thoughts might be a shopping list. He adored Abe Lincoln, the former Whig who stepped forward to declare slavery a wrong, and loathed his famous rival, "Steve Douglas the chameleon." With tensions between North and South at breaking point in the last year of his life, he raised a small militia, the Zouave Wide Awake Artillery Corps, drilled his worker-soldiers on the Yonkers Green and won permission to fire his little cannon "between the hours of seven and 8 o'clock with as little annoyance as possible." His most impassioned comments in his diaries concern slavery and abolition: "The old men of the revolution were fanatical and sophisticated enough to believe all men created free and equal," he wrote, "and I believe the same. Slavery is a crime against the life and soul of man." As for his inventions, he was philosophical. One note reads, "Machines the tools of liberty." It is a good epitaph for him and his age.

LEWIS TAPPAN Every day billions of transactions rely on the innovation pioneered by this evangelist crusader — credit rating

1788–1873

A MAN ON A WHITE CHARGER led a roaring mob to Lewis Tappan's new home in Rose Street, New York, on the sweaty night of July 7, 1834. The marauders broke in, ransacked the house, threw out every piece of furniture, bedding and pictures and window frames and made a bonfire of it all in the street. The word had been that they were going to tar and feather their prey, but he had heard the rumor and taken his wife and children to safety. Another mob laid violent siege to the Pearl Street store he ran with his brother, Arthur, who had barricaded himself inside with 30 or so clerks. Arthur passed out guns as they listened to the thud of a battering ram against the front door. "Steady, boys, fire low," he ordered. "Shoot them in the legs, then they can't run." Just as the mob broke in, the mayor arrived with troops, having belatedly placed the city under martial law.

The rioters were after Lewis and Arthur Tappan (1786–1865) because the brothers wanted to end slavery. New York had no slaves, only 14,000 free emancipated Negroes, but it was hardly the citadel of liberalism. As the nation's leading exporter of cotton, it was filled with businessmen who had a vested interest in protecting the "southern way of life," and they were not alone. Many in the city's work force saw the Tappans as smug middle-class reformers who did not know what it was like to be a workingman living with the fear of losing a job. White porters, laborers, draymen and butcher boys were led to believe abolition would mean an influx of Negroes ready to work for a few cents. The fears were not entirely groundless. Even before the recession that began in 1837, jobs were scarce: One in eight of the city's labor force was unemployed.

To say the brothers opposed slavery is a massive understatement. They were on

A crusader against slavery and debt, Tappan strode the wharves of New York in fine clothes, proffering Bibles and religious tracts.

fire with the evangelical fervor of Calvinists who knew God was on their side. The Tappans came from a line of early English settlers in rural Massachusetts, and there was little in their family tree to suggest they would move mountains, still less that one of them would introduce a revolutionary innovation in the conduct of the nation's business. They were brought up as strict Congregationalists in a family of ten children, their father an unambitious country storekeeper in Northampton, and only Arthur and Lewis became abolitionists. The formative influence on them was their intense mother, Sarah. She was from a stern Calvinist family but she was more sharp minded and better read than the average country Puritan and she had an exalted view of her Christian duty: not simply to explain Christ's truths to the depraved world but to enforce them. "Think not a moral character sufficient," she cautioned Lewis when at 15 he left home in search of his fortune. He took his mother's idea of Gomorrah, and eight dollars and a Bible, on the sleigh stage to Boston for an apprenticeship with a dry-goods merchant. In five years, he learned the newfangled double-entry bookkeeping, wrapped himself in a blanket in his chilly room to read history, religion and trade manuals, and so impressed his master he was able to borrow from him to start his own store in Philadelphia in 1809.

When the war with Britain broke out in 1812, the 25-year old Lewis was lucky enough to have in stock a collection of English goods. Their inflated prices yielded him windfall profits of $75,000 when an artisan would have had to work a whole year to earn $5,000. He bought a hardware store in Boston, married 18-year-old Susan Aspinall from an aristocratic Boston family (and in time had six children) and plunged into raising money for the deaf, indigent boys and the insane. His brother Arthur had been ruined in Montreal selling blankets for the Indian trade, and Lewis lent him $12,000 to try his hand in New York in 1815 as an importer of silks.

A decade later it was Arthur's turn to help. Lewis, self-described as "foolishly young and headstrong," had gone bankrupt; he had invested in woolen and cotton mills and a nail factory, but his timing was off: A recession in 1826 and British dumping of iron products and cheap cottons had forced him to close his factories. In 1827 Arthur took him in as his partner in his now thriving store on Pearl Street, New York, one of the largest in the city. Lewis supervised the shipments in and out of cheap silks, straw hats and ladies' stockings, umbrellas, fans and feathers, and he dealt with the country merchants who came into the city in spring and fall to buy stock for their own all-purpose stores. The Tappans ran their Pearl Street store on Christian principles. Clerks had to be as reverential as the brothers. They gathered in the loft every morning for prayers. Bachelors were required to live in religious boardinghouses, abstain from drink and be in bed by 10 p.m. The only Sunday activity tolerated was a walk to church. Arthur's example of fortitude was to take as lunch a single cracker and a glass of water without pausing in his cluttered office. The brothers carried their

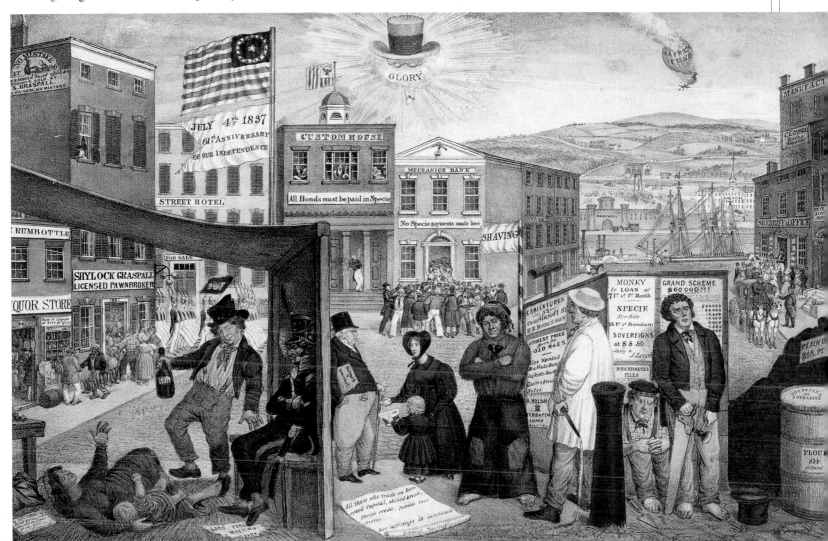

The Panic of 1837, a New York engraving from the period, is suggestive of the conditions that ruined Arthur Tappan.

Christian principles into the daily transactions of trade. Selling for cash was one Christian principle; there was no taint of usury. Another was to treat everyone equally. Other merchants negotiated prices according to the eagerness and status of the buyer. With the Tappans, every customer was offered the same fixed price, an innovation in which Arthur anticipated the retail pioneers A. T. Stewart and John Wanamaker of Philadelphia.

The brothers were not content to do the Lord's work only in their business. Awed by

The Tappans' campaign against sin in all its forms had limited appeal to Tammany Hall, the Democratic machine in the city, which cherished the time-honored trinity of political patronage in drink, prostitution and gambling. Yet the brothers were made of sterner stuff than the city's middle-class mosquitoes who buzzed wordily in the bosses' ears. The Tappans set up Christian spy cells to report infractions of neglected city ordinances for taverns and gaming houses. Arthur, borrowing an idea from the Magdalene Asylum in London,

up on good business practices. Lewis had less to give away at this stage, but both of them expiated with exceptional energy. Lewis strode the wharves of the East River in his fine clothes, proffering Bibles and religious tracts; over time, he and his other distributors reckoned they had given out six million morally improving pages. He allowed himself rare pleasures denied in his youth. He might drink a glass of wine and occasionally attend the theater and dances; he read novels sparingly. He derived most fun from all kinds of formal occasions, parades, ceremonies, gun salutes and debates.

In June 1830 a singular young man entered the Pearl Street store, dressed like a dandy, Lewis noted, but with a face alight with "conscious rectitude." The unknown visitor asked if he might meet Arthur, not a privilege freely granted. Arthur was a testy man who kept no chairs in his office where callers might sit and waste time, and this visitor was soon in and out of the inner sanctum, though not out of their lives. He was William Lloyd Garrison (1805–1879), destined for fame as the bravest and most prophetically brilliant agitator to free the slaves without delay. At the end of May, Arthur had read an eight-page essay in which Garrison described how he had landed in Baltimore Jail with a six-month sentence for penning a denunciation of a Baltimore businessman's perfidious connection with the slave trade. Arthur had straightaway sent him an unsolicited $100, which paid the fine the visitor had refused to pay as an alternative to six months in prison. He emerged from Arthur's office still beaming and with another $100 in his hand. He spent it on a lecture tour dramatizing the cruelties of slavery and looking for a city to start a newspaper. *The Liberator,* launched in Boston in January 1831, was the outcome.

Arthur soon committed himself to the cause, though he came to be concerned that the unrelenting Garrison was a bit too revolutionary. Garrison's vision, well described by his biographer Henry Mayer, was not limited to freeing Negroes. He campaigned for a society of equals; he embraced all the oppressed, immigrant workers on starvation wages, women treated as chattels; and he castigated the enemies of free speech in

The old merchant's sign summed up the feelings of traders with loads of bad debt. Until Lewis Tappan arrived, there was no way of knowing who could be trusted.

their mother's spirituality, they sought salvation by the conversion of sinners and immersion in the good deeds that could save souls—evangelical activities in which Calvinists could join reform movements with Methodists, Presbyterians, Baptists, Congregationalists and members of the Dutch Reformed Church, but not with Catholics and Masons. The brothers were effective because they were as munificent with their money as with their zeal. They founded the *New York Journal of Commerce,* which carried religious and economic news but no ads for the theater or liquor, except a "nonalcoholic" Burgundy from France they hoped would replace Communion wine.

opened a boardinghouse at Five Points, one of the roughest spots in a brawling city, where regular meals, Bible readings and the prospect of decent work awaited fallen women persuaded to seek new lives. Though plagued by headaches and weak vision, he himself went boldly into brothels, "the recesses of Satan," to pluck prostitutes "from the roaring lions who seek to devour them." His brother felt about vice, said Lewis, as he would a toad in his pocket. Arthur was prolific in his generosity, endowing many causes, often secretly. Among others, he founded the Mercantile Library Association of New York, where traders and young men starting careers could read the churches and political parties. Still, Arthur went along, and after three years Lewis, too, became an abolitionist. He outdid his brother in flamboyant zeal. He took a brace of slave whips into his public addresses and cracked them theatrically as he lashed northern complicity in the "ungodly" institution. He had a universally melodramatic view of the South, unrelieved by any personal knowledge, and instilled the stereotype of noble Negroes and villainous white southerners into his Sunday school classes. He carried around copies of *The Slave's Friend,* a little magazine with pro-Christian, antislavery stories, drawings and poems, and gave out copies in Sunday school, beseeching the children to get their playmates to read it, too, and "pray to God to break the rod of the oppressor and let the oppressed go free."

Not surprisingly, the proslavery forces denounced both brothers as "leaders in the crusade against white people." The Bible Society, founded in 1809 to spread the Christian gospel, shrank from reelecting Arthur to its board. When the brothers called an antislavery meeting on October 2, 1833, a large angry mob descended on Clinton Hall. Luckily for the Tappans, the mob had the wrong address; the hall's trustees had objected to an abolitionist meeting so the campaigners had adjourned to the Chatham Street Chapel. While the rioters boiled over into Tammany Hall, the Tappans formed the New York Anti-Slavery Society, with Arthur as president and the chief financial sponsor of the society, and a newspaper, *The Emancipator.* They gave the great Presbyterian revivalist Charles Grandison Finney his own church in New York in 1832 and in 1836, the year he became a Congregationalist, established him in the Broadway Tabernacle. Finney refused Communion to slaveholders and his preaching against slavery moved a widening circle of young abolitionists, notably the fiery Theodore Weld.

It was a feverish and a fearful time. When Lewis's home and the store were attacked in 1834, a couple of church deacons and merchants were among the crowd watching and applauding the rioters, as did all the city's daily newspapers except the Tappans' own *Journal of Commerce.* Black homes, schools and churches were gutted,

too, with the Tammany democratic mayor, Cornelius Lawrence, doing nothing to protect them. Antislavery meetings were disrupted and newspapers pilloried the brothers, commending the rowdies for their civic spirit. Lewis received a package containing a Negro's severed ear. In the South, a price was put on Arthur's head and he was burned in effigy in Charleston. A rally in Louisiana pledged $30,000 for his kidnapping and a South Carolina minister raised the stake to $100,000. Arthur responded with bravado: "If that sum is placed in a New York bank, I may possibly think of giving myself up."

The business community and the conservative aristocracy associated with the Episcopalian Church stood aloof. They had a gift for disassociation. Insofar as they were troubled by slavery in the South, and embarrassed by Britain's abolition of the institution throughout the Empire in 1833, they were rather more alarmed by anything that might affect their social status or interrupt the cotton trade. Condescendingly critical of the southern treatment of blacks, they were nonetheless repelled by the idea that a black might attend their own white churches or dine in their company. And indeed Arthur Tappan and Company had no Negro clerks, no black in any position above porter. There were few, in any event, who had the requisite education and fluency for commerce, but the Tappans rarely mixed with any of the Negroes in the city. Even Finney seated blacks in the gallery in his church.

Christian evangelicals in the abolitionist movement distinguished between the political equality of emancipation, which they demanded as a national priority and a constitutional right, and the social equality of integration, which even the more radical appreciated might take generations to achieve. Lewis was ready to press for swifter progress to ending discrimination, Arthur was more of a gradualist, but both were aware that the black community suffered first when anything was done to inflame the ugly racial animosities of the city. The brothers were not saints and certainly not liberals in the modern sense. They more and more resented, as both ethically in error and tactically imprudent, Garrison's radical determination to campaign for

human rights for everyone. They especially shared the repugnance of the conservative clerics at Garrison's espousal of the admittance of women as equals in the antislavery movement; Lewis said they were likely to bring "unnecessary reproach and embarrassment to the cause of the enslaved," since women in politics was "at variance with general usage and the sentiments of this and all other nations." In May 1840 the brothers headed the "schismatics," who broke away on this issue from the American Anti-Slavery Society and founded their own American and Foreign Anti-Slavery Society, in which women were specifically denied the vote.

Still, if the brothers could be bigoted, arrogant, stubborn and opportunistic, they undoubtedly had more genuine moral imagination and empathy for the Negro than their peers. Bertram Wyatt-Brown, a historian of the evangelical war against slavery, points out that Lewis's view of the South might have been unsophisticated, sentimental and abstract, but he and the other abolitionists did much to advance the recognition that deprivation, rather than genetics, explained the Negro condition. "Respectable" opinion, North and South, was skittish if not downright reactionary. Arthur gave money to set up a Negro training school not far from Yale, but the scholars did not support it and the community took fright in the wake of Nat Turner's slave rebellion of August 1831, when 55 whites were killed. Rowdies, presumed to be Yale students on the rampage, smashed a window at Arthur's home in New York and yelled curses on him and his family.

The Tappans' crusade had many cascading effects. Their help in setting up Ohio's Oberlin College in 1835 created an abolitionist nerve center. Finney, on his appointment to the faculty, insisted that free Negroes should be eligible for entry, and he attracted a large body of faculty and students from Cincinnati's Lane Seminary. The downside for the Tappans was the vulnerability of their business activities. The proslavery forces organized a boycott of their store. Merchants all across the South ceased buying from them. The strain induced Arthur to relent on cash-only trade. He gave credit for longer and longer periods without increasing his prices to compensate for the inevitable bad debts. In the panic of 1837, when 600 banks went out of business, Arthur owed $1.1 million on goods he had not been able to sell. The size of the debt—equal to hundreds of millions in modern dollars—stunned Lewis and the public alike. On May 1, 1837, Arthur suspended payments to his creditors. Within 18 months, by personal austerity and cost cutting, the ailing Arthur had repaid all his 100 creditors with accumulated interest at from 9 to 15 percent a year. It was a mark of his rectitude, but the store was never the same again.

By 1839 Lewis was distraught to find that Arthur was drifting back to easy credit. "The crediting system," he wrote, "is ruinous to the habits and morals of so many and detrimental to the community and the country." He was preaching against the tide. There was too little cash and too much need to trade. National currency then was gold and silver specie, with state-chartered banknotes but no federal greenbacks, and there was simply not enough money in circulation to finance the growing of America: Everyone needed credit to bridge the period between planting and harvesting, between building a factory and receiving cash for its output, between buying in New York and selling months later thousands of long, troublesome miles away. The expansion of the American empire and the opening of the West meant the big city merchants who wished to make hay in the good times could no longer confine themselves to dealing with a small web of kin and friends known to be trustworthy. The general custom among all those who facilitated trade with retailers—wholesalers, manufacturers, jobbers, commission agents—was now to be liberal with credit and pray for the soundness of the letters of recommendation the country trader brought with him from clergymen, bankers and lawyers back home. Creditors had little means of verifying the information on debtors, and in boom times they were tempted into blind trust. Why risk losing a sale when your competitor down the street was ready to trade? In 1827 a group of New York wholesalers had sent Sheldon P. Church to travel around collecting information about the viability of merchants,

but his annual reports had not been timely enough or comprehensive.

"At 53," Lewis wrote to a cousin in 1841, "I found myself worth nothing." Yet out of disaster came inspiration. Lewis began to screen applicants for credit at the store more rigorously. He kept files of creditworthiness judgments. Jobbers aware of his expertise started to drop by to ask what he knew about a saloon keeper in San Francisco and Lewis would be able to say the man's building was of trifling value and he drank too much of his own stock to be a safe debtor. There was probably a moment

One of Tappan's early spies, Abraham Lincoln, lawyer

when some creditor, rescued from being taken for a ride, expressed his relief in a way that stuck in Lewis's mind. He never elaborated, but his eureka moment must have come when he realized that his inquirers were ready to pay for the kind of information he had—even more for the kind he could collect if he organized a network of informants. And he already had a network! There were all his abolitionist allies and his friends in the churches across the country, and some of them were lawyers, too. He could engage them as correspondents for little or no expense in return for the promise that he would recommend them for hire by aggrieved creditors.

Lewis was so bullish on his idea that he decided to leave Arthur's faltering business and set up an institution that would monitor the credit-worthiness of as many mer-

chants as he could. It would profitably exploit the economies of scale, since he could sell the same information to a large number of creditors, and—of course—in pursuit of honesty it would serve God. Arthur tried to dissuade his brother from this new venture, predicting it would soon fail. Lewis was sure his scheme would attract thousands of firms. They were all at risk from bad debts and could not afford to run their own investigations. He had worked out his total expenses at $21,000. His business plan was to ask $50 a year from a small firm and up to a maximum of $300 a year from larger firms, so with a few hundred subscribers he would be in profit. He would also have a business invulnerable to the business cycles, so savage then. He wrote excitedly to a nephew, "In prosperous times they will feel able to pay for the information and in bad times they feel they must have it."

Thus in the summer of 1841 was born the Mercantile Agency. It was to become the world's first large-scale and successful network of credit reporting.

The originality of the scheme has been questioned. Financial historian James D. Norris has suggested that Lewis borrowed the details of his new business from a company of lawyers called Griffen, Cleaveland, and Campbell formed six years earlier. In 1835 it claimed to have organized lawyers throughout New York State to report on merchants, with 100 subscribers to its service. Lewis knew about this company; he referred obliquely to its "errors" in his prospectus and in 1842 he bought its files. But none of this much dims his innovative achievement, no more than Fulton or the Wright brothers or Henry Ford are diminished by earlier experimenters in steamboats, aviation or automobiles.

Whatever the sources of his inspiration, Tappan was the innovator because he made the idea work. He developed and managed a complex organization, he brought to it what he had learned in 30 years as a merchant, he infused it with his integrity, he stuck it out when the going was rough and he ensured that it did not die with him. The Mercantile Agency succeeded, as indeed James Norris writes, because Lewis "combined his visionary reform sentiments with hardheaded Yankee practicality."

The first call on Lewis's skill and determination was finding and training the correspondents. His diary of August 6, 1841, notes: "For the last two months I have been very busy—working generally 10 hours a day. My eyes have suffered some." He was cold-shouldered in the South, but in the East and the Midwest his network of abolitionists gave him a good start. He was able to recruit Ellis Gray Loring of Boston, Roger Sherman Baldwin of Connecticut, James G. Birney of Michigan (a presidential candidate for the Liberty Party in 1840 and 1844) and Salmon P. Chase in Ohio, later secretary of the treasury in Lincoln's wartime cabinet and the sixth chief justice of the Supreme Court. Lincoln himself, as a young lawyer in Springfield, Illinois, was an early correspondent. (Of one grocer, he reported that he had a rat hole that "would bear looking into.") On the day after Lincoln's assassination, a clerk wrote a eulogy in the margins next to the entry on Lincoln's law firm. The anonymous employee drew a picture of a weeping willow arching over a tombstone in the shape of a cross and wrote, "This office has had the honor of having Old Abe as a correspondent." Later correspondents included three future presidents: Ulysses S. Grant, Grover Cleveland and William McKinley. (Grant is on record reporting that a young lawyer had nothing but an office chair, a barrel with a board and a fine young wife, but reckoned his ability and ambition compensated for his lack of capital.)

Lewis wanted to know a subject's estimated net worth and general business prospects, but he emphasized character as the most important criterion. Drunkenness and gambling were spelled out as danger points; adultery was seen as unimportant unless a debtor strained his income by setting up a mistress in a separate house.

Once the Mercantile Agency was launched on August 1, 1841, by an advertisement in the New York Commercial Adviser, Lewis had to keep up his spirits amid the slings and arrows that the inert and the jealous customarily shower on innovators. He was first and foremost denounced as an invader of privacy. James Gordon Bennett's New York Herald sneered that Tappan's was "an office for looking after everybody's business but his own." A southern lawyer sounded off in New York's largest newspaper, the antiabolitionist Courier and Enquirer, saying he had been asked to spy and "even a slave would scorn such a job."

The attacks made it painful for Lewis as he trudged around lower Manhattan trying to sign up subscribers. He had told his diary that he had committed his plan "to Him in whose hands are the hearts of men," but he was unhappy about the Lord's handiwork. Five months after starting, he had only 11 firms when he had reckoned he would have hundreds. In January 1842 he

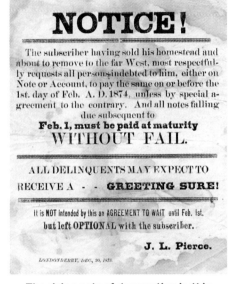

NOTICE!

The subscriber having sold his homestead and about to remove to the far West, most respectfully requests all persons indebted to him, either on Note or Account, to pay the same on or before the 1st. day of Feb. A. D. 1874, unless by special agreement to the contrary. And all notes falling due subsequent to

Feb. 1, must be paid at maturity WITHOUT FAIL.

ALL DELINQUENTS MAY EXPECT TO RECEIVE A - - **GREETING SURE!**

It is NOT intended by this an AGREEMENT TO WAIT until Feb. 1st. but left OPTIONAL with the subscriber.

J. L. Pierce.

LONDONDERRY, DEC. 20, 1873.

The rising note of desperation in this creditor's appeal-cum-warning reflects the chaos endemic in the system before Tappan's innovations.

attracted 33 more by lowering his rate, but he did not make his first-year budget of $9,000. One day, disheartened in the course of being rejected several times, he wondered why "the Lord should have directed my mind to this new employment, and then disappointed me in my expectations." But he drove himself hard and at a pace that was the despair of his clerks. He kept knocking on doors, "brought into sweet submission to His will. I think it was the desire of my heart to make a fair trial and perform any duty and then quietly submit to what would appear to be the act of Him who cannot err." Luckily, he did not have to undergo an agonizing reappraisal of his relations with God: The next merchant he solicited readily subscribed "with very kind and flattering words. . . . I applied to another, and he, after some persuasion, subscribed. This elated me much. I was surprised at my former unbelief and the present revulsion in my feelings. I almost felt guilty. My gratitude was now raised high and I expressed it in ejaculatory prayer. The Lord this day has been better to me than my fears. Blessed be His holy name." By early 1843 his faith was fully restored. "The Mercantile Agency is now quite popular. . . . It checks knavery and purifies the mercantile air." By mid-1844 he had 280 clients and 300 correspondents, though none yet in the South. The agency earned him $15,000, $5,000 more than the year before, and he rescued Arthur by giving him a share of the business and a job monitoring the accounts. Spurred by competition from imitators impressed by his success, Lewis opened branches in Boston, Philadelphia and Baltimore.

As an innovator, Lewis is intriguing for the way he was able to reconcile his core beliefs with commercial opportunity. He refused to sell services to a distillery. He offended the ribald by refusing to laugh at dirty jokes. These attitudes did him no harm; they enhanced his reputation for probity. But in the biggest decision he had to make about the agency, he stunned his abolitionist friends in 1846 by making his heir apparent a man who was an inflexible defender of slavery and the states' rights Lewis had spent years fighting. Benjamin Douglass (1816–1900), Lewis's chief clerk, was a steely Scottish Presbyterian, then aged 30. He was not only an able and imposing figure; he knew the South, having been a merchant in Charleston, South Carolina, and New Orleans, and the South had been the Achilles heel of the Mercantile Agency. Arthur objected strenuously to the promotion of Douglass to a third share of the agency. Lewis, who avoided confronting Douglass on his beliefs, focused only on the business when he responded to Arthur's objections, "I have never had the slightest discordance with him." And again: "He is as smart as a steel trap." As Bertram Brown-Wyatt writes, when Lewis's ambition outran his stern code "he did not always engage in a show of repentance. He could be perfectly unreflective and perhaps that selective lack of retrospection had something to do with his decisiveness and dynamic earnestness." Certainly, the pragmatic Yankee deal maker prevailed over Don Quixote.

Douglass effectively ran the business from 1849. He bought out Arthur in 1854, so he owned the business outright, and he rode the booming economy of the 1850s with skill and bravery. He opened branches in the South and West. By 1851 he had 700 big subscribers in New York alone, serviced by 2,000 full-time paid correspondents preparing reports from all states of the Union. Their letters gave a picture of life in various regions. From Leadville, Colorado, the correspondent wrote that "men walk down the main street swinging their revolvers by a chain attached to their wrists, ready for use on the slightest provocation. Shooting scrapes in a gambling joint next to our office are regular occurrences."

Some of the reports are pithy. Of Leland Stanford and Charles Crocker, builders of the transcontinental railroad, the correspondent summed up: "Are making money fat. Shrewd wide awake men." A more typical report is this from Nevada Territory: "Carson. Louis Epstein. Restaurant and Hotel. Mar 9/72. Jew 40 married. In business 7 or 8 years. Rents house. Owns furniture and value. Does a small business apparently only makes a living has good habits but doesn't pay his debts and is an inveterate liar. Is in debt in San Francisco. Could make nothing out of him, has no credit here, lately pawned his watch to buy a bill of groceries." Thirty men in New York copied and indexed the reports to read them to subscribers who came into the office. (The telephone was not available, Alexander Graham Bell being only four years old.)

Douglass was a man of iron will—significant for the future of credit reporting. As agencies proliferated, the idea of so much snooping, as it seemed, made people uneasy, especially when correspondents were sloppy. In 1849 Mercantile had to pay

$10,000 in libel damages to two brothers, John and Horace Beardsley, of Norwalk, Ohio, but the ramifications of *Beardsley v. Tappan* were more damaging still. The agencies had always taken comfort in the doctrine of privilege, that they could not be sued for what transpired in the privileged relationships between agency and subscriber. The judge ruled otherwise. Credit agency information, he declared, lost its privileged protection once it was copied for dissemination. Information could be safely imparted only by individual to individual, not by "an establishment conducted by an unlimited number of partners and clerks." Agencies responded by making still more elaborate disclaimers of legal liability when they gave out information, but one good thing came out of this first big case: The precedent set by the courage of Douglass. He spent 20 days in jail for refusing to reveal the identity of the agency's reporters. It was a sacrifice, he said, that "aided greatly in establishing the agencies in the confidence of the public because men saw that they could give information to the agencies and that these would not betray the confidence reposed in them." The business community applauded. Lewis Tappan was busy now trying to make Texas a free state, but he defended Douglass in an article in the New York *Evening Post*.

It took years of court cases to have the restrictive ruling reversed, and it occurred in 1858 in *Ormsby v. Douglass* just as Douglass was handing over the business to Robert Graham Dun (1826–1900), his brother-in-law. Dun, who had joined the company in 1850 when he was 24, would be the central figure for 42 years until his death at the end of the century. It was a climactic moment for the business of credit reporting. The legal, political and public hostility Dun faced was perhaps most colorfully expressed by an editor in Baton Rouge: "Talk of Vidocq or Fuche police-Japanese espionage-damnable leechers and hireling bloodsuckers; it is all honorable—legitimate—and proper in the place of this most villainous inquisition." The attacks reached a high point in 1876 with the publication of an indictment of the quality of reporting by a former employee, Thomas F. Meagher. Dun moved to replace inexpensive local

correspondents with trained traveling reporters. His second response, over time, was to change the focus of reporting. Lewis had emphasized character. Dun emphasized capital. In his *Reference Book* in 1864, Dun used statistical ratings and provided estimates of the capital worth of corporations; by 1875 firms were being asked to submit their own financial statements. Subscribers were now able to make valid comparisons among applicants for credit. The business was no longer simply credit reporting. It was credit rating and credit ratios.

Just as big a challenge to Dun was a failed dry-goods retailer who in 1855 invaded New York from Cincinnati. Forty-year-old John M. Bradstreet cut deeply into the profits of Tappan and Douglass by publishing credit ratings in a book, rather than by having subscribers visit the office to have reports read to them. Volume 1 of Bradstreet's *Book of Commercial Reports* (1857) listed 20,000 merchants, each described with a numerical code that could be understood only by looking up the numbers in a second volume. Douglass had always refused to put Mercantile's reports in print for fear that general access to the information would render it actionable. But he made the decision to imitate Bradstreet and publish *The Mercantile Agency's Reference Books* in February 1859, each volume equipped with a lock and key.

Dun was a quiet, assiduous family man. His father, who had moved from Scotland to Chillicothe, Ohio, was a business failure who died when Dun was still young. Dun and his three impoverished sisters were brought up by a wealthy uncle. He was not a crusader like the Tappans; his God "intended the Negro to be the servant and slave of the superior race." Nor was he as bold. In 1876 he had a chance to establish a near monopoly by offering a daily reporting service on selected firms that his rivals were in no position to duplicate; he shied away from it when some subscribers grumbled about his rates. It is not surprising that Dun was sensitive to risk; at his back, he could always hear competitors hurrying near. Scores sprang up after the Civil War: 41 agencies went bankrupt in a few years in the 1870s. But if Dun was not a classic entrepreneur of the Gilded Age, or an innovator in the stamp of Tappan, he was representative of an important new phenomenon in American business, the professional manager. He was among the first to see the potential of the typewriter. Diligently, carefully and always soliciting advice before acting, he grew the business from a small credit reporting firm with fewer than 20 branches to a large international organization with more than 135 branches ready to furnish timely credit reports on any business in the world. In 1933, thirty-three years after his death, R. G. Dun and Co. merged with Bradstreet, its only large-scale rival. Dun and Bradstreet became the largest credit-reporting agency in the world.

For many decades, there was nothing like the credit rating system elsewhere, and successive commentators suggest how long it took the world to catch up. The *Times of London* in 1852 commented on the "novel system of protection that had arisen out of the peculiar position of traders in the Union, their go-ahead spirit of speculation and the wide extent of their commercial transactions." P. R. Earling, an authority on credit, wrote in 1890: "It is paradoxical, for the free citizen of the United States is the only one on the face of the globe who tolerates it. It is a purely American institution and flourishes only on American soil." In 1913 Professor James Hagerty of Ohio State University noted that in "no country other than the United States do the mercantile agencies go so far in investigations of private business." Two British banks trading in America at the time, Baring Brothers and Brown Brothers, did institute a system of credit reporting but only for their own protection; it was not open to everyone. Credit reporting in Britain itself was then performed by mutual protection societies, closed groups again. And European credit reporting systems were only regional in scope; run as nonprofit protection societies, they did not develop the resources to go national. The long-run economic consequences of Lewis Tappan can be judged by the fact that America achieved its single continental market in the middle of the 19th century, something that eluded Europe until the 1990s.

AMISTAD CAMPAIGNER

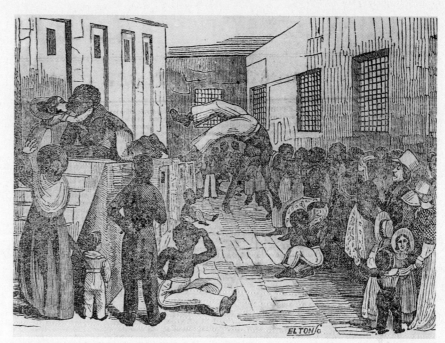

A LITTLE MOCKERY FROM NEW YORK: The *Morning Herald* newspaper headlined its drawing "Teaching Philosophy to Lewis Tappan & Co. in the prison at Hartford."

IN DECEMBER of 1841, the year the Mercantile Agency opened its doors, Lewis Tappan came to Staten Island to say goodbye to 50 or so black men on the way to their homes in Africa. In June 1838, when they were being transported to slavery in Cuba, they had mutinied and seized their slave ship, the *Amistad*. Arrested by the U.S. Coast Guard for piracy and murder, they were held in jail in New Haven. After years of captivity, they were released only by order of the Supreme Court, an order accompanied by a long public argument in which Tappan, even as he brooded on his credit agency, had been a doughty campaigner, as well as a spiritual counselor to the prisoners.

The Mercantile Agency, like the *Amistad* affair, was a vindication of Lewis Tappan's faith and perseverance in the face of prejudice and opposition. In 1865 he published *Is It Right to be Rich?*, a pious attack on the materialism of the age and a warning against the use of credit and paper money. When he died aged 85 in Brooklyn Heights on June 21, 1873, Henry Ward Beecher spoke at the funeral.

Cinque, leader of the *Amistad* uprising; Magru; and Banna: expressions of gratitude to Tappan

THEODORE DEHONE JUDAH

Everyone thought he was a little crazy, but it was his brilliance and tenacity that gave us the transcontinental railway

1826–1863

THERE WERE two Theodore Judahs, or "Ted" as he was sometimes hailed in encounters on the dusty sidewalks of Sacramento in the 1850s. One was a sunburned adventurer on horseback leading his pack-mule toward the forbidding Sierra Nevada who would not be seen again for weeks at a time. The other, more a Theodore than a Ted, was a grave young man in a long black frock coat buttonholing senators in Washington's Capitol building, bureaucrats in their departments, newspaper correspondents at their desks and the president in the White House.

The uniting thread of Judah's double life was his passionate conviction that a transcontinental railway could be built—and must be built. He was by no means the first to come up with the idea. When America rushed to build short-run railways in the 1830s, various individuals and townships advocated a railway from the Great Lakes through Oregon to the Pacific. The most fervent and effective campaigner was Asa Whitney, the young trader of our acquaintance from the introduction to this book grown to wealth in his middle age. Grief-stricken by the death of his wife, he swore to devote the rest of his life to mankind and in 1844, after visiting China, began a crusade for a railway that would link the East Coast of America to the West as a means of saving the immigrant millions from "the tempting vices of our cities" and of spreading America's civilizing influence among the "starving, ignorant and oppressed" teeming millions of Asia. He argued that the railway would pay for itself by the sale of land adjacent to the right-of-way if only Congress would set aside a 60-mile strip for the track. The redoubtable senator Thomas Hart Benton told him to

his face he was quite mad since science was unequal to overcoming the Allegheny Mountains and it was humbug to talk of scaling the Rocky Mountains, which were four or five times as high. (Later, Benton changed his mind and became a leading proponent of a Pacific railroad.)

Whitney's response to the engineering problems was to spend 26 days on his knees in a canoe surveying the Missouri wilderness region in 1845, which did nothing to answer Benton's point but did enable him to challenge the dismal view of the West of that determined East Coaster Daniel Webster: "What do we want with this region of savages and wild beasts and deserts, of shifting sands and whirlwinds of dust, of cactus and prairie dogs?" Many people even in midcentury regarded California as an aberration that should be left to enjoy its isolation; its population even after the gold rush in 1849 was less than 250,000. "Our rich possessions west of the 99th meridian have turned out to be worthless," declared the *North American Review*.

Whitney gave up after seven years, but he got the issue on the nation's agenda to the extent that by 1853 Congress instructed the secretary of war, Jefferson Davis, to have the army survey four possible routes. The 12 elaborate volumes published in 1855 painted an appealing picture of a West ripe for settlement, but aggravated an underlying tension that had only become tauter since Whitney began his campaign. Davis, the future president of the Confederacy, favored the most southerly route, an advantage to slave states and New Orleans in particular. The free states, particularly the empire builders of Chicago, could never countenance this. It was stalemate.

Theodore Judah had a transcontinental railroad on his mind from the day he arrived in California mid-May of 1854 at the age of 28. He had been engaged by a speculative company to build a line 22 miles up the Sacramento Valley, a modest venture over easy ground along the American River to the putative new town of Folsom. It was not by itself enough to pull a very ambitious young man away from the railway manias of the East, but this Sacramento Valley line was causing a stir locally. It was to be the very first railroad west of the Rockies, whereas 18,000 miles had been built in the rest of the country. (Within five years it would be 30,000.) The 22-mile line Judah was to build would save a full day for people and freight on the backbreaking haul from Sacramento to the mines, but to him it was a first step in a grand scheme to cross the supposedly impassable Sierras into Nevada and Utah and beyond.

Judah loved to contemplate the impossible. He had made a name for himself at the age of 26 by accepting a commission to run a railway down the narrow gorge of the Niagara River. "Gentlemen," he had told the anxious directors, "raise the money and I will build your road." He did it by cutting a scary track into the cliffs high above the boiling rapids. Years later his wife, Anna, reflected: "Did it not show what was in the man then, to grasp the gigantic and the daring?" He had risen fast from a difficult start. The early death of his clergyman father ended his hopes of attending a naval

VISIONARY: A daguerreotype of the well-dressed Judah as he was seen in the halls of Congress in the late 1850s. He was alert to the new technology of portraiture. He gave his artist wife, Anna, another daguerreotype taken as early as 1848.

academy and he had become a surveyor's assistant at 13, without any engineering training. He got that later by attending the Troy School of Technology (later Rensselaer Polytechnic), New York, and learning on the job under the renowned A. W. Hall on the Schenectady & Troy Railroad, the first in New York State. At 18, he helped to build a line from Springfield, Massachusetts, to Brattleboro, Vermont, then did engineering work for the New Haven, Hartford and Springfield Railroad, leading to his appointment as assistant to the chief engineer for the Connecticut Valley Railroad when he was only 20. He was the chief engineer connecting the Buffalo and New York railroads when there came the invitation to go to California at short notice.

He was so ambitious he had schooled Anna always to "have the right pairs of gaiters on" for unexpected opportunities. He promptly apologized for taking the California job by telling her he saw it as an opportunity to "know that country and help build the Pacific railroad. It will be built and I am going to have something to do with it." Working for 18 months to build the Sacramento Valley Railroad within daily sight of the looming mountain walls that closed off California, he became so obsessed with a transcontinental railroad he was known everywhere as "Crazy Judah." It was an affectionate title; he was respected, and popular opinion considered that a Pacific (i.e., national) railroad would be splendid for California—they just wished he could talk about something practical for a change. To build a railway that would have to rise 7,000 feet up into the Sierra Nevada in fewer than 20 miles from western base to summit (to say nothing of crossing the peaks and deserts of Nevada and Utah) was clearly a costly fantasy, locals declared, even if those folks in Washington could be persuaded to release the land and find the money. Look at the 25 years it had taken for the Baltimore and Ohio Railroad to get through the Alleghenies, mountains so much smaller than the Sierras with nothing like western snowfalls.

Judah made some forays into the foothills of the Sierras for sponsors of the very different proposition of a wagon road, but realized that if he were to advance the national railway he must first understand the mechanisms of the federal government as well as he understood the capacities of a steam locomotive. Anna got her gaiters on for the long journey back east by way of the Isthmus of Panama in April 1856 and a longer stay in Washington than either of them expected: He became so absorbed in the legislative maneuvering in committee and cloakroom, they were still there into the fall and winter as the clouds of secession and disunion gathered. It was shrewd of Judah to linger, though it was a strain on his finances; he developed a keen appreciation of the sectional jealousies over the route for the railway, city against city, township against township, county against county, to say nothing of the increasingly bitter antagonisms between free and slave states: "Rather than not have their own route, they will not have any." On New Year's Day of 1857 in Washington, he promulgated his solution, a pungent 13,000-word pamphlet he called *A Practical Plan for Building the Pacific Railroad*. It was both hardheaded and idealistic, and very much an engineer's document.

Judah offered a glimpse of special express trains traveling across the prairies at 100 miles an hour with passengers at ease in restaurants, smoking and reading rooms, and ladies in partitioned sleeping berths, all with plenty of room because for these express specials he would connect two 11-foot cars side by side to run on two tracks. He wrote that he heard "a hundred scientific gentlemen start up with their objections," so he dealt with them one by one, calculating the effect of 90-ton engines and 14-foot driving wheels.

They would get nowhere arguing, said Judah, whether the route should be the central line favored by the California convention, a northern line favored by the Oregonians or a southern route insisted on by the South. The railroad should be taken out of politics because "a house divided against itself" would never agree. There was only one way to decide and that was by engineering criteria. Government had spent so much money and time on so many routes it had as yet never got around to a proper survey of any one of them. Judah was caustic about the irrelevance of previous surveys: "When a Boston capitalist is invited to invest in a Railroad project, he does not care to be informed that there are 999 different varieties and species of plants and herbs, or that grass is abundant at this point, or Buffalo scarce at that; that the latitude and longitude of various points are calculated to a surprising degree of accuracy, and the temperatures of the atmosphere carefully noted each day." This was not a fair comment on the previous surveys. Boston capitalists and any other investors were entitled to ask the questions earlier surveys tried to answer, in particular whether the regions were attractive for settlement. Whether grass—and water—was abundant was very much an issue. A railway line, however brilliantly engineered, was not going to be viable if nobody wanted to settle in its vicinity. Where Judah was more on the mark was in spelling out the detailed questions the railroad engineer must answer. How many cubic yards of the various kinds of excavation and embankment had he calculated section by section? How many tunnels and how long would they be? How much masonry and where was the stone to come from? How many bridges, river crossings and culverts were to be built and on what foundations? What would timber and fuel cost and where would it come from? What was to be the length, number and radius of the curves? What was the answer to avalanches and Indian war parties? How did all this add up in cost per mile and what were the potential passenger receipts?

Judah's *Practical Plan* advanced a scheme for financing the national railway that looks unrealistic today but was a brave attempt to end the stalemate. It would cost $200,000, Judah concluded, for a qualified railway engineer to report in eight months. He should be left to get on with it as an independent professional and his estimates of the cost of building the line (but not its direction) could then be scrutinized by six of the most eminent railway engineers. Elite opinion was very divided. The relatively new Republican Party, spokesmen for expansion and industrialization, favored government subsidy. The Democratic Party, fearful in those days of Washington interference with slavery, considered federal support to be unconstitutional. The influential Henry Varnum Poor, editor of the *American Railroad Journal,* was still

THE BIG FOUR

COLLIS POTTER HUNTINGTON
(1821–1900)

He was a Promethean figure, a natural bully but crafty, and certainly the key one of the Big Four in raising the money, negotiating purchases, harassing bankers and lobbying in Washington. He was without pomp; he worked most of the six years on the railway from a grim little office on Wall Street and let Stanford be the public face of the railway. Adversity was always just an invitation to rebound; he was like a jujitsu artist deriving energy from opposition. On his way from New York to the goldfields in 1849, he was delayed for three months in Panama, and walked across the poisonous isthmus 24 times buying and selling. Assailed by the dominant but squabbling Democrats as a "nigger lover" because he was outraged by the brutal treatment of fugitive slaves, he helped to found the state Republican Party in March 1856, defying hoodlums and not-so-empty threats of a public hanging. He sup-

ported the vigilantes in San Francisco when they took to the streets and hanged individuals they considered corruptly protected by the Democrats.

Huntington's initial railway fortune, like that of his associates, derived mainly from the kickbacks in construction. Less is known of the Central Pacific dodges than those of the Union Pacific because "Uncle" Mark Hopkins "accidentally" burned all the papers. The Central Pacific whetted Huntington's ravenous appetite and he became a dominant figure in railroads, buying up and consolidating lines in California for an empire depicted as the all-powerful Octopus. As a merchant, he made a point of meeting every obligation, but otherwise, to quote an unknown epigrammatist, he was scrupulously dishonest. He never hesitated to bribe a congressman to "do the right thing." In Frank Norris's novel *The Octopus,* Huntington is reviled as the villainous Shelgrim, but the historian Eugene Huddleston draws attention to the solid accomplishment at the end of the melodramas: "He left a transportation system that unified East with West by shrinking the vast distances of the American continent and he was instrumental in transforming the vision of the West from Great American Desert to the Golden West."

He met death as other reverses: head-on. He simply announced, on August 14, 1900, "I am very, very ill," and promptly died.

CHARLES CROCKER
(1822–1888)

The construction boss whose innovation was hiring Chinese crews. He was worth $20 million when he died, and left not a cent in philanthropy—a stark contrast to Stanford. His name was eventually that of one of California's most powerful banks (see Giannini, page 258).

MARK HOPKINS
(1802–1887)

Huntington's store partner was the keeper of the Central Pacific books, an austere vegetarian of few words who left one of the largest fortunes in the country.

The celebrated hotel in San Francisco is named after him.

LELAND STANFORD
(1824–1893)

Stanford, governor and later senator, was the political fixer for the Central Pacific, but his more important contribution to the making of America was the money he left to found Stanford University on the site of his 900-acre horse farm at Palo Alto. The bequest was in memory of his only son, Leland Jr., who died just shy of his 16th birthday in 1884. Stanford came from a prosperous landowning family near Albany, New York, and was admitted to the bar in 1848. He hung out his shingle in Port Washington on the shore of Lake Michigan without much luck, and was attracted to California by the better prospects of his five brothers in a chain grocery business. He swept the floor of the Sacramento store, took it over in the mid-1850s—and then struck gold. Two merchants who owed him money gave him 76 of their 93 shares in the slow Lincoln mine, which then became very productive and is said to have earned Stanford half a million dollars. He was the most articulate and cultivated of the four associates—though a bloviator of the Warren Harding school of oratory—and the most concerned to use modern technologies in the building of the railway that Crocker and his conservative construction boss resisted. One of Stanford's racehorses achieved immortality in Eadweard Muybridge's famous photographs freezing the motions of a gallop, which later inspired Edison with ideas for his movie camera. Huntington squeezed Stanford out of the presidency of the Southern Pacific in 1890.

opposed to subsidy in 1857, while conceding that private capital would not be forthcoming for a project he regarded as essential to the national interest. Judah shared the view that giving money to "moonshine speculators" would rapidly have the railway surrounded by a body of thieves. Nor did Judah favor raising money by bond offerings that could impose inflationary interest charges on a risky venture. Instead, he envisaged that every cent of the $150 million needed at $75,000 a mile could be raised for a people's railway by a massive national campaign: "There is not a man in the whole community who has hands to labor with who cannot afford to take one share of $100, and pay $10 a year per year, or three cents a day, upon it . . . there is no retail merchant, doing even an ordinary business, who cannot afford to take ten shares and pay in his $100 a year, or thirty-three cents a day. If they have children what better heirloom can they leave them than shares of this stock?"

Judah was an adroit publicist. He had made a friend of Lauren Upson, the editor of the *Sacramento Union,* then the leading newspaper of the coast. The *Practical Plan* was given full treatment and together editor and evangelist were instrumental in having the California State Legislature summon a Pacific Railway Convention. The hundred delegates meeting in San Francisco on September 20 had had their fill of rhetoric about Manifest Destiny and voted to send Judah to the nation's capital as the state's accredited agent with three specific proposals: that the federal government should guarantee a 5 percent return to whichever company built the road; grant right-of-way over whatever route the builder deemed the best; and the pledge that California would finance part of the road "within her limits." Once again, Anna and Ted set off for the long journey

east, paying excess baggage charges on trunks stuffed with documents. On the same steamer leaving San Francisco were Congressman-elect John Burch and General Joseph Lane, the former governor of Oregon and now a senator for the newly

FAITHFUL ANNA: She was the daughter of a well-connected merchant in Greenfield, Massachusetts. In half a dozen years after her marriage to Judah, on May 10, 1847, she made a home in 20 different places, and was steadfast at his side in all his adventures in California. Sometimes she went with him on his mountain expeditions to make sketches and collect fossils.

admitted free state, whom Judah immediately inundated with facts and figures on every conceivable point of a possible bill. Lane was noncommittal, but Burch enlisted

in the cause. "We became immediate and intimate friends," said Burch. "His knowledge was so thorough, his manners so gentle and insinuating, his conversation on the subject so entertaining, that few could resist his appeals." Anna worked her charm, too. At dinner one night she told the congressman she had packed all Judah's charts from the San Francisco convention and paintings she had made of the mountains, along with fossils and ores from the mountains, and would it not be splendid if they could all form an exhibit on Capitol Hill?

Her idea bore fruit. Theodore Judah was afforded the rare privilege of a room of his own, the old vice president's capacious quarters on the same floor with the Hall of the House and Senate; packed with his maps, reports, specimens, surveys and papers of every kind, it became known as the Pacific Railroad Museum and the battle headquarters for a Pacific Railroad Bill. His first political call was on the faltering Democratic president James Buchanan, with Burch and California's Democratic senator William Gwin, a southern sympathizer. Buchanan was preoccupied keeping the country together. "Forget the railroad, young man," he told Judah, "until we see what is going to happen to the nation." Still, at the end of the audience, he reversed himself on federal involvement, promising his support, albeit for the southern route Gwin favored.

There were, in fact, four bills for different routes. Judah agreed that it was sensible to back the one that emerged from a House Select Committee, chaired by Iowa's Samuel C. Curtis, for the single central line Judah favored from Iowa-Nebraska to Sacramento; it sanctioned land grants and a government loan of $60 million, about half the cost of the project, to be paid out as per-mile subsidies to designated corporations.

From January 1860, Judah was intimately involved in months of labyrinthine committee compromises for the bill. He was quietly assiduous in cultivating good relations with the sponsors of a railway from the east to meet the railway from the west; they were less concerned about the route than getting their hands on the money. To the last minute, it looked as if an East-West Republican coalition would see the bill through with a concession to the southern congressmen that a southern route would be included in an engineering survey. It was not to be; sectional tempers were too frayed. In March 1860, the bill was tabled until December, ostensibly for press of business, but thanks in large part to Judah's tireless advocacy and alliances it had become part of the Republican platform for the elections due in the fall when the presidential candidate would be Abraham Lincoln, the rail-splitter.

The question congressmen visiting Judah's museum asked most was just why he was sure it was possible to run a railway through the highest mountains in America. Truth to tell, he had no evidence to present. He made up his mind he would never go to Washington again until he had explored the Sierra Nevada thoroughly enough to identify and cost the most feasible passage over the summit. "With facts and figures," he exclaimed, "they cannot gainsay my honest convictions, as now!" Three weeks after they got back to Sacramento, he carefully rolled up his instruments in chamois bundles and trekked on horse and foot over several known passes north of Lake Tahoe. After the smoky Washington committee rooms, he relished the sparkling air and magnificent vistas (while keeping an eye out for mountain lions), but the crests and ravines he surveyed, crossing the summit 23 times, were an engineer's nightmare. The Sierras Judah traversed, as the western historian Helen Hinckley noted, were really two impenetrable walls with a trough between them, but no river passage cutting through the walls. He was immediately intrigued on a return visit to Anna some weeks later when she gave him a letter from Daniel Strong, a druggist everyone called "Doc," who wanted a surveyor's help in checking a path for a wagon route he believed would bene-

fit his little mining town of Dutch Flat, 55 miles east in the hills above Sacramento. Judah was on the stagecoach to Dutch Flat the next day and the following morning he and Doc, immediate friends, saddled up to check out Doc's prospect. Doc led him along a disused trail to a long ridge between two deep river valleys. The region had ominous echoes—they could look down on Donner Lake, where 41 members of the snowbound Donner party starved to death in the winter of '46—but it was good news for Judah. He could see how his railroad could ascend to the ridge by an almost continuous incline, a ramp more or less to the Donner Summit. The ridge itself had problems, but the descent eastward to the Truckee River valley did not look forbidding, and then the line would be in the Truckee Meadows and heading for open desert.

"Dr. Strong is entitled to the credit for the suggestion," wrote the engineering historian John Debo Galloway, "but it needed the trained eye of a practical engineer to determine in a preliminary way the merits of the location." Judah's eye could see it would be challenging. Instead of crossing two parallel ridge lines as everywhere else in the Sierras, his railroad would be able to ascend and traverse only one, though the one ridge would need a number of long tunnels and many bridges over wide ravines. Still, he reckoned no gradient was more than 100 feet per mile, and he knew the locomotives being developed could do that. It was a thrilling moment of vindication for his faith. On the way down in foul weather, the pair had a foretaste of the conditions that had doomed the Donner party and a hazard the railway would have to combat. A snowstorm developed quickly while they were asleep and threatened to engulf them. They struck their tents in the middle of the night and stumbled down a canyon in darkness—"and none too soon were they," said Anna. An exhausted Judah "could not sleep or rest after they got into town and the store until he had stretched his paper on the counter and made his figures thereon." There and then Judah worked out the profile of the route and also drew up the Articles of Association for a Central Pacific Railroad Company, anxious to have a corporation in place before Con-

gress picked up the Curtis bill again. By California state law, the pair had to raise $11,500 to incorporate: 10 percent of the capitalization of $115,000 ($1,000 for each of the 115 miles from Sacramento to the state line).

To raise the money for a more thorough detailed survey of their route, Doc went canvassing the townships for $100 subscriptions, $10 down, while Judah prepared an explanatory map and pamphlet, *Central Pacific Railroad to California.* The partners were short only $4,500 when they went into a meeting with San Francisco businessmen on November 14. Judah was brimming with confidence and hope; after all, the new (and short-lived) Pony Express had just come through with the news that on November 2 Lincoln and the Republicans had triumphed in the election, improving the prospects for the Pacific bill. And an incorporated railway company with a defined route in mind stood the best chance of winning the federal and state subsidies and land grants.

Anna recalled his dejection when he came back to their hotel: "Weary and disappointed, his words to me were these, 'Not two years will go over the heads of these gentlemen I have left tonight, but they would give all they hope in their present enterprises and business to have what they gave away tonight.'" The San Franciscans had their doubts whether the line could be built in the seven years Judah claimed, but their fundamental lack of enthusiasm lay in the money they were making in businesses that would be hurt by a national railroad (steamships, express, stage and wagon companies).

Judah hastened to Sacramento. With the help of the *Sacramento Union,* he unrolled his charts before 30 or so men in a room at the St. Charles Hotel on J Street. The rich jeweler James Bailey and the merchant Lucius Booth and a few others were keen, but their subscriptions were not enough to close the gap and finance a survey. At the end, a burly black-bearded man, who had sat watchfully motionless through the presentation, paused on his way out of the hotel and said to Judah: "You are going about this thing the wrong way. If you want to come to my office some evening, I will talk with you about the railroad." This was

Collis Potter Huntington, 39 years old, a tinker's son from rural Connecticut, who had come for the gold rush in 1849, then made himself a prosperous merchant selling axes and shovels to the miners for as much as he could get; his biographer, Oscar Lewis, tells us that in later life he looked back on the half day he had once spent shoveling gravel from a creek bed as one of his major mistakes in judgment. Huntington was the most prominent of a number of forty-niners who had concluded that selling to people trying to find gold was easier than finding it yourself. He had put money into a wagon road from Placerville to Carson City and invested in the telegraph line even then on its way to Sacramento.

His loyal partner, Mark Hopkins, was the opposite of Huntington in physique and temperament. If it was fair to say of Huntington in later life that he was a cheery old man with the soul of a shark, "Uncle Mark" had the soul of an auditor. He was 10 years older than Huntington, politely soft spoken and thin, very thin by the side of his 220-pound partner: They would have been the Laurel and Hardy of their day if there had been any fun about them.

The first meeting with Judah was most likely in the office of the Huntington-Hopkins store at 54 K Street. Huntington put him through his paces, unimpressed with his democratic $10 sales pitches, and not convinced about the railway either. "The wire Mr. Judah could pull on these 'far-seeing wise men,'" as Anna put it, was that the preliminary work on the railway would enable them to build and own a wagon road over the pass from Dutch Flat. They could charge a toll and would have a near-commercial monopoly in the money-no-object booming silver towns of the Nevada plains, where folks were desperate, impatient with the long delays in machinery and foods coming up the snakelike tracks from Sacramento. "They grasped that," said Anna, "and were led to do what he asked—simply through that." Another of Huntington's biographers, David Lavender, suggests "the born plunger" was more excited by the railway. "Think of it," he imagines Huntington saying to the reluctant Hopkins, "a transcontinental railroad in our grasp for only 1,500 dollars!" Huntington assured Hopkins he would commit

only to a detailed survey, then he raised $1,500 each from two more Republicans, both sumo-wrestler shopkeepers over 220 pounds—Leland Stanford and Charles Crocker—who completed the quartet that became known as the Big Four. Stanford, a lawyer by training who had joined his brothers' wholesale grocery, was such a slow speaker listeners often felt stranded in midstream; he was a slow thinker, too, but he loved the limelight as much as Huntington hated it (and him) and at 38 was on his way to being governor. "Bull" Crocker, at 40, was the noisiest and most profane of the quartet. He had trekked the whole continent to end up measuring calico for pioneer ladies visiting his dry-goods store, but he was happy so long as he made money.

With other smaller investors and Bailey, Judah's Central Pacific now had its capital. Huntington told Judah the quartet would subscribe his $1,500 and put him on the payroll as chief engineer at $100 a month and they would pay for the proper precise survey. Stanford became president, Huntington vice president, James Bailey, secretary, and Mark Hopkins, treasurer. As a gesture to the counties where the road would go, Huntington shrewdly added Doc Strong and Charles Marsh of Nevada City.

It was clear from the start that Huntington meant to control the enterprise. He threw his weight about on everything, demonstrating that energy in the service of ignorance is no virtue. He challenged Judah's engineering expertise to the extent of insisting on a trip into the mountains to look at other routes before allowing a precise instrument survey of Judah's Donner Pass route. The beefy Huntington edging miles along a canyon wall of the Feather River with a Chinese servant carrying a blanket was a happy spectacle, but a waste of time and money to confirm what Judah already knew, that the route would be 65 miles longer than the Donner Pass, even if the river could be commanded not to flood, and would require 30 to 40 expensive tunnels. The other directors, except Bailey and Strong, all wanted to play railway engineers, and it was early in January 1861 before Judah was given permission to begin the instrument survey. On a crisp morning, as state after southern state was pulling out of the Union, Judah walked in on Anna at

Vernon House and exclaimed: "If you want to see the first work done on the Pacific railroad look out of your bedroom window, I am going to work there this a.m. and I am going to have these men pay for it." The survey he began on muddy Front Street, looking east along the American River, took him with 10 men and Doc Strong into "the most difficult country ever conceived of for a Railroad." He edged along steep hillsides above the 1,500-foot Bear Gorge, "the worst place I ever saw," trying to pound in survey sticks to mark the line to be graded, cheerfully writing Anna, "the river is 1,200 feet below us and the top of the ridge 700 ft above, in places so steep that if you once slip, it is all over." He narrowly escaped death when Doc Strong's horse bolted and Judah's own mare panicked, throwing him headfirst into the flailing hooves of the stampeder. But the survey he lived to complete in late July was a brilliant piece of work, a "stupendous leap forward in transcontinental data" in the judgment of David Haward Bain, the discerning historian of the line. On an amazing 90-foot map and five 20-foot maps, it delineated a navigable route to the summit and a descent by gentle arcs into the Truckee Valley almost to the state line, albeit with 18 tunnels through granite and a first estimated cost in the region of $88,000 per mile.

Judah was ready for the frowns he expected from the quartet. He calculated that every day the railway would handle 50 tons of outward freight for $2,500, 100 tons of return freight for $2,500 and 25 passengers each way at $1,250, yielding $1,956,250 for a year of 313 days. (He was further ingenious in spinning gold from flax: "If we allow that 300,000 acres, or two thirds of this land, contains only ten trees per acre, from which can be cut six logs twelve feet long per tree, averaging twenty-four inches square. This gives 3,400 feet board measure per tree, and the total quantity amounts to ten thousand million feet of lumber, which delivered at Sacramento at, say, $15 per thousand, amounts to one hundred and fifty millions dollars.")

On October 11, Theodore and Anna boarded the steamer yet again for Washington with his precious maps and 470 pounds of baggage, including a demijohn, two feather pillows, a bunch of hats, a

BIRTHPLACE: Judah sold the concept of the transcontinental railroad to Collis Huntington and Mark Hopkins in a meeting over their hardware store in Sacramento, some 35 miles downriver of the first gold strike in 1849. It was a rough town of about 6,000 when he arrived in 1854 and his moral standards proved too high for his founding partners.

French grammar and a medicine chest. This time, also accompanied by James Bailey, he was going as the chief engineer and accredited agent of the Central Pacific Railroad Company, and this time to a Washington at war with all the secessionists who had for so many years obstructed the project. And once again, Judah had a useful political companion for the journey; the newly elected congressman from Nevada City, California, turned out to be a friend, the 34-year-old newspaper editor Aaron Sargent. He made another in California's senior Democratic senator, James A. McDougall, chairman of the Senate's Committee on the Pacific Railroad, who

was impressed by the memorial Judah had written and sent to 1,000 key people, including President Lincoln, no slouch on railway questions. Briefed to the hilt by Judah, who had reopened his "museum," Sargent made an impassioned speech on a dull day in the House and was appointed chairman of the House subcommittee charged with breathing life into the old Curtis bill. In an extraordinary sleight of hand, which would not be approved today, the clerk of Sargent's committee, the secretary of the full House committee, and the secretary of the Senate's railroad committee, was one and the same person: Theodore Judah.

The bill that finally became the Pacific Railroad Act, signed by President Lincoln on July 1, 1862, a day of dark defeat for the Union Army in Virginia, was basically written by Judah. It designated two companies to benefit from 200-foot grants of right-of-way to build and operate a railroad between the Missouri River and Sacramento, the Central Pacific from the west and the Union Pacific from the east. The companies were given five alternate sections of land per mile on each side of the road within the limits of 10 miles. Once they had built an approved 40 miles of railroad on easy land, 20 on difficult, they would be entitled to government loans,

DONNER PASS BEFORE: Judah stood on this ridge in 1860 looking toward Donner Lake and saw a passage for his railway.

$16,000 for every mile on flat land, $32,000 for the foothills and $48,000 in the mountains. The loans would be in the form of U.S. 30-year bonds which the railway companies would have to sell but on which the United States would pay 6 percent interest. If the railway was not finished by July 1, 1876, the companies would forfeit the property.

Forty-two congressmen and 18 senators signed a testimonial paying tribute to Judah's "indefatigable exertions and intelligent explanations" that had convinced Congress of the practicability of the line and ensured the bill's passage. For his part, he used the newly strung Pacific Telegraph to send a message to his Sacramento partners: "We have drawn the elephant. Now let us see if we can harness him."

BACK IN SACRAMENTO, Judah soon ran into granite. He had taken the precaution while in the East of placing orders for locomotives, rails, freight cars, six first-class passenger cars, switches, turntables and rails—a wise move in view of the war's inflationary impact, but Huntington was scrambling to find the money. It was urgently needed after ground was broken on January 8, 1863, in a patriotic little ceremony where Crocker and Stanford (but not

Judah) orated. Coming back from directors' meetings, he lamented to Anna: "I cannot make these men, some of them, appreciate the 'elephant' they have on their shoulders. We shall just as sure have trouble in Congress as the sun rises in the east if they go on in this way." He was particularly distressed that Crocker had been awarded the first construction contract; he suspected that the Big Four intended to cream construction profits and he was right. On Judah's protest, Crocker resigned from the board, but his place was taken by his cunning, smooth-talking brother, E. B. Crocker, a judge and the company's attorney. Judah worried, too, that money due the railways was being diverted to the wagon road; the Big Four had excluded him from the separate company they set up for that. For his part, Huntington was furious that Judah held out against a stratagem to exploit the Act's provision of extra money for hillier country. Governor Stanford sent the politically pliable state's geologist to help out here, sending him with Crocker to Arcade Creek, where the official, standing on ground that stretched flatly for miles, solemnly certified that the reddish soil indicated he was standing at the point of ascent. Other political appointees, the state's surveyor general and the U.S. sur-

veyor general for California, endorsed the statement, Aaron Sargent presented it to President Lincoln and he signed. It is said the president remarked, "Here is a case in which Abraham's Faith has moved mountains"; some accounts attribute the joke to Sargent. In any case, it was, as Judah said, "a fraud," and he, Bailey and Strong would have nothing to do with it. To Huntington, Judah's high-mindedness was intolerable since the move yielded the company an extra $240,000 in government bonds (at $16,000 a mile) at a time when they needed every cent: Huntington and his friends, for all their apparent wealth, had undertaken a stupendous risk since they were collectively worth only around $120,000, according to the sworn testimony of Stanford in 1862 (gossip put Huntington's wealth at a million). Lavender defends the Big Four: "Since Judah had estimated that the cost of construction could cost more than $100,000 a mile, against which only $48,000 in subsidies were available, a compensating game in the flatlands seemed fair." But not to an idealist like Judah, who in his *Practical Plan* had inveighed so much against theft—a man so scrupulous he had robbed himself, charging only $40 for expenses against the $2,500 it had cost him to represent the California Convention in the capital.

The original directors' board regularly split four to four (Huntington, Stanford, Hopkins and Marsh against Judah, Strong, Bailey and Booth), but two new directors answered to Huntington. The tensions came to a head in May. "Oh, some of those days were terrible to us!" wrote Anna. "He felt they were ungrateful to their trust and to him . . . it is all too personal and wrings my heart." In July Huntington provoked a showdown. Judah had positioned the railway where it could receive steamer cargo from the wharf. Huntington vetoed this sensible plan, tearing up what had been done and moving the line several blocks on grounds of cost. Then he demanded that all the directors pay up fully for the stock they owned. Judah, Bailey and Strong did not have the money to do that, Huntington stopped all work, and then challenged the four dissident directors, now in a minority: Sell out to us or buy us out.

The others sold. Judah exchanged his 500 Central Pacific shares for $100,000 in Central Pacific bonds, but was determined to exercise, as soon as he could, an option to buy out each of the Big Four for $100,000 each. He aroused interest in the East, but sensibly played his cards close to

the chest before he boarded the steamer *St. Louis* in San Francisco for Panama on October 3. Anna writes: "He had secured the right and had the power to buy out the men opposed to him and the true interests of the Pacific railroad at that time; everything was arranged for a meeting in New York city on his arrival—gentlemen from New York and Boston who were ready to take their places." Though stories were circulated at the time that Judah had given up on the railroad and gone to Europe with his wife, they were false ("cruel as death," said Anna); there is every indication he was in earnest and might well have succeeded in his plan. "New York" is most likely a reference to the steamship and railway millionaire Cornelius Vanderbilt (1794–1877), who had once contemplated building a canal between Nicaragua and the Pacific, and in 1862 was on a spree buying, acquiring and consolidating railways: He was the kind of tough and efficient buccaneer capable of taking on the Central Pacific. The "Boston" reference is probably to Oakes and Oliver Ames, the shovel company owners who became very active and rich running the Union Pacific from 1866. Both Vanderbilt and Ames had capital resources out of all proportion to Huntington, who was build-

ing the railway hand to mouth and at this stage would very well have been glad to take his profit.

Judah never met his new prospects. As the steamer received passengers at Aspinwall on the Caribbean side of the Panama isthmus, the rain was torrential. Instead of watching it all from their stateroom, he went out to help the unaccompanied women and children, shielding them under a big umbrella he had just bought. He was bitten by a mosquito carrying the yellow fever virus. Anna nursed him for eight terrible days and nights. "Anna what cannot I do in New York now?" he exclaimed one feverish night. "I have always had to set my brains and will too much against other men's money. Now with money equal, what can I not do!" A week or so later, on November 2, at the Metropolitan Hotel, he died in her arms. Seven years later, on May 10, 1869, when the final spike was driven into the rails at Promontory Point to complete the railway, the western route of his devising, it was the 22nd wedding anniversary of Theodore and Anna. "It seemed," she wrote, "as though the spirit of my brave husband descended upon me, and together we were there unseen, unheard of man."

DONNER PASS AFTER: The view from the ridge nine years later when telegraph poles and snowsheds marked the route of the Central Pacific.

A hundred inventions a day became commonplace in the second industrial revolution

TURN ON THE LIGHT, tune in the radio, phone home, catch a movie, watch TV, cash a check, take a photograph, comparison shop automobiles, and now, exhausted with all that, fly somewhere to the sunshine. The innovators who made all these activities possible were prolific in the second half of the 19th century and into the 20th. The graph of American inventiveness from the Civil War years onward is much the profile of the mountain chains the builders of the first transcontinental railway had to ascend: 3,773 U.S. patents in the foothills of 1863, more than 12,000 in 1869, rising to 24,000 by the end of century, then 100 every week and by the '30s, 1,000 a week.

There was an accelerating feedback between discovery and invention, on the one hand, and innovation on the other—between what-knowledge and how-knowledge, to adapt historian John Mokyr's terms. In this Part II, we categorize the innovators in three broad groups:

•INVENTORS who did not simply have a bright idea, but adapted and improved their inventions to change society in a significant way.

•DEMOCRATIZERS who enabled the common man to enjoy uncommon benefits.

•EMPIRE BUILDERS whose innovative energies created industrial empires on a scale never seen before.

We open Part II with the men who dared to run a railway across a continent, the single most important innovation of the 19th century, speeding a trading nation into an industrial colossus.

INDUSTRIAL MIGHT: The Mexican artist Diego Rivera (1886–1957) based his murals, now at the Detroit Institute of Arts, on what he saw at the River Rouge plant in Dearborn, Michigan, in 1932. Rivera painted at the invitation of Edsel Ford, who wanted to celebrate the new era in automobile technology.

Adventurous Men

*I*nch by inch along the cliffs, yard by yard across the rivers, mile by mile through the desert: how they built the first transcontinental railroad

The two railway lines that linked at Promontory Summit, Utah, on May 10, 1869, were the supreme feats of management and engineering of the 19th century. "My little road," as Theodore Judah described the Central Pacific (CP), conquered the Sierras from the west, with rail laid 690 miles through the mountains along the line he had prescribed, and it famously met the Union Pacific (UP), which had pushed 1,086 miles westward across the desert plains from Iowa's little river settlement of Council Bluffs and surmounted the Rockies.

The project called forth an array of determined adventurers, some with the disciplined passions of the engineer, some with the calculating spirit of merchants focused on making money, some who proved able to assemble and motivate vast forces of workers and streamline their labors, some who could navigate the halls of legislatures and Congress to win an unprecedented scale of government action and others who could persuade bankers to steer their money to a project that looked not just risky, but insane — truly a Mission: Impossible.

Private enterprise in the form of Theodore Judah and the Big Four was the father of the Central Pacific Railroad Company. The Union Pacific Railroad Company, on the other hand, was an orphan of government and, in the words of the railway historian Albo Martin, "blighted in its development for a generation by congressional sodomy of the worst kind." The Pacific Railroad Act of 1862 charged 163 commissioners with organizing a company called Union Pacific to build westward from the 100th meridian. It was the first corporation chartered by the national government since the Second Bank of the United States in 1816, and Lincoln, the one-time railway lawyer, made it happen. Lincoln had long seen the transcontinental as a way of holding the Pacific region to the Union and, more recently, as a military necessity in wartime. In 1864, when the UP had not been able to sell a single share, he signed another, more generous bill doubling the amount of land granted to both companies and permitting them to sell bonds with a first claim on assets ahead of the government mortgage. Had he not, the UP would have certainly foundered and so might have the CP.

The Union Pacific looked a much easier proposition than the Central Pacific, but the deceptively simple line on the map reached out from the unbridged Missouri into 700 miles of uninhabited and unmapped land swarming with numerous hostile Indian tribes before crossing three mountain ranges at the highest elevation yet attempted. On the plains, there were no supplies of any kind, no hardwood for the ties and bridges, no machine shops, no masonry, little water for much of the way for 10,000 men and as many horses, mules and oxen. The immense mass of iron, tools, ties, spikes, bolts and lumber had to be brought in by steamboat from St. Louis on a river navigable only four months a year; for the rest of the year supplies had to be hauled by wagon over dirt roads from the end of the track of the Chicago–Rock Island Pacific Railroad inching across Iowa toward Council Bluffs on the Missouri. Water had to be supplied by aqueducts feeding holding tanks along the route and by water trains.

On both lines, the innovative practical intelligence of the surveyors, engineers and managers, and the fortitude of the work crews — indeed their heroism — remains undimmed, but the UP started as the kind of stock swindle foreseen by Judah — rapacity tempered by incompetence, in the phrase of the journalist Ambrose Bierce. The man who managed to take control of UP a year after the Pacific Railroad Act was the stealthy Wall Street gambler and railway promoter Doc Thomas Durant, who at the age of 23 had abandoned teaching surgery for moneymaking, acquired a straggly Vandyke beard and shed whatever scruples he had. "I had rather have a man about me," he said, "that did all his enemies claimed, [who] had pluck and energy and resources within himself to accomplish his work, than to have a dolt, though he might be as honest as the sun." Durant was the central figure in the complex scheme by which the originators raked in excessive construction and stock profits through a front company they cutely called the Credit Mobilier of America after an equally fevered speculation in France. His most important contribution to actually building the railway lay in his recruitment of two engineers, both generals in the Civil War, Grenville Dodge

Unite a Continent

(1831–1916) and "Jack" Stephen Casement. Dodge's professional integrity rescued the UP from endless meddling by Durant, who wanted to cheapen the construction and lengthen the mileage, for federal per-mile dollars.

When the company's first chief engineer quit in disgust at the end of 1864, saying working with Durant was "like dancing with a whirlwind," UP had yet to lay a single rail, while the Central Pacific was running regular passenger services 18 miles into the foothills from Sacramento. Durant needed Dodge so badly he had to agree that as chief engineer Dodge would have full control of where the line went and how it was built. Durant was always trying to wriggle out of that. It required no less a figure than General Ulysses S. Grant to rescue Dodge from the machinations of Durant (see Showdown, page 141).

Dodge deserved his vindication. During the Civil War he had been preoccupied fighting Confederates and the Indians after Appomattox, but he had for some years agitated for a Pacific railroad through Nebraska's Platte Valley and had speculated in land it might require. He was like Judah, infused with the earnestness and discipline of a trained engineer, provoking Huntington to complain: "If you should see Dodge you would swear it was Judah. The same low cunning that he [Judah] had. Then a large amount of that kind of cheap dignity that Judah had." Lincoln esteemed Dodge. They had met in August 1859 when Lincoln was speaking as a Republican candidate at a Council Bluffs hotel, and on the porch he had asked the 28-year-old railway engineer where the best place was to start the eastern half of the transcontinental. "Here from this town out the Platte Valley along the 42nd parallel," said Dodge, and proved it in theory to Lincoln's satisfaction there and later when Lincoln summoned him from the front lines to the White House. (He feared he was going to be reprimanded for arming freed slaves.) He proved it in practice by organizing work on the railway as a military campaign. His tiny but dynamic construction bosses, Casement and his brother Dan, were Union Army veterans as were thousands of the men they engaged; many were Irish, but there were also Scandinavians from Chicago, Mexicans, several hundred Negro mulewhackers, some "Galvanized Yanks" from the South, and 5,000 Mormons who graded the Utah section.

The Casements were hard as nails but inventive. They devised a work train, two locomotives pulling up to 22 cars bearing a forge, a carpentry and machine shop, a saddler's, iron, timber, tools, water, bunks for 144 men, kitchens and a baker's car and dining tables for 200. Fresh beef was provided from a herd of Durham Shorthorn and Galloway cattle trotting alongside the train all the way through three winters from Nebraska to Utah. Up to 200 miles ahead of Casement's train, preparing the ground for the track (of which more in a moment), Dodge and his construction boss, Sam Reed, had 3,500 graders. They cut deeply through hills and took soil forward to fill canyons. With pick and shovel, they dug out millions of tons of earth and rock in wheelbarrows and hauled it away in one-horse dump cars. Several thousand lumberjacks cut the hardwood timber ties on which the track would rest. A thousand engineers were still further ahead, designing and erecting bridges and re-erecting them when flood and storm brought them down. At Dale Creek, they erected 700 feet of wood in a stupendous trestle bridge over a gorge 130 feet deep.

Dodge had also to contend with lethal interruptions by Sioux, Cheyenne, Crow and Arapahoe. "Our Indian troubles commenced in 1864," he writes, "and lasted until the tracks joined at Promontory." Twenty-five thousand warriors were reported to be out on the plains, enraged by the invasion of their hunting grounds and punitive reprisals for massacres of settlers and train wrecks. Dodge wanted 5,000 cavalrymen, and the War Department did not have the funds for anything like that number. Only 200 mounted men and 600 foot soldiers were available between Omaha and Denver; "it's awkward as hell," said a trooper, "for one soldier to surround three Indians." In his memoir Dodge writes, "We lost most of our men and stock while building from Fort Kearney to Bitter Creek." There were very few deaths, but much fear. Most vulnerable to ambushes were Dodge's small advance survey parties of 18 to 20 men finding a way for the line through untracked grassland hills and semidesert with perhaps six cavalrymen to guard them. Dodge gave out carbines and pistols to every grading, tie-ing, surveying and bridging crew accompanied by a simple order: "Never run when attacked." He wrote, "I do not know that the order was disobeyed in a single instance." Indian raiders, as it happened, were fortuitous for Dodge. In his Indian-fighting days in the army at the end of the Civil War, the projected railway had never been far from his mind. During the Powder River campaign against marauders in September 1865, he had taken six men with him to explore a route into the Black Hills, a major obstacle to a railway. At noon, they were spotted in a valley by more than 100 Crow warriors in a position to cut them off. Dodge hurried his men to a high ridge between Crow Creek and Lodgepole Creek, lit a signal fire to alert the main body of cavalry below and settled down behind boulders to shoot at Crow who got too close. It was nearly nightfall when Dodge's signals were seen, but when the rescuing cavalry troop escorted his party down the mountain he discovered that the ridge that had been his little party's refuge led down to the plains without a break. "We saved our scalps," said Dodge, "and I believe we have found the crossing of the Black Hills." And so he had. When he returned with his rail crews in 1866–67, the route proved ideal. He named it Sherman Pass, after the general in charge of protecting the line from the Indians. Eventually, it ran over Wyoming's Sherman Summit, at 8,242 feet becoming the highest point crossed by any railroad on the continent. (The narrative here is based on Dodge's 1910 memoir. Professor Wallace Farnham disputes it and credits the English-born engineer James A. Evans with the discovery.)

In the west, the "mad bull" Charles Crocker discovered in himself depths of managerial skill he never knew he had. He made no profession of engineering knowledge. "I could no more measure a cut than I could fly," but he was a driver and motivator of men. His major innovation was to make railway builders of thousands of Chinese laundrymen, chefs, errand boys, gardeners, street merchants, fishermen and scrabblers in half-abandoned gold workings. The 60,000 Chinese in California were much despised and abused; they paid $2 million in discriminatory taxes but were denied the vote, denied access to courts, barred from public schools, restricted to the "tailings" in gold country and supposedly banned from coming into California at all after 1858. Typical of the prejudice was the depiction of them as "the dregs" and "that degraded race" in campaign speeches for the governor's mansion by CP president Leland Stanford. The CP's bullying construction boss James Strobridge, fearsome with a black eye patch and a pick handle he exercised on difficult workers, told Crocker he would not have Chinese on the railway on any account. He regarded all of them as four-foot-nothing pigtailed weaklings whose employment would cause the whites to walk out.

Stanford changed his mind about the Chinese when the CP could not get white men in 1865; out of 1,000 transported free to the railhead, 900 slipped away within days to head for Nevada's silver towns. Crocker imposed a trial 50 Chinese on Strobridge. "They built the Great Wall of China, didn't they?" yelled Crocker. "Stro" was astonished that the little pigtailed men had tremendous endurance and discovered in due course they were also his best hope of getting the better of the granite walls of the Sierras. "Crocker's pets," as they came to be called—his name for them was Celestials—asked if they could be allowed to work on the precipice known as Cape Horn. The only way to pin a track along the cliff wall was to suspend a worker by rope from the cliff top 1,000 feet above the American River gorge so that he could hack at the rock with hammer and chisel and explode black powder charges from time to time. The available white gangs worked awk-

wardly from a bosun's chair. The Chinese came up with the idea of suspending themselves in waist-high baskets they wove from reed. Inch by inch through the fall and winter of 1865–66, the Chinese basket men chipped away, each aiming for eight inches of depth a day, until by summer there was a track for either one of the line's two engines, C. P. Huntington and Theodore D. Judah. The Chinese were quiet, methodical workers who organized their own teams, and by the end

SHOWDOWN AT FORT SANDERS: On the sweltering day of July 26, 1868, the generals and the railwaymen met in the officers' club at Fort Sanders, Wyoming. The rascally Thomas Durant, hands clasped to the right of top-hatted General William Harney, had appealed to General Ulysses S. Grant, Republican nominee for the presidency, to have General Dodge fired as chief engineer. Dodge, in peaked cap, stands at far left, about as far from Durant as he can get, and next to him is Union Pacific director Sidney Dillon. Behind Durant's catalog of complaints about Dodge was the financier's simple desire to get a longer line and hence more federal dollars. Dodge replied to the charges: "If Durant or anyone with the Federal Government changes my lines, I will quit the road." Grant (center in straw hat, in front of birdcage) broke the ensuing silence. "The United States government," he said, "expects the Union Pacific to meet its obligations and it expects General Dodge to remain in control of the location of the line." Durant backed off, but never stopped conniving behind Dodge's back. The other generals at the showdown in this detail from Andrew J. Russell's photograph included Philip Sheridan (third from left) and William Tecumseh Sherman (framed in the doorway), who was in charge of defending Dodge's tracklayers and bridge builders from Indian ambushes.

CITADEL ROCK: Casement's tracklayers were moving so fast only the temporary bridge (right) had been thrown up by the time they reached the crossing of the Green River at Citadel Rock late in 1868. The buttresses of the permanent masonry bridge are visible on the left.

AT LAST! "General Jack" Casement's crews take a break at the paymaster's card at Promontory Summit, the meeting point of the Union Pacific and the Central Pacific. Andrew Russell's photograph does not identify him, but Casement is probably the cross-legged, bearded man fourth from the left.

of 1865 Stro was happy to be directing 7,000 of them along with 2,000 whites, and asking for fresh imports from China (whatever the California law said, Stanford could take care of that little detail).

The CP crews had no mechanical power worth speaking of and nitroglycerine was not available until 1867. Swarms of men worked night and day in three shifts of eight hours, making cuts at the approaches to the summit tunnels in 1865: In two miles of line they had to cut their way through no fewer than seven walls of solid granite. The power that drove a drill into rock so that it could hold a charge of black powder was the muscle wielding an 18-pound sledgehammer. When monstrous snowstorms struck in 1866–67, the tunnelers made tunnels *in the snow* to get to work on the granite tunnel beyond; some of the snow tunnels were big enough to have a two-horse team remove the excavated rocks. In one of the scores of storms, a log cabin was crushed and buried under 15 feet of snow; hours later, someone noticed the house had vanished and 13 of the 16 occupants were dug out alive. In 1868, an astounding 40 feet of snow fell in Summit Valley. Twenty Chinese workers died in an avalanche. Ted Judah had been oversanguine that ordinary snowplows could keep the line open; Crocker and Stanford put 2,500 men to work building 37 miles of snowsheds.

The summit tunnel at the head of the Donner Pass was finally opened in November 1867, at 1,659 feet the longest of 15 tunnels. In doing all this, Crocker had to work against the frustrations of a ponderous supply line. Nobody on the West Coast could supply what he needed. Locomotives and work trains and thousands of tons of rails and spikes—made in America as the 1862 Act stipulated—had to be shipped from the East to San Francisco on a four-month passage by boat 18,000 miles around Cape Horn. Thirty vessels might be at sea at any one time. From San Francisco Bay, the heavier machinery was winched into steamers and barges for 130 miles of tricky river navigation to Sacramento and then sent to the work sites by horse wagon and flatcars. Heavy hoisting machinery was taken apart and reassembled at the summit.

The competition between CP and UP found its truest expression in the business of laying track, a matter of intense interest to both sides as they came closer together with every mile worth so many tens of thousands of federal dollars. Indeed, Casement's work train crews had perfected assembly-line techniques and time and motion studies years before Frederick Winslow Taylor invented them. Rails were loaded on a light car drawn by a single horse. As it came forward, two men seized the end of a 30-foot rail and moved forward. Two other men took hold of another section, and then two more, moving forward until the rails were clear of the car. On the command "down," they dropped the rail in place on the bedded tie, one every 30 seconds. While the iron men grabbed the next rail that moved toward them, spikers drove ten spikes into the ties for each length of rail. Bolters thrust bolts through the

PAPILLION VALLEY: The temporary wooden trestle and 95-foot fill across the Papio, or Papillion, Valley, near Omaha, Nebraska. In 1908 the permanent bridge was wide enough to accommodate the Union Pacific double tracks.

fishplates to join rail length to rail length. Track liners with crowbars made sure the rails were in perfect line. Levelers shoveled ballast of broken stone under each rail. Hundreds of tampers with shovels and tamping bars beat the track fully into the bed. Four hundred rails made a mile.

Three miles of tracklaying a day was exceptionally good going, but as the lines raced toward each other in April 1869, the iron men of Casement and Crocker were determined to outdo each other. Casement's men laid six miles in one day. Crocker responded with seven. Casement invited guests to see his men get up at sunrise and by dusk finish seven and a half miles, short of a few rail lengths. Crocker wagered he could lay 10 miles, and got a telegram from Durant saying, "Ten thousand dollars that you can't do it before witnesses." Crocker bided his time until the lines were so close the UP had only nine miles to lay and the CP fourteen, so there would not be time for a counterblow. He rehearsed every move with Strobridge and his teams: "No man stops, no man passes another." On April 28, 1869, he had 5 trains ready-loaded with iron, 41 teams of horses pulling 16 flatcars, 848 Chinese laborers, and 8 Irish rail handlers—Thomas Dailey, George Elliott, Patrick Joyce, Michael Kennedy, Edward Killeen, Fred McNamara, Michael Shay and Michael Sullivan. At sunrise, 7:15 a.m., on the signal, the Chinese assaulted the cars and had cleared them all in 16 minutes. Two handlers with tongs picked up the front of the first 560-pound iron rail, two others took the rear together and ran it forward, dropped it on the ties on command and stepped aside for the spikers. One spiker drove down one spike with three blows and moved on, while the fishplate men joined the rails and then stepped aside for the crowbar men straightening the line, who themselves stepped aside for a line of Chinese ballasters and tampers. Strobridge blew the whistle for lunch at 1:30 p.m. They had covered six miles. In the afternoon, they lost the better part of an hour because there was a curve in the line ascending the west slope of Promontory Mountain and the rails had to be bent into the right shape. But at 7 p.m., when Strobridge shouted, "Lay off!" the Central Pacific was 10 miles and 200 feet closer to the Union Pacific. The handlers had lifted and positioned more than 2,000 tons of iron in 3,520 lengths of rail, advancing exactly 4 feet, 8 and a half inches apart. The rails were spiked to 25,800 ties by 55,000 spikes, joined by 7,040 fishplates and 14,080 bolts, all hectically supplied by the Chinese laborers. It was an incredible ballet of iron and bicep, all minutely choreographed by Crocker. That night a locomotive ran over the stretch at 40 miles an hour.

With accomplishments like that, it did not matter much that the last ceremonial spike on the last rail on that flawless spring day in the middle of nowhere was missed by Governor Stanford's silver hammer, and then by Durant's, so that the embedded telegraph wire was not activated. The transcontinental railway instantly liberated the American imagination. Railways would go anywhere!

It was a new epoch in which anything was possible.

The first transcontinental had its limitations—the track in parts was poorly laid, and the CP and UP never concerted their schedules and rates—but its completion marked an accelerating burst of railway building. When ground was broken in Sacramento in 1863, there were some 33,000 miles of railroad track in operation. By 1897, trains were running on nearly 200,000 miles of railroad. In little more than 20 years, the single line that could barely be built in the 1860s was joined by four other transcontinental railroads. Thanks to the machinations of Collis Huntington, the Central Pacific spawned the Southern Pacific, a transcontinental from San Francisco to New Orleans via the 32nd parallel, and at Deming, New Mexico, it formed a thriving connection to Kansas with the new Atchison, Topeka and Santa Fe—that magic name!

The East and Midwest had enjoyed the dynamic effects of railways two decades before the transcontinental, but there was much waste and duplication because the lines continued to reflect the parochial jealousies of their conception. Many were short and transfer from one line to another was a pain. The iron-maker Erastus Corning (1794–1872), who began his working life as a clerk in a hardware store in Troy, New York, had the right vision; perhaps there is something in the air in Troy, where Crocker was born and Judah learned his engineering. In 1853 Corning had promoted the consolidation of 14 short railways yielding, in the New York Central, one continuous line between Buffalo and Albany, an innovation characterized by Albo Martin as perhaps the first great corporate merger in American history. With the completion of the Lake Shore Railroad and the Michigan Southern Railroad, the throbbing link of New York and Chicago was effected. The last of the five 19th-century intercontinentals, the Great Northern, was constructed without federal dollars by James J. Hill (1838–1916); he was another 200-pound railway baron, "not always suave," according to one of his more ironic workmen, but a superb and honest manager. He built the 900 miles from Minneapolis through the Rockies in subzero temperatures to reach Seattle in 1893—the same year another great professional railway manager, Edward H. Harriman (1848–1909), rescued and revived the Union Pacific. As the management authority Alfred Chandler Jr. has demonstrated so well, the more superior railways became universities of modern management techniques. Canada, too, guaranteed its future as a nation with the transcontinental Canadian Pacific, completed in the 1880s with the help of men who had cut their teeth on the American lines.

All these railway developments were achieved two decades before the Trans-Siberian Railway (1916), and with a dispatch that confounded history. It had taken 200 years for America to expand over the Alleghenies and into the Ohio Valley from a few isolated villages on the Atlantic shore, and most people had thought it would take another two centuries to occupy the vast lands acquired by conquest and purchase.

WESTWARD HO

In the emigrant trains, the journey west was less comfortable than in George Pullman's palaces on wheels. The novelist Robert Louis Stevenson, who made the crossing in 1880, marveled at how an emigrant could be borne in less than two weeks from the Atlantic to the Golden Gate for some 12 pounds sterling, but he grumbled stylishly about "the worthlessness of rest on that unresting vehicle." The wood-burning trains carried people and freight from Omaha to Sacramento in four and a half days, at an average speed of 22 miles an hour. The sleepers lay in uneasy attitudes, "roughly shaken on narrow wooden benches . . . with scarce elbow room for two to sit and not space enough for one to lie." He would have enjoyed rather less the alternatives before the transcontinental railroad: from New York, two days by rail to St. Louis and then a 25-day bone-shaking odyssey by stagecoach; or the overland ordeal of four months by wagon train where the problem with sleep was not space but the Sioux, and the ghosts of the migrants who died starving in the snows of the Rockies. There was a circuitous route by steamship, 26 days from New York to Panama, then another ship to San Francisco or a hazardous trek through the pestilential jungles of Central America.

Instead, the new railways quickly settled hundreds of thousands of immigrants on the virgin lands west of the Missouri and in California, especially within radius of the golden magnet of San Francisco, its bay area businesses and the farms, orchards and vineyards of its valley tributaries (Los Angeles was a backwater). They carried the settlers' livestock, tools and plows and came back to fetch their grain, cows and sheep for St. Louis and Chicago. Along the way, they grew trading posts into townships bustling with sawmills, furnaces, machine shops, saddleries, warehouses, grain elevators, land agencies, newspaper and telegraph offices. No longer were California and Oregon just gold-crazed islands on the Pacific. They joined the national economy, and remained part of the American scene. On the sleepy banks of the Missouri, Omaha rose at the eastern end of the terminus, and on the shores of San Francisco Bay, Oakland flourished.

With the integration of the transcontinentals, the trunk line railroads of the Northeast (especially the New York Central and the Pennsylvania) and the so-called granger roads running through the nation's breadbasket, the great landmass of America—and all its riches—was finally open to the fullest exploitation by an exceptionally energetic people. The new railways moved mountains of coal for steam power and heat and gas lighting in the cities; hundreds of tons of chemicals and copper; convoys of iron ore and coke for Sir Henry Bessemer's revolutionary new steel (which in the hands of Andrew Carnegie and his rivals provided steel rails); oceans of petroleum from Pennsylvania; forests of Washington timber; perishable fruit from California; all the potato fields of Idaho, and all the wheat fields of Kansas; a cornucopia of products noted by the economics historian Harold Underwood Faulkner as the work of small manufacturers who were "often a combination of engineer, inventor or scientist, and businessman, and who were the original entrepreneurs or builders of American manufacturing." The immense potential of all the western grasslands could at last be realized. The number of farms tripled between 1860 and 1900, from 2 to 6 million, and grain production tripled, too (with help from Cyrus McCormick). Nebraska's population of just over 100,000 rose to more than a million by 1890, the Dakotas to 500,000 from a mere 15,000 in 1870. Not only were colossal resources opened up to exploitation, but scenes of spectacular beauty drew tourists from far away. They came to see "the new America."

Everything speeded up. The railways by 1870 enabled the U.S. mail to deliver to Chicago a letter posted in New York the day before. And where the railway line went, so went the telegraph pole. Rochester's Hiram Sibley (1807–1888), a one-time shoemaker and sawyer, manufacturer and banker, was very early to recognize that it made no sense to have short distance telegraphy; it had to be national to realize its utility. When his business colleagues failed to share his enthusiasm, he went ahead on his own, advocating an intercontinental telegraph line, and won the ensuing federal contract. In October 1861, when the wires hummed between New York and San Francisco, Sibley had laid the foundation for Western Union, the first national monopoly and a company with the resources to bet on the inventive talents of Elisha Gray and Thomas Edison (page 151).

A retinue of satellite innovations attended the railways in all manner of engineering, construction and agriculture; out of enlightened self-interest, various railroads promoted improved farming methods and introduced new crops and livestock. Farmers in remote areas learned of likely changes in the weather from semaphores or flags flown by trains moving along the track. The weather data came from the Weather Service, which was first established in the Army Signal Corps around 1880; collated local information was telegraphed to Washington from around the country, then predictions were sent out via railroad and telegraph. George Pullman's Palace Car Company, incorporated in 1867, marked the advent of the transcontinental by applying for patents on a hotel car with sleeping berths, dining rooms and staterooms where families could gaze on the plains in comfort and hold hands while descending the precipitous twists in the Sierras. Pullman (1831–1897) won the sleeping car contract for the Central Pacific, and was by 1876 operating 700 luxurious palaces on wheels in the manner of special trains built for Napoleon Bonaparte. It was well said that he reproduced the privilege of royalty for the middle-class American. Philip Danforth Armour (1832–1901) and Gustavus Swift (1839–1903), having developed assembly-line techniques in the killing yards, shipped dressed beef from Chicago in special railcars refrig-

erated at first by ice blocks and, not many years later, by machines. The House of Armour's sales in 1893 alone topped $110 million, derived in part from exploitation of every animal by-product for companies making leather, glue, gelatin, fertilizer, margarine. A certain Joseph G. McCoy (1837–1915) of Springfield, Illinois, persuaded the Union Pacific to run a spur into Kansas, and at Abilene, he built stockyards and a hotel and printed handbills telling Texans he would pay high prices if they drove longhorns for shipment on the hoof to Chicago. In 1867, some 35,000 were driven across the Red River and up along the traders' trail pioneered by Jesse Chisholm. The cowboys who drove cattle into the first cowtown of Abilene, then Newton and Dodge as the railroads moved west, became part of the American legend, along with Wild Bill Hickock, hired by McCoy as his mayor of Abilene to keep order in the saloons. Thus it was the railways gave birth to a whole new industry based on the legends of the West in pulp-fiction romances, novels, movies and TV series.

The railroad barons became black-hat bad guys themselves in the latter part of the 19th century as they gouged the farmers, suppressed labor unions and bribed congressmen to do their will. Along with banks, merchants and landlords, they were one of the focal points of the rebellion by tenant farmers and planters that led to the rise of the Populist Party. Still, the ribbon of iron running through the continent from 1869 was a marriage of private and public energies, a symbol of indissoluble political union, but it was above all a permanent way for the locomotive of the new American economy. By the end of the century, railways had created and now served a gigantic new national mass market, the progenitor of all the innovations of mass production, of mail-order catalogs and advertising and consumer magazines, and the Populist movement, triumphing within the Democratic Party, was the seed of the reformist Progressive movement that came to fruition in both parties at the opening of the 20th century. Even now the role of these Pacific railroads is not finished. Containers packed with products from newly industrialized nations like South Korea can, in essence, flow to New York on two paths, either westward by water through the Strait of Malacca and the Suez Canal, or eastward across the Pacific in ships to Los Angeles or Seattle and then by rail to the East Coast. The two journeys each take a bit more than 20 days. The western railroads also figure in the most gigantic transfers of bulk goods ever. This is the flow of ever longer trains, pulled by ever more powerful engines, which are filled with low-sulfur coal from the Powder River in Wyoming of Dodge's adventure to electric power plants all over the East. The minimum friction in steel wheels on steel makes the railway impressively energy-efficient, 10 times moreso than road, even today: A single locomotive with two men can haul the freight requiring 70 drivers and 70 modern semitrailer truck rigs.

Well did Asa Whitney and Theodore Judah contrive!

INVENTORS

Only a tiny fraction of the thousands of patents ever pass into society as practicable, useful innovations. Very often, when they do, the complex process is left to others—to entrepreneurs, improvers, financiers, corporations. In these first narratives of Part II, we open with inventors who did their own innovating. The Wright Brothers had a hard time doing that because of their own secrecy and universal skepticism. The African-American Garrett Morgan was penalized by racial discrimination. Leo Baekeland and Edwin Armstrong moved with reluctance from invention to innovation, Baekeland because manufacturers had trouble making Bakelite, Armstrong because the corporation that employed him wanted to sabotage his invention.

Young Thomas Edison, seated like a king among his associates at Menlo Park (center, hands on knees), was proud to call himself an inventor and to depict Menlo Park and his much bigger laboratory opened in 1887 in West Orange, New Jersey, as engaged on invention. He did himself less than justice, a rare event. We can now see that "inventor" was an inadequate appellation and a misleading one about the nature of Edison's distinction. It was a correct title for him only in the sense that he invented a system of inventing, but he did not leave it at that. The inventor's patents were but the first steps in Edison's scheme of things; to him that was the easy part before the "long laborious trouble of working them out and producing apparatus which is commercial"—in short, innovating.

THOMAS ALVA EDISON He gave us light, heat, power, music and the movies. He made innovation from science and made a

1847–1931

It sits in isolation on a slope in the middle of a cow pasture, a two-story white clapboard house surrounded by a picket fence and silence. Approached from the front, it looks like an ordinary family home with high sash windows, a gracefully arched porch ascended by sagging wooden steps and a little balustraded balcony above. The first surprise is how far back the house extends. From the modest 30-foot façade, it stretches at least 100 feet to the fringe of a virgin forest.

It is late on a winter's night in 1876. There is snow on the ground. Nobody is about, but wood smoke curls from two brick chimneys. When we enter the house and climb the uncarpeted stairs in darkness we find ourselves in a big bare-boarded room lit by gas jets and kerosene lamps. The room runs the building's full 100 feet, its ceiling laced with wire and piping, its walls lined floor to roof with jars of liquids and hundreds of bottles of powder of every color. There is a rack in the center of the room stacked with galvanic batteries, and every other nook and surface is covered with bits of copper, brass, lead and tinfoil; and crucibles and vials and small darkened panes of glass; and microscopes, spectrometers, telegraph keys and galvanometers; and rubber tubing and wax and small discs of some obscure material. At scattered workbenches and heaped-up tables there are a dozen young men too engrossed or tired to break off what they are doing. A bearded pair of them observe a spark jumping from an electromagnet to a metal lever; another is boiling a smelly chemical; another has his ear to some kind of telephone receiver; another, chewing tobacco, has his head down, frowning at the needle on an instrument. In the far corner of the room, stretched out amid a score of opened books, is a pale young man with a mop of hair over his forehead and acid stains on his hands, entirely lost to the world because he is concentrating on making a new one.

This is Thomas Alva Edison, at 31, in his invention factory at Menlo Park, New Jersey. If we can stay long enough, we will see him uncoil his shabby 5 feet 8 inches and, stooping slightly, move slowly among the workbenches, cupping an ear to listen to observations on the night's work, reaching over to tweak an instrument, joshing one fellow and then another, breaking out in laughter as one of them makes a riposte at his expense. His black frock coat and waistcoat are dusty, and a white silk handkerchief around his neck is tied in a careless knot, falling over the stiff bosom of a white shirt rather the worse for wear, but what stands out is the extreme brightness of his eyes.

Around midnight he and his comrades in discovery will settle in front of a blazing fire for pie, ham, crackers, smoked herring and beer, and nonstop talk and banter. There is as likely to be a competition in mocking doggerel or crude cartoons as a debate on the proper expression of Newton's law of gravitation. Someone, maybe Edison himself if he has had a good day, will blast out a melody on a huge pipe organ at the end of the big room and they will raise the rafters singing sentimental (and censorable) ditties. Then they will all go back to their benches and books until the early hours while down the hill in Edison's farmhouse home, Mary Edison, his wife and the mother of two of his children, will have given up on him and gone to sleep with a revolver under her pillow. One late night soon a disheveled Edison will forget his keys, climb onto the roof and let himself in through an open bedroom window. Mary, ever fearful of intruders, will nearly shoot him with her .38 Smith & Wesson. In the words of his journal, he will again "resolve to work daytimes and stay home nights," but he cannot keep a promise to himself when his head is filled working out the complexities standing between a panoramic vision and the steps to its realization.

Thomas Edison, remembered by everyone as the inventor of the incandescent lightbulb, was America's most productive inventor in the 19th century and as such remains so into the 21st. The 1,093 patents in his name are by no means the proper measure of the man, for reasons to be explored, but a summary of his major inventions gives a glimpse of the peaks in a mountain range (number of patents in parentheses):

- Telegraph inventions (150 patents): The **repeater** played messages at slower speeds for easier transcription. The **printer** translated dots and dashes into roman letters for stock tickers. The **perforator,** the opposite of the printer, turned roman letters back into Morse code, which could then be transmitted. The **quadruplex** sent two messages and received two more simultaneously over one telegraph wire
- The Edison **mimeograph** (5), an electric motorized pen for stenciling hundreds of copies
- The **incandescent bulb** with manifold electrical connections (389)
- The **phonograph** (195) recorded and reproduced speech and music for the first time
- The carbon-button **telephone transmitter** (34), dramatically improving audibility: the heart of the modern telephone
- The **tasimeter** electrically measured the minutest variations of temperature
- The **ore separator** (62), a system of gigantic electromagnets that sorted iron ore in mining
- A **cement works** (40) produced cheap prefabricated housing and factories
- The **kinetoscope** (9), an early projector of motion pictures
- The rechargeable alkaline **storage battery** (141) for electric cars and trucks
- The **Edison effect,** a discovery he did not follow through but which became the basis of the vacuum tube, the foundation of radio and electronics (and also the first coast-to-coast telephone call in 1915 through the telephone repeater invented by Lee De Forest and reengineered by Harold Arnold of AT&T)
- And **Menlo Park** itself, the forerunner of the industrial research laboratory

But Edison's greatness lies not in any single invention, not even in the whole panoply, but in what he did with his own and other men's cleverness. The single invention for which he is most remembered, the incandescent bulb, is emblematic. The idea of the bulb itself had been around for 50 years; many other inventors had tried to heat filaments to incandescence: Ira Flatow, the science reporter, gave up counting at 23. The light Edison made was a marked advance—and we will put that invention in context—but the piercing vision, and it was Edison's alone, was how he would bring light and power to millions of homes and offices. The technology historian Ruth Cowan writes that Edison from the beginning understood that he wanted to build a technological system *and* a series of businesses to manage that system. Edison created innovation from science, but made a science of innovation—the exploitation of invention to build entire industries. Nobody had done anything like that before.

By the time he applied for any patent, Edison had already envisaged how his machine shops could translate the invention, once perfected, into a tangible, commercial product and how it might be financed and marketed; indeed, he would not begin the research otherwise. This came from experience. His first operative invention as a freelance inventor and his first recorded patent (on June 1, 1869), when he was 22, was an electromagnetic machine by which politicians in assembly, at the press of a button, could vote sitting down. But the officials were horrified at anything so scientific reducing the mysteries of party maneuver. A hundred-dollar investment wasted, Edison thereafter resolved to invent only things for which there was a surefire demand and one he could himself service. "Anything that won't sell, I don't want to invent. Its sale is proof of utility and utility is success."

Edison was a classic innovator. More, he was an impresario of innovation. Almost every individual featured in this book had one basic innovation, one plotline. Edison had scores where he more or less successfully married the art of invention with the business of innovation. "Only Leonardo da Vinci evokes the inventive spirit as impressively," writes the historian Thomas Hughes, "but, unlike Edison, Leonardo actually constructed only a few of his brilliant conceptions." Purists might respond that Leonardo was on his own whereas Edison had clever men at his beck and call—but what a sensible notion that was! One man could hardly hope to keep up with the efflorescence of knowledge in the sciences and the profusion of new techniques and new materials. In the decades after 1870, when industrialization in manufacturing superseded the machine shop culture, it

was quite brilliant to finance and focus multidisciplinary research in an organized manner with the deliberate intention of manufacturing from the results. The momentum by which the United States surpassed Britain as the greatest industrial power near the turn of the century was in significant part due to the culture of research and development. In the year Edison was born, 1847, only 495 inventors won patents; in the year of his 40th birthday, he

EDISON AT FOUR: Little Al to his mother and oldest sister, Marion

had more than 20,000 lesser mortals for company.

What were the sources of Edison's gifts? Little Al, as he was called then, did not do well at school. At the age of eight, he burst into tears hearing a teacher describe him as "a little addled." His father thought he was stupid: "Teachers told us to keep him in the streets for he would never make a scholar." Edison himself recalled, "I was always at the foot of the class. I used to feel that the teachers did not sympathize with me." All his life, he was scathing about formal education. Part of the trouble was that he missed years of public school lessons through a series of infections, one of which seriously damaged his hearing, and his father fell behind in paying the fees at little private catch-up schools. He was also of a temperament that rejected any kind of rote learning; he could reach understanding only by doing and making. It was this predilection for experiment that found

him one morning in 1853 standing in the cobbled public square in his birthplace of Milan, Ohio, to receive a whipping by his father. "Just to see what it would do," he had set a little fire in the family barn and burned it to the ground.

His father, Sam, was a handsome jack-of-all-trades of Dutch-American extraction, a political iconoclast who had had to flee Canada in the big rebellion against the Crown in 1836–37. Sam Edison was variously a sailor, tailor, carpenter, innkeeper, miller, and became a lighthouse keeper on moving his family of three, diminished by death, to a big rented house in Port Huron, Michigan, in 1854. He had endless schemes for getting rich that never quite came off; nevertheless, while it suits the mythology to portray Alva as growing up in poverty, the little family was comfortable by the standards of the day, if erratically in debt. The public whipping suggests a cruel man, but he was not. Errant boys were regularly beaten by their fathers—and their mothers. Sam was, in fact, a fairly easygoing father, but not surprisingly he found it hard to understand a son who varied between the mischievous high spirits of a Tom Sawyer and the total self-absorption of a monk: He was baffled and ashamed when Alva at five watched a friend vanish while they were swimming in a creek and said nothing about it when he came home while the whole town was out with lighted lanterns looking for the drowned boy. In his later reflections, Edison himself always found his conduct hard to understand. Empathy was not his strong suit as a husband and as the father of six aimless children caught in the glare of his reputation.

The salvation of Al was his very protective mother, Nancy, a devout Presbyterian who always dressed in black in memory of three children dead in infancy. Al was basically brought up as an only child—big brother and sisters were in their teens—and she divined that he had a visual imagination and unusual powers of reasoning. She had been a village schoolmistress before bearing seven children, Alva being the last, and she made it her business to take him out of the little school that found him defective when he was eight. She was a severe disciplinarian; she did not spare the rod herself. But Nancy read classics like

Gibbon's *Decline and Fall of the Roman Empire* and Sears's *History of the World* to her son, and when he kept asking questions she could not answer (What is electricity? What is pitch made of?), she put into his hands, at the age of nine, R. G. Parker's *A School Compendium of Natural and Experimental Philosophy*. It illustrated simple home experiments in chemistry and electricity and Al attempted every one of them, setting him on the trajectory of his life. He rigged up the liquid batteries and a crude sender for a homemade telegraph and ran through the woods stringing the wire to a pal's house. There was nothing remarkable about that—Samuel Morse had inspired hundreds of boys—and when Alva, a boy with a large head and jutting jaw, left school for good at 13, he was "dead set on being an engineer of a locomotive."

His first job, secured by his father, was to climb aboard a train at Port Huron at 7 a.m. with a bundle of copies of the *Detroit Free Press* newspaper to sell to passengers on the three-hour journey to Detroit and back. In the daylong layover he read in a Detroit free library. He attempted Isaac Newton's *Principles* unaided in the reading room there for several days and came down the mountain gasping for breath: "It gave me a distaste for mathematics from which I have never recovered." Matthew Josephson, in his classic biography, remarks that it was a thousand pities Edison was not properly educated in the way of James Clerk Maxwell and Lord Kelvin, future masters of electricity in England; on the other hand, some of Edison's scientific breakthroughs derived from "wild" ideas trained minds rejected.

He certainly studied capitalism 101. He persuaded the train conductor to let him store berries, fruit and vegetables, as well as sandwiches and peanuts, and deputed two other boys to sell the food for him. He produced probably the world's first train newspaper, writing, handsetting and printing 24 issues of the *Weekly Herald,* a sheet of local news for which he got 500 subscriptions from passengers at 8 cents a time. He made a cheeky habit of walking into the composing room of the *Detroit Free Press* to find out what the next day's headlines would be and a year into the Civil War, on April 6, 1862, he scored a coup. He saw

a proof of next day's sensational front page reporting that as many as 60,000 might be dead in a battle at Shiloh (actual deaths were 24,000). He had enough money to buy only 300 papers to sell on the train and wanted more. The distribution manager refused, so Edison talked his way into the sanctum of the fierce managing editor, Wilbur F. Storey, and persuaded him to part with 1,000 copies on credit. He had already bribed officials at the railroad office to telegraph the fact that there had been a battle to every train station on the way back to Port Huron. He was mobbed at the first stop, raised prices every station thereafter and ended with a sell-out auction in his home station, and the princely sum of around $150. "It was then," he said later, "it struck me that the telegraph was just about the best thing going. I determined at once to be a telegrapher."

His luck was in. Late that summer, waiting at Mount Clemens station, he plucked a three-year-old boy from the path of a boxcar rolling toward him. Edison made light of it in his recollections, but *Scribner's Magazine* featured a drawing of the heroic 15-year-old train boy dashing to the rescue, and the grateful father, the railroad's stationmaster, James McKenzie, offered telegraph lessons as a reward. Five months later, Al changed his boyhood name to Tom and wandered middle America as one of the hundreds of young "tramp" telegraph operators (preceded by such budding stars as Andrew Carnegie and Theodore Vail). The "tramps" were fond of gambling, cursing, drinking, smoking, playing jokes and carousing with women. Edison was no prude. He chewed tobacco ceaselessly, gambled a little and played practical jokes, but he spent most of his spare time reading in lonely boardinghouse rooms and fiddling with telegraph equipment in railway stations on his preferred night shifts. In Indianapolis, working for Western Union, he struggled with press reports coming in on the sounders. He was able to hear the sharp dot-dash clicks, but not write them down in longhand at the speed they were being sent. His answer was to find an original Morse machine where the dots and dashes were indented on a paper tape at whatever speed they came in, and then have his companion feed the tape into a second

Morse telegraph at a speed convenient for transcription. "They would come in at the rate of fifty words a minute," Edison wrote, "and we would grind them out at the rate of twenty five. Then weren't we proud. Our copy used to be so clean and beautiful that we hung it up on exhibition." When the manager found out about the repeater, he suspected it was a dodge to cover incompetence and shut it down.

The end of the Civil War brought an upsurge in telegraph traffic. In ravaged Memphis in 1866—"the whole town was only 13 miles from Hell"—18-year-old Edison got part-time work with the military, where he experimented with ideas for doubling the carrying capacity of the wires. He was fired for his pains. "Any damn fool," yelled the officer in charge, "ought to know that a wire can't be worked both ways at the same time." The repeater and the duplex experiments were just the beginning of Edison's multiple inventions in telegraphy, in biographer Robert Conot's words, "the fountainhead of the most productive inventive career in the history of the world."

By the time he arrived in Boston in 1868, Edison was a haunted man. The little sleep he got was populated by a plethora of polarized magnets, springs, cylinders, rotating gears, armatures, batteries and rheostats, all dancing intricate patterns with labyrinthine strands of wire to make the most marvelous advances in telegraphy, and all vanishing as soon as he awoke. He worked the night shift at Western Union, expertly transcribing press messages (and reducing the numerous office cockroaches by electrocution). In Boston, then the leading center of American science and technology, he found a backer to put up a few hundred dollars and rented a corner of Charles Williams Jr.'s instrument workshop. (This was the same workshop where Alexander Graham Bell encountered his collaborator mechanic Thomas A. Watson.) Here Edison tried to render his nocturnal phantoms into models of printing telegraphs, duplexes and facsimile recorders. He improved on the standard stock telegraph tape printer, and went into business with a number of other telegraphers to sell his machine and a stock-and-gold quotation service to local banks and traders. Edison clambered over the roofs of houses looping wires from the Boston gold exchange to the 30 or so customers they acquired. The month he started the service, January 1869, he resigned from Western Union and announced in a trade paper that Thomas A. Edison "would hereafter devote his full time to bringing out his inventions." It must have sounded very pretentious.

Still, he was on his way and more complicated science began to inform intuition. In a secondhand bookshop on Cornhill he bought *Experimental Researches in Electricity,* three volumes of exploration unimpeded by mathematical abstraction by the self-taught English genius Michael Faraday. He repeated Faraday's experiments and was inspired to begin his own lifelong journal. It proved a way of reconciling his free imagination with his determination not to invent something nobody wanted. He formally recorded steps in all his inventive processes, for protection in patent disputes, but he also jotted down any wild idea that came into his head: "Dot and Dash and Automatic Printing Translating System. Invented for myself exclusively and not for any small-brained capitalist."

There were too many ideas, and not enough money for them was forthcoming in Boston. Edison abruptly abandoned his partners for New York. His arrival in Manhattan in June 1869, at the age of 22, could not have been better timed. He had no money and the city was awash with it in an inflationary boom for anyone who could speed up financial intelligence, especially fluctuations in the price of gold: With the end of the Civil War, the country had returned to the gold standard and gold prices affected everything. Edison was in the offices of Dr. Sam Laws's Gold Indicator's wire service as a piecework assistant when its machine broke down and bedlam ensued. Hundreds of brokers' messengers fought at the door for the information while Wall Street came to a stop and Dr. Laws and the experts responsible for transmission worked themselves into impotent rage. Edison fixed the machine.

He was the golden boy. Already noted in telegraph circles for his Boston stock ticker, and someone whose can-do zest attracted others, he formed a series of short-lived partnerships to manufacture and market his inventions in the dizzily evolving telegraph world in which rival companies were eager for ways to avoid the embrace of the 900-pound gorilla, West-

NOTE TO HIMSELF: He sketched hundreds of ideas in his journal. Here he describes "a new system of telegraphy using neither dots nor dashes, but receiving the message by a puncher . . . I may improve it in time."

ern Union, the near monopoly dominated by the railroad baron Cornelius Vanderbilt. Western Union, for its part, was determined to stay ahead; it was representative of a new style of company that was ready to invest heavily in unproven technology. Thus Edison was knocking at an open door when he went before the directors of a Western Union subsidiary, Gold and Stock, to present a device that ensured stock tickers in outside offices would always remain in alignment with the central station. "What did he want for it?" they asked. In a memoir, he tells us he had thought to ask

for $5,000, was ready to settle for $3,000, and then bit his tongue. "Suppose you make me an offer," he said, and was asked: "How would forty thousand dollars strike you?" For that, they wanted the right to all his future inventions in the form of stock tickers. In fact, according to the records, the $40,000 Edison remembered was only $30,000, but it was still a substantial sum. It enabled him to make the transition from employee-operator-inventor to independent inventor-manufacturer; his confidence, already sublime, came to border on the reckless.

On behalf of partners, and then on behalf of himself, he boldly contracted to deliver private telegraph machines and electrical equipment as well as 1,200 speeded-up stock tickers for Western Union, manufacturing them from 1870–76 with a machinist in Newark named William Unger. By working 16 hours at a stretch, living on coffee and apple pie and cigars, he delivered all the machines, though his bookkeeping mixed up the accounts of rival companies. He rebuked one aggrieved client, "You cannot expect a man to invent and work night and day and then be worried to the point of exasperation about how to obtain money to pay bills." Unger had never been in a cyclone before. Once the 1,200 printers were finished, he announced his intention to close the factory and sell the machinery. It might have been a serious setback for Edison, leaving him with no base, but without drawing breath he bought out Unger with $2,500 in cash, an IOU for $5,000, and borrowed working capital at 18 percent interest on a two-year note.

Edison was now his own man. He acted as foreman of 50 or more pieceworkers in the four-story Newark factory, but the industrial arts were a secondary preoccupation. He set up a laboratory on the top floor, equipped with the latest scientific equipment, so that he might continue his inventing. He took the ferry from Newark for a course in chemistry at the Cooper

THE DEAF INVENTOR'S PHONOGRAPH

EDISON yelled, "Halloo" into a telephone diaphragm. The sound caused an attached embossing point to move across paraffined paper, indenting the vibrations. Then he pulled the paper through again so that the marks actuated the point of another diaphragm. As he later recounted, "Batchelor and I listened breathlessly. We heard a distinct sound which a strong imagination might have translated as 'Halloo!'"

That was at Menlo Park in July 1877. His original idea was to record and reproduce sound coming over Bell's telephone. In November, he gave to his mechanic John Kruesi the sketch of a little machine, a cylinder wrapped in tinfoil with two diaphragms, each with a stylus. The most he hoped for was that it might possibly record a word. Into one of its two diaphragms, he shouted the nursery rhyme "Mary Had a Little Lamb" while turning the handle of the shaft so the indentions on the tinfoil were recorded at different places. He wound the cylinder back, replaced the recording diaphragm with what he hoped was a reproducing diaphragm, and wound the cylinder forward again. "I was never so taken aback in my life," Edison said later. "Mary Had a Little Lamb" came so clearly in Edison's high-pitched voice, Kruesi paled and exclaimed, "Mein Gott in Himmel!"

News of the talking machine was an overnight sensation, but it was only a beginning. Hundreds of experiments lay between Mary and the few hand-cranked phonographs Edison put on the market for $30 in March 1878. He tired of the little lamb and got to uttering, "Mary has a new sheath gown, it is too tight by half. Who cares a damn for Mary's lamb, when you can see her calf?" His greatest joy was to speak into the diaphragm pretending to be two men talking, using a deep bass for one and a high shrill falsetto for the other. "Such a conversation coming back to us over the tin foil," said Francis Jehl, "sent us into spasms of laughter. Mr. Edison himself laughed like a boy while the tears ran down his cheeks."

Edison was too busy with the incandescent light to do more, and when he returned to making a viable commercial product of his baby, as he called it, he had competition. Alexander Graham Bell's associate Charles Tainter had made a vast improvement with a "graphophone," a foot-treadle machine which substituted wax for tinfoil. Edison rejected an overture from them for a partnership and raced to perfect an electric phonograph, also with a wax recording surface, which he manufactured in a new works at West Orange. He saw it as a dictating machine, rather than part of the nascent entertainment industry. Though he outsold the graphophone 50 to 1, not enough businessmen adopted it for dictation, nor did he have the resources to market it effectively. The business was a disaster. What did succeed beyond expectations was the nickel-in-the-slot musical phonograph in penny amusement arcades. Edison stopped making dictating machines and quickly switched to selling recordings. By 1907, he was selling no fewer than 20 million records. Then he successfully reentered the business market with electrical office systems led initially by three separate machines, one for dictation, one for transcription and one to shave old cylinders for reuse. The Ediphone in the '20s and the Voicewriter in the '30s became as familiar in offices as the typewriter. Once again, Edison's business strategy was to introduce new technology not as individual product but as part of a system.

How did the increasingly deaf Edison listen to music? The same way he had tested acoustic material researching his "speaking telegraph": by vibrations. With the telephonic devices, he clenched his teeth on a metal plate; auditioning pianists he bit into the grand piano itself and listened. He always refused to reply to pleas to invent a hearing aid. He said the quiet enabled him to think: "Most nerve strain of our modern life comes to us through our ears."

TEAMWORK: Edison (seated in the center) displays his wax-recording phonograph. Fred Ott is seated at left; standing behind Ott is William Dickson, important in developing Edison's movie camera; and the indispensable Charles Batchelor is standing on the far right.

Union in Manhattan. He did prodigious research. One of his associates described seeing him go through a five-foot-high pile of chemical journals from Europe, eating and sleeping in his chair over six weeks, writing a volume of abstracts and conducting hundreds of experiments. Most important, in the early 1870s he recruited three men who would be crucial to his graduation from inventor to innovator: Charles Batchelor, an English textile machinist; John Kruesi, a Swiss clockmaker; and Edward Johnson, a voluble railroad and telegraph engineer. Edison's intellectual partnerships were more durable than those with investing and manufacturing partners. All three were to stay loyally with him for years. Batchelor would render a rough Edison sketch into a precise drawing, Kruesi would make a model that could be entered into an application for a patent and Johnson would organize patent applications, contracts and payroll.

Edison's inventiveness was perfectly complemented by his instinct for the kind of people he needed to stimulate and service his fertile imagination, and the right people were drawn like moths to his creative flame. His journal of February 1872 had more than 100 sketches; with the help of Batchelor and Kruesi, he won 34 patents in that single year. If he hit a block working on an invention, he would "just put it aside and go at something else; and the first thing I know the very idea I wanted will come to me. Then I drop the other and go back to it and work it out." The same February there is amid the technical drawings a poignant little scribble: "Mrs. Mary Edison, My wife Dearly Beloved Cannot invent worth a Damn!" Two weeks later on Valentine's Day of all days, he reiterates: "My Wife Popsy Can't Invent." It is not entirely clear what 24-year-old Edison hoped to find in the gentle and demure Mary Stilwell he had married on Christmas Day 1871. She was 16, a tall beauty with golden hair, the working-class daughter of a Newark sawyer and inventor. She had been employed punching perforations in telegraph tape for Edison; he said his deafness helped him— "it excused me getting quite a little closer to her than I would have dared"—but though he was affectionate and generous, he could not keep his mind off his work

for very long. On his wedding day he scurried off to the factory to attend to some troublesome stock tickers, returning by one account around dinnertime and another midnight. His notebook entry no doubt reflects his disappointment to find that he could not discuss rheostats every night over dinner. Poor Mary soon became lonely and plump, gorging on chocolates, while the Newark quartet worked into the small hours oblivious of the clock.

The five years Edison spent as a freelance inventor and manufacturer extended his range and reinforced his determination to control his own destiny. The climactic event was his invention of the quadruplex, by which two messages could be sent in one direction and two in the other. His commonest method of inventing was metaphorical, to build or visualize something he did understand to enable him to "see" the abstract solution. For the quadruplex, based on introducing four different modulations of current, he built a model hydraulic apparatus in which a pump forced fluid back and forth through pipes and valves "in the pattern of the wires and controls planned for his quadruplex system."

To test his elaborate circuits, Edison needed access to Western Union's secret experimental quarters, and this meant making himself beholden to the corporation. Its president, William Orton, had decreed that Western Union would not be host to a "flood of capricious inventions;" it seemed to Orton that every telegraph operator was stealing company time to experiment. But Orton was well aware of Edison's menacing talent and deployed a little tactical cunning, granting Edison access on the understanding that Western Union's chief electrician, George Prescott, would collaborate and that Western Union would have prior claim on the result. Prescott's nominal involvement was no more than a ruse to have a Western Union claim on the invention. Edison notes that he spent "one hundred nights" away from home working at Western Union. He slept on the marble floor—fully dressed, his usual sleep mode; he was at Western Union when Mary gave birth to a daughter on February 18. His quadruplex—and it was Edison's and Edison's alone—was worth a fortune to Western Union, $20 million by Matthew

Josephson's calculation, and Orton put it into operation straightaway while dickering with Edison over its value. Edison had spent every cent he had on research and test runs. He was desperate for money to pay his 120 Newark workmen and his mortgage, and the sheriff was knocking at Mary's door for unpaid grocery bills—shades of the late Charles Goodyear, whose lament *The Trials of an Inventor* was still fresh in everyone's memory. Edison's request for $25,000 in two parts, with royalties, or $10,000 a year for 17 years, the life of the patent, was reasonable, especially since he had agreed to share it with Prescott. But over Christmas 1874, when Western Union had already been operating the quadruplex for five months, Orton went away to the Midwest for several weeks, still without agreeing on a settlement.

Edison had been stalked by Jay Gould, the owner of the Atlantic and Pacific Telegraph Company, and Western Union's bogeyman. It is clearer now than in those heated times that Gould was a brilliantly innovative manager (of railway systems in particular), but he was feared and reviled on Wall Street as a corporate raider and the creator of chaos in the gold markets on "Black Friday" (September 24, 1869). Having suborned President Grant's brother-in-law, Gould and Jim Fisk cornered the $15 million worth of gold in circulation and squeezed up its price, threatening the whole country's currency system until Grant realized what had happened and on September 24 ordered the sale of government gold. Gould, in short, was the stereotype of the unscrupulous robber baron, and it took nerve for Edison to do what he did. He went secretly through the servants' entrance into Gould's plush Fifth Avenue mansion to do a deal: country boy meets city slicker. Gould told Edison he would "save him" from an outrageous fraud by Western Union. For Edison's personal half share of his patent, he offered $25,000 in cash, shares valued at $75,000 in his Atlantic and Pacific Telegraph Company and the job of chief electrician. Edison demurred and they closed on $30,000 in cash and the shares. Edison told the apoplectic Western Union he was now involved with real businessmen—"men that sleep with their boots on."

In the resulting telegraph war between Gould and Western Union, played out in rate wars and long legal hearings, Edison was portrayed by Western Union as having "basely betrayed" his indulgent patrons. One of the counsel who cross-examined him, Grosvenor Lowrey for Western Union, came to be Edison's own counsel and friend. Another company, to whom Edison had also half-promised his invention, testified that he was "the professor of duplicity and quadruplicity." Maybe so, but the dean of duplicity was certainly Gould. Early in 1875, in a stock shuffle, he cheated Edison of the shares, worth $250,000 by then. Edison was sore, but philosophical. There must be something wrong with a person like Gould who put money before the excitement of building an enterprise, "a strain of insanity somewhere." And to Edison, Gould was not much worse than others. "His conscience seemed to be atrophied, but that may have been due to the fact that he was contending with men [of Western Union] who never had any to be atrophied." Edison spent more time ventilating another grievance: "I tried several times to get off what seemed to me a funny story, but he failed to see any humor in them. I was very fond of stories and had a choice lot . . . with which I could usually throw a man into convulsions."

In the end, it was against the interests of both Edison and Western Union to prolong their quarrel. Orton accepted Edison's first terms for a half share and Edison was able to pay all his debts, put $20,000 in the bank and reach some conclusions about the direction of his life. He summoned his 71-year-old father to New York and gave him an assignment. Sam was an old billygoat. At 67, a few weeks after Nancy's death, he had married a girl of 17 and fathered two children Edison never acknowledged. Sam had as lively an eye for property—he had built the home in Milan, Ohio, with his own hands—and it was he who found the pasture in New Jersey and oversaw the building of the curiously shaped house where Edison set up his laboratory in March 1876.

MENLO PARK

Thomas Hughes describes Menlo Park as a cross between Camelot and a monastic cloister. Edison installed his chief knights, Batchelor and Kruesi, in two rented houses in the village close to his own farmhouse and invited Mrs. Sarah Jordan, a distant relative, to open a six-bedroomed boardinghouse for the young bachelor squires the mechanics called "muckers" who followed Edison from Newark. They were all exhilarated by the multifaceted inventions they listed for conquest—a speaking telegraph, a duplexed cable for the Atlantic, an electric pen and mimeograph, an electric sewing machine, electric shears and an artificial flying bird, and curiosities such as an artificial perfumed rose and a refillable cigar to be marketed by his American Novelty Company. Every downstairs room in the lab had a needling quotation from Joshua Reynolds: "There is no expedient to which a man will not resort to avoid the real labor of thinking." Every clock had its spring removed to show that the place would not be a slave to time as measured by a machine; the length of the days would be fixed by Edison, who would often work for 24 hours, with tiny naps stretched out on floor or bench, and then sleep for 18 hours.

Menlo Park is unsurpassed in the annals of invention. Here Edison and his men made Alexander Graham Bell's telephone really audible and commercially viable. Here he invented the phonograph and the incandescent bulb. But more significant than any single invention was his conception and execution of systems by which inventions would come to a full flowering. He had won money and acclaim for his telegraphic inventions and that would have satisfied many men, but Edison found it frustrating to consign his inventions to a corporation over which he had no control. He was no longer content to play first violin; he was intent on conducting the whole orchestra and to a symphony of his own composition.

His happy band of brothers knew something fresh and big was brewing at the end of August 1878 when a well-tanned Edison bounced into the lab wearing a big black sombrero. His exuberance was so different from July when, sick and exhausted, he had gone off by himself to the Rockies for a vacation watching the total eclipse of the sun with a group of scientists. He had stayed away a worryingly long time, refusing to come home even for a telegraphed "return at once" appeal from Mary's doctor who told him she was suffering from "nervous prostration" while carrying their third child. Edison considered Mary a hypochondriac; he had sometimes doodled her maiden name of Stilwell into "Stillsick." On his return to Menlo Park, he paused dutifully to take Mary for a buggy ride and spin western yarns to a press that now referred to him as "the wizard of Menlo Park," but as soon as he decently could be, he was back in the lab talking expansively with Batchelor.

One of the scientists on the Rockies trip, Professor George Barker of the University of Pennsylvania, had enthused about a system of lights the electrical inventor Moses Farmer had installed at the brass and copper foundry of William Wallace in Ansonia, Connecticut. Four years earlier, Wallace had been coinventor of the first American dynamo. The lights at his foundry were arc lights, so called because the light was an arch of elongated sparks reaching across the gap between two carbon electrodes. Arc lights were as bright as searchlights, as much as 4,000 candlepower compared to eight for an ordinary house gaslight. They had been familiar since the '60s in British and American lighthouses and a few places of public assembly, but were too blinding for domestic use and the sparks could be a fire hazard. It happened that Barker's advocacy coincided with the arrival at Menlo Park of a file of papers on the same subject from Edison's friend and lawyer Grosvenor P. Lowrey, still general counsel of Western Union. Lowrey urged Edison to think seriously about lighting in view of the way various American inventors were following up the success of the émigré Russian military engineer Paul Jablochkoff, who had illuminated half a mile of the Avenue de l'Opera in Paris with arc lights from an alternating current generator. He had managed to make his "candles" less glaring and they lasted longer. John Wanamaker's department store in Philadelphia had adapted the Russian candles, though they were costly to run and really suitable only for such large-ceilinged open spaces or streets. They were also wired in series, which meant that when one went off they all did (an irritation familiar to anyone who has strung up Christmas tree lights).

Edison himself had experimented with a battery-powered arc light in 1875 but then moved on to other things. When he took the train to Ansonia with Barker and Batchelor on Sunday, September 8, it was not so much the line of eight big arc lamps at the foundry that excited him as the system he examined that morning: electric light without reliance on batteries. Wallace generated the electricity with a primitive little 8-horsepower dynamo and wired the current a quarter mile to the foundry. Edison had a double epiphany. Here he was seeing for the first time practical proof that electric power could be sent a distance—and it could be subdivided between lamps. That seems obvious now, but it was not then. His next very Edisonian question was whether it could be done at a profit. A reporter for Charles Dana's *New York Sun* who had come along, writing about the unfamiliar object of a dynamo as "an instrument," captured the moment of realization: "Edison was enraptured. He fairly gloated over it. . . . He ran from the instrument to the lights and from the lights back to the instrument. He sprawled over a table with the *simplicity of a child,* and made all kinds of calculations. He estimated the power of the instrument and of the lights, the probable loss of power in transmission, the amount of coal the instrument would save in a day, a week, a month, a year, and the result of such saving on manufacturing."

Edison's intuition was to think small. Instead of sending current to create a leap of light between the electrodes of big arc lamps, useless for domestic lighting, why not send it along the wire and into a filament in a small incandescent lamp? It was well known, following the observations of James Prescott Joule (1818–1899), that if electric current could be passed through a resistant conductor it would get white hot and the heat energy would turn to luminous energy only very briefly before melting or burning out unless oxygen could be excluded from the bulb. Nobody in the world in 50 years of experiment had been able to keep an incandescent light alive for more than a few moments. The chemist Joseph Swan in England had given up in 1860 after 12 years of effort. Arc lighting looked much the most promising way forward, but once Edison had envisaged gentle incandescent lights linked in a network he could not let the vision be ruined by the little local detail of an invention that had eluded him and everyone else. His competitive juices and thrill of discovery coalesced to produce the less than gracious departing remark to his hosts: "I believe I can beat you making the electric light. I do not think you are working in the right direction."

Back at Menlo Park he worked euphorically through two nights. "I discovered the necessary secret, so simple that a bootblack

FOUR YEARS ON: By 1880, the original lab building (with steps) is part of a complex. Edison has added a machine shop at the rear, a glassblower's shed jutting out diagonally, and a library-office (middle foreground). Just discernible at right is his little electric railroad. He liked to drive his locomotive at 40 miles an hour along a track of two and a half miles, but was too busy with his lighting system to carry this invention much further.

MENLO PARK GALAXY: Edison's inventive assistants around the central sun—he is the clean-shaven man with the short hair, center. Batchelor is behind Edison's left shoulder, Upton is the tall, bearded man two men to Batchelor's left.

might understand it," he wrote. "It suddenly came to me. The subdivision of light is all right. I am already positive it will be cheaper than gas, but have not determined how much cheaper." His emphasis on the economics of electric lighting should be noted; he regarded the imperatives of cost as directing, rather than frustrating, his subsequent research. This would have been the time, one would think, for stealth; an indiscretion might well help a rival—William Sawyer of New York City was already on the trail of using carbon for an incandescent lamp—but Edison went public only a week after his visit to Ansonia. The former editor of the *Weekly Herald* knew how to exploit the "human interest" appetite of Charles Dana's *New York Sun* and James Gordon Bennett Jr.'s *New York Herald*. Edison's spicy quotes got full play

when he told reporters that he would supplant gas lighting. He had not only found the way to create an incandescent bulb but would be able to light the "entire lower part of New York" with one 500-horsepower engine and 15 or 20 Wallace dynamos: "I have it now! With a process I have just discovered, I can produce a thousand—aye, ten thousand [lamps] from one machine. Indeed, the number may be said to be infinite. The same wire that brings light to you will also bring power and heat . . . with the same power you can run an elevator, a sewing machine, or any other mechanical contrivance, and by means of the heat you may cook your food."

It was hot air. The bootblack "secret" was something he had visualized but not realized, a thermal regulator to cut off current to the filament before it melted or

burned out. He had done that sort of thing with electromagnets and switches in his telegraph inventions, but simply assumed the same approach would work for incandescence. The Edison scholars Robert Friedel and Paul Israel underline the audacity: "For Edison, the search for a practical incandescent light was a bold, even foolhardy, plunge into the unknown guided at first more by over confidence and a few half-baked ideas than by science. To suggest otherwise is to rob the inventive act of its human dimension and thus to miss an understanding of the act itself." A mocking chorus reverberated across the Atlantic. Professor Silvanus Thompson of London called Edison's forecast "sheer nonsense." Sir William Preece, electrical consultant to the British Post Office, smote with Latin: "A subdivision of the electric light is an ignis fatuus" (foolish fire). To talk about cooking food from the same electricity used in a lamp was "absurd," said the English authority John T. Sprague. In America, William Sawyer predicted "final, necessary and ignominious failure." A Newark arclight manufacturer called Edison's plans "so manifestly absurd as to indicate a positive want of knowledge of the electric circuit and the principles governing the construction and operation of electric machines." Three imponderables excited the derision. Apart from the fact that nobody had stopped a light from burning out, there was the issue of generating enough power with the crude dynamos of the day, and then distributing power over a large area. Other experimenters in both arc and incandescent lighting had used a great deal of current, and Edison's critics assumed he would have to use even more to compensate for the energy losses of distribution over the distances he envisaged. This meant he would have to enlarge the cross-section of his copper conductors, at great expense. But that was not all. Independent switching, Edison promised, could be achieved only if the lamps were wired in the newfangled system of parallel or branch-line wiring—consuming exponentially more current requiring still fatter expensive copper cabling, so that the whole project was scientifically stupid and economically hopeless.

Edison was not even on the fringes of being able to resolve these dilemmas in the early fall of 1878 when his sage counselor, Grosvenor Lowrey, who had encouraged Edison to fly his colorful kite in the press, moved adroitly on his behalf in New York's banking parlors. It has been customary to portray Edison as a scientific visionary having to contend with ignorant capitalists, but at this initial point it was the scientific community that was myopic and the capitalists associated with Vanderbilt's Western Union and the banker John Pierpoint Morgan who sponsored innovation, though taking a bigger gamble than they realized at the time. In October Lowrey swiftly raised $300,000 to form the Edison Electric Light Company with $50,000 in cash for Edison and 2,500 of the 3,000 shares in return for his yielding his electric light patents and any improvements he might

make for the next five years. In comparison, William Sawyer's funding from the speculator Albon Man around the same time amounted to no more than $4,000. Edison already had the finest research laboratory in the country and progressively he enriched it with a machine shop and university-trained engineers and scientists. The most significant appointment walked across the muddy lane from the railway stop on Friday, the 13th of December, one Francis Upton, disciplined mathematical graduate of Bowdoin College and Princeton, two years Edison's senior. The diffident Bostonian was a pianist of some note; Edison nicknamed him "Culture."

Some weeks before Upton's arrival, Edison had struck out in a new direction, having utterly failed with the thermal regulator. He had used platinum as a filament (or "burner") because it did not oxidize and had a relatively high melting point, but under the intense heat required for incandescence it melted or broke or gave only a faint and brief flickering light. Edison had gone back to basics, studying in detail what everyone had tried before. In November, Batchelor had drawn his attention to someone else's design that had "enormous large conductors owing to [the] small resistance" in each lamp. "Small resistance" was a key phrase. Edison had a hunch that he could flow a low amount of current (now called amps) through a thin copper wire if he proportionately raised the voltage (pressure pushing the current along the wire) and increased the resistance of the filament. "That conclusion is easily reached by an elementary application of Ohm's law," writes the electrical authority Harold Passer. "But in Edison's time it was an important achievement which placed him ahead of other incandescent light scientists." Ohm's law—that the flow of current is directly proportional to the voltage and inversely proportional to the resistance—had been propounded in 1827 by the German physicist George Ohm in the formula voltage = current (amps) x resistance (ohms), but it was imperfectly understood. Edison himself said later, "At the time I experimented I did not understand Ohm's law. Moreover, I do not want to understand Ohm's law. It would prevent me from experimenting." This is Edison in his folksy

genius mode. Understanding the dynamic relationship between voltage, current and resistance was crucial to the development of the incandescent light, and he understood it intuitively even if he did not express it in a mathematical formula. The piquant fact is that if he had not contemplated serving a large area he would not have been propelled in the direction of thin-wire high resistance. Upton was routinely astonished by Edison's insights. "I cannot imagine why I did not see the elementary facts in 1878 and 1879 more clearly than I did," he said later. "I came to Mr. Edison a trained man, with a year's experience at Helmholtz's laboratory . . . a working knowledge of calculus and a mathematical turn of mind. Yet my eyes were blind in comparison with the eyes of today; and . . . I want to say that I had *company*."

Upton's arrival affected a marvelous marriage of the intuitive and the instructed. Upton bought instruments to measure resistance and current and calculated what would happen if they tried to light 100-watt lamps by sending just one ampere of current at 100 volts on multiple circuits. Answer, the resistance of the filament would have been as high as 100 ohms. In this way incandescence could be maintained with a low expenditure of energy—and a saving in the dimensions of copper cabling. It was the scale of the saving that astounded them both: They would need only one-hundredth of the weight of copper as was needed in a low-resistance system.

Edison patented a high-resistance lamp in April 1879. All they had to do now was find a filament that did not oxidize but had a high resistance, then heat it up to incandescence in a bulb as close to airless as they could get to hinder oxidization. (All!) Filament and vacuum proved more elusive than Edison had hoped. He had discarded carbon because it burned up so readily. Platinum wire in short strands offered only low resistance but did not oxidize and therefore seemed to offer the best prospect of staying alight. They worked on making long spirals of thin platinum, to increase the resistance, but it was delicate and dangerous work. At one point a glowing platinum filament burst on Edison's head, the explosion temporarily blinding him. From

10:00 p.m. until 4:00 a.m. he suffered "the pains of hell." A massive dose of morphine finally allowed him to sleep.

The other headache for Edison, always prickly about any doubting of his genius, came from the investors pressing to see what they had got for $50,000. In mid-April, Lowrey led a group of them into the darkened lab. Twelve platinum lamps on the walls were linked in series. Edison told John Kruesi to turn on the current slowly. Francis Jehl, an assistant, recalls: "I can see those lamps rising to a cherry red and hear Mr. Edison saying, 'A little more juice,' and the lamps began to glow. A little more . . . and then one emits a light like a star after which there is an eruption and a puff, and the machine shop is in total darkness." Batchelor replaced the dud lamp; the same thing happened a few minutes later. Only Lowrey's eloquence and the steadfastness of 42-year-old John Pierpont Morgan held the group together.

What they did not see was the grim condition of the Wallace dynamos, which overheated. There was no dynamo anywhere that could generate constant voltage current efficient enough for high-resistance lights. To Wallace's anger, Edison told the world he was designing a new dynamo. It turned out to be made up of two five-foot cylindrical bipolar magnets with a coil between the legs, so reminiscent of a recumbent female it was dubbed the "long-legged Mary Ann" (edited for the public into long-waisted Mary Ann). It proved to have 90 percent efficiency, twice as good as the few crude dynamos in service, but Edison still could not keep a lamp glowing more than "an hour or two." Too much air was left in the sealed glass bulbs after tedious hand-pumping.

The romantic story, retailed by Marshall Fox in the *New York Herald,* and no doubt encouraged by Edison as mythmaker, is that at this frustrating moment he received a gift from the gods. "Sitting one night in his laboratory . . . Edison began abstractly rolling between his fingers a piece of compressed lampblack mixed with tar for use in his telephone. For several minutes, his thoughts continued far away, his fingers meanwhile mechanically rolling over the little piece of tarred lampblack until it had become a slender filament. Happening to glance at it the idea occurred to him that it might give good results as a burner if made incandescent." A few minutes later, lampblack—carbon—was tried and hey! Presto! They were on the road to success.

WHAT IS GOING ON? Charles Batchelor looks into one of the first lamps, provoking Edison to observe: "Just consider this: we have an almost infinitesimal filament heated to a degree which it is difficult to comprehend, and it is in a vacuum under conditions of which we are wholly ignorant. You cannot use your eyes to help you, and you really know nothing of what is going on in that tiny bulb. I speak without exaggeration when I say that I have constructed 3,000 different theories in connection with the electric light, each of them reasonable and apparently likely to be true. Yet in two cases only did my experiments prove the truth of my theory."

Alas, the laboratory notebooks suggest cerebral rather than celestial inspiration: It was the prospect of evacuating most of the air from the bulb that allowed Edison to consider carbon again. He had put a classified advertisement in the *New York Herald* for a glassblower and found himself greeting an 18-year-old in a little red German student cap. The Menlo Park mockers were amused by the dainty Ludwig Boehm and his pince-nez, but he blew a better bulb to Edison's design and he helped the vacuum team work out a new way of evacuating a bulb by infusions of mercury. Edison combined two different types of pump to create the most effective vacuum pump of the time. It was laborious, frustrating work— many bulbs shattered—but in September,

after weeks of effort, they achieved a vacuum of one-hundredth of an atmosphere. Edison urged them to keep trying, but he discovered that even at this level they had so reduced the oxygen in the bulb that a carbon stick did not burn up quickly and it gave a better light than platinum ever had. That was the good news; the less good news was that resistance to this particular piece of carbon was only around two ohms (which would mean more current, more copper). Resistance could be raised by shaping a tiny filament in a small spiral, to reduce radiating surface, but the filament would have to be no thicker than fifteen-thousandths of an inch. It was counterintuitive to think that such a frail element could survive in superhigh temperatures, but Edison set everyone in a frenzy trying to roll carbon into reeds no thicker than cotton thread. Day after day, night after night, the spiral reeds kept breaking.

After two sleepless weeks, Edison relieved the carbon rollers. His new idea was to bake the carbon into a length of plain cotton thread. It was heartbreaking work trying to attach the delicate thread to the lead-in wires. On the ninth attempt, at 1:30 a.m. on October 21, the dexterous Batchelor held his breath while carrying a tiny thread bent into the shape of a horseshoe to Boehm's house for insertion in a bulb Boehm hoped to evacuate to one-millionth of an atmosphere. "Just as we reached the glass blower's house, the wretched carbon broke," Edison recalled. "We turned back to the main laboratory and set to work again. It was late in the afternoon before we produced another carbon, which was broken by a jeweler's screwdriver falling against it. But we turned back again and before nightfall the carbon was completed and inserted in the lamp. The bulb was exhausted of air and sealed, the current turned on, and the sight we had so long desired to see met our eyes."

Thread No. 9 lasted until 3 p.m. —13 hours without faltering, whereupon Edison added a stronger battery to boost the light to 30 candles, or three times gaslight. Edison, Upton, Batchelor, Kruesi and Boehm watched the tiny filament struggle with the intense heat. The light continued for 60 minutes. It was a crack in the glass that turned the room back into darkness—a darkness lit by the cheers of exhausted men. The measured resistance was an agreeable 113–140 ohms. Edison, after examining the charred filament under a microscope, launched another search for an organic fibrous material, some form of cellulose that might yield even more resistance than cotton. In the following weeks Batchelor carbonized cedar shavings, hickory, maple, cork, twine, celluloid, coconut hair, flax, paper, vulcanized fiber, flax, fishing line and cork. By November 16 they settled on a piece of common cardboard. Edison records: "None of us could go to bed, and there was no sleep for any of us for forty hours. We sat and watched it with anxiety growing into elation. The lamp lasted about forty-five hours, and I realized that the practical incandescent lamp had been born. I was sure that if this rather crude experimental lamp would burn forty-five hours, I could make a lamp that would burn hundreds of hours, and even up to a

thousand." It was a lamp, moreover, that would be economical in current. Encouraged by the new methods of obtaining a vacuum, Joseph Swan in England had returned to the race with a functioning lamp, but he used a low resistance stick of carbon that consumed a hundred times more current than Edison's.

The distinctive characteristic of Edison as inventor is that he never stopped cohabiting with Edison as innovator. Even while the frenzied trials of the filament charged the lab with expectation and anxiety, he was preparing to establish electric beachheads in New York, Paris and London. He got Western Union to run temporary overhead wires from the lab to the railway stop, pathways and six houses. Marshall Fox of the *New York Herald* broke the story on December 21, writing of seeing "a bright, beautiful light like the mellow sunset of an Italian autumn." The lab staff worked frantically making bulbs by hand, one by one, so that on New

UPTON'S CARTOON:
"I shed the light of my countenance for $15,000 per share."

Year's Eve, when Edison opened Menlo Park to a public exhibition, he had around 300 bulbs in stock. Some 3,000 people came to gaze and put questions to the great man wandering among them without an overcoat: "How [did you get] that red-hot hairpin into that bottle?" Still the experts in America and England refused to be dazzled. In January 1880, Professor Elihu Thomson (1853–1937), a brilliant English immigrant, came to Menlo Park from Philadelphia, where he had started an arc-lighting business with his original professor, Edwin J. Houston. Edison graciously made him a present of a bulb, but Thomson told the newspapers he thought little of it. There was no future for it because setting up parallel circuits for many lights would need "all the copper in the world." Dr. Henry Morton of the Stevens Institute, who had been on that Rockies expedition with Edison, was not to be cheated of his position as the leading naysayer. Edison, he charged, was

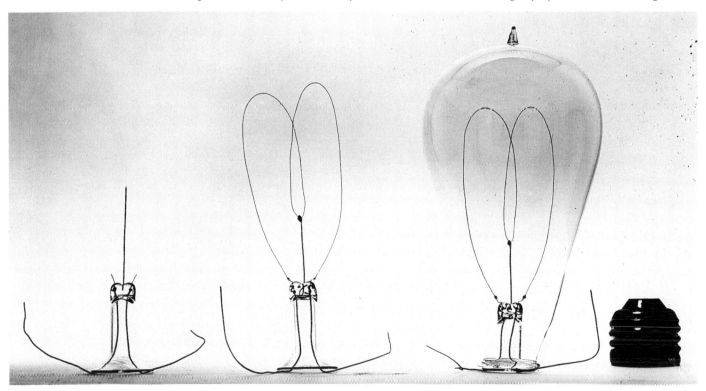

MAKING A LAMP: Edison was constantly trying to improve the manufacture of his lamps. Here are the steps from the stem to completed mount, and the exhausted bulb and base of a Gem lamp. The screw socket is at right.

perpetrating "a fraud upon the public," provoking Edison to make another promise: He would erect a statue of Morton at Menlo Park and shine an eternal electric light on his gloomy countenance.

THE BOYISH EDISON had kept up his spirits and those around him in the face of "granite walls" that the graybeards declared could never be scaled. What he attempted next can only be characterized as awesome, as if having climbed Everest he sprouted wings and flew from the top. "There is a wide difference," he correctly said, "between completing an invention and putting the manufactured article on the market," but marketing an electric lightbulb was the least of it. He had to invent the electrical industry. He had to conceive a system down to its very last detail—and then manufacture everything in it. To spell out the enormity of the task, he had to build a central power station; design and manufacture his own dynamos economically to convert steam power into electrical energy; ensure an even flow of current; connect a 14-mile network of underground wiring; insulate the wiring against damaging moisture and the accidental discharge of electrical charges; install safety devices against fire; design commercially efficient motors to use electricity in daylight hours for elevators, printing presses, lathes, fans and the like; design and install meters to measure individual consumption of power; and invent and manufacture a plethora of switches, sockets, fuses, distributing boxes and lamp holders.

For all this activity, Edison had to put up most of the money himself. Luckily, he was worth about half a million dollars by then; Western Union had made big payments for his telegraph and telephone patents. The directors of the original Edison Electric Light Company put $80,000 into the Edison Electric Illuminating Company mainly for the central station, but they were very scared of manufacturing—and without manufacturing there could be no lighting. "Since capital is timid, I will raise

EDISON THINKING: Most images of Edison present the youth or the white-haired seer of old age. This portrait of him in midlife and midcareer, around 50, captures his thoughtfulness.

it and supply it," Edison declared. "The issue is factories or death!" Subsequent profits justified his confidence in manufacturing: By midsummer 1882 he had sold all 264 new sixty-light dynamos, and textile factories were clamoring for bigger dynamos because the risk of fire was less than with gas lighting. But this still left him without enough cash to retain a controlling equity in the founding Edison Electric Light Company. When the capital of the company was tripled, he had to sell most of his 2,500 shares.

Shuttling between Menlo Park and his grand new headquarters in a double brownstone mansion at 65 Fifth Avenue, Edison the industrialist organized a coherent group of companies between 1880–81, the progenitors of the modern Con Edison and General Electric. He successfully designated the day-to-day management of them to his associates, rewarding them with equity. He sent Batchelor to Paris and Johnson to London to sell Europe on the Edison system. He charged Upton with manufacturing bulbs and installing lights in the *Columbia* steamship. When it sailed into the Atlantic on May 9, 1880, it was the world's first incandescent installation on a vessel and a glittering advertisement for electrical power.

To win approval for digging up New York streets for conduits over the objections of the gas lobby, he put on his best clothes and invited the mayor and aldermen of the notoriously corrupt Tammany City Hall to see night turned into day at Menlo Park around Christmas 1880; at the suggestion of Lowrey, who precisely calibrated the appetite for science, the bright lights were swiftly succeeded by a spectacular champagne banquet in the upstairs lab catered by Delmonico's with waiters in tails and white gloves. Edison took it on himself to prowl dingy areas of downtown New York looking for a site for a central power station designed to pump current into premises spread over a half square mile at the tip of Manhattan; he bought a couple of dilapidated warehouses at 255–257 Pearl Street within sight of the high towers of the unfinished Brooklyn Bridge and just to the east of City Hall (the same street where the Tappan brothers (page 118) endured a mob assault on their shop). In

December 1881, he began to dig up cobblestoned streets for conduits radiating symmetrically outward from Pearl Street. The city fathers had decreed that the work had to be done between 8 p.m. and 4 a.m., which he did not mind in the least. He was often down in the trenches in the raw early hours checking the connections made by the wiring runners. It took six months to do the work.

Sunday was normally the one day of the week reserved for Mary and the children, but Sunday, September 3, 1882, was different. All day and into the night Edison was on Pearl Street rehearsing every part of the operation with his engineers for the system's debut due on Monday afternoon. "If I ever did any thinking in my life," he reflected later, "it was on that day." So much might go wrong when he gave the orders for the steam to flow, spinning the dynamos and flowing the current beneath city streets and into offices, homes and restaurants. "The gas companies were our bitter enemies, ready to pounce upon us at the slightest failure. Success meant worldwide adoption of our central station plan. Failure meant loss of money and prestige and setting back of our enterprise." When the chief electrician pulled the switch on schedule at 3 p.m., only one of the six dynamo sets worked and the steam engine was wobbly. Edison had walked over to the Broad Street offices of Drexel, Morgan and Company, dressed in a black Prince Albert coat, high derby and white starched shirt, ready for the big moment when he would ceremonially connect the 106 lamps there. They all came on! They came on, too, at the offices of the *New York Times,* "in fairy tale style," said the paper, 52 filaments appearing to glow stronger as the night drew in. A gleeful Edison meanwhile had left the celebrations at Drexel, Morgan to mend a blown fuse in an underground safety-catch box; a couple of days later he was amused by an urgent summons to deal with dancing horses at Ann and Nassau streets where current leaked into a puddle of water.

There were fewer customers for Pearl Street power than Edison had hoped—450 lamps in 85 premises—and the station had cost nearly three times his estimate, but it was at once a historic vindication and an incitement. Sawyer and Man challenged the incandescent patent granted to Edison in 1879, though the clever Sawyer had been drunk too often to work out a distributing system. Elihu Thomson, who had declared Edison's system impractical, had no compunction stealing what he could when Thomson-Houston Electric opened for business in Lynn, Massachusetts, in 1883. The inventor Hiram Stevens Maxim (1840–1916) poached the glassblower Boehm, intimidated by horseplay in the lamp factory, and had some Edison-style lamps on display in New York in 1880 (four years later, in London, he invented the first fully automatic machine gun). George Westinghouse (1846–1914), the formidable inventor of railway safety brakes, had been shown Edison's system at Menlo Park and blandly announced he would make lamps having "profited by the public experience of others."

Edison had a robust attitude to "the lies of these infamous shysters." He would not sue, he would out-invent, undersell them all. Until 1884, when the competition started to hurt, he resisted the entreaties of his backers to seek legal redress in America and England. Of the English challenges, he said, "the British have a beautiful system of patent law invented I think by King Canute." The divergence between Edison and his corporate backers was a question of philosophy. When the bankers heard the word *competition,* they reached instinctively for writs and balance sheets with a view to consolidation to keep prices up; Edison went back to the lab with a view to keeping prices down. Between 1881–83, he won no fewer than 259 patents, mostly related to electric light and power. He was constantly seeking cheaper, quicker, safer ways to do things. He invented a three-wire system to improve the reach of direct current distribution and designed new tools and machinery to fabricate the lamp filament and fix it to the base. Lamps that had cost $1.21 to make in 1880 came down to 30 cents by 1883, 28 cents in 1889 and 15 cents in the 1890s. Dissatisfied with the lamp-life

of 300 hours and irritated by legal claims of precedence, he simply invented another bulb. Looking up from his microscope in 1880, he adjured his staff: "Somewhere in God Almighty's workshop, there is a dense woody growth, with fibers almost geometrically parallel. . . . Look for it." The faithful Batchelor spent months looking at rags, wild grasses, flour paste, leather, old carpets, macaroni, bast, bamboo, sassafras,

WIRE RUNNER: One of Edison's teams laying cables under New York City streets, 1882. He would have been furious about the exposed connection that killed a woman in 2004.

pith, cinnamon bark, eucalyptus, turnip, ginger root. A carbonized whisker from a lab assistant's beard glowed a rich vermilion. A thread of spider web turned pink and glowed a green phosphorescence. They found bamboo to be best, giving 1,200 hours of a reddish light, but Edison exercised his flair for publicity by dispatching a variety of Harrison Fords to search the jungles of the Amazon and China and India for rare fibers, redundant adventures romanticized by the press. Except for Edison's blind spot over the challenge from alternating current systems, there is validity in the observation of the Edison scholar Paul Israel that "Edison consistently sought technological solu-

tions to business problems. He saw continued innovation as the best means of defeating the competition." You are only as good as your last invention.

When Pearl Street went online in 1882, no fewer than 200 companies across America had already signed up with the Edison Company for Isolated Lighting, using 45,000 lamps a day: companies like Cyrus McCormick's Harvesters and Marshall Field's dry-goods store in Chicago, George Eastman's Photographic Company in Rochester, the Stetson Hat Company in Philadelphia and Dillard's Oregon Railway and Navigation Company. Overseas, Edison's electrical evangelists had done their work well. In fact, Edison had preempted himself in London, where Edward Johnson opened the world's first power station early in 1882, easily stringing wires under Holborn Viaduct for the English licensees. It was not to flourish, stymied by depression and government interference, but at the Paris Exposition in 1882 and a London show at the Crystal Palace, Edison's system outclassed all the rivals. A London newspaper summed up the acclaim: "There is but one Edison." William Preese, who had been so scathing, recanted: "Mr. Edison's system has been worked out in detail, with a thoroughness and mastery of the subject that can extract nothing but eulogy from his bitterest opponents." Joseph Swan, who had done such genuine pioneering work himself, honorably conceded that Edison's system was superior when the two companies merged in 1883 in the Edison and Swan United Electric Company. On the Continent by then some 158 isolated plants were providing light in Manchester, Berlin, Paris, Bordeaux, Munich, Bologna, Rome. Edison's candles lit the world.

The critics who had assailed Edison's original scheme as absurd because of the cost of copper cabling had been routed by his adoption of high resistance. Still, it remained a fact that there were limitations to Edison's system of direct current (DC), carrying electrons moved directly along copper wire at low voltage or pressure. Low voltage meant that energy levels could be maintained at a distance only by costly thickening of increasingly expensive copper cables. Edison improved the radius with a three-ring system, but from the mid-80s he was challenged by a new system emanating from Europe, in which the current constantly reversed itself, aptly called alternating current. The superiority of AC was that it could be stepped up thousands of volts, and such high voltages could be transmitted very long distances cheaply and efficiently. The downside was safety: High-voltage alternating current could be lethal.

In 1886 Edison turned down the chance to acquire a version of AC because of not infrequent fires and electrocutions in arc-light AC circuits of about 2,000 volts. He wrote: "I cannot for the life of me see how alternating high current pressure mains — which in large cities could never stop — could be repaired." High voltage cables did indeed occasionally fall and fry men and horses in city streets. But George Westinghouse was not deterred. At the instigation of the New York inventor William Stanley, he began to offer AC systems, and gained momentum. Edison was unshakable in his conviction that AC was too dangerous, but his mind was on other adventures.

EDISON WAS VINDICATED at the end of the long patent battles in America and England. On July 14, 1891, federal judge William A. Wallace forthrightly rejected counterclaims by Westinghouse and Thomson-Houston. Edison's priority, the judge declared, was complete. His great legal victory was matched by the roaring success of the factories for which he had not been able to raise a cent from his corporate backers. To meet the hunger for electricity toward the end of the decade, the Edison companies needed more capital while Edison himself was drifting away from the lighting business, more and more preoccupied by other inventions, especially an electromagnetic machine that would separate iron from low-grade ore. Henry Villard — Civil War correspondent, financier, railway tycoon — had been a friend and staunch supporter of Edison's ever since he joined the rubberneckers in Menlo Park when the lights went on. He persuaded Edison

THE TWO MRS. EDISONS

MARY: Edison was distraught about the darkness that overtook the mind of Mary in 1884 when she was only 29. "Send trained man who is not afraid of person out of her mind," he telegraphed his assistant Samuel Insull (page 318) in March. "Send as soon as possible." The Edisons were not long back from what had become an annual vacation in St. Augustine, Florida, where Mary picked strawberries and orange blossoms and Edison fished and hunted. The eldest daughter, Marion (nicknamed Dot), recalled: "Mother was never happier than when she went to Florida, for then Father belonged entirely to her." Mary had returned to their home in Menlo Park while Edison went back to day-night experimentation with batteries in Bergmann's factory in the city. He was summoned by telegram just before she died at 2 a.m. on August 9; Marion found him at the bedside "shaking with grief, weeping and sobbing so he could hardly tell me that Mother had died in the night."

MINA: Edison at 37 was a widower with three motherless children, aged 13, 8 and 6. He fell in love with Mina, the 19-year-old daughter of the inventor Lewis Miller. He taught his beloved the Morse code so they could flirt with each other secretly while in company, tapping each other's hands with dots and dashes. That was how he asked her to marry him; she tapped back yes. All his children suffered from his emotional distance, but notably those of the first marriage. His first daughter, Marion, had a tense relationship with her young stepmother. Tom Edison Jr. earned his father's fury by exploiting the name Edison and marrying a gold digger. He became an alcoholic, wandering the streets of New York with homeless men. Later, he adopted the name Burton William and committed suicide in 1935. William Leslie incurred his father's wrath by asking to borrow money. The children of the second marriage, Madeline (1888), Charles (1890) and Theodore (1898), fared better. Mina made sure that Charles took over Thomas Edison Enterprises. Later, he served in President Franklin Roosevelt's National Recovery Administration and was the Democratic governor of New Jersey for three years from 1940. Theodore remained in research. On his mother's death in 1947, he gave $1,260,000, half of his inheritance, to social welfare for 2,700 employees.

to let him merge his companies into Edison General Electric. Edison did not like the bureaucracy in the new company, in which he had a 5 percent stake, but it yielded him $1,750,000 in stock and cash. The real killing was made by the original Drexel-Morgan group. They had put in $779,000 and realized $2,700,000, a profit of more than 350 percent.

But such was the prodigious growth of the electrical industry. Millions more in capital investment were required. The moneymen worried that price wars were going to cut into profits. Villard therefore proposed that Edison General Electric merge with Thomson-Houston; it was financed

with big Boston money and vigorously led by a hard-charging one-time shoe salesman, Charles Coffin, the Jack Welch of his day. Edison was philosophically opposed to merging with anyone, and hated the idea of sleeping with the enemy — "They have infringed upon every patent we use." The way to win was to cut prices 50 to 75 percent, and he could do it if he were freed from "the leaden collar" the new corporate structure had imposed on him. He told Villard: "If you make a coalition my usefulness as an inventor is gone. I can only invent under powerful incentives. No competition means no invention. It's the same with the men I have around me. It's not money

HOW EDISON FIRST MADE MOVIES

EDISON ENJOYED himself when Eadweard Muybridge came to West Orange, New Jersey, in February 1888. Muybridge, famous for sequential photography (and shooting his wife's lover), used numerous cameras and fast shutter speeds to take stop-action photographs of horses, baseball players, birds, boxers and fencers, but mostly of innumerable naked women (some things never change). Muybridge thrilled audiences with the illusion of movement by mounting the more seemly images on the rim of a spinning circular glass plate he called a Zoopraxiscope.

The idea of making movies—for such it was—had been somewhere in Edison's cavernous mind at least since 1878, when Batchelor clipped a fanciful newspaper story suggesting Edison might next create "magical moving pictures." But Muybridge's visit was critical. Edison bought plates from the photographer's *Animal Motion* and told one of his English immigrant assistants, a talented photographer called William Dickson, to develop a motion picture mechanism. In October, Edison filed a caveat, a declaration of intent to apply for a patent.

Only after meeting the French physiologist Etienne-Jules Marey at the Paris Exposition in 1889 did Edison bring together the three essential elements leading to the movie camera: Muybridge's stop-action photographs, Marey's continuous filmstrips and thin nitrocellulose film from George Eastman. These were other men's work, but Edison was yet again the innovative catalyst.

He called the camera a kinetograph, as clumsy a word as the camera itself—but it was the first camera specifically designed to film motion pictures. In 1893, he spent $638 housing it in the world's first movie studio, a tar paper shack mounted on a circular track with a hinged roof that opened up to catch sunlight. Staff called it "the Black Maria" for its resemblance to a police paddy wagon. Edison hired filmmakers to shoot prize fights, acrobats, knife-throwers, strongmen, French ballet dancers, cockfights, barroom brawls; he took himself backstage at Daly's Theatre in New York City to get the Gaiety Girls to come and dance, and he also filmed Buffalo Bill and Annie Oakley. The general public got its first sight of the first 30–40 second films in amusement arcades in April 1894, peering through eyepieces in an upright coin-in-the-box kinetoscope: The film was advanced past the eye by an electric motor at some 46 frames a second. The kinetoscopes were an immediate hit. Edison sold nearly a thousand of them at $200 each to a syndicate; his $24,000 on experiments had by February 1895 returned $177,847 on machines and films. He took the first step on

CARMENCITA (1894): One of the first films shot in his Black Maria studios

the long, long road to synchronized talking pictures, which would require other adventurers, when he installed a phonograph in a peep-show machine.

The big future, of course, was in projecting film onto a screen. Edison was too happy with his manufacturing profits to give it priority in the lab (and Dickson had gone to a rival company). It took something for the Old Man, as he was now called, to suppress the pride of the inventor in favor of the enterprise of the innovator, but he did just that when Thomas Armat and Charles F. Jenkins came up with a projector in advance of anything else. He agreed to manufacture it in quantity and the inventors agreed that it made marketing sense to use Edison's name on their Vitascope, the first commercial projector mass-produced for the American market. Its debut was on April 23, 1896, at Koster and Bial's Music Hall in New York City, where an enraptured elite audience lapped up short chasers of vaudeville skits, ballet dancers, burlesque boxers and the like.

The magic of the movie as a storytelling medium was still to come. Edison had the good sense to leave that to people like Edwin S. Porter, who ran his studio for 11 years and in 1904 released the 10-minute epic melodrama *The Great Train Robbery,* shot in New Jersey and along the Erie and Lackawanna railway line. So long as his studio and independents made "good, clean pictures" encapsulating hometown virtues, Edison was happy enough to count the takings.

Edison's 10-minute epic, the first Western

they want but a chance for their ambition to grow."

Villard persisted. A merger made sense in technical consolidation. Edison companies led the way in incandescent lighting, electric motors and isolated power stations. Thomson-Houston was ahead in arc lights for streets and its alternating current installations were winning the battle for central power stations. The climax of all the corporate Kabuki was J. P. Morgan and the Vanderbilts deciding that the better merger was a takeover of the larger Edison by the smaller Thomson-Houston (and the departure of Villard!). Coffin had walked over to 23 Wall Street and convinced Morgan his company was better managed. He made much of the fact that he had a profit of $11 million on sales of $2.7 million whereas the Edison companies made only $1.4 million on sales of $11 million.

Edison had no part in the merger creating General Electric in February 1892. His secretary Alfred Tate wrote that news first reached Edison from a newspaper: "I had never before seen him change color. His complexion was naturally pale, but it turned white as my collar." The popular newspapers played it as lone inventor betrayed by Wall Street. Edison himself said, "This is not a game of freeze out. . . . Electric lights are too old for me. I simply want to get as large dividends as possible from such stock as I hold. I think I was the first to urge consolidation." His close associate Samuel Insull maintained that Edison expressed his disenchantment with the deal by insisting the corporation drop his name from its title. General Electric gave him a seat on the board, but after one directors' meeting in August 1892 he never went to another.

"Something died in Edison's heart," writes Tate. "His pride had been wounded. He had a deep-seated, enduring pride in his name. And this name had been violated, torn from the title of the great industry created by his genius through years of intensive planning and unremitting toil." On the other hand, he could count on an infusion of cash to pursue new excitements by selling his 10 percent of stock in the new company. He had millions of ideas for research and manufacture in the great laboratory he built at West Orange, New Jer-

sey, and he was convinced he would find gold and iron by means of his magnetic separation machinery. "It will be so much bigger than anything I've ever done before," he said, "people will forget that my name was ever connected with anything electrical." Sporadically between 1892–99, he roughed it in the Musconetcong Mountains near Ogdensburg, New Jersey, directing 200 or 300 men dwarfed by their steam shovels and rockcrushers, joining with them in the maelstrom of dust and noise. Biographer Neil Baldwin observes that Edison "drew sustenance from his regressive descent into the workers' world he once inhabited and still desperately craved."

It was expensive sustenance. His "Ogden baby" consumed millions of dollars, unable to compete against the newly discovered rich, near-the-surface ore in the Mesabi Range in northeast Minnesota owned by John D. Rockefeller, who also owned the railroad and the fleet of ships to bring the ore to eastern Lake Erie, where the railroads ran straight south to Pittsburgh. Edison shut down operations in 1899 and went back full-time to West Orange. "What would my GE shares be worth today?" he asked, as they boomed in the market. "About four and a quarter million dollars," he was told. He beamed: "Well it's all gone, but we had a hell of a good time spending it."

A MAN OF EDISON'S STATURE attracts myth-makers and debunkers. In the words of the historian Daniel Kevles he became very early on "the paragon of the self-made man, a mythic, unschooled, inventive genius." He was an ideal public hero for a nation of fixers and an era where progress was synonymous with invention. This was the Gilded Age, so dubbed by Mark Twain in 1873, in which capital was seen as corrupt and exploitative. Edison, setting up his first factory at the age of 24, joked to his mother, "I am what 'you' Democrats call a Bloated Eastern Manufacturer," but nobody ever saw him that way when he had 20 factories. His ascent was the rags to riches American dream personified by the homespun boy next door, and everybody could be happy in his success because they shared it. He delivered good, practical things to enrich and entertain—electric light and an

audible telephone and moving pictures for a start—whereas the creative mergers in steel, railways, oil and electricity were incomprehensible, vast and vaguely menacing. Americans saw themselves on an express gathering speed from small-town Arcadia to an industrialized metropolis where Andrew Carnegie and John D. Rockefeller crushed the unions while mysterious corporate "trusts" called the shots. Edison was the bridge between nostalgia and realism. By contrast to the titans of the age, he was a doer in the hands-on democratic American tradition, accessible, unaffected, plainspoken, competitive, joustingly humorous, a pithy country store populist in his social views and optimistic—that was the thing: Confident optimism was mother's milk to Americans. On starting up Menlo Park in 1876, this cocky youth with the jutting jaw and broad, irrepressible smile promised "a minor invention every ten days and a big thing every six months," and that was just what people wanted to hear as their good times too often collapsed into bad.

He did not keep all his promises, of course, nor was he quite such a paragon. His tenacity shaded into downright obduracy; he was convivial among men and of kindly disposition, but he was an indifferent father and a thoughtless husband. In his own field, he found it impossible to acknowledge the achievements of rivals. His pride in the phonograph blinded him to the potential of radio, which he dismissed as a fad, a craze. He was at his worst in the war of the currents. Edison was genuine in his conviction that AC was too dangerous, but he went beyond the bounds of decency, even for those crueler days, in allowing an anti-AC campaigner to use his West Orange lab for the electrocution of cats, dogs and horses in 1888. Then he let his name be used to justify the introduction of the high-voltage electric chair, suggesting it was a painlessly effec-

TESLA'S DRAMA: Nikola Tesla, the Serbian-born genius who transmitted electricity from Niagara Falls. He is pointing to a photograph of one of his fabulous displays of artificial lightning.

tive way of carrying out the death penalty. He did not believe in capital punishment, and he had no authority for the medical statements he made; he just hoped that Westinghouse would be associated with death. The world's first electric chair execution, on August 6, 1890, contrived by Edison, was a grisly spectacle. But he maintained this conservative position long after the time grounding and insulation techniques had rendered very high voltage AC relatively safe and transformers had been invented to step down the voltage of current fed into homes. It took him 20 years to concede. Meeting William Stanley's son, he said, "Oh, by the way, tell your father I was wrong."

Science and business have tended to converge in a pincers movement on Edison's reputation. The template for a certain scientific disdain was fashioned by Nikola Tesla (1856–1943), the brilliantly eccentric discoverer of the rotating magnetic field; his polyphase alternating dynamos were adopted for Westinghouse's epic generation of electricity at Niagara Falls in 1895 and its transmission 25 miles to run Buffalo's streetcars. "If Edison had to find a needle in a haystack," said Tesla, who had been at Menlo Park, "he would proceed at once with

the diligence of the bee to examine straw after straw. I was a sorry witness of such doings, knowing that a little theory and calculation would have saved him ninety percent of his labor." Edison himself encouraged the notion that he was just a tinkerer. "A scientific man busies himself with theory," he declared. "He is absolutely impractical. An inventor is essentially practical. They are such different casts of mind I do not think the two can very well co-exist in one man." A necessary corrective, as biographer Matthew Josephson points out, lies in an examination of Edison's notebooks, which show he had "a wide understanding of the principles of science as known at the time and tremendous faith in the method of scientific experimentation." Joseph Henry, the preeminent pure scientist of the period, called Edison "the most ingenious inventor in this country or any other," and Henry was fastidious in his judgment (as Samuel Morse had discovered). It does not vitiate Henry's endorsement to keep a perspective on Edison's inventiveness per se. Strictly speaking, he invented from scratch only one device—not our incandescent lightbulb, of course, but the phonograph. Hundreds of his 1,093 patents were for

improvements on inventions already in operation—but that is the essence of the innovative process. Tesla missed the larger point about Edison—that if his methods of invention seemed haphazard from time to time, his method of innovation, the creation of new industries, was systematic and complete. An apparently more damaging disparagement was made by Francis Jehl, who fell in and out of love with his master. In his negative mode, he said: "Edison is in reality a collective noun and means the work of many men." Indeed, the image of the solitary inventor, memorialized in the famous 1888 picture of a Napoleonic Edison slumped over his phonograph, is the antithesis of Edison in full flower. For most of his inventive life, he was exuberantly at the head of a battalion of chemists, scientists, mathematicians and engineers and intuitive men like himself; he paid them and directed them to research, tinker, improve and invent. Other photographic frames on the same occasion celebrating the phonograph depict seven helpers posing with Edison (page 156), but the group pictures are rarely published, such is our fixation on inventors as lonely men crying "Eureka." The names of collaborators did not go on Edison's patents, as they would if they emerged from a corporate research laboratory today. He rewarded research assistants generously with shares and royalties, but he was jealous about sharing public credit. Yet none of the inventions would have emerged as they did without Edison's original mind, probing, directing, challenging. None of the principal longtime inventive associates (Batchelor, Upton and Dickson) produced anything of significance once released from Edison's magnetic orbit.

While Edison may have had somewhat excessive adulation as a lone inventor, he has had less than justice as a businessman. It is part of the received wisdom that he was too preoccupied with invention to manage anything well. Henry Ford was a friend of the older Edison, so in awe of him that he always called him "Mr. Edison," but he also remarked that Edison was the world's greatest inventor and the world's worst businessman. That was an epigram in search of a victim. The modern management guru Peter Drucker has said much the same thing, comparing Edison to the

THE EDISON QUIZ

Edison had an encyclopedic memory and expected job applicants to have a similar knowledge. The test he administered to every job seeker had 150 questions, each test tailored for a specific position. Some of the things college graduates were expected to know: (Answers upside down)

1. What city in the United States is noted for its laundry-machine making?
2. Who was Leonidas?
3. Who invented logarithms?
4. Where is Magdalena Bay?
5. What is the first line in the *Aeneid*?
6. What is the weight of air in a room?
7. Where is Korea?
8. Who composed *Il Trovatore*?
9. What voltage is used on streetcars?

CABINETMAKERS HAD TO KNOW

10. Which countries supply the most mahogany?
11. Who was the Roman Emperor when Christ was born?

MASONS HAD TO ANSWER

12. What is the weight of air in a room 10 x 20 x 30 feet?
13. Who assassinated President Lincoln?

CARPENTERS WERE ASKED

14. Name 20 different carpentry joints.
15. What are the ingredients of a Martini cocktail?

Edison did not demand perfect scores, merely 90 percent, which had been likened to having an IQ of 180. Out of 718 college men Edison tested for jobs, only 10 percent got a "fair," or passing grade. Edison said, "Only two per cent of the people *think,* as I gather from my questionnaire." Magazines, which loved running stories on Edison's employment test, gave Edison pop quizzes with similar questions on a variety of subjects. He averaged 95 percent.

ANSWERS

1. Newton, Iowa, 2. Spartan general who died at Thermopylae, 3. John Napier, 4. Baja California, 5. *Arma virumque cano, Troiae qui primus ab oris,* 6. 48.42 pounds, 7. Asian peninsula between China and Japan, 8. Giuseppe Verdi, 9. 600 volts, 10. Brazil, Bolivia, 11. Augustus, 12. Air at 0.075 pounds per cubit foot x 6000 = 450 pounds, 13. John Wilkes Booth, 14. Bridle, butt, biscuit, box, corner halving, dado, dovetail, dovetail halving, dowell, finger, housing, lap, half lap, mitred butt, mitred, mortice and tenon, scarf, spline, rabbet, tongue and groove, 15. Gin and dry vermouth.

---

dot-com spendthrifts, reinforcing the stereotype of the absentminded inventor with no head or care for figures. Ford and Drucker are heavy bookends to displace on any question of business, but the pronouncements are tendentious.

Various Edison companies failed, but he was continually pushing the envelope, starting adventurous new businesses on the basis of scientific research—and market research, too, though the tools at his disposal were primitive by comparison with modern times. In Robert Conot's vivid analogy, he was a discoverer of new continents where prevailing opinion held none existed. If sometimes he sailed off the edge of the world, that was part of the risk of exploration. But even in simple accountancy terms, Edison succeeded much more than he failed. His business blunders are chickenfeed to the business vision that delivered fortunes in electrifying cities: At the time of his death the industry based on his vision and executed by him was worth $15 billion. There was no businessman anywhere in the world then who could have done what Edison did in his secret economy trial of January 28, 1881: work through to a decimal point what it would cost in capital, equipment, labor, energy losses and coal to run 425 lights for 12 hours. The answer, that Edison's light could be run at a fraction of one cent, was the bottom line on which the world was electrified. Biographer Matthew Josephson dryly notes that while Edison enjoyed life more than the obsessed moneymen, even so, as a part-time avocation, he managed to create a "family business embracing about thirty different endeavors whose gross annual sales in the closing years of his life amounted to $20-$27 million." By 1914 some 3,600 workers were employed in his factories around West Orange.

Indeed, a detailed contemporary examination of Edison's business record by the technology historian Andre Millard in 1990 concludes that "the same energy and ingenuity that he [Edison] brought to his experiments were applied to his business affairs." In his factories, he was swift to employ mass-production methods to progressively lower the cost of making lamps and other equipment. Far from being a careless bookkeeper, in his prime he was a stickler for detailed, accurate records; he knew full well that to keep winning financial backers, as he did, he had to produce proper accounting. His well-publicized jokes about the futility of bookkeepers were not evidence of eccentricity so much as the cunning cultivation of the image, useful in the constant litigation, of the innocent inventor set upon by scoundrels. He appointed good managers (even if he often second-guessed them). He was a pioneer of diversifying business based on industrial research. From the 1890s onward, when one business fell foul of depression, another took the strain. When he died, he turned over a resilient business based on five product lines: musical phonographs, dictating machines, primary batteries, storage batteries and cement. His estate was worth $12 million. Thomas A. Edison Incorporated, formed in 1911, was not an enterprise on the scale of Ford or U.S. Steel. No doubt if he had concentrated on one innovation like electricity, he would have approached that scale, but we can be glad he was the starting point for at least three industries—electricity, motion pictures and musical entertainment—each generating billions of dollars.

The real bottom line is our modern civilization.

LEO HENDRIK BAEKELAND
The cobbler became a chemist and his formula became plastic

1863–1944

T HE 13-YEAR-OLD apprentice at the cobbler's last took another hobnail out of his mouth and tapped it into the leather sole. He was learning the trade at the insistence of his father, Karel, a shoemaker in Ghent in Belgium. His mother, Rosalia, had other ideas. She had always thought the boy was gifted. She was a housemaid who saw how the rich valued education, and borrowed books for him. She set about wearing down the father's objections that more schooling was a waste of a good shoemaker, and they could not afford it anyway.

The master cobbler had to relent when Rosalia won their son a scholarship from the City of Ghent. The boy abandoned the shoemaker's last for lessons at the Royal Athenaeum, a government high school, where he very soon vindicated his mother's persistence. His abilities and enthusiasm dazzled the teachers. He was voracious. After a day's immersion at the Athenaeum, he spent the evenings at a vocational school. He became intrigued by photography, which meant learning how to mix chemicals for developing and printing. He

did it consummately well. Before he was 21, he won a summa cum laude doctorate from Ghent University; by 24 he was an associate professor of chemistry and physics at Bruges University.

The year was 1887 and the boy, Leo Hendrik Baekeland, was poised to help transform the twentieth century.

History would have probably been different if Leo had not fallen in love, twice over. Once with the idea of the American dream. Among the books his mother secured for him was Benjamin Franklin's autobiography; he read it when he was eight. The second infatuation was with a young woman in the lab of his senior chemistry professor. He was so dazzled by her beauty he dropped two beakers at her feet. Celine Swarts, he liked to say, was his most important discovery at university, but she was the daughter of his distinguished boss, Professor Theodore Swarts, and Swarts did not favor the liaison. He was also irritated at the way Leo was diverting his energies from teaching by getting mixed up in Ghent's manufacturing of photographic plates and emulsions. The boy genius was moonlighting as the head of a wobbly little company marketing his first invention, a simpler plate that was proving tricky to manufacture. It was not so much the attraction of making a little extra money that seduced Leo to commerce. He dutifully wanted to repay his parents, but the driving force was his ambition to make something of use. Throughout his life, practice supplanted theory.

The tension was resolved only when Leo resigned his academic job and provoked an extraordinary three-way deal involving the minister for education, anxious to keep him in university life, the Swartses and Leo. He withdrew his resignation and gave up his frail little business in return for an associ-

ate professorship at Ghent, and the hand of Celine. It was unlucky for the Belgians that Leo then went on to win a traveling scholarship awarded by Belgium's four universities. He married Celine on August 8, 1889, and two days later sailed from Antwerp to America, never to return.

He was 26 years old when he stepped off the boat in New York City with his bride. The honeymoon was over. His infatuation with photography took him to the Camera Club, where he met Richard Anthony, whose company later became Agfa Ansco. Anthony introduced Leo to his chemistry consultant Charles F. Chandler, a professor at Columbia, who talked expansively of the opportunities in New York. Within a few days, the new husband was explaining to his wife why they would not be going home: He was trading the security and prestige of an associate professorship at Ghent University for the risks of working in American business at Anthony and Company.

Why did Belgium so quickly lose one of its stars? It cannot be said the country failed to recognize his promise. Ghent was famous for its prowess in chemistry research, and Leo was assured of support for experiment. The answer is that he was excited most of all by applied science, and the psychological proximity of academic research and business enterprise, a significant element in America's rise to economic supremacy. Later in life, he added the stultifying role of the Catholic Church in his country's intellectual life to his disenchantment with academia's condescending attitude to commerce.

DRESSED UP: Baekeland has got himself suited up more like a bank president for the posed picture in his lab at Yonkers. Most days, he refused to wear formal clothes and pottered about in sneakers.

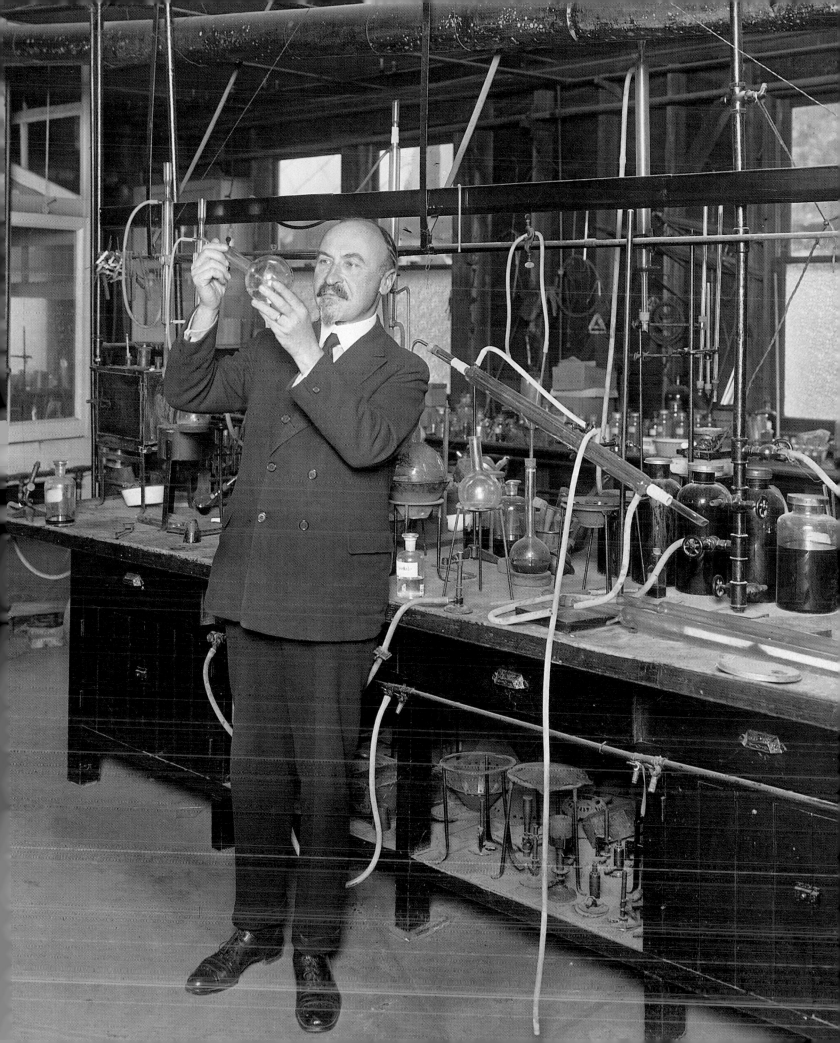

Leo certainly arrived in New York City at an exciting time, exchanging the cloistered calm of ancient Bruges for a teeming metropolis where everything was happening at once. He was one of around half a million immigrants who landed in 1889, the year America hurtled into its second century, the American Century, as it was to be called. The Statue of Liberty he sailed past was only three years old and the soaring Brooklyn Bridge only six. He saw his and New York's first skeleton steel skyscraper, finished in 1889, and no sooner was it topped out than another taller building began to rise. Most of the young country's 62 million people were crammed into the cities of the East. The immigrant poor in New York City, speaking no English, lived in gaslit tenements more densely packed than in Dickens's London, without running water or indoor toilets, but the unskilled could earn in three months as much as they earned in a year in Europe, and the West was opening up: 1889 was the year the homesteaders stampeded across the Indian territory line to create the state of Oklahoma. The country throbbed with energy and invention. Anything seemed possible to a brilliant and confident young man. Fortunes were being made overnight and the rich were ostentatious in their wealth. It was the Gilded Age of Andrew Carnegie, John Rockefeller, John Pierpont Morgan, of Thomas Edison—and the young tycoon of photography, George Eastman (page 250), who was to play a pivotal role in Leo's life.

The year the Baekelands arrived was the year Eastman brought out his first Kodak camera using film. Amateur photography was taking the country by storm. The cash-flush Eastman was fighting to get control of supplies of photographic paper; he did a deal with a raw-paper cartel in Leo's Belgium. It would make a nice fairy story for Leo to arrive fresh-faced in Eastman's office flourishing the innovative photographic paper his father-in-law had frowned on in

Ghent, but it didn't happen that way. Celine, soon pregnant, went home to have their daughter. Leo was so preoccupied with the problems of perfecting his paper he did not cross the Atlantic to see his daughter. Celine, hurt by his insensitivity, stayed away for two years while her brilliant husband, having given up his job at Anthony's for

THE SUITOR: He is a new doctor of chemistry, teaching at the Normal School in Bruges—and falling in love.

work as a consultant, faced isolation and ruin. He had few clients; he tinkered with this and that beyond the point of exhaustion and fell seriously ill. It was a turning point. "While I was hovering twixt life and death, with all my cash gone and the uncomfortable sentiment of rapidly growing debts . . . it dawned upon me that instead of keeping too many irons in the fire, I should concentrate my attention upon one single thing which would give me the best chance of the quickest possible result." The concentration, once he slowly recovered, led him to experiment with several hundred silver chlo-

ride emulsions and eventually the invention of a photographic paper that enabled images to be developed by artificial light rather than daylight. He called it Velox and by 1893 he was marketing Velox for a little company in Yonkers, Nepera Chemical, headed by Leonard Jacobi. Amateurs began to use Velox for making contact prints, in preference to Kodak's Solio paper. The legend is that Baekeland offered Velox to Eastman for $100,000, got a no, and then was surprised to be summoned by telephone to Rochester to discuss a deal. Baekeland himself says he spent a sleepless night on the train upstate wondering what price to ask and was stunned when Eastman straightaway said, "Hendrik, I will give you $1 million for that damn paper, and not a penny more." Had his knees not been locked, said Baekeland, he would have fallen flat. The trouble with the story, repeated in a *New York Times* obituary, is that Nepera was not Baekeland's to sell. Eastman bought it in June 1899 in Yonkers after three days of bitter negotiation with Jacobi, who stuck out for his price of $750,000, provoking Eastman unattractively to write, "Mr. Jacobi, who is of Semitic origin, with all the characteristics of his race, evidently thinks the combine must purchase his business at his own price. . . ." Eastman yielded; Nepera was an essential part of the deal he had made with the Belgian cartel to control the American market.

Baekeland's share of the $750,000, the equivalent of $25 million in 2000—made him a rich man, just nine years after landing. Eastman admired the immigrant "Doc," and offered him a consultancy. Baekeland probably would have stayed on as chief photochemist and remained a minor figure in history, but Eastman gave that job to someone else, so Baekeland, somewhat disappointed, took himself off to the Paris Exposition of 1900. Two years later, still only 37, the new American citizen who christened his son George Washington came home to "retirement" in Harmony

Park in northern Yonkers, away from the clamor of the city. He lived at Snug Rock, a turreted mansion high above the Hudson where Celine (now called Bonbon by everyone) grew prize roses, painted in oils and entertained on a grand scale.

Baekeland had no appetite for such elegancies. His diaries express a constant yearning for "a simpler and unemotional life." He wrote: "My wife cannot live without some so-called society, a stupid conventionalism and the unwarranted cause of all our complicated living. What do we want such a large house for and why all the servants? Why all the complicated trash of unnecessary furniture?" He indulged himself with one of the earliest automobiles, a 4 horsepower vehicle, coughing its way through the quiet streets of Yonkers, and he pondered its engineering. He made all kinds of wine and he was prolific in writing letters and articles on photography, but he was most happy mixing chemicals in a little lab he fashioned from an old stable across the back lawn. He had become what would now be called a young fogy. He disliked new clothes, ate out of tins and preferred sneakers to shoes even as a reluctant guest at Celine's evenings. He frowned on her playing the piano and deplored makeup on women. He exulted in American ideals, but his temperament was patrician and aristocratic. "The doctor," the two children learned, must never be interrupted. He loved his family, but the flame closest to his heart was a Bunsen burner.

Baekeland did not have much initial success as the gentleman chemist of Snug Rock. At the turn of the century, a number of chemists in Europe and America were galvanized by the awkward habits of the female *Laccifer lacca* beetle. Found only in India and Burma, the beetle's excretions were the only source of shellac, initially valued as a hard varnish for wood products but in escalating demand as an insulator of electricity. Without shellac, electrical devices were near impossible to build, but 15,000 female lac beetles required six months to produce resin for a pound of shellac and they died in the process. A dozen chemists scattered over Europe were trying to find a substitute. It was a chaotic scene. Stimulated by the vivid reaction of mixing phenol and formaldehyde, Germans, Englishmen and Frenchmen were trying hit-or-miss tactics of mixing in various solvents and condensation agents, in vacuo and at atmospheric pressure, and with and without heat. They ended up with smelly sticky syrups that could not be applied as an insulating lacquer or dye, or took ages to dry, or with unmanageable

NEWLY RICH: A stereoscopic view of Leo and Celine Baekeland on the porch of their grand new home in Yonkers around 1900 with daughter, Nina Rose, and son, George. Baekeland was a loving but autocratic father.

PIONEER: He was a very early motorist, intrigued by the inside of his 4-horsepower vehicle, proud to take the family on noisy outings.

solids that looked like frozen beer. In 1902, Baekeland took up the trail and followed it intermittently for the next five years, with the assistance of Nathaniel Thurlow. Baekeland's notebooks record failure after failure. He interrupted himself to try to make interesting foods out of soybean, gave up and toyed with the effect of X rays on organic compounds, tried to make an improved nitrocellulose base for movie film, researched production of the electrolytic cell for Elton H. Hooker and came back again to phenol and formaldehyde.

When he failed to find a solvent for the unmanageable solids that emerged, as every chemist before him had failed, he set about systematically mapping the roles of all the variables in the reaction. This was a different approach from the dye researchers who identified products by chemical analysis, or the Celluloid engineers who just moved on from one solvent to another.

In 1906, following the lead of others, Thurlow produced a feasible shellac substitute they called Novalac. Thurlow was keen to explore its potential as a hardening varnish; Baekeland came to think it a dead end because they couldn't find a way to make insoluble products with it. In June 18, 1907, when Thurlow was away, Baekeland returned to the lab with an idea he says

Thurlow had discouraged as eccentric ("He looked at it too much from a one-sided standpoint. He wanted soluble products"). Baekeland was indeed to try something rather odd: Mohammed had tried to walk to the mountain, now he would make the mountain come to him. "I reasoned that if nothing could be done with the substance when it was once produced in flask or any other vessel, I should attempt to carry out the reaction so as to produce the substance right on the spot where I wanted it." Instead of creating a mixture, and then applying it to the object, Baekeland soaked two five-inch pieces of wood in equal volumes of the two intriguing chemicals and heated the impregnated blocks. The other phenol experimenters had worked at temperatures of under a 100°C because the heat speeded an uncontrollable reaction. Baekeland set his oven at 140°–150°C.

He tried seven different ways of varnishing his bits of wood that June 18 at Snug Rock and didn't go to bed. The result was one of those glorious failures like Alexander Fleming's discovery that his growth of bacteria had been spoiled by the penicillin mold. On the next day, Baekeland looked at the wood again and stopped. His diary reads: "The surface of wood does not feel hard although a small part of gum that has oozed out is very hard." Many clever men would not have bothered with that excrescence. The German chemist Adolf von Baeyer of the Bayer aspirin, who was then famous for synthesizing indigo, had noted that mixing phenol and formaldehyde threw off "a colorless resin." He discarded it as worthless because it did not crystallize; without crystallization it was not possible to establish accurately the chemical composition for large-scale production of a dye. Another German chemist, W. Kleeberg, also gave up because his product, being insoluble, defeated chemical analysis. Baekeland, like Edison, was less interested in the chemical structure of a substance than its potential usefulness. Could anything useful be made out of the "gum" that had oozed out from his failed attempt to create a varnish in situ? The Eureka moment was only the spark of a long exegesis for the fastidiously methodical Baekeland. What would happen if the fast reaction produced by the high tem-

perature was stopped before it turned the wood soft? What would happen at temperatures of 300°–400°C? Or pressures as high as 100 pounds to the square inch? What would result from incremental variations of the mix of phenol and formaldehyde? His notes have no record of sleep for the four days and nights he spent in his little lab. He was wholly absorbed boiling the chemicals in varying proportions, trying out all the reactions with different filler materials in both closed and sealed tubes. He discovered that dividing his "cooking" was a key. Before a wood specimen was pulped by the mixture, he transferred it to a "Baekeliser," a sealed vessel like a large egg where he could apply heat and pressure; the idea of applying high pressure to chemicals had been tried by a Russian experimenter only in 1903. Baekeland ended up excited by several different products which he labeled A (the liquid condensate first produced); B (a soluble rubbery product); C (an infusible, insoluble hard gum); and D (insoluble in all solvent and does not soften). Of D, he wrote: "I call it Bakelite and it is obtained by heating A or B or C in closed vessels."

Thirty-three pages of notes follow that sentence. He called what he had produced *oxybenzylmethylenglycolanhydride.*

It was to become so much more than a varnish or substitute for shellac insulation, or the perfect material for molding. It was the first true synthetic, the first material made by man and so superior in its chemical, mechanical and physical qualities, so manifold in its uses, so swift and inexpensive, it was swiftly to transform modern living. It was not, of course, the first plastic material suitable for molding. Alexander Parkes in England made Parkesine from pyroxylin and oil in 1855, and there were others. But the most notable of them, Celluloid, invented by John Wesley Hyatt and marketed in 1872, was derived from existing materials and had severe limitations. Its thermoplastic virtue, meaning it softened when heated, made it easy to mold, but rendered it vulnerable to stress or heat. A lighted cigar applied to one of Hyatt's celluloid billiard balls would at once result in a serious flame. He wrote: "Occasionally the violent contact of the balls would produce a mild explosion like a percussion cap.

We had a letter from a billiard saloon proprietor in Colorado, mentioning this fact and saying he did not care so much about it, but that instantly every man in the room pulled his gun."

Bakelite, as it was developed by Baekeland in succeeding experiments, would not catch fire, melt or break, it would not conduct electricity, and it was quick and cheap to make. It was the first thermoset plastic; once poured into a mold as a liquid it could

name Bakelite was promoted. He was thrilled to meet "the ever smiling" Charles Steinmetz at General Electric and interested the company in Bakelite insulators. In December, he sold 100,000 insulator pots to the New York Central Railroad to replace porcelain. In March 1908 he got the congratulations of John Hyatt for making a perfect billiard ball.

Billiard balls and insulators were the merest beginnings. In 1908 a New Jersey

that could be produced after a few hours in a mold under high temperature and pressure.

Automotive engineers fell on a product that could be molded as they wanted for magneto couplings and timing gears, and keep its shape despite heat and physical stress. The resourceful Charles "Boss" Kettering used Bakelite for his "Delco" ignition and starting systems because it solved the tough problem of short-circuiting.

Mary Pickford, a beneficiary of Baekeland, as were all women who found cranking a car too much for them. Bakelite was essential for Charles Kettering's electrical self-starter in automobiles.

Henry Ford, another beneficiary of Baekeland's inventiveness, demonstrates the toughness of the plastic deck lid (trunk panel) on a 1941 Ford.

be molded into all sorts of different shapes in a minute or two and once molded under pressure retained its shape forever. It was revolutionary material, triggering the search for other thermosetting synthetics. It was the harbinger of the Synthetic Century.

Baekeland did not know the chemical structure of his discovery that June day. He was a disciplined researcher and his studies were the foundation of modern polymer science, but his innovative genius was pragmatic. "I have not slept well . . . Almost every day since I invented Bakelite I have been thinking about the best method for developing this into a substantial business. . . ." His first instinct was as a marketer. He patented his invention in broad terms, then went around the country trying to excite people in no fewer than 43 industries. He offered the liquid form at 25 cents per pound in large lots on condition the

company found it could use Bakelite to make bobbin ends molded to a tolerance of plus or minus one-thousandth of an inch. It meant that for next to nothing Bakelite could be used to make millions of parts in millions of products. It was the essential element in the evolving era of interchangeable parts, and it was a crucial moment in the birth of the electrical and automotive industries. The many new electrical devices needed cheaply made parts that would not conduct electricity or catch fire. Soon, Bakelite was everywhere adding innovation to the innovative industries. It was in toasters, washing machines, electric irons, vacuum cleaners, shavers, ventilators, lamp sockets, headphones. It was on the subway tracks as the third-rail insulator (1910); it covered meters (1914); it was the circuit breaker insulation for elevators. When America entered World War I, Bakelite engineers perfected plastic propellers

Designers loved Bakelite for fashioning steering wheels, door handles, instrument panels, gearshift knobs and radiator caps.

Design and Bakelite formed a marriage of mutual convenience. In the '20s, with the emergence of industrial design as a profession, Bakelite was more than the material of choice for Art Deco or Moderne. It set the streamlined aerodynamic style, partly because it was more difficult to mold Bakelite into sharp angles. The writer Stephen Fenichell called it "the signature medium of a new hardboiled, hard-drinking 'lost generation,' the sophisticated urbane set that devoured the original *Vanity Fair* and giggled at the quirky cartoons in *The New Yorker*." Those elegant cigarette holders were Bakelite, the millions of futuristic radios were Bakelite inside and out, the Kodak Brownie of the early '30s was Bakelite, and Henry Dreyfus won the Bell Telephone competition for a telephone that

THE NEW STYLE: The advertisement in Fortune in 1939 suggests the versatility of Bakelite. Coco Chanel launched a line of Bakelite accessories in Paris in 1925 and Bakelite costume jewelry retains its appeal.

would combine the speaker and microphone in one handset with his classic black rotary dial phone in the "rugged, durable, phenolic resin" specified by Bell, i.e., Bakelite. Parker Pen adopted Bakelite and in 1927 invited a crowd to watch its Duofold brand with a Bakelite barrel flung to the pavement from 23 stories. "Picked up unbroken!"

The brilliance of Baekeland's invention, of course, was that it was nothing and it was everything. Its lack of specificity was its miracle strength. It could be used for almost any purpose.

Baekeland was a zestful inventor but a reluctant innovator. He had no wish, he wrote, to become "one of those slave millionaires in Wall Street." Instead of running a business himself, he planned to grant licenses. That did not work. Manufacturers kept making mistakes in production. Ironically, one of Baekeland's business innovations stemmed from these disasters. He wound up selling not just the product, but a host of consulting services. The knowledge and advice he sold became as important as the Bakelite itself. He was one of the first to realize the power of scientific expertise and profit from it.

By 1910 he had no choice but to lead his own manufacturing and distribution corporation and it grew rapidly. In 1913 he sold 700,000 pounds of Bakelite; 8.8 million pounds were sold in 1922. He stipulated

that he would do "no office routine." He was imperious; he expected deference and he got it. But he remained adventurous. The "Doctor," one employee recalled, "would go down to the laboratory and if the technical people had some crazy ideas they wanted to try, or they wanted to buy some tool we didn't have, he would tell them to go ahead. He would ride round the rest of the brass and deal with the fellows in the lab himself." For a decade, he had to throw himself into legal battles to defend his hundreds of patents against "pirates," as he called them. It strained him to the point where he wrote that he had come "to hate the whole Bakelite enterprise." He won every legal case, but he was adept at turning enemies into friends, offering partnerships and affiliations to the infringers who had something to offer. He indoctrinated his whole company with this attitude. The Bakelite Corporation flourished. It survived the Great Depression because he had insisted on liquidity: "I congratulate myself for having rebuffed the solicitation of bankers who advised me to put our stock on the market. I was also able to prevent our employees, directors, etc from being seduced by speculation on our stock. I know a large number of cases in which such speculation led to the ruin and the disorganization of certain otherwise well founded companies."

Why did Baekeland succeed where so many failed before him? His personal strength was an open, ever-questioning mind. Nothing was sacred. He accepted no assumptions; he tested. He not infrequently analyzed ingredients brought from vendors to make sure they were pure. He was not a grand hypothetical thinker, but he could visualize the potential value of research. He was methodical, taking step by step; his slogan was "commit your blunders on a small scale, and make your profits on a large scale." But he had courage. He was ready to risk derision for trying something new. L. V. Redman, the brilliant Canadian who challenged the Bakelite patents and became a collaborator, wrote: "In his research he was unorthodox, was guided by original thought, not precedent nor prevailing style. All his successes had their origin in clashes between observed fact and accepted theory." His scientific epiphany with Bakelite was to attempt the opposite of the obvious in speeding up a chemical reaction everyone else had been trying to slow down. All innovators have to develop antibodies to the contempt and jealous cynicism of the established. Baekeland had endured it as a young man when professional photographers trying Velox denounced him as a charlatan while the amateurs, who troubled to read the instructions, embraced it with joy. Science, and mankind, stands to gain much from the "ignorant" newcomer who does not know enough to shy away from lateral thinking.

Baekeland also nurtured formidable powers of concentration. He could switch from subject to subject and back again, but this was a facility that could mislead. During his gentleman-chemist period, he for a time forgot the lesson he'd taught himself when he was a struggling new immigrant in New York in the 1890s trying to survive as an independent chemical consultant when few knew what one was. He flitted between all sorts of "half-baked" ideas ranging from an electrolytic process for extracting tin to a safety explosive without developing any of them.

But personal qualities are rarely a full explanation of an innovator's success. Innovation does not spring fully formed from a vacuum. It emerges from the vibrations of a body of professional knowledge. Baekeland, for all his seclusion at Snug Rock, was not a monk. His career is a reflection of the social and economic matrix in which science and industry came together in America more readily and swiftly than anywhere else. Wiebe E. Bijker, the Dutch professor of technology and society at the University of Limburg, Maastricht, has analyzed how Baekeland's agility as an experimenter is traceable to the interlacing technological and social frameworks of his life, his cross-pollinations, the biologist would say. At first, Baekeland began his phenolformaldehyde work by following the same map as the Celluloid engineers, trying different solvents; they all failed. But then he was able to break away from that tradition and explore by a new map, by "new ways" that were really old ways for him. His early work as a photographic chemist informed his decision to test all the variables one by one. His experience as an embryonic entrepre-

neur in photochemistry and later as a chemical engineer prodded him to think beyond the lab to strategies for large-scale production. His awareness of the work for the Celluloid researchers gave him insights into the process of finding solvents. In his patent application for the Bakelizing process, he conceded that it was "similar in some respects to the vulcanization of

THE LONER: Holding lion cubs in the Berlin zoo when he was 60. In his later years, he spent more and more time away from his family.

rubber." He derived unconscious benefit, in short, from the fluidity and very American approachability among social and professional groups of photochemists, electrochemists, pioneering motorists and innovators. Many of the other famous innovators of the time were his correspondents, including the Wright brothers, Thomas Edison, Henry Ford, Willis Whitney of General Electric, the Du Ponts, Alexander Graham Bell. When Elmer Sperry attended Baekeland's presentation to the American Chemical Society on February 5, 1909, he presciently suggested that the engineering

potential of the discovery might be greater than the chemical.

In 1926, as Baekeland's first patents began to expire, a flood of new phenolformaldehyde products rushed into the market, including Duranoid, Durea, Lacanite, Lennite, Makaloi, Neolith and Textolite. Baekeland, though selling direct to industry, had always marketed his product to consumers, emphasizing the prestige and quality of the brand just as Intel was to do decades later. The strategy paid off for a while; consumers continued to insist on genuine Bakelite. By the late '30s, however, "the father of plastics" realized that new plastics were challenging and the future lay with large vertically integrated chemical powerhouses that could convert raw materials into a broad range of products for many markets. They all had large research and service divisions, and he probably saw that to compete Bakelite would need a large infusion of capital or a merger; George Washington did not want to follow in his father's footsteps. So in 1939, when he was 75, Baekeland sold the company to Union Carbide for stock valued then at $16.5 million. It still makes Bakelite eclectically for pot handles, clarinet mouthpieces and fashionable jewelry.

The reluctant innovator had been drawn away from his lab for thirty years but at last in retirement in Florida on the former estate of William Jennings Bryan he found the secluded simple life he had longed for, sailing a yacht along the intercoastal waterway, tending to his exotic fruits and flowers, playing with children on the beach, writing pungently about the importance of heredity and eugenics. He slept in a Spartan room with no furnishings except a plain white-cast iron bed. He ate off a Bakelite tray, serving himself split-pea soup from a can, with a little added seawater, or tinned sardines. By day he dressed entirely in white, like Mark Twain, whom he knew and admired. On very hot days, he demonstrated he had retained his grasp of scientific principles to surprised visitors by calmly walking fully clothed into his swimming pool or the ocean, then resuming a conversation exactly where he had left it. "The evaporation," he explained to anyone who dared to ask, "is what keeps you cool." Practical to the end.

WILBUR AND ORVILLE WRIGHT
The modest brothers who gave the world wings

1867–1912   1871–1948

B Y WHAT FREAK OF CHANCE was the mystery of flight, the perplexity of great minds for centuries, finally mastered by two young bachelor brothers growing up in a religious household in heartland America, one of them apparently handicapped for life, and neither of them with a college education? How bizarre was it that the great Age of Aviation should have its gen-esis at the oily workbench of a small-town bicycle repair shop?

Chance had nothing to do with it. The invention of the first powered airplane by Wilbur Wright and his younger brother Orville in 1903 was a triumph of reason and imagination, of hundreds of small painstaking deductions, risky experiments and giant conceptual leaps. It was also a celebration of empiricism and a vindication of human values of courage and brotherly love. They argued the bewildering variables of heavier-than-air flight with a vehemence that shook the family home but never their trust in each other. "I love to scrap with Orv," said Wilbur. "Orv is such a good scrapper."

Biographers have been divided on who was the real Mr. Wright. Now and again,

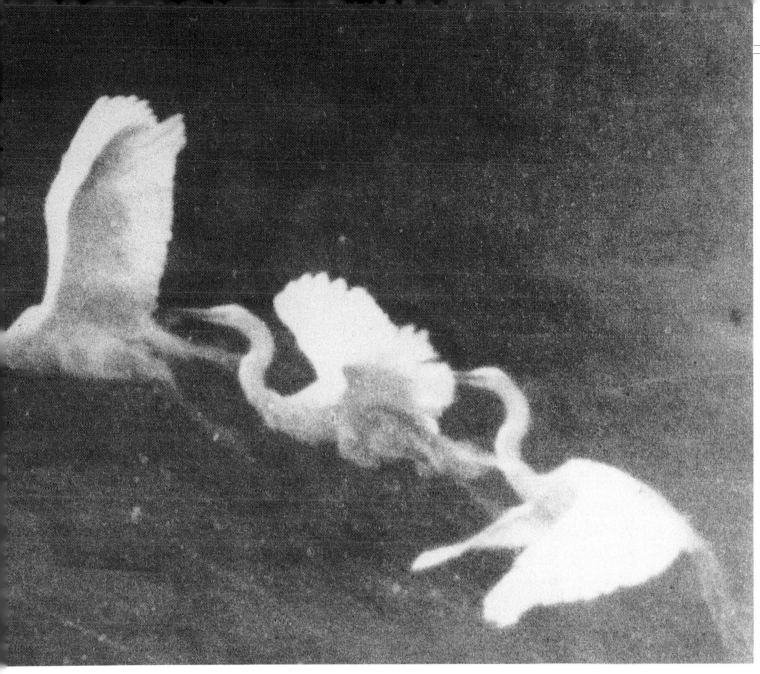

WILBUR'S BIRDS: A photograph of a heron in flight by the French physiologist and photographer Etienne-Jules Marey. The first thing Wilbur did was take down Professor Marey's book *Animal Mechanisms* in the home library.

Orville had to remind Wilbur to say "we" when he said "I." Wilbur offered only generalization when he wrote about their partnership shortly before he died at the age of 45: "My brother Orville and myself lived together, played together, worked together, and in fact thought together." Three decades later, in 1943, Orville approved an official biography by Fred Kelly, an Ohio journalist, in which he allowed himself to be identified as having "a little edge on Wilbur in the importance of suggestions offered," a con-

clusion that provoked a 1975 biographer, John Evangelist Walsh, to accuse Orville and Kelly of covering up Wilbur's primary role.

Two volumes comprising more than 1,000 pages of correspondence and notes released by the Library of Congress in 1953 suggest that individually both men contributed important insights, that their partnership was truly creative and sustained through tensions and temptations that would have broken up conventional corporations, but that Wilbur's was the originating, controlling intelligence. Sickness denied him a university education, but a thirst for knowledge, a devotion to reading and instincts of honesty and self-reliance led him to preside at a marvelous marriage of the theoretical and the practi-

cal. Orville was more daring as a pilot and prolifically inventive, but it was Wilbur who first had the certain vision that they might fly at a time when many scientists believed flight to be impossible, when aviation experimenters were routinely killing themselves and when the efforts of the previous 100 years seemed to be leading nowhere. Yet it is unlikely that Wilbur, without Orville, could have resolved the complexities, a judgment that remained the settled conviction of their most important mentor, their father. The aviation specialist and biographer Fred Howard made the point well: "Wilbur and Orville were among the blessed few who combined mechanical ability with intelligence. One man with this dual gift is exceptional. Two such men

ORVILLE: More mischievous as a boy than his older brother, and more outgoing as a man. He was the first to show business enterprise—though Wilbur was originally the initiator of the great project.

whose lives and fortunes are closely linked can raise this combination of qualities to a point where their combined talents are akin to genius."

Far from being semiliterate mechanics who stumbled on the secrets of aviation by random trial-and-error tinkering, the brothers were disciplined researchers who sustained each other in pursuit of a singular vision. From the start, they made a decision of breathtaking simplicity but one that escaped so many: They would learn to fly before they built a machine that could fly. In the excitement over the invention of the world's first powered airplane, it tends to be forgotten that it also required someone brave enough to be the world's first test pilot. Both brothers had raw courage, but they were not reckless. They weighed the odds as best they could, but inescapably they risked their lives when they committed themselves to "the uncertain embrace of the air," in John Walsh's phrase. They had to fly or crash. Orville Wright reflected years later, "I look with amazement upon our audacity in attempting flights with a new and untried machine."

They fashioned with their own hands all the gliders in which they learned to fly, but they read much, observed much and pondered more before assembling their bits of wire and wood, screws and cloth, and then reassembling them again and again. In all their work, with airplanes as with bicycles, they translated their perceptions by methodical and meticulous craftsmanship, sensitive to the feel of materials and always ready to graft an idea from an apparently unrelated technology. They moved inch by inch but in a straight line.

The other virtues the brothers shared were curiosity, industry, ethical sensitivity, modesty and patient resolve. They were impressively stoical when they came to be defamed as liars and fakers. They knew who they were; the emotional security of their childhood was the thread that never broke. As befitted the dutiful sons of a bishop, they were formal and very proper. (Out of respect for their father, they never worked on a Sunday.) They had tempers, but impressed their chief mechanic by being able to vent without profanity. They wore suits to their bicycle shop, Orville more dressy with sleeve cuffs and an apron of blue-and-white ticking, and they always emerged immaculate from the grime. On every practice glide and flight in the wilds of Kitty Hawk, they wore a fresh stiff white celluloid collar and necktie.

They were hard to read. Wilbur was a man of few words, but behind the poker face was a mind brimming with ideas; his writing was witty and erudite. Orville was more approachable and more of a talker, the tease and practical joker of the family. He hated to put pen to paper, but when he wrote letters home they were warm and jolly. He grew a walrus moustache, unthinkable embroidery for Wilbur, whose bare cerebral dome was emblematic of his intellectual rigor. They were self-possessed young men, but both noticeably nervous around young women.

As inventors, they were astoundingly swift, three years from kite to success at Kitty Hawk; as innovators, translators of invention into an operating business, they seem in retrospect astoundingly slow, but they emerged in a world that was astoundingly unready to recognize and cherish one of the most transforming acts of the 20th century.

THE EFFECT OF A DEATH

In later years Orville said the brothers owed much to the "exceptional advantages" they had growing up. Their advantages were love and literature. The Wrights were not rich, but comfortable on the salary of their father, Milton Wright, who earned rather more than the median income as he rose from circuit preacher to bishop of the Church of the United Brethren and the editor of its weekly newspaper. With his long white beard, he looked like a repressive patriarch, but he was a forthright, open-minded intellectual who in his youth before the Civil War had been a passionate crusader against slavery. His church was a movement that had grown away from Calvinism and emphasized the individualistic and democratic spirit of the frontier. His upstairs library in their home from 1884, a white frame house in a streetcar suburb of Dayton, Ohio, included books by agnostics, and mother and father encouraged their four sons and only daughter, Katharine, to read widely and think deeply. In the downstairs library, there was

WILBUR: "I think there is a slight possibility of achieving fame and fortune from it," Wilbur wrote his father. A British reporter remarked on Wilbur's "extraordinarily keen, observant, hawklike eyes" in a fine-drawn face.

Gibbon's *Decline and Fall of the Roman Empire,* Plutarch's *Lives,* sets of Sir Walter Scott and Washington Irving and two encyclopedias.

Their mother, Susan Koerner Wright, inherited and bequeathed a more technical DNA. She was the daughter of a German wagonmaker who migrated to the United States in 1818 because he was outraged by Prussian militarism. At a time when few women went to college at all, she studied science and literature at Hartsville College in central Indiana (where Milton Wright had been a teacher) and was the top mathematician in her class; "Mother," said the family, "could mend anything," but she was more than a fixer. One winter she made a sled that she promised her first two sons, Reuchlin and Lorin, would run faster than anyone else's because its narrow shape and closeness to the snow reduced air resistance.

The initial interest of the bishop's boys in flying was sparked, according to Orville, when he was 7 and Wilbur 11. Their father came home from one of his frequent church trips with what we would now call a toy helicopter, a flying object made of paper and bamboo and powered by a twisted rubber band. It was the creation of a brilliant but short-lived French experimenter, Alphonse Pénaud. Wilbur tried to make larger versions but none of them flew well. (Later in life he learned that doubling the dimensions cubed rather than doubled the power required to make his machines fly.) They fared better selling Orville's homemade kites to friends. From early childhood, they were exceptionally good with their hands. Wilbur was only 14, and Orville 10, when they made a foot-powered lathe while living near Grandfather Koerner's farm in Richmond, Indiana. When the family moved to 7 Hawthorn Street, Dayton, Ohio, the two teenagers fashioned symmetrical curves on the woodwork forming a spacious wraparound porch. Technology and science interested them both—they developed their own plate-glass photographic negatives—but Wilbur, the more assiduous scholar, was headed for the church and Orville for commerce.

We owe the invention of the airplane to what seemed at the time a disastrous setback, an accident in a hockey game on a frozen pond in 1886 that smashed in 19-year-old Wilbur's face and left him with chronic stomach pains and nervous palpitations. He had been among the best athletes in the county, a daring balancer on the horizontal bar and a fine figure skater, but after the accident he was too much of an invalid to continue high school or go on to Yale Divinity School. A writer in *Popular Science* later reported: "It seemed to everyone that the boy was handicapped for life." For four years he was a frail ghost about the house. "What will Wilbur do?" Lorin asked, writing from college. "He ought to do something. Is he still cook and chambermaid?" What Wilbur did was devotedly nurse his mother when she fell sick with tuberculosis. Milton was much away with his church in turmoil. It was Wilbur who every day carried her tenderly from her bed to the downstairs parlor until the day she died at the age of 58 on July 4, 1889. In the same summer Milton lost a battle in the church against elements who wanted to compromise its hostility to Freemasons and their secret lodges. Milton led a breakaway group and Wilbur helped him in the legal battles over disputed church property. He was also reading, reading, reading.

Orville was the first to show business initiative. While still a teenager, he built a printing press. At 18, about the time of his mother's death, he started a little printing business and a weekly newspaper. It became a four-page daily when a revived Wilbur joined him in April 1889 as a crusading editor; the older brother argued for votes for women and against American expansion overseas. Within 16 months, they were crushed by the bigger Dayton newspapers, but by then they had become fascinated by a new technology sweeping the country: the replacement of the wobbly high-wheel (penny-farthing) bicycles with European-style "safety bicycles," which had same-size wheels and a chain drive. (See Pope, page 101) Orville fancied himself as a racer, but both brothers had the self-awareness to recognize they were better mechanics than athletes.

Wilbur opened a bike shop in the spring of 1893 and Orville joined him. They sold, rented and repaired bikes, then in 1895 began an auspicious partnership in manufacturing, handcrafting their own brand.

Business at the Wright Cycle Company boomed in the bike craze. Dayton was full of such little workshops; with a population of only 60,000 but more patents per head than any other city in the United States, it called itself the "city of a thousand factories." Orville wanted to branch out into making automobiles, as their friend and part-time employee Cordy Ruse had done with the first horseless carriage in Dayton. According to Orville, Wilbur's response was, "To try and build one that would be of any account, you'd be tackling the impossible. Why, it would be easier to build a flying machine!"

Thwarted from his chosen career, ever more conscious—without vanity—of his mental abilities, aware even in 1894 that the bike craze was waning, Wilbur felt, at 27, that life was passing him by. He brooded about making a late entry into college. He wrote his father, "I do not think I am specially fitted for success in any commercial pursuit." It was in this mood, searching for meaning in his life, that in September 1894 Wilbur looked in some awe on a few pages in a new illustrated magazine, *McClure's* (page 185). Photographic reproduction of motion had been achieved only in the preceding few years, as had the ability to reproduce half-tones, so the impact of what Wilbur saw was amplified by the technical virtuosity. The photographs were headlined "The Flying Man" and they froze a moment in time when a red-bearded gentleman in knickerbockers dangled from a hang glider of his construction, having launched himself into the wind from a hilltop. The brothers were thereafter hooked on the adventures of the intrepid aviator, a 46-year-old mechanical engineer by the name of Otto Lilienthal, who made more than 2,000 glides in monoplane and biplane creations and managed to stay aloft for up to 15 seconds. Then on August 9, 1896, their hero's glider stalled at 50 feet and the ensuing plunge killed him. In the following month, Orville, too, lay close to death, infected with typhoid.

The two calamities were ever after associated in their minds. As Wilbur and their 22-year-old sister, Katharine, sat in Orville's hot little bedroom through six weeks of pain and delirium, Wilbur was on fire himself. If God spared Orville, the two of them could dedicate their lives to taming the forces that had killed Lilienthal. There was nothing mystical about the way they tackled the problem of flight. The first thing Wilbur did after Lilienthal's disaster was to go to their home library and take down a book he had read several times: *Animal Mechanisms* by Professor Etienne-Jules Marey, the French physiologist and photographer of bird flight. As Orville recovered, Wilbur went off to a wild location in the Pinnacles in Dayton where he lay on his back for hours training his spyglass on birds swooping and soaring. Over the next two or three years, he plowed through every relevant book in the Dayton libraries, and when he had exhausted every source, he wrote to the Smithsonian Institution. "I wish to avail myself of all that is already and then if possible add my mite to help on the future workers who will attain success." On June 6, 1899, a package arrived at the Wright home packed with pamphlets, Smithsonian publications, writings on aerodynamics and a list of suggested books. The brothers read avidly between two schools of thought, the pioneers like Lilienthal who believed in graduating to powered flight from unpowered gliders, and the proponents of power who concentrated their efforts on propulsion. Samuel Pierpont Langley (1834–1906), an astrophysicist and mathematician who was secretary of the Smithsonian no less, had built six model planes powered by steam engines. In May and November 1896 he had successfully catapulted two models he called Aerodromes from a barge on the Potomac, the last traveling 4,200 feet in 1 minute, 45 seconds. War with Spain in 1898 prompted the army to give Langley $50,000, in secret, to develop a plane capable of carrying a man.

The Wrights were not scared off by Langley's apparent momentum with his Aerodromes. They found themselves more in tune with a friend of Langley's, Octave Chanute (1832–1910), a railway engineer who in middle age had dedicated himself to aviation. Chanute believed that Langley was mistaken in putting too much emphasis on launching a plane and too little on keeping it in the air. At the age of 60, Chanute put his theories to the test, venturing into the air by proxy. He persuaded three young men to try out a variety of his

THE PICTURE THAT STARTED IT ALL

It was a photograph of the German experimenter Otto Lilienthal (above) that first triggered Wilbur's interest in flying. Lilienthal launched himself from 50-foot hilltops and might glide for 1,000 feet borne along on 20-foot wings of woven muslin. He eventually broke his neck. (Below) The American inventor Hiram Maxim (1840–1916) believed that the key to flight was generating power. In England, on July 31, 1894, he launched a four-ton machine with a steam engine generating 300 horsepower, with himself and two passengers aboard. They rose only a few inches from the ground, along a guide rail, then an outrigger broke through the guide rail and Maxim cut power lest the behemoth crash. The machine did not fly again, but it was the first time a heavier-than-air machine had lifted off the ground under its own power.

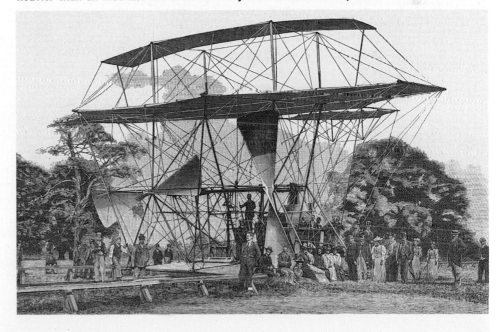

GOOD TIMES: Orville (standing) loved to play practical jokes when friends came around to 7 Hawthorn Street—their home from 1884–1914. The group is amused by photographs taken by Orville and Wilbur, courtesy of George Eastman. Katharine (inset) kept house and taught Latin at high school, but Orville cut her out of his life when at the age of 52 she married an Oberlin professor. In 1929, when she was stricken with pneumonia, Orville refused to visit her, then relented on the insistence of his older brother Lorin and went to her bedside before she died.

two- and three-wing hang gliders with braced wings and a fixed tail. The longest they stayed up was 10.3 seconds. One of the engineer-fliers, Augustus Herring (1867–1926), an overambitious Georgian the same age as Wilbur, claimed he stayed up for 14 seconds—when nobody was looking.

Wilbur and Orville were scathing about the literature, but they absorbed more than they realized at the time from the long line of experimenters before them, certainly the conclusion that wing-flapping contraptions were futile and that fixed wings with a curve had better lift than flat wings. It was Wilbur who made the conceptual breakthrough. Man had been kept earthbound by three areas of difficulty—lift, propulsion and control. The literature suggested the aerodynamics of lift were in some sort of soft focus, and propulsion would soon be possible as the new internal combustion engine supplanted steam. But the third area, control, was in darkness, and

to Wilbur it was fundamental. How was the operator of a flying machine to become a pilot rather than a projectile? Until they learned how to balance and steer an aircraft in flight, he wrote, "all other difficulties are of minor importance."

The challenge Wilbur set himself was to build an aircraft in which control was integral to every stage of the design. From his reading and deep thought, he reasoned this meant he had to enable the pilot to exert coordinated control of a craft in all three axes of motion—its pitch, nose up or nose down; its balance, the lateral roll of the wings; and its yaw, the side to side movement of the nose. None of the other experimenters had appreciated the essential interrelationship of these three movements. Others who tried to make winged aircraft, like Langley and Chanute, had done everything they could to keep the wings straight, stabilizing the aircraft on the horizontal plane so that there was less

danger of being flipped over by a clumsy move or wind gust. Turns—if the putative pilot ever stayed up that long—were to be achieved by working a rear rudder, yielding the wide flat turns typically produced by working the rudder of a boat.

Dramatically, Wilbur took the opposite tack. He would not seek stability, but a dynamic balance: His aircraft would roll, but it would be a roll controlled by the pilot for the purpose of achieving a banking turn—the kind of banking turn made by a cyclist. It was a stunning piece of lateral thinking rooted in Wilbur's experience as a bicycler. The bicycle is inherently unstable, yet totally controllable. Small adjustments could achieve equilibrium—or destroy it. Balance in motion was the key both to cycling and flying, with the awkward difference that flying required balance in three dimensions and the penalties for failure were apt to be final.

But how could an airplane be a bicycle? Turns depending on movements of the pilot's body weight (Lilienthal's method) would limit the size of the craft—and put a premium on obesity. What could be achieved by the dynamic interactions of airflow and wing? While other aviation experimenters concentrated on the energies that might get some kind of craft off the ground, Wilbur asked the simple question he would ask of a cyclist trying one of the bicycles he made: Once you are moving, how do you control this inherently unstable machine?

Wilbur said inspiration came from watching birds over Ohio's Great Miami River, a romantic idea to which Orville in later life added a dash of skepticism. "Learning the secret of flight from a bird," he wrote, "was a good deal like learning the secret of magic from a magician. After you once know the trick and know what to look for, you see things that you did not notice when you did not know exactly what to look for." Biographer John E. Walsh has suggested that while the rapid flutterings of a bird's wings may indeed be too hard for the human eye to discern, many hours of observation stirred something in Wilbur's imagination. Wilbur himself said he came to realize that the oscillation of a bird's wings—the tilting from side to side—was occurring so fast as not to be explained by the shifting of the bird's body weight. "Some other force than gravity was at work," Wilbur wrote. "The thought came that it had adjusted the tips of wings about a lateral transverse axis as to present one tip at a positive angle and the other at a negative angle." Simply put, Wilbur concluded that if the underside of one wing of an aircraft could be exposed at an increased angle, the wing would rise because of increased air pres-

sure; likewise if the topside of the opposite wing were simultaneously exposed, that wing would drop. An entire flying machine could be made to turn as a result, banking as one wing lifted and the other dropped.

Wilbur called this effect "the torsion principle." But how could it be introduced into the design of an aircraft whose fixed wings had to be strong enough to hold up pilot and plane against onrushing winds? It was a baffling question. In the third week of July 1899, Wilbur was alone in the bicycle shop when a customer came in with a flat tire, unable to cope with the new technology of the day—an inner tube. Wilbur sent him happily on his way with a tube in his tire, and was about to throw out the 2-inch-by-10-inch box in which the replacement tube had been packed when an idea struck him. Without its end flaps, the box resembled the wings of a biplane. He tore off the ends and played around, squeezing the two diagonal corners of the box in such a way as to effect his torsion principle. By twisting the box in a slight helical motion, one "wing" dropped, the other simultaneously rose, just like the buzzard's wings. He had discovered wing warping, a dazzling concept. If a pilot could warp the wings of a real plane in this way, he could, on Wilbur's theory, make tight and safe banking turns.

For the next week, Wilbur did little aside from making a biplane kite with a five-foot wingspan. On the evening of July 27, 1899, he took it to a field. Watched by a group of small boys, he stuck four pegs in the ground with cords leading from each peg to the wingtips. He wanted to see if the kite banked when he tugged on the appropriate cord to warp the wings. It worked brilliantly. One flight, however, ended so precipitately the boys had to throw themselves to the ground. What Wilbur had by chance reproduced was the crash that killed Lilienthal and was nearly to kill him at Kitty Hawk. The fatal flaw was that he had shaped his wings on Lilienthal's reasoning, following the arc of a circle and not the curve of a parabola. The difference was crucial, but Wilbur didn't know it yet.

The next step was to build a glider capable of carrying a man. It took a full year. The brothers had some idea of the aerodynamic requirements from calculations Lilienthal had published of the lift produced by wing area, curvature, angle of attack (also called angle of incidence) and velocity. "From the tables of Lilienthal," wrote Orville, "we calculated that a machine having an area of a little over 150 square feet would support a man when flown in a wind of sixteen miles an hour." Still, they felt they needed more guidance, so between repairing cycles Wilbur took up his pen and on five pages of the Wright Cycle Company's lower-decked blue paper wrote to Chanute. "For some years I have been afflicted with the belief that flight is possible to man. My disease has increased in severity and I feel that it will soon cost me an increased amount of money if not my life." He asked Chanute for advice. "The problem is too great for one man alone and unaided to solve in secret." Chanute was responsive. It was the beginning of a long dialectical relationship in which Chanute was the encourager rather than the educator: He was a generous man, but soon enough found it hard to adjust to the reality that the brothers he always regarded as pupils overtook the master.

The glider the brothers built was deceptively simple in appearance, light at 52 pounds unmanned, but strong and easy to adjust. Control wires ran through pulleys to connect the end of each wing to a foot pedal. The pilot's pressure on the pedal would change the shape of the outer panel and put the aircraft into a controlled turn. Wilbur thought even a small deflection would enable the pilot to bank the aircraft.

Once in the air, balancing a flying machine was simple—in theory. The equilibrium of straight level flight is achieved when the

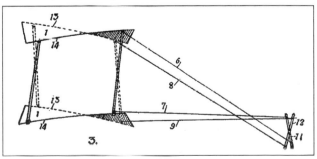

WING WARPING: How Wilbur tested the principle with a kite, lines and sticks in the ground

center of air pressure against the wing coincides with the center of gravity—think of it as the center of balance of the aircraft. In practice, as Wilbur later wrote, "there seems to be an almost boundless incompatibility of temper which prevents their remaining peaceably together for a single instant." Coping with the variables that produced a constantly roving center of pressure was to be the principal conundrum of flight. Everything the Wrights had read said that the center of pressure—the point of focus of all the forces pushing against the wing—moves steadily backward as the plane veers upward. It does move progressively with a flat surface, but not over the surface of a wing with a curved surface called a camber, an arch or curve in which the center is higher than the ends. The smaller a camber the less the wing is curved. On a cambered surface the center of pressure travels very sharply to the back as the nose of the plane lifts, playing havoc with the plane's balance. Wilbur knew nothing of this when he fashioned his glider wings with a camber close to 1-in-22, a shallower camber than the 1-in-12 favored by Lilienthal. He reasoned the relative flatness would slow the dance of the center of pressure along the surface of the wing. For the same reason, he put the deepest curve three inches from the front of the wing.

At the front of the plane, they built a horizontal stabilizing surface, a pair of small movable wings the pilot could control so as to keep pace with the roving center of pressure. Wilbur called this device a "horizontal rudder," more commonly described as an elevator. For straight flight, a pilot would keep the elevator horizontal. To take off and climb, he would deflect the elevator upward, and move it down for a descent. With this, theoretically, Wilbur had made the world's first flying machine a pilot could control both longitudinally and laterally. It had cost $15 in materials. He had no idea if it would work.

The answer lay in the winds of the North Carolina skies. Wilbur needed a long windy stretch of ground and the U.S. Weather Bureau suggested Kitty Hawk on the 100-mile Outer Banks of North Carolina. Its beach was about 1 mile wide, and ran for nearly 60 miles with few trees or hills—and the sand offered a soft landing.

There was no hotel or house to rent, but Kitty Hawk's fisherman-postmaster, Bill Tate, promised a cheerful welcome. Wilbur left Orville in charge of the bicycle shop and on September 6, 1900, set off by train to Old Point, Virginia, and then by steamer from Hampton Roads to Norfolk, along with a large trunk carrying tools, metal fittings, spools of wire and a bale of French sateen for the wings. All that was missing were 18-foot spars for the edges of the wings, which Wilbur reckoned to buy at a lumberyard in Norfolk. Alas, on his arrival he found the longest spars the yard could supply were 16 feet and in pine, not white spruce. With total surface area of the wings reduced by 35 square feet, stronger winds would be required, making flying more dangerous.

Kitty Hawk was the next problem. When he arrived in Elizabeth City, 40 miles across Albemarle Sound from the Outer Banks, nobody had heard of Kitty Hawk. He hunted around for 36 hours and fell gratefully on Israel Perry, a local salt who had a decrepit flat-bottomed fishing schooner, *Curlicue,* and thought he knew where Kitty Hawk was. As the boat entered the sound just before dark, carrying Wilbur, Perry and a boy deckhand, a crosswind rose and *Curlicue* began to be pushed off course. Waves grew choppier and higher. They took on water. In the blackness of midnight, five hours out, and everyone still bailing frantically, winds reached gale force. Perry said he was going to make for whatever lee he could find. Suddenly the topsail ripped away. A few minutes later the mainsail tore loose. Only the jib sail remained. If that went, too, they would be helpless. Perry headed into the wind, risking capsize. For every minute of a desperate shuddering hour, their lives and the soaring future in Wilbur's trunk lay hostage to the tempestuous present. Perry found the shelter of a point after 1 a.m. and the exhausted party hid out there the rest of the night and morning. The following afternoon, having patched up the sails as best they could, they set out again, making landfall at sunset on September 12. Kitty Hawk was populated by no more than 60 people scattered in shacks in the sandhills. Nobody was about. The store was darkened. All three sailors spent a miserable night on the boat. Except for a jar of jam that his sister had packed for him, Wilbur hadn't eaten in two days.

The next morning Wilbur found his way to the sparsely furnished home of 40-year-old Bill Tate and his wife, Addie. Door to door, it had taken him a week, but the welcome was as warm as promised and Wilbur's material was undamaged. He set up a canvas lean-to and began obsessively assembling his glider in 100-degree heat. It took him 10 days to make the wings, borrowing Mrs. Tate's sewing machine to resew the white sateen coverings to the new wing length. Orville arrived on September 28 and the glider was finished in the first week of October. He had brought a large tent, the luxuries of coffee, tea and sugar and his mandolin for campfire melodies; Wilbur had a harmonica. The brothers set up camp on the dunes half a mile away from the Tates' and a 300-yard walk to fetch fresh water in a bucket. Orville cooked, Wilbur did the dishes. Their tent was roped to one of the few trees around, but they were up often enough in the middle of the night holding it down in blinding sandstorms.

The precise date they tried out the glider is uncertain, probably Wednesday, October 3, when winds were about the right velocity. With Bill Tate's help, they floated the glider as a kite, letting it rise 20 feet or so while they tried out maneuvers by pulling on cords. The performance was indifferent, but Wilbur could not resist trying the first manned but tethered test. He climbed inside the 18-inch cutout in the lower wing and lay facedown. No other experimenter had lain prone. With Orville at one side and Bill Tate at the other, each holding a rope attached to a wingtip, they ran forward. They released the ropes inches at a time, and the glider lifted. Fifteen feet off the ground, it bucked wildly, with Wilbur yelling, "Let me down! Let me down!" Orville, hauling his brother to safety, was disappointed. Wilbur explained: "I promised Pop, that I'd look after myself." They had all heard that the English glider pilot Percy Pilcher, who had fallen to his death in 1898, had been trying to fly a glider with exactly the same 165-foot lifting surface as theirs.

Wilbur calmed down and tried again. This time the ropes were held tightly, so the glider did not rise much above head-height. Gingerly, Wilbur tested the controls. He was able to make the nose rise and fall. He practiced for about half an hour and said he was ready to try another tethered flight but this time starting from a 15-foot wooden derrick they had erected on the sand. Wilbur soared, held only by the rope attached to the tower, the wind rushing at his half-closed eyes. Yet something was badly wrong. To stay up, the glider had to be angled up into the wind at about 20 degrees when it was designed to fly nearly parallel to the ground, at a 3-degree angle. They tied one of the tethering cords to a grocer's spring-weighing balance and with a little trigonometry were able to work out what lift and drag they were experiencing: They were getting only about half the lift they had calculated on the basis of Lilienthal's well-regarded formula. Recalled Orville, "When we got through Will was so mixed up he couldn't even theorize."

There were several days when they could not try anything in the absence of wind. They went back to observing seabirds. Kitty Hawk people remember the brothers running around the beach with arms outstretched. "They would imitate every movement of the wings of those gannets," recalled John Daniels. "We thought they were crazy." When the wind came, it was not helpful. It picked up the stationary glider and hurled it 20 feet so that it cracked and splintered. Five days were lost on repairs, two more in storms. It was at this moment, Orville wrote home to Katharine, that they "began to consider getting home." The sight of buzzards, sea chickens and bald eagles swooping in the air no longer heartened his brother. "Will is most sick of them."

They ended up changing the position of the elevator and managed to have the kite as glider stay up and level for an encouraging distance of 30 feet at about man-height. Having spent six weeks in on-off experiments, Wilbur was determined to try another manned glide—and this time not tethered to the derrick. It was a big risk: His glider's wings were six feet longer than the span Lilienthal and Chanute considered prudent for the pilot to have any hope of control, but on Saturday, October 19, the brothers loaded the glider on Bill Tate's wagon and lugged it four miles to the highest of three sand dunes known as "Kill Devil Hill." Wilbur climbed onto the lower wing for his first-ever attempt at a free glide. The wind was blowing 12 to 14 miles per hour. Orville and Tate ran down the hill for 100 yards before letting go of the wingtips. The glider rose about a foot above the ground and traveled about 50 yards before Wilbur tested the elevator for descent. It worked like a dream. He traveled another 30 feet on the sand after a gentle touching down. This first glide lasted about half as long as Lilienthal's longest, but it was perfectly controlled. It was also the first time someone had flown lying down. (Chanute had warned they might plow the ground with their noses, but Wilbur thought it worth the risk to reduce air resistance by 60 percent.)

Wilbur made about a dozen exultant glides that day, several of them up to 20 seconds in duration and more than 100 yards, surpassing Lilienthal. They seemed to have mastered the fore and aft balance that had defeated so many experimenters. Still, six weeks on the Banks had yielded about three minutes of practical flying rather than the hours of practice they had hoped for, and they were still baffled by the inadequacy of lift. It was time to go home. The brothers gave the glider to the Tates, who used the wood for their fireplace and the sateen for new dresses for their daughters. Wilbur and Orville promised they would be back with a new machine.

THE DAYTON HOME of the close-knit Wright family was now managed by their clever and beguiling sister, 26-year-old Katharine, who had graduated from Oberlin College and was teaching Latin in high school. With Katharine looking after them and their father, the brothers built a new glider in the workroom behind the bike shop in the first six months of 1901. In June they hired a clever machinist, Charlie Taylor, to help in the bike shop, income from which they very much needed. Their main concern in the new glider had been to get enough lift in low winds. They increased the wingspan from 17 feet to 22 feet, altogether yielding 290 square feet of wing surface. It was the same basic design but bigger—bigger indeed than anyone had ever made—and at 98 pounds twice as heavy as Glider No. 1. They also increased the wing camber to Lilienthal's 1-in-12 and moved the steepest part of the curve farther back in the camber.

The celebrated Chanute visited them in Dayton on June 26 and 27, every inch the professional engineer, his formal air enhanced by his white hair and trim goatee. Chanute was a passionate believer in the dissemination of knowledge. He wanted approval to tell the world what the Wrights were up to. They did not divulge details of construction but they told him much about their system of lateral and longitudinal control. He also wanted them to welcome two of his assistants to Kitty Hawk, Edward C. Huffaker and George Spratt, in the hope that the brothers would help with a new glider of his own design. Wilbur was not eager for anybody else to see what they were up to at Kitty Hawk, but Chanute had given them two instruments—a French handheld anemometer to measure the force and direction of wind and wing velocity, and a clinometer to measure the angle of the glider's descent or the angle of sand dune—so when he wrote assuring the brothers of the discretion of his assistants, Wilbur penned a remarkably open—and naïve—response. "We have felt no uneasiness on this point, since we do not think the class of people who are interested in aeronautics would naturally be of a character to act unfairly."

The brothers packed their glider in boxes and on July 7, 1901, headed back to Kitty Hawk and into the greatest storm in its history. The rain was still lashing the oceanfront when they lugged their equipment four miles over the soft sand to a spot at the base of the largest Kill Devil Hill dune, which they called Big Hill. Without fresh water—before they drove a pipe 10 feet into the sand—they survived the day on rainwater collected in a kitchen pan as it ran down the side of the tent they struggled to erect. The biblical rains were followed by a comparable plague of mosquitoes. Orville wrote to Katharine, "The sand and grass and trees and everything was fairly covered with them. They chewed us clear through our underwear and socks." No sooner had they settled in than they were assailed by Huffaker, who arrived a

MYSTERIES: The 1901 glider failed to live up to their expectations. Instead of rising steadily, it stalled or nose-dived after a few yards. Wilbur and Orville tested it as a kite and noted how the leading edge dipped down in strong winds. They retrussed the wings to a less sharp camber, then tried again and again.

few weeks in advance of Chanute himself with the bits and pieces of a three-decker glider he had made for Chanute out of paper and tubes. (Chanute's glider instantly collapsed, though not his self-esteem.) Huffaker's ability to squirt tobacco juice into a spittoon several yards away and to wear the same shirt for a week were not attributes the Wrights rated. It was an uneasy camp until another of Chanute's young men arrived, Dr. George Spratt from Pennsylvania, whom Chanute had designated for Kitty Hawk because he thought the brothers might need Spratt's medical training.

Each attempt to fly on Saturday, July 27, ended in a glide of only a few feet and a shower of sand. The hope had been that with Chanute's clinometer they could plot the undulations of the dune and fly safely just a few feet higher, but that was academic. There was just no lift in the machine. Wilbur moved his position back an inch at a time in the hope he would find the point at which the center of balance and center of pressure coincided. On the ninth attempt, when he was a full foot behind from where he started, the glider rose a few feet. He sailed a bumpy 315 feet in 19 seconds. Orville and the others ran up excitedly to find Wilbur shaking his head. The elevator, he told them, was not as responsive as it had been in the first glider. Everyone was for calling it a day while they worked out the problem—everyone except Wilbur. There was only one way to learn to fly and that was to fly again and again. This time Wilbur found himself carried up 30 feet or so, which was never the intention since they had all agreed it was prudent to fly only at heights where a fall would be bearable. Spratt ran under the glider yelling and waving his hands to make Wilbur aware of a danger he was too well aware of. The glider was stalling. It was, Orville wrote Katharine later, "precisely the fix Lilienthal got into when he was killed." Wilbur had the elevator fully down, but he still rose. He did the only thing possible, heaving himself forward toward the leading edge, hoping his weight would make the difference. The glider flattened out, and at about 20 feet headed down. At this point, Wilbur raised the elevator to get maximum lift in front. The machine parachuted gently to the ground, maintaining its horizontal position almost perfectly. He was proud of his performance, relieved that the front elevator had proved its value, and puzzled. Why was the pitch so volatile?

At this point, Spratt and Huffaker earned their baked bean suppers. Something was happening to propel the center of pressure away from the center of balance and convulse the fore-aft balance. That something, they suggested, was the new deep camber. It was Wilbur's genius to devise a test of a theory he was reluctant to accept. They took off a wing and flew it like a kite. In light winds the wing assumed

FRUSTRATION: Will this thing ever really fly? Wilbur stays lying on the sand after a short ride in the 1901 glider. "To the person who had never attempted to control an uncontrollable flying machine in the air this may seem somewhat strange, but the operator on the machine is so busy manipulating his rudder and looking for a soft place to alight that his ideas of what actually happens are very hazy."

an angle of six degrees to the horizon and pulled upward against the retaining rope. This meant the center of pressure was in front of the wing's center of balance. In medium winds, the wing stayed in horizontal equilibrium, but in strong winds, the leading edge of the wing dipped sharply down. Instead of continuing to move forward, the center of pressure had moved backward from just in front of the center of balance to just behind it—with devastating results. The brothers retrussed the wings to a less sharp camber of 1-in-19.

On Thursday, August 8, with considerable courage, Wilbur tried gliding with the new curvatures in a wind of 18 to 22 miles an hour. "At first," he wrote, "we felt some doubt as to the safety of attempting free flight in so strong a wind with a machine of over 300 square feet and a practice of less than five minutes spent in actual flight." He found he was able to make glide after glide, flying in a straight line but gracefully

following every bump and hummock and sometimes sailing high. That day and the next he made 30 glides, in one of them staying up for 17.5 seconds and traveling 390 feet.

The lift was still problematic, but the next day Wilbur was encouraged enough to try a turn. Watched by Chanute and the others, he tried his first turn at 12 miles per hour about 10 feet above the sand. He warped the wings slightly and the plane gently dropped on the left and the right wing rose. The glider started to turn left exactly as he had said it would. Ten seconds into the turn, he felt something vaguely unstable. He leveled the wings and quickly landed. Making a second attempt, he turned smoothly but was alarmed to find the descent too steep. He yanked on the elevator control. Too late, beginning a spin, the glider hit the ground with its left wing, bounced and twisted. Wilbur was thrown headfirst into the elevator ribs. When the others reached him

he was sitting still with his head in his hands. He was all right, with just a bruised nose and a black eye. He was utterly preoccupied with what had happened.

He stayed on the ground for a week, thinking. On August 16 he figured out what the birds could not have told him. It was the discovery of what is now called adverse yaw. As the machine banked in the direction of the lower wing, the other wing rose, as intended, but then increased drag on this upper wing with a greater angle of incidence retarded its advance, causing the machine to lose air speed and tumble out of control. The discovery that the relative velocities of the right and left wings bore a very important part in lateral equilibrium was a fact, Orville wrote, "never before considered by any investigators." Wilbur had diagnosed the problem; he had no idea how to correct it and his mood blackened. On the long journey back to Dayton Wilbur brooded. He reportedly said to Orville that if man ever did fly, "it would not be within our lifetime . . . not within a thousand years!"

Many people were saying the same thing, a skepticism that stirred Wilbur out of his depression. Invited to address the Western Society of Engineers in Chicago on September 18, he realized that he and Orville knew more about flying than anyone in the world. They worked hard on the lecture, which Wilbur delivered dressed in one of Orville's suits—he drew the line at wearing a tuxedo. The speech was well received, but when they came to prepare it for publication they decided they must go beyond the mere assertion of suspicion about Lilienthal's tables on which aeronautics everywhere were based. They began by riding a bicycle around the streets of Dayton. This apparently aimless activity was in fact the beginning of a scientific odyssey, an exploration of how the invisible eddying currents of air respond to intrusion.

The brothers took a spare wheel and positioned it horizontally in front of the handlebars so that it would spin in the wind as they cycled. On the rim, they clamped two small surfaces perpendicular to the airflow. One was a tiny model wing shaped to the 1-in-12 curve used by Lilienthal, and the other was a flat square metal plate representing drag. They set these surfaces half a wheel apart from each other. Air flowing over the curved plate would produce lift and turn the wheel clockwise. Air hitting the flat plate would send the wheel in the other direction. The question was whether the airflow striking the cambered surface generated enough lift to balance exactly the pressure on the flat plate so that the bike wheel was held motionless. Lilienthal's tables suggested a five-degree angle of attack of the curved plate would produce enough lift to equal the pressure on the flat plate. The cycling Wrights found that at the small five-degree angle of attack, the lift on the airfoil was inadequate to balance the plate. Balance was achieved only when the airfoil was set at an angle of 18 degrees. Ergo, they had been misled in the calcula-

AERODYNAMICS 1: The bicycle Wilbur and Orville rode around Dayton to test the effect of airflow over a curved plate mounted on a spare wheel

tions of lift they had used to design their gliders.

That was suggestive, but it was a crude test and it did not tell them just how variations of wing angle, shape, wing area and velocity affected the ability to lift a given weight, assuming a constant value for air pressure. In a more sophisticated groundbreaking experiment, beginning just before Thanksgiving, they made a six-foot wind tunnel in a gaslit room over the bike shop. The tunnel itself was unexceptional: a box 6 feet long, 16 inches square, with a fan propelled by a one-cylinder illuminating-gas engine and an observation window. The ingenuity lay in the delicate homemade instrument they placed inside, two balances fashioned out of wheel spokes and hacksaw blades. On a top axle, they suspended, one at a time, a variety of small wing shapes. When the airstream struck the wing, it produced a movement of the axle, the precise degree indicated on a pointer at the base. On a lower axle, they permanently suspended four flat plates to measure drag. When the airstreams struck the airfoil the lift was indicated on a dial. The degree of lift was then offset by the amount of drag recorded when the airstream hit the flat plates.

Over two months, they stood stock-still in the same place, held their breaths and turned on the fan to test wing after wing with a wide range of design variables: They tested 200 wings of subtly different varying cambers, wings with pointed tips, rounded ends, tapering sharply or slowly, wings with sharp leading edges and wings whose front

edges had been thickened with layers of wax or soldered tin, wings in monoplane and biplane configurations. They also tested aspect ratios, the span of the wing in relation to its chord (width), the straight line measurement from leading to trailing edge. They discovered an astonishing and utterly unpredictable range of responses to variations. Wings of the same area produced different lifts: Wide short wings were less efficient than long narrow wings. And Langley had been wrong to assume that a sharp-edged wing would meet less resistance than a rounded leading edge.

The brothers narrowed down the models to 38 and over three weeks carried out 43 tests on each, determining to within one quarter of a degree the angle to the airstream at which lift began. In calculating their 1900–1 wing design, they had plugged in too high a figure for air pressure, hence the disappointing lift, because they had accepted the common coefficient suggested by an English engineer, John Smeaton, in 1759.

By Christmas of 1901, the brothers now had the most complete and accurate data in the world on the design of wings and a new figure for the coefficient of air pressure. They accepted Chanute's offer to do the mechanical drudgery of turning the wind

AERODYNAMICS 2: The more sophisticated balances they tested in their 1901 wind tunnel

tunnel findings into tables while they made bicycles for sale in the shop and gave some attention to this basic business. Of course, there was also the small matter of translating their new knowledge into the handcrafting of a third glider they would make in parts in Dayton and assemble in Kitty Hawk. Over the winter months, they designed a glider with much slimmer wings than Glider 2. The wing camber was 1-in-24, the shallowest curve yet, and its deepest part was pushed even farther back from the front edge of the wing. The chord (width of the wing) was two feet narrower than Glider 2 to reduce drag. Glider 3 was a much bigger machine with a 32-foot wingspan, 10 feet longer than Glider 2, and about twice the length of Glider 1. In an attempt to make the elevator more sensitive, the Wrights also designed it to be shaped more like a wing and reduced it by three square feet.

But the biggest innovation they sketched in February, after days of pondering the 1901 bad experiences, was something entirely new, a fixed vertical tail at the rear of the glider—two fins six feet high. Wing warping for turns worked because it generated unequal amounts of lift, but the wing rising to initiate a turn presented a higher angle of incidence and slowed in relation to the lower wing having a smaller angle of incidence, causing the glider to spin. Therefore, they reasoned, the answer was to slow the forward speed of the lower wingtip. The fixed tail presenting itself to the airflow on the side of the low wing would retard its forward movement so the velocities of right and left wings would be in equilibrium.

Save for the precisely curved wingtips and ribs, contracted out to a local carriage maker, the brothers made the entire Glider 3 by hand in the spring and early summer of 1902. The most exacting job was making the muslin skin for the wings; they took over the first floor of their home. Katharine wrote to their father: "Will spins the sewing machine around by the hour while Orv squats around marking the places to sew. There is no place in the house to live."

The brothers returned to their base at the foot of Kitty Hawk's Big Hill on August 28. They drilled a 16-foot well for fresh water, and built a living-room/kitchen addition to the shed. The photographs show a kitchen to delight an army quartermaster: serried ranks of boiling and baking pans, tinned food, and dishes, and eggs marked up in order of freshness. Two sheets of heavy burlap were strung up in the rafters for beds. On a new bicycle they'd designed with gears for the soft sandy road, they found they could get to Kitty Hawk and

ORDER PREVAILS: The 1902 Kitty Hawk kitchen lit by kerosene lamps, with serried ranks of tinned tomatoes, peaches, beans, eggs and cans of Chase and Sanborn coffee. Orville cooked, Wilbur washed up.

back in one hour instead of three. It took them two weeks to unpack the crates and assemble their spars, outriggers, crosspieces, and sheathe the wings with the skin they had stitched in Dayton. The result was the most graceful of the gliders; it looked like a real airplane though it weighed only 15 pounds more than Glider 2. The Wrights were buoyed by the expectation that they had beaten the problem of lift, but they had imposed an extra burden on themselves by changing the control system. The elevator control bar was in a new position, and to gain altitude the bar had to be pushed down—whereas in Glider 2 elevation was achieved by pushing the bar up. Wing warping was no longer effected by foot levers but by wires attached to a pivoted hip cradle. The pilot now had only to shift his hips an inch or two to warp the outer sections of a wing. That was an improvement, but pilot-writer Harry Combs is right to emphasize such changes meant they had

to relearn to fly the instant they left the ground. "It is extraordinarily difficult for an airman to fly the instant he leaves the ground—well, I would have used the term impossible, except that the Wrights did accomplish the impossible. When they left the ground it was either fly or crash—nothing in between. It was total commitment and nothing less." Altogether the brothers developed nine different control systems.

Glider 3 was ready for testing on September 19. They tried it as a kite that day, noting it could stay up at low angles of attack—a big improvement in lift. The next day the Wrights and Bill Tate sweated the 115 pounds up Big Hill and Wilbur climbed in for his first glides. They went beautifully. He tried a longer glide from higher up the hill. A gust hit the side, raising the left wing. Wilbur executed the hip movement to bring the right wing up, and neatly moved the elevator control bar to the full down position, intending to come lower. Trouble was he had acted out of instinct: He had forgotten that they had reversed the position so his intention to descend became a command to ascend. "Almost instantly," he wrote, "it reared up as though bent on a bad attempt to pierce the heavens." Without the cradle Wilbur would have rolled out to his probable death as the stalled glider tumbled out of the sky. The right wingtip hit the sand and the machine spun around it. Wilbur held on tight. It was the closest call yet. By the end of the day he had mastered the controls, making 25 low glides without accident, but that night, as a precaution against side gusts, they modified the trussing to make the wingtips droop.

It was time to give Orville a chance. He proved a natural pilot. On September 23, when they had made about 75 flights, Orville rose 30 feet on the last glide of the day. This was alarming; it was always their intention to glide not much higher than 10 feet so that glider and pilot would not come down with too much of a bump. Tate and Wilbur ran alongside the glider, yelling about the danger. In the howling wind, Orville didn't hear them. When he tried to level off, the top wing moved higher and the nose lifted. The glider fell sideways and backward as it slowed to a complete stop in the air, then the wind hurled Orville violently backward into the hill. The spars cracked, the fabric sundered, the machine rolled down the hill. Orville stepped out of "the heap of flying machine, cloth and sticks" without a mark, cross that his suit had ripped.

On the morning of September 29, watched by the newly arrived George Spratt and Lorin Wright, the brothers each

potentially lethal aeronautics. Orville drank cup after cup of coffee as they talked, and when the others finally went to their burlap hammocks, he stayed fretfully awake all night. The next morning at breakfast he expounded the cause of the new side-slipping and the solution. Piloting error and crosswinds were not to blame. It was the new vertical tail fins. They had worked too well, solving one problem only to create its diametric opposite. The tail retarded the lower wing speed, as they hoped, but it speeded up the higher wing. The increased

were really halves of one whole." Orville set to work with hardly a pause to make a rudder 5 feet high, 14 inches deep, and fix it to the glider.

It was frustrating at this exciting moment to receive two more visitors — Chanute and his assistant Herring, with two Chanute-Herring hang gliders they wanted to test. Orville and Wilbur did their best to help, but the guests' multiple-wing gliders were duds. The best run was a hop of 50 feet. Who knows with what mix of emotions the visitors watched the Wrights

made a series of long, successful glides in the rebuilt machine. In the afternoon the wind picked up to 18 miles an hour. Wilbur tried a turn. Again, the glider slipped sideways toward the lower wing. Wilbur was prepared this time. He threw his weight toward the opposite side of the cradle and with the help of the elevator was able to level the machine for a soft landing. Orville experienced the same frightening slippage, seeing his lower wing hit the ground while the rest of the plane spiraled around it, digging a hole in the sand. It was something another generation of fliers would call a tailspin; they called it well-digging. Over the next two days they made 40 glides, some as long as 25 seconds, but on average one out of every three attempts to turn would end up with the pilot fighting tailspin.

The brothers and Spratt had stayed up late on October 2 arguing about these

speed of the high wing gave it still greater lift and the decreased speed of the lower wing gave it a lesser lift, so that it dropped and the higher wing went still higher, leading to a disastrous tailspin crash. The answer, said Orville, was to make the two fins into a single rudder and give the pilot control of it. With a mobile rudder, the pilot could increase or decrease the speed of the low wing on a turn to match the velocity of the upper wingtip and so maintain a controlled turn without sideslip or spin. According to Lorin, Wilbur surprised everyone that celebrated morning: He didn't argue. He demurred only in feeling that the pilot already had too much to do with the elevator and the warping. Let's combine the rudder control with the wing warping, he said. It was striking that they conceived this together, remarks biographer James Tobin, "as if their two minds

LONGER, LONGER: Wilbur pilots the big 1902 glider from Big Hill. He achieved a distance of 622 feet, Orville 615 feet. They tried never to ascend too high. In the final frame here, taken on October 10, 1902, the figure running alongside is Dan Tate, Bill Tate's half-brother, who helped Orville with the launch.

take out their big white machine to begin a halcyon period of gliding. The working rudder, a balancing not a steering component, was the triumphant completion of the three-dimensional system of control the brothers would seek to patent. Chanute found solace in France telling everyone about the three-axis system of control his "pupils" had devised. Herring, who had also worked with Langley, went straight off to Washington to offer Langley all of the Wrights' secrets in exchange for a job at the Smithsonian. That honorable man refused to see him, though

Langley himself tried, and failed, to get an invitation to Kitty Hawk.

Over the next two weeks, happily on their own, the Wrights honed their skills from dawn to dusk. They made 250 glides in two weeks setting all kinds of records for flying in the highest winds (30 miles an hour). On October 23, they had a little contest to see who could travel farthest. Wilbur's best was 622 feet. His kid brother was just behind at 615 feet.

Two days later they packed up and started out for Dayton. Winter was com-

anything to offer. The technology of the time could have produced lightweight engines, as Langley's assistant, young Charles Manly, showed in 1903, but the engine makers did not see a commercial demand. The Wrights would have to make their own engine.

They went about it in characteristic fashion—step by step. Orville studied engine design, then their bike shop mechanic-manager Charles Taylor made a skeleton model. The outcome was a 12-horsepower, 4-cylinder engine in which the carburetor

Charlie Taylor said, "I think the propeller was the hardest job Will and Orv had." The discussions were fierce. "I don't think they really got mad, but they sure got awfully hot," said Taylor. "One morning following the worst argument I ever heard, Orv came in and said he guessed he'd been wrong and they ought to do it Will's way. A few minutes later Will came in and said he'd been thinking it over and perhaps Orv was right. First thing I knew they were arguing it all over again, only this time they had switched ideas." Dinners at home in

ing and they walked four miles over the beach through a cold rain. Now they needed an engine.

WILBUR AND ORVILLE had initially thought an engine would be the least of their worries. Now it began to dawn on them just how much work remained. A powered aircraft would have to be bigger than the glider: 520 square feet of wing area would be needed to lift a pilot of 140 pounds, engine and transmission of 220 to 225 pounds and airframe of 260 pounds, a total of about 625 pounds. In camp, where they did these calculations, they specified an engine of 8 to 9 horsepower as a satisfactory permutation of weight and thrust, with the propellers at 330 revolutions per minute producing 90 pounds of thrust pushing from the rear. Not one of a dozen companies in the embryonic auto industry had

was fashioned from a tomato can, the fuel line was a speaking tube, and the ignition was from a dry battery turned on by a light switch bought at a hardware store.

They also discovered that aerial propellers didn't exist. They had planned to borrow from marine propulsion theory, but discovered there was none. Shipbuilders, trying this and that, got by with propellers with 50 percent efficiency. A marine propeller has simply to displace water, but an aerial propeller has to cope with aerodynamic forces of great complexity; even the laconic Wilbur called designing their own propellers "a very perplexing problem." Orville wrote: "With the machine moving forward, the air flying backwards, the propellers turning sideways, and nothing standing still, it seemed impossible to find even a starting point from which to trace the various simultaneous reactions."

Hawthorn Street were alternations of rapid-fire exchanges and angry silences. Voices were raised so high one night, Katharine yelled, "If you don't stop arguing I'll leave home!"

They wrestled with the perplexities not by trying different propellers, but by envisaging the propeller as a rotary wing required to produce lift. The upshot of symbols and figures scrawled at all hours on walls and fabric and crammed in nine notebooks was a pair of propellers eight feet in diameter, made of three laminations of spruce, with more than 60 percent efficiency. "Isn't it astonishing," wrote Orville to George Spratt in June 1903, "that all these secrets have been preserved for so many years just so that we could discover them!!"

Chanute was on their case all this time, questioning the power of the engine, but

THE TRICKIEST PART: In September 1902 turns too often ended in frightening slippages or tailspins. They found the answer in having a movable single rudder, and here is a rear view of Wilbur using the rudder and wing warping for a smooth banking turn.

badgering them to beat competitors in the St. Louis World's Fair of 1904. Wilbur refused to be rattled. He wrote to Chanute: "The newspapers are full of accounts of flying machines which have been building in cellars, garrets, stables, and other secret places, each one of which will undoubtedly carry off the $200,000 at St. Louis. They all have the problem 'completely solved,' but usually there is some insignificant detail yet to be decided, such as whether to use steam, electricity, or a water motor to drive it. Mule power might give greater ascensional force if properly applied, but I fear

would be too dangerous unless the mule wore pneumatic shoes." He had to take more seriously the rumor in the summer of 1903 that Langley and Manly were ready to launch an airplane built with $50,000 from the United States military: The Wrights had spent only about $1,000 of their own money from the bicycle business and felt confident, but Chanute had been talking indiscreetly all over the place about what the Wrights were up to. It was a worrying time for the brothers. They had applied a second time for a U.S. patent, having been rejected in March 1903 by officials

who regarded all aviation applications as the work of fantasists. If their ideas were stolen or Langley succeeded, they would say goodbye to all their hopes.

Through July, they worked hard at the Flyer, as they called their new machine. On August 8, Langley unveiled not a completed airplane but a quarter-sized, unmanned model with a three-horsepower, five-cylinder gasoline engine designed by Manly. Projected over the Potomac from a catapult on a barge, the model was in the air for 27 seconds, traveling 1,000 feet before smashing into the water. Langley claimed a

success—it was the first flight of a heavier-than-air machine powered by a gasoline engine. The army approved the launch of a full-scale version.

The brothers made the long journey to the coast of the Outer Banks at the end of September. The Flyer was no longer something they could carry along in a trunk. While they waited for crates to arrive, they practiced with the 1902 glider, setting new records and suffering a few spills in gusty winds. They licked their wounds cheerfully enough: News came through that on October 7, Langley's full-scale Aerodrome, piloted by Manly, had fallen straight into the Potomac only a few yards from the launch barge. The *New York Times* wrote: "At no time was there any semblance of flight." The *Washington Post* said the Aerodrome was as incapable of flight as a dancing pavilion floor. A *New York Times* editorial surmised flight might be possible if mathematicians and mechanics spent the next 1 million to 10 million years working together on the problem. Soon afterward, Chanute sent them a newspaper report that the army was nevertheless likely to approve another test by Langley.

The last parts of the new Flyer arrived on October 9. The Wrights had their work cut out to fly before winter: They had to assemble plane and engine, complete a 60-foot monorail from which to launch and then test the new machine as an unpowered glider before risking it with power. With the engine and a pilot, the weight came to 700 pounds, 75 more than their estimate. Chanute calculated they would lose so much power in the chain transmission from engine to propeller the plane would not get off the ground. It all depended on whether the propeller could deliver 100 pounds of thrust compared with the design thrust of 90 pounds at 330 revolutions per minute. The work ran into November, when they were normally back home in Dayton. Spratt finished laying the 60-foot rail on November 4. He was shocked when he came into the shed to be told by the brothers that they had decided to forego any tests of the machine as a glider. He protested. They would be risking their lives in an untried machine. How did they know they could handle a wholly new machine with the added weight and thrust from the pusher propellers? The brothers were adamant. It was a thoroughly uncharacteristic gamble; throughout they had tested and tested, taking a step at a time.

Before they carried the Flyer to Spratt's rail, the brothers ran tests of the engine. The propellers were a triumph, delivering an ample 132 pounds of thrust at 350 revolutions per minute. But they had found it hard to tighten the sprockets, and when the engine misfired the vibrations shook the screws out half a turn. The shafts were badly damaged. They would have to go back to Dayton for repair. Spratt volunteered to put them on an express train from Norfolk, but it would be 10 days before they would be back, and Spratt decided he would not be coming back. He was sure they had missed their chance that year. The brothers spent the time repairing the engine fault (a magneto), and testing. They ran short of food, not for the first time. Their tempers were not improved by the arrival of Chanute, grumbling about the cold and their cuisine. Orville wrote to Katharine, "We had come down to condensed milk and crackers for supper, with prospects for coffee and rice cakes for breakfast." They had never stayed on the shore of the Atlantic so close to winter. Wilbur described the progressive stages he went through to try to stay warm at night: "5 blankets and two quilts. Next come 5 blankets, 2 quilts and fire; then 5, 2, fire and hot water jug. This is as far as we have got so far. Next comes the addition of sleeping without undressing, then shoes, hats, and finally overcoats."

It was all too much for Chanute. He left on November 12. "He doesn't seem to think our machines are so much superior as the manner in which we handle them," Orville wrote Katharine. "We are just of the reverse opinion." Chanute's estimate of their chances was manifest in his farewell offer. If (and when) they failed, said he, they might want to become his assistants and help him perfect his own gliders. He further cheered them up by mailing them a picture of Langley's machine with the information that Langley's engine of about the same weight had four times the power of theirs.

"Day closes in deep gloom," Orville wrote. When the shafts came back, the sprockets still rattled loose. Had they not been cycle mechanics, the experiment might well have ended then, at least for the season, but Orville remembered something. "Arnstein," he pronounced, "will fix anything from a stop watch to a thrashing machine." Arnstein, though sounding like a wizard mechanic, was, in actuality, a glue. The brothers used it in Dayton to fix tires. Perhaps it could fix propellers.

It could. Now the problem was weather. It grew worse: more rain, biting cold, snow flurries, high winds. On November 28, as they readied to take advantage of a weather window, they were hit with another calamity: There was a hairline crack in a propeller shaft. They would have to switch to shafts made of solid steel. Orville rushed to Dayton to make them, with the weather window narrowing all the time. The brothers vowed to each other, hail or hurricane, to stick it out until they had at least tried to fly once.

Orville was back on December 11 with the solid steel shafts—and news of Langley. He had tried again to fly on December 8. His airplane, the *New York Sun* reported, "slid into the water like a handful of mortar." Pilot Manly had nearly drowned in the icy Potomac and the press howled about the waste of public money. The army cut off its subvention.

Now the weather again frustrated the men on the beach. After weeks when they might have flown, the gusts dropped, so that when they were finally all set to go, on Saturday, December 12, there was not enough wind for starting from flat and not enough time to go to the hill. The next day, the weather was perfect, with good, steady winds—but the Wrights would not fly on the Sabbath. Monday morning, December 14, had weak winds, again no more than eight miles per hour. Still, they decided to try in the afternoon. They raised a flag that could be seen by the four men at the nearby lifesaving station, a prearranged signal to come and be witnesses. At 1:30 p.m., in biting cold, the two brothers and their helpers slowly moved the 600-pound Flyer up the dune on a dolly. Picking up the end of the rail, and tediously again laying it in front of the Flyer, made for a tough hour.

Just before 3 p.m., on December 14, Wilbur flipped a coin to see who would be the first to fly. Orville called. Wilbur won.

He crawled facedown alongside the racketing engine, his hips in the cradle, his hand on the small elevator lever to control climb and dive. Orville held the right wingtip. One of the witnesses held up the left wingtip. Orville pressed his stopwatch the moment the Flyer left the rail. It rose—for four seconds. Some 105 feet out from the rail, Wilbur lost speed. He had pushed the elevator too sharply forward. "As the machine had barely speed enough for support already, " he wrote, "this slowed it down so much that before I could correct the error, the machine began to come down

though turned up at a big angle." The Flyer thudded to the sand, snapping the skids underneath and smashing the elevator.

Four seconds after so much effort! The Wrights decided it hardly merited being called a flight. It was basically a stall, winds too low to help lift a craft leaving the rail at only about six miles per hour. The next day and a half were devoted to repairs. They were ready to try again on the afternoon of the 16th, but the wind was again too weak. Thursday, December 17, was a gray, cold day. Few birds were out. The wind had risen, whipping up the white beach sand at 27 miles per hour. That was several miles an hour faster than they calculated to fly the powered plane—they had glided at wind speeds up to 37 miles per hour—but they decided they had to seize the moment. It was Orville's turn and he was eager. At 10 a.m. they hoisted the signal flag again to summon the lifesavers. At 10:30 a.m., the Flyer was warming up on the launch rail. Orville set up a tripod and camera and instructed one of the lifesavers, 28-year-old John Daniels, how to put his head under black cloth to take a photograph. Orville and Wilbur talked quietly next to the buzzing engine. Daniels said later, "After a while they shook hands, and we couldn't help notice how they held on to each other's hand, sort o' like they hated to let go; like two folks parting who weren't sure they'd ever see each other again."

This time it was Orville who clambered through the wire struts at 10:35 a.m., cap pulled down, eyes half-closed against the wind and eddies of sand. Wilbur told the witnesses, said Daniels, "not to look sad, but to laugh and hollo and clap our hands and try to cheer Orville up when he started." Wilbur held up the right wingtip. The cheering started and Orville slipped the restraining wire. The Flyer started moving with Wilbur running alongside. Toward the end of the track Orville put the elevator in positive and the Flyer rose, tilted a bit to the left and cleared the end of the track at a height of about 3 feet. It went up to 10 feet, down to 4 or 5, and rose and dipped several times, then stayed down. The Flyer had traveled 40 yards (120 feet) and stayed aloft for 12 seconds.

Those 12 seconds represented the first controlled mechanical flight in history,

defined as taking off on one's own power and landing at a place no lower than the point of takeoff. That was good enough for posterity, but not for the Wrights. Once everyone got warmed up over their improvised stove back at the hut, a cracked elevator strut was bandaged and the plane dragged up the hill again. Orville noted that if they had tested the machine as a glider first they would have found that the elevator was more sensitive than that of their previous craft, the slightest movement liable to pitch the craft up or down. Wilbur climbed into the cradle at 11:20 a.m. He flew for a bobbing 13 seconds, finding the elevator as tricky as Orville had, but at least he had an easy landing. Orville was off again at 11:40. This time he flew straight for 15 seconds, then the right wing lifted and the plane started tilting. Orville overcompensated and came down at once. At noon, Wilbur was launched. He undulated wildly at the start, at 200 feet coming within a foot of the sand. At 300 feet out, the Flyer steadied. At 400 feet out, Wilbur was flying level about 10 feet above the ground. To the awe of those watching, he receded into the distance, 500 feet, 600, 700. Wilbur could see houses and trees approaching. At 800 feet out, he tried to get higher, and the machine started bucking again. It came down and pitched forward, cracking the elevator. His flight had taken him nearly 300 yards, and kept him up 1 second short of a full minute. He stopped the engine and for minutes alone in the vastness stood next to his great white bird in silent communion with the wind, adversary and ally in his long adventure. No man since the dawn of time had done what he had done. He was 36 years old, the world was at his feet and he had only 9 more years to live.

They headed for camp for a break before resuming. But the Flyer had flown its first and its last. The wind suddenly gusted to 30 miles per hour, and in a flash the frail machine was turned upside down. The valiant Daniels, who tried to hold it down from inside, was trapped, tangled and screaming in the wires, as the wind rolled the Flyer over the sands toward the ocean. He broke nearly every wire and upright getting out; the chain guides were bent out of shape and the legs of the engine frame broke off. The world's leading airmen were too buoyed up to be more than briefly upset. They went over to the Kitty Hawk weather station to telegraph their family: "Success four flights Thursday morning all against 21 mile winds from the level with engine power alone average speed through air thirty one miles longest 59 seconds inform press home Christmas."

DANIELS'S EPIC PRIVATE SNAPSHOT of the Flyer leaving the rail on December 17 has epitomized for all time man's conquest of the air: Wilbur's slim black silhouette nervily outlined against the sand, pilot Orville prosaically memorialized by his shoe leathers. The Flyer was a triumph of human ingenuity and determination, but it was not a practical airplane. To achieve that, they would have to produce a machine that could fly more than fitfully, that would assuredly take off, rise above obstructions, go safely where they wanted it to go, carry more than the pilot and land safely. The photograph is the pure poetry of invention. Now, while perfecting the invention, the brothers had to move to the tricky geometry of innovation. Determined to keep control of powered flight and to become rich and famous from it, they would have to finance and market their invention, which meant exposure and publicity, while all along the way protecting their unpatented ideas from predators, which meant stealth and secrecy. Extrapolating their genius for aeronautics, they would have to master patent law, negotiation, politics and public relations in a world at once skeptical and jealous.

Wilbur described the choice that confronts every inventor. "We found ourselves at a fork in the road. On the one hand, we could continue at playing with the problem of flying so long as youth and leisure would permit but carefully avoiding those features which would require continuous effort and expenditure of considerable sums of money. On the other hand, we believed that if we would take the risks of devoting our entire time and financial resources we could conquer the difficulties in the path to success before increasing years impaired our physical activity." They took the fork of high risk, cutting their involvement in the bicycle business and budgeting new experiments against the

$4,900 or so they had accumulated, largely the result of their father's sale of a farm. They hired a patent lawyer to work on language that would cover not so much the nuts and bolts of the Flyer as their three-axis system of control.

Wilbur seemed a little mixed up by the prospect of becoming an innovator. His ethical instincts and training were aquiver. He wrote frankly enough that he wanted to make his family rich, then added, "without exploiting the invention commercially or assuming any business responsibilities." His resolution of the contradictions in that position was to aim at selling a developed machine at a fair price to government, preferably to the United States, but, if need be, a foreign government. In that way, he reasoned, they could have a sure return, escape tedious lawsuits and get back to scientific exploration. Thieves and rivals were thick on the ground. Herring had the gall to write from Washington to say he already had a patent that predated the glider he saw at Kitty Hawk. He was willing not to sue the Wrights if they formed a partnership with him. As the "true originator" of the glider, he would take one third. Wilbur told Chanute they would ignore his former associate's "rascality."

In 1904, the gliding phase over, they transferred operations from the wilds of the Outer Banks to a cow pasture outside Dayton called Huffman's pasture. Kitty Hawk was no place to test a plane with an engine. In terms of discretion, the new location was a wash: It reduced the romantic excitement attached to the location of their early efforts, but multiplied the risk of exposure by proximity to the city. Interest in December 17 had been perfunctory after a sporadic initial flurry of fanciful reporting. The Dayton newspapers, in common with the rest of the world, confused airship and airplane, attaching more significance to the long drift of a flabby gasbag than a short heavier-than-air controlled flight. Subsequently, the brothers withheld the Daniels photograph and refused to elaborate on details, but on May 25, 1904, they invited every newspaper in Dayton and Cincinnati to attend a test flight, the conditions being no cameras and no sensationalism. Flyer No. 2 was similar to the 1903 machine but with a horsepower increased from 12 to 18,

slightly less camber in the wing and the rudders redesigned.

The Wrights took this airplane out of its shed in front of about 40 spectators. Wind and rain forced several hours of delay. In late afternoon, Wilbur moved down the track only to bump to the ground at the end. He explained he was having engine problems. Not until Thursday was another demonstration attempted. Only half a dozen of the reporters showed up. This time the best Orville could do was rise 8 feet and land about 30 feet down the field, cracking the landing skids. The *Chicago Tribune* wrote it up as a success, the *New York Times* as a failure with the headline "Fall Wrecks Airship." The newspapers lost interest after that. The Flyer was a flop at Huffman's field through the long hot summer of 1904, failing to rise from the track or quickly sinking to the ground. On August 24, Orville escaped death by inches when he crashed the nose and the upper wing spar broke in two instead of striking his back.

The resilience of the brothers was extraordinary. In this long, trying period in the cow pasture, unknown to the legend, they were continually heaving the now-900-pound plane over soggy ground, sewing canvas, sanding wood, fiddling with pump and carburetor, moving radiator and fuel tanks back and forth to improve pitch, nursing their wounds from crashes, waiting for the wind to rise, waiting for another chance to hazard their lives. Wilbur was not to know then that, by contrast with Kitty Hawk, they were attempting to fly in air of high humidity yielding, in Harry Combs's calculations, a difference in density altitude of 4,700 feet that robbed plane and engine of the little margin of power.

Toward the end of summer, they built a catapult. Helpers pulled on a rope raising a huge weight (1,600 pounds) to the top of a 16-foot derrick. The 350 pounds of force generated by its release pulled a line that went under the track, wrenching the plane forward into a surefire launch. The catapult, and cooler air for flying, made all the difference: They flew half a mile. On September 20, a gray day promising rain, a writer witnessed Wilbur execute a perfect circle of 400-yard diameter. It was the first complete circle ever flown by an aircraft

and the observer's report was history's first published eyewitness account of manned flight. The observer, who turned up at Hawthorn Street and intrigued the Wrights, was Amos Ives Root (1839–1923), a wealthy iconoclast who had made his fortune inventing a beehive that made it possible to harvest honey without destroying the colony of bees. He had devoted his later years attempting to reconcile Christians to technological progress: He reached Dayton by driving one of the new automobiles across the state from Medina. Root wrote that what he saw on September 20 was "one of the grandest sights, if not the grandest sight, of my life." Readers of *Gleanings in a Bee Culture,* where he published his report, no doubt took his declaration that it "outrivaled the Arabian nights" as a taste for exotic simile rather than nostalgia, but Root of the purple prose had more prescience than the learned editors of the *Scientific American.* They weren't interested in printing his silly story about a plane flying in a circle, no doubt smiling at the flowery fantasy of the apiculturist's prediction: "No

living being can give a guess of what is coming along this line, much better than any one living could conjecture the final outcome of Columbus's' experiment when he rushed off through trackless waters." Nobody else followed up Root's scoop.

Orville caught up with his brother, making his first circle on October 14, then Wilbur capped him with four circles. He flew for five minutes, traveling a total of three miles. "The strain upon the human system of a single individual involved in the navigation of an aeroplane," he wrote later, "was excessively great. Control of the equilibrium, the vertical steering, the horizontal steering . . . kept the mind and body under continuous stress." On December 1 Orville made his own five-minute flight and they packed up for the winter unaware that their inadequate engine would work better in colder weather.

By the end of 1904, they had made some 105 flights and accumulated an hour's flying between them. Their patent application was still pending, but they felt ready to offer their invention to the U.S. government. On

January 3, 1905, Wilbur called at the Dayton home of their local congressman, who volunteered to deliver a proposal personally to the new secretary of war, fellow Ohioan William Howard Taft. Since the Wrights are sometimes faulted for being vague and unbusinesslike in their approach to government, the essence of the letter is worth quoting.

"The series of aeronautical experiments upon which we have been engaged for the past five years has ended in the production of a flying machine of the type fitted for practical use. It not only flies through the air at high speed, but also lands without being wrecked. During the year 1904, one hundred and five flights were made at our experimenting station on the Huffman Prairie, east of the city; and though our experience in handling the machine has been too short to give any high degree of skill, we nevertheless succeeded, toward the end of the season, in making two flights of five minutes each, in which we sailed round and round the field until a distance of about three miles had been covered at a

CONFERENCE TIME: The world lost interest after the first powered hop of December 17, 1903, but the brothers—pictured here in May 1904—did much the more serious flying in their base on a cow pasture in Dayton.

speed of thirty five miles an hour. The numerous flights in straight lines, in circles, and over 8-shaped courses, in calms and in winds, have made it quite certain that flying has been brought to a point where it can be made of great practical use in various ways, one of which is that of scouting and carrying messages in time of war." Then the Wrights spelled out two options: "If the latter features are of interest to our own government, we shall be pleased to take up the matter on a basis of providing machines of agreed specification, at a contract price, or of furnishing all the scientific and practical information we have accumulated in these years of experimenting together with a license to use our patents; thus putting the government in a position to operate on its own account."

Taft never saw the letter. The congressman fell ill and the proposal was batted from the War Department to the Board of Ordnance and Fortification the same military men who had burned their fingers on Langley. A sort of catch-22 ensued. The Wrights had written to say they had brought a plane to the stage of practical operation without expense to the United States, and the War Department said it would be pleased to hear from them when they had brought a plane to the stage of practical operation without expense to the United States.

Wilbur and Orville were annoyed at receiving a negative echo, clearly a form letter for cranks. As a result, they felt free to do a deal with the British. The connection had been made in October when the British Army's Lieutenant Colonel John B. Capper visited Dayton. Fresh from the St. Louis World's Fair competition for the Grand Prize for Aeronautical Achievement between airships, bamboo-winged ornithopters, silk sky cycles, inflated saucers, dirigibles, tetrahedron kites and pterodactyl hang gliders, he had arrived thoroughly jaundiced. "It is no use pointing out something to an ordinary American," he complained, "they are all so damned certain they know everything and so absolutely ignorant of the theory of aeronautics that they only resent it." But the visit to Dayton had converted the dyspeptic colonel into an apostle of the Wrights. They had not flown for him, but they had convinced him with explanation and photographs of the Flyer in flight. In February, London promised to send a military attaché to investigate the Wrights' claim to be able to supply a machine carrying two men 50 miles.

THE BROTHERS turned back to perfecting the Flyer. The one major anxiety remaining was the tendency still to sideslip on some turns. The pilot might overcome it, they concluded, if they went back to Orville's original proposal to give the pilot direct, sensitive control of the rudder independent of the wing-warping actions. The pilot would now have two levers, one on the left to operate the elevator, one on the right to move the rudder. For insurance with vertical stability, they put two vertical semicircular vanes between the elevator wings.

Once again, the changes called on the pilot to unlearn and relearn. When they returned to the field in a rain-drenched summer, Wilbur had trouble with the vertical rudder. Takeoffs were botched. Orville made a turn at 40 miles per hour and was thrown against the upper wing and into the crumpled elevator. Wilbur fell 10 feet to the ground, smashing the plane's ribs, skids, spars and engine legs. In August they improved takeoff by increasing the twin elevator surfaces from 50 to 80 square feet and positioned them 12 feet in front of the wings. They enlarged the rudder from 20 to 35 square feet hoping to ease the steering.

Successful flights increased. Wilbur flew four circuits of the field. Orville flew a figure eight. When the rains intensified in the third week of September 1905, they took the opportunity to make even more adjustments.

On Tuesday, September 25, Wilbur stayed up until he had emptied the gas tank, flying circuits totaling 11 miles at about 34 miles per hour. On Thursday, Orville made a climactic flight that was nearly his last. He had been eight minutes circling the field just above the treetops when he shifted his hips to the right to raise the left wing and effect another turn. The left wing stayed down. Caught in the familiar menacing stall-slip they hoped they had fixed, Orville was headed straight for a nasty 40-foot honey locust tree. In a flash he decided on an emergency landing, risking a crash. He threw the elevator into full negative, causing the plane's nose to go down—and to his astonishment the left wing began to rise, turning the plane to the right, away from the menace, catching only one branch. He landed easily.

Orville and Wilbur finally figured out what was happening. In a turn, the centrifugal force was an added load for an underpowered machine: Even for a straight flight, it had only a marginal surplus of power. As a turn tightened, the centrifugal weight the plane had to overcome

increased rapidly. Approaching the locust tree, there had just not been enough energy to lift that left wing. Putting the Flyer in a dive, as Orville had done, added the impetus of gravity, so air speed resumed and the stall ended. On earlier troubled flights, they had never been high enough to resort to a dive—and dive is an extreme verb. "When we had discovered the real nature of the trouble," wrote Wilbur, "and knew that it could always be remedied by tilting the machine forward a little . . . we felt we were ready to place flying machines on the market."

The big wide world outside Dayton was still unaware of the breakthroughs at Huffman's pasture. The Wrights crated their machine and were not to fly again for three years, until 1908. Their energies would now go into marketing. What did fly in that long hiatus was rumor and skepticism supercharged with envy. They had played their cards so close to the chest hardly anybody believed they had an ace.

ALL THESE YEARS, the Wrights had earned nothing from their invention. The month they stopped flying, they prodded the sleeping beast in Washington. This time they wrote directly to Taft to say they had got nowhere in January with the Board of Ordnance and Fortification. Taft's office deftly diverted the missive back to the same board, which reflected on its own competence and found itself fully satisfied.

A month after the War Department snub, on November 15, 1905, the brothers reported their achievements in letters to aviation journals in Germany and France, and the Aeronautical Society of Great Britain. The French could not countenance the idea that a couple of uneducated Americans had intruded into airspace historically theirs since Montgolfiers' balloon ascended over Annonay in 1783. The claim to have flown more than half an hour was dismissed as typical American bombast, *le bluff americain*. If the Wrights had done what they claimed, why had it not been headlined in the American press? The turn of the century was, of course, a larger world unconnected by radio, television and international newsmagazines, but the same theme was taken up in America by the prestigious *Scientific American*: "Is it possible to

believe that the enterprising American reporter, who, it is well known, comes down the chimney when the door is locked in his face, would not have ascertained all about them and published broadcast long ago?"

Only in Britain were the Wrights given any credence. The War Office woke up to the fact that all these months its military attaché had been too busy in Mexico to go to Dayton. He was instructed again, but his instructions, he told the Wrights, were to see a flight. Nothing less would do. When the Wrights refused, the British broke off negotiations. Though London was alive to the prospects of the Wrights and Washington was not, both capitals as it turned out were on the same negotiating page. They were both saying they'd believe it when they saw it. The Wrights were saying you'll see it when you believe it.

Various writers have criticized the Wrights for refusing to fly even a secret demonstration, but the judgments are too harsh. Not until June 1906 did the Wrights have a U.S. patent, and events showed they were entirely right to think that pirates were at large. The immediate evidence followed their direct attempts to sell to the French what they had failed to sell to the Americans and British.

IN APRIL 1903 Chanute had toured France, lecturing on the Wrights' accomplishments with their glider. He had also written a long article in *L'Aerophile,* describing in detail the Wrights' 1902 glider and explaining how its wing warping and elevator worked. An explosion of chauvinistic rivalry followed. Ernest Archdeacon, a wealthy patron of French aviation, financed attempts to copy the Wrights. Captain Ferdinand Ferber, a wily artillery captain, modeled a glider on what he thought was their 1901 machine and managed nine meters. Even as he tried to sweet-talk his way into the camp at Kitty Hawk, Ferber called on patriotic Frenchmen not to allow the Americans to fly first.

In Christmas week of 1905, after a year of fruitless negotiation, the Wrights subdued their suspicions of Ferber and received his friend, a Frenchman named Arnold Fordyce, who came to Dayton saying he represented a syndicate of French businessmen. They were patriots who wanted to buy the Flyer and present it to the French government. It sounded like a tall story, but it was genuine enough. They were ready to pay a million francs ($200,000) for an option to buy a plane and exclusive access to its secrets for three months if the Wrights could make a trial flight of 50 kilometers before August 1, 1906, train pilots and forgo other negotiations. If the deal had not been confirmed by April 5, the French would forfeit 25,000 francs ($5,000) to the Wrights.

Fordyce and the Wrights swiftly signed a contract on December 30. The French War Ministry, anxious to keep an eye on red-hot German ambitions in North Africa, endorsed the proposed deal, put the $5,000 in an escrow account in New York and dispatched to Dayton a five-man secret commission headed by Commandant Henri Bonel, chief of engineers of the French General Staff, famously skeptical of flying machines. The brothers met them every day for two weeks of heated argument in a room above the old bicycle shop. Bonel became convinced, but a faction in Paris double-crossed him, changing the terms to 12 months exclusivity and a higher altitude of 1,000 feet. Three days before the expiry of the deadline, the Wrights called in Chanute in the hope his superior French would smooth the way.

They were still deadlocked when the option ran out. Bonel and his men were recalled to Paris. Now, at five past midnight, the brothers recanted and agreed to all the terms. The French preferred to pay the forfeit. Ironically, as historian Fred Howard remarks, the first money earned from the Flyer was not by selling but by not selling a flying machine.

The $5,000 was more than enough to cover all of the expenses Wilbur and Orville had incurred in all their efforts since 1899, but after 16 months they were back where they had started. Chanute wanted to press the brothers' case directly with President Theodore Roosevelt. It was a sound idea. Roosevelt seemed like the man to envision the power of an Air Force as he had embraced and put into effect the big-Navy concept of Admiral Mahan. The brothers demurred. In a fastidious reply to Chanute, Wilbur wrote: "It has for years been our business practice to sell to those who wished to buy, instead of trying to force

goods upon people who did not want them. If the American Government has decided to spend no more money on flying machines until their practical use has been demonstrated in actual service abroad, we are sorry, but we cannot reasonably object. They are the judges." Chanute argued that if they flew before a large crowd, governments would come knocking at their door. Wilbur said they would not make "a Roman holiday for accident loving crowds."

On the heels of the departing French, the wealthy Cabot family of Boston initiated a third attempt to rouse Washington through its rising star, Senator Henry Cabot Lodge, one of Roosevelt's closest friends. The Wrights had declined a very early offer of investment by the family, which wanted to carry coal by air, but it was in a spirit of outraged patriotism that the Cabots took up the cause. Cabot stormed the citadel himself, traveling to Washington to call on General William Frazier. The general said they would give careful consideration to any proposal from the Wrights, to which Wilbur responded, "We are ready to negotiate whenever the Board is ready, but as the former correspondence closed with strong intimation that the board did not wish to be bothered by our offers, we naturally have no intention of taking the initiative again." It was petulant, but knowing what we know, we can share the Wrights' exasperation. They had finally been awarded a patent, No. 821,393, on May 22, 1906, three years and two months after their first application.

In its way, the hiatus is a reflection of the incredibility of the Wrights' accomplishment, if not an advertisement for their bureaucratic effectiveness. But part of their pace came from Wilbur's belief that it would take anyone else five years to catch up. Was it a dangerous complacency? Chanute was sure it was; he knew of six derivative machines with propellers more or less ready to make attempts in Europe in 1906. "Are you not too cocksure," he wrote, "that yours is the only secret worth knowing and that others may not hit upon a solution in less than . . . five years? It took you much less than that and there are a few (very few) other able inventors in the world."

Wilbur kept his nerve when France went wild in October 1906 over the "conquest of

the air" by Alberto Santos-Dumont, a rich young Parisian-Brazilian. He flew 200 feet in a machine with 33-foot wings made of "box-kite cells," a wheeled undercarriage and a 25-horsepower motor, and surpassed that on November 12, flying 700 feet before a cheering and near-hysterical crowd. It was the longest flight Europe had seen, and the European press seized on the event to say that the first successful flight had taken place not in America, but in Europe. Santos-Dumont, whose machine was just an adaptation of the Wright configuration, was quoted as saying the Wrights were nothing. Some American editors wanted to come out slugging in their defense, but the brothers kept their poker faces. They would say nothing and demonstrate nothing until they had made a deal with someone.

But who? It was fast approaching three years since they had last flown, nearly two since their first approach to the War Department. They sensibly accepted a proposal from the New York–based Flint Company to help them get a deal with a foreign government. Charles Flint, the founder, was known as an armaments middleman and industrialist. He had dabbled in warships and submarines and was close to Theodore Roosevelt, the Rothschilds and J. P. Morgan (he later seeded what became IBM, page 356). His company would get a commission of 20 percent and the Wrights could bill him up to $10,000 in expenses. Wilbur went to Paris to meet their new agent, Hart Berg, who reported back that he saw in Wilbur's eye "that peculiar glint of genius." Wilbur expended it writing draft contracts, waiting in hotel rooms for a summons to the powerful, suddenly realizing he was underdressed for that kind of work. Selling an airplane in Europe required a dress suit and a Prince Albert. "They will cost about a hundred dollars. Please have Orville send me an American Express money order for $150. I cannot hobnob with the Emperor when I go to Berlin without some clothes."

In Dayton, Orville fretted that they might be cheated. "If we organize a [European] company, we must take precautions against being frozen out. If they have control of the business, they could dispose of us in a very short time." But Orville was usefully on hand in Dayton when, for the fourth time, somebody thought he could drag the Board of Ordnance and Fortification into the 20th century. This time it was U.S. congressman Herbert Parsons of New York, brother-in-law of the president of the Aero Club of New York, who met Wilbur just before he sailed off to Europe. Parsons sent Roosevelt an article in which *Scientific American* had finally done justice to the Wrights' achievements. The adventurous Roosevelt smelled Progress and no one ignored TR on military matters. The generals were moved to ask the Wrights if they had anything on their mind. On June 15, 1907, Orville proposed to sell one machine for $100,000. The board wanted to know if the offer was exclusive to the United States. Orville had to say that the recent contract (with Flint) precluded this. He heard no more, the still recalcitrant board having found a fig leaf if TR should ask what was happening.

Orville joined Wilbur in Paris. The big brother did most of the talking in the on-again, off-again talks, with Berg translating. The French, like the British before them, kept thinking one of their own pioneer aviators would relieve them of putting up francs for the Flyer, and the Germans, pen raised to sign a contract, got cold feet about whether Wilbur could do what he said he could. The European press was openly hostile, calling the Wrights "*bluffeurs,*" while even the Paris edition of the *New York Herald* headlined an editorial "Fliers or Liars." The Wrights still had no deal when significant flights began to be made by three French aviators—who had all borrowed much from the Wrights, thanks to the blabbermouth Chanute. On November 5, a Parisian artist, Léon Delagrange, flew for 40 seconds, traveling 1,500 feet, in a biplane built by Charles and Gabriel Voisin. Four days later the British-born but French-domiciled Henry Farman, a former automobile and cycle racer, flew half a mile in just over a minute, and the papers said, "It was the most wonderful flight ever made in such a contrivance." It wasn't of course, but the Huffman pasture was on another planet. And in America their monopoly of flight was challenged by a brilliant and wealthy man: the inventor of the telephone, Alexander Graham Bell. He had long dreamed of making a machine

that would fly like a bird and experimented with giant tetrahedral kites that might lift a man. At 60, in October 1907, he formed the Aerial Experimental Association with four exceptional young engineers (Glenn Curtiss, Casey Baldwin, Thomas Selfridge and Douglas MacCurdy) who committed themselves to developing a series of biplanes.

Chanute's forebodings seemed about to be realized. At this point, an unlikely body came to the rescue of the Wrights' peace of mind: the Board of Ordnance and Fortification. The Signals Corps had formed an Aeronautical Division in August 1907 and assigned to it was a young balloonist, Lieutenant Frank P. Lahm, whose father, another balloonist, had met the Wrights and believed in them. The chief signal officer, Brigadier General James Allan, was a member of the Board of Ordnance. He responded to young Lahm's advocacy by getting his three colleagues to finally reply to Orville's proposal of June to sell a Flyer for $100,000. The bad news was they didn't have the money without first going to Congress, which would take months. The good news was they were at last really listening. Orville said they would make every concession, and they did. Wilbur stopped off in Washington on his way home for Christmas, met three of the board and settled for $25,000 to supply an aircraft capable of carrying two men at 40 miles per hour, staying in the air for at least one hour and landing without serious damage.

About the time the army contract was agreed, Hart Berg finally got a French syndicate together to buy the French patent rights. It would form a company to make, sell and license Flyers in France, and train fliers. The Wright Company was capitalized on March 3, 1908, at 700,000 francs (about $140,000), with the Wrights owning most of the stock and royalties on every machine sold. The deal, and a signing bonus, was contingent on four demonstration flights for which they would receive $16,000.

The brothers were now committed to nearly simultaneous debuts in 1908 on both sides of the Atlantic. Rivals screwed anxiety levels another notch. Farman and Delagrange outdid each other in Europe. Delagrange flew 4 kilometers in 6.5 min-

utes in April, then stayed aloft for 16.5 minutes in June, whereupon Farman flew 20 minutes, 20 seconds covering 20 kilometers and was offered $25,000 for a U.S. demonstration at Brighton Beach near New York.

The Wrights remained confident that imitations of the Flyer were inferior, not a real airplane a pilot could control. Wilbur had noted that Farman's acclaimed flights were made up of wide jagged quarter-turns produced by yanking on the rudder. The French engines were superior, more powerful pound for pound, but the brothers trusted their new 35-horsepower motor, so much so that they declined the gift of a fine 50-horsepower engine from Glenn Curtiss. They were relentless in improving on their creation—adding to their own piloting difficulties. In the redesigned Flyer, the pilot sat upright, with a passenger on his right. The wing warping and vertical rudder controls were now in one lever between the two seats, with the front elevator rudder on the pilot's left. It was effectively a new plane, and both pilots were rusty after three years on the ground.

In April 1908, five months before their big test, the brothers went back to the wilds of Kitty Hawk to practice with their new plane. The camp was a wreck, the old living quarters under a foot of sand and the press was camped out on the periphery. Wilbur slept in the rafters, Orville on boards flung across the ceiling joists. Both fell sick. The heat was unbearable and the flying was tricky. On the first flight, Wilbur was so occupied with the right lever that he failed to maintain elevation with the left hand and landed after 22 seconds. Over succeeding days, they recovered their touch. Orville made a circling flight for 3 minutes, 20 seconds. Charles Furnas, a Dayton mechanic, became the world's first air passenger for 22.6 seconds with Wilbur and 4 minutes with Orville. Wilbur vanished behind West Hill on a long solo flight. At seven and a half minutes into the flight, the only noise Orville and Furnas base camp could hear was the screaming of the crows and gulls: The chattering, clattering sound of the crossed chains driving the Flyer's propellers had ceased. Orville and Furnas ran with pounding hearts across the sand and up the dune. Wilbur was a still figure inside the wreck. He was stunned. He was cut and bruised about his face, shoulders, arms and hands. He had pushed the elevator forward when he meant to pull it back and with the following wind had hit the ground at 41 miles an hour and flung against the wires and upper wing.

At this point, a telegram arrived from the Flint Company. It was imperative that within the week one of the brothers return to France. The successful flights of Farman, Delagrange and Louis Bleriot, trumpeted in the press, were giving the Wrights' French backers cold feet. Wilbur hurried over to Le Havre, arriving in June to a hail of ridicule from the French press, which was more than ever convinced he was a fraud. Orville stayed in America preparing for the army test. The brothers were to be separated for the most important moments of their lives. "We must take things as we find them," wrote Wilbur.

WILBUR had two nasty shocks. The first was on June 6 when he opened the crates shipped from Dayton. The disassembled Flyer was a wreck. Wood was broken and jumbled. Cloth wings were ripped and ribs broken. The radiator was mashed, the coils torn up, the aluminum tubes and a propeller axle bent. Wilbur sent a furious Big Brother rebuke to Orville. "I never saw such idiocy in my life," he raged. "If you have any conscience it ought to be pretty sore." He had to apologize. The damage was the unaided work of French customs officers rummaging through the mysterious crates.

Wilbur worked 6 weeks, 10 hours a day, reconstructing the Flyer in a corner of an auto factory in Le Mans, 25 miles south of Paris, where the trial was to be held. He ate his lunch out of a pail alongside the auto workmen, who were amazed at how unaristocratic the American aviator was compared to airmen like Santos-Dumont. Wilbur, hardworking and laconic, came to typify the average American for the many French people who met him.

The second shock came on Independence Day. On Saturday, July 4, when he was in rolled shirtsleeves correcting work on the engine the French mechanics had done, the radiator hose snapped and jetted boiling water on his chest and exposed fore-

arm. He stumbled out of the room in fierce pain. He was stoic about the four-inch blisters, while admitting to his father that the scald over his heart had "more dangerous possibilities" in the light of his troubles at the age of 18. On the same July 4, Glenn Curtiss in his *June Bug* biplane won the *Scientific American* trophy for a flight of one kilometer or more (achieving a kilometer and a half in 1 minute 42.5 seconds). He was soon to team up with the perfidious Augustus Herring. Orville wrote to Wilbur: "Curtiss et al are using our patents, I understand, and are now offering machines for sale at $5,000 each, according to the *Scientific American*. They have got good cheek!" Toward the end of the month, Orville became worried that Henry Farman looked to be stealing their thunder with an exhibition flight over Brighton Beach. He urged Wilbur to hurry with the French trials and come home.

The Flyer was not restored until August 4, then it was moved under cover of darkness to Les Hunaudieres racetrack, near Le Mans where the big test was to be staged. He set up his cot, chair and washbasin on the dirt floor of the little hangar housing the Flyer; he had a gas stove, and the richest man in Le Mans sent in "the finest sardines, anchovies, asparagus, etc." Storms delayed a demonstration and scattered the crowds for the next two days. It gave his arm more healing time; it was still so sore he did not know how it would affect his ability to handle the controls. On Saturday, August 8, the weather cleared and Wilbur wrote Orville, "I thought it would be a good thing to do a little something." It was. The French press, mocking the story of the wounded arm, was calling Wilbur a coward and his plane a phantom. The grandstands, where crowds had been cheering on the French fliers, were sparsely occupied.

Everything turned on the flight. He was about to fly a plane he had never flown before, with new controls and over an unfamiliar course. He had only a little more than an hour's total experience with a powered plane, most of it going back three years. His most recent flight at Kitty Hawk had ended in a crash. A show of nerves would have been in order, but he seemed calm. His careful check-check of everything was routine. He was up at 7 a.m. doing that, and

not until 2 p.m. was the Flyer brought out into the sun. The folded tail was opened up to its full 10 feet. As the machine was being wheeled to the catapult, the man holding the right wingtip let the underside drop. It caught and tore on a stump in the ground. Wilbur stopped, took out a patch, needle, thread and glue, and unhurriedly repaired the six-inch tear. Once the airplane was in place, Wilbur returned to the shed, took off his overalls and returned to the machine dressed in a gray suit with a high starched collar and tie. His cap was turned backward so as not to catch the wind. At 5 p.m., he checked the engine and found a short in one wire. At 6:30 he climbed into the pilot's seat. One witness wrote, "He called to one of his mechanics who was standing at the back of the machine asking him whether some quite small last-minute adjustment had been made on the motor. The man replied promptly that it had. At which Wilbur sat silent for a moment. Then, slowly leaving his pilot's seat, he walked around the machine just to make sure, with his own eyes, that this particular adjustment had, without the slightest shadow of a doubt, been well and truly well made."

He tested the controls and then suddenly, with seemingly no warning, took off. A reporter for the *New York Herald* wrote, "In a flash the catapult had acted. Mr. Wright has shot into the air, while the spectators gasp in astonishment." The plane quickly rose to 30 feet. European pilots had never been able to take off so suddenly and go so high so quickly. The people in the stands—officials, press and experts—were stunned. Then Wilbur tried something else none of them had ever before seen. He started banking. The cry went up that the plane was falling out of the sky. Women screamed and covered their eyes. The sight was "terrifying," said spectators. Wilbur banked in an easy circle around 60 yards in diameter past the viewing stands. He circled the field a second time, and the screams turned to cheers. He landed after two minutes. This was just his warm-up run, but the French fliers were amazed. "Monsieur Wright has us all in his hands," said Louis Bleriot. "Compared to the Wrights," said René Gasnier, "we are as children." For Wilbur the two minutes were

FRENCH TEST: A rare photograph of "Le Hangar" at Cap d'Auvours, where Wilbur readied himself and the plane for the crucial flight.

pleasing balm on years of anxiety. One reporter said he was flushed with pleasure, another that he was pale with emotion. But all he said when a reporter asked him if he was pleased with his flight was, "Not altogether. While in the air I made no less than ten mistakes." He reserved his exultation for his family. "You never saw anything like the complete reversal of position that took place, after two or three little flights of less than two minutes each."

It was too late for another flight that day and the Flyer was rolled back into its shed. Wilbur was invited by the other aviators to go into Le Mans to celebrate. He excused himself. He had to check on his machine; he could not do it the next day because it was a Sunday. He was toasted in absentia. The next morning the French press and almost all the doubters apologized. The flight, they conceded, was "not merely a success but a triumph." Wilbur was "the birdman."

On Monday morning at 11 a.m. Wilbur was ready again at the catapult, this time in front of 4,000 people. All the great aviators were there. Léon Delagrange had rushed back from Italy. Wilbur shot out of the catapult and curved past the stands. He made a wide circle and misjudged how long he could fly straight. He was now heading into the trees, which were too high to fly over at the last second. An aviator in a French biplane would have hit the trees or

crash-landed. Wilbur made a sensational turn of only 30 yards diameter, leveled out and touched down. The jubilant crowd could not be contained by the police line. That afternoon Wilbur made a figure eight. No one in Europe had seen that either. He showed nothing of his exultation but did not conceal it from his family. He wrote home: "Blériot and Delagrange were so excited that they could scarcely speak, and [Henry] Kapferer could only gasp and could not talk at all. You would have almost died of laughter if you could have seen them." Delagrange just threw up his hands and said, "We are beaten! We just don't exist."

The next day Wilbur weaved in and out between the trees. His mastery was absolute, and the crowd shouted itself hoarse. He was the hero of the hour. Wilbur wrote to Katharine, "I cannot even take a bath without having a hundred or two people peeking at me. Fortunately every one seems to be filled with a spirit of friendliness and this makes it possible to deal with them without a fuss." A song composed in his honor, "Il Vole" [He flies], was an instant hit."

On Tuesday just before sunset, with the excited crowds spilling out of the viewing stands, Wilbur flew above the treetops at an altitude of 70 feet. No one else had ever flown so high. The point of the demonstration was stability and control in flying circuits over a small course so that the

crowds could see, but he made longer and longer circuits of seven and eight minutes. On August 13, he became confused performing a steep bank. The right wing was rising too high and he pulled the wrong wire. The left wing hit the earth at 30 miles per hour and the machine skidded across the grass, skids and elevator damaged. If anything, the crash enhanced his reputation. "Mr. Wright," said the airship designer Edouard Surcouf, "is as superb in his accidents as in his flights." At a gala dinner in his honor in Paris, the story was relayed. Wilbur was told his response to the toast should have been longer, and he replied, "I know of only one bird, the parrot, that talks, and it can't fly very high."

WILBUR'S MAIN ANXIETY now was not being with Orville for the army trials due at Fort Myer, Virginia, across the Potomac from Washington. "Do not let yourself be forced into doing anything before you are ready," he advised. "Be very cautious and proceed slowly in attempting flights in the middle of the day when gusts are frequent. . . ." And again: "Be careful of your electrical connections."

The trials began on September 1 when Orville demonstrated how the machine could be dismantled and carried on a combat wagon. On his first flight on September 3, he was confused by the new controls and headed for one of the tents. The emergency landing did a little damage to the skids and rudder braces. Although Wilbur had shown the Europeans just how maneuverable the sophisticated Flyer was, Delagrange was still being acclaimed just for staying up. On September 6 near Paris the Frenchman flew 30 minutes one day and 31 minutes the next. Orville was determined to show the amateurs. On September 9, he circled Fort Myer for 57 minutes. It was a triple retort, not simply duration but height and aerobatics. He flew at 110 feet and in impressive circles. In the afternoon, he flew for more than an hour. In fading light in the evening, he took up a passenger who truly deserved the honor, Lieutenant Lahm, whose entreaties had got the Board of Ordnance to move. They were up six minutes, breaking the record set by Orville himself at Kitty Hawk on May 14, and he broke it again three days later taking up Major

Squier for nine minutes. The same day he flew for 1 hour, 14 minutes and set a new altitude record of 310 feet. And he did it all with a flourish, performing two figure eights. Four days, nine world records, a worldwide sensation. From Paris, Wilbur sent a mock rueful message: "A week ago I was a marvel of skill, now they do not hesitate to tell me that I am nothing but a 'dud' and you are the only genuine skyscraper. Such is fame!"

A third army officer was waiting to be a passenger: Lieutenant Thomas Selfridge, a member of the Aeronautical Board but also a 26-year-old pilot with Alexander Graham Bell's Aerial Experiment Association and bitterly regarded by the Wrights. They suspected he had abused their trust by apparently innocuous inquiries for information that the AEA had then exploited commercially in violation of the Wright patent. Orville not unreasonably regarded both Selfridge and Curtiss, who attended the trials, as spies. "I will be glad to have Selfridge out of the way," he wrote. "I don't trust him an inch. He is intensely interested in the subject, and plans to meet me often at dinners, etc. where he can try to pump me."

Selfridge's eventual flight with Orville on September 17 was tragic. After four circuits, Orville heard a tapping noise behind him. He was about to turn off the engine and glide to a landing when a propeller blade split, flew off and broke the vertical rudder upper stay wire. Orville crashed from a height of 100 feet at an angle of 30 degrees. Lieutenant Selfridge suffered a broken skull and became the first person killed during a powered flight. Orville broke his leg, ribs and hip. He was in the hospital for seven weeks. Wilbur kept the flag flying in France while his brother recovered. In the three months after Orville's accident, Wilbur made almost 100 flights and broke every record for altitude and speed. On December 31, 1908, he stayed up 2 hours, 20 minutes, and traveled 75 miles in a cold rain. It earned him the Coupe Michelin and 20,000 francs. Then he flew demonstrations in Italy.

Katharine nursed Orville from crutches to cane, looked after the

DO IT YOURSELF: Wilbur checked and checked again.

80-year-old bishop, worried about the family income and finally was rewarded when everyone agreed she and Orville could join up with Wilbur in Europe. She was swept up with her brothers in a whirl of social life, taking breakfasts with kings and premiers; on February 15, 1909, in the French spa of Pau, Wilbur took her up for a seven-minute spin. The brothers, the *bluffeurs,* were the men of the hour, world celebrities. When they returned to America, laden with gold medals and garlands and worth about $200,000 from the French contract and prizes, they were feted across the land, honored in the White House by newly inaugurated President Taft, eulogized in June in gala parades in Dayton, the latter much against their will. All of this took them away from preparing for the renewed army tri-

FAMILY RIDE: Orville (left) wishes Katharine well as she prepares for her first flight, with Wilbur at the controls, in Pau, France.

als. The contract required Orville to stay in the air more than an hour carrying a passenger (he did it in a record 1 hour, 12 minutes) and fly at 40 miles an hour with a bonus of $2,500 for each mile per hour faster. He averaged 42.8, so the army paid $30,000 for the machine. Orville went to Germany to fly more demonstrations. Wilbur agreed to perform a flyover in September for festivities marking the 300th anniversary of Henry Hudson's entry into New York Harbor and the centennial of Robert Fulton's voyage to Albany. He was to be paid $15,000 for any flight of his choice of not less than 10 miles. The city hired Glenn Curtiss in case something went wrong with Wilbur; he was to be paid $5,000 if he could fly from Governors Island up the Hudson to Grant's Tomb, a total journey of about 23 miles. Curtiss was among those accused by the Wrights of breach of patent, but the two were civil enough when they met at the Governors Island Army post, where a field had been set aside for launching the rival planes.

The exhibition venue was treacherous, with crisscrossing air currents above the harbor, volatile weather and the difficulty of somewhere to land. Curtiss made it clear he would only fly over the river. "I wouldn't fly over the buildings of the city if they deeded me everything that I passed over." The winds were judged too strong on Tuesday, September 28. Early the next morning Curtiss's plane was seen moving along the ground into mist, and it was back in a few minutes; whether it took off at all became a matter of dispute, but it was not seen in the air. At about 10 a.m. Wilbur announced he would make a short flight. He had attached a red canoe to the bottom wing of the white plane so he might float if he had to come down in the water, though it changed the aerodynamics. A reporter said, "He looked toward the Statue of Liberty and made a significant nod, meant only for his head mechanic." Five hundred thousand people clogged the Brooklyn shores, binoculars and telescopes at the ready. Similar crowds jammed Lower Manhattan around Battery Park. Wilbur went up in bright sunshine, circled Governors Island and headed straight for Lady Liberty. Boats throughout the harbor hooted. Twenty feet from the statue's waist he banked and disappeared behind her arm, which was about the length of the plane. For a few moments he was feared to have crashed. He reemerged to the hooting and tooting of boats throughout the harbor, having put a band of hope around Lady Liberty's waist. He flew under the raised torch and banked one more time around her waist. Coming back, he flew over the new ocean liner the *Lusitania* on its way to Europe and, according to the *New York Herald*, looked down to see "everywhere on her decks whirlpools of handkerchiefs, hats, umbrellas, and even wraps and coats that the passengers had stripped from their backs and were waving in delirious joy." The ship tooted its deep horn twice in a salute "from the Queen of the water to the King of the air," as the reporter put it. On landing, Wilbur "put his hands in his pockets and looked just a trifle pleased." According to the *New York Times,* he told friend and machinist Taylor, "Goes pretty well, Charlie." Taylor replied, "Looks all right to me, Will." The American papers, like the French, loved portraying Wilbur's nonchalance.

The next few days brought windy, cloudy weather. Curtiss, a very fine pilot, tried to get up, but circled only once and landed, saying, "I did not like the way the air was boiling." Winds remained too powerful for either flier. On Monday, October 4, Curtiss departed, never having made a long flight. Wilbur announced he would attempt the 23-mile flight upriver to Grant's Tomb and back, considered a riskier adventure than Louis Bleriot's crossing of the English Channel that year because of the hot gases rising from the armada of the world's warships assembled for the celebrations, and the winds that would come bowling down the long streets between Manhattan's skyscrapers. He took off into a gray day at 9:53 a.m., two Stars and Stripes fluttering from his elevator struts. He headed upriver against the wind in a cacophony of whistles, foghorns, bells and cheers from the million spectators crowded in the streets and screaming and waving from windows and rooftops. A gust howling down Twenty-third Street knocked him sideways. He leveled up. He kept hugging Manhattan's curving shore at 40 miles per hour, turning slightly at Thirty-fourth Street and occasionally descending as low as 20 feet to give spectators a better view. At Grant's Tomb, 20 minutes into the flight, he banked gently around the British cruisers *Drake* and *Argyll,* and turned to head back down the Hudson, seeming, said one spectator, to float and run. He hugged the Palisades on the Jersey shoreline, buffeted by the thermal updrafts as he passed the warships flying their flags for the festival. The crews cheered and danced jigs on the decks. A German officer presciently and chillingly told a reporter, the plane's performance was "another indication of the important part the aeroplane will play in the next big war."

For almost everyone that day it was their first sight of an airplane flight. For Wilbur, it was his last public flight.

ONE OF THE WATCHERS that memorable day was a boyish 24-year-old by the name of Clinton R. Peterkin, a junior partner at the investment banking firm of J. P. Morgan. He summoned up the nerve to call on the hero pilot at his Manhattan hotel, and said he would like to spearhead the formation of an American company to make and sell Wright planes. The Wrights had discussed doing this with Russell and Fred Alger of the Packard Motor Company. Wilbur was amused at Peterkin's pretension, but said he could try. Only about a month later, the Wright Company was incorporated with big-name investors with a capital stock issue of $1 million, offices on Fifth Avenue in New York and a factory projected in Dayton. For their rights and expertise the Wrights received $100,000 in cash, one-third of the shares, plus a 10 percent royalty on every plane. And the new firm would bear the expense of prosecuting the many patent infringers.

Some of the brothers' time over the next few years was spent training other pilots. (Orville taught the Canadian pilot Roy Brown, who half a decade later would shoot down Captain Manfred von Richthofen, a.k.a. the Red Baron.) But far too much of their time was spent defending their patent and their honor. There was a sad little quarrel with Chanute, who thought he deserved more credit and said things, out of a naive candor rather than malice, that were not helpful to the Wrights in their patent case. Most companies bowed out of the patent battles and paid for a license, but the Curtiss Company fought on, partly because it thought that Augustus Herring had patents

PIONEER PILOTS: The Wrights with some of the pilots they trained for hugely popular prize exhibitions in competition with Glenn Curtiss fliers and others from Europe. Two Wright pilots died flying exhibitions—Ralph Johnstone, a former trick cyclist and clown, and Arch Hussey. More than a hundred pilots and passengers had lost their lives by the end of 1911.

RECORD: Most of the early flights by the Wright brothers were low altitude. This is Orville establishing an unofficial altitude record of 500 meters at Bornstedt Field, Germany, on October 2, 1909.

that predated the Wrights': He hadn't, he was a liar and a cheat. The Wright Company won all its patent battles, but the cost was the exhaustion of Wilbur. He was often on the stand being deposed and consulting with his lawyers. Wilbur wrote in April 1912, "It is rather amusing, after having been called fools and fakers for six or eight years, to find now that people knew exactly how to fly all the time." He was in this weakened condition when he became sick after eating seafood in a Boston restaurant. He died of typhoid fever three weeks later, early on May 30, 1912, surrounded by his family. His father's tribute was touching: "An unfailing intellect, imperturbable temper, great self-reliance and as great modesty, seeing the right clearly, pursuing it steadily, he lived and died."

Orville did not have the energy or interest for business. Having won a legal monopoly of flight he could now have made millions by snapping up competitors, but he despised the hassle and longed to spend his time in the research laboratory he built at the Wrights' new family home, a neoclassical mansion in Dayton. He was endlessly inventive at the house and in the air.

The huge crowds feared the worst when they lost sight of Wilbur, but he was safe, just flying around the waist of Lady Liberty.

The domesticated Orville, who designed a perfect toaster and an automatic vacuum-cleaning system for the home, was the same daredevil pilot-inventor who on the snowy New Year's Eve of 1913 put on goggles, fastened bicycle clips around his trouser legs and climbed into a Wright plane to fly seven circuits without his hands on the controls. For that feat, performed with the pendulum-based automatic stabilizer he had invented, he won the Collier prize, awarded by the Aero Club of America. He improved the lift of the hydroplane with a split-flap airfoil he patented. Neither of these inventions conquered the world. (Elmer and Lawrence Sperry took the autopilot market with their more practical gyroscopic stabilizer and Handley Page's slotted wing was preferred for hydroplanes.) Orville's most successful invention among many was a toy called Flips and Flops, which featured miniature clowns on a trapeze. It expressed his playful personality and became a big hit. That gave him more joy than balance sheets.

In the fall of 1915, Orville bought up the Wright Company and then sold it at a profit of $1.5 million to a syndicate of eastern businessmen. Ten months after the sale, the Wright Company merged with Glenn Martin's company to become the Wright-Martin Aircraft Corporation in August 1916 with $10 million in capital. When America entered World War I in 1917, the United States enforced a pooling of patent rights; Wright-Martin and Curtiss Aeroplane each received $2 million. Orville remained the éminence grise at aeronautical functions. He shied away from any formal public speaking, but was always ready to reaffirm to reporters his touching faith in the airplane as a peacekeeper. In the 1930s, he suggested it had made war so awful nobody would start another; in the 1940s he suggested the *Enola Gay* might have ended war for all time.

For the rest of his life, Orville was passionately concerned to defend what the brothers had achieved. It was a savage irony that his bitterest fight was with the Smithsonian, whose literature had set the Wrights on their way. Its secretary, Charles D. Walcott, betrayed the institution by lending its great name to a shameful conspiracy against the Wrights. The Aerodrome, "Langley's Folly," had clearly failed to fly, but Walcott, eager to honor his predecessor, installed it in an honored place in the museum with the weasel words "The first man-carrying aeroplane in the history of the world capable of sustained free flight." Glenn Curtiss, losing his patent battles with the Wrights, calculated he could undermine them in the courtroom if he could prove that the Aerodrome robbed the Wrights of primacy as aviation pioneers. Albert Zahm, custodian of the remains of the Aerodrome, was an oily character with a grudge against the Wrights. These three—Walcott, Zahm and Curtiss—

then contrived for the Smithsonian to release the Aerodrome so that on May 28, 1914, Curtiss could fly it from Lake Keuka as a gala re-creation of history. The plane on the lake may have traveled perhaps 150 feet—but it wasn't the Aerodrome. It was a fake, the old Aerodrome fitted with a new more powerful motor, radically altered to take account of the Wrights' discoveries. Zahm reported that the Aerodrome had flown "without modification." Before putting it on display at the Smithsonian, Walcott concealed the fraud by having the exhibit restored to its original unflyable condition.

In 1925 Orville finally got the right kind of national attention when he sent the 1903 Flyer for display in the London Science Museum. The impasse was broken only during World War II when Walcott's successor, Charles G. Abbot, finally conceded that Curtiss had changed the Aerodrome, and made an overdue public apology to the first men to

fly. Orville relented. On the 40th anniversary of the first flight, it was announced that the national treasure would be coming home from England after the war, and so it did, just shortly before Orville died in Dayton at 77 on Friday, January 30, 1948. The inspiring creation of the brothers, refurbished but real enough, was suspended from the ceiling of the Smithsonian in December 1948. One hundred years after that chill morning at Kitty Hawk, it is there for all our imaginations, so sublime in its simplicity, so frail but so prodigious: one large step for mankind.

GARRETT AUGUSTUS MORGAN
He came to the rescue with his gas mask

1877–1963

THIRTY-TWO MEN gasp for air in a tunnel 282 feet under Lake Erie. An explosion has filled the tunnel with carbon monoxide. They are five miles out from Cleveland, at work on a project to draw freshwater from under the lake, and the only way to reach the trapped men quickly is from a rig in the middle of the lake where an elevator shaft has been sunk into the tunnel. It is 9:30 p.m. when the alarm is raised on July 24, 1916.

Tugs ferry scores of firefighters and policemen through the darkness to the rig, where a rescue team of a dozen or so forms up under the plant superintendent, John Johnson. The mayor and police chief watch Johnson lead the party down the elevator shaft. They are quickly overcome by the fumes; only Johnson and another man breathlessly reemerge. Two hours later a second search party of 10 is led down by another superintendent, Gus (Gustav) Van Dusen. Only 4 of the eleven make it back, without Van Dusen. Nobody else is willing to risk a third attempt. It seems the men in the tunnel are doomed.

Around 3:30 a.m. guilty desperation triggers simultaneous recall; a marine officer, a detective and a policeman named John Chafin all suddenly remember that there is a man in Cleveland who has invented a gas mask. The inventor, 39-year-old Garrett Morgan, is out of bed in a flash with the first of the three phone calls, running to his automobile without shirt or shoes. He throws 25 of his masks into the trunk, collects his brother Frank, and in less than 30 minutes is at the pier and on a fire tug to the rig. By this time, everybody there has pretty well given up hope. Morgan urgently explains his mask to the rescuers. It is a curious contraption, a large canvas helmet with eyepieces that encloses the head and has two breathing tubes attached; one of them is lined with absorbent material and trails down to the ground. It is not designed for heavy pressure, says Morgan, but should give protection from fumes for 15 minutes or so. There are no takers. Morgan volunteers to go down himself with his brother, then two other men come forward, Tom Clancy, who says his father, Gus Dusen, is one of the trapped men, and another called Thomas Castelbery.

The police chief warns them it is at their own risk. The mayor of Cleveland, Harry L. Davis, takes Morgan's hand as he enters the shaft. If anything happens to him, he should rest assured the city will take care of his wife and children.

At the bottom of the elevator shaft, the barefoot Morgan leads the way through an iron door and into the smoke-filled tunnel, crawling at first on hands and knees. The mask is working. At 20 feet or so, he comes on an apparently lifeless body. While the others are putting this worker on a truck car, Morgan goes forward in the darkness to investigate a noise, which turns out to be a groaning man. Clancy comes up with the light; it is his father. There is no time to do more. They bring both men to the surface. Clancy stays on top to attend to his father. The Morgans and Castelbery replenish the air in the canvas sack and descend again.

Electric wires lie exposed in the blackness. Morgan steps on one, lighting up the tunnel in a flash, but he is unharmed and presses on into the smoke, breathing through the tube trailing behind him. They find three more lifeless bodies, then Morgan comes upon another worker who is alive. He carries him out on his shoulder and up the elevator shaft to the cheers of the crowd on the rig. Now rescue workers rush to put on the Morgan masks. They go down the shaft without mishap and extricate 29 survivors. Most of the 19 dead could have been saved, too, if the masks had been available those five hours before.

Garrett Morgan was a hero—for those few minutes on the rig on July 24, 1916. He was not mentioned in the press reports. When Mayor Davis set up an investigation into the accident, testimony was taken from everyone except Morgan. He sat in court for three days; the city attorney refused to call him. The official report made no mention of his name at all, still less his role as inventor-to-the-rescue. Clancy, who followed Morgan on one trip into the tunnel, and three other men who did not even go down, were given $500 and awarded medals from the Carnegie Hero

MORGAN'S MASKS: "Four good men equipped with our helmet, at the total cost of $100, can accomplish more in the first 10 minutes of a fire than the entire company in the next hour and a half."

TRAFFIC COP: Morgan's traffic signal, patented in 1923, saved lives on chaotic streets.

Fund Commission for voluntarily risking their own lives "in saving or attempting to save the life of a fellow-being."

The difference between Morgan and those honored was this: Morgan was a black man. The prejudice against African Americans was that real. When Morgan took his invention on traveling road shows after winning a patent in 1914, he felt he had to ask a white friend to pose as the inventor to give the mask credibility. Morgan cast himself as the inventor's docile guinea pig, outfitting himself as a Canadian Indian by the name of "Big Chief Mason." The October 22, 1914, *Times-Picayune* in New Orleans described a typical demonstration: "One

of the spectacular shows of the day was given by the National Safety Device Company of Cleveland. A canvas tent, close flapped and secure was erected and inside the tent a fire started. The fuel was made up of tar, sulfur, formaldehyde and manure, and the character of the smoke was the thickest and most evil smelling imaginable. Charles P. Salan, former director of public works of Cleveland, conducted the tests. Fitting a big canvas affair that had the appearance of a diver's helmet on the head of "Big Chief" Mason, a full-blooded Indian from the Walpole Reservation, Canada, Mr. Salan sent the Indian under the flaps into the smoke filled tent. The

smoke was thick enough to strangle an elephant, but Mason lingered around in the suffocating atmosphere for a full twenty minutes. He came out of the tent 'as good as new' and a little later gave another exhibition of the salamander business."

The Ohio *Alliance Leader* described a similar test in a sealed, fume-filled room for the benefit of Fire Chief Stickley: "The demonstrator, Mr. Mason, entered the building with the hood over his head and remained in these poisonous gases for about twenty minutes. The test was very satisfactory to Chief Stickley who will recommend the purchase of several of the hoods for the department."

Morgan's mask was based on the simple observation that gas and smoke rise, leaving a layer of less polluted air at ground level. One of two long tubes from the hood enabled the wearer to breathe in this air, filtered through a sponge soaked with water. When the ground-level air was contaminated, the wearer plugged the tube and breathed in air from the cavernous helmet and big sealed sack at the back. There was enough for 15 minutes' exertion. The oxygen masks of the day offered longer protection, but they were cumbersome, with heavy gas tanks, valves, neck bindings and buckles, whereas a fireman could don Morgan's 3.5-pound National Safety Hood and Smoke Protector in seven seconds and move easily with both hands free. The nearest mask comparable to the Morgan, R. S. Dart's National Smoke and Ammonia Helmet, sold for $109, the Morgan for $25, and the Morgan performed better. The *Buffalo Courier* reported on November 16, 1913, that in a test done by the Yonkers, New York, fire department, a fireman wearing Morgan's helmet lasted 25 minutes in a smoke-filled room, while someone wearing the Peerless Helmet, built by S. F. Hayward and Company in New York City, lasted only 14 minutes.

The Morgan won the first grand prize at the Second International Exposition of Safety and Sanitation, sponsored by the American Museum of Safety and the Grand Central Palace in New York. Masks on display were said to have been used by New York City firefighters rescuing victims of a subway disaster. The New York City fire department adopted the Morgan and over

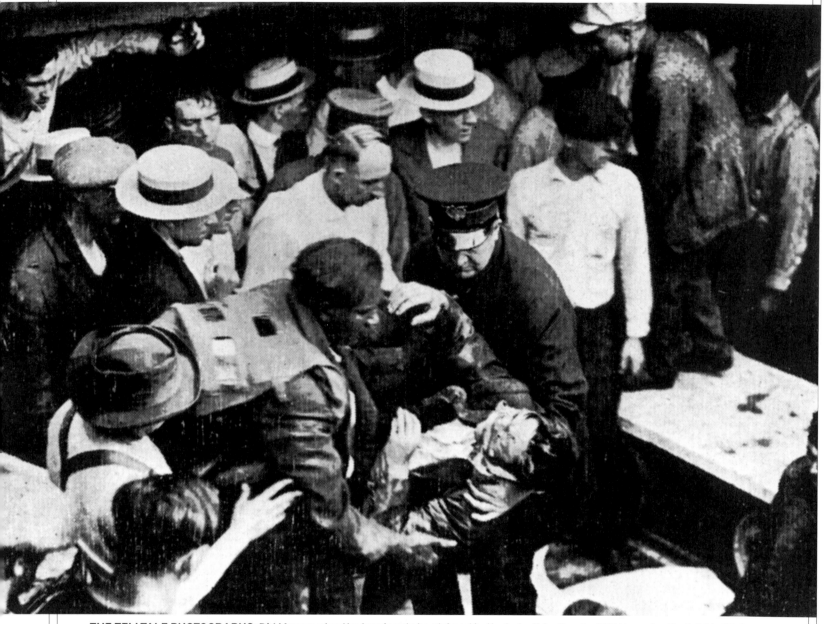

THE TELLTALE PHOTOGRAPHS: Did Morgan play the heroic role he claimed in the Lake Erie disaster? Did he actually risk his life to save others? Had he even been there? For years officials and the press in Cleveland ignored or scoffed at him. There are photographs that document the injustice to the inventor-hero. This one shows Morgan carrying a body off a boat; his antigas hood is draped over his shoulders. The photographs were never published at the relevant time or referred to in the years of controversy. (One of them shows his bare feet, corroborating his story that he had been awakened in the middle of the night and raced to the disaster, with his masks but without putting on his shoes.)

time 500 cities followed. The U.S. Navy bought helmets after Morgan showed them off in a submerged submarine, and the army issued an improved version to troops in World War I.

From early on, Morgan had good sales in the North. Orders from the South were problematical, prone to cancellation if the buyer learned that the inventor was a Negro. When the mask won an award from the International Association of Fire Chiefs in New Orleans in 1914, Morgan still thought it prudent to lie low. He had a white friend assume his identity to receive the prize. Morgan sat in the segregated audience.

What pride mingled with bitterness he must have felt. He was a prolific inventor, recognized in Cleveland for his early inventions directed at African Americans, but he longed for—and deserved—the acclaim of a wider community. All his life he had to contend with the ambiguities and absurdities of race. Both his parents were former slaves. His mother, half Indian, half black, was the daughter of a Baptist minister, and devout. His father, half white, was the son of Confederate Colonel John Hunt Morgan, leader of Morgan's Raiders in the Civil War. Garrett Morgan looked like the Indian chief he pretended to be. The large family with nine children lived on a dirt-poor farm. At 16, with a sixth-grade education, Garrett left home for a job sweeping the floor in a textile factory in Cleveland. He made it an education. He watched how the power sewing machines worked, and being good-looking, cheerful and engaging he had little trouble getting the white machine adjusters to show him what they did. He became a Merlin of the machines, working his magic on electrical and mechanical faults, inventing improvements as he went along. He was the only Negro in Cleveland to be employed as an adjuster.

In 1907 Morgan set up his own shop to repair and sell sewing machines, and the following year, when he was 30, he took as his bride a seamstress from Bohemia, Mary Anna Hassek. It was one of Cleveland's first interracial marriages and a fruitful partnership. The two of them started a tailoring business, with 32 workers making coats, suits and dresses, and then opened Morgan's Cut Rate Ladies Clothing Store. Morgan's enterprise was the talk of his community—he was the first African American in Cleveland to own a car—and he had only just begun.

Morgan experimented in his home workshop concocting a polish that might prevent fast-moving needles from scorching the thread of woolens. Called to dinner by his wife, he wiped his hands on a bit of pony-fur cloth and noticed on coming back to it that the fibers on the cloth were now standing straight up. The polish had the same effect on the hair of an Airedale, to the consternation of the neighbor who owned it. Morgan, ever sensitive to the physical characteristics that marked African Americans, marketed G. A. Morgan Hair Refining Cream as a human-hair straightener. He was on a roll. He followed up with the first curved-tooth comb for straightening women's hair and G. A. Morgan's Bleechcen ointment "to lighten the color of the skin." It sounds somewhat exploitative of black insecurities, but Morgan considered that making money by relieving anxiety was the Lord's work. He was an active member of the NAACP, and helped to finance three African-American colleges. And he put his own hair on the line. He was one of his own best customers for Morgan's Hair Refining Cream.

The insult of his invisibility in the Cleveland rescue became too much for Morgan. He set out to collect evidence and testimony for belated official recognition. White businessmen had taken up the $10 shares in his National Safety Device Company when the African-American community failed to subscribe, thwarting his desire to have an all-black company. Some of these businesspeople, led by Victor Sincere, who had been a backer of Morgan's gas mask company, wrote to the Carnegie Hero Fund. Rescue workers and police signed affidavits. John Chafin, the policeman who first called him out, verified Morgan's account of the rescue. Morgan wrote to the mayor, to the press and to the Carnegie Fund. He told the fund: "I never charged one cent for what I did and would not think of doing so, although Mr. Danzenberg, the Water Commissioner, asked me to send in a bill for seven livrs [sic] and I would not do so. I was told that he did this so it would bar me out of the Carnegie hero fund . . . God is a just God, and I know that I will get [recognition] in the next world."

The city took no notice. Carnegie disqualified him for the lack of official evidence. He sought consolation in black causes. In 1920 he started the *Cleveland Call*, a black newspaper later called the *Cleveland Call and Post*. He bought a 121-acre farm and turned it into an all-black country club. He kept on inventing. Shocked at seeing an automobile run into a horse carriage at a busy intersection, he put his mind to traffic safety. The roads were chaotic, a free-for-all in which automobiles shared the same hectic street grids with horses, street cars, carriages and bikes, the hazy concept of right-of-way dependent on the occasional cop. Morgan devised a mechanical traffic signal, a T-shaped pole unit whose arms displayed one of three instructions: Stop, Go and an all-directional Stop position to stop traffic in all directions so that pedestrians could cross. Other inventors experimented with traffic signals, but Morgan was the first to apply for and acquire a U.S. patent. It was granted on November 20, 1923, and he patented it in Canada and Britain. The first signal, operated by a policeman, stood in the town of Willoughby, Ohio, the second in the heart of downtown Cleveland. Morgan sold the rights to the General Electric Company for $40,000 and his signal was used throughout North America for many years until replaced by the system of red, yellow and green lights of today. In the 1950s, he came up with the idea of capping cigarettes with a plastic-wrapped water pellet that would extinguish a cigarette accidentally dropped.

But the offense of the Lake Erie disaster rankled through the '20s and '30s. When the Great Depression struck, Morgan lost most of his money and sought some form of monetary compensation for the nightmares he said he suffered. He came to haunt City Hall. His persistence was not endearing. The city council in 1929 debated giving him $2,000 in compensation. The city manager urged some sort of award for moral if not legal reasons. White groups were vehemently opposed. Why, they protested, when there had been no mention of Morgan in the official investigation or the press. Newspapers began to refer to him as a "self-styled hero." The *Cleveland Gazette* portrayed him as a half-ridiculous, half-scheming village idiot "prowling round City Hall." The older he got the more he embellished details, making the story more dramatic but his own real achievement less credible; his family and friends had him saving 32 people.

Morgan never got his Carnegie medal, but gradually some recognition came. A citizens group organized by Victor Sincere presented him with a diamond-studded medal. He was honored by the Cleveland Association of Colored Men. The International Association of Fire Engineers presented him with a gold medal and honorary membership. Western Reserve University awarded him an honorary degree. The city of Cleveland awarded him a watch in 1962 and paid the funeral expenses of his brother Frank. His hometown of Claysville, Kentucky, originally named for the Civil War hero Samuel Clay, was renamed Garrett Morgan Place. And when he died, aged 86, on July 27, 1963, the *Cleveland Plain Dealer*'s obituary began, "Garrett A. Morgan, a hero of the waterworks crib explosion in Lake Erie. . . ." After almost 50 years, Garrett Morgan was no longer labeled "a self-styled hero."

EDWIN HOWARD ARMSTRONG Every time you hear a clear sound on your television or radio or make a cell-phone call you are indebted to this man, the father of modern radio

1890–1954

The silence, first of all, is awesome. It has none of the pin-prick interruptions of "silence" accompanying the moment a radio is tuned to a station, the random static that sounds like a mouse investigating a potato chip. This is real silence, and its purity puzzles the audience of radio engineers who have just seen the lecturer turn on a radio receiver in their clubhouse on Thirty-ninth Street in New York City. The speaker on the stage this night of November 5, 1935, is one of the members of the Institute of Radio Engineers, Edwin Howard Armstrong. He is a tall, phlegmatic man of 47 who looks the part of a professor of electrical engineering at Columbia University. His high-domed head, devoid of a wisp of hair, is shaped like a perfect melon. He speaks with slow, deliberate emphasis. "Now suppose," he concludes softly, "we have a little demonstration."

The listeners' ears are attuned for the crackle customary as a radio is tuned to a station, and they know that some is inevitable in the experimental broadcast Armstrong has just introduced. There is no such noise. Armstrong's radio ham friend, Randy Runyon, announces that he is speaking from W2AG at Yonkers, which is just a name for his parlor and backyard antenna. He is so crystal clear the audience can hardly believe he is not in the same room with them. It is no trick, merely a prelude to the dramas conceived by Armstrong for his "little demonstration."

> Hear water poured into a glass!
> Listen to the crumpling of a piece of paper!
> Hear the striking of a match!

The listeners are stunned. The sensation lies in the fact that these mundane

THE WESTERN FRONT: In World War I, the American Expeditionary Force had little or no radio communication until Captain Armstrong arrived. He insisted on being flown over the battlefront to test how well his airplane radio worked.

sounds are clear and precise, not the thunderous "Niagaras" and noisy "forest fires" transmitted by routine low-fidelity broadcasts. There is a short Mozart piano piece, every note clear, and a guitar solo, and a phonograph of a brassy Sousa march, and perhaps most dazzling of all, a tap on an Oriental gong with rapid dissonance in the upper register. "The shimmering afterglow," a listener reports, "hangs in the room with uncanny, lambent clarity."

The effects were not by reason of some momentary harmony in the electromagnetic universe. The listeners were hearing the first public demonstration of transmission on a broadband carrier wave by the modulation of very high frequencies—frequency modulation or FM as we know it today. Armstrong's invention was rather more than a matter of making the world nicer for aesthetes. It enriched the whole culture—and frequency modulation was but the crowning achievement of Arm-

strong's life as the inventor and innovator who did the most to make any kind of radio broadcasting possible at all. He built on the work of others; that is the commonplace of uncommon achievement. John Ambrose Fleming, Reginald Aubrey Fessenden and Lee De Forest are three who made important contributions for the continuous-wave transmission of sound, as distinct from the stuttering impulses of wireless telegraphy invented by Guglielmo Marconi. But it was four basic discoveries by Armstrong, developed in thousands of experiments over 40 years, that extended the potential of human communication to the ends of the earth and beyond the planet. The shade of Armstrong's genius prevails whenever we turn on a radio or television, hear an announcement over a public-address system, listen

DAREDEVIL: Armstrong in a balletic pose above Forty-second Street, New York, having climbed the radio tower on top of the Aeolian Hall

to a stereo concert, or when an astronaut converses with ground control, a ship's captain with the Coast Guard, a taxi dispatcher with the drivers, a president with a nation, the Pentagon with a tank commander in a foreign desert, a fire chief with a rescue team or umpteen millions of us with a cellular phone when we feel it necessary to assure someone we have not been abducted by space aliens.

Armstrong's story is darkened by the frustrations of innovation as well as illuminated by the thrills of invention. It began with excitement roused in a 13-year-old boy by a book brought back from London by his father, John, in 1904. It was the *Boy's Book of Inventions* by the famous American journalist Ray Stannard Baker, which included an account of Marconi's sensational transmission of Morse code from Cornwall across the Atlantic to Newfoundland on December 12, 1901. Baker's collection of romantic stories of inventors was followed within a year of the boy's enthusiastic reception by a second, Russell Doubleday's *Stories of Inventors: The Adventures of Inventors and Engineers.* Armstrong's father, publisher of the American branch of Oxford University Press, and ex-schoolteacher mother, Emily, worried whether their only son and eldest child of three could catch up on his education. At nine, he had been afflicted by St. Vitus Dance (chorea) after a bout of rheumatic fever. They had kept him out of school for two whole years, summoning uncles, aunts and cousins in the extended Armstrong-Smith family, strict Presbyterians all, to tutor the boy at home, a brownstone on Manhattan's Upper West Side. When he was 10 concern for his health led both families to move upriver to the leafier Yonkers, where they lived next door to each other in large Victorian houses overlooking the Hudson.

Howard entered Yonkers High School in 1905. He was a lanky adolescent with an occasional St. Vitus twitch in neck, mouth and shoulders, but strong: His tennis serve was a killer. He did not immediately excel in his schoolwork, but the inventor stories had stirred his imagination. His heroes were the Italian Marconi, with his youthful optimism, and an Englishman, Michael Faraday, a blacksmith's son whose discoveries founded the science of electromagnetism and inspired the boy with the thrill of discovery by deduction. Armstrong could never recall what made him finally decide to be an inventor in wireless, then as vast and mysterious a terrain as confronted Lewis and Clark in 1804. Perhaps it was the image of the youthful Marconi, only 23 when he sent his first signal nine miles across the Bristol Channel; perhaps it was Marconi's predictions that greater discoveries were at hand.

"They were uncannily correct," Armstrong remarked 50 years later, and so they were in telegraphy, though curiously Marconi did not foresee modern radio broadcasting. His method of generating radio waves through sudden bursts of electricity was incapable of transmitting anything other than dots and dashes. The later development of continuous radio waves inspired in Armstrong a passion for sending words and music over the air for anyone to tune in and hear. Marconi saw no future in voice transmission without wires.

Armstrong, with his youthful optimism, saw further.

Radio signals, like light, are waves of electromagnetic energy moving through space. Every changing current radiates electromagnetic energy, and a moving electromagnetic wave stimulates a current in a circuit. Early radio pioneers like Marconi developed rudimentary circuits for generating and detecting these waves. Building home versions of these sets soon became a national fad. Armstrong began as a radio ham, one of the thousands of schoolboys as much hooked on telegraphy as the 1980s generation was on video games and computers. His high third-floor room under the cupola at 1032 Warburton Avenue was filled with crystals, Leyden jars, coils, coherers and condensers, and for an antenna he ran thousands of feet of wire along an embankment. He was known as "Buzz" Armstrong. Day and night, headphones on, he strained to hear dots and dashes from transmitters as remote as Nova Scotia or Key West, and tapped out Morse code himself to a circle of friends in Yonkers, one of them the same Randy Runyon who would be the point man on the big night in 1935.

Armstrong had a more accessible mentor than Marconi. At the top of the hill from home lived the inventor Charles Underhill, who had devised a form of teletype machine to convert Morse code into words. A great-uncle in Howard's pedagogic extended family introduced the teenager, and he took to cycling up the hill to Underhill's house after school to bombard the kindly man with hours of questions. He got patient answers and pieces of equipment to embellish his hobby. Schoolboy Howard transcended all the local hams in the range of his ambition and the reach of his signals, even before, at 19, he built a wooden tower to fix an antenna 125 feet above the ground and 300 feet over the Hudson. It was a feat, using two-by-fours and guy ropes, that took months with the help of only his youngest sister. He hauled himself to the top in a bosun's chair, indulging an intoxication with heights that never left him. The trajectory of his whole life, in fact, is discernible in those teenage years.

As important as the equipment Underhill gave him was the intellectual attitude Underhill fostered. The teenager told Underhill he was puzzled because he had connected a spark transmitter wrongly, according to one of his wireless books, but it had performed better than the "correct" connection. Underhill rebuked him, "What do you care about what's in the book? You're an original thinker!" In innumerable experiments later, when Armstrong tried to do something everyone knew was impossible, his mantra had become the words of Josh Billings: "It ain't ignorance that causes all the trouble in this world. It's the things people know that ain't so."

By the time Armstrong entered the electrical engineering school at Columbia University in 1909, whizzing there on a red Indian motorcycle at daredevil speeds, a human voice had been transmitted for the first time without wires—not by Marconi but by a Canadian-born polymath, Reginald Aubrey Fessenden (page 219). On Christmas Eve, 1906, Fessenden generated a continuous electromagnetic wave—something Marconi thought impossible—and hitched a ride on it to give history's first ever radio broadcast of music and voice. Crews on a few ships of the United Fruit Company in the Atlantic Ocean and the Caribbean Sea equipped with receivers designed by Fessenden heard a recording of Handel's "Largo," and also the inventor himself

THE MAVERICK

REGINALD AUBREY FESSENDEN (1866–1932)

THE ENGINEER who first transmitted intelligible speech without wires was a big bear of a man, a bearded and voluble cigar-chomper variously described as "irascible, choleric, demanding, bombastic, arrogant, vain, egotistical, domineering, combative." He was a maverick, always challenging famous certainties from the eminent. It is to mankind's benefit that he had the self-confidence to brush aside skeptics, however illustrious. He worked for a time in the 1880s in Thomas Edison's lab and asked the great man what he thought of the possibility of broadcasting voices. Edison's answer: "Fezzie, what do you say are men's chances of jumping over the moon? I think one as likely as the other."

Less than 20 years later, on December 23, 1900, "Fezzie" leapt the technological moon with a question into a microphone: "Is it snowing where you are, Mr. Thiessen? If it is, telegraph back." His voice, carried on a spark-induced wave more continuous than was usual, radiated from a 50-foot antenna on Cobb Island in the Potomac River, Maryland, to another 50-foot antenna not quite a mile away. The sound was rough, but Fessenden had proved it could be done. Professor Thiessen heard well enough to telegraph back that it was indeed snowing.

Fessenden was a Canadian-born prodigy who worked most of his life in America. He discounted Marconi's conviction about the way radiation was transmitted through the ether by the invisible electromagnetic waves first identified by Heinrich Hertz's spark-gap generator in 1886. Marconi and everyone else were sure the waves were stop-go, a kind of whiplash effect. Fessenden threw a rock into Chemung Lake in Ontario in 1897. "See how the waves circle out where the rock hit?" he remarked to a skeptic. "If they are going to carry the whole range of voice sounds, the Hertzian waves must radiate like that from the antenna at the transmitting end, and they must keep going in a steady stream until they encircle the antenna at the receiving station. They must never let up even for a split second." He was right and

Marconi wrong. But two problems took years of thinking and experimenting as he moved from professor's chairs at Purdue and Western University of Pennsylvania to the U.S. Weather Bureau and a business partnership with two Pittsburgh millionaires. The first was producing a continuous wave of electrical vibrations at the high frequency required to carry sound. Fessenden was not able to do this, so in 1904 he turned to General Electric, where there were two more brilliant immigrants: Charles Steinmetz (1865-1923), from Germany, and Ernst Alexanderson (1878-1875). In 1906 Alexanderson succeeded in making a generator that could throw off 100,000 cycles a second. The question then was what happened at the receiving end to convert these high-frequency waves to audible sound. The human ear cannot detect waves that pass a given point more than 20,000 times a second.

Fessenden's Christmas concert from Brant Rock in December 1906, mentioned in the profile of Edwin Armstrong, was based on Fessenden's method of converting the high-frequency waves to a lower audio frequency. It was a landmark triumph, and a key to the technology of radio communication. He called his discovery the heterodyne principle—a semantic witticism marrying two Greek words to express an electromagnetic marriage of two frequencies, *heteros* meaning external and *dynamis* meaning force. The Christmas song "O Holy Night" was transmitted on high-frequency radio waves, and at the receiving end the incoming sound was married to a steady signal that acted as a form of subtraction. The fluctuations between the two signals were detectable and recognizable as audible sound: Think of it as removing the radio wave and leaving the audio. Modulating the amplitude of the radio wave into the shape of a sound wave was a breakthrough, the origin of broadcasting by amplitude modulation, or AM.

Fessenden was the first to achieve two-way voice transmission by radio across the Atlantic (Marconi having achieved one way and only in Morse code). Fessenden did not maintain his momentum in radio after 1912,

leaving it to Armstrong to carry through the revolution. But he remained prolific. When we flick on a light switch, the room does not catch fire as it did in J. P. Morgan's mansion after he had installed Edison's new system. Fessenden devised wiring with rubber insulation in galvanized tubing. He sharply reduced the cost of making lightbulbs, enabling George Westinghouse to light the great Columbian exhibition in Chicago. He developed the silicon steel that preserves the iron cores of transformers and electric motors. In the wake of the *Titanic* disaster of 1912, his sonar system enabled ships to detect icebergs miles away and his Fathometer accurately to ascertain the depth of the ocean.

He conformed to the stereotype of brilliant men of science in being vulnerable to exploitation. He had to sue and sue again to collect any money from his many wireless inventions; of the $500,000 RCA was forced to pay him in 1928, some $200,000 went into legal bills.

He summed up his own life this way: "My parents despaired of me. They saw my future as a church minister or teacher, but when I closed my eyes I dreamed. I saw an inventor who could send voices around the world without using wires or cables. 'There's no future in that,' my mother told me, and she was both right and wrong. Despite all my hard work I lived most of my life near poverty. I fought years of court battles before seeing even a penny from my greatest invention. And worst of all I was ridiculed by journalists, businessmen and even scientists for believing that voice could ever be transmitted without wires. But by the time of my death, not only was I wealthy from my patents but all of those people who laughed at my ideas were twisting the dials of newly bought radios to hear the latest weather and news."

singing Gounod's "O Holy Night" and playing a violin, and his wife nervously attempting passages from the Bible. All this was beamed from a shack on Brant Rock, a seaside community near Plymouth, Massachusetts, where Fessenden erected a 400-foot tower to transmit continuous high-frequency electromagnetic waves provided by a powerful alternating generator devised by Ernst Alexanderson, a Swedish immigrant working with Charles Steinmetz at General Electric and destined for fame as one of America's greatest inventors. There were practical difficulties in the size and cost of the generator and the stability of the signal, but the principle by which Fessenden made the breakthrough was to be a source of one of Armstrong's greatest achievements.

That year of 1906 was indeed a watershed year in the development of radio. Fessenden played his violin. An international conference in Berlin officially changed "wireless" to "radio" from the Latin word *radius* for ray or beam of light. Lee De Forest, a 33-year-old telegraph enthusiast (see panel), found a way to exploit a power far subtler and far more effective than Alexanderson's huge machine: the infinitesimal subatomic particle called the electron. The genesis of the discovery, and what happened to it thereafter, is central to the titanic struggle that enveloped the earnest Armstrong and the flamboyant De Forest.

The electron, at least a thousand times smaller than the smallest atom, was discovered in 1897 by the British physicist Joseph John Thomson (1856–1940). Its existence might have been realized a half-decade earlier. Thomas Edison, experimenting with his lightbulb, had noticed that particles of his heated carbon filament had somehow been transferred from the middle of the bulb to the inside of the glass. He put a metal plate inside his bulb to prevent the transfer. One of his technicians connected it to the positive terminal of a battery and noted that an electric current flowed across the space between the lamp's hot filament and the plate. This came to be called "the Edison effect." Edison did not deduce the eventual explanation—that the carbon was propelled by millions of electrons boiling off the surface of the filament. Nor did Edison realize that his

device might be put to use in the detection of electromagnetic waves. He saw his two-element lamp as simply a way of measuring the flow of current in the main lamp circuit. He patented it as he patented everything else and then did no more.

John Ambrose Fleming, a young English scientist working for the Edison Electric Light Company in London, was intrigued by the "Edison effect," but he, too, gave up on it in 1896. He went to work for Marconi; it was Fleming sitting in a power station on the Cornish coast on December 12, 1901, who tapped out the letter *s* in Morse code received in St. John's, Newfoundland. In 1904, no longer associated with Marconi, he was looking for a way to detect and measure alternating currents. Then he remembered the Edison effect. He found a two-element Edison bulb gathering dust in a cupboard. He curved the cold metallic plate (the anode) into a cylinder around the central heated filament (the cathode) to receive the myriads of electrons it emitted. Then he connected the cylindrical anode to a source of alternating current. The rapid alternation of crests and troughs of the incoming current caused the voltage in the anode to flip between positive and negative. Only when it was in positive mode did the anode attract a current of the negatively charged electrons from the filament. In negative, the plate repelled the electrons. The device thus transformed the incoming alternating current into a one-way direct current leading out of the anode.

Fleming secured a patent for his invention, which he called an "oscillation valve." He assigned the rights to the Marconi Company as part of an arrangement that reestablished his relationship to the firm. Then, like Edison, he moved on to other projects. Marconi realized that Fleming's "valve" would detect and transform not only an alternating current but also an incoming radio wave. The resulting direct current could then be sent through another wire to actuate a faint sound in a headphone. Radio signals became electrical signals that became sound.

The valve made no immediate impact in wireless, mainly because the Marconi Company found it technically deficient as a detector, but it remained of interest to

researchers. It was left to De Forest to exploit Fleming's discovery. He read Fleming's report to the Royal Society in 1905 and simply copied the invention in 1906. That was De Forest's way; that very same year, he lost a patent suit for misappropriating an invention he had seen in Fessenden's lab. But about a month after his little bit of Fleming piracy, De Forest transcended himself. In the most original act of his life, he introduced a third element into the tube, a piece of wire bent in a zigzag, and placed it between electron-emitting filament and electron-absorbing anode. A weak signal coming into this mesh or "grid," as De Forest called it, would control the stronger stream of electrons flowing between the cathode and the anode. The signal would thus be changed into direct current and could in principle be amplified to the level of the cathode/anode current. The tiny intrusion, the creation of a three-electrode valve (or triode) he called the Audion, produced only a slight improvement in sound in a receiver, but it was to be the seed of modern electronics, computers and the Internet: Until the advent of the transistor in the 1940s, developments of the triode valve were at the heart of all electronic equipment.

The word "developments" is important. No glimpse of that dazzling future was perceived by anyone in 1906 or many years after that and for two very good reasons: Nobody completely understood how the Audion worked, not even its inventor, and it didn't work very well. Radio signals could be detected but heard only with earphones pressed painfully tight; it helped if the listener held his breath. At some frequencies, the tube emitted only a faint whistling sound. The effort of fiddling with the five-dollar Audion seemed hardly worth it. Nine years after putting the Audion on the market, De Forest made a point of stressing he could not guarantee how any particular bulb would perform. It varied, he remarked, "to an astonishing degree . . . what may appear to be a fixed law for one bulb may not hold for another." He was frank in his application for a patent that he had been "unable to explain this action of the Audion."

For six years, radio remained pretty well where it had been for a decade. Some 99.9

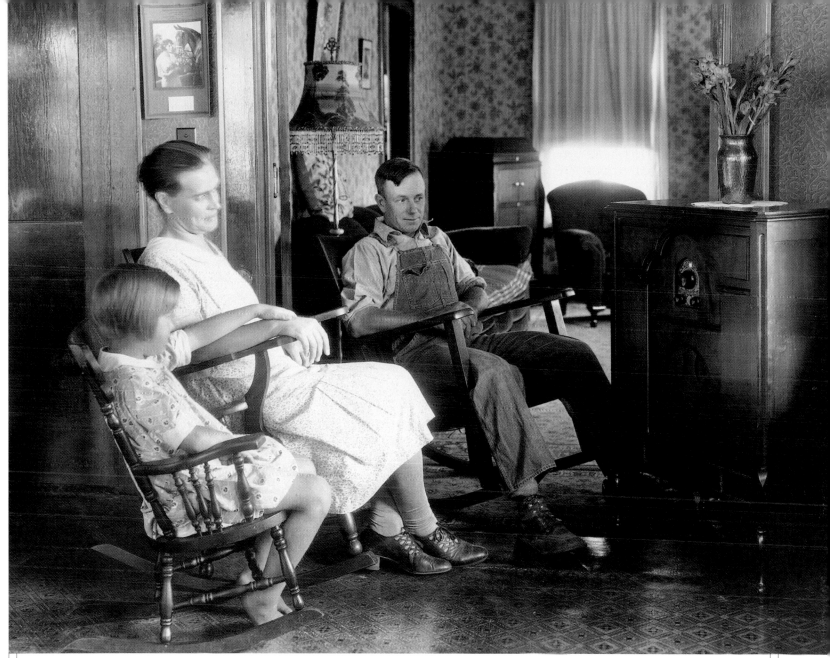

RADIO TAKES OFF: Westinghouse's purchase of Armstrong's superheterodyne receiver, his Paris brainstorm, enabled it to start America's first radio station, KDKA in Pittsburgh, in 1920, and a whole new culture. It is a fair guess that like millions of others this farming family in the late 1920s is listening to one of the phenomenally popular daily 15-minute episodes of *Amos and Andy*, the comedy show that opened evening radio from 1928 and ran to 1960. Most Americans heard of Pearl Harbor on their radios.

percent of radio communication was still dot-dash in 1912, carried on with spark transmitters and rectifiers (crystal and Fleming's valve) or magnetic detectors. Only one page out of 3,000 in the leading radio textbooks mentioned De Forest's invention. It looked like a dead end—until the 20-year-old undergraduate Armstrong picked up the Audion and became determined to find out exactly what went on inside it.

This was an investigation requiring a tricky combination of science and diplomacy. "My main object in life just then," he wrote, "was supposed to be obtaining the degree of Electrical Engineer at Columbia University and the professors could not be relied upon for the necessary charity mark of 6 unless a certain so-called reasonable amount of time was devoted to their particular courses." He offended a number of the faculty who were ready to see him depart without a degree. He seemed arrogant because he was impatient with glibness. On one occasion in the lecture theater, he made a fool of a supercilious visiting professor from Cornell who warned the class that grasping the end of a coil emitting sparks might be fatal. Armstrong did just that, without harm; he was confident that seizing the grounded end second created a safe circuit: After a shock in his attic lab, he had worked it out that electricity went through the body seeking a ground. The professor tried it but made the mistake of first seizing the apparently safe grounded end. "Before he got within six inches of the terminal, the spark jumped to his hand," Armstrong told Underhill with rather too much glee. "He pulled most of the apparatus off the table before they turned the current off." In fact, Armstrong's little trick was a piece of reckless bravado that might have misfired. There was no spark when he grabbed the grounded end of the coil second because most of the electrical energy was dissipated in the resistance presented by his body, leaving very little voltage available across the gap between his hand and the grounded end. But if the voltage in the apparatus had been high enough, or if his hands had been sweaty, allowing more of the current to flow through his body rather than, as was likely, on the surface of the skin, he could have electrocuted himself by grabbing the two ends of the coil, no matter the order in which he grabbed either end. He was lucky.

The small incident, as historian Tom Lewis suggests, typified Armstrong's approach. If something looked dangerous or right or wrong, it did not mean it was. You had to prove it was and then find out why. Carl Dreher, who visited Armstrong in the 1950s for a *Harper's* profile, remarked that his mental processes were abnormally purposive. "Watching him with a radio circuit spread out on a laboratory table, or just talking with him, you felt the intensity, the preoccupation, the dogged resolution." In all of his career Armstrong proceeded as he did as a young man with the Audion tube in his hand. Forget all the theories about how it worked. Practical experiment came first, explanation after nature had stopped coming up with surprises. Only when there were no surprises was it time for theory. A colleague summed up the technique: "Listen, look and measure."

Armstrong was fortunate that there were a handful of teachers at Columbia who could appreciate the stubborn originality of his intellect. His principal teacher, the immigrant inventor Michael Pupin (page 225), was also a believer in discovery through practical work. Another professor, Morton Arendt, realized that Armstrong knew more than he did about telegraphy and encouraged him to go it alone. John Morecroft, in the alternating current lab, helped Armstrong investigate the Audion with his department's sensitive equipment for electromagnetic detection and measurement. Earphones on, day after day when everyone else had gone home, Armstrong listened intently for pattern in the current from the Audion's "grid" plate, interposed between the anode and cathode. In early 1912 after endless hours of experiment, Armstrong discovered that some current leaving the anode plate was

alternating. Theory held that the current should be nothing but direct. Alternating current, on the other hand, could be tuned. He knew how the circuit connecting the antenna to the "grid" plate was tuned to the incoming radio waves. Now he wondered what would happen if he also tuned the "wing" circuit connecting the anode and headphones to this alternating current and fed it back into the grid. To put it simply, he asked himself if he could devise a feedback circuit. Armstrong hoped to create an endless loop of high-frequency oscillations. Sound orders of greater magnitude might result: The Audion could be transformed from a mere detector of signals into an amplifier. The potential for radio and for telephone signals was enormous. Others came up with this idea almost simultaneously, but Armstrong was the first to build and understand such a circuit.

He kept his hunch pretty much to himself while he tried out various circuits in his attic bedroom in Yonkers. He has said that the breakthrough idea came to him on a mountain in Vermont in August, "an idea suggested by the fundamental axiom of radio, 'wherever there are high frequency oscillations, tune the circuit.'" He rushed home to see what would happen if he tuned the Audion's plate circuit by means of an inductance. His sister Ethel recalls what happened on the night of September 22, 1912: "Mother and Father were out playing cards with friends and I was fast asleep in bed. All of a sudden Howard burst into my room carrying a small box. He danced round and round the room shouting, 'I've done it! I've done it!'" The dots and dashes he made her listen to through his earphones were loud and clear. His regeneration circuit, shuttling electrons back and forth many thousands of times a second, built up the strength of signals by several hundred times, as much as a thousand times for a weak signal.

Thomas Styles, a fellow student and trusted friend, was invited to the attic soon afterward and remembers his astonished excitement on being let in on Armstrong's still-secret discovery. Armstrong showed how with a very small antenna he was now able to receive reliable signals from the navy shore stations on the Pacific Coast and the Marconi station at Clifden in Ireland. Nor

was it necessary any longer to wear headphones. Sound could be amplified through loudspeakers.

Armstrong was still not satisfied. At a certain point when the amplification was boosted, a station would disappear, "and in its place was a loud hissing tone, undeniably the same station, but recognizable only by the characteristic swing and the message transmitted." Where had it come from? It was not an intruder. The hiss came from within the equipment itself. Anyone else might have shrugged that it was a pity and moved on. De Forest's only thought when he detected this hiss and howl was how to eliminate the nuisance.

Armstrong determined to wrestle with this fundamentally new phenomenon, as obscure as the operation of the Audion itself. He went back to the lab and begged some more instrument time. Several months later, he concluded that he had opened up an entirely new field of practical operation. Beyond maximum amplification, the triode began to oscillate on its own, meaning it produced high-frequency radio waves. The hiss marked the point at which the system passed from simple amplification to generating in radio circuits its own continuous electromagnetic waves capable of carrying voices and music as spark transmitters were not. He now had a tube that was not only a detector and amplifier of radio waves but also, most crucially, a generator of them. It was a simple, elegant circuit that could replace Marconi's clumsy spark-gap machines and Fessenden's massive AC generators. Armstrong had many refinements still to make, but his first biographer, Lawrence Lessing, is justified in writing, "With this dual-purpose circuit, still the basis of all radio transmitters, modern radio was born."

There was a downside to Armstrong's propensity for secrecy. It may have been due less to stubborn self-confidence, which he certainly had, as to his imperative to shut up until he had a full and perfect understanding. In any event, Armstrong excluded his teachers from knowledge of his two achievements, even those who had helped him most. Nor did he make contemporaneous notes; he preferred to keep everything in his head until he was sure and ready to disclose. At Columbia, the first

THE NEMESIS

LEE DE FOREST (1873–1961)

LEE DE FOREST was a clever Ph.D. from Yale, a frugal parson's son who went on his knees to pray for business success, so ambitious that when he was only 28 he set up a national telegraph company to compete with Marconi's, so careless in business affairs he had it stolen from him by stock manipulators who went to jail. De Forest was acquitted of fraud, rightly, but he was a reckless businessman. In his career, he founded or was associated with 25 different companies that went into bankruptcy. He was also a scientific kleptomaniac. The responder in his first telegraph company was something he borrowed from a German magazine; he stole an important idea from Fessenden when he visited his lab; Fessenden sued and De Forest had to pay damages. De Forest's lasting achievement, the Audion, was based on John Fleming's vacuum tube; Fleming sued him and won, a decision that, in this case with good reason, was later overturned.

It was legitimate, indeed admirable, to adapt other men's ideas—that is much of the history of progress—but De Forest was too much in love with the idea of being an inventor to share the glory and too restless in his imagination to push one scientific idea to its limits. He moved on from his Audion without understanding it. Armstrong, who did, was the recipient of his undying hatred.

In a tumultuous life, De Forest was involved in lawsuits galore as plaintiff and defendant, made and lost three fortunes, married four times and up into his seventies was a "shotgun" scientist, forever seeking patents for one visionary scheme or another—a "radio knife" for surgery, a self-guiding night bomb, a scanner of eventual importance to radar, a portable machine for diathermy, a radiophone, a form of television called telefilm. In 1921 he showed the nascent film industry how to add synchronized optical soundtrack to silent movies. Hollywood chose a different and inferior system for *The Jazz Singer* in 1927 in which it recorded Al Jolson's songs. The De Forest Phonofilm Corporation collapsed in 1925, and only later did Hollywood come back to De Forest's method, acknowledging its ill treatment of him with an Oscar in 1959.

What De Forest ached for most of all was acclaim as the "father of radio," the title of his autobiography. That was a reach—Armstrong is certainly the technical father of radio, Fessenden the grandfather—but De Forest was a pioneer of wireless telephony. He saw it, as he always did, as a way of making a fortune by challenging the wired telephone in two-way communication, but he was fired up also by an elevated vision of public service broadcasting to a general audience. Just a year after Fessenden's solo, De Forest sent classical music by wireless telephone to the sailors on the U.S. Navy battleships assembled in San Francisco Bay to begin the "Great White Fleet's" historic trip around the world ordered by President Theodore Roosevelt. De Forest used the same means, a Poulsen arc generator for transmission and receivers tuned to the appropriate frequency, for a half-dozen opera experiments. He broadcast Caruso live from the Metropolitan Opera House, though the great tenor sounded as if he was being strangled in a shower. In 1916, now using his Audion as an oscillator for the radio telephone, he made the first news broadcast from New York City, sending out the *New York American*'s hour-by-hour reports of the Hughes-Wilson cliffhanger presidential race between Woodrow Wilson and Charles Evans Hughes. The broadcast was heard by 7,000 wireless operators in the radius of 200 miles of the city, an achievement that lost a little of its glow because the paper called it prematurely and wrongly for Hughes.

De Forest was disappointed in his dream of bringing culture to the masses, especially his beloved opera. He expressed his frustration with satisfying vehemence. In 1933 he denounced "uncouth sandwich men" whose advertisements had come to dominate radio. "From the ecstasies of Beethoven and Tchaikovsky, listeners are suddenly dumped into a cold mess of ginger ale and cigarettes." He was still in anguish in 1946. He told broadcasters they had sent his "child" into the street "in rags of ragtime, tatters of jive and boogie woogie, to collect money from all and sundry, for hubba hubba and audio jitterbug."

mention he made of his bedroom invention was on December 7, 1912, when he told a college friend that he made a connection for "intensifying sound." When word reached Morton Arendt, he at once urged Armstrong to take out a patent. He did not have the $150 to do that. He sold his motorcycle but still did not have enough. One of the most poignant elements of the Armstrong story is that his father, who had so devotedly nurtured his son, refused to give or lend him the money. He could have it if and when he graduated the following spring. The father's impatience with his son's distractions was an error of judgment but understandable. What were all these howls and squeals from incomprehensible circuitry compared to a good degree? Lost days and weeks were to cost Armstrong grievously for the rest of his life. Only on January 31, 1913, on the advice of the helpful uncle, did he invest 25 cents in establishing his priority. He dropped into a real estate office and had a quick drawing on tracing cloth stamped by a notary named John Goodwin. Only in June did he demonstrate the regenerative receiver to an experienced patent lawyer, William Davis, several months after more urging by Arendt. Only on October 29, 1913, did he follow Davis's advice to make a formal application. And he told neither Davis nor Arendt that his system could generate radio waves as well as receive them. This was the most significant aspect of what was effectively a single invention, but Armstrong, having spent extra months working on transmission, thought of it as a separate invention. Davis, on November 14, urged him to include all the features of his invention in a single patent application. He did not. Only on December 18, 1913, his 23rd birthday, did he apply for a patent covering the tube's capacity to transmit. His biographers attribute these misjudgments, mutated by history into blunders, to youthful inexperience, naïveté, obduracy and a characteristic distrust of anyone he did not know intimately.

He was, in any event, flying high in this period. He had survived the skepticism of the more unfriendly examiners, won his degree and was at once invited to stay on at Columbia as an assistant to the great Pupin. He gave papers to the Institute of Radio Engineers and the Radio Club of America explaining his inventions with definitive lucidity, and a minimum of mathematics, to the acclaim of the professionals but very much to the fury of De Forest since the explanations made nonsense of his own theories. In March 1914, De Forest tried to catch up with an application for a patent for an "ultra Audion" receiver. He claimed it would do what the Armstrong circuit did, both as receiver and transmitter, but without recourse to feedback or regeneration. De Forest repeatedly disclaimed any regenerative features. There is no doubt that in 1912 De Forest invented a circuit that happened to oscillate, but the evidence strongly suggests he had no idea that this oscillation was of any importance. Because the ultra Audion claimed properties similar to Armstrong's circuit, the Patent Office launched interference proceedings, blocking De Forest's patents for 10 years. On October 6, 1914, the Patent Office awarded Armstrong patent 1,113,149. De Forest's claim dangled in limbo.

Pupin proudly showed off his student's invention to engineers from American Telephone and Telegraph, who evinced little interest, and the Marconi Company. Three men from Marconi were led into the lab by their "chief inspector," who was to be a leading figure in Armstrong's life. He was a kid, not yet 21, a fast-talking, streetwise, Jewish refugee from Russia by the name of David Sarnoff. He might as well have been a different generation from Armstrong, who was only 23 himself but a reserved, laconic middle-class Wasp who towered over the chief inspector. They hit it off as opposites sometimes do. Sarnoff invited Armstrong to bring his circuit to Marconi's station at Belmar on the New Jersey coast where they might hear how well it received telegraph signals by comparison with Marconi's standard equipment. They shivered together in the porous Marconi shack on a bitterly cold January 30–31, 1914, listening excitedly to Armstrong's system outclass Marconi's in volume and quality of reception from Honolulu and Ireland. Sarnoff urged his bosses in London to negotiate. As Armstrong said later in life when he was a millionaire many times over, "I was hard up in those days. If anybody had said, 'Here's $10,000 and a job at $75 a month,' I'd have sold out so fast!" Marconi made no offer. The managers in London were too suspicious of American patents and the vision of Marconi himself was limited. Only two years later did the Marconi managers open one eyelid and offer Armstrong $500 a month for a license. In the meantime, Lee De Forest had started the process of challenging Armstrong's priority, a move that was to embroil 30 judges in 21 years of court battles in 13 different courts.

When the United States entered World War I in April 1917, Armstrong was an early volunteer and the army at once recognized his value. He was dispatched to France with a captain's commission in the U.S. Signal Corps. He installed radios in the warplanes, and insisted on testing them in the air, a conscientious effort that also gratified his taste for speed and height. But his major challenge was to pinpoint enemy radio activity. Neither the American Expeditionary Forces or the French could effectively monitor German short-wave radio traffic on the battlefronts, even across no-man's-land, so they were without an important clue to the location and movements of the enemy. The Germans had devised a way of sending messages over very high frequency waves of from 500,000 to three million cycles a second. The problem, of course, was not the speed of the signals since high-frequency notes do not travel any faster than low-frequency notes; both travel at the velocity of light. But high-frequency waves have a shorter range than low-frequency waves so they are harder to pick up from a distance.

Happily, there was at Marconi in Britain a radio engineer by the name of Henry Joseph Round who was very much on the same intellectual wavelength as Armstrong. Delayed by bad weather on passing through Britain, Armstrong had used the time to call at Marconi House in London. Round's work there was shrouded in as much secrecy as the city was by fog, but he recognized Armstrong's genius and let him know what he was up to. At anchor in their bases, the German captains communicated with each other by low-powered radio sets on a high-frequency wavelength of 200 meters (1.5 million cycles). They were confident they could not be heard more

THE TEACHER

MICHAEL PUPIN (1858–1935)

ARMSTRONG'S TEACHER at Columbia was a Serbian goatherd who arrived alone in America at the age of 16 with a red Turkish fez on his head and five cents in his pocket. He knew only a few words of English, but he had learned how Benjamin Franklin had flown a kite in a thunderstorm to find proof of electricity in nature and he admired Abraham Lincoln. Tradeless and penniless, the words "Franklin" and "Lincoln" supposedly helped him get past the immigration inspectors. His father had died and his mother had insisted he try to continue his education in America. Both parents were illiterate peasants, but they had always sacrificed for his schooling. In America, he drove a mule team hauling manure on a Maryland farm, painted in the docks at Hoboken, New Jersey, hauled coal in New York and took a job in a cracker factory, but he spent every minute he could in the Cooper Union Library in New York City reading *Scientific American* and the *New York Sun*'s science supplement with the help of a pocket dictionary. He won a scholarship to Columbia University, studied at Cambridge University, England, and the University of Berlin. His work on X-ray radiation made it feasible for medical diagnosis by shortening the time a patient would be exposed from nearly an hour to a few seconds. He helped to make long-distance telephone calls cheaper and clearer by a system of inductance coils for placement at predetermined distances along underground telephone cables. AT&T bought out his patents in 1901. His autobiography, *From Immigrant to Inventor*, won a Pulitzer Prize in 1924.

"Knowledge," he liked to say, "is the golden ladder over which we climb to heaven."

than a few miles. Round, however, had devised a vacuum tube that could pick up such weak signals, and by coupling as many as 130 of these tubes he had been able to tip off the Royal Navy in May 1916 that there had been a 1.5-degree change in the bearing of the German flagship 300 miles away at Wilhelmshaven. This suggested the fleet was putting to sea. The Royal Navy was able to intercept the enemy fleet and fight the Battle of Jutland. But Round still found it impossible to tune in and listen to signals with frequencies higher than 1.5 million cycles. Did Armstrong have any ideas?

Armstrong was not able to come up with anything immediately, but this meeting with Round was one of the three elements that led him to his second great invention, the superheterodyne, the basic circuit in 98 percent of all radio and television receivers today. The second element was his knowledge of an experiment by Reginald Fessenden at Brant Rock, and the third was a spark in his mind on a dark night in March 1918 in Paris when he was walking to his apartment and heard the antiaircraft batteries firing futilely at German bombers they could not see. "Thinking of some way of improving the methods of locating the positions of the planes, I conceived the idea that perhaps the very short waves sent by the motor ignition system might be used. . . . All three links of the chain suddenly joined up, and the super heterodyne method of amplification was practically forced into existence. Not one link in the chain could have been dispensed with. This, I think, is the only completely synthetic invention I have ever made." Years later he said that if dropped blindfolded into Paris he could go straight to the street of inspiration.

Fessenden's Brant Rock experiment married an inaudible high-frequency radio wave with a steady signal at the receiving end, the difference between the two signals being detectable and recognizable as audible sound. Armstrong's flash of insight was that he might use the same principle to bring down the elusively weak high-frequency waves to a lower frequency and then enter a process of amplification. An incoming high-frequency wave of, say, 3 million cycles a second would be mixed with a wave of 2,900,000 cycles produced in the

receiver by an oscillating vacuum tube. A circuit would then detect the difference of the two wave frequencies. The resulting "beat" wave of 100,000 cycles would still be inaudible, but this wave could be amplified several thousand times to the point where it could be picked up by the regeneration circuit and converted to direct current for relay to earphones or loudspeaker. Converting this hypothesis to wiring and circuits proved, in Armstrong's words, "a task of the greatest difficulty." The basic system used eight tubes. It took more than a year and the assistance of a doggedly clever country boy, Sergeant Harry W. Houck. Armstrong worked like a demon and always inspired those around him to do the same, but as his confidence developed so did a certain explosive impatience, vented not in the work but in minor frustrations. Told by a junior officer he could not have the army pool car he wanted, he punched the man to the floor.

The war was over in November 1918 before the superheterodyne receiver could be put to use to direct antiaircraft fire. Its golden future was in broadcasting. Soon after the Armistice in November 1918, Armstrong patented his invention in France and the United States, the seminal event of the Age of Radio. He came home in triumph. The French awarded him the ribbon of the *chevalier de la legion d'honneur,* the U.S. Army raised him to the permanent rank of major, a title adopted for the rest of his life, and in America the Institute of Radio Engineers awarded him its first medal of honor for his invention of the regenerative circuit.

He was thrilled, but kept his head. Still only 28, he was surprisingly deft in playing off the competitive ambitions of Westinghouse and the newly formed Radio Corporation of America, where Sarnoff, as general manager, was a fast-rising advocate of commercial broadcasting. Westinghouse paid Armstrong $335,000 in 1920 for rights to both his heterodyne and regeneration patents, with another $200,000 if he was successful in defending his regeneration patent against attack by Lee De Forest. Its pioneering radio station, KDKA, began broadcasting in October 1920. By the end of 1922 there were 580 commercial broadcasting stations in America, a million listeners and hundreds of companies making receivers. Sarnoff, determined to beat Westinghouse in broadcasting, was galvanized when he heard that Armstrong had made yet another advance in technology. This time it was a patented extension of regeneration he called super-regeneration. Something less effective but similar enough to cause a problem had been patented in England by John Bolitho. RCA's London office was told to buy out Bolitho so as to weaken Armstrong's hand in Sarnoff's negotiation. To their astonishment, Bolitho responded: "See a fellow named Armstrong." Armstrong had tracked Bolitho to the Egyptian Sudan and bought him out. The circuit was to prove more useful in military, police and ship to shore communication than in broadcasting, but Sarnoff ended up paying Armstrong with $200,000 in cash and 60,000 RCA shares. The deal made Armstrong RCA's largest shareholder. "Arriving in England on Saturday," he cabled H. J. Round, "with the contents of the Radio Corporation's safe."

Another 20,000 shares came his way the following year when RCA urgently needed to simplify for mass consumption its manufacture of radios on Armstrong's principles. He and Harry Houck produced a best-selling radio with two knobs for tuning and one for volume. Armstrong turned

up for lunch at Morey's Bar and Grill in lower Manhattan and took Houck's breath away by writing him a check for some $100,000.

Armstrong had a fondness for gestures. He stunned RCA's top men in 1923 by entering an apartment for a negotiation carrying a very large radio going full blast. It was the world's first portable. It was also a token of love. Armstrong, who had never been known to have a date in his life, was hot in pursuit of a tall, vivacious young secretary he had met in his frequent visits to David Sarnoff's RCA offices and the portable was made for her. Marion (Minnie) McInnis was 22, 10 years his junior. Armstrong wooed her with the intense versatility he brought to wiring circuits. He came back from a vacation in Europe with a dashing Hispano-Suiza and drove her at 100 miles an hour on the Vanderbilts' private motor parkway in Long Island. On the afternoon of May 15, 1923, with Minnie in mind, he climbed to the top of a new antenna tower on top of Aeolian Hall on West Forty-second Street, and hung by his knees from a cross arm. At dusk, half an hour before the dedication ceremonies for the WJZ station, he was back up the open steelwork in a stiff breeze, this time vertiginously straddling an ornamental globe

on the very top of the antenna some 400 feet above the street. The next day he sent Minnie and Sarnoff a series of the daredevil photographs. An angry Sarnoff banned him from the tower: "If you have made up your mind that this mundane universe of ours is not a suitable place for you to be spending your time in, I don't want to quarrel with your decision, but keep away from the Aeolian Hall towers or any other property of the Radio Corporation." But Minnie succumbed. They were married on December 1, 1923, a few days before his 33rd birthday, with Sarnoff in attendance, and photographed on the sands of Palm Beach, Florida, with the improbable portable between them, the very first beach boom box. Up north storm clouds gathered.

Armstrong wrote later that he had luck in his invention of the vacuum tube as both the regenerative receiver and regenerative oscillator, or transmitter. If so, it was the 1 percent crowning a 99 percent of thoughtful and imaginative experiment. He certainly had little luck in the long-running sequel when a furiously jealous De Forest tried to claim regeneration as his own.

Armstrong returned from his honeymoon in 1924 to be plunged back into a war of patents he thought he had won long ago.

After failing with his 1914 application for a patent for the ultra-Audion receiver, De Forest had recast his challenge. In September 1915 he had applied for a patent for an "oscillating Audion," now claiming that he was the first to discover regeneration and both the receiving and transmitting features. This tactic exploited the mistake Armstrong had made in applying for two separate patents. It had halted the issuance of his second patent for the transmitting function. The conflict had come to the federal court of the Southern District of New York, in May of 1921. The presiding judge, Julius M. Mayer, had not been impressed with De Forest's "faulty memory" or his argument. The documentary evidence showed that in 1912 De Forest was working with an assistant named Herbert Van Etten in an effort to use the Audion as a telephone relay and amplifier and did not then know how to produce a radio frequency. Van Etten, said the judge, failed to corroborate De Forest as to the existence

BIRTH OF THE BOOM BOX: Married on December 1, 1923, Howard and Marion (McInnis) Armstrong honeymooned in Palm Beach—where the bridegroom presented the bride with the world's first "portable" radio.

of feedback. Neither man "had any realization of what is now the invention in suit." De Forest's sole evidence of priority was a circuit diagram and entries from a notebook kept by Van Etten, dated August 5 and 6, 1912. But two items from the original notebook had been omitted from the document presented to the court. Both suggested De Forest might have produced a regeneration circuit of audio, but not of high-frequency waves. In fact, De Forest had damned himself. In his angry retort to Armstrong's Institute of Radio paper in March 1915, De Forest had been so eager to refute Armstrong's caustic analysis he argued that regeneration did not produce oscillations. In finding for Armstrong, Judge Mayer had noted how the calendar made a mockery of De Forest's claims. If August 6, 1912, was indeed the date he had conceived of feedback, why did De Forest, a prolific and precipitous filer of patents, delay making any application until September 1915—two years after his alleged discovery, one year after Armstrong's patent, and six months after Armstrong had explained how the triode with feedback really worked? And in the meantime, the judge noted dryly, De Forest had filed 30 other patents. On March 13, 1922, three judges of the federal circuit court of appeals had unanimously rejected De Forest's appeal.

Thomas Manton, writing the unanimous decision, said Armstrong's "wholly novel idea" of feedback was "a great advance in the wireless art," and "a radical modification of an instrument that was little understood."

That should have been the end of it. By that time Armstrong had also been upheld by the Patent Office in three claims of interference. He had heard himself highly praised from the bench for having a modesty that belied his extraordinary ability, his "thoughtful and imaginative mind." Armstrong was all of that, but he was also proud and obstinate and he seemed to think of people only as either positive or negative. De Forest was, in his mind, a char-

latan who had never understood the Audion, a man of dishonor who had defaulted on the award of damages to Fessenden in 1906 and a thief and liar who lied to rob him of the fruits of his labors. Armstrong's generous nature was corrupted by his anger. He taunted De Forest by flying from his first aerial on Warburton Avenue a huge flag bearing the white numerals 1,113,149 so that De Forest might see it from

MORE MADNESS: Armstrong climbed the Aeolian Tower as part of his campaign to woo Marion McInnis. Not satisfied that the photographer had a good image of his daylight climb (page 217), Armstrong did it again that night, straddling the ornamental globe 400 feet above the street.

his home in the Bronx. Armstrong refused to waive the assessment of costs to be fixed by a special master against the nearly bankrupt De Forest and he refused to let De Forest license a regeneration circuit of his own. In the Alice in Wonderland world of patent litigation, dragging things out was a fatal error. The long delay arguing the assessment meant no final decree was entered. In the spring of 1924 De Forest struck back by appealing the adverse Patent Office decisions in the District of Columbia circuit court, the ruling body over the Patent Office. Armstrong hired a detective

to go undercover into De Forest's company; he suborned a secretary so he could search for evidence. De Forest, in fact, had no new evidence, no new arguments, but he had sold himself to American Telephone and Telegraph, and AT&T's lawyers had been very clever. They had persuaded an examiner in the Patent Office to change a single vital word in De Forest's application. He had claimed his Audion produced oscillations of high frequency, meaning radio waves. The lawyers got that changed to the broader "electrical oscillations," which could be construed as either low-frequency audio waves or high-frequency radio waves. De Forest was not thinking about radio transmission in 1912, but the important change in wording enabled the lawyers to claim he might have been thinking about it. The trick of verisimilitude was to make the language of the patent application seem older rather than newer. "Electrical" was something he could have written in 1912. "High frequency" obviously was not. It's the technique of forgers of old paintings who age their canvases and avoid pigments manufactured in the 20th century.

Judge Josiah Van Orsdel, the associate judge who gave the opinion in the District of Columbia, did not see through the subterfuge. He wrote that he was judging De Forest's applications "as broadly as their language will permit." He failed to distinguish between the generation of lower, audio-frequency electrical waves—noise!—to which De Forest might lay claim, and the generation of higher, radio-frequency electronic waves to which he could not lay claim, but did. Orsdel disregarded the reasonings and the conclusions of the two informed decisions of the New York courts for Armstrong as never having been decreed final, and he disregarded the three Patent Office rulings. On May 8, 1924, 10 years after Armstrong's award of the patent, Judge Orsdel took it away from him and awarded the patent

Out of this tangled mess in his lab, Armstrong created radio that could for the first time broadcast the full range of sound detectable by the human ear.

rights for regeneration and oscillation to De Forest.

The nightmare for Armstrong was generated by three "electrodes"—his own original obstinacy in disregarding his counsel and in refusing to waive his costs; the vested interest of those corporations who had a financial stake in a De Forest victory; and judicial precedence set by Orsdel rather than Mayer and Manton. These elements prevailed through 10 more years of litigation right up to the Supreme Court, and there not once but twice. On the first occasion, in 1928, finding for De Forest, the Court would consider only the judicial precedent, not the scientific facts. On the second, in 1934, when it did look at the science, it had to contend with a resounding 1933 revalidation of Armstrong as the inventor by a three-judge verdict in the Second Circuit appeals court. Justice Cardozo, who wrote the opinion for the Supreme Court, was a clever man unwilling to admit he was out of his depth. He accepted De Forest's assertions in a way that demonstrated, in the words of historian Lewis, "a fundamental lack of understanding of the circuit." Cardozo's reasoning and opinion upholding De Forest stunned the community of radio engineers and scientists, raising again the question of whether a lay

court is competent to adjudicate the issues of modern science. Professor Michael Pupin spoke for them in 1934: "The De Forest record and his testimony on the witness stand tell two different stories which do not agree. The court does not seem to understand the story of the record and therefore accepts the testimony. The scientific mind understands the record and rejects the contradictory testimony."

The radio community came through for Armstrong. After the Cardozo blunder, Armstrong went along to the annual convention of the Institute of Radio Engineers in Philadelphia that May. He strode to the podium—a bulky lumberjack of a man who walked like a sailor, said a *Fortune* reporter—with a speech in his hand. He had decided to say he was returning his gold medal. He did not get to say it. The board of directors had heard of Armstrong's intention. They were going to have none of it. The chairman announced that they "strongly reaffirmed" their institute's original citation honoring Armstrong's "engineering and scientific achievements in relation to regeneration and the generation of oscillations by vacuum tubes." Nearly a thousand engineers in the hall cheered and cheered the decision. Armstrong accepted the medal the second time with tears in his eyes.

The long litigation was traumatic, time-consuming, wearying to the soul and expensive for Armstrong personally after 1928 when Westinghouse and RCA pulled out of backing him. Westinghouse's financial interest expired with its patent in 1931 and an RCA deal with AT&T had incidentally given it an interest in the longer-running De Forest patent. Through the late '20s and into the '30s, however, Armstrong was buoyed up by a further flowering of his imagination.

He had taught the world how to amplify signals, advancing the form of radio known as amplitude modulation (AM). "Modulation" in the vocabulary of radio takes a low-frequency audio signal from a microphone, converts it to electrical energy and superimposes that signal on a radio wave of higher frequency; an AM radio receiver detects the waves and uses electrical circuits to remove the carrier wave and convert the modulating signal back into sound. To picture an AM wave, imagine a series of basketballs, tennis balls and softballs stuffed through a rubbery Slinky, expandable in its cross section. The varying height, or amplitude, of the Slinky carries information. The trouble was that the circuits also amplified natural interference from other transmissions, among them auto ignitions, storms and solar flares. Prewar and postwar, Armstrong and Pupin had tried devising filters to clean up the sound, removing static and improving fidelity. So had others, notably at the AT&T Bell Labs, where John Renshaw Carson was the brilliant chief mathematician-physicist. Everyone failed. Some engineers began to think of experimenting with the other variable, frequency rather than amplitude: hold the strength of the carrier wave constant, but change its frequency to match the pattern of imposed sound. Imagine our Slinky lying flat but one end moved back and forth at varying speeds. The frequency of the waves passing through the coil now carry information. Armstrong explained frequency modulation this way: "When the speaker talks or shouts, the wave widens. If he whispers, the wave narrows. This enables a differentiation between what goes through the microphone and the natural disturbances such as static."

All very well in theory, but making FM work proved beyond the ability of the radio industry. Westinghouse, General Electric and the Radio Corporation of America pooled resources in an attempt to devise a system of frequency modulation in which the signal occupied very little space on a long carrier wave—narrowband FM. There were two reasons for focusing on narrowband. The first was to relieve congestion of the airwaves, since sharply tuned narrowband takes less of the spectrum. The second was to reduce static. Wide bandwidth yields a superior range of audio, but by the same token offers a bigger target for interference. The narrower the band of transmission, the better the so-called signal to noise ratio, just in the same way, in technologist Gary Frost's metaphor, that shorter books harbor fewer typographical errors.

RCA managed to employ a form of FM to avoid fading the signal in a telephone relay broadcast of the Max Schmeling and Young Stribling fight in July 1931, but nobody could make narrowband FM practical in terms of high fidelity and low noise. Carson delivered the FM funeral oration in 1928, an irrefutable series of mathematical equations that showed narrowband FM solved neither congestion nor static. He was right on that. He was wrong on this: "As the essential nature of the problem is more clearly perceived, we are forced to the conclusion that static, like the poor, will always be with us." Certainty of this kind, born of abstract reasoning, was a red rag to the empiricist Armstrong. "You don't make inventions by fancy mathematics," he remarked crossly to an RCA engineer, "but by jackass's storage batteries around a lab."

He had begun to ponder FM in 1925, but in the wake of his Supreme Court reverse in October 1928, he embarked on prodigious experiments with manic energy. With his own money—he accepted only $1 a year from Columbia—he hired two assistants, his childhood friend Tom Styles, and John Shaughnessy, an Irish radio inspector from the New York Navy Yard. Every morning for five years, seven days a week, Armstrong went down to the Marcellus Hartley Lab in the basement of Philosophy Hall at Columbia University to pursue will-o'-the wisps. There were more of them, he said,

than "I ever thought possible." He stopped only at lunchtime for a single cheese sandwich and a glass of milk, often worked into the night and had no time for any conversation that was not about the immediate frustration. Shaughnessy survived making the occasional joke only by being very good at making models of circuit boards. Armstrong's genius for visualizing spatial arrangements translated into boards of increasing complexity, spreading from table to table until they occupied two whole rooms. He used 100 vacuum tubes in scores of thousands of tests and made scores of thousands of measurements. Three years later, none of the circuits was working as he had hoped.

In 1931 he seized conventional wisdom by the throat. If John Carson was proving right about narrowband FM, what about broadband? Carson's equations were unassailable on the assumptions he made, but he had assumed that broadband FM would behave as broadband AM. Everyone knew that the noise admitted into a circuit is proportional to its bandwidth—but what if there was some way in FM of limiting those interruptions? If that were feasible, a listener would enjoy high-frequency broadband's capacity to transmit the full range of frequencies audible to the human ear. An Everest of "ifs," but two years later, Armstrong had conquered it. His special transmitter varied the broadband frequencies to mirror the signals from the microphone. It was altogether tougher developing a receiver that would accept and amplify the FM signals without also amplifying static admitted into the broadband. Armstrong's answer was a form of electronic colander. His special vacuumtube circuit called a *limiter* strained off all amplitude noise (static), allowing only the pure frequency modulated current to pass through to a second circuit called a *discriminator.* This converted the original FM variations into amplitude variations ready for the loudspeaker. The limiter and the discriminator were integral to a wholly new form of radio, high fidelity able to play all the notes Mozart wrote without intervening buzzsaws—and Armstrong's FM inhabited the relatively untenanted ultra-short waves. Its range was limited to not much more than the horizon, but with the signal's amplitude held constant it was able to function on a fraction of the power of AM radio.

Armstrong was as excited as ever he could be on Saturday, December 23, 1933, when he invited his old friend Sarnoff to the Columbia basement to see what he had secretly contrived and already patented. The memory Armstrong cherished of the boss of RCA was of the voluble young supporter who had shivered with him in Marconi's radio shack at Belmar, New Jersey, in 1914. Sarnoff had gone on to wield great corporate power, which involved him in decisions inimical to Armstrong, but Armstrong persisted in seeing him through a bifocal lens. There, close up, was the friend he took coffee with on visits to RCA, who had come to his wedding, who had even sent him a private telegram of warm congratulations when he briefly triumphed over De Forest and RCA in the Second Circuit court in 1933. Fuzzily in the distance was the Sarnoff who on the following day did his corporate duty by authorizing a press release rejecting the very same circuit court decision. In the trauma of the Supreme Court hearing in 1934, Armstrong tried to fuse the two Sarnoffs. He appealed to him directly to tell Justice Cardozo that Armstrong was the true inventor of regeneration—"in all friendship to yourself" and then he added: "and to the Radio Corporation." The intention was to warn the corporate Sarnoff that he risked Armstrong's future cooperation with RCA. It was all to no avail.

Sarnoff, in doing the supergeneration deal, had secured for RCA the right of first refusal to Armstrong's next invention, which happened to be FM. Sarnoff was as impressed in the Columbia basement as he had been at Belmar, but this time he inwardly recoiled. What he had seen and heard, he told his friend, was not so much an invention as a revolution. That was a two-edged compliment. RCA was heavily invested in making AM radios and broadcasting AM through its National Broadcasting Corporation, founded in 1926. In the middle of the Great Depression, Sarnoff saw no future in asking Americans to spend more for better radio. The industry as a whole had $75 million invested in equipment and there were 40 million radio sets that would be obsolete. Sarnoff was sure enough, anyway, that radio was dead. The possibility of television obsessed Sarnoff and he called the shots now at RCA. He did not turn Armstrong down flat. Instead, he stalled, asking for tests and then more tests. They were conducted from RCA's experimental television station on top of the Empire State Building. Armstrong broadcast to a house 70 miles away in Westhampton Beach, Long Island, and then to a house 80 miles away in Haddonfield, New Jersey. The sound remained sweet and serene, while storms blacked out more powerful AM stations. In November 1934, he proved the capacity of FM broadband by simultaneously transmitting music, telegraphy and a high-quality facsimile front page of the *New York Times* on a single FM carrier wave. (Radio transmission of facsimiles had been achieved between Europe and the United States by the German scientist Arthur Korn in 1922, but despite big investments the fax machine as we know it did not become popular until the 1960s.) The following April, still without RCA's endorsement, he went public with the announcement that staticless radio was here. "After ten years of eclipse, my star is again rising," he wrote. Still, Sarnoff held back. In July Armstrong was told he had to move his equipment out of the Empire State Building to make way for more trials of television. It meant, it seemed, that he had no means of going public with a demonstration.

Armstrong was not so easily thwarted. He recruited three recent Columbia graduates to help and the result was the broadcast to the Institute of Engineers from Yonkers in November 1935. One of his new men, a clever experimenter called John Bose, who went on to make a name in Bose high-quality audio equipment, remembers Sarnoff's private visit the day after the club meeting: "He arrived with Fedora, cane, spats and whatnot. To an engineer this was pretty creepy." Once again Sarnoff tap-danced around a commitment. Armstrong at least realized what he was up against: His precious invention was consigned to corporate oblivion. He made a momentous decision. He would himself introduce FM broadcasting to the world; he would be an innovator as well as an inventor. For the

first time in his life, he would not look to a large corporation to convert his inventions into commercial operations. With the sale of a hunk of RCA shares, he had $5 million in cash. He was now as rich as he was brilliant, but it was a big gamble. This was no longer a question of supplying superior components to existing systems. He had to launch and manage a whole new broadcasting system. He had to find ways of demonstrating FM to the world to convince broadcasters and listeners that FM was the future; he had to design and build transmission equipment. He had to design radios to receive FM and have them manufactured. He had to expect, and duly did receive, sabotage from RCA, its affiliate NBC, the Columbia Broadcasting System and the rest of the radio industry. And he had to grapple in the murky world of licensing, where the premier science was not electromagnetic radiation but logrolling.

The controlling body, the newly enfranchised Federal Communications Commission, reflected RCA's hostility to FM. There was, historian Tom Lewis records, such "bureaucratic collusion" that innuendo propagated by RCA was swiftly recycled by ever-willing engineers at the regulatory agency; the door between RCA and the FCC was a revolving one. Armstrong's routine application for an experimental license to test FM was rejected in January 1936. It took him six months, counsel and a new FCC chief engineer—the earlier rejectionist having taken a job with RCA—before he was allocated the 42.5–43.5 megahertz band. It was less than he wanted but enough for five stations and enough for him to beam programs into New York if he could set up a station and find a suitable site for a high transmitting tower. In the meantime, Sarnoff was busy trying to steal his invention. He allowed RCA attorneys to object to Armstrong's patent on the false grounds that an RCA engineer had thought of it first. The interference case was ultimately resolved in Armstrong's favor but it took time, money and emotional energy. Armstrong, never given to hyperbole, finally spoke out about the campaign against FM by "intangible forces" originating in "vested interests, customs, habits and legislation." The friendship was over.

Armstrong found the site for his tower on 11 acres of the Palisades in Alpine, New Jersey, just across the river from the Yonkers attic of his boyhood. Alone, with only Marion at his side, he ceremonially chopped down the first tree in April 1937 with "a little three dollar axe." They saved the first wood chip as a memento and toasted the enterprise with champagne. These were heady months for Armstrong. As the steel tower rose to its full 425 feet, he was up and down in snow and ice, swinging high above the ground in a bosun's chair as he had done when he built his first wooden tower. The tower's blood-red signal lights started up for low-power tests in June of 1938. His friend Runyon concluded each demonstration by dropping ice into a glass, pouring in scotch and dousing it with a siphon.

Those who heard the "highball" tests were intoxicated. They included some enterprising broadcasters, John Shepard of the Yankee Network in Boston, and Franklin Doolittle, who opened a station in Meriden, Connecticut. But their enthusiasm in a small corner of the United States would have counted for little if the public could not tune in with radios capable of receiving FM. They were certainly not going to be made by RCA, which was busy advising the FCC and everyone else that Americans had tin ears; they did not care about high fidelity and would never pay for it. But there were other lions in the jungle. Armstrong had shrewdly commissioned the giant General Electric Company to build 25 FM radio sets for him at a cost of $400 each. Realizing how good a product it was, General Electric negotiated a license from Armstrong to manufacture "golden tone" radios at $70, paying Armstrong a royalty of 2 percent. To his precious mementos, he added, uncashed, the first royalty check from GE for $22.66, the first money he had collected in the five years since his invention.

Armstrong's mission gathered momentum. Alpine W2XMN went on the air at 35,000 watts of power on July 18, 1939, relaying classical music flawlessly from WQXR in New York. In January 1940, the Yankee Network, and Doolittle and Runyon, joined Alpine in a test Lewis rightly describes as having monumental implications for long-distance communication. A

TO THE RESCUE ON 9/11: Armstrong removes icicles from the 425-foot radio tower he built in 1937–38 in Alpine, New Jersey, for America's first FM radio station. Sixty-four years later major New York City television broadcasters rushed to install antennas on his tower (among them WNBC, Ch. 4; WABC, Ch. 7; WPIX, Ch. 11; and WNET, Ch. 13.) They lost their signals when the antenna on top of the north tower of the World Trade Center collapsed in the 9/11 outrage.

program from Runyon's little station in Yonkers was picked up by Alpine, which sent it to Meriden, which in turn sent it to Worcester, Massachusetts, from where it went to Mount Washington and then Boston—825 miles in a fraction of a second, and of a quality never before achieved by costly relays through the telephone system. It was a satisfactory punch on the nose for AT&T's famous Bell Labs, which had declared such long-distance relay impossible: "The song of the whirling electrons sets a natural limit and will always be with us." By then 40 experimental FM stations were broadcasting and 109 more were pending. Armstrong's dream of national high-fidelity broadcasting no longer seemed a fantasy. The FCC was ready to discuss permanent frequency allocations. Sarnoff did his best to set the clock back. He led the networks (then his company's NBC affiliate and CBS) in propaganda and crafty maneuvers against high fidelity. He cam-

paigned for wavelengths for television that had the incidental attraction of ensuring FM would not have room to grow. Armstrong, no longer an ingenue in the world of commerce, led a delegation to Washington from the newly formed FM Broadcasters Association in January 1940 with a little bombshell in his briefcase: copies of RCA engineering reports favorable to FM that the corporation had withheld from the FCC.

The FCC, subject to dizzying oscillations in its staffing, now had a chairman who was not captive to the industry, a New Deal activist lawyer by the name of James Lawrence Fly. He was shocked by RCA's deceit, then angered when Sarnoff, overplaying his hand, tried to pressure the FCC with a heavy advertising campaign for television sets before the channels had been allocated between FM and television. Sarnoff, the champion gymnast of business politics, performed an immaculate

back flip. RCA suddenly testified to the commission in favor of FM. Armstrong emerged from the hearings with wavebands of 40–50 megacycles for FM broadcasters, enough for 40 channels that could carry up to 2000 stations. FM was already coming to be favored as the mobile radio system by police, military and emergency services. And the commission ruled that FM was to be the standard for television sound.

Armstrong found himself courted by the corporation that had done so much to undermine him. Through intermediaries, Sarnoff finally sought a license to manufacture FM radios. Armstrong, still pained by RCA's support of De Forest, still angry at the way Sarnoff had toyed with him, dragged out the talks. In June 1940, he rejected an offer of $1 million cash, no royalties, for a nonexclusionary license. He insisted on royalties. Historian Lewis judges this "an astounding decision that demonstrated once again Armstrong's inflexible

FIGHTING FOR FM: Armstrong testifies at a Federal Communications Commission hearing in early 1940, when RCA was doing everything it could to stop the spread of FM. The commission allocated permanent frequencies for FM and specified it as the standard for television sound.

nature," but that is harsh. Armstrong was offering the enemy the same terms of a license against royalties accepted by General Electric, Zenith, Freed-Eisenmann, Scott and Stromberg Carlson. It was Sarnoff who was stubborn—and stealthy. In his mind was a scheme to circumvent his old friend's monopoly.

World War II frustrated Armstrong's prediction that the AM broadcasting system would be superseded within five years. Everything was on hold for five years. During this time Major Armstrong waived his royalties in favor of the army and navy and adapted his FM researches to radar: On January 10, 1946, U.S. Army Signal Corps engineers bounced the first radio signal off the moon and did it from Belmar, New Jersey, where 34 years before Armstrong and Sarnoff had heard his regenerative circuit pick up signals from around the world. The

moon shot proved FM waves communication could penetrate the ionosphere so radio communication through space was feasible—provided it was FM, not AM.

These postwar years should have been the pinnacle of Armstrong's career. He was universally esteemed in the engineering and academic communities; the Franklin Institute brushed aside the legal confusions by awarding him its medal for "pioneer work in Regeneration and the Oscillating Vacuum Tube Circuits, the invention of the Super heterodyne Circuit, the Super regeneration and a system of wide-swing Frequency Modulation." But once again, corporate chicanery and a pliant bureaucracy conspired to cheat him of his rewards.

The FCC, led now by Charles R. Denny, became captive to Sarnoff, who was the mastermind of a series of FCC decisions that crippled FM for 20 years. Making

technical judgments that Armstrong amply showed were fallacious, it needlessly removed FM from its hard-won frequencies to the 88–108 band, rendering obsolete 50 stations and half a million FM radios, and robbing Armstrong of millions of dollars of royalties expected in a postwar boom that he needed to recoup his long investment and continue research. After his rulings adverse to FM, Denny was hired as vice president and general counsel of NBC, RCA's subsidiary at the then high salary of $30,000 a year. Too late, Congress passed a law against regulators taking jobs from the industry they were supposed to regulate.

When Armstrong rejected the terms of Sarnoff's bid for a license in 1940, the RCA negotiator Gano Dunn had warned him he would be making a big mistake if he thought he could fight "anyone so powerful as the RCA." RCA did not make an FM radio until 1946 but when it did it was by the simple expedient of ignoring Armstrong's patents. They were ignored also in RCA's installation of FM for television. Sarnoff's ploy was as simple as it was brutal. RCA would claim it had invented its own system of FM independently of Armstrong. If challenged in legal actions, it would easily outspend and outlast a lone inventor. His royalty revenues were due to run down on expiration of his patents from Christmas Day, 1950. If RCA could drag out the retrial process of discovery of documents and testimony, it could in that time at least avoid a court ruling preventing its manufacture of FM equipment.

An exasperated Armstrong filed suits in the federal district court in Wilmington, Delaware, on July 22, 1948, charging RCA/NBC with willfully infringing five of his basic patents and having encouraged others in violation. Sarnoff's lawyers were ready—not ready for trial, but ready for a rearguard action against trial. They excelled themselves in the discovery process. World-class procrastinators, they kept Armstrong in a chair for a year with petty, irrelevant and vindictive questioning. The court finally ordered that stopped, but the filibuster and the fog in its wake persisted as the depositions slipped into 1950, then 1951, then 1952, then 1953, without any prospect of a trial. "They will stall this along," Armstrong remarked, "until I am dead or broke." He eviscerated the testimony of RCA's expert witnesses, but the emotional cost to him of proving that he was the inventor was wrenching. He directed every aspect of the case. It took priority over his beloved research with John Bose in the Columbia lab. His wife was in third place; she had no children to comfort her. He had installed Marion in a grand apartment in the River House, midtown on the East River, but he had also converted it into his office and she agonized as he wilted before her. He would get out of bed in the middle of the night to read and reread transcripts and make notes for the next day. St. Vitus twitches became more pronounced. He had his stomach pumped out one night convinced he had been poisoned. The stress of living with a man obsessed put Marion into psychiatric care. She tried to jump into the East River. She spent months in a mental hospital. Armstrong seemed oblivious to everything except the injustice being done to him.

There was a brief moment of exultation in March 1953. He and Bose announced they had succeeded with multiplexing. It made it possible to transmit two or three different programs over the same FM band at the same time—or to bring stereophonic reality from a concert hall. But Armstrong was indeed being poisoned—by the case. One day in a lawyer's office in 1953 he watched Sarnoff himself loftily answer questions. He was smooth. "We saw each other frequently either in my office or my home," Sarnoff said of Armstrong. "We were close friends. I hope we still are." But as the questioning went on he showed his teeth. "I had as many technical advisers," he snapped, "as a dog has fleas." He had been advised by his patent attorney that RCA engineers, not Armstrong, discovered the basic law of FM, and then he let forth: "I will go further and say that the RCA and NBC have done more to develop FM than anybody in this country, including Armstrong." A lawyer present says Armstrong's eyes fixed Sarnoff with a look of pure hatred. Only a saint could have borne it with equanimity and Armstrong was not a saint.

One of the last social commitments he recognized was an annual Thanksgiving dinner with radio friends and their wives. This night, when the guests had gone, he told Marion for the first time the case had ruined them. He was nearly bankrupt. He asked if she would consider giving back to him some of the money he had given her. She demurred. It was money put aside for their retirement. Perhaps he should accept a settlement that had been vaguely mooted. Rage overtook him. Wildly, he reached out for a poker and swung it. Her forearm took a glancing blow. She rushed out of the River House to a doctor and her sister's home in Connecticut. Armstrong was alone in the empty rooms. She never came back in his lifetime.

Armstrong was alone over Christmas and the New Year. The proposed RCA settlement had proved derisory—RCA made an offer—$200,000 and an additional million maybe and only if other man-

ufacturers agreed. On the last Wednesday of January, he stirred himself to call on Bose at the lab. He "looked like hell," said Bose. "He was always telling me what to do—Do this, do this, try this, but this time he had no interest." Sunday, January 31, 1954, was the 40th anniversary of the night young Edwin Howard Armstrong had proved the power of his first invention to Sarnoff in the hut in Belmar, New Jersey. He took up a yellow legal pad and wrote to Marion:

Dearest Marion,
I am heartbroken because I cannot see you once again. I deeply regret what has happened between us. I cannot understand how I could hurt the dearest thing in the whole world to me. I would give my life to turn back to the time when we were so happy and free. My estate is solvent, especially if RCA comes through. Also, the Telephone Company should pay something, for they have been using my inventions.
God keep you and may the Lord have mercy on my soul.

Ed

He put on his overcoat, with scarf and gloves, climbed outside his 13th-floor bedroom and jumped to his death.

DEMOCRATIZERS

The innovators in this section could hardly come from more different fields. What is there in common between the automobile and the camera, between banking, beauty treatments and the Boeing 747? The short answer is that Henry Ford, George Eastman, Amadeo Giannini, Madam C. J. Walker, Martha Matilda Harper and Pan Am's Juan Trippe all profited from the mass markets they created and served. But time and again we find the innovators moved by the desire to be remembered as public benefactors, and their effect on society was more profound than simply feeding the engines of consumption. They also served the American dedication to democracy—to equal rights and freedom. Eastman's Kodak gave everyone equal access to memory; Giannini opened banking doors, and opportunity, long shut against the less affluent; Walker and Harper offered women a chance of independence; Trippe put the world at the feet of the common man. There was risk in it. It was by no means sure when they adventured in this manner that the masses would buy cars or open bank accounts, or take to intercontinental air travel.

Of course, the democratization of consumption is not the same thing as the democratization of political rights, but the two seem to march in tandem in American history. The U.S. Constitution created expectations of the good life, and these innovators fulfilled them.

RACING HENRY: Henry Ford loved fast racing cars. This is Ford (left) in his 999 racer apparently dueling it out with Harry Harkness in a Simplex. Since photography was then not capable of freezing motion at high speed, Ford had the photograph retouched—but it is a fair representation of his style and this is the model he drove on the ice of Lake St. Clair in 1904, at a top speed of 91.37 miles an hour, a world record.

HENRY FORD He gave practical reality
to the rhetoric of American democracy
by fighting for the people's car

1863–1947

IT WAS NOT LOVE at first sight. In the early years of the twentieth century, the mass of Americans resented the automobile as an offensive symbol of the gulf between the pretentious rich and the working poor. Literally, only one American in a million could afford a car. "Nothing has spread socialistic feeling in this country more than the use of the automobile, a picture of the arrogance of wealth," declared the president of Princeton University in 1907, one Woodrow Wilson. The future president of the United States was at the crest of a tide of public opinion. "Our millionaires and especially their idle and degenerate children have been flaunting their money in the faces of the poor as if actually wishing to provoke them," warned a writer in the *North American Review,* evoking images of the Paris aristocracy before the guillotine fell. In the first six months of 1906, fumed the magazine's contributor, rich people speeding through the streets in their big cars killed more Americans than the enemy did throughout the entire Spanish American war in 1898. A farmers' magazine, *Breeder's Gazette,* indicted motorists as "a reckless, bloodthirsty, villainous lot of purse-proud crazy trespassers." To thwart and frighten the noisy intruders on Arcadia, country people dug ditches on roads, put up log barriers, scattered shards of jagged glass and took pot-shots with their rifles. In one assault in 1909, farmers near Sacramento, California, forced 13 cars into the ditch. Several counties in Pennsylvania and West Virginia banned cars from country roads. Vermont compelled every motorist to have "a person of mature age" walk ahead bearing a red flag.

A country boy from Michigan put all that into reverse. While much of the nation fulminated about the ravages of the rich man's plaything, a middle-aged Henry Ford sat in a rocking chair in his Detroit workshop perfecting a counter-revolution.

THE MAN YOU WERE LOOKING FOR

Did you spot the genius in the group portrait of workers taken about 1893 at the Edison Electric Illuminating Company, Detroit, Michigan (pages 2–3)? It is chief engineer Henry Ford.

Ford launched his Model T in October 1908 with the proclamation "Even You Can Afford a Ford." Despite a common notion that he invented the automobile, Ford was in fact slow to pursue the idea he claims to have had as a boy. As a carmaker, he was preceded by others in America and he was preceded in his hometown of Detroit. More than 3,000 car companies were formed in America in the decade between 1895 and 1905, and hundreds of them actually put cars on the market. So Henry Ford's singular achievement was not being first. It was in liberating the common man. His inexpensive and rugged Model T, by which America came to have a love affair with the car, was a perfect unostentatious instrument for a restless people inhabiting a vast continent where each man considered himself as good as the next. In a statement that is believed to have been made as early as 1903, he promised, "I will build a motor car for the great multitude. It will be so low in price that no man making a good salary will be unable to own one—and enjoy with his family the blessing of hours of pleasure in God's greatest open spaces."

His aspiration would have been empty piety without his genius in organizing the technological and social means of mass production, and his lonely heroism in the face of legal warfare waged by monopolists intent on keeping prices high. Just as democracy had never been tried on such a scale as in the United States, so had democratic industrialization never been brought to such a pitch as Ford brought it. As the art critic John Kouwenhoven observed, the Model T was also an unabashed expression of the no-frills vernacular tradition in America, just like its modern successor in popular affection, the honest-to-God army jeep.

Two great managerial innovations of the twentieth century were at the heart of Ford's achievement: the assembly line and the $5-a-day wage. He cannot be said to be the sole inventor of either of these, but his vision was the fulcrum. This faith, and inspired leadership in its execution, was Henry Ford's supreme innovative contribution to the making of a new America. It would be for another generation to resolve the paradox inherent in the Ford system, that a car for the multitude had to be made in factories that were about as authoritarian as one could get.

FORD'S FATHER, William, was 21 when he arrived in America, one of a family of poor County Cork tenant farmers, Anglicans in a Catholic community, who fled the potato famine in 1847. They settled near relatives in Michigan where William became a carpenter with the Michigan Central Railroad and was enchanted to discover, as his immigrant father before him did, that if he saved he could be a landowner. This, said his daughter, Margaret Ford Ruddiman, was to him "the great miracle of America." Henry Ford absorbed this sense of romance

growing up on the prospering family farm in Springfield Township, 10 miles west of Detroit, but he liked the concept of land-owning democrats much more than farming itself. There was too much physical labor in it for little reward, so he remembered his father's acquisition of a McCormick reaper as an epochal event. The labor-saving reaper no doubt intensified his interest in the way machines worked. He was in the Eli Whitney mold of the schoolboy tinkerer and as fascinated by steam as Oliver Evans. He plugged a kettle to see what would happen and suffered a wound from exploding shrapnel. His favorite pastime was dismembering watches. A neighbor remarked, "Every clock in the Ford home shuddered when it saw him coming."

He learned little in his years in a one-room schoolhouse in Dearborn beyond simple arithmetic. He was a poor reader and his scattershot spelling rivaled Sam Colt's, but in an unformed hand he jotted down in a small blue spiral notebook aphorisms from the McGuffey Eclectic Reader series. They were emotionally engraved in his heart by his revered mother, Mary. "Life will give you many unpleasant tasks to do," he remembered her reciting to him. "Your duty will be hard and disagreeable and painful to you at times, but you must do it. You may have pity on others, but you must not pity yourself. Do what you find to do and what you know you must do to the best of your ability." Ford later said McGuffey was the man most responsible for teaching "industry and morality" to America.

Mary died in March 1876 giving birth to her eighth child. She was only 37, and 13-year-old Henry was inconsolable. In his autobiography, he wrote, "The house was like a watch without a mainspring." Even in his poetic moments, Ford's metaphors were mechanical.

Soon after her death, he was excited to see a steam engine moving down a road on its own power. Stationary engines were common, but this farmer had run a chain to the wheels of his wagon. According to Ford's autobiography, this was the moment he determined to devote himself to making a vehicle. And the rest is history, as they say. History, when it is not bunk, is actually a good deal more complex than a eureka moment and the straight-line progression implied in the standard Ford story. At the very least, if we take Ford at his word, he went about his mission in a roundabout manner; he certainly did not soon manifest the obsessive single-minded drive characteristic of many of our innovators. Through much of his life, his personality alternated between bouts of laserlike focus and bouts of abstraction.

At 16 he was fired after six days as an apprentice engineer in Detroit. He found and finished other engineering apprenticeships. At 20, he was back on the family farm, dutifully walking four miles to church every Sunday. Then he took a job with Westinghouse, traveling to southern Michigan to fix farmers' steam-traction engines. He made a workshop at home, but the night-school lessons he took in the city in this period were generalized—mechanical drawing was sandwiched in with accounting, typing and business administration. At 25 he married Clara Jane Bryant on her 22nd birthday; she was a small, vivacious, sensible young woman from a neighboring farm whom he'd met at a square dance and taken for romantic rides in a pretty horse-drawn sleigh. He seemed to be reconciling himself to a country life on 80 wooded acres given him by his father.

All the while in Europe an automobile industry had already sprung up. Gottlieb Daimler had built his gas-powered four

wheel car in 1886, and the French had swiftly followed, bequeathing us *automobile* and *chauffeur*.

Ford was hardly galvanized by these European breakthroughs. Clara designed a charming balustraded house with a verandah and Henry built it, cutting most of the lumber with a steam-powered saw. In 1890—four years after Daimler had put a car on the road—Ford, on a service errand for Westinghouse, had his first sight of one of the four-stroke gasoline engines called Ottos (after the 1876 inventor Nikolaus Otto) installed in a soda plant in Detroit. He was familiar with steam engines, but the action of explosive gas ignited in a cylinder excited him. The legend is that back at the Square House, as they called home, he took a sheet of music Clara had been playing at the piano and sketched for her how he would adapt a "silent Otto" as the heart of a vehicle. He set himself to design a gasoline engine. He had trouble understanding how the Otto's ignition worked—but again he did not seem to be in any hurry to find out. Both Henry and Clara independently affirmed that they decided to leave Dearborn so he could learn more about electricity, but not until September 1891 did they abandon their rural idyll for a $10-dollar-a-week apartment in clamorous Detroit, where 200,000 people were in the grip of a cruel recession. A night-shift supervisor transmitting power for the Edison Illuminating Company in Detroit had been killed on the job, and Ford stepped into the dead man's shoes (in the same year, Samuel Insull started electrifying Chicago). Ford kept the lights glowing without a flicker on city streets and in a few hundred homes, and in rapid promotions he was raised to be chief engineer of the main plant, overseeing a staff of 50 at the handsome salary of $1,000 a year. He was 28 but looked

Roadster $590

Torpedo $590

Town Car $900

Delivery $700

NEW YORK STREET SCENE, 1908: By 1914 every second car on the road was a Model T.

older, with a full mustache, and holding his lean frame erect he looked taller than his 5 feet 8 inches. He was a good boss. He earned respect by the way he organized the plant and affection by his readiness to laugh and joke with his coworkers; in the group photograph shown on page 160, he contains a smile, clearly at ease among them, and when they break up he will no doubt indulge his pastime of challenging someone to race him back to the plant.

Yet Ford was never "one of the boys": He could turn the lights off within himself. There was always about him a sense of difference, manifest in his bearing, his ambitious imagination and his candor, as much as in his abstemious eating habits and his abhorrence of alcohol and tobacco. Contradictions raged within him, "a conflict that makes one feel that two personalities are striving within him for mastery," the Reverend Samuel S. Marquis, head of Ford Motor Company Sociological Department, later said. "There are in him lights so high and shadows so deep that I cannot get the whole of him in the proper focus at the

same time." Ford could easily downshift from amiable to aloof, charming to plain vicious. He was always intriguing. "Henry had some sort of a magnet," said an Edison plant worker, Frederick Strauss, who had been an apprentice with him. "He could draw people to him; that was the funny thing about him." People routinely said they could feel when he entered a room. His engaging spirit had the same effect on the night-school students when he taught machine-shop practice in his spare time in the winter of 1892–93 (at the very same institution where the train boy Thomas Edison had spent hours in the library). The night owls were thrilled when he enlisted them in his little private projects. At the power plant, too, where Ford set aside a room with a lathe and electric coils for his own work, engineers and mechanics were eager to lend him a hand making his big ideas a reality. When he turned up at night for pie and milk at their favorite dive, the Night Owl Lunch Wagon, an eatery in a converted trolley, shift workers taking a break would cluster around him, asking to

see any bits of a gasoline engine he might have in his pocket. According to Strauss, it was in the power plant that Ford (and his collaborators) finished the model of an engine. Strauss says, "It took us about six weeks to get this little engine built . . . we had an awful time with the ignition."

Six weeks! Since it was ready for testing in December 1893, this means Ford did not start work on it until September, nearly three years after his sketch on the piano music. What finally got him moving? On September 20, 1893, in the town of Springfield, Massachusetts, J. Frank Duryea drove a four-cylinder vehicle designed by him and his brother, Charles; the speed was five miles an hour. On November 6, Clara gave birth to their son, Edsel, their only child. Biographer Robert Lacey suggests a baby in the home concentrated Ford's mind. It may also have been around this time that Ford heard the sermon he reported to his sister Margaret in which the preacher urged the congregation to "hitch your wagon to a star." All his life he honored simple aphorisms and exhortations as a way of honoring his mother. He told Margaret, "'Hitch your wagon to a star' is what I am going to do."

On Christmas Eve 1893, while Clara cooked dinner for visitors, Henry rigged up his model one-cylinder engine on the kitchen table. When he eventually got it going, Clara's kitchen was filled with black smoke.

He was on his way at last, but when he started in earnest in 1894–95, he was well behind other experimenters in America—who, in turn, were two or three years behind the Europeans. In 1893, Ford could have simply ordered an engine. This is what another Detroit inventor did. Charles Brady King was a Cornell-educated draftsman with a Michigan railroad car company who had grown up in an army family in the West. Fascinated by watching the transcontinental lines stretch westward, he had invented a brake beam for railroad cars and a pneumatic hammer and exhibited them both at the Columbian Exposition in Chicago in 1893. There he saw—and ordered—a Sintz gasoline engine then being made in Grand Rapids. Contriving the best way to make the engine propel a carriage was still a conundrum for him when he met Ford in the winter of 1894, probably through a King employee, Oliver Barthel, one of Ford's YMCA disciples. King, five years younger than Ford and familiar with the European technical literature, talked over automotive problems with Ford and they became friends. "I used to go over to Henry's place at night and help him tinker with a little one-cylinder experimental engine which he'd clamped into place on the kitchen sink," he told Brendan Gill of *The New Yorker* in 1946. "Mrs. Ford used to be cross with us. She had baby Edsel sleeping in a crib in the next room and she was afraid the gasoline fumes would poison him."

Ford and King shared an interest in psychic as well as practical issues. Biographer Robert Lacey writes that some of Ford's own thinking was shaped by King's theory that human inventiveness was a special, undying essence preserved in "the immortal mind." It reinforced Ford's belief in reincarnation; born at the end of the month that started with the battle of Gettysburg, Ford reckoned he was continuing the life of a Union soldier killed in the battle.

"Time was no longer limited," he reflected later. "I was no longer a slave to the hands of the clock. There was time enough to plan and create." Every experience in this life was worth having because it would be passed on to the next life.

It was in King's office, said Barthel, in January 1896, that Ford caught sight of a November 1895 copy of the *American Machinist* containing diagrams of an internal combustion engine that could be made with ordinary machine tools. "I want to build one of these," said Ford, but by the time he had put everything together his friend King had beaten him to produce the first car in Detroit. King had the benefit

Ford, at the age of 43, tries to impress Clara with his cartwheel, on vacation in Atlantic City. Little Edsel in the background is too busy to notice.

of a wooden chassis from Emerson and Fisher Company of Cincinnati. He installed his four-cylinder engine, a foot operated accelerator, a muffler and gasoline and water tanks, and on the evening of March 6, 1896, drove his vehicle through cold fog along St. Antoine Street at five miles an hour with a serious looking Ford cycling alongside. In the same year, Charles Duryea launched the first American company to manufacture and sell gasoline-powered vehicles; he sold 13 high-priced cars.

Ford was now in an uncharacteristic frenzy to finish a vehicle he had begun in a shed next to half of a brick house he rented at 58 Bagley Avenue. Again, it was a cooperative venture with three of his

power-plant wizards—Jim Bishop, George Cato and Edward S. "Spider" Huff—running on Ford's high-octane enthusiasm. A blacksmith named Daniel Bell, hired to forge the metal parts, said, "I never saw Mr. Ford do anything. He was always doing the directing," but in fact Ford did work himself to a frazzle in this final phase. He was inventive—his design for the 1897 carburetor won a patent—but the early car was a multiple of inventions. There was nobody to supply a magneto to carry electricity to a spark plug, and nobody to supply a spark plug. Everything had to be invented and then fabricated from scrap metal. So, too, with the multiple components like valves, camshafts, piston rings, pushrods, transmissions and radiators. The generous King gave Ford some valves and guaranteed his credit at the hardware store for 10 feet of drive chain.

Three months after King's triumph, in the early hours of June 4, 1896, with Clara looking on, Ford finished his own inaugural horseless carriage, having worked nonstop for 48 hours. Of course, as everyone likes to point out, he commemorated human fallibility by failing to notice that the shed door was too small for his ceremonial entrance into the street. At 4 a.m., having demolished the doorway, and with his arms still covered in grease patches, he steered his aptly named Quadricycle into the darkness and drizzle with Jim Bishop cycling ahead to warn carriages: The Quadricycle, his engine conjugated with four bicycle wheels, had no steering wheel, just a tiller, no brakes and no reverse gear. What it had was speed. Ford was an also-ran in terms of historic firsts, but as we have seen time and again being first does not guarantee innovative success. His curious latecomer was four times faster than the carriage cars made by Duryea and King. Ford's car weighed only 500 pounds, against King's 1,300 pounds, and its two-cylinder, four-horsepower engine propelled the driver at 20 miles per hour. King went along for a spin, and soon afterward Ford drove the eight miles to see his family in Dearborn, Clara sitting at his side, baby Edsel on her lap, on the board that passed for a seat.

The question looming for Ford was whether to start a business, to make the risky leap from inventor to innovator. King decided not to follow the Duryea brothers into the competitive melee. The magazine *Horseless Carriage* estimated that between July 1 and November 1, 1895, no fewer than 300 motor vehicles were under construction (by bicycle mechanics for the most part). Few of them saw the light of day, but to make a serious entry into business, Ford would have to give up a secure, well-paying job with fine prospects. It was also still unclear whether the internal combustion engine would become the dominant technology. Of 4,000 cars manufactured in America around the turn of the century, three-quarters were electric or steam.

Thomas Edison made the difference. In August 1896, Ford met the man he had idolized from youth as the guest of Detroit Edison's president, Alexander Dow, at the Edison Illuminating Company's 17th annual convention on Coney Island. Dow introduced Ford to Edison as the "young fellow who has made a gas car" and the insatiable Edison pounced. Ford recalled, "He asked me no end of details and I sketched everything for him, for I have always found that I could convey an idea quicker by sketching it." Edison, then 49, pounded his fist on the table and said, "Young man, that's the thing! You have it. Keep at it. Electric cars must keep near to power stations. The storage battery is too heavy. Steam cars won't do either, for they have a boiler and fire.

Your car is self-contained, carries its own power plant, no fire, no boiler, no smoke and no steam. You have the thing. Keep at it!" This was an endorsement remarkable for its disinterest since Edison himself was experimenting with electric cars. The encounter was the final push that led Ford into business: "That bang on the table was worth worlds to me," he later wrote. He told Clara on his return, "You won't be seeing much of me for the next year!"

He kept his job, but set up a machine shop of his own, spent $100 on machine tools—and moved the uncomplaining Clara to still cheaper lodgings. (Over 20 years in Detroit he moved his family once every two years.) He was the Pied Piper, attracting ambitious young men to help him in a backyard barn. A little resentment simmers in a comment by his old friend Fred Strauss—"Henry never used his hands, to tell the truth. He never came to work until after nine, either"—but how much Strauss himself was spellbound is indicated by the fact that he put in working stints for Ford without pay. Out of these combined efforts emerged the second-generation Ford car early in 1898, and then a third in 1899, a bigger, shinier car with brakes. These were all handmade, one at a painful time. On August 5, the city's business elite, led by the mayor and family friend William Maybury, put down $15,000 so that Ford could start multiple production in the $150,000 Detroit Automobile Company, the city's first auto company with a factory on the rural outskirts. Ford quit the Edison power plant, forsaking a splendid salary of $1,900 a year as general superintendent for a small stake in the new enterprise.

What happened next is strange, though it chimes with Ford's willful character: He did not continue the line he had begun. On January 12, 1900, he unveiled a heavy delivery wagon; it prompted an ecstatic fanfare in the *Detroit News-Tribune* following an ingratiating ride Ford gave the reporter, but the company lost $250 on each of the dozen or so produced. Ford became the invisible man in his own workplace. He showed up less and less. He went off for hours into the nearby woods, leaving word that if his investors asked for him his associates were to say he was out of town. In his autobiography, he suggests his heart was

not in the company because his backers were not interested, as he was, in building cars for the mass market. "The whole thought," he wrote, "was to make to order and get the largest price possible for each car. It was merely a money-making concern." (And not ever that. It lost $86,000.) But as biographer Douglas Brinkley writes, "It was Ford himself who didn't hold up his end of the deal." He made a few cars as well as the delivery vehicles, but he spent much of the investors' money building a race car.

The Detroit Automobile Company was dissolved in February 1901, but the directors indulged Ford, letting him stay in part of the plant to finish his racer. Most of the key men had to take other jobs, but they moonlighted for Ford. Childe Harold Wills, whose bulk belied his delicacy in draftsmanship, flitted between Ford and his desk at the Boyer Machine Company; Barthel, the clever designer who worked with King, sacrificed his evenings; Ed Verlinden, a lathe operator; Charlie Mitchell, a blacksmith; and Spider Huff, the electrician, all found time for Ford. The atmosphere was go-as-you-please. A young Alfred Sloan Jr., who then headed the nearby Hyatt Roller Bearing Company, dropped in on the unheated workshop one winter and found Ford and Wills flailing at each other with boxing gloves. It wasn't a fight, just their jokey way of getting warm enough to work at the drawing board. Ford liked nothing better than catching workmates with exploding cigars and electrified doorknobs. His concentration on a race car sounds like another jape, an irresponsible diversion from his proclaimed ambition to build a people's car. Though unfair to his backers, it might actually have been the shrewdest thing he did. He learned more about achieving speed and reliability under stress than he would ever have done in the dogged one-by-one duplication of a car he had already designed. Work on the race car put him in the forefront of automobile technology: Spider Huff invented a spark coil that Barthel took to his dentist for insulating in porcelain, the genesis of our modern spark plug.

In October 1901, it was announced that the celebrated Alexander Winton, a Cleveland carmaker, would race all comers 25 times around the new one-mile oval dirt track at the Grosse Point resort east of Detroit. Winton in his 40-horsepower Bullet had beaten everyone in a Chicago event, averaging 38 miles an hour. No one had yet hit 60 miles per hour, and the mile-a-minute mark was a threshold to beat like the sound barrier a half-century later. Henri Fournier, a French speed demon who held the world mile record of 1 minute, 14.2 seconds, failed to show up on the big day at Grosse Point and two other racers pulled out. Winton consoled the crowd by driving an exhibition lap in which he clipped a full second off Fournier's record. There was nobody left to race against except Ford, a late entrant unknown in the racing world. The organizers cut the distance to 10 miles to spare spectators the boredom of watching Winton the pro lapping the amateur who was driving a two-cylinder car producing only 26 horsepower. Ford was allowed two practice loops and found it hard to follow the tight curves. Spider Huff volunteered to act as a counterweight to the centripetal forces by standing on the running board so he could lean out in the curves, holding on for dear life as he did so. This was the kind of dedication Ford attracted—and expected.

Winton knew how to take curves and in the race he quickly left Ford choking on his dust. Three miles into the race, Ford got the hang of things and actually started to gain on the champion. The crowd of 7,000 cheered wildly. On the seventh lap Winton's engine spouted smoke and Ford overtook him to finish. Clara wrote to her brother afterward, "Henry has been covering himself with glory and dust . . . you should have heard the cheering when he passed Winton. The people went wild." The hometown boy won the race in 13 minutes, 23.8 seconds. He had averaged a promising 45 miles per hour, but standing in the winner's circle with friends, he vowed: "Boy, I'll never do that again. That broad fence was right here in front of my face all the time. I was scared to death."

BACK TO NATURE: The solitary ice-skater is Ford at 55. He loved to escape the tumult by getting out in the countryside. Watching birds was his favorite pastime. He bought up farmland and forest around Dearborn to create bird sanctuaries, then built a 56-room mansion on a 1,300-acre estate called Fair Lane. It is now a National Historic Landmark.

Clara wrote, "That race has advertised him far and wide," and she was right. Investors came back, including some who had been burned in the first fiasco. A new company, called the Henry Ford Company, was incorporated on November 20, 1901, dedicated to building a lightweight car for $1,000—or so they thought. Once again, Ford secretly concentrated the talents of his team on a race car, this time a monster called 999 that was nearly 10 feet long with four huge cylinders yielding at least 70 horsepower. (It was to set a record of five miles in 5 minutes, 28 seconds on October 25, 1902, with Eli "Barney" Oldfield at the wheel.) Nine months of Ford's masquerade

was too much for the investors. They brought in 57-year-old Henry M. Leland, a graduate of the Springfield Armory and the Sam Colt factory in Hartford who had developed machine tools with a tolerance of 1/100,000th of an inch. Leland was not interested in the mass market. He wanted to build big, finely engineered, highly priced cars. In March 1902, Ford resigned, or was fired, with $900 compensation and the right to reclaim his name. The Henry Ford Company was renamed the Cadillac Motor Company, after Detroit's founder, Antoine de la Mothe Cadillac. (When General Motors was formed in 1909, Cadillac would be its top seller.)

Once again Ford was on his own—but not for long. He took a decisive step toward the people's car the day he gave Harold Wills the commission to design a mass-market model. Ford told his attorney, John Anderson, what he had in mind: "The way to make automobiles is to make one automobile like another, to make them all alike . . . just as one pin is like another when it comes from the pin factory." Racing experience persuaded Ford and Wills that they would get more power and less vibration if they placed the two cylinders upright, instead of horizontally, an innovation of lasting importance. Detroit's dominant coal merchant, a Scottish immigrant in his thirties by the name of Alexander Malcomson, had seen the 999 race and offered to finance a prototype of the Ford Wills model that would become Model A—a bold act of faith at a time when there was much distrust of the automobile. In November Ford and Malcomson signed an agreement to manufacture cars in premises on Mack Avenue under the rubric of the Ford Motor Company. They would jointly own a 51 percent majority if investors could be found to subscribe $150,000. As biographer Douglas Brinkley writes, "The founding of the Ford Motor Company ranks among the most significant events in twentieth-century U.S. industrial history,"

ARMATURE ASSEMBLY LINE, RIVER ROUGE, 1934: It was tough, concentrated work. Spotters jumped on talkers. But River Rouge was a model factory, way ahead of its time. Ford installed elaborate dust and climate controls, and 2,900 drinking fountains. His nutritionists ensured box lunches never contained fewer than 800–900 calories. Five thousand men did nothing all day except keep the place spotless.

his consolation prize. (He later became a two-term senator.) Couzens brought immediate sense to the accounting and flow of cash and parts. He conceived of a distribution network of high-quality dealers (by 1905 he had 450 of them). He kept a tight hand on Ford's perfectionism. Ford the dominator accepted the abrasive Couzens's rulings; it was one of those disparate partnerships that mark so many innovative enterprises.

Five years into car manufacturing there were now suppliers of components. Ford farmed out much of the engine building to two redheaded tearaways who ran the best machine shop in Detroit: Horace and John Dodge. The Dodge brothers sent engines and transmissions by horse and carriage to Ford's shop to be married to bodies from C. R. Wilson and wheels from the Prudden Company of Lansing. It was touch and go for Ford and Couzens in the first few months of production in 1903. The $150,000 had not been fully subscribed, they had no advance orders, the bank balance dropped to as low as $223.65, and until a car was sold they could not pay the two Dodge brothers, who credibly threatened to beat up Ford. The great day was July 23, 1903, when a Chicago dentist, Dr. Ernst Pfenning, paid the full cash price of $850 for a Model A.

The Ford Motor Company roared into profits. It made $150 on a single car, and 215 were sold in two months. By the end of the first year Ford had sold 1,000 cars and employed 125 people. The company was profitable so quickly that within 15 months investors got 100 percent dividends on their stock, and Henry Ford, with $25,000 for his zero investment, bought a dress suit and thought about moving out of a cramped rented house near the workshop. In late 1904, the Ford Motor Company moved into premises on Piquette Avenue. The space was so big, a mechanic told Ford he doubted the company would ever grow enough to fill it all. Ford's response was: "Let's run it!" He raced the mechanic from one end of the building to the other and then back, more than 220 yards; into his old age he enjoyed challenging people to footraces. Twelve-year-old Edsel was able to cycle around the empty building—but not for long. It soon filled with machinery

but it was hard selling the stock and it was not until June 16, 1903, that the company was incorporated with 12 stockholders. Ford vetoed a thirteenth investor as bad luck. (It was clearly bad luck for the investor: By 1919 his spurned $500 would have been worth $1,750,000.)

Ford made light of his two bad starts: "Failure is the opportunity to begin again, more intelligently." His antenna for talent was the principal force behind the bare survival and then the brilliant success of his third company—not just engineering talent but managerial. Malcomson deputed his factotum and cashier James Couzens to help Ford. Couzens, nine years younger than Ford, looked a stiff when he turned

up in a derby and well-polished shoes, peering through a wire-framed pince-nez as he acknowledged he did not know the first thing about automobiles. Growing up in Chatham, Ontario, he had taken hard knocks from his salesman father and vowed he would never be dominated by anyone again. He had worked to improve himself, getting a job in Detroit checking railcar inventories for the Michigan Railroad Company, and coming to the attention of Malcomson by the way he hounded coal companies for late shipments. Couzens liked to complain to his mother that, as Canadian-born, he could become neither king of England nor president of the United States. Running Ford's empire would be to meet the demand for Model A, then B, and AC, C, F, K, N, R and S, all from the collaboration of Ford and Wills (whom Ford had promised 10 percent of his own stake).

As these nine models raced erratically through the alphabet between 1903 and the debut of the Model T in 1908, it was Ford's habit to roam the factory floor. One of Couzens's men, George Brown, recalled, "God! He could get anything out of the men because he just talked and would tell them stories. He'd never say, 'I want this done!' He'd say, 'I wonder if we can do this, I wonder.' Well the men would just break their necks to see if they could do it. As far as I can remember, Mr. Ford always wore a business suit. I can't ever remember Mr. Ford in coveralls, not even when he was working on machines."

Ford was still not producing the cheapest car. That distinction belonged to the Oldsmobile, $150 cheaper, produced by Ransom Olds until he was pushed out by partners determined to go upmarket. This was the trend throughout the industry and Malcomson favored the same strategy. More money was to be made unit by unit on cars costing $1,300 and beyond. In 1904, Malcomson led other directors pressing for a radically larger, more powerful, four-cylinder deluxe vehicle, the Model B, at $2,000. Ford protested quietly. Historian Roger Burlingame writes, "It is reasonable to suppose that Ford, far from being as sure of himself as so many have painted him, was slowly feeling his way, setting up brackets of cost, searching for the answer to large-scale production." He must also have nurtured some yearning to race again, despite his vow to stay alive. On January 12, 1904, as a prelude to launching the car, 41-year-old Ford climbed into a Model B on the frozen Lake St. Clair and roared off across the ice and snow. He broke the world record, traveling a mile in 36 seconds, with a top speed of 91.37 miles per hour. His entire team was treated to a muskrat dinner. Ford was twice vindicated: Despite the board of directors' hopes, the Model B did not sell as well as the cheaper models.

The experience reinforced Ford's original convictions just at the point Malcomson decided he wanted Couzens fired so he, Malcomson, could share the driving seat with Ford. In the showdown, in the summer of 1905, the stockholders split evenly but Ford sided with Couzens (and let him buy 11 percent of the stock). Malcomson, his old friend and backer, was forced out and Ford was at last master of the company bearing his name. That day he told a mechanic, Fred Rockelman, who drove him

HENRY AND CLARA: Norman Rockwell's romantic painting to celebrate the 1896 birth of the Quadiricycle

home, "Fred, this is a great day. We're going to expand this company, and you will see that it will grow by leaps and bounds. The proper system, as I have it in mind, is to get the car to the multitude."

The revolution was born in a tiny room, 12 feet by 15 feet, an aerie at the top of the three-story Piquette Avenue plant; this time Ford made sure the door was wide enough for a car. Early in 1907, he squeezed in six inventive engineers and assistants. The plant had drill presses, lathes and spare parts, but the most important tool was the blackboard where he wrote and drew his ideas and engineers drew theirs: It was Menlo Park on wheels. Much of the time he rocked in his mother's "lucky" old chair, watching in his detached way and indulging his perfectionism to the last detail in the hundreds of elements making up a car. He was, in John Reed's words, "a slight boyish figure with thin, long sure

hands incessantly moving." Unlike Edison, he rarely got those hands dirty; he could not read a blueprint, but he had an uncanny knack of figuring out what worked and what didn't just by watching. "Charlie," he said, "the trouble with the plate is that we have not insulated it properly." He was talking to the Danish-born tool and die maker "Cast-iron Charlie" Sorensen about a magneto to carry electricity to the spark plugs Spider Huff and Oliver Barthel had devised in 1901. Huff's latest invention—a flywheel with 16 copper coils and magnets—was the first to produce sparks in the cylinders of an inexpensive car without a battery but it inexplicably kept failing. The next day Ford came up the stairs with big kettles used for boiling maple syrup. He and Sorensen converted his kettles to pressure cookers, basted the magnetos in heavy varnish, cooked and then baked them for six hours. The magnetos emerged perfectly insulated. "Mr. Ford and I worked about forty-two hours without letup," recalled Sorensen. (Around the same time in Yonkers a chemist born the same year as Ford was inventing a material to solve all insulation problems, page 172.)

Ford insisted on two seemingly contradictory features in the embryonic Model T. It had to be big enough for five people but light for speed and light on the wear and tear of tires. The trick was to find a strong ultralight material. Few in the auto industry had bothered about metallurgy; they just took whatever steel came in from Ohio and Pennsylvania. There are two versions of how Ford found his magic metal. The romantic one told by Ford and favored by biographer Robert Lacey is that at a race meeting Ford picked up a little valve-stem strip from the wreckage of a French car and it proved to be vanadium steel—much lighter than ordinary steel yet ten times stronger, but not then made in the United States. The more prosaic version, favored by the historian and biographer Douglas Brinkley, credits Harold Wills with picking up the idea from experts in the Pittsburgh steel industry who watched what the French were doing. Whatever, there is no doubt the driving obsession with lightness with strength was Ford's, and his uncanny instinct for people once again paid off. When Wills suggested that they hire a university-trained metallurgist, Ford indulged his prejudice against higher education. He told Wills to train John Wandersee, who had recently graduated from sweeping floors to minding machines. Wandersee flourished to become Ford's chief metallurgist, and vanadium the key material.

Vanadium steel (vanadium, a transition metal, alloyed with ordinary steel at very high temperatures) was used in more than half of the car; the total weight of 1,200 pounds was the same as the Model N preceding it, but the N was flimsy and carried only three people. Taking a Model N into the countryside was an adventure. Ford designed the Model T with a 100-inch wheelbase and a high frame for the rutted wagon tracks of rural America; only 20 percent of American roads had any paving. It was more than a vehicle. It was a powerhouse: The sturdy 20-horsepower engine could be hooked up to saw logs, pump water and churn milk to cheese. But Ford took what seemed to be an odd decision in moving the steering wheel from the then customary right side to the left. On the right, a driver was well placed to keep an eye on the ditch. Ford, instead, was looking straight ahead. He foresaw the day when a driver's biggest concern would be oncoming traffic. As for the ditch, the Model T driver could just rock out of it by hitting one pedal to drive the car forward and another pedal to drive it back. One of the test drivers in 1908 reported: "Mr. Ford, the rougher you were with them, the better they run." The laconic Mr. Ford agreed: "I think we've got something here."

The "something" was, of course, a runaway success—18,664 sold in 1909–10, 34,528 in 1910–11 and doubling again the following year and the next. It was not perfect. There was no self-starter as yet and cranking was a problem for women; the whipback was known to break a wrist or two. But the Model T was full of innovations and an amazing value at $850: "No car under $2,000 offers more," the advertisements proclaimed. In town and country it was soon regarded as a reliable friend, as "Lizzie," a genuinely American character.

Into Lizzie, as Roger Burlingame writes, went a part of Ford's own character—his contempt for wealth and show and servants, his rustic-bred pride in independence, his unarticulated belief in equality of opportunity, his toughness of will. E. B. White invested Lizzie with mystical powers: "As a vehicle it was hard working, commonplace, heroic; and it often seemed to transmit those qualities to the persons who rode in it. My own generation identifies it with Youth, with its gaudy, irretrievable excitements." John Steinbeck, who put the dustbowl Joad family in an overloaded Model T on the road to California in *Grapes of Wrath,* made a grocer's Model T truck central to the plot of his *Cannery Row.*

When Ford first started preaching that he hoped to produce 1,000 Model Ts a day, the response was typically scornful, but in 1908 he had already bought a 60-acre site for the world's largest auto plant, financed, to the displeasure of his investors, out of profits. His new factory was on the grounds of a racetrack in Highland Park, Michigan, about six miles from downtown Detroit, and he stuffed it with machinery to cut costs and sell a low-price, high-quality car not to hundreds of thousands but to millions. As Ford noted in his autobiography, his competitors could not have been more pleased by his announcement. In 1908 and 1909, the question on everyone's lips, Ford wrote, was "How soon will Ford blow up? . . . It is asked only because of the failure to grasp that a principle rather than an individual is at work, and the principle is so simple that it seems mysterious."

The principle was low-cost, standardized mass production, small unit profits on high volume. Ford's way of achieving this, which became known as Fordism, was a mix of the commonplace elevated to a science, and innovation transmuted to an act of faith. Douglas Brinkley defines it as corporate development through unceasing improvement—"the restless approach to management that would sweep the industrialized world during the next fifteen years and that has remained the business norm ever since." Ford's investments in machinery raised productivity by 50 percent and more, but all the auto companies were doing that and making similar gains. He cut his dependence on outside suppliers; cost apart, it was a logistical nightmare to coordinate deliveries and check engineering tolerances. Again, there was nothing innovative about this, though at the River Rouge plant in the twenties and beyond Ford was to push vertical integration to its limit: He came to own the forests from which he got the wood for the chassis, glassworks for the windshields, sixteen coal mines in Kentucky for his steam, a 2.6-million-acre plantation in Brazil for his rubber, a fleet of Great Lakes boats to deliver his ore. Again, he was an apostle of interchangeable parts, and some authorities believe the increases in productivity achieved by this were greater than those from the more visible assembly line. But once more he was only in the mainstream of the American System from Eli Whitney and the national armories, through Sam Colt and Isaac Singer.

The distinguishing innovation was the assembly line for manufacturing. There had been assembly lines before—we saw Oliver Evans automate a flour mill—but as Ford pushed the concept it was new and revolutionary. To say Ford "invented" the assembly line is to misunderstand the collaborative nature of the advance and the way he worked. It does no disservice to Ford's importance as an innovator to get the facts right. It makes his contribution that much more magical. It was not something Ford ordered overnight; it happened over seven years and it sprang from the spirit of enterprise he encouraged and instantly rewarded rather than from any directive. Henry Ford said let there be innovation and there was. The assembly line was an outcome of his long-term vision that the basic product should not change so much as the methods used to produce it. The Model T changed little over the 19 years it was made, but because of that constancy the means to produce it were more easily transformed.

Ford's men are known to have been inspired by a visit to Chicago's stockyards, where they saw the Swift "disassembly line." William C. Klann, a foreman, came back and more closely studied one worker make a magneto, sitting at a bench and dipping into a box for parts. In the spring of 1913, Klann broke the job down and had workers stand by a conveyor belt where each man performed just one or two of the 29 different manual operations to assemble a magneto. With this first assembly line, the time for a magneto dropped from 15 minutes to 13 minutes, 10 seconds. Further mechanized and improved, the time fell to 7 minutes and then 5 minutes.

It was a start, but there were still 1,500 other parts in a Model T. Different parts slowly started being assembled faster and faster as the system swept through the factory: the transmission assembly, the rear axle, the radiator, the engine and the chassis. The man who placed a part did not fasten it. The man who put in a bolt did not put on the nut; the man who put on the nut did not tighten it. The bottleneck became completion of the chassis and engine. A stopwatch examination in the month of August showed that it took 250 assemblers and 80 parts carriers working 9 hours a day for 26 days to complete 6,182 chassis and motors — an average of 12.7 man-hours for each chassis. Ford engineers conceived of slowly winching a chassis along 250 feet of rope. Six assemblers trotted alongside it picking up parts in storage containers along the way and putting the car together as it moved. The average number of man-hours per chassis soon fell to 5 hours and 50 minutes. The next stage was to have the assemblers stand still, subdivide the work even more, and move the unfinished chassis along at waist-height. Assembly time now

fell to 93 minutes. Ford said, "Every piece of work in the shop moves; it may move on hooks, on overhead chains . . . it may travel on a moving platform, or it may go by gravity, but the point is that there is no lifting or trucking. . . . No workman has anything to do with moving or lifting anything. . . . Save ten steps a day for each of 12,000 employees, and you will have saved fifty miles of wasted motion and misspent energy."

Speed of assembly was such that Ford doubled production every year for the next decade and cut the retail price two-thirds, to $440. By 1914, just about every second car on the road was a Ford, yet there were only 13,000 workers compared with 66,000 at other plants. By May 27, 1927, when the last Model T came off the line, 15 million had been bought; and Ford assembled cars in 21 countries.

Ford loved the assembly line. He had his supervisors time each improvement with stopwatches. One joke told of a worker who had been fired for incompetence. What had he done? He dropped his wrench, and when he looked up again he was 16 cars behind. "Time loves to be wasted," Ford frequently said. "From time wasted there can be no salvage. It is the easiest of all waste and the hardest to correct because it does not litter the floor." People had to become machinelike to keep pace with the machines, a conceit illustrated in movies like Fritz Lang's *Metropolis* and Charlie Chaplin's *Modern Times*. Ford was not moved. He remarked on one occasion, "A great business is really too big to be human." His justification for subjecting men to the dronelike nature of the work was that some people were different. "Repetitive labor is a terrifying prospect to a certain kind of mind," he said. "It is terrifying to me. I could not possibly do the same thing day in and day out. But to other minds, perhaps I might say to the majority of minds, repetitive operations hold no terrors. In fact, to some types of minds, thought is absolutely appalling."

The workers begged to differ. By the end of 1913 they had thought with their feet, only 100 staying for 964 hired. Relations between workers and capitalists were strained everywhere. Violent strikes were common. Socialism and communism were becoming attractive to many.

Ford's masterstroke—a social, moral, political and managerial masterstroke—burst on the world on January 5, 1914. Three reporters only were called to the plant, where Couzens, with Ford silent by the window, read a statement: "The Ford Motor Company, the greatest and most successful automobile manufacturing company in the world, will, on January 12, inaugurate the greatest revolution in the matter of rewards for its workers ever known to the industrial world. At one stroke it will reduce the hours of labor from nine to eight, and add to every man's pay a share of the profits of the house. The smallest to be received by a man 22 years old and upwards will be $5 a day." The reporters were momentarily speechless. Had they heard right? This was at least an instant doubling of pay. Ten thousand job hunters descended on Highland Park in the midst of a harsh Detroit winter. In the ensuing riot, the police hosed the mob with freezing water.

At the time nobody in his right mind believed the economics could possibly turn out to be as sound as they did in reducing turnover, giving workers an incentive and creating a whole new class of customers who could afford to buy the cars they made. The publisher of the *New York Times,* Adolph S. Ochs, remarked, "He's crazy isn't he?" The *Wall Street Journal* called Ford a criminal. It saw the $5 day as "the application of spiritual principles where they don't belong" and a threat to organized society. The *Journal* was wrong in prediction, but right in diagnosis.

Moral principles had penetrated a citadel of industrial capitalism. Who first had the impulse is a matter of contention. Henry Ford recalls his dismay just before Christmas 1912 when 20-year-old Edsel, walking with him through the factory, witnessed two men beating each other insensate: Were they brutalized by the way they worked? Sorensen says soon afterward he and Ford worked out the $5-a-day concept and presented the idea to Couzens and other executives. Ida Tarbell, who interviewed Couzens for a book on Ford that she never published, traces the genesis to an evening in December when Couzens read a magazine "of socialist tendencies." Tarbell wrote, "An idea flashed through his head. Why shouldn't the Ford Motor

Company take a decided lead in paying the highest wages to its workers, thus enabling them to enjoy better living conditions." Couzens was known, like Ford, for his sympathy for the working class. When he became mayor of Detroit in the 1920s the work relief program he started was a model for the New Deal. He donated much of his own wealth to charity. He later said, "We had been driving our men at top speed for a year and here we were turning them out to spend the Christmas holidays with no pay. The company had piled up a huge profit from the labor of these men; the stockholders were rolling in wealth, but all that the workers themselves got was a bare living wage." He said he was appalled by the "gross injustice of all this."

In this version, Ford initially thought the full $5-a-day raise would be too risky, but became hooked on volunteering some kind of raise. Couzens would not be put off. He told Ford, "If we talk for more than forty-eight hours, we'll never do it." Ford then decided to make wages $3.50 a day. Couzens balked. "No, it's five or nothing." Ford countered, "Then make it four." "Five or nothing," Couzens said. Ford agreed that factory workers would get the wage, but not everyone. Couzens stood firm. George Brown wrote, "The only friction that I knew of between Mr. Couzens and Mr. Ford was when the $5 Day went into effect. That was what I could really say was the first friction. Mr. Ford figured it was just the men at the machines who were entitled to it. Mr. Couzens couldn't see that. He couldn't see that that was on the level, and he and Mrs. Ford had the same idea, that what's good for one is good for the other. Between the two, they fought Mr. Ford."

Ford never could say no to Clara, but he was torn. It was the sort of moment that comes inevitably to every head of a large organization, the sort Thomas Watson Jr. would face when he bet the company on the digital computer. It must have been a daunting, even terrifying, moment. Ford had never paid more than the going rate. He would probably never have taken the risk he did had he not been exposed much earlier than once thought to the philosophical teachings of Ralph Waldo Emerson. Robert Lacey found a slender, worn blue volume of Emerson heavily marked in 1913 with Ford's spidery exclamations of approval. Ford read and reread Emerson's essay on compensation, marking several passages, including this one: "He is great who confers the most benefits. He is base—and that is the one base thing in the universe—to receive favors and render none. . . . Beware of too much good staying in your hand. It will fast corrupt and worm worms."

Ford, in later life, took pleasure in making it sound as if his $5-a-day plan was just a matter of "efficiency engineering . . . one of the finest cost cutting moves we ever made." It was a neat way of getting back at the fellow capitalists who had denounced him for "the most foolish thing ever attempted in the industrial world." In fact, as Lacey concludes, there was "true generosity and not a little rashness" in the initiative taken by Ford and Couzens, or Couzens and Ford.

In 1919, Ford bought out the other stockholders and the Ford Motor Company became a wholly family-owned business. We have a picture of him at this point from the novelist and one-time muckraker Upton Sinclair: "Henry was now fifty five, slender, grey-haired, with sensitive features and quick nervous manner. His long thin hands were never still, but always playing with something. He was a kind man, unassuming, not changed by his great success. Having less than a grammar school education, his speech was full of peculiarities of the plain folk of the middle west. He had never learned to deal with theories and confronted with one, he would scuttle back to the facts like a rabbit to its hole. What he knew he had learned from experience, and if he learned more, it would be in the same manner."

Five years later, in 1924, the 10 millionth Model T rolled off the assembly line, and 15 million had been bought when the last Model T rolled off the assembly line on May 27, 1927.

So ends the greatest of Ford's innovations. He would continue being involved in his company, making key decisions, bringing his business to near ruin and then allowing it to catch up to its competitors. But by then the character of Henry Ford was much changed, the story of Henry Ford, innovator, is over, and the tale of Henry Ford, the persona, has begun. He sloughed off the men who had helped him most, Couzens and Wills. He became a name, a national symbol, a cracker-barrel philosopher who spent part of his fortune promoting anti-Semitic theories and a worldview nearly as regressive as his innovations had been advanced. He involved himself in politics, in square dancing, museums, farming, aviation and newspapers. He still moved the world, but from this point onward it was through his money and power, less through innovation and technology.

The great irony of Henry Ford's story is that he succeeded because he understood the American character, but his success changed the very nature of America. By the middle of the twenties, the country was moving from the mass era to what Alfred Sloan Jr. called the "mass-class" era: Americans, wealthier than the generation before, wanted nicer, better cars, and choice. Sloan's development of "a car for every purse and purpose" hooked consumers, who no longer wanted cars that came in any color, as long as it was black. General Motors under Sloan took advantage of this trend, pioneering installment paying for a range of cars. Ford refused to see the writing on the wall. His son, Edsel, unlike Thomas Watson Jr., did not have the strength to battle his father.

If the democratization of consumer goods is based on ever greater choice, the era of customization is a natural outgrowth of Henry Ford's innovations. The long reign of General Motors began because he became stuck in an earlier era. He did so much to boost the wealth of the nation, through his car and through his example, that Americans gradually became accustomed to luxury. Ford never really understood this great transition. America was changing; he refused to change with it, nearly destroying his company. Ford Motor Company had to be saved by his descendants.

Ford and his son, Edsel, in 1933, in front of Ford Motor Company's huge River Rouge plant, characterized by Diego Rivera's mural (pages 136–7), at a time when the company was facing keen competition from Chevrolet.

FORD'S LONELIEST HOUR
He saved the people's car from the monopolists

Henry Ford navigated as many dangerous corners on the road to the people's car as he ever did on the racetrack, then he ran into a solid steel barrier reinforced by the law of the land. Judges ruled that he was in breach of patent law: He had no right to manufacture cars without paying royalties to a trade ring that would not in any event allow him to market an inexpensive model.

Spies reported cars that did not have this brass plate from the monopolists. Lawsuits followed.

The case was full of nonsense that ensured it a long life. George Selden, who had never actually made a car, managed to stymie the entire American auto industry for more than a decade with a patent for a car that didn't work. Born in 1846, he was an inventive young man, but his real talent was the law: By 1878 he was an expert on patent law. (He was George Eastman's first patent lawyer.) He was genuinely early in appreciating the internal combustion engine and with the help of a Rochester machinist, William Gomm, he labored in his workshop to scale down the two-stroke engine invented by George Brayton. It was 1,160 pounds in a 10-foot frame. He got it down to 360 pounds and small enough to fit in a relatively light vehicle. The only snag when he tested it in May 1878 was he could not get it to run for more than five minutes. He nonetheless filed his application for a patent in the spring of 1879, powerfully assisted by the chaos in the Patent Office then. He did not submit a proper working model of his car, as required, simply saying he hoped that the Patent Office would accept his outline of the general features of the car.

Selden had two distinctions: The first was that nobody in the world applied before he did for a patent covering gasoline cars as a single combination of the parts. The second was that the "prince of procrastination" contrived to delay scrutiny of his claim for no less than 16½ years. The lawyer Hannah Yang wryly observes that in addition to "inventing" the automobile Selden invented "the submarine patent." At the time he applied, an inventor was allowed two years to complete his application—in complete secrecy—but Selden got around that requirement by filing a series of continuations and amendments. Selden wrote more than 100 changes into his patent, and all 19 of his original claims were emended to incorporate later advances in automotive technology made by others. William Greenleaf, the author of the definitive study, writes: "It is one of the paradoxes of the American system that such sweeping claims were awarded to an individual who made no practical contributions to the art of the motor car during its most vital phase of experimental development."

Selden timed his final acceptance of a patent perfectly, just as the gasoline carmakers began to rev up; four years after Patent No. 549,160 was granted on November 5, 1895, sales of gasoline cars approached 1,000. The ailing Electric Vehicle Company, headed by William Whitney, a New York lawyer, and Colonel Albert Pope (page 101) bought the patent on November 4, 1899, at first as a defensive measure, then as a blunt instrument for demanding a royalty on every gasoline car made in America; Selden himself stood to collect $15 per car. The irony is all too clear—a failing electric car company using an obsolete patent to control the fast-rising gasoline car industry.

The outcome of huffing and puffing by Whitney and 10 leading automobile manufacturers—excluding Ford—was an alliance between them to exploit the patent and control entry to the auto trade for the life of the patent (to November 1912). Whitney and the privileged group of automakers formed the Association of Licensed Automobile Manufacturers (ALAM) in March 1903, a few weeks before the incorporation of the Ford Motor Company. Automakers granted a license under the patent paid a 1.25 percent royalty on every car sold; unlicensed automakers were liable to prosecution. Ford's application for a license was immediately rejected. He later told a dealer he could have had a license if he had been prepared to charge $1,000 a car and make no more than 10,000 a year, but the grounds given for rejection were that he was a mere fly-by-night assembler; it was spurious since most of ALAM's members and 80 percent of all car companies were assemblers. As Hannah Yang points out, it was a catch-22. You cannot build a car without a license, and you cannot get a license, because of the Selden patent, without showing you can build a car. The intent was to ruin Ford and limit competition altogether.

The Association put out bloodcurdling announcements of what would happen to anyone who made or bought a car without a license. Ford defied the Association by continuing to sell. There was a long silence at a meeting of his stockholders in September 1903 when they heard that if they were sued the costs of fighting might be $40,000—at a time when gross profits were around $36,000—but they resolved to "throw down the gauntlet and not join the Association come what may." The Ford Motor Company issued a stinging open letter denouncing the "mendacious" monopoly and challenging the ring to sue. It did. It took six long years to record depositions for the trial. Greenleaf describes Ford in court as reticent in manner but plucky in spirit. "His tone was pitched at the level of a small-town mechanic talking of shop matters."

In the meantime, the Association was relentless. It installed brass tags with Selden's patent number, "No. 549,160," in all ALAM cars and engaged networks of spies to inspect parked cars. William J. Moore, a New Yorker with an unlicensed Martini car, fled rather than pay damages. He was later reported dead in Texas, but ALAM detectives tracked him down to Albany and served papers as the "dead" man had a drink at the bar of the Ten Eyck Hotel.

Ford's response to this campaign was to put up a $1,000 bond for everyone who bought a Ford. But the odds did not look good. Ford had little cash reserve at the start, whereas the hosts arrayed against him had a capitalization of $70 million.

Oral presentations finally began on May 28, 1909, in the New York District Court of Judge Charles Merrill Hough. Selden and ALAM had been making strenuous efforts to build a car to the Selden patent (called Exhibit 89), with dismal results, and in the midst of arguments chance presented the judge with a snapshot of the underlying realities. A transcontinental auto race was starting in City Hall Park, right under the courthouse windows. Judge Hough called a recess so everyone could watch. Young Fred Coudert, for Ford, piped up. "Your Honor," he said with feigned surprise, "there is something that puzzles me. I see a Ford car, two Ford cars, but I see no Selden car!" The judge is reported to have joined the laughter, but he had no patience with the technical issues in dispute. On September 15, 1909, he found for the monopoly on the grounds that while Selden did not invent the different components that made up his car, he was the inventor of the ensemble.

William Durant at General Motors gave in and started paying ALAM back royalties of about $1 million. ALAM went after

unlicensed carmakers with renewed vigor. Mass numbers of unlicensed carmakers started clamoring for entry into ALAM. Ford was almost completely isolated now in his opposition to ALAM.

ALAM offered Ford major concessions if he would join, but his blood was up. He sent telegrams to all his dealers and to editors across the country: "We will fight to a finish." He posted $12 million in bonds through the National Surety Company for

GEORGE SELDEN: Smoke and jets of boiling water, and a journey of 3,450 feet, was all Selden's exhibit 89 could achieve—but his defeat led the Supreme Court in 1912 to adopt new rules of equity revising the patent law he had exploited.

buyers who were threatened with lawsuits. Ford was putting a brave face on things. Privately, he was not so sanguine. In autumn 1909, Ford and Couzens were ready to sell the company to Durant for $8 million, a deal that fell through when Durant's bankers jibbed. They told him, "We have changed our minds. The Ford business is not worth that much money."

Ford and Couzens sat through every day of the appeal in a Manhattan courtroom from November 22, 1910. Walter Chadwick Noyes, the youngest of the three judges, took the lead. Coudert lambasted Exhibit 89, arguing, "If men who had that vehicle in charge for ten months could not, at the end of ten months, make it do better than 1,309 feet in one hour and twenty minutes, will anyone say that Selden showed a vehicle with any radius of action?"

Coudert was nursing a coup. A few weeks before the trial, he had been

kept waiting half an hour in the office of ALAM's counsel, Samuel R. Betts. There was a table piled high with books and magazines and to kill time he began reading a set of proofs of a new book by the Scotsman Dugald Clerk, an expert witness on the gasoline engine who had testified for ALAM at the first trial. Clerk had changed his mind. There in cold print he wrote that Selden was not entitled to a pioneer patent since he was now satisfied the basic work had been done by Otto, Daimler and Benz.

Coudert started by asking Betts in court if their case was largely based on Clerk's testimony. Betts said, "Yes." As soon as he realized where Coudert was going, he started to object, but Judge Noyes overruled him. In his book Clerk in essence contradicted the testimony he had given in trial. The appeals judges were swift with a verdict for Ford and some French automobile interests battling alongside him.

Noyes wrote that Selden's patent had contributed nothing of social value and that Selden's patent in 1879 did not cover any novel combination of elements. Further, Selden had not been the first to use a Brayton engine for propulsion; it had been used prior to 1879 to run boats and streetcars.

Within 24 hours of the decision 1,000 telegrams flooded Ford's office. Charles Duryea wrote, "It was a plucky, hard and above all HONEST fight. I hope the great American public will awake to the result and not fail to appreciate the champion of their rights against trust methods." Coudert wrote, "Your indomitable courage and tenacity in fighting the case through when practically all other American companies had abandoned you was certainly admirable." Even Richard Joy, of the Packard Motor Car Company and ALAM, wrote, "While my interests are on the other side of the fence I cannot but admire your determination to fight the matter out to the bitter end." Ford and Couzens were invited to the annual ALAM banquet in New York, and Ford was greeted with cheers of "FORD! FORD! FORD!" when he entered the hall. He bowed but said nothing. While everyone else was in fancy eveningwear, Ford wore only a simple suit.

The case made Ford into a folk hero.

GEORGE EASTMAN The bank clerk who democratized photography with the Kodak

1854–1932

EASTMAN moved at 1/1000th of a second. In November 1877, when he was a slim and neat 23-year-old bank clerk in Rochester, he bought his first camera, and set out for Lake Huron, Michigan, to photograph the natural bridge on Mackinac Island. The camera was five by eight inches, and with it he took a tripod, a darkroom tent and bottles of chemicals. In the darkness of the tent on site, he coated a glass plate with a thin solution of egg white, then laid on an emulsion of gun cotton and alcohol mixed with bromide salts; when the emulsion was set but still moist, he dipped the plate in a solution of nitrate of silver and shielded it from light as he put it in the camera.

He had no chemistry training, but he was determined to make photography simpler: "One ought to be able to carry less than a pack-horse load." A little over two years later, he had not only invented an emulsion that enabled exposures to be made with a dry plate, he had also designed a machine to coat the plates in quantity, installed his machine in a little room over a music store and was selling plates as fast as he could make them. In between, he had worked through nights in his mother's kitchen to test hundreds of versions of his emulsion of ripened gelatin and silver bromide (itself based on a discovery by English originator Charles Bennett), sailed to London to patent his machine there because it was the photographic capital of the world and returned to Rochester to form the Eastman Dry Plate Company in an enduring partnership with Henry Strong, a buggy-whip manufacturer.

Within four years, he had replaced glass plates with paper treated with a photographic emulsion and patented a roll-holder for a camera (with William Hall Walker). In May 1888, he launched a handheld camera with a light-sensitive roll of the treated paper sealed inside; he called it a Kodak because he liked the strength of the letter *K* and reckoned it was a word that would

EASTMAN AT 31: A portrait made on a dry glass plate, mechanically coated by a machine of Eastman's invention, before he perfected film

be pronounced the same in every language. In 1889, he fitted his Kodaks with transparent nitrocellulose film, a major breakthrough in photography. It was a brilliant stroke to make picture-taking easy, and brilliant again to separate the taking of a picture from its very complicated processing. In the first year, 13,000 people paid $25 for a Kodak; they each took 100 pictures, returned the camera and within 10 days Kodak sent back the prints and camera with film for another 100 pictures. Eastman's slogan—"You press the button, We do the rest"—entered everyday vocabulary.

He fretted that the succeeding models of the Kodak were becoming too complicated, limiting the heart of his business, the sale of film. (He often likened his camera to King Gillette's safety razor, which was practically given away at a fifth of its cost to sell blades at five times their cost.) In 1900, his leading camera craftsman, Frank Brownell, finally delivered Eastman's dream of "a film camera pure and simple"—the Brownie, the camera soon to confound the 1897 prediction of the great photographer Alfred

Stieglitz that the Kodak was "a fad well nigh on its last legs, thanks to the bicycle craze."

The Brownie went on sale at $1, and within a year 250,000 people were Brownie-snappers. This was the camera that truly democratized photography. In one form or another, it stayed in production for 80 years. Eastman had originally designed it as a camera a child could use—called a Brownie after the elves in Palmer Cox's illustrations—but its appeal was universal. The shutter speed could be set anywhere from 1/25th to 1/50th of a second, and depth of field (an f/14 aperture) was so good pictures were in focus from a few feet to infinity.

The photographic elite disdained Eastman's little fixed-focus cameras. "The button pusher," wrote Ernest Beringer, "becomes as much a photographic artist as the winder of a barrel organ becomes a celebrated musician." But the Brownie rescued photography from being a hobby that might be superseded by another hobby. It offered millions of people instant celebration—no ceremonial event has truly happened until the button has been pressed—and it gave us timeless memories: memory of a loved one, a baby's first steps through to graduation, the wedding day, the countryside under snow, the vacation, the cat.

Eastman was a zealous acquirer of patents, always restless for the next best thing—though, curiously, while happily supplying his friend Thomas Edison with film, he regarded the movies as a fad. For more than 40 years in command of his business, he reinvented himself every five years: young inventor, dry-platemaker, film manufacturer, camera-maker, master of self-financing through the London markets, all of it managed from a little desk side by side with his omnicompetent secretary. He was a very early pursuer of color prints. In 1912, he bet on the abilities of C. E. Kenneth Mees, a

Strip film made it possible to take passport-style photographs like the 16 on the following page (ca. 1925).

E 9 8 7 6 5

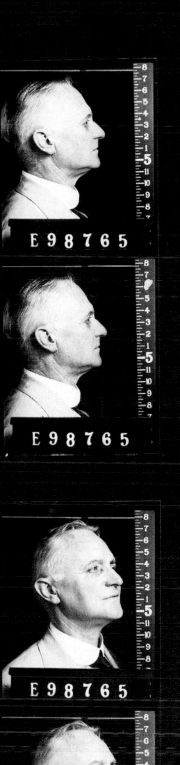

E 9 8 7 6 5

E 9 8 7 6 5

E 9 8 7 6 5

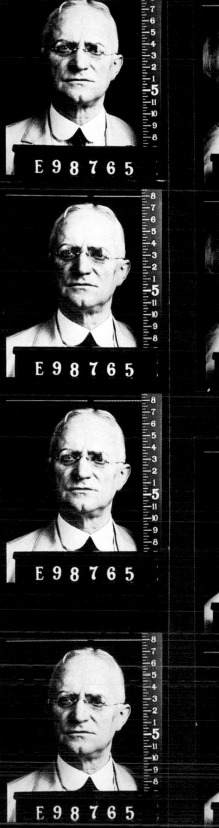

E 9 8 7 6 5

E 9 8 7 6 5

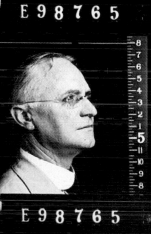

E 9 8 7 6 5

30-year-old English employee, and gave him the simple instruction "Your mission is the future." It was left to Mees, two years after Eastman's death, to announce his legacy in color: Kodachrome slides and prints invented at Kodak by the professional musicians and amateur photographers Leopold Mannes and Leopold Godowsky.

Eastman was a meticulous organizer of men and money. His energy was focused. He ran up stairs two at a time, but he had formidable concentration. His biographer Elizabeth Brayer describes him standing motionless, feet apart, hands in his back pockets, chin thrust up, apparently oblivious of commotion around him. But the main reason for his eventual dominance was his original perception of the snap as a precious family document. His first advertisement was of a father on one knee, snapping a picture of his little daughter. Eastman claimed that no man was more astonished than he was at the popularity of his "press the button" slogan, but in fact he had a gift for communication with the masses. He was very modest—he lived with his mother and never married—but he appreciated sex appeal. "A picture of a pretty girl," he told his advertising agency, "tells more than a tree or house." He invented the Kodak Girl, a girl in a striped dress holding a camera; sedate in her early years at the turn of the century, she later learned to syncopate with the twenties flappers. Eastman was also extremely generous. He paid small salaries, but gave large stock options. On making his first million in 1898, he distributed a third to staff; altogether he gave away more than $100 million for education, and racial advancement.

For himself, he built a splendid mansion and installed an organist, so that he might breakfast to Bach; his organist, Harold Gleason, varied the programs to match Eastman's mood. In 1930, Eastman called out, "Harold, please play my 'funeral march.'" When Harold had finished Gounod's "Marche Romaine," Eastman shouted, "We'll give them hell when they carry me out the front door."

Two years later, on March 14, 1932, suffering from spinal pain, he made his room tidy, then shot himself in the heart. He left a note: "To my friends: My work is done, why wait?"

Eastman Family Album
(as he might have written it)

On board the steamer to England in 1890, I pulled the cord, turned the key and pressed the button in order to snap my patent attorney, Frederick "Chappie" Church—while Church snapped me.

My beloved mother, Maria, on our trip to California. (My father died when I was only seven.) One of the reasons I liked the letter K for Kodak so much is that my mother's maiden name was Kilbourn.

THE
KODAK CAMERA.

ANYBODY CAN USE IT.

Photography reduced to
Three Motions.

*Silver Medal, Minneapolis Convention, P. A. of A., for most Important
Invention for the year.*

1. Pull the Cord.   2. Turn the Key.   3. Press the Button.

And so on for one hundred pictures.

One Hundred Shots Before Reloading.

Size of Camera, 3¼ x 3¾ x 6½ inches.
Size of Picture, 2⅝ inches diameter.
Weight of Camera, 25 ounces.

PRICE, Loaded, - - - $25.00.

Extra spools, film for 100 pictures, . . . $2.00
Amateurs can finish their own pictures, or the
exposed film can be sent to the factory, by
mail, to be developed and pictures finished.
Price for 100 finished pictures, including spool
of 100 films, for reloading, ... 10.00

THE EASTMAN DRY PLATE AND FILM CO.,

115 Oxford St., London.   ROCHESTER, N. Y.
(x)

*I wrote my own copy for advertisements,
and the first instruction booklet, because
I wanted everything simple and direct.*

On the edge, at the Grand Canyon, 1930.

Lunch with the ladies at my favorite picnic site south of Rochester.

My rhino, 1928.

SARAH BREEDLOVE WALKER

She came from nowhere to be the role model for the self-made American businesswoman

1867–1919

Sᴀʀᴀʜ ʙʀᴇᴇᴅʟᴏᴠᴇ, later known as Madam C. J. Walker, was not the first black woman to make a significant business out of hair care—that distinction belongs to Annie Turnbo Malone (1869–1957), who trained her, outlived her and was as generous a philanthropist. But nobody among all our innovators emerges from such swamplands of ignorance, squalor, disease, injustice and prejudice, and with such bravura and style, as Sarah Breedlove.

See her at birth, at the age of seven, and ten, and twenty, and into her thirties, and she is clearly doomed to a life of misery. She is born into utter destitution to freed slaves in a one-room sharecropper's cabin in steamy, pestilential Delta, Louisiana, in the uneasy days of Reconstruction, when thousands of "uppity" blacks are murdered. She does not go to school; she picks cotton and does not learn to read. She is six when her mother dies, and soon her father dies, too, so she is an orphan at seven. At about the age of nine, she treks with her 21-year-old sister, Louvenia, across the Mississippi to Vicksburg, where there is much to fear: an epidemic of yellow fever, the brutalities of the man who gives them shelter, the revival of the Ku Klux Klan.

The illiterate barefoot waifs knock on doors asking for dirty laundry, then haul water from the levee for daylong boiling, pounding, scrubbing and pressing with heavy flatirons. At 14, she marries a laborer, at 17 she has a daughter she christens Lelia and at 20 she is a widow.

She moves to St. Louis with her baby, where there are three Breedlove brothers;

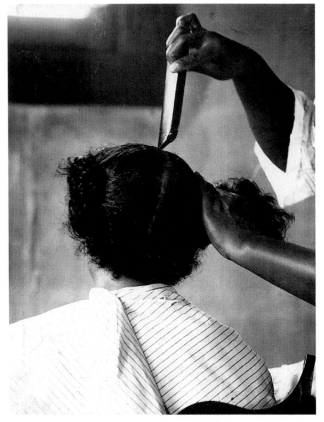

MADAM WALKER'S SKILLFUL HANDS: Her work was distinct from that of the much criticized (and mainly white) retailers of hair straighteners.

her older brother, Alexander, has a barber shop, a community intelligence center for blacks. She finds that the St. Louis Colored Orphans Home will take Lelia on a day-to-day basis as a half-orphan, so Sarah can work, but she has no hope of being anything but a washerwoman. She has taught herself to read after a fashion, and in any event good jobs in stores and offices are closed to blacks. Flash forward a full fifteen years and Sarah Breedlove at 35 is still doing backbreaking laundry for $1.50 a day (and her hair has fallen out).

What was there to break the apparently inexorable decline?

First of all, there was the church, the most important institution for oppressed blacks from slavery through the Jim Crow decades and the civil rights movement. It was segregated, of course, but that was its virtue. The church was a sanctuary where blacks could talk without fear, help each other and sustain their secret hopes. In St. Louis, Sarah attended the city's most respected church, the vibrant St. Paul's African Methodist Episcopal Church, where her great-great-granddaughter and biographer, A'Leila Bundles, says she derived much from the association with middle-class women, some of them teachers. Around the age of 19, she—the victim of circumstance—was inspired to an act of impulsive compassion. It is a quality of significance in her destiny. She read in the *St. Louis Post-Dispatch* how a poor blind man struggled to care for his blind sister and invalid wife. That Sunday she summoned up the courage to address the matrons of the church and ask for money for him. She succeeded in that, and in doing so she also discovered a gift within herself: She could speak; she could command an audience.

The second escape hatch for Sarah Breedlove was another black woman, Annie Turnbo Malone (Annie Turnbo Pope at the time). In many accounts of Sarah's life, there is no mention of Malone, which is unjust. Malone had a far better start in life as the daughter of a landowning farmer in Metropolis, Illinois. Though both her parents died early, like Breedlove's, she was nurtured by older brothers and sisters—she was the 10th of 11 children—and

attended high school in Peoria for a year. From her chemistry lessons there, she learned enough to realize how much harm was done to follicles and scalp by the mixtures of goose fat, meat drippings and coarse soaps black women used to style their hair. Hundreds of products were on the market, but she broke through with her Wonderful Hair Grower, a rinse most likely made with a mixture of sage, egg, other herbs and perhaps a little sulfur. She also perfected a hair-pressing iron and hot wide-toothed comb for frizzy hair. She made her shampoos with help from her sister, Laura Turnbo Roberts, and sold door-to-door in Lovejoy, an all-black Illinois town, then in 1902 took her wares to St. Louis in good time to sell to the thousands attracted to the 1904 World's Fair.

She was brave, a 33-year-old single woman living in a honky-tonk district of the big city, starting a business on her own in an era when respectable women, black or white, were expected to stay where they belonged—over the stove and washbasin. She did well enough to employ and train three commission agents, and one of them was Sarah Breedlove, whose hair she treated when, as Sarah said later, it was "less than a finger's length." Sarah's early accounts of how she became a hair specialist go like this: "One night I had a dream, and in that dream a big black man appeared to me and told me what to mix up for my hair. Some of the remedy was grown in Africa, but I sent for it, mixed it, put it on my scalp, and in a few weeks my hair was coming in faster than it had ever fallen out. I made up my mind to sell it."

In the words of biographer Beverly Lowry, this is hogwash, and later in life Sarah spoke less and less of the dream, until it vanished, and she attributed her rise to the virtues of patience, thrift and the acquisition of practical skills, then being preached by and practiced by the former slave Booker T. Washington (1856–1915), who founded the Tuskegee Institute. If Sarah's dream was the work of a poetic imagination, it was nonetheless a shrewd marketing ploy, as A'Lelia Bundles recognizes: "By invoking Africa she invested her

Madam Walker portrait taken around 1914 by Addison Scurlock, a talented African-American photographer

A MADAM WALKER CLASS: Agents were screened, asked to wear white shirtwaists and given a long list of rules for hygiene, safety and deportment, anticipating by decades the health regulations for modern beauty salons.

potion with the magical power of herbal medicine still practiced by some of her potential customers." It was also, Bundles notes, "a virtually unchallengeable defense to fend off a rival who claimed noticeable similarities between the hair products they both manufactured." More credibly, Sarah told a *New York Times* reporter in 1917: "I was at my tubs one morning with a heavy wash before me. As I bent over the washboard and looked at my arms buried in soapsuds, I said to myself, 'What are you going to do when you grow old and your back gets stiff? Who is going to take care of your little girl?'"

It is possible Sarah invented her own shampoos and rinses with the money she earned on commission. It is more likely that she adapted Malone's formula. Zenobia (Peg) Fisher, who worked many years with Sarah, is on record saying that a druggist named Edmund Scholtz offered to analyze Malone's mix in a period when Sarah was cooking for him, telling her, "Then you can make your own." This was in Denver, where she moved in July 1905 as a Malone commission agent. She was successful selling for Malone, and vehement in warning off imitators, but gradually during the year there was an evanescence: Sarah Breedlove disappeared and emerged as Madam C. J. Walker, an independent hairdresser and retailer of cosmetic creams. The game-

keeper turned poacher. Mrs. Annie Turnbo Pope accused her of "fraudulent misrepresentation," but, instead of suing, Annie copyrighted a new trade name, Poro, engaged more agents and developed a mail-order business.

The degree of inventiveness displayed by Sarah Breedlove in the matter of a secret shampoo is less important than her invention of herself. Her managerial skills were undoubted, her marketing concepts were clever, her fortitude phenomenal, but her real contribution was as a role model for the self-made American businesswoman. She made a celebrity of herself, prefiguring Oprah Winfrey in our time, and in doing so she liberated millions of women. If she could come from nowhere, prosper and do good works, so could they.

The man she married at the age of 38 in Denver, Charles Joseph Walker, was an advertising and publicity man and he was undoubtedly helpful with raising money and promotion. He got her to adopt the more resonant name; he drafted and placed "before-and-after" illustrated advertisements of what Madam Walker's treatments had done for her; they became adept at orchestrating tease campaigns announcing that Madam was leaving the city shortly and people had better rush. They did; "shortly" proved very stretchable.

But it was the genius of Madam Walker to make herself a figure of mythic appeal to women. She was an empathetic salesperson on her early door-to-door pitches in Denver and then exhausting barnstorming journeys through several states, traveling hundreds of miles for weeks on end to demonstrate and lecture in homes, schools, black community centers and churches. She was more dramatically dressed than most of the women she addressed, a big bosomy figure in fine silks and fancy feathered bonnets, with a zest for life her audiences found infectious. As her business flourished, she put her now-glamorous daughter Lelia in charge of a mail-order campaign for some five skin and hair beauty products, and took on hundreds of agent-operators on commission. All of them had to be able to tell Madam's personal story "in an intelligent and emphatic way, watch his or her face, note what statements impress most. AND THEN DRIVE THE NAIL HOME, ALWAYS REMEMBERING YOU ARE THERE TO SELL."

Madam made a point of making a grand entrance at social functions in the city, notably at a church for the elite she had once again discovered, the Shorter Community African Methodist Episcopal Church in Denver, where she conspicuously attended a lecture by the wife of Booker T. Washington. Madam Walker was giving employment to black women as beauty culturists and agents at a time when there were few jobs for them other than domestic service and manual labor, but Washington kept her at arm's length for years. Though he counseled living with Jim Crow ("We can be as separate as the fingers, yet one as the hand in all things essential to mutual progress"), he recoiled from the idea that black women might want to look like white women, overlooking the point that many of them just wanted to look like women.

In 1910, after flirting with Pittsburgh, she settled on Indianapolis as her corporate headquarters for sales, training, research and manufacturing. She swept into the city heralded by the advance press notices of visiting royalty. She hired a classy young business manager and attorney, Freeman Birley Ransom, cool and ever loyal, and took private lessons in elocution and penmanship from Miss Alice Kelly, a Louisville business schoolteacher who became her confidante and forelady of her Indianapolis factory. She made a stir buying a spacious $10,000 house at 640 North-West Street she intended to make glow with good taste, and notoriously buzzed around the city in a $1,600 Pope-Waverly Electric runabout. (She would soon enough hire a chauffeur.) By this time, C. J. Walker had passed his sell-by date; he felt that Madam was going too fast, and he himself was going too fast with other women. They were divorced in 1912.

Madam expanded her vision; now she talked about the Walker System, about women's independence and budgeting as well as grooming. She was a careful planner, but it was impulse again, as in the speech for the blind man, that carried her on a new trajectory. Julius Rosenwald, having made his fortune with Sears, Roebuck, had prescribed a course of radical philanthropy for himself, not just making straight gifts but offering large sums on condition the recipients themselves raised the rest of the money to "cure the things that seem to be wrong." In the fall of 1911, he pledged $25,000 for Indianapolis if within 10 days it would find the other $75,000 for building a black YMCA. The challenge to the pride of the city produced a frenzied campaign to raise $60,000 from the white community and $15,000 from the black. At the launch meeting, white businessmen pledged $15,000; two prominent black men—George Knox, the editor of the *Indianapolis Freeman,* and Dr. Joseph Ward—offered $250 each and Madam L. E. McNairdee, a well-off black clairvoyant and boardinghouse keeper, took the audience's breath away by doubling those $250 pledges. Madam Walker got to her feet to say they should one day create a colored girls' association—and then surprised everyone, including herself, by saying she would double the doubled pledge and give $1,000. She was ever afterward introduced as the First Woman in the World Who Gave $1,000 to a Colored YMCA. Indianapolis got its YMCA. (So did St. Louis, where Annie Malone followed suit with a $1,000 donation.)

Editor Knox had led the campaign for the YMCA. He expressed his gratitude to Madam in a glowing profile of her in his newspaper, starting on the front page. The National Association of Colored Women invited her to address 400 delegates. She funded scholarships at Tuskegee, but Booker T. Washington continued to refuse to include her curriculum in the college—or indeed acknowledge her in any way. It all came to a head in front of a packed convention of the National Negro Business League in Chicago in August 1912, when for two days Washington patently ignored her petition to speak about her efforts to elevate her race (while affording time to

She sold beauty products but more and more sold pride.

two competitors). In the closing hours, unable to contain her frustration, she acted from impulse again, jumped up and challenged the chair, "Surely you are not going to shut the door in my face. I feel that I am in a business that is a credit to the womanhood of our race. . . . I am not ashamed of my humble beginning. Don't think because you have to go down in the wash-tub that you are any less a lady!" To ringing applause, she itemized her success: revenues of $18,000 so far in 1912 from a start in business on $1.50 a day. She vowed "with the help of God" to build an industrial school in Africa. Washington took his time coming around, but on July 13, at the dedication ceremonies for the new YMCA, he accepted an invitation to stay at her home and the services of her uniformed chauffeur driving the splendid seven-seater open Cole

touring car she had just acquired. It was an endorsement she had long cherished.

Her sales soared to a quarter of a million dollars, making the Walker Company the largest black business in the United States. More and more she married philanthropy with business. She and Malone competed in how much they could give away; Malone eventually founded Poro colleges in 30 cities. Madam Walker invested in building racial pride with gifts to churches, educational institutions and the principal activist groups of the National Association for the Advancement of Colored People and the National Association of Colored Women: Hers was the largest single donation to the NACW to preserve the home of Frederick Douglass, and her $5,000 for the NAACP antilynching campaign was the largest in its 10-year history. Organizing her 20,000 agents into Walker Clubs, she rewarded those who raised the most money for charity along with those who registered the highest sales. She moved as vigorously as her failing health would allow into political activism, joining a delegation to Washington to protest lynchings and President Woodrow Wilson's segregationism.

Madam Walker had only three years to live when she left Indianapolis for a final flourish in New York City and a Hudson Valley estate. She built an Italianate villa at Irvington-on-the-Hudson, which she called Villa Lewaro, as an example of what someone of her race could achieve. She gathered vegetables, plucked weeds and presided over political conferences and soirees—cheek by jowl with the estates of John D. Rockefeller and Jay Gould. The villa's name was an acronym from her daughter Lelia's acquired name, A'Lelia Walker Robinson, and it was Lelia who inherited it and maintained its celebrity as a focus for the Harlem Renaissance when Madam died in 1919.

The commonplace acclaim for Madam Walker is that she was the first black millionaire. Her federal tax returns for business and real estate suggest it was more like $600,000 about $6 million today—but that is not a measure of her legacy. As Bundles writes, it was the promise she bequeathed to future generations, that they might realize greater successes and dream more elaborate dreams.

AMADEO PETER GIANNINI
The big man on the side of the little man: the people's banker

1870–1949

NOBODY ENJOYED an earlier introduction to innovation than Amadeo Peter Giannini. When the epochal transcontinental railway started services to California after the line's completion in May 1869, Amadeo was among the very first passengers. His newlywed father and mother, Luigi and Virginia, boarded the majestic Union Pacific express in Omaha, Nebraska, for a journey of four and a half days to Sacramento, the last stage of their emigration from Italy. For $100 they could spare themselves the crowded third-class coaches with narrow wooden benches and enjoy the landscapes from the plush seats of the glitteringly ornamented Pullman car. Amadeo had the snuggest journey of all; he was ensconced in the womb of his plucky mother.

Had there been no transcontinental railway when the Gianninis set out, it is by no means sure they would have arrived unharmed in California. The bone-shaking stagecoaches to the West were a risky ordeal for a pregnant woman. If the Gianninis had gone by sea from New York via the Isthmus of Panama they would have been exposed, among other miseries, to malaria, smallpox and cholera. Luigi himself had contracted smallpox on his way home to Italy from the California goldfields. It was a return made specifically to secure a bride he had never met. One night around a diggers' campfire, another miner had read out a letter from his sister, Virginia, and Luigi decided there and then, sight unseen, she was the woman he would marry.

Virginia was 15. She was mature for her years, but she was a risk-taker then and throughout her life. She left the bosom of her prosperous family farm in Chavari, near

Giannini was determined to serve all the people. He found a way to give banking facilities to Japanese interned in World War II.

Genoa, and tied her destiny to a man of 22 she knew for only six weeks before their marriage. Luigi had arrived at the farm with 20 gold pieces in his money belt and lyrical descriptions of the life they might have growing fruit in California's fertile Santa Clara Valley. It is nice to think that as the young immigrants crossed the Rockies, their adventurous spirits somehow crossed the placental barrier. Amadeo, born to them on May 6, 1870, in a cheap hotel room in San José, was a visionary whose innovations in banking built the state of California and transformed the finances of the common person.

The couple, struggling to learn English, made a good start in San José, 50 miles from San Francisco at the southern end of the bay. On the main street in a white clapboard house, they leased and ran a 22-room "Swiss Hotel," mainly for Italian immigrants, and

did well enough to sell up and buy a 40-acre ranch (as Californians call farms) in the town of Aviso, between San José and the bay; they grew fruits and vegetables for sale in booming San Francisco. In 1877, Luigi's luck ran out on the ranch of his dreams. He was gunned down by a workman in a dispute over two dollars, and six-year-old Amadeo saw it happen. Virginia, aged 22, was pregnant with a third child. She decided to run the ranch by herself. She sent A.P., as Amadeo Peter came to be known, to a one-room schoolhouse in Aviso where he mixed with a polyglot group of children, native Americans with immigrant Italians, Portuguese, French, German, Armenian, Spanish and Japanese, an experience that gave him a tolerance and understanding of other cultures that was to play a decisive role in his life. Sometimes Virginia took him with her on the boat to San Francisco to sell her fruit. As biographer Felice Bonadio describes it, a steamship captain remembered the two Gianninis as "a pleasant, attractive widowed woman and her small son sitting quietly together on the crowded deck of his boat as it moved through the cold, predawn darkness of the bay." In 1880, the widow married Lorenzo Scatena, a 26 year old Italian immigrant who had worked his passage to America and acquired a team and wagon. Eventually she had three more children with him. He was a quiet, gentle man, adored by his three stepsons. A.P. was happy to call him "Pop."

Virginia was the force in the family. In 1882, she decided that they would be better off buying and selling in San Francisco than growing in Aviso. Scatena was reluctant to give up the ranch, but Virginia insisted. He took a job in a commission house, buying in bulk for sale on slim margins to retailers, and did so well, working 16 hours a day, he

GIANNINI IN HIS PRIME: An open door and an iron fist

Fast-growing Giannini did not have much taste for the social rituals of the Italian community. The transition to dealing in money was more or less an accident following his 1892 marriage to Clorinda Agnes Cuneo.

was able to come proudly home one evening with the news that his monthly salary was more than doubled, from $100 to $250. "Say no," Virginia instructed, "ask for $300." When Luigi was refused his request, Virginia told him to quit and start his own firm. Once again, she was vindicated. By the end of the year Scatena was thriving as an independent wholesaler. They

bought a house in North Beach, the city's Italian district; Giannini shared a bed with his more scholastic brother, Attilio, who was later to graduate in medicine, but his most vivid memory was the family's pride in having a house with a bay window.

Graduating to the Washington Grammar School, A.P. concentrated on his arithmetic and penmanship. He had good enough grades, but what he lived for was Saturday midnight when he boarded his stepfather's wagon and rolled down the dark streets to the wharves, where he would wait tensely for the farmers' boats to come in. It was rough on the dock, jostling with other buyers, settled Americans and an immigrant medley of Italians, Chinese, Portuguese, Irish and Syrians, all yelling their competitive bids in a babble of tongues for the best buys of beans and potatoes, corn and peas, grapes and cantaloupes, cherries and plums, which they would then rush to sell to retailers before coming back for more. These Saturdays, A.P. sneaked quietly out of the house in his stocking feet because Virginia was always fretting about her boy being pushed around. Again, it was a learning experience. "The old waterfront commission business was a pretty stiff school for men," A.P. reflected later. "I used to study them down there and I suppose I picked up the knack of sizing up men."

He did more. Scatena was surprised one day to start getting offers from farmers whose business he had not solicited. It turned out A.P. had put his penmanship to use. The 12-year-old had written to growers all around San Francisco promising "honest prices on the barrelhead and quick service" if they made L. Scatena and Company their exclusive middlemen. No one before had challenged the big commission houses in this way. At 14 in the eighth grade, A.P. dropped out of school and took a five-month course at Heald's Business College. Virginia did not want him in the business. She told his stepfather to discourage him. Scatena set a trap. Knowing that oranges were hard to come by, he told A.P. that if he could buy a carload from a new supplier he would be so impressed he would give him a gold watch. The expectation was that the boy would be demoralized by the experience. Three weeks later he had secured not one but two boxcars of oranges, and

his stepfather felt obliged to give him a full-time job. Years later, A.P. recalled, "I still have the gold watch Pop Scatena gave me for the letter that brought that business. It reminds me that the only pleasure I ever wanted as a young boy was the reward and pleasure of a successful transaction."

By the age of 15, "young Scatena" was six feet two and a half inches tall, and 170 pounds, able to hold his own in the occasional waterfront fight. He had inherited his mother's dominating nature and a volatile temper with it, but he had his father's instinct for people. "No one could bluff, intimidate or out-general him," a rival commission salesman remembered. "He had an extraordinary faculty for gauging just how long the other fellow could stand the gaff. No such salesman was ever known on the waterfront." He made friends of the company's bookkeeper, an old Irishman named Tim Delay, learned how to read the ledgers and worked 18 hours a day, shunning teenage parties and dances: "I decided what I wanted and then went after it hammer and tongs." He was ready to try anything to make a deal. Noting that early peas fetched a good price in the San Francisco market, he didn't wait for the next round of wharf-front battles. He stuffed bread and cheese into his pocket and rode out to ranchers in the valley to offer them a quick sale if they would plant more and pick early. He went into the Sacramento and San Joaquin Valleys hundreds of miles on foot, by horse and buggy, stagecoach and riverboat. He pursued one large grower from his doorstep to Sunday church and up the aisle; on another occasion he waded a swamp, holding his clothes above his head, to get ahead of a rival. In short, he was a pest. He was insatiable to know who was who and what affected their trade. He made a point of straight dealing, disingenuously revealing his profit margins. The strategy, which he was to carry into banking, was low margins, high volume. He came to concentrate on the big growers, but he wanted to be every rancher's best friend. When he saw a new cultivation technique, he passed it on. He deposited something else in his memory bank: how farmers needing trees and irrigation were reliant on reasonable credit. He lent them money, on behalf of Scatena and Company.

Other commission houses made advances, too, but as the Bank of America historians Marquis and Bessie James observe, they were not in the habit of allowing a teenager decide who could be trusted.

At 22, A.P. was a full partner in his stepfather's business and the cynosure of every female eye as he strode through town in top hat, Prince Albert, gloves and cane. Scatena and Company was on the way to becoming the top firm in their trade west of Chicago. It was mainly A.P.'s doing. In the words of a family member, he was "crazy with ideas . . . he had a new one every time he moved." In 1901, when he was 31, with an additional monthly income of $250 from real estate investments, he sold his half interest in Scatena and Company to ten employees for $100,000. His explanation was that he was doubly bored; there was nobody to fight anymore and he had no interest in accumulating great wealth. "I don't want to be rich," he insisted in one of his many remarks over the years hammering the same point. "No man actually owns a fortune; it owns him." A bank associate recalled, "He could remember the balance sheets of scores of banks from year to year but he never knew how much money was in his own bank account." Many entrepreneurs affect a similar asceticism. It was bred in Giannini's bones. Only the call of family duty plunged him into the business of banking, where billions passed through his hands and there, too, he made sure relatively little stuck to him.

A.P. had married Clorinda Agnes Cuneo in 1892. Her father, another Italian immigrant, had found his fortune in real estate rather than the gold that had originally drawn him to California. He died in 1902, leaving a widow and 11 children but no will. There were sons older than A.P., but it was the shrewd son-in-law to whom the family turned, asking him to manage Cuneo's ownership of more than 100 parcels of real estate. A.P.'s dabbling in property had given him what biographer Gerald Nash describes as a "phenomenal" knowledge of the market. He took over with relish, and one propitious day assumed his late father-in-law's seat at the directors' table of the Columbus Savings and Loan Society in North Beach. Board meetings conducted with murmurous solemnity under the gaslit

ALL DRESSED UP: Amadeo Giannini (front and center) with his stepfather Scatena (top right) around 1910

Lorenzo Scatena's fruit and vegetable depot, where Giannini first made his mark

chandeliers rapidly escalated from polite assent to debate to argument and then shouting matches that reverberated through North Beach. Columbus had been the first bank in the community, built in 1893 by another failed Italian-American goldminer, John F. Fugazi, on the strength of having one of the few safes offering security and interest payments to people whose notion of a savings bank was a tin hidden in a sock. Fugazi and his associates lent money to the favored few, established home-builders and merchants; in the words of the Jameses, they were in the banking business first of all to make money for themselves. Like almost all the banks at the time, they showed the door to anyone needing to borrow $100 or so; the loan sharks were welcome to such time-wasters.

Giannini earned Fugazi's furious protest that he was "a young ambitious hotshot . . . infatuated with big plans and crazy ideas" because he campaigned for making such small loans and to more people. To Fugazi and most of his associates, the newcomer's advocacy of more aggressive and popular banking was a gloss for gambling. To Giannini, money lying idle in the safe was lost opportunity. He argued, "We should loan the bank's money right up to the hilt." He urged they go out and solicit loans, a heresy even to bankers more adventurous than Fugazi. The regular confrontations ended with A.P. storming out with four other directors in his slipstream. "I might never have gone into banking," he remarked, "if I hadn't gotten so damn mad." He marched straight into the American National Bank, where Scatena and Company had their account, and to his good friend James F. Fagan, a vice president, he blurted, "Giacomo [Italian for James], I'm going to start a bank. Tell me how to do it!"

Giannini's game plan was to create a bank anchored in the Italian community by liberal lending and by spreading the ownership widely to make people feel it was their bank. He split the $300,000 of capital, half paid up, into 3,000 shares of $100 each. No director was to hold more than 100 shares; 620 shares were divided among 143 people who had four shares or fewer. A.P. told the world: "We don't want a situation in which any individual can promote his own interests over those of the

bank." He was faithful to this concept all his life; he and his family never owned more than a trifling percentage of stock. As so often with Giannini, though, an appealing purpose was also a good tactic. With no large stockholders, his authority was enhanced; he was a general in an army composed largely of privates. His Bank of Italy was owned by people with vowels at the end of their names, among others, by a fish dealer, a retired saloonkeeper, a drayman, a druggist, baker, accountant, restaurant owner, plumber, housepainter and barber, giving the bank a more popular democratic cast than the names associated with the big "American" banks in San Francisco such as the Crocker-Woolworth or, more so, the "foreign banks," Lazard Freres and Rothschild. Antonio Chichizola, a Columbus defector, was the president and Giannini a vice president, but even before the doors opened for business it was clear Giannini would run the show by the force of personality and his pocketbook. Armando

BUSINESS AS USUAL: Giannini's baritone voice offering loans for rebuilding boomed across the devastation following the San Francisco earthquake and fire, 1906.

Padrini was a charismatic assistant cashier at the Columbus renowned for bowing and kissing the hand of female customers, European-style. Giannini poached him by doubling his salary without consulting his board-in-waiting. The directors protested that Padrini could hardly be worth twice as much for bowing and kissing for them. Giannini retorted that women were crazy about him, then added, "and he gives the

man in overalls as much attention as a big depositor." When populism failed to convince, A.P. pledged that if the bank's profits did not justify the appointment, he would pay Padrini himself until they did. The populist had a mailed fist.

Looking for a prime site for the bank, he persuaded the owner of a saloon in the Drexler Building on Montgomery Avenue to sell his lease, and then he negotiated with

the principal leaseholders for the entire building, which just happened to house the Columbus Bank. One morning Fugazi discovered that a team of carpenters was turning the saloon next door into a rival bank that had stolen a key man from him, that the despised hothead Giannini was now his landlord and that overnight this new landlord had tripled his rent. The Columbus moved across the street, where Fugazi and his sons could glare at people going into the Bank of Italy at No. 1 Montgomery Avenue. It cost less than $10,000 to transform saloon into bank, with $750 for a safe Giannini described as "little more than a tin box without a top." The day the teller Victor Caglieri opened the Bank of Italy doors, on Monday, October 17, 1904, for both savings and checking accounts, and lending, 28 people came in to deposit a total of $8,780. One of them was Virginia with $1,000. The tellers huddled with the new customers, patiently helping them write deposit slips; many could not read English. The climax of the big day was shortly before closing when a bunch of North Beach fish merchants, friends of one of the directors, marched through the doors in their hip boots and leather aprons. Only the little local Italian paper noticed the opening; if it was mentioned at all in commercial San Francisco, it was "that little Dago bank in North Beach."

A.P. practiced what he had preached at Columbus. Nobody was safe from his beaming candor, his coaxing hand on an elbow, his formidable remembrance of names and family circumstances. At night, A.P. and young Charles Grondona, another director, performed door-to-door duets on the joys of interest. Thousands and thousands of dollars in gold and silver were flushed out of hiding places. A.P. advertised. He kept the bank open longer hours and on weekends to fit into working people's schedules. He went into the countryside beyond North Beach, peddling loans at 3 percent. Giannini was ready to do business in real estate loans as small as $375, personal loans as little as $25. His bank risked money financing homes for rent. For the first time in San Francisco, in the words of historians Marquis and Bessie James, the small man in need of a small sum was provided a place where he could borrow at bank rates.

WE WANT GIANNINI: Citizens of little Yorba Linda, population 2,000, gather in a coffee shop to express support for a branch of the Bank of America.

loans exceeded deposits by $200,000, the highest ratio in the state. Some of them had no collateral. It could all have gone wrong early on. So much depended on assessing the integrity of the borrower that A.P. called these transactions "character loans." He believed that prospects could serve as adequate collateral and cheerfully lent money to pay doctors' bills for delivering a baby. In a radical departure from typical banking, he had the bank give up its investments in bonds and stocks in preference to real estate and personal loans. The liberal lending had a multiplier effect in the community in good times, increasing both trade and the bank's popularity. In July, by which time his stepfather, Lorenzo Scatena, had become president, A.P. declared a 5 percent dividend. "Throughout his career," writes Nash, "he believed in the payment of dividends whenever possible as a means of retaining a loyal following." By December 1905, all was going right. Assets had reached $1 million. Giannini gave raises to all six bank employees and the bank relieved him of paying Padrini. In February 1906, it was decided to increase the capital by $200,000 to $500,000 and the day fixed for distribution of 2,000 additional shares at $105 each was Wednesday, April 21.

For 28 seconds, at exactly 5:13 a.m. that Wednesday, there occurred the event that gave Giannini a chance to exploit his genius for inspiring trust. He was thrown from his bed at Seven Oaks, San Mateo, by the shock of the earthquake; his chimney collapsed, taking part of the roof with it. He calmed a panicked Clorinda, who was now expecting her eighth child, took a commuter train part of the 17 miles between home and the bank and ran, walked and hitched the rest of the way to arrive at noon. Padrini had opened the undamaged bank for business at 9 a.m., having taken a horse and buggy to collect some $80,000 in gold, silver, nickel and copper coins and paper money issued by the federal government and national banks; they had previously been in the habit of depositing their cash overnight with the Crocker-Woolworth National Bank. A.P.'s journey into town, against the tide of refugees streaming out of the city with everything they could carry, had given him a sense of

The financial community thought all this vulgar and unethical. When the ex-vegetable king applied to join the Pacific Union Club in elite Nob Hill, he was swiftly blackballed, a further stoking of paranoia latent in his business isolation, a Roman Catholic of Italian extraction in a predominantly Anglo-Saxon Protestant America. "They thought I was undignified," he said. "I could never figure that out. I always thought that if business was worth having, it was worth going after. How can people know what a bank can do for them unless they're told?" Giannini was not the first banker to think of the "little guy." Andrea E. Sbarro, who came from Italy in 1852, started the Italian-American Bank in 1899,

offering credit for small as well as large homebuilders, and thereby gaining on the Columbus, for which A.P. had criticized Fugazi. But if not the first, Giannini was the foremost. He was to push popular banking to a quite different order of reality; successful innovators, we often find, are entrepreneurs of the exponential. They squeeze the lemon until the pips make music. Lewis Tappan and Robert Fulton succeeded for the same reason.

Giannini's baby bank learned to crawl slowly. Deposits by the end of December were $134,413, a bagatelle compared with the Columbus or Italian-American, which had around $2 million each. But it was adventurous for its size. After a year and a half, the calamity still impending: fire spreading from the devastation of the flimsy houses around the bay. The army was out with orders to shoot looters. "I reckoned," he said later, "we had about two hours to get out of there. I realized that no place in San Francisco could be a safe storage spot for the money." He got two teams and wagons from his stepfather's company that were loaded with orange crates, hid all the bank's gold and paper money underneath the crates and waited for nightfall. "We didn't have any guards. All the police and soldiers were busy fighting the fires. It was extremely difficult to disguise the load we were carrying, and I thought I saw would-be robbers on every street corner." It took all night on clogged roads to reach Seven Oaks, where he and brother Attilio hid the money in the ash trap of the living room fireplace and stood guard until daybreak with two other men upstairs as lookouts. "The idea of the crates worked out," said A.P., "but for weeks after the bank's money smelled like orange juice."

It might have been a debilitating moment the next forenoon to find his baby bank a mass of charred rubble. One third of the city had been burned down. Most of the residents of North Beach were among the 250,000 people whose homes and businesses had been destroyed. The Bank of Italy had no building and only $80,000 in cash to cover deposits of $846,000. How could it survive a panicky run? Giannini did not fret. What he saw was not a disaster to mourn, but an opportunity to exploit: He would promote the Bank of Italy as the savior of the community. According to biographer Felice Bonadio, the story told by A.P. and his bank, and implied in the scholarly history by the Jameses, gilds the facts just a little. The most heroic version has it that A.P. was the first banker to reopen in the ruins, lending so people could rebuild. He was not, in fact, alone, but once again he triumphed in reality and perception by thoroughness, by his speed, memory, showmanship and driving competitiveness. The bigger banks were handicapped by loss of records and vaults too hot to open immediately, and they were more scrupulous in observing a government moratorium on banking; the bank holiday lasted more than a month. Giannini meanwhile was down

Bank of America loans enabled Starr Baldwin, editor of the weekly *St. Helena Star*, to buy presses for the *Star*, three other weeklies and a monthly, and photoengraving machines to serve other small California newspapers.

on the Washington Street wharf close to still-smoldering North Beach, his baritone booming across the desolation. Where he had bargained over for potatoes a few years ago, he was doing banking business on a plank laid across two barrels with a hand-made banner advertising his readiness to act. For withdrawals and loans, he dipped into a conspicuous bag of money retrieved from the Seven Oaks fireplace. He told everyone, "We're going to rebuild San Francisco, and it will be greater than ever." Loan seekers at the wharf and the office he opened on April 27 at his brother's home were embraced as heroes. In fact, the booster would give only half of what people said they needed. He had a good idea how much gold was still hidden away and sensed, correctly, that the hoarders were now more scared of fire and robbery and destitution than the mysteries of banking. If he could get the hoarded money into circulation, it

would find its way into rebuilding and a fair chunk into his bank. Six weeks after the earthquake, deposits were exceeding withdrawals. He started looking up steamship captains he knew to shove money into their hands, saying, "Get up north and get lumber." As a result of Giannini's exuberant initiatives, North Beach was rebuilt faster than other areas, to the acclaim of the city's newspapers. Two months after the fire, Giannini went ahead with his $500,000 stock offer and used the money to build a new permanent office. By the end of the year, deposits had soared from $706,000 to $1,355,000. The thrill of helping to rebuild the city convinced Giannini that banking was what he wanted to do for the rest of his life, but just around the corner was another disaster that was to determine him to do it in a revolutionary manner.

The event was the financial panic of 1907. A.P. sensed storm clouds were gathering in

April during a working vacation in the East with Clorinda. The confident lender returned in May as the rigorous retrencher. He raised rates for real estate. He personally reviewed every loan. He told tellers not to make payments in gold. His most knowledgeable director, James Fagan, his ally in starting the Bank of Italy, was now cashier of the big Crocker-Woolworth National Bank. Fagan thought Giannini was carrying things too far. North Beachers were gold fetishists, he believed. They would never accept paper. But they did; they had faith in *their* bank. Giannini put the gold aside for the rainy day and added to it by using paper to discharge debts to other banks for cleared checks; normally, the payments cleared through the Crocker included a proportion of gold, but nobody asked and Giannini did not tell. When banks began to fail by the hundreds from east to west in October—6 banks and 4 trust companies failed in New York, 16 in California—the Bank of Italy had $100,000 in gold above normal needs. Giannini piled it high in the teller cages. Other banks limited withdrawals, asked for advance notice and paid only in Clearing House scrip—"funny money." But Bank of Italy tellers were now told to give gold promptly to anyone who asked for it. In a grim year, the big eight banks in San Francisco all lost deposits, the great Crocker, $3 million. The little Bank of Italy ended up with $300,000 more in deposits and was able to help out Crocker with gold; in return, at the peak of the crisis, it was accommodated with a high overdraft.

It was all a testament to Giannini's acute understanding of the psychology of confidence. When the Bank of Italy opened a pleasing new nine-story building, Giannini refused an inner office and placed himself at a desk in the middle of the big, open floor. He was brisk with staff and would sometimes shout impatiently to a clerk clear across the room, but he was genially available to all and sundry who came into the bank. "The people who come to see me tell me what's going on," he told a reporter. "I figure I get at least five times more business done than I could handle by other methods." His personal charm was considerable. He would entertain a couple of toddlers while their mother filled out a

CATCH THEM YOUNG: Children with their piggy bank coins were made especially welcome by Giannini. The scene is a Saturday morning at the Santa Monica branch in 1949.

deposit slip, congratulate a father on an addition to the family by reaching into his pocket for a $5 gold coin. "For the little fellow, and remember that a savings account in his name can be started for one dollar." He beamed his approval of dads who walked across the floor and did just that. He had time to talk because he was a good picker of subordinates and left details to them. He kept them on their toes by throwing out one of his "crazy" ideas to see whether they'd duck or take him on. At those who agreed with him too eagerly, he'd yell, "Are you yessing me?" His day began at 5 a.m. and ended, he would say, "in sleep,"

which he assented to only because he could subliminally plan his next day without interruption. Once up, work was the only thing on his mind. Two or three executives were with him talking banking every morning when he was shaved in the barbershop; looking in the mirror at himself was, like all leisure and social activities, "a damn waste of time." He demanded from his staff a monastic devotion to banking and purity of soul. Those tempted to find relief from bookkeeping in San Francisco's exciting nightlife were likely to find themselves reported on around the clock by private detectives. Yet for all his self-righteous rigor, Giannini aroused devotion. He demanded nothing of his enemies except total surrender, but he was not mean among his staff. If intimidating, he was fair. If hard-driving, he was generous. He was one of the first American businessmen to promote employee ownership and profit sharing, and early with pension schemes.

The question haunting the obsessive Giannini after the panic was how he could reconcile the need to be big enough for safety but small enough to retain local affections. Perhaps the answer lay in branch banking. He had one branch up and running on Mission Street in August 1907. Once again, he was not the first into branch banking of that kind—no more than "another teller's window up the street," in the Jameses' phrase. In Los Angeles, seven of the banks had one or more extra teller's windows in the city, and a few large country banks also had rural branches for petty transactions. Two big banks in San Francisco had such branches. This, however, was one-dimensional branch banking, limited in scope, size and reach. Giannini's ambitions were in 3-D, panoramic in their geography, daunting in their perplexity and potentially ruinous in their call on resources. He dreamed of setting up a state network of branches for all the variegated regions, then a national and international system. It was a vision utterly alien to American banking philosophy. The random nature of frontier development, the early difficulty of communication and the ever-present American fear of concentrated power had produced a proliferation of independent "unit" banks. They did not for the most part share resources of information. Two-thirds or more were private banks that had received their charters from the state; others were "national" banks with charters from the Treasury and more assets generally. Membership of the Federal Reserve System was automatic for national banks; state banks had to apply separately.

The tradition of the small-town banker was so firmly rooted in American culture that only 12 states had laws permitting branch banking, nine prohibited it altogether and national banks were generally forbidden to open branches. When California enacted a tough new law regulating banks after the panic, it was with a nega

tive attitude to branches: No new branch could open without satisfying the superintendent of banking that it would promote "the public convenience and advantage"; $25,000 additional capital was required and a bank would be disqualified if it had loans on the books to officers or employees. State and national bankers assembled in two conferences in 1908 resented as scare talk the sage words of President McKinley's former treasury secretary, Lyman J. Gage, that in a crisis the 10,000 or more isolated banks in the United States would always fly apart competing with each other to survive. At the American Bankers Association, small bankers froze when they heard the president of Princeton Univeristy, Woodrow Wilson, urge rich big-city banks to set up outposts in credit-starved small towns and villages in their states. They could do without the competition. It was a reasonable fear; if the small banks lost their deposits they would not be able to lend. But that, of course, was the rationale for the branch banking system. One community with inadequate savings could be nurtured by another.

Wilson's analysis of how communities were hobbled when confined to local credit was something Giannini had seen in his horse and buggy foraging for fruit and vegetables in the Santa Clara Valley. It happened that soon after Giannini had heard Wilson's speech, California's newly appointed banking superintendent found a weakness in the financial structure of one of the grandest of the Santa Clara banks, the Commercial and Savings Bank back in Giannini's birthplace of San José. The Commercial's directors needed funds and it was to the Bank of Italy they came, hat in hand. It was a moment of intense pride for Giannini. As the Jameses write, "For all its backing by the grandee families of the valley, here was aristocratic old Commercial asking the son of an obscure immigrant to save it." Giannini, with assets now of more than $3 million, was able to help out and happier still to follow up with an offer to buy the Commercial for conversion into a branch of the Bank of Italy. Giannini's fluent young attorney, Jim Bacigalupi, navigated various legal obstacles with California's banking superintendent, who was glad to see his problem with Commercial

resolved so quickly, and approved Giannini's application as being to the public convenience and advantage.

There was sense as well as sentiment in the acquisition. The Commercial had concentrated on serving the larger landowners. Giannini could see the potential in the smaller orchards, farms and vineyards run by people like his mother and father. He was not best friends with his brother Attilio, but he asked him to give up his medical practice and take over this first branch. Determined to identify the bank with all elements of the community, A.P. mixed old families, large growers and immigrants on an advisory committee he created, the first time new immigrants had had such a voice. The watermark of the Giannini approach was his retention of many of the Commercial's staff. Familiar Bank of Italy salesmen, driven around the clock by their boss, were at weddings, baptisms, community picnics, children's fairs, church services, retirement parties; they were expected to find jobs for customers, translate documents into English, visit the sick, pay grocery bills and sometimes settle domestic disputes. Giannini knew that if the Bank of Italy branch was to do more for the community than the independent banks, there was a tremendous amount to learn about local people, and the rhythms of debt and redemption for various kinds of produce.

This makes it all the more surprising, reckless even, that he next took the plunge of invading Los Angeles, the suburb in search of a metropolis exploding across its valley in a tumult of enterprise and immigration. It was like going straight from an afternoon's punt on a millpond to exploring the Amazon in a canoe under a rain of arrows from hostile tribesmen. Los Angeles was not merely a good day's journey away—it was in another business universe. It was into movies and oil, speculative subdivision housing on semiarid land, glassmaking, shipping and food processing, and the growth industry of fads and frauds. The population was overwhelmingly Protestant, American and xenophobic, with only 3,802 foreign-born Italians to give the Bank of Italy a start (compared to 16,918 in San Francisco). When, in 1913, A.P. bought and converted the first of several L.A. banks

into a Bank of Italy branch, the *Los Angeles Times* ran the headline: "Italians take over Park Bank." The powerful L.A. banker Joseph Sartori tried to have Giannini barred from the city, protesting to the state superintendent that the man was "a serious menace to the banking profession" and "too grandiose for his own good." Just as he had been in the fruit business, Giannini was called a secret agent of the pope, a member of the Sicilian mafia, a player in a secret international organization bent on overthrowing America.

The defamation of "the dago huckster" had effect. For the first time in his banking career, Giannini suffered losses. He went out in the midday sun to dusty real estate tracts, pressing loans on homeowners even as they carried in their furniture. He advertised. Deposits remained slow. He sat up at night with Padrini wondering what else they could do, fuming about the plots to ruin him. Most of his colleagues on the Bank of Italy board were sure what to do: pull out. They saw no return on the $2 million investment. An exhausted Giannini, returning to San Francisco, was unable to still the criticisms, especially when he told them he intended moving the main branch to a better location in the heart of downtown, at a $60,000 lease, four times the current lease. The board vetoed the move and Giannini began a long game of brinksmanship. In January 1914 he stunned them by saying he was so tired he would retire at the end of the year, then left for a three-month vacation. In his absence, under new leadership, Los Angeles lost another $200,000 in deposits. When Giannini returned in the spring, a reenergized 44-year-old, he flung himself into work in L.A. He sacked "a bunch of seat warmers," collected on dubious debt, sold more small loans and started making a few thousand in monthly profits despite diminished deposits. The critics were unrelenting. Rumors circulated that the bank was quitting Los Angeles. In November, Giannini came back to San Francisco to read his

DAY AND NIGHT: Giannini was very conscious of how inconvenient traditional banking hours were for workers. This branch at One Powell Street, San Francisco, was open six days a week from 10 in the morning till 10 at night.

directors a put-up-or-shut-up memorandum. They should either close Los Angeles right away or move into the new quarters, for which he had now negotiated a better price of $50,000, and if they were closing the branches in Los Angeles he would buy them both there and in San Mateo, too. For his valediction, he gave them a pep talk: "The institution has never known and should never know the word 'failure' in any matter, large or small," he scolded, "nor will 'cold feet' ever bring it enduring or any sort of success. Our flourishing San José and Market Street branches are pertinent illustrations of what 'boosting' and constant optimistic demeanor accomplished for us in the face of trying and at times disheartening odds."

Giannini was barely in his seat before the dissenting directors gave in. They voted to negotiate for the site he had selected and at a meeting on January 12, 1915, passed a unanimous resolution rejecting his resignation. In April, the L.A. headquarters moved to its new location, celebrated with a punch party. A.P. worked the crowd with a big smile. The Los Angeles branches were in the black and another was opening. He could renew his dream of a coast-to-coast banking empire.

Having survived his urban ambush, Giannini moved into agricultural communities from 1916 on in a frenzy of buying banks and making them Bank of Italy branches. Farmers got to know his black Packard, racing along dirt roads on Sunday scouting missions with his family. "Travel on those dusty, bumpy roads was difficult then. The car was always breaking down. We had to plead with our father to stop," his daughter, Claire, remembers. "By then it was usually late at night and we would be forced to share a single room in some godawful place. All of us would complain, but my father never did." When they pulled in one night, he could not stop talking about the chauffeur's idea of the bank issuing traveler's checks. He lived so much for the bank, Clorinda's refrain was, "For God's sake, Amadeo, can't you talk about something else?" Family life was warm, but stalked by sadness. The boy who saw his father murdered saw only three of his eight children to adulthood and two of these, Lawrence Mario and Virgil Thomas, were

hemophiliacs who lived with crippling pain and frequent trips to the hospital. Virgil died at 38. Mario, who became his father's most trusted aide and successor, worked 15 and 16 hours a day and outlived his father by only three years.

In his headlong drive to establish branches across the state, Giannini was fearful of being turned to stone by any one of three gorgons: enemies at local, state and federal levels. If he moved too aggressively and stirred up local resistance, he would destroy the very foundation of the Bank of Italy's success as "the people's bank." He made himself invisible when necessary, buying shares quietly in a closed corporation in the shrewd conviction that all families eventually fought over money. After a purchase, he reappointed staff where he could, and insisted on risk-taking for local needs. Rural banks typically charged up to 12 percent interest, five points above the national average. Old loyalties tended to wane when farmers discovered that the "foreigner" who had bought the Main Street bank had lowered their interest rates to 7 percent. In one year in Madera, 75 percent of small orchards and vineyards signed up with the Bank of Italy. Giannini could point with pride to the fact that in a number of agricultural communities his branches had lent out double the amount of their deposits, something no unit bank could do, even borrowing to the limit from the Federal Reserve or from correspondent city banks.

The more constant worry was the double-headed nature of regulation. When state regulators sanctioned a new branch, federal regulators might not, and vice versa. Giannini's ability to extend branch banking always depended on the eddies of politics, changes in membership of the Federal Reserve Board and personal chemistry. He had winged sandals, but while he was hastening here and there two antibranch lobbies with the zeal of Holy Rollers were at work in Washington and Sacramento: the United States Association Opposed to Branch Banking and the California League of Independent Bankers. Giannini won state approval for a branch in Santa Maria, for instance, after an unusually bruising two-year struggle, only to have it vetoed two years later by an ally of the lobbyists,

President Harding's comptroller of the currency, Henry M. Dawes from Chicago. Dawes called branch banking "essentially monopolistic" and "destructive of sound banking principles." (One would have thought the monopolies charging 12 percent fitted that description.) In 1923, Dawes got the Federal Reserve Board to rule, in a four-to-three vote, that state banks outside the corporate limits or contiguous territory would not be admitted to the system. His argument was that branch banking would mean the destruction of the national banks and thereby the destruction of the Federal Reserve System. It was specious.

In Santa Maria, the small business owners and immigrant workers who wanted a Bank of Italy had been threatened with violence and denied credit everywhere in town. The establishment portrayed Giannini as Mussolini. (That was not totally unfair; he visited Mussolini and admired him for his strong hand, as did many at the time, including Franklin Roosevelt.) But Giannini could play rough, too, and his aides were messianic. He had a department keep tabs on the moral and financial standing of every living Italian in California, a secret intelligence network that was the steel in his proclaimed "faith in the honesty of the common man." The network's agents would bribe rival bank employees to get the names of Italians who deposited money at other banks and then pressure them, their families and employers to make the erring Italian see light. The intelligence also gave Giannini political clout in helping various political leaders he favored, notably Roosevelt. Biographer Felice Bonadio reports that small-town bankers reluctant to sell were bribed or browbeaten. One aide remembers driving with A.P. to a small bank whose owner had dug in. They parked across the street from the building. "Giannini had me get out of the car, walk back and forth and pretend I was measuring distances with my feet. When the owner of the bank came out to ask what was going on, I told him the man seated in the back of the car was A.P. Giannini and that he was planning to open a branch of his bank on the corner." They got the bank. According to the testimony of a cashier to the House Committee on Banking, disputed by Giannini, the Bank of Italy sent officers through Santa Maria county buying up passbooks, and then took 85 of them into the Bank of Santa Maria, laid them on the counter and demanded the money. "The next day one of the Bank of Italy men returned and asked how the bank liked that 'wallop.' He said, 'The Bank of Italy had many such more wallops to administer.'" Bonadio writes: "Giannini was always eager to publicize the dirty tricks of his opposition, but when it came to getting what he wanted, he had no difficulty tolerating excesses by his own employees."

Dawes's 1923 Federal Reserve edict did not survive long intact, but even as Giannini's bank grew, he insisted on it remaining the friend of the small borrower; a man in a regular job could get up to $300 just on his signature. Anybody could call the bank, ask for Mr. Giannini and be put right through without a moment's hassle. He would sometimes drop in on a branch manager, ask to see the list of loan rejections and want to know why in each case; but he would also look over every nonperforming loan.

Throughout, Giannini was a tireless innovator. He discovered that bankers' acceptances—drafts guaranteed by the federal government—would be accepted as the basis for more liberal lending. He used the 1916 Federal Farm Loan Act to provide long-term credit for farm improvements at low interest rates. He organized his banks into "foreign" departments to cater to ethnic groups: Yugoslav, Russian, Portuguese, Greek, Mexican, Spanish, Chinese. In San Francisco, he got the board of education to name the Bank of Italy the "official" bank for schoolchildren; by 1918, the bank had 22,000 children's accounts with deposits of $1 million. Others before him had focused on women. A.P.'s distinction was in the follow-through. His Women's Bank opened in 1921 with a staff of 12 women and their own floor. In two years, 10,000 customers deposited $1.5 million.

Following an initiative by his brother, Giannini was a pioneer in financing movies. Wall Street was sure they were a passing fad; there may also have been some anti-Semitism since many of the earliest filmmakers were Jewish. As early as 1919, Jesse Lasky, Samuel Goldwyn's brother-in-law, got $50,000 to start a Hollywood studio, and Charlie Chaplin, $250,000 in 1921 to make *The Kid* with Jackie Coogan. A.P. authorized a $3 million loan to Darryl Zanuck and Joseph Schenck to start 20th Century-Fox in 1930. Some of Hollywood's classic hits might never have seen the light of the projector without the Bank of America. The Gianninis lent Cecil B. DeMille $200,000 to make the original *Ten Commandments,* and backed Selznick's *King Kong* at Paramount. Selznick ran out of money trying to develop a Civil War melodrama on his own; A.P. visited the set in the fall of 1935 and was impressed enough to authorize a $1.5 million completion loan for *Gone With the Wind.* He backed Selznick in *A Tale of Two Cities, The Prisoner of Zenda, The Adventures of Tom Sawyer, A Star is Born.* By the 1940s, the loans to filmmakers totaled $306 million. Some of those dollars put Walt Disney on the road to fame. He was half a million dollars over budget and out of cash in making his first full-length animated film, *Snow White and the Seven Dwarfs.* Attilio, playing Grumpy for a change, turned him down. Disney went up to San Francisco hoping A.P. would play Happy, and he did, overruling his brother. Giannini family members say the decision had less to do with his artistic insight than a desire to overrule his little-liked younger brother. Still, *Snow White* needed nerves. In the fall of 1937, Disney was back asking, and getting, another $1.2 million. A bank executive wrote: "Had Snow White not succeeded, the loss would have bankrupted Disney and hurt the Bank of America enormously." But *Snow White* delivered, making $22 million on its first showing, and A.P. went on to finance *Fantasia, Pinocchio, Peter Pan, Cinderella* and *Dumbo.* He also financed Disney's new $1.5 million studio in Burbank, and Disneyland.

Giannini took nary a breath when he achieved the first statewide banking system in the United States. It was the close of 1918, when he had 24 branches in 18 cities, with combined resources of $93 million. He created a network of vaguely linked banking institutions—the Bank of Italy, the Commercial National, Bank of America of Los Angeles, and Liberty Bank—then danced between the interstices of state and federal regulation. Where one bank group was thwarted in branch expansion, another could take its place. When the McFadden Act of 1927 allowed a national bank to absorb another, and preexisting branches in certain circumstances, Giannini executed a dazzling series of rescues and mergers. From the spring to the winter of 1927 he bought 100 new banks with assets of $200 million. He consolidated his Bank of Italy in the national system. It meant a sad parting with the title of 26 years, but the trade-off meant he then had a Bank of America of California with 276 branches in 199 localities, far outstripping state competitors, and still growing fast. Bank of America was the third largest bank in the country, with assets of $750 million. It was an extraordinary achievement, but Giannini was irredeemably hypermetropic: Nothing would induce him to take his eye off nationwide banking. He argued the national interest: "Under a nationwide system a section distressed through crop failures, floods, unemployment, or for any other reason, would experience no diminution in its local financial support for any legitimate purpose. A new factory, for instance, would be financed as well in a distressed section as in a prosperous one. The amount of distress would be lessened and recovery speeded up."

The argument was sound, but unlike a department store or telephone company a bank was not allowed to do business in more than one state. So how was nationwide banking to be achieved? The only answer, and a clumsy one, was a house-that-Jack-built, a holding company that would own the stock of separate bank groups in each state. In 1928, Giannini formed such a holding company, Transamerica Corporation, which was watched with great suspicion by federal regulators and competitors, none more formidable than the financial emperor of New York, John Pierpont Morgan Jr. (1867–1943), who also had effective control of the Federal Reserve Bank of New York. When Giannini set out to build an East Coast network of banks, it involved him in a conflict with Morgan that dogged him for the rest of his life. Sent to scout New York, his longtime associate Leo Belden reported that a wrangle among stockholders gave them a chance to buy the proud old Bank of America in New York with 47 branches. The deal needed

FAIRY TALES: Walt Disney's groundbreaking *Snow White and the Seven Dwarfs* was saved by the Bank of America. Walt Disney (left) in a 1949 meeting with Bernard Giannini, BoA vice president, and Roy O. Disney, Walt's brother.

the blessing of "the Corner," Wall Street idiom for the House of Morgan, and Giannini came east to seek it. His arrival at Grand Central Station, with his wife and daughter, created excitement in the press. Among the pack of reporters, the *New York Times* man described him as "a regular knockout of a personality, with a titanic head, a face like a rock and a voice like a howitzer." The *New York Sun* referred to "a modern day Medici." Giannini was as cunning. He was "as meek as a lamb" when he met "Jack" Morgan, so called in order to distinguish him from his look-alike famous father. He did not argue much when Morgan demanded a veto on New York board appointments and the sacking of the Bank of America's Jewish chairman; even Gian-

nini, who could blurt vile remarks about Jews, was taken aback by the vehemence of Morgan's anti-Semitism. At the end of an unpleasant session, Giannini owned the third largest bank in New York, though it was to be run by a Morgan man, Edward C. Delafield. It was an unhappy marriage, split between the Italian-Americans from California and the native blue bloods. The Italian-Americans felt Delafield condescended to them; and to the fury of all the Gianninis—Attilio, Mario and the absent A.P.—it was symptomatic that Delafield snubbed a man wanting to open a $200 savings account. As the Jameses dryly observe, the idea of strengthening an economy with the protected savings of workers, small tradesmen and small farmers was not part of the Morgan conception of banking.

The showdown finally came when Morgan, through the Federal Reserve of New York, decided to stop Transamerica from owning Bank of America stock. It would have been the end of Giannini's coast-to-coast strategy. He held fast, and Transamerica acquired the stock without hindrance, but he was stressed out. Away in Rome, fussing over an Italian bank he had bought, he suffered a crippling and potentially fatal bout of polyneuritis, and possibly a stroke; he could hardly walk. Not quite ready to hand over everything to Mario, he set about finding a Wall Street man he could trust who would pursue the strategy of assembling banks, and other industries, under the umbrella of Transamerica and know how to fend off Morgan and his friends at the Fed. He found him in Elisha Walker, a tall, well-starched, neatly moustached banker of 49 educated at Yale and MIT who presided over an investment bank. On January 16, 1930, Giannini handed control of Transamerica to Walker and his associate, a young Frenchman named Jean Monnet, later to be the architect of the European Community. Giannini went to European spas for treatment, but the California connection was maintained with Mario as president, Jim Bacigalupi as vice president in New York and Attilio "Doc" Giannini in San Francisco.

Walker and Monnet seemed committed to Giannini's basic strategy to achieve nationwide banking, then they were unnerved by the severity of the Great Depression, with hundreds of banks failing, and Transamerica's stock in free fall. Still, that was no reason to keep the Gianninis in the dark about what they were doing. Walker humiliated Mario, making him sit outside his office for hours and then sending word he was too busy for him. When Mario quit and Walker put the cool Bacigalupi in his place, Giannini smelled a rat. "You folks may think I'm daffy," he wrote to Mario and Bacigalupi, "but still insist Wall Street gang behind the whole thing. No great effort has been made to boost Transamerica's earnings." Bacigalupi made soothing noises, but as Transamerica's stock continued to fall—down from 67 in 1919 to 2 in 1931—Giannini was con-

vinced it was the victim of a conspiracy, a massive bear squeeze concerted by Morgan with the help of "that contemptible traitor" Leo Belden. Giannini went public, demanding a congressional investigation without naming names. *Time* magazine was one mouthpiece for the Corner, telling readers they should "pay no heed," because "Mr. Giannini, 60, retired, is no longer official spokesman for Transamerica, having been succeeded by the astute Elisha Walker . . . The Giannini outburst was merely florid sales talk by an ageing master of finance." In fact, Giannini's paranoid instincts were right. Congressional hearings later in the thirties revealed that at the time Morgan and Giannini were in contention Giannini's Belden was being paid by Morgan; he was on their "preferred list" of special clients who were sold stock below market price. It also came out that Walker and Monnet had secret meetings with Thomas Lamont, one of Jack Morgan's senior partners, to discuss having "new interests" take over Transamerica to bring it into "fresh and changed alignment." The full extent of these talks has never been disclosed, but, as Felice Bonadio writes, the evidence is that "Walker's conduct as the corporation's chief operating officer could be traced to policies set in motion by senior Morgan partners like Lamont."

Transamerica was undoubtedly in great trouble under its new management. In the first six months of 1931, the bank lost deposits in California twice as fast as its rivals; it had always been the other way around. The national bank examiners marked down $15 million in bad debt. Walker told his board privately in June that they had no choice but to liquidate. Goodbye, Bank of America. Goodbye, transcontinental banking. Goodbye, all that Giannini had built in 25 years. Betraying his old patron, Bacigalupi put his eloquence at Walker's service to persuade everyone Walker was right, including Giannini's early brave appointment, the sinuous Armando Padrini. There was no resistance from other directors who had been part of the bank's ascent. Mario told his father they had been bought off with higher salaries. Walker then started a smear campaign to discredit the absent Giannini before the sell-off was due to be announced in September. Word was

leaked that A.P.'s income and fortune were much greater than he acknowledged. It turned out that his wealth was less than half a million, exactly what he said it was. "Hell," growled Giannini, "why should a man pile up a lot of goddamned money for somebody else to spend after he's gone?"

On September 22, 1931, the day of the crucial directors' meeting at Transamerica, a man with heavily hooded eyes, who did not look as if he had long to live, came through the doors and striding to the table announced, "If this board goes along with Walker's program, it will have a fight on its hands." It was Giannini, secretly back from Europe, and back from the dead, it seemed, with a vigor that belied his pallor. Propelled by a fury at the Walkerites, he had abandoned his treatment and come back into the United States, sailing first to Quebec under the name of S. A. Williams. His old lieutenants, caught in guilty surprise, were silent as he threw down his formal resignation and strode out again.

Wall Street rejoiced at his departure. The *American Banker* wrote, "A meteor in the banking firmament disintegrates with A. P. Giannini's loss of control. . . . Giannini's ambitions were excessive and the shortening of his shadow in the banking world will be good for banking in a future that should look askance at too large financial egg-baskets." *Time* magazine sniffed that the Gianninis were "echoes." On October 21, 1931, Walker started his disinvestment program by putting Bank of America of New York on the block.

In California, there was an immediate grassroots revolt. Charles Fay, a former postmaster of San Francisco, rallied stockholders; the city newspapers were detached but most of the rural papers spoke up for Giannini, a hero to the farmers. The only recourse against dismemberment of the bank was to persuade a majority of the 20,000 stockholders with 20 million shares to vote Walker out, and there was less than six months before this could be done at the annual meeting in Wilmington, Delaware. The reinvigorated Giannini began his campaign with a whirlwind tour of California, visiting 35 towns and cities in less than two months. In San Francisco, he filled the Dreamland Auditorium with 10,000 stockholders. The theme of the frankly populist campaign was the West versus the East. One speaker cried out that the issue was "whether we are to be dictated to by the golfers of Long Island or the cocktail sippers around some mahogany bar in the Bronx." The more A.P. traveled and spoke, the more he glowed. He was buoyed up by the surge of affection and support from bank employees and officers. The Walkerites told the staff that on pain of dismissal their 1,500,000 shares must be pledged to management; they gave notice to debtors that their loans would have to be repaid at once if they did not commit their 1,600,000 shares to Walker; Jim Bacigalupi appealed to his longtime associates to understand that only "an intimate knowledge of the facts" had persuaded him to desert "our old Chief and Friend." Walker and Monnet were confident they could defeat demagogy. They had only to win over the 7,000 holders who controlled more than 13 million shares, so the big holders were entertained at private lunches, and federal heavyweights were secretly enlisted to exert pressure, notably President Hoover's comptroller of the currency, John Pole, and John U. Calkins, the head of the Federal Reserve Board in San Francisco. They favored Walker's policy of drastic write-offs and liquidation, but there was also personal animus. Calkins had been a cashier at a small bank bought by A.P. in 1910 and had vowed he would never work for "some damn dago."

On February 15, 1932, all the forces converged on Wilmington. A.P. and his thousands of supporters took the lower floors, Walker and his allies, the upper floors. All waited while the proxies were counted. Just past midnight, the winner was declared: Giannini had 63 percent of the proxies. At 3 a.m. Walker conceded defeat, got into a limousine and was driven back to New York. A.P. had a victory breakfast, crying for perhaps the second time in his life: His daughter Claire says the first time she ever saw her father cry had been a few months before when he raced back from Europe too late to say goodbye to his beloved stepfather, who died in the middle of the row with Walker. Giannini told his jubilant supporters he wanted a quiet return to San Francisco, but the celebrations were spontaneous. Hundreds of people turned out to hail him at the station. The *San Francisco Chronicle* called the scene "victorious Caesar returning to Rome" and the "greatest Wall Street defeat of all time."

The symbolism of the new beginning was openness. The partitions Walker had installed in the San Francisco offices came down, and in February 1932 Giannini made himself available once again in the middle of the big, open office. But symbolism would not suffice to rescue the bank, still less restore it to its preeminence. Because of Walker's uninspired leadership and the gathering depression, it was losing deposits at the rate of $3 million a day—$138 million in the last six months of 1931. Walker's defeatism faithfully reflected the mood of Hoover's Washington and Wall Street. What was truly remarkable about Giannini was his preemption of New Deal pragmatic optimism, his insistence that recovery would never come unless California agriculture and industry was put to work rather than shut down and sold off. He did not join the line of lemmings, marching over the cliffs as they intoned the slogans of the balanced national budget; Giannini, like John Maynard Keynes, saw that deficit spending to put money in people's pockets was imperative in a depression characterized by lack of demand. It sounds obvious but then it was not obvious to all. Indeed, the received wisdom was typified by Treasury Secretary Andrew Mellon's advice to "liquidate stocks, liquidate the farmers, liquidate real estate." Giannini went into overdrive to sell what he sold best: confidence. He went back to his beginnings, to mix with the early morning vegetable dealers on the wharf. He set out to rouse all the state's 410 branches, all 6,000 employees, to join him in a deposit-soliciting campaign. One of his entourage winced at the memory: "We never stopped from morning to night, in a few instances until 3 o'clock in the morning, and then on to another city before the branch manager arrived in the morning." Giannini put $300,000 of the bank's money into a "Back to Good Times" poster campaign, exhorting "Keep Your Dollars Moving." He gave weekly booster talks on radio. When voters approved a $32 million bond issue for the Golden Gate Bridge, he bought the entire bond issue to restore the Bank of America's image. Forty-one days after taking control, Giannini had reversed the negative flow of deposits and repaid $7 million in debt. By the end of 1932, deposits were up by $100 million with 220,000 new customers. Many of the loans that the jittery national examiners had wanted to write off proved to be viable; they had taken no account of the factors Giannini rated highly, character and prospects.

Once again, Bank of America was sound, sounder than hundreds of other banks. But it still had enemies. Like every other bank in the country it observed the emergency four-day bank holiday decreed by President Roosevelt on Tuesday, March 7. The San Francisco banking establishment ran the local Federal Reserve and so resented Giannini's buoyancy that the head of the Federal Reserve, John Calkins, advised the treasury secretary he was not going to allow the Bank of America to reopen along with other banks when Roosevelt's bank holiday was due to end on Monday, March 13. Roosevelt's response was that Calkins would have to come clean and tell the public it was his personal decision to keep the Bank of America shuttered. Calkins gave in.

Giannini was a Roosevelt man from then on. He seized every opportunity put forward by New Deal legislation after 1933. The Federal Housing Act guaranteed bank loans to modernize and improve homes. Other bankers shied away, leaving borrowers in the hands still of the 17,000 finance companies charging between 10 and 30 percent interest. Giannini interpreted home improvement broadly. He created an installment-plan credit with the title Timeplan, which enabled hundreds of thousands of Californians to buy stoves, washing machines, vacuum cleaners, refrigerators and other household appliances. Timeplan loans went from $22 million in 1935 to $95 million in 1937 and $313 million in 1939. The whole electrical industry was borne up as a result. But Giannini did not stop there. He and Mario, made president, pushed the concept of "home" improvement to cover car loans. Most banks would not touch them. Giannini made them at 6 percent, and at a profit, to the fury of the finance corporations. A typical ad exulted: "Every five minutes another Bank of

America financed car." The bank became the second only to General Motors in financing cars in California. Again, the Gianninis jumped in with the Housing Act, which guaranteed long-term loans for homes. In 1938, the bank financed 163,000 new homeowners. After the war, while other banks were wondering what to do about the GI Bill, the Gianninis were in the forefront of paying out money for veterans to go to college—$600 million by 1948.

Not only did the lion in winter save his beloved bank, and get it back on the transcontinental high road, he revived California in time for the war and the boom that followed. No man could do so much good without being maligned. It was said he wore the mask of populism to create a dangerous instrument of personal power and personal wealth. The truth is in the paradox that the man whose life was money had no interest in it. He turned down frequent salary increases. He never took the frequent bonus increases voted by the board. He refused all gifts. No Bank of America employee could make an overdraft against his deposit account, borrow money from a client or buy securities on inside knowledge: Giannini preceded the Securities and Exchange Commission in banning insider trading. Shortly after leaving the chairmanship in 1945, when he found himself "in danger" of becoming a millionaire, he set up the Bank of America–Giannini Foundation and gave it half his personal fortune. His estate on his death was appraised at $489,278, which would be worth about $3,690,000 today. If he was an autocrat in administration, he was democratic in his capitalism. On his death, no less than 40 percent of the bank stock was owned by his employees.

As for politics, it is true he did not stand aloof. He put his bank behind the Democratic underdog Clement Young for governor of California in 1926. He was one of the few bankers to support the New Deal. He earned the renewed enmity of New York bankers by supporting FDR's Banking Act of 1935, reforming the Federal Reserve system, which had proved itself more concerned with short-term banking profits than the public interest in sustained economic activity. There was a benefit to the Bank of America in the reform, which took control of currency and credit from the hated Wall Street and 12 Federal Reserve banks and put it firmly in the hands of a public authority, the Federal Reserve Board, but Giannini supported it as giving the United States a real central bank for the first time, able to control credit and currency in the national interest. All this was in the public eye. There is no evidence that Giannini sought more than he campaigned for openly—liberal bank legislation and a fair deal for California. Close investigation, as Fred Carstensen concluded, does not bear out the charges popularized by Carey McWilliams in *Factories in the Field* that he was somehow complicit in the evils of the migratory labor system and maintained a near stranglehold on the farmlands through mortgage holdings. On the contrary, he did much to relieve rural poverty. In Felice Bonadio's phrase, by providing easy credit to the working class, the Bank of America "expanded the boundaries of life for millions."

Of course, the sheer size of the empire Giannini created raised legitimate fears. Its decision-making was more and more decentralized as it grew, but "financial octopus" and "hydra-headed banking empire" were common labels. The more successful an innovator, the more he or she is able to take over the market, until the next innovation comes along. The Bank of America in Giannini's lifetime had 504 branches and 4 million depositors. It controlled 40 percent of California's savings accounts. The inherent irony of Giannini is that he had to get big and powerful—his bank had to assume monopolistic proportions—in order to liberate people from the petty tyrannies of small-town bankers and the condescension of Wall Street. His bank could never have been so liberal had it not been protected by the size of its resources. It could never have "deeply impressed" the conservative banker Eugene Meyer by the service it rendered California homebuyers, in which the vast majority of $65 million in real estate mortgages in the thirties were under $2,500. It could never have come to the rescue of bean farmers in 1919, the raisin growers in 1922, finance half the state's cotton crop in 1929, nearly all the winemaking in 1938, keep water flowing for irrigation projects, nor could it have accelerated California's new industries and lent it huge resources across the nation. The bank that approved $2,500 mortgages for little people put millions into Lockheed, Dupont, GE, ATT, Bechtel. It was Giannini's personal support, and introduction to President Roosevelt, that enabled his like-minded friend Henry Kaiser to transform shipbuilding. "What is good for California is good for the bank," Giannini told his man in charge of refinancing desperate water authorities, and he was consistent in that attitude. He saw California as a symbol of the promise of America—"California hasn't started yet" was his mantra—and he was proudest of all of having freed the West from the yoke of eastern capital so that it could go its own idiosyncratically adventurous way. This was the source of much of the suspicion and hostility Giannini generated in New York and Washington. The East was long uncomfortable with the radical notions of the Wild West, reasonably so in the early days. Surely there must be something fishy about a bank for ordinary people that could grow so fast? Roosevelt's Treasury Secretary Henry Morgenthau Jr. was very much in this mold. He did his exasperating best to nail Giannini one way or another, and failed, and time has shown that it was right he should fail.

Of course, in his never-ending defense of his people's bank, Giannini did not behave like one of the Catholic saints for whom he regularly lit candles. When Morgenthau was harassing him, Giannini called the secretary at midnight, 3 a.m. Washington time, and shouted into the telephone, "You Jew son-of-a-bitch, I'm going to tell the world about you." He lost Louis B. Mayer and Joseph Schenck as board members when he railed about Morgenthau, Meyer and Fleishacker as "a damn pack of Jews" in conspiracy against him. This was one conspiracy too far. Giannini believed in racial stereotypes but it is unclear to what extent he was anti-Semitic. In late 1948, an Irish lawyer named Bartley Crum asked A.P.

about securing loans for the new state of Israel. East Coast bankers had all turned him down on the grounds that the loans would be too risky. A.P. listened to Crum's request and asked, "What security have they got?" Crum said, "They've got a lot of land." A.P. replied, "We'd have a fine time trying to foreclose on that. Any other security?" Crum couldn't think of anything and was starting to despair, when Giannini said, "You have overlooked the best possible security of all, the character and integrity of the Jewish people." He approved a loan of $15 million.

At 75, Giannini received the big news that Bank of America had become the world's biggest private bank, passing Manhattan's Chase National. He had become a colossus, but as Marquis and Bessie James testify, he personally would still discuss a $50 loan as earnestly as a $5 million loan. When he died on June 3, 1949, a few weeks after his 79th birthday, his funeral drew hundreds of ordinary people to San Francisco's cathedral. The *San Francisco Chronicle* reported: "On side aisle and in balcony they sat, men in hard brown hats, the pants legs hitched to reveal high black shoes. Women dressed in stern black and with black felt hats pulled down lower over their foreheads. Just as he had not forgotten them, they had not forgotten him on this day." In keeping with his wishes, all Bank of America branches remained open that day for business.

IN 2002, THE BANK OF AMERICA merged with NationsBank of Charlotte in a $48 billion agreement and the following year merged with FleetBoston Financial Corporation, with almost $1 trillion in combined assets. It thus became the first bank in the United States to have branches coast to coast, the fulfillment of the visionary who, 98 years before on a Monday morning at No. 1 Montgomery Avenue, San Francisco, had called out, "Vic, you may now open the front door." Amadeo Peter Giannini crossed as many deserts, scaled as many mountains, forded as many rivers and survived as many battles as the epic builders of the transcontinental railway who bore him to California in 1869.

CONQUEROR: Giannini at 78, head of the world's largest private bank, but still interested in the little man

MARTHA MATILDA HARPER

The servant girl who became the mother of America's first retail franchise network

1857–1950

HAVE BROOM, WILL TRAVEL FAR: Martha Matilda Harper as a maid in Rochester

MARTHA MATILDA HARPER, a 31-year-old maid in Rochester, New York, made the beds, cleaned the house, did the shopping, served tea, dressed her employer's hair and lent a sympathetic ear to the older woman's anxieties. When she got a spare moment in a long day Martha went back to her room to mix hair tonic and skin creams to a "secret" formula. Secret formulas were boasted every day in newspaper advertisements, but Martha's was more scientifically based, given to her by a kindly doctor when she was the maid in his family's house in Ontario, Canada. He had also taught her the elements of physiology. For a young woman who had known nothing but domestic service since her inept Canadian father bound her out to service at the age of seven, Martha had learned much, and by 1888 had saved enough ($360) to plan opening Rochester's first public hairdressing salon, a risky venture since well-to-do-women expected hairdressers to come to their homes.

At this unpropitious moment, Martha collapsed from exhaustion. It was the making of her. Mrs. Helen Pine Smith, a healing practitioner of the Christian Science faith, was summoned to her bedside; the two women prayed and Martha recovered. But that was only the beginning of a lifelong devotion that was to be decisive in her business life. Jane R. Plitt, the biographer who in 2000 rescued Martha from the oblivion to which history had consigned her, writes that Martha's soul was calmed "and her ambition was propelled towards a distinct model influenced by Christian Science values." Her association with Christian Science also introduced her to a benevolent network of people supportive of independence for women: This was a time when Rochester-based Susan B. Anthony and Elizabeth Cady Stanton were cam-

paigning against the taboos that kept women out of the voting booth and denied them property rights without a husband's support. In 1890 only 17 percent of the paid workforce was female.

Martha had little education but she was a lively and warm young woman, determined to become financially independent by running her own business. She was sensible enough to invest some of her precious dollars in engaging a prominent local attorney when the owner of Rochester's most prestigious premises refused to rent her a room: A woman in business was bad enough, but a woman carrying out hairdressing and skin care in a public place, rather than the privacy of a home, surely would involve his building in scandal.

She won her room and proudly displayed on the door a photograph of the barely 5-foot Martha as Rapunzel, with hair down to her feet and glowing with good health. Her hair was to be an effective trademark advertisement, but customers were few until a music teacher moved in next door and Martha kindly and cleverly offered him her salon as a room where mothers could wait for the end of the lessons; naturally, they killed time with hair and beauty treatments.

The Harper Method, as Martha came to call her offering, was as much about the soul as a haircut. In the therapeutic serenity of her salon, she taught that every person could glow with beauty if spiritually whole and physically obedient to "the laws of cleanliness, nourishment, exercise and breathing." She was practical about it. She designed the first reclining shampoo chair (and foolishly did not bother to seek a patent). Susan B. Anthony came to her salon, as did more and more celebrities, and visitors from out of town urged her to set up a salon in their cities.

This is where Martha's ethical sense inspired her crowning innovation. Instead of commissioning agents (as had Cyrus McCormick, Madam Walker and Annie Malone), from 1891 she developed what we now know as the franchise system. She installed working-class women like herself in salons exactly like hers, dedicated to her philosophy and her products—but not as salaried employees (as in Isaac Singer's branches). The women in what finally became a satellite network of 500 salons in America, and then Europe, Central America and Asia, owned their Harper salons so long as they bought Harper and followed Martha's rules. They benefited from group insurances, retraining and inspections by Martha. They prospered worldwide from her advertising and promotional campaigns down the decades. What was good enough for Susan B. Anthony, Woodrow Wilson, Calvin and Grace Coolidge, Jacqueline Kennedy, Danny Kaye, Helen Hayes and Lady Bird Johnson must be good enough for the rest of the world. Martha revised her techniques and was unafraid to share them with the industry that she dominated. By the 1920s, when beauty became fashionable, she opened the Harper Training School as a separate facility and a Harper laboratory. She was rich; she accepted her good fortune with joy and encouraged "her girls" to be proud of theirs.

Martha was 63 and did not look it when in 1920 she married a 39-year-old, an army captain named Robert McBain, to the alarm of her staffs. She stepped aside from day-to-day operations when she was 78 and handed control to McBain. Today, only one Harper salon remains, the Harper Method Founders Shop in Rochester, but her legacy is manifold. Franchising accounts for more than half of retail sales in America; her methods have been adopted, and health and beauty are related as they were not before Martha. But as she wrote in Miss Harper's Personal Message, the greatest achievement of the Harper Method was not the daily dollars or the number of salons or the scientific perfection of her treatments or wholly in the treatments. "The Great Achievement of the Harper Method is the women it has made."

SPRING IN EVERY STEP: Graduates of the Harper Method

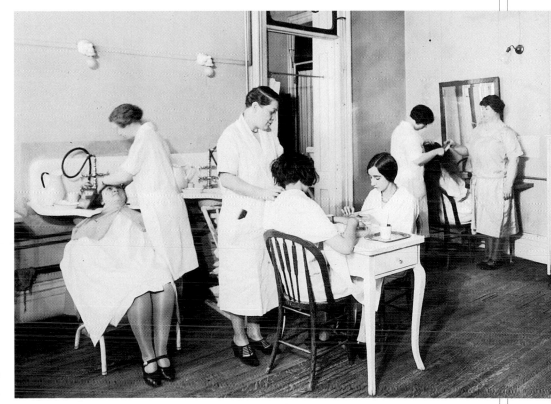

A Harper salon with a client taking advantage of Martha Harper's invention, the reclining shampoo chair

RAYMOND "PAPPY" INGRAM SMITH
The F. W. Woolworth of chance

| 1887-1968 |

PATRIARCH: Raymond I. Smith in 1953, flanked by his sons, Raymond A. Smith (left) and Harold (right), who gave his name to the seminal Reno club.

IN THE BIG CITIES in the twenties, there were plenty of popular diversions. Babe Ruth's home runs, Charlie Chaplin's *Gold Rush,* Harry Houdini's serial escapes, the Ziegfeld Follies, Alvin "Shipwreck" Kelly, who did nothing but survive for 23 days and 7 hours perched on a flagpole. On the plains of the Midwest, the farmers and their families, dressed up in gingham and Sunday suits, got in their Model T flivvers and mule-drawn wagons and went to a country carnival. Guess the pig's weight. Catch a goldfish for a box of chocolates for Ma. Knock down a pyramid of leather bottles with a baseball and win a fluffy doll for Sis. Bet a dime for a dollar on a jackknife wheel.

At such a county fair in Kansas, Raymond Ingram Smith risked his livelihood putting virtue to the test. He invited bets on where a wheel with numbers on its rim would come to rest. Winners got a jackknife that they exchanged for cash, the standard dodge in communities where cash gambling was illegal. He was in his mid-thirties, a rangy fellow with a barker's baritone, but he was new to the game and quietly nervous. On either side of him were two competitive wheels run by sharp and seasoned grifters. They followed the conventional practice of their trade. They rigged the wheel so that nobody betting had a chance. Smith looked like a hill farmer and he had a gritty Yankee honesty. He let chance decide where the wheel would stop and soon found himself parting with cash. The competitors sniggered at the sucker. But not for very long. When the farming families cottoned on that you could win on Raymond's wheel and he actually paid out dollar bills, they played his stand to the neglect of the others. In the late afternoon, when the sheriff reckoned the folks had had enough fun, he shut down all the gambling games, but Raymond, unlike the others, had earned enough money to pay his fine and get out of town. That afternoon at the Kansas carnival, according to Raymond's son Harold, was the genesis of the family's fortunes. Their policy would be to play fair in their gambling and expect no more than a fair return.

Raymond's family never had a cent to gamble. He grew up on a 200-acre farm in Vermont, where the family of four lived on $500 a year. His father died when he was seven. At 14 his mother sold the farm and

he was on his own. He spent much of his youth traveling the country, working where he could digging ditches, herding cattle, peddling milk, mixing cement, selling newspapers, clerking in a store; one time he was the "peanut butcher" handing out the nuts on the Denver and Rio Grande Railroad. He never touched a drink or a cigarette and kept very much to himself, a man who would stand aside when everyone else was raising hell.

By the time he tried his hand at the wheel, he had two sons, Raymond A. Smith, born in 1907, and Harold, in 1910, but he was abandoned by his wife, Dora, a chuck-a-luck dealer, who ran off with an Englishman. She, in turn, was abandoned by her new lover, but she brought the boys up in a 23-room gaslit boardinghouse she ran in Cleveland, where they scrubbed the floors and washed windows. Raymond, who was out on the road, sent Dora $5 a week for the boys. He had ambitions for them. Harold at 14 decided he wanted to be a violinist and summoned up the nerve to ask his roaming father for $150 for an instrument. "He didn't think $150 for a violin was frivolous," Harold wrote in a memoir. "He sent it immediately. Later when he learned that I was studying hard and practicing hard, he was pleased. Daddy could always appreciate something if you did it by hard work. It is a strange fact of my life that every productive quality I have developed, from self-confidence to decisiveness, has come from trying to please him."

Raymond was not an easy father to please. In 1924, when he was in Chutes-at-the-Beach in San Francisco with his wheel games, he sent for Harold and Raymond to help out. Harold never forgot the day at the fairground his father asked him to find him a hammer. He came back to say he couldn't find one. His father snapped: "I won't stand any can't-me guys around. When I say get a hammer, get one!" Harold found one finally at another operator's booth. When Harold was running a booth, his father would walk by and bark, "You stand right there at the center of that counter and sell every person that walks by." And: "Speak up to 'em, son. Speak loud.

WINNERS: There were more of them at Harolds Club machines in 1945 than anywhere else in Reno.

They ain't going to hear your pitch if you're whisperin'. An' smile. Show your teeth. These folks are here for fun and recreation." And: "Don't be a smart alec. Don't josh ladies old enough to be your Mama. They're havin' the fun—let them be the comics; you're workin'." And: "They're walkin' by you. Get out in the pike and turn 'em. Get thirty feet out front and direct the traffic to your stand."

Harold reflected later: "After that year on the beach at San Francisco, I couldn't have run a sloppy, unprofitable or tricky game if I tried." He learned to cut his losses. For Chicago, his father had devised a goldfish game. They put 10,000 goldfish in a small tank and invited players to catch them with a hoop covered with glazed paper. It was an easy game—and a big flop: "Even a nineteen year old soon comprehends that the way to go broke is to try selling the American public something it doesn't want." But it was his father's Kansas example that led him to success. Operators of a nail game invited players to drive a nail into wood with three blows of a hammer. They made it tough to win, with hardwood, and they were niggardly with the prizes. Harold decided to use a medium-soft wood, charge 15 cents a try and offer attractive prizes. He practiced so often his wrists started swelling, but he got good enough to drive in the nail with just two blows. "And suddenly I had my pitch. I would sell this game on the ease of it. Since I was a smalling lad of 143 pounds, I would concentrate on big men, particularly those strolling with girls, and challenge them to surpass my two-stroke performance. I wanted them to win of course." On opening night, Harold began shouting in the best Raymond style, "If I can drive this nail . . . in the block with two strokes of the hammuh . . . you ought to do it in three strokes and WIN A PRIZE." One husky man said, "Let's see you do it." Harold did. Then the man tried and didn't quite make it. He paid another 15 cents for a second try and won a box of chocolates. He came every night, carrying off cartons of prizes. The significance of all this tomfoolery was the insight into human nature. The nail game appealed to a man's ego, and he was not humiliated. He carried away a tangible reward. It was quid pro quo, something for something, not caveat emptor as in so many amusement park games.

The stock market crash of 1929 catapulted the Smiths into applying this fairground psychology in ways that transformed gambling in America. Fifty miles north of San Francisco, Raymond found in the resort town of Rio Nido on the Russian River a bingo game that had gone under. The owner of the booth had seen bingo as simply a means to make a quick buck; he didn't bother to build a partition to keep propelled the Smiths to Nevada, where gambling was legal.

Shortly thereafter, Harold found himself wandering down the lower main street of Reno, a typical decrepit end of a western town close enough to the railway tracks to be Skid Row. There was a cut-rate restaurant, a loan office, a cleaners, and three bingo parlors jammed under a walk-up hotel. Harold went into the middle bingo parlor, not much more than a hole in the wall, and hurried Raymond to see it. The vertiginously from the moving hand of a public clock. They had chanted "Har-wold Lloyd! Har-wold Lloyd!" It struck Harold that his own name was auspicious. He had named the bingo game on the Russian River "Harold's." Raymond and he decided to use the same name and call their new place Harolds Club—without the possessive apostrophe. Raymond hated the word "casino"; he always called the club his "store" or "shop," where people could buy a thrill as they might buy a kettle.

HALF A MILLION ON A ROLL OF THE DIE: Robert Carnahan, a casino operator from the Midwest, tried to break the bank at craps, two dice on the green baize, double odds. He would not quit and kept raising the stakes. After 33 hours, still at the table in his brown fedora, he owed the house $348,000. An enormous crowd had gathered. He wanted to go to half a million. Harold suggested peeweeing the dice, that is, using one die: Winner takes all. If Carnahan lost he would owe half a million. If he won, he would win the difference—$152,000—between his debt and his stake: His debt would be reduced to $196,000. Harold rolled first. He got a one. Carnahan now had 5/6th of a chance of winning. He also got a one. Tie. Harold's next roll was two. Carnahan got a three. The die roll put Harolds Club in the big league. It was then the highest-stake payout in the history of legalized gambling in Nevada.

DOES A ROULETTE WHEEL HAVE A MEMORY?: Two mathematics students from the California Institute of Technology showed up in Reno with pads of paper and pencils and a "system" for winning at roulette. They started at a smaller casino and then moved to Harolds, where their nonstop wins became the talk of the town. Observers saw that they kept betting between $3 and $13 on number nine, skipping a few rounds occasionally. After 48 hours, they had made between $10,000 and $15,000. Their stunt was all over the news. Mathematics professors opined on whether there could be a scientific system or whether a Reno croupier was more on the ball with the remark, "A roulette wheel has got no memory."

Raymond did not in the least mind the losses. He treated the students like royalty, plying them with free meals, drinks and cigarettes. The attention to Harolds, he told the press, was worth "at least $150,000."

night breezes from his clients, mostly older women, and his prizes were cheap geegaws. "He didn't treat his customers as if they were real ladies out to buy enjoyment with their money," said Raymond. "I did and prospered." Harold ran the booth for his father, knocked up a little place to sleep behind it and lived on about $2.50 a day. By the end of the summer, they had earned $4,500. The next stop was Florida, where Ray for seven summers ran a new game called Fascination—a sort of electric bingo. In winter, they moved to Modesto, California. Gambling was still illegal in the state. The cops turned a blind eye to bingo, recognizing the Smiths' honesty, but in 1934 the Smiths risked adding a penny roulette. They were arrested, given a 90-day suspended sentence and fined $500 for running a gambling game. As Raymond put it, he had backed the wrong man for district attorney. It was a lucky break. It owner of three months, still in debt for lumber, rent and prizes, had convinced himself Reno was too sour a place to make money. Harold, only five minutes in the parlor with Raymond, recalls an electric tingle, "a hunch, extra sensory perception or what you will." He urged his father, "'This is it. We must have this place.' Daddy stared thoughtfully at the ten lonely customers, half of them shills. He was learning to give my hunches more than passing thought. We took the place."

They pooled their money to get it for $2,000 and paid an extra $500 to settle claims stemming from the seller's bad debts. The Modesto court had allowed the Smiths to keep their gaming equipment. All they needed was a new name. In San Francisco's Chinatown once, Harold had been mistaken by a group of Chinese for the silent-movie comedy star Harold Lloyd, famous for risking limb if not life hanging Reno had grown up along the Truckee River in the mining boom of the 1870s. Its quickie marriage and divorce laws had made it the place to come for both. It had banned gambling in 1909, but as Gilam Ostrander describes it in his 1966 book on the state, this led only to gambling going underground, like the liquor trade, so that it fell under the influence of criminals who bought off corrupt cops. Scandals forced a modified repeal, then in 1931, when Nevada's monopoly of revenue from marital easy-come-easy-go was threatened by other states, it responded to pressure from constituents to legalize gambling completely. The old Libertarian Western tradition, in short, made a quick marriage with new states' rights avarice. Commercial consummation was delayed a surprisingly long time. "Most Nevada gamblers," writes Ostrander, "continued their dear old ways throughout the thirties. They operated in

THEY MADE AMERICA

darkened and secluded halls, the dealers in shirtsleeves and green eyeshades, dealing to the same customers as in former days." Gaming managers were as conservative as the bankers whom Amadeo Giannini outraged by advertising the services of his Bank of Italy (page 259). Oscar Lewis in *Sagebrush Casinos* suggests the cautious habits formed during the years of secret undercover operations had become ingrained. Such publicity as anyone in gambling did was limited to tiny street signs, and the few ads in newspapers were apologetic in tone, usually describing things that could be bought at the bar with the main business limited to the single word "gaming" tucked away in small type.

Raymond Smith's business genius, a characteristic of many innovators, was to recognize an entire new market, and to have the courage and imagination to serve it. He did not gamble himself, though he was a crafty bridge player, and he shared the revulsion of ordinary people for the smoky, dark casinos associated with professional gamblers and wild debt. Harolds Club was located between three dingy bingo joints, but Raymond transformed the space by introducing the festive atmosphere of his carnival days. He opened it to the street, lit the interior with bright lights, and altogether tried to make it inviting to passersby. And he set out to implement his heretical notion that it was good for the house to lose and the punter to win—just as long as the house won slightly more often.

None of this went down well with competitors. One hour before the club was set to open on Harold's 26th birthday, February 23, 1936, he stood in front of the new "store," giddy from anxiety aggravated by altitude sickness. They had $600 set aside for opening costs, but adjoining bingo parlors had started a price war that threatened to be fatal. The Smiths decided not to enter the fray; bingo, Raymond decided, could be left to his competitors. Instead, he suspended from the ceiling an electric eight-foot roulette wheel from Modesto and a gigantic mirror. This "flasher" wheel had places for 43 players who could watch the wheel light up numbers, hopefully theirs. It was the first of its kind in Nevada, but the more significant innovation was Raymond's decision that anyone could play for

a penny. This was at a time when 25-cent bets were the lowest permitted anywhere else in town.

Harolds Club was packed the first night and people kept coming back to a place where they won more often than anywhere else. Raymond rented two battered slot machines which he soon threw out because they paid out only once a year. He installed new machines, reduced the stake money to 5 or 10 cents a shot and fixed them so that they paid out often. Next to the slots, he posted the number of winnings from the previous 24 hours, something like:

Dollar jackpots:	25
50-cent jackpots:	17
25-cent jackpots:	34
10-cent jackpots:	66
5-cent jackpots:	172

It was low-margin business, but they hoped it would be high volume. In the mid-1930s, the average house take in Reno was as high as 20 percent. Raymond slashed the house cut to 4 percent. On some games, like the slots and craps, he cut the house percentage to 3 percent. In other words, people got money back 97 percent of the time. His 11 craps tables gave the house a mathematical edge of 1.4 percent, the 29 blackjack tables a 2.5 percent advantage. It was viable because the games were so quick; the profit came in the high turnover of small change. Other gaming houses were furious. To stay competitive, they had to cut their own take. Raymond's leadership, in fact, was to create the modern industry standard.

That first year Harolds Club lost $2,000. The Smiths worked 90 hours a week, keeping the club open seven days from noon to midnight. Struggling in the Depression, the rest of the family turned up to help: Harold's brother, Raymond A. Smith; Dora, Pappy Raymond's ex-wife and Harold's mother; one of Raymond's brothers, Harry, a Jehovah's Witness who dealt 21 by day and sold the *Watchtower* at night. Many around Reno just knew the crazy club could not last, deriding it as Smith's Honky Tonk, Harold's Penny Arcade, Gamblers' Five and Ten, which pleased Raymond. To be thought of as the dime store joint, with its echo of F. W. Woolworth's, was just the image he wanted.

There were other challenges. Their first bouncer stole the currency box from their first crap table. His girlfriend, who was dealing at a table, hid takings in her handkerchief and handed them over in the bathroom. Sharpies managed to mark cards. And cheaters tried to get hired so they could let conspirators break the bank. "We were obviously the most innocent operators in the land," writes Harold, "and every cross-loading son-of-a-bitch in Nevada wanted to work for us. We hired most of them, too, I think." The local mob controlled a number of casinos in Reno at the time. The first attack on Harolds came when a corrupt councilman, controlled by the notorious Chicago gangster Baby Face Nelson, brought the rest of the council to examine the big penny roulette wheel. A city ordinance imposed a tax on every wheel in every casino. The councilman tried to convince the rest of the council that Harolds should pay the tax not just on its wheel but also on each of the 43 gaming places. It would have ended the club there and then. The councillors demurred. A few days later, Harold heard that hoodlums were going to wreck the place. He and his brother, Raymond, got ready. They were alone in the unopened club when at 10 a.m. seven men swaggered in and started knocking things over. They approached Raymond, who was behind a crap table. Harold pulled his loaded .38 from the roulette counter. His hand on the gun was clammy with sweat, but he said coolly enough: "You're not going to shoot any dice, so just turn around and walk out of the door." He was resolved to put a bullet near the foot of the first man to move and a bullet into the body of the second. He stared at them for 1,000 years. They stared at the gun. He knew that if he wilted, every hoodlum in the area would clobber them, but if the men retreated before a gun, the psychological advantage was with Harolds. The sinister seven backed up, talked in a circle, then turned around and walked out. Harold continued to carry a gun when leaving the casino. He writes: "The eventual success of Harolds Club, Harrah's and several others was all that saved legal gambling from the clutching talons of those racketeers." He had always to be on his guard. Sometimes he would enter his room and find a

woman sitting there, soon to be followed by a man. He started hiring a bodyguard less for protection than as a witness for any attempt at blackmail.

Raymond persisted with the friendly approach and in fairly short order, he created a Disney World of Gambling before Disney started his first theme park. His bet was that his unconventional approaches would be rewarded in the long run. The babysitting service he provided in the casino obviously enabled more women to gamble. The same spirit of enlightened self-interest lay behind the signs he put up throughout the club and on slot machines: "Gambling is a game of chance. We advise you to wager no more than you can afford to lose.—The Management." Dealers had standing permission to advise players when they were making a dumb bet, such as a self-defeating blackjack play, something that would have got them an automatic firing at other places. One of Raymond's most novel innovations, which he allowed to be publicized, was his practice of giving travel fare ("once only") to gamblers who went bust. A man approached Smith one time crying that his wife was in the hospital, he had rent and car payments due, three children to feed; he had come to Reno in the hopes of making money and lost everything instead. Harold checked out the story and ordered every cent of the man's money refunded with a few extra dollars. Cynics may be unimpressed, but the fair play and generosity does seem to have been more than an occasional stunt. There were years when Harolds Club refunded up to half a million dollars, all of it after paying the state's take of 5.5 percent, on three conditions—that the loser did not tell anyone what he had received; that if they came back to Reno they would come to Harolds; and they would tell people that Harolds was an honest place to play. "You'd be surprised," said Harold, "how this bread cast on water has returned. Daddy believes it has brought more business in the long run than our familiar billboards."

All the time Raymond invested in the future. In the early days, he did not have enough money to advertise much, but he had a talent for ballyhoo. Without fail every night, he would tour the club, walk up to the tables and place double bets for every-

one. It cost the casino $500 a day, but word got around.

One day as Raymond stood by the doorway, a woman came in, took two or three steps toward the first game and stopped short. "There are no women here!" she cried and fled. Raymond badly wanted women in the club. He came up with an idea that enraged the professional gamblers and disturbed Harold: He would hire women dealers. His hypothesis was that women would feel more comfortable gambling with other women and men would come to look. Oscar Lewis says the industry diehards, furious at the breach of man's last secure bulwark against the encroachment of womankind, predicted that unless this move was nipped in the bud, it would wreck the industry and make Nevada the laughingstock of the country. Raymond quieted Harold's misgivings and ignored the hostile mutterings. In 1939, he ran ads that read, "Girl dealers. Pleasant appearance and friendly personalities essential. Wages $15 to $17.50 a day." He dressed the women in yellow sateen blouses with the name Harolds Club embroidered on front and back, with black pants, white belts and white shoes. He gave them extensive training before allowing them on the floor, not

just on how to handle a deck of cards but how to sympathize with losers, discourage players out of their depths and handle rowdies. Every girl who made change was told to tell the gambler "Good luck!" As Raymond predicted, the women dealers were such a novelty that crowds came just to see. The scoffers in other casinos started hiring women, too; they proved more honest than the run of men. Harold later admitted that he had been wrong. "The veneer of society rubs a little thin around a gambling house. Sometimes inhibitions do slip temporarily. If it weren't for the ladies, it could get rough. Their presence helps us all maintain our perspective." He credits the advent of women with the introduction in Nevada casinos of carpeted rooms, floor shows and free daiquiris.

Until 1941, Harolds Club was largely dependent on locals. That winter Reno began to boom; people came pouring over the mountains with their money and the club expanded upwards, downwards and sideways on Virginia Street. The club had enough income now to mount a major advertising campaign. Raymond began with billboards in a 500-mile circle from the casino, all the same, with a covered wagon and the slogan "Harolds Club or Bust."

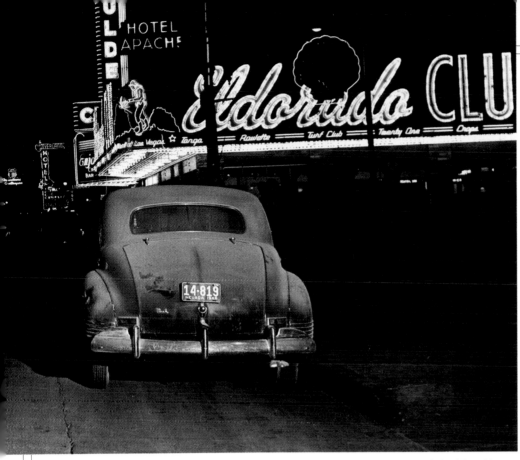

NIGHTLIFE, RENO, 1945: Harolds Club was the model for the then nonexistent Las Vegas.

Eventually he had more than 2,400 across the country. National magazines ran photographs of Eskimos above the Arctic Circle holding "Harolds Club or Bust" signs. People clamored to have club signs erected on their property, tempted by the rumor, which Raymond did nothing to contradict, that the club would give $100 in free chips to anyone who put up a Harolds billboard. By the '50s and '60s the advertising was more likely to tout Nevada's landscape and historic mining towns. Raymond's theory was that once enticed to within 50 miles of Reno, tourists would be unable to resist a visit to Harolds.

He was forever inventing new games, new stunts. Harold returned from travel appalled to find that his father had hired a mouse and its "trainer" at $100 a week. The mouse had replaced the ball on the spinning disk of the roulette wheel. Players bet on which of the numbered holes it would run into. Harold had seen a similar version in Florida and saw that nobody could win. The operator there kept very quiet just until the mouse started sniffing around a hole, then he called loudly, "Anyone want a Coke?" The mouse would then scamper into a losing hole. The Great Reno Mouse Roulette lasted one week only, but the pub-

licity was gigantic across the country. It drew thousands to Harolds Club; 25 years after its debut people were still asking to see the mouse roulette.

Raymond was also one of the first to add attractions to a casino: a Saturday night fireworks show in the summer. A Western museum of frontier days. A gun museum. He relented only very late in teetotal opposition to serving drinks, and was against serving meals, but he was early in setting the model for the then nonexistent Las Vegas by bringing in famous entertainers. Bing Crosby would sometimes play the tables, as would Jimmy Durante, Eddie Cantor, Dean Martin and Edward G. Robinson. James Stewart came to parties.

By the 1950s, Harolds was claiming to be the largest casino in the world. It is hard to verify that conclusively, but the numbers do point to extraordinary volume. Typically, 19,000 came through the doors over a 24-hour period; the daily attendance was greater than the typical weekly custom at the famous casino in Monte Carlo. On weekends and holidays, it was common to have 10,000 at any one time. Annually, more than 1.5 million people came to Harolds, about 10 times the population of Nevada at the time. The Smiths never dis-

closed how much money they made. In the year 1953, according to *Time* magazine, they took in more than $100 million in bets; the *New York Times* in 1962 reported a gross of about $12 million annually. They were big contributors to charities. By the 1960s, Raymond was giving about $100,000 a year to the Boy Scouts of America, the Community Chest and just about every church in Nevada, and $90,000 to finance full college scholarships for some 160 students.

Harold, alas, did not take the advice plastered around the club. After his first wife left him, he took to drink, as much as four quarts of whiskey a day. He gambled millions of dollars, ending up only with an insight in that "We'll stand longer losing than winning." Raymond put a limit of $50,000 on his wayward son's gambling. Harold kicked his addictions after 10 years, though he still was wont to ride into the club on a horse. By then Harrah's was starting to eclipse Harolds, and Las Vegas was starting to shoulder Reno aside as the center of gambling in America. The Smiths sold to a New York syndicate for $16,675,000 in 1962; eventually Howard Hughes bought Harolds after acquiring six Las Vegas operations.

In the end, of course, Las Vegas overtook Reno. Its proximity to Los Angeles counts, as does the popular appeal of the Strip, with its glitzy hotels offering entertainment and dining. But contemporary Las Vegas is Smith's model, only more so. It was Smith who first democratized gambling in a way that attracted millions of people to Nevada. He invented a new type of casino, far more plebeian than the snooty European type but far more wholesome than the casinos inhabited by cutthroat professionals. His emphasis was all on the fun of the carnival: recreational betting for the masses, honestly conducted and far removed from the dark joints where the dice were loaded. Smith showed something else: that good profits can lie in fair play. The industry, now Nevada's largest, earns far more than the fixers, cheaters and corner cutters ever thought possible.

Neil Cobb, Raymond's son-in-law, put his finger on the source of Raymond's success: "Pappy was a wonderful visionary. He knew what the common people wanted because he was one of them."

JUAN TERRY TRIPPE

He was an air taxi pilot with a single-engine seaplane who flew the whole world into the jet age

1899–1981

AMONG THE AWESTRUCK crowds cheering Wilbur Wright, when he flew around the Statue of Liberty on September 29, 1909, was a 10-year-old who would make aviation history himself over and over again: Juan Terry Trippe. As an adult, he hated the Juan in his name, suggestive of Latin American blood, and for several years signed himself J. T. Trippe. There was, in fact, some Latin American blood in him, Venezuelan mixed with racy Irish on his mother's side, and he grew up to have the dark good looks of a Rudolph Valentino, albeit given to plumpness. But the reason he recoiled from "Juan," he explained years later, was that everyone pronounced it "Wang," unappealing to a boy whose father was proud to tell him the family could trace its English roots back to the Norman Conquest.

Both parents doted on him. His Baltimore-born father, Charles, was a Wall Street broker and banker; his mother, Lucy, speculated in real estate. They had money, though nothing like as much money as the people they preferred to mix with, the Rockefellers, the Vanderbilts and the Morgans.

At Yale in 1917, there was already an outline of Trippe's destiny and style of operating. He was a well-knit six-footer and football star, but he failed the 20/20 eye test to be a World War I naval aviator, so his father had a word with the assistant secretary of the navy—Franklin Roosevelt no less—and another test was arranged. He got through by committing the bottom line to memory and all his life could recite A-E-P-H-T-I-Y. He just missed getting into action in Europe, having qualified as a night bomber pilot in October 1918 in Pensacola, Florida, but he romanced his future wife, Betty Stettinius, with adventurous tales of dramas from the Florida war front. Many cadets were mysteriously plunging to their deaths in their trainee biplanes, he told her, then it was discovered that a German saboteur on the base had daubed acid on the aircraft control wires. The spy, he said, turned out to be a German teacher from his boarding school, the Hill in Pottstown, near Philadelphia, and Ensign Trippe told his sweetheart that he was assigned to the firing squad. It must have been entirely credible to Betty, fixed by her suitor's brown, almond-shaped eyes and radiant smile, but throughout Trippe's life people kept appearing who said they regretted being taken in by his charm: "He'll talk to you in his suavest, most deferential manner," said Frank Russell, an investment banker. "You'll think there is no one like him and believe every word he says, and all of a sudden he's stolen your pants." In her sweet personal diary, Betty admiringly records of J.T., "When he graduated he had a report card of all A's" in the Hill School. This statement was contradicted in the well-researched 1982 biography by Marilyn Bender and her husband Selig Altschul: "Juan, nicknamed Tripe or Trippy by his classmates, did not cover himself with scholar's laurels. A failing mark in German nearly kept him from graduating." German? It seems Betty was asked to believe that Trippe, in the name of duty, shot his teacher.

Whatever else he learned in youth, Trippe thoroughly absorbed the advice of the upperclassman character called Le Baron in Owen Johnson's bestseller *Stover at Yale* to mix with the right crowd, the crowd that mattered: "You may think the world begins outside of college. It doesn't.

OCEAN BREAKER: On March 26, 1939, Captain Harold Gray flew a Boeing B-314 Yankee Clipper across the northern Atlantic on a test. This was the plane that went on to complete the world's first commercial round-trip flight across the Atlantic.

CHAMPIONS: Fifty-six fliers in Curtiss training planes competed in a 25-mile race among 12 colleges in 1920 from Mineola, Long Island. Pilots Trippe (left) and George Horne of Yale won by 10 seconds. Below: He was right guard on the freshman football team. A kick in the spine in a pileup in 1920 damaged three vertebrae and ended his football career.

It begins right here. You want to make the friends who will help you, here and outside." One of our first innovators, Yale's Eli Whitney, had exploited his Yale network (page 50), and so did many other Yalies (including Trippe's contemporaries Henry Luce and Brit Hadden, who went on to start *Time* magazine).

Still, one has to admire the dash and vision of Trippe at 24, when, two years out of Yale, and just as the stock market boom was revving up, he gave up an easy life as a bond broker, bought seven junked navy training seaplanes in an auction for $500 each, then went knocking on doors of wealthy men, saying, "I'd like to show you my figures." Some of these were introductions from his blue-chip friends — Cornelius Vanderbilt (Sonny) Whitney, William Rockefeller and Harvard's dashing wounded air ace John Hambleton — but Trippe did much of the dirty work for the company that emerged, Long Island Airways.

He stripped out the 90-horsepower Curtiss OX5 engines and substituted 220-horsepower Hispano-Suizas with smaller propellers so that the pontoon biplanes could carry two passengers. In the summer, he and his six ex–military pilots gave rides at the beaches from Coney Island to Fire Island, then he rented a barge near the Statue of Liberty as his pickup point for air taxi trips at $5 a time to the Hamptons, Newport and Atlantic City. He swept the hangar and kept the books; and he appeared in court when one of his pilots sued for $470 in back wages. He was in court another time for flying just above the sidewalks of lower Broadway for a movie company filming a chase: case dismissed because there was no law then about airplanes on city streets.

Nothing seemed to faze Trippe. As his little airline drowned in red ink, he presented himself as an expert to a congressional subcommittee on commercial aviation that converged at the Waldorf-Astoria Hotel in New York. In the high, earnest voice of youth, he said he knew of "one hundred possible buyers of aviation equipment who are being held back because there are no suitable landing facilities around New York." He predicted a surge in air travel if only there were the air-

fields and navigation facilities — New York's La Guardia was an amusement park, Idlewild (later Kennedy) was known for its golf course.

The Boston Brahmin grandee Godfrey Lowell Cabot led the applause on behalf of the assembled aircraft manufacturers and bankers. Congress was then on its way to relieving the U.S. Army of the duty of carrying mail by air, opening the way for private operators to do it for a fee. Even as Representative Melville Kelly, a Republican from Pittsburgh, steered the enabling bill through Congress in 1924–25, Trippe was acting on two fronts — in Washington to lobby Kelly, and in Boston to form an alliance between his friends and a group of solid New England businessmen. Together, the two groups created Colonial Air Transport. They won the license to carry airmail between Boston and New York, and "J. Terry Trippe" became vice president in charge of operations.

The service was not due to start until July 1926 and Trippe did not have board approval when he pledged $75,000 for two three-engine planes from a new friend, Anthony (Tony) H. G. Fokker, the Dutch designer of Germany's ace fighter plane, who had emigrated to the United States. Trippe rightly guessed that the directors would object to such large, expensive aircraft, especially when the post office contract specified single engines, so he organized a campaign to publicize the virtues of the Fokker; press and public in New York City were invited to take tea aloft and try out Fokker's pride, the first airborne toilet.

Trippe and Hambleton meanwhile plotted how they could win contracts for more routes, and especially break into the business of carrying mail overseas, still restricted by Congress. With an introduction from the head of President Coolidge's air board, Dwight Morrow of J. P. Morgan (the future father-in-law of Charles Lindbergh), Trippe got to see and captivated the man in charge of post office airmail as second assistant postmaster general, New Jersey Republican W. Irving Glover. Then he took a trip with Fokker to Cuba and through a Yale connection secured an audience with the new president, Gerardo Machado. Somehow, Trippe emerged with

MAILMAN: Thirty-three-year-old Trippe, as managing director of Colonial Air Transport, greets the arrival of the first airmail flight from Boston to New York on July 1, 1929.

a secret license for exclusive landing rights at Cuba's national airport outside Havana.

The Yankee directors of Colonial were not amused by Trippe's high-flying initiatives. Their minds were not on such fanciful exploits as flights to Havana but on the railways of the North-East corridor and how they could capture their mail traffic. Nor did it help that Colonial's service lost money on the Boston–New York service from the day it began on July 1, 1926, and that Trippe had very different views about how they might win the New York–Chicago franchise. The expansionist-minded young New York aviator blue bloods and the conservative Yankees were oil and water. Trippe had told the attorney setting up the company, Robert G. Thach, another wartime flier friend, to "fix it so that our crowd is in charge," but it was Trippe who was booted out. Sigmund Janas, who became president of Colonial, was caustic: "The trouble with Trippe is that he just can't tell the truth." Evasiveness would have been a fairer charge. Throughout his career, Trippe was gripped by such an urgent certainty about the future of air travel that he could never bear to risk it being impeded by second-guessers.

Trippe and his friends lost the money they had invested in the Colonial venture, but what rankled more with him for years was the thought that "a bunch of old fogies" had robbed him of the chance to build the largest domestic airline (Colonial was absorbed in 1930 into a predecessor company of American Airlines).

After the debacle, Trippe concentrated his ambitions on licenses for carrying mail overseas, starting with the 90-mile Key West to Havana route, as soon as the post office invited applications. He formed the Aviation Corporation of America with $300,000 in capital, put up mainly by the original trio of Trippe, Vanderbilt and Hambleton plus William Rockefeller (all $25,000 each) and Sherman Fairchild ($5,000) and other Yalies, on the understanding that the airline would be run by Trippe. They faced competition from two other groups, which also had money and famous fliers. The favorite to win the license was one led by a former navy pilot, Captain John Montgomery, with Richard Bevier, in association with Major Henry "Hap" Arnold (who would create and head the army air force in World War II). The other was led by the adroit 39-year-old investment banker Richard Hoyt, who had merged the Wright and Curtiss aircraft companies and chaired Curtiss-Wright. The flying star here was America's top gun in World War I, Eddie Rickenbacker, who had founded Florida Airways, bankrupt for want of people willing to fly.

For the post office, Assistant Postmaster General Glover urged the partnerships to merge. Trippe and Hoyt were willing to tango, the Montgomery-Arnold group was not, but when the music stopped Trippe's group had the floor. Montgomery's group had to withdraw altogether when Trippe quietly revealed the exclusive landing rights deal he had made in 1925 with President Machado: A U.S. license to fly mail was useless if there was nowhere to deliver it. Hoyt's group had to be satisfied with a minority stake in Trippe and company's Aviation Corporation. Hoyt became chairman of Aviation, with Trippe, at 28, taking over as president and general manager of the operating company, a little puddle-jumper that nobody had heard of: Pan American Airways, Inc.

On October 19, 1927, Cuba's postmaster rowed out into Havana's harbor to collect seven bags with 30,000 letters just brought in by a single-engine seaplane flown by Cy Caldwell from Key West. Trippe had borrowed the seaplane from West Indian Aerial Express because his Fokkers were not ready. Nine days later, on October 28, the first regularly scheduled flight began of the first enduring American international airline, seven bags of mail flown in by Captain Hugh Wells and Ed Musick piloting the Fokker trimotor *General Machado,* named after the first but not the last dictator to be garlanded for his cooperation with aviation, the twentieth century's glamour technology par excellence. Freshly painted on the fuselage was the blue PAA legend.

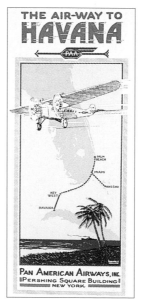

THE AIR-WAY TO HAVANA

PAN AMERICAN AIRWAYS, INC.
PERSHING SQUARE BUILDING
NEW YORK

Pan Am thrived on Trippe's almost clairvoyant grasp of the future, his determination to find aircraft to fit that vision, his laser eye for the small print, his discernment of talent, his ruthless extermination of competition and his stealthy lobbying for U.S. mail licenses: As biographer Robert Daley wrote, shame was not part of his makeup. He would do anything to get a license to carry mail because it was the essential source of revenue when there were so few passengers or planes to carry them in. Trippe carried out his lobbying in secrecy, a characteristic that was a mixed blessing in the actual operations of the company. He vanished for weeks at a time without anyone knowing where he was. He was never on time for a meeting. In the early months, he always locked his rolltop desk whenever he left, even for a moment, the little office he shared with the "the engineering department." The department was then one man (or rather one superman), namely Andre Priester, whom he had appointed as chief engineer.

By most accounts Trippe was a cold, stingy and distant employer. Perhaps he was not so much rude as preoccupied. Roger Lewis, who served many years on the Pan Am board, described him as "the politest and least compassionate man I have ever known." All the same, Trippe selected and for years kept the loyalty of a core of fine professionals ready to work devotedly with an intelligent pirate and system builder. Top talent appreciated that while Trippe set the strategy, he gave the executives he trusted the room to breathe.

The tiny Andre Priester, a Dutch compatriot of Fokker's, was a fanatic for safe flying, an attribute appreciated by Trippe's other inspired selection, chief pilot Ed Musick. Musick had been a daring birdman for carnival crowds but one near-death experience had made him a meticulous planner with no time for idle chatter. He and Trippe had that in common. When Pan Am moved offices with a rapidly growing staff, Trippe had no casual

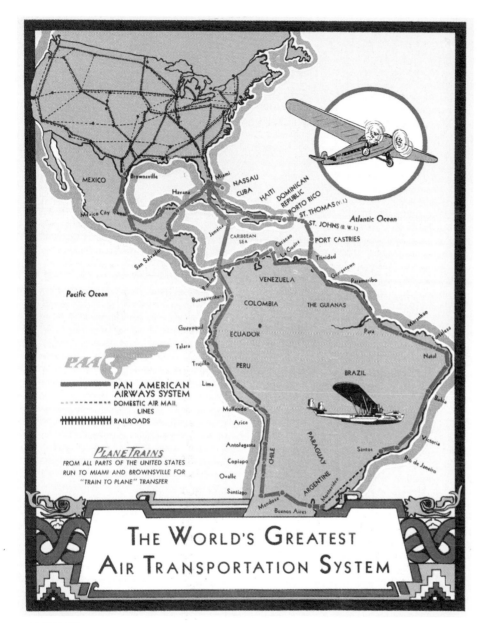

DOMINANT: By 1930 Trippe was running the world's largest airline, with a monopoly of services through North, Central and South America from gateways in Texas and Florida. He had also absorbed the internal airlines of Cuba, Columbia, Mexico, Brazil and Peru.

words for anyone, no welcome, no memory for anyone's name, no thanks after a long, hard day. But he did not curse people out, either. His strongest profanities were "Darn" and "Oh, gosh." His marriage to Betty was long, affectionate and faithful. His romantic feelings were otherwise reserved for exotic journeys.

Secreted in his desk in these frantic early days, the high-velocity Trippe had a pair of seven-league boots, a big, bold plan carefully worked out by range of aircraft and potential mail contracts to win foreign landing rights ahead of anyone else. He virtu-

ally camped out on Capitol Hill for three months before Congress voted on March 8, 1928, to let the post office pay up to $2 a pound for airmail carried overseas by private operators. Basil Rowe, pilot owner of West Indian Aerial Express, was sure he had the best claim to Foreign Air Mail Contract no. 5 for the route to Puerto Rico and Trinidad via Cuba, Haiti and the Dominican Republic. (Rowe had lent Trippe his seaplane for that first Key West–Havana airmail run.) When he lost to Pan Am, he became the first of a long line of airline operators to note, "While we had been out

developing our airline in the West Indies, our competitors had been busy on the much more important job of developing a lobby in Washington." He spoke more truly than he knew. Biographers Bender and Altschul revealed that Irving Glover let Trippe examine Rowe's bid so he could amend his own. Pan Am soon took over Rowe's airline and enlisted him as a pilot.

Whenever he moved offices in Manhattan, Trippe gave pride of place to a gigantic old globe of the world and bent over it with a piece of string, measuring distances. He planned his moves in aviation very much as he had learned the modern game of American football, invented by his coach at Yale, the great Walter Camp (1859–1925). Before Trippe put the ball in play at a meeting with Washington officials, with Hoyt and the holding company board or with his growing staff, he knew exactly where he wanted it to go and how to run interference. He would then shed his laconic style and wear people down with talk.

"We're going to fly over there to the South American mainland," Trippe casually announced at a staff meeting in the Sevilla Biltmore Hotel in Havana one afternoon in July 1928. The plan to carry mail and passengers around the Caribbean and to the mainland countries had been in his mind for at least a year, but only now did he tell the staff about services they were supposed to start in six short months. They were already stretched, preparing for the daily nonstop flights from Miami to Havana in just two months' time. Angrily, they asked, what was the hurry? Trippe replied quietly but relentlessly: German and French operators are infiltrating our hemisphere. Priester, the man who most needed to know in advance what was planned, was reduced to swearing in Dutch. He pointed out that their Fokkers had only a range of 100 miles. Ah yes, said Trippe, that was why he had ordered a Sikorsky S-38 amphibian for delivery in the fall. The company lawyer, Robert Thach (of the Colonial Airways misadventure), pointed out that they could never start in the fall because it would take months if not years to get permissions from all the foreign governments. True, said Trippe, which was why his agents were at that moment in the field backed up with foreign airmail licenses he had already privately secured from the U.S. Post Office. All very well, said James M. Eaton, the traffic manager, but there was no market for passenger flights. Well, said Trippe, they would make one and his colleagues would be pleased to know he had arranged to solicit custom among passengers traveling south on the trains to Florida. Americans who found Prohibition hard to bear were to be urged: "Fly to Havana and you can bathe in Bacardi rum two hours from now." (One of Pan Am's early customers was mobster Al Capone, not because he needed a drink, but to have an alibi when his hit men carried out the St. Valentine's Day massacre in 1929.)

Trippe exhibited exquisite patience and let everyone talk for eight hours without a break, until they began to feel ashamed of not sharing the sublime fortitude of their patriotic boss. Around 10 p.m., capitulation was signaled by Priester saying he was willing to go ahead "on the understanding that the job was impossible anyway."

At that point, it was. The whole project would have ended in disaster had Priester not spotted and Trippe not recruited Hugo C. Leuteritz, a 31-year-old engineer working at the new Radio Corporation of America (RCA). Pilots flew by the seat of their pants. They had no navigation aids, no communication with the ground; they flew visually at first, with maps clipped from old Spanish geography books. When Caribbean squalls brewed, they bounced around in black clouds with no idea where they were or what to do. At RCA, Leuteritz had asked for $25,000 to design a radio transmitter and receiver that was light enough for aviation but had been turned down by David Sarnoff, the creator of RCA. On his own, Leuteritz designed a 10-watt transmitter for single-engine planes and a 100-watt transmitter for a trimotor. At Priester's urging, Trippe had invited him to test it both by telegraph and by voice on the Key West–Havana run. On one flight with Leuteritz and a passenger, the pilot lost his way in a storm. He circled a ship and dropped asking for directions, but had to ditch in the sea. The passenger was drowned, Leuteritz suffered a broken pelvis and shoulder—and Pan Am lost half its fleet.

PALS: Trippe and Lindbergh in front of a Fokker F-10 at the dedication of Pan Am's 36th Street Airport in Miami, January 9, 1929. On the trips, the two men were like a pair of schoolboys, engaging in pillow fights on stopovers. Lindbergh ensured that Trippe made the first entry at receptions by jabbing him in the behind with a pin stuck in a pencil eraser.

Incredibly, Trippe still managed to hypnotize Leuteritz into joining Pan Am. By the time of Trippe's Sevilla Biltmore meeting, the brilliant Leuteritz had designed an airborne radio telegraph and defined the frequencies for transmissions up to 150 miles (requiring a Morse code operator to be added to the flight crew). Then he moved on to devise a fixed-loop direction finder, based on Morse signals from the pilot and triangulated with the shore and ships at sea. The pilots hated it, arguing night after night on the porch of the La Concha Hotel until Leuteritz was finally able to show them how riskily far off course they often strayed.

Priester left the stormy Sevilla Biltmore meeting, blessing the day he had heard of Leuteritz. A labor of Hercules began. The 1,400 miles of new airway he had been asked to set up needed radio links as well as weather stations and refueling strips. Pan Am was the first American airline to use radio and direct-finding communications; eventually, by agreement with 27 countries, 93 radio and weather stations were spread across the continent and Pan Am became the largest private radio network in the world. Opening up the Caribbean and Central and South America was truly a prodigious enterprise, of which technology was only one essential. To find the refueling strips, survey pilots flew over mountain ranges more precipitous than the Rockies and over jungles and swamps, dropping sacks of flour to mark likely spots in the matted green below. Construction crews came in by burro and canoe, mixing bundles of five-and-dime-store trinkets and candles with their shovels and machetes: The Indians in the jungle had to be persuaded first to put aside their poisoned arrows and then to carry five-gallon tins of fuel from the nearest seaport. All of this had to be financed as well as organized by Pan Am without aid from taxpayers at home or local governments. (For the domestic airlines, the U.S. government footed an infrastructure bill of more than 50 million dollars.)

Trippe always had something up his sleeve. The moustached man barely discernible beneath dark glasses and pulled-down hat who had visited Trippe's apartment at 100 East 42nd street for long hours in 1928 was not really "Mr. James Stewart" as announced, but Charles Lindbergh in disguise. Early in 1929, Trippe revealed that Lindbergh was joining Pan Am as technical adviser (for $10,000 a year and shares). To hire the hero of the hour—the hero of the decade—was a coup, carefully plotted by Trippe with the help of Hoyt, Hambleton and Whitney. They had given Lindbergh space to think after the mad frenzy following his nonstop solo flight to Paris in May 1927. Lindbergh's presence was a powerful aid to Trippe in convincing officials in Washington that Pan Am should be what came to be called the "Chosen Instrument" of the United States, a monopoly strong enough to prevent French, German and British airlines from dominating areas of importance to American interests. Trippe's case was that what was good for Pan Am was good for the country (as "Engine" Charlie Wilson would argue in the Eisenhower fifties on behalf of General Motors). No political party in America favored having state-supported airlines, like Britain's Imperial Airways and Holland's KLM, so both Republican and Democratic policymakers bought the argu-

ment that a well-organized private monopoly was preferable to free-for-all competition that might result in weaker and less safe services in a capital-intensive business.

What made the difference to Lindbergh in deciding in 1929 to ally himself with Pan Am was Trippe's passion for flying the flag all around the world, starting with the Western Hemisphere but extending to Europe and Asia. They lost no time about it. On January 9, 1929, Lindbergh and Trippe, with their brides—Anne Morrow and Betty Trippe, both daughters of J. P. Morgan partners—took off from their new airport in Miami for a 7,000-mile tour of the Caribbean aboard a Sikorsky S-38 amphibian. They were carrying the U.S. mails; their reception throughout could not have been more ecstatic had they been scattering gold dust. Flag-waving multitudes lined the streets of 16 major cities all the way to Paramaribo, Dutch Guiana (now Suriname). At Barranquilla in Colombia, the mobs flanking the runway made it impossible to land. Lindbergh ran out of gas circling, but he had spotted an uncluttered stretch of water and brought them down on what Anne called "the sweetest little lake"; she was amused at how discomfited the prudish Juan Trippe was when they were rescued by Indian men in G-strings.

After the tour it was down to business. On March 10, 1929, Juan and Betty went aboard one of the first all-metal planes, an eight-seater trimotor built by Henry Ford, known as "the tin goose." Lindbergh was the pilot for the inaugural mail and passenger daily round-trip service between Brownsville, Texas, and Mexico City, flying the 470 miles in 5 hours and 18 minutes. ("The clouds  were so thick," Betty wrote in her diary, "it was like flying in a milk bottle. The air was so rough, the airplane sometimes dropped hundreds of feet in air pockets.")

The Brownsville–Mexico City run was made under U.S. post office Foreign Air Mail Contract no. 8. Irving Glover, still at the post office, aroused wrath when he awarded the route to Trippe, whose bid at the maximum rate of two dollars a mile was twice as costly as losing bids. The post

office had a good defense: Because of Mexican law, Pan Am was the only company that could guarantee all-round performance of the contract. The postrevolutionary Mexican government of the day permitted no foreign airline to carry Mexican mail on its territory; and the rights to that had been awarded to a Mexican company called Compañía Mexicana de Aviación or CMA. The year before, as it happened, Trippe had flown in a shaky little single-engine Fairchild over the forbidding Sierra Madre to Mexico City and while there had bought 100 percent control of the little CMA line.

Trippe's de facto acquisition of U.S. mail licenses was critical to his imperial ambitions. In January 1929 the licenses he had already secured (in advance of scheduled operations) were the fulcrum that enabled him to extend his reach down the west coast to Chile and Peru through an unavoidable marriage of convenience with the W. R. Grace Corporation of New York. Pan-American Grace Airlines (PANAGRA) was an uneasy fifty-fifty arrangement, facing competition from the French Aerospostale Company and the German-controlled SCADTA line based in Colombia. Trippe never told Grace when in 1931 he secretly acquired financial control of SCADTA from Peter Paul von Bauer. (In a May 1940 putsch, with the help of Colombian troops, he removed von Bauer and all the German immigrants on the phony pretext that they might be a threat to the Panama Canal, then took majority stock in the nationalized Avianca.) Meanwhile, to Trippe's fury, the countries along South America's far more prosperous and potentially lucrative east coast, notably Argentina and Brazil, had given domestic mail concessions and rights to land in their great city harbors to a swashbuckling airline entrepreneur in Trippe's class: the World War I flying ace and engineer Ralph O'Neill (1896–1980). Associated with Trippe's old rivals Montgomery and Bevier, O'Neill founded the New York, Rio and Buenos Aires Airline (NYRBA) in 1929, backed by $8.5 million from a consortium led by James Rand, head of Remington Rand, the aircraft manufacturer Reuben Fleet and Ford's W. B. Mayo. The dashing Spanish-speaking O'Neill deployed diplomacy and adventurous flying to start a variety of scheduled services within the region from August 21, 1929, climaxing in the great day of February 30, 1930, when his fleet of Commodore and Sikorsky flying boats began a relay of regular flights between Buenos Aires and Miami. O'Neill's NYRBA, with Ford tri-motors and then Commodore flying boats built by Reuben, was much better equipped than Trippe's Pan Am—but O'Neill had an Achilles' heel. He assumed that his fleet, unsurpassed in the world, would guarantee him the license to carry U.S. mails. He reckoned without Trippe. While O'Neill was wooing the Latin American presidents, building his bases and training crews, Trippe was consolidating his Washington connections. The U.S. postmaster general, Walter Folger Brown, duly refused O'Neill a license. Aviation historian R. E. G. Davies writes that O'Neill's mistake was to underestimate the extent to which "Trippe would resort to Machiavellian cunning to achieve his objectives by fair means or foul." Without the license, a heartbroken O'Neill was forced to sell out to Trippe in August 1930. For a bargain price, Trippe acquired a fleet superior to his own to "fly down to Rio": 11 Commodores seating 22 to 26 passengers each.

The other side of Trippe's politicking was that he knew how to run an airline. Between 1932 and 1934, the percentage of scheduled flights completed by Pan Am was 99.46 percent—flying long distances over dangerous mountain ranges, often enough in bad weather. Priester's pilot-training programs were rigorous. Betty Trippe remembers how bitterly the veteran pilots resented hearing about blind flying from a kid just out of the army, 23-year-old Harold Gray. "They were finally convinced," she wrote, "when they were over Key West, having flown with curtains covering the windows of the cockpit, and all they could see was the instrument panel."

Trippe's appointment of Lindbergh yielded so much more than the buzz of his celebrity. Lindbergh was the bravest of the brave, test-flying routes that no one had ever flown before. In July 1931, with his new bride Anne as conavigator and radio operator, he took off from Long Island in a Lockheed single-engine Sirius monoplane to explore the Great Circle route via Alaska to the Orient, made a forced landing in fog

in Japan's Kurile Islands and then crossed the East China Sea to Shanghai and Hankow. The mutual respect of Lindbergh and Trippe was great not just for Pan Am but the whole future of air travel. Perhaps the aviation historian Clive Irving had it right that it took one enigma to understand another.

By the early '30s, Trippe's deft footwork in Washington and his appetite for buying out or undermining anybody who looked at all competitive had made Pan Am the world's largest airline and America's sole international carrier. Every milestone of Trippe's advance in Central and South America was marked by festivals with speeches and brass bands. Cash for expansion was flowing. In due course, he came to own or control most of the internal airlines of the region (and of China, too).

Pan Am consistently received preferential treatment in the Departments of State, Commerce and War throughout the years of Republican ascendancy in the Roaring Twenties and into the Democratic New Deal of the '30s. Trippe's hand may have been the invisible ink between every line of legislation, but after the absorption of NYRBA, Pan Am really was the only airline, which enabled diplomacy to be integrated with commerce. "Pan Am," writes Roger Lewis, "gradually came to be recognized as a symbol, not of aggrandizement but of progress. The air transport expertise which Pan American pilots, engineers and administrators brought to Latin America contributed in no small measure to the social well-being of millions of people."

Trippe was in perpetual quest of power on the two levels of politics and engineering. A remarkable fact of his whole career was that he was always ahead of the makers of the machines, always pressing them for advancements in performance. He understood the purpose and use of technologies even before the men who developed them did: It is perhaps his distinctive trait as an innovator. His Pan Am was the first airline to order and purchase aircraft built to its own specifications. It was quite a scene in Trippe's office when he chewed a cheap cigar while Lindbergh leaned his long frame over drawings and specifications produced by another legend-in-the-making, Igor Sikorsky (1889–1972). The mystical pioneer of multiengine aircraft had fled Russia for America in 1919, inspired, he said, by the works of Edison and Ford. Trippe had once again identified a crucial talent.

The S-40 flying boat that Sikorsky delivered in the fall of 1931, with Pratt and Whitney engines, was America's first four-engine transport, the largest plane so far built in the United States. It seated 40 passengers, and could carry them 700 miles. Lindbergh was not wild about the plane, but he agreed with Trippe that he would captain it on his first commercial flight, at the age of 29, and he would take Sikorsky along with him. They left Miami on November 19, 1931, on the maiden scheduled service 1,500 miles to Panama. High over the Caribbean, Lindbergh could not contain himself. He handed the controls to copilot Basil Rowe and went back into the cabin where Sikorsky was making another record of sorts, consuming the first hot meal prepared over water in an American aircraft galley. Lindbergh took the menu, not to order from the uniformed steward but to sketch his ideas for the plane Sikorsky should build next, an even greater and more luxurious flying boat, capable of crossing the Atlantic in two stops. Lindbergh kept sketching on the menus through a stopover dinner in Kingston, Jamaica.

Trippe meanwhile had been on an apparently less glamorous mission, climbing the steps of the New York Public Library. He spent days there until he found the buried treasure, the original logs of the famous tall-masted China tea trader sailing ships that clipped so much time off voyages

THE MEXICAN CONNECTION: A Mexican company called Compañía Mexicana de Aviación (CMA) had exclusive license to carry Mexican mail within Mexico. The elegant solution of Trippe—here in Mexico in the mid-1930s—was to buy control of CMA.

across the Atlantic and Pacific Oceans in the 19th century. There was still no plane that could cross the ocean as the original clippers had. He studied their routes and their stopping points—and he knew precisely what he had to demand next from the makers of new aircraft he would in the future call Clippers. He pitted Sikorsky against Glenn L. Martin, a Baltimore manufacturer of large military aircraft, inviting them both to build three airliners for new North and South Atlantic routes. Harder than the aircraft procurement was the political negotiation to put the projected new planes into service on the Atlantic. Even before the first airframe blueprint, he was in negotiation with the authorities in Britain, Newfoundland, Ireland and France for permission to land. None of the European flag carriers had any capacity to fly the Atlantic in two stops, as Trippe expected he would have, but by the time Sikorsky delivered his giant flying boat S-42 in 1934, with a range of nearly 3,000 miles, all the effort proved futile. The British, in particular, would not afford landing rights at intermediate bases in Newfoundland and Bermuda until their Imperial Airways, too, could cross the ocean, so there was a stalemate in Anglo-American aviation understandings. Trippe and Lindbergh kept up the pressure to demonstrate that they were ready. Instead of essaying the Atlantic, Lindbergh left Miami for Buenos Aires on August 16, 1934, flying Sikorsky's luxury *Brazilian Clipper.* He cut the flying time from eight to five days.

The headlines were ecstatic, but Trippe was not done. When he gathered Priester and Leuteritz around his map table one evening that summer, with two other executives, he found satisfaction again in the cries of alarm he evoked. His Clippers, he told them, were prevented from flying the Atlantic pond, so they were going to fly mail and passengers right across the Pacific—the Pacific!—from California to China. How could he even think it? That was 8,700 miles of ocean visited by tumultuous storms, its wind patterns never charted. Even the big new Martin with the latest navigation systems would have a range of only 3,200 miles. Yes, they could reach Honolulu (2,400 miles), or Midway (2,800 miles) but what then? Trippe's days in the New York Public

Library had given him the answer. One of the places where the clipper ships had stopped was a dot barely visible on the biggest maps, an uninhabited bit of coral called Wake Island. (Few even knew it existed then, seven years before it became the scene of an American stand against conquering Japanese.) Trippe had already been to Washington to seek long-term leases for refueling bases on Wake and also on the stepping-stone islands of Midway and Guam.

At that time, Trippe was not wildly popular with the Democratic New Deal administration. "Juan Trippe," said President Roosevelt, "is the most fascinating Yale gangster I ever met." James Farley, the new Democratic postmaster general, was plagued by congressional complaints that Pan Am had received favored treatment from Republican predecessors and had then ungraciously overcharged for carrying the mails. There was little doubt that bills had been padded, so for nine months Farley kept Trippe in a state of exhausted suspense about the renewal of Pan Am's licenses while Trippe struggled to pacify his board about his considerable spending in advance of a Pacific airmail contract. His luck held. Franklin Roosevelt, that avid philatelist, was less worried about the price of an airmail stamp than the vulnerability of U.S. naval installations from Honolulu to Manila in the event of war with Japan.

Trippe got the airmail contract (with a small reduction in the top $2 per pound fee) and the vital leases to the Pacific outposts from the U.S. Navy. There were secret understandings that Pan Am's radio and weather stations, and a little tactful spying, would be in the national interest.

Lewis calls what followed "the most efficient single program of preparatory work ever accomplished in starting a new air route." Careful survey flights were made of the airways to Hawaii, Midway, Wake and Guam by an S-42 stripped down to carry more fuel; 14 solo fliers had previously lost their lives attempting that crossing (and the celebrated female flier Amelia Earhart was to vanish in the middle of the Pacific in July 1937). Then bases had to be built. On March 27, 1935, Pan Am's cargo ship *North Haven* steamed into the ocean through the Golden Gate Bridge with materials for 2 complete villages, 5 air bases, 40 long antenna masts

for Leuteritz's direction finders, 250,000 gallons of fuel, 6 months' supply of food and vegetable seeds, motor launches, landing barges, electric generators, windmills and water storage equipment, plus 44 airline technicians and 74 construction workers. In 55 days after landing, they built air bases on Wake, Midway and Guam and improved the ones at Honolulu and Manila.

Six days before Thanksgiving, on November 22, 1935, an apparently calm Trippe stood amid a crescendo of excitement around San Francisco Bay. Millions were tuned in by radio for the debut of the world's first commercial service across the Pacific to the Philippines (extended later to Hong Kong), listening as they would listen and watch the moon landing 34 years later. Trippe had organized a spectacular send-off from Alameda, across the bay from San Francisco; 25,000 crowded the dockside and 150,000 people awaited the big moment at every vantage point in the city. Trippe's gleaming *China Clipper,* the graceful Martin 130 flying boat, bobbed at anchor in the afternoon sunshine. Seven uniformed officers, in crisp navy-and-white uniforms, had gone ceremoniously aboard. So had a full ton of mail. In the cockpit, Captain Ed Musick awaited the order from Trippe as goodwill messages were broadcast to the crowds from the Clipper's Pacific destinations and Pan Am island bases. "Captain Musick," Trippe finally announced, "you have your sailing orders. Cast off and depart for Manila in accordance therewith." The *Star Spangled Banner* sounded, the crowds of people roared, sirens screeched, hundreds of car drivers honked, ships whistled, flags waved, and at 3:46 p.m. *China Clipper* bounced down San Francisco Bay, rose slowly—and headed straight for the catwalk cables strung between the towers of the unfinished San Francisco–Oakland Bay Bridge. At the last second, the flying boat dived under the bridge. Everybody cheered the maneuver as part of a great afternoon's show. In fact, the phlegmatic Musick realized that he was rising too slowly to clear the cables, so in a flash he put the Clipper's nose down and threaded his way through the construction cables. "We all ducked and held our breaths until we were in the clear," said Second Engineering Officer Victor Wright.

China Clipper arrived in Manila on November 29, in 59 hours, 48 minutes' flying time with its 111,000 letters; fighting against ferocious headwinds, it was back on December 6 with 98,000 letters. Scheduled passenger services started on October 21, 1936; they were only for the rich, at $850 one way from San Francisco to Manila, about $10,000 in today's money. When the Clipper stopped to refuel on the formerly barren atolls, the adventurers were met by attendants in crisp white uniforms and led through lawned and landscaped gardens on Midway, and paths of crushed coral on Wake, to Pan Am's hotel rooms, tastefully decorated by Delano and Aldrich. Trippe had style.

The Clippers flew on to Hong Kong from Manila in November 1936 and to New Zealand in 1937, landing rights having been refused by Australia. The planes were the symbol of glamour in the grim days of the Great Depression. Newsreels and magazine photographs offered millions the vicarious excitement of the Orient, dinner en route in the elegant Clipper lounge with notables such as Jack Dempsey and heiress Doris Duke. Backstage, Trippe and his associates were doing everything they could to reduce the risks, which were real. The image was Humphrey Bogart in the Warner Brothers movie, *China Clipper,* battling a typhoon to arrive spotlessly in time. The actuality for Ed Musick and his crew was that cruising the landless 1,200-mile lane between Midway and Wake could be a nightmare. Charts were sometimes as much as 100 miles off. Making calculations in unpressurized cabins, navigators never dared to take a rest, much less sleep, for 18 or more hours. In typhoon areas the Clipper might drop 1,000 feet or go up 1,000 feet. On January 11, 1938, Musick and his crew died in the South Pacific when the S-42B *Samoan Clipper* blew up in flames while surveying a route through Pago Pago to New Zealand. Six months later, the Martin *Hawaiian Clipper* vanished between Guam and Manila.

Two such dreadful accidents might have ended another airline. Left with only two transocean flying boats, Trippe had to cut scheduled services in half, and public confidence was collapsing anyway. Once again, he rescued himself by imaginative forward

planning. Always sure that the aircraft manufacturers could be pressed to perform miracles, he had opened talks with Boeing in 1935, even while Martin built the 310. Bigger, faster, better was always Trippe's refrain, and from Boeing he had demanded an aircraft with twice the power of the Martins, able to carry 77 passengers at least 3,500 miles at close to 200 miles an hour. It would be the largest civil aircraft in service.

He had signed a contract for six Boeing 314s on July 21, 1936, and the first test flight was two years later on June 7, 1938. When its teething troubles were over the whale-

shaped B-314 was probably the finest flying boat ever to go into service, certainly the first real transoceanic airliner with a payload that made economic sense. But Trippe encountered turbulence. Anxious about the delays, his board was exasperated that he did not simply hold the cards close to his chest; he never took them out of his pocket. A State Department official summed it up: "Up to the spring of 1939, Trippe had been the three dimensions of his company. He wrote policy with his right hand, executed it with the left, and saw to it that neither hand knew what the other was doing." He also drove the lawyers crazy with his second, third and even tenth thoughts. "He was always nibbling for a little more," said his attorney. "He managed to close his contracts just before he lost them."

His old friend Sonny Whitney, owning $4 million of stock, led a board revolt that demoted Trippe and promoted Whitney to chief executive officer. He took Trippe's office on the 58th floor of the Chrysler Building, pushing him to the other end of the corridor, and presided at meetings where Trippe glowered. He bit his tongue, but it was unbearable for him that Whitney was in charge, preening himself on Trippe's feathers, when the Boeing 314, the *Yankee Clipper,* took to the air from Port

Washington, New York, on May 20, 1939, a historic day in aviation—and political—history. Captain A. E. LaPorte carried almost a ton of mail to Marseilles, via the Azores and Lisbon; Captain Harold Gray opened the northern service to Southampton on June 24, 1939. Four days after that, Captain R. O. D. Sullivan inaugurated the first regular passenger service from New York to Southampton via Newfoundland, all according to Trippe's plans. These flights were commercial milestones, but they also marked the beginning of the end of the relative isolation of the United States from Europe, and they were an augury of the massive air ferrying operations that would help the Allies to prevail in World War II.

Whitney did not have long to savor the triumph. He could not fathom what was happening or supposed to happen in what was now a large and far-flung company whose fortunes turned on complex technological decisions. As the war clouds gathered in Europe, the board realized that only Trippe, for all his maddening ways, could navigate Pan Am—could decide for instance whether to order more B-314s or press Boeing for still longer ranges, more passenger space, more engine power. Within months, Trippe was restored to power (and pressing Boeing again) on condition that he make weekly reports. He was still only 40 and had great things to do.

The military airlifts across both the great oceans, with Pan Am in the lead, were the precursors of the first world airline network. Pan Am flew the troops to war and back. Churchill enlisted his help for a trans-African supply route to Egypt for General Montgomery. President Roosevelt celebrated his 61st birthday in 1943 aboard the *Dixie Clipper* on the way back from the Casablanca conference with Churchill and Stalin. Trippe, like Coca-Cola, was eager to brand the world with Pan Am, but his international monopoly could not endure in the postwar years (when land-based planes succeeded the Clippers). Permission to operate across the Atlantic was given to a man who would have been a serious threat to Trippe personally and to his airline if he had managed to preserve his sanity: Howard Hughes, the flamboyant billionaire moviemaker, flying boat inventor,

Trippe was lavish in his interior decorations. The Consolidated Commodore, carrying 22 passengers, had real alligator skin on the walls.

round-the-world pilot, and owner from May 1939 of what became Trans World Airlines. In December 1945, after President Truman had approved the entry of several U.S. domestic airlines into overseas markets, TWA flew its first overseas service from Washington to Paris; regular services to London and Frankfurt followed in 1950. Trippe never stopped trying to reinstate his international monopoly while trying and failing to acquire domestic routes. The chairman of the new oversight Civil Aeronautics Board at the time was Jim Landis, a longtime New Dealer and former dean of Harvard Law School, and in 1949 he reflected ruefully on what it was like saying no to Juan Trippe. "If Trippe had been a woman, about 50 B.C., her name would have been Cleopatra. A great seductress, a person of elegance, considerable charm, keen intuition. . . . He never relaxed the pressure. When I tried to shut the door on his men, they came through the windows." Stanley Gewirtz, a young lawyer who was Landis's executive assistant and subsequently an executive at the American Airport Association, found Trippe's monologues so taxing he often left the open phone vibrating on his desk. "He was completely insistent and completely consistent." Ironically, Gewirtz became a Pan Am executive.

In testimony to Congress, however, Trippe performed surprisingly poorly; he stumbled so badly, it was suspected that the man who was so persuasive in private

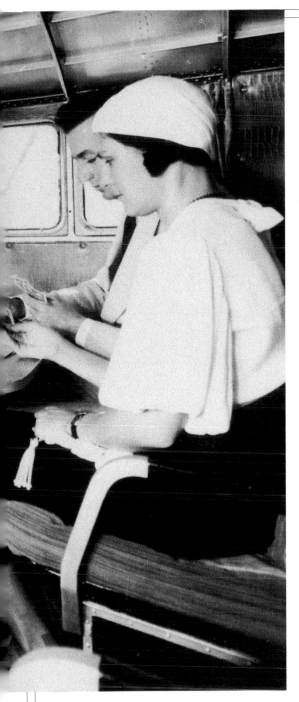

not have Trippe's vision soon enough about the future of commercial aviation. He bought RKO Studios (having made a wartime star of Jane Russell in *The Outlaw*), whereas Trippe's focus was unswerving. There was nobody in the airline business with his drive and imagination. His critic Landis remarked, "What vision. Some people look ahead six months or a year. Juan Trippe is thinking about the next decade, maybe two or three." Trippe was so sure that there would be an explosion in air travel, he put money and managerial energies into creating decent hotels wherever Pan Am might touch down, starting with the Intercontinental chain in 1946. On June 17, 1947, Trippe launched the first round-the-world scheduled service, a 250-miles-an-hour Lockheed Constellation seducing globalists with pressurized cabins at 20,000 feet and comfortable sleeping quarters.

But that was not where his heart now was. That kind of air travel was for the elite and Trippe was consumed by the notion that air travel was for the masses. In this, he was alone among the airlines belonging to the International Air Transport Association (IATA). Trippe provoked outrage at the cartel's 1945 convention in Montreal by announcing a round-trip "tourist class" fare between New York and London of $275 when the European airlines, which did not have anything like Pan Am's carrying capacity, had just agreed the minimum fare should be $572 *one way*. Britain actually stopped Pan Am flying into London with tourist-class passengers. Trippe had to divert to Shannon, Ireland.

Trippe proved his point with $75 flights from New York to San Juan, Puerto Rico—every flight was packed—but it took him four years to batter down the international cartel, by which time other countries were on a technically even footing. In 1952, IATA finally accepted that tourist fares were legitimate and feasible and there was surprised gratification that lower fares produced a rise in bookings of as much as 30 percent. But that was only the start of Trippe's postwar revolution.

In 1945 Lindbergh had gone to Europe to look into the new technology of the jet engine, deployed in fighter planes by Britain and Germany in the final days of the war. In Germany, he had come across the renowned inventor Willy Messerschmitt, sitting in a cow barn adjacent to his house, now occupied by the American army. Messerschmitt told Lindbergh that within four years he could have built a big jet airliner. Lindbergh reported his findings to the engine-makers Pratt and Whitney and to Trippe. Nobody else in the American airline business wanted jet engines; they were considered to be noisy, heavy on maintenance (a fallacy), too big for most airports and too expensive. The tough Texan who ran American Airlines, Cyrus Smith, was speaking for domestic flights, but he expressed the attitude of all the airlines: "We can't go backward to the jet." Indeed, all the airlines were thrilled with the new generation of turboprop aircraft. They felt vindicated when the world's first jet passenger plane, Britain's 470-miles-an-hour Comet, was withdrawn in 1953 after three catastrophic midair explosions from metal fatigue. But the Comet had proved, contrary to widely held opinion, even in the highest levels of aeronautics, that jet engines could be adapted for commercial use.

Trippe was undeterred. He remained determined that Pan Am would skip the turboprop generation. Bigger planes meant more passengers, which in turn meant airlines could make profits with lower fares. But to make economic sense, the jetliner he set his mind on would have to carry something like 180 passengers, seated six abreast on a nonstop crossing of the Atlantic, and none of the big three airframe manufacturers—Boeing, Douglas and Lockheed—was

was deliberately playing the country boy. This ploy did not always work and the charismatic Hughes was often at his throat. He gave off-the-record interviews to accuse Trippe of bribing senators (in an interview in 1949 he said, "Once Trippe gets an Atlantic monopoly, he'll sit back on his fat ass and stick both the public and the government as he did during his South American monopoly"). Trippe was outwardly serene but he pressed his monopoly argument too far and too long.

Hughes was no mean innovator himself when it came to airplanes, but he just did ready to attempt such a plane. There were risks on both sides. If they invested millions trying to meet his demands, other airlines might not buy, such was the antipathy to the jet. If Trippe did not get the plane he wanted in time, he would lose passengers to the faster and quieter turboprops.

A battle of wills ensued. Lockheed dropped out. Boeing built and Douglas designed a jetliner to seat 100 passengers, respectively, the 707 (based on the Dash-80 prototype built with Boeing's own money) and the DC-8. The trouble was that the J-57 Pratt and Whitney engine both proposed to install was fine for coast-to-coast journeys at nearly 600 miles an hour—but not all the way across the Atlantic, and it was too weak to climb quickly to 30,000 feet. So Trippe, the czar of international travel, thanked them very politely and declined. Both makers told him the designs were frozen. He played the rivals against each other, Boeing's adventurous lawyer-chairman Bill Allen against the more cautious MIT-trained Donald Douglas. When it seemed that Trippe might make a deal with Douglas, Boeing undertook a costly redesign to yield a payload of 147 on its putative 707 (with the swept-wing technology of the KC-135 tanker developed with Pentagon money). Most of Trippe's staff thought he would call that a victory. Allen waited for the applause. He was appreciated for his effort, but again came that quiet, smiling insistence, "We are not going to take any plane," said Trippe, "that can't do the Atlantic nonstop."

Through the summer of 1955, as he negotiated to a stalemate, Trippe could hear a continuous roar, the remembered sound of a monster engine installed in an isolated concrete building in East Hartford, Connecticut. Trippe had been told in December 1953 that Pratt and Whitney was working on a secret air force contract for an engine called the J-75, rumored to have nearly twice as much power as the J-57, and Lindbergh had managed to see it. (Lindbergh's reputation had been dimmed by his thirties flirtation with the Nazis but he was still a god in aviation.) Trippe had staffed Pan Am with top flight engineers, led now by John Borger, an engaging big bear of a man who understood engines: The most technically complex reports did not have

to be explained to him. Trippe badgered Pratt and Whitney mercilessly for a J-75 for a possible jetliner. Visiting East Hartford, he got a swift no from company leaders Fred Rentschler and William Gwinn. They told him: "We haven't flown it yet. We just don't know how good it's going to be." He asked them to a power lunch at the Cloud Club atop the Chrysler skyscraper in Manhattan. Still no, and no again and again. Come back in two years, they told him. Trippe tried another tack. Word filtered back to Pratt and Whitney that he was meeting Britain's prestigious Rolls-Royce company with a view to acquiring its new Conway engine. Rentschler summoned his technical staff. "Are you sure we can't do it?" They began to think they just might. Over another lunch, Trippe put $40 million on the table. Pan American would pay $250,000 each for 25 of the J-75s, he told Rentschler. "How soon can I have them?" The answer was 1959.

Trippe had committed himself to buy engines for which he had no airframe. He still had none when he emerged from a long meeting with Bill Allen in Seattle. It was the 707 as it was, said Allen, take it or leave

it. Trippe left it and moved down the coast to see Donald Douglas. It was easier for Douglas; his DC-8 had been designed but not built. As Robert Daley writes, "All Douglas had to do was crash it in the wastepaper basket." He accepted Trippe's order for 25 DC-8s with the J-75 engine.

This is where one comes to appreciate Trippe's capacity for three-dimensional chess. His deal with Douglas was conditional on secrecy. Having concluded it, he went back to Boeing and seemed to be gracefully accepting defeat with an order for 21 of the small short-range 707s. On October 13, 1955, Trippe threw a party at his Gracie Square apartment overlooking the East River for the heads of foreign airlines in town for IATA meetings. Very casually,

as he moved from group to group, he mentioned how happy he was to have just invested $269 million in 45 big jetliners—45 big jets! Everyone realized at once that they were sunk, their shorter-range turboprops rendered obsolete almost overnight; they would have to buy American jets. It was a bad evening for them but a worse morning for Boeing's Bill Allen reading the news the next day in the *Wall Street Journal*. His European salesmen were telling him soon enough that they could not sell the short-range 707. The big long-range DC-8 was going to dominate the market—and Boeing would be squeezed out of it. There was a swift reshuffle in the Pan Am–Boeing contract. The order for 21 short-range 707s was converted to 15 long-range big jets yet to be designed, and with 6 of the short-range to be delivered in the fall of 1958. Pan Am introduced a one-stop Atlantic jet service on October 26, 1958, just beaten by Britain's BOAC Comet 4 on October 4, 1958.

The following year Trippe's big jet came into its own: New York nonstop to Paris in 6 hours, 35 minutes, half the time of the fastest piston plane, San Francisco to Tokyo in 12 hours, 45 minutes instead of 25 hours. Trippe's jets carried twice as many people as the propeller-driven Stratocruiser and at 32,000 feet they enjoyed pampered tranquillity not associated with moving at 600 miles an hour. Richard Branson, the innovator of Britain's cut-rate Virgin Airlines, notes that Trippe exploited the glamour of his first celebrity jet-set passengers, but he was also calculating the new jet-age math—"what we call in our business 'bums on seats'"—the seat-mile cost. Trippe changed the five abreast seating to six abreast, cut fares and attracted hundreds of thousands more passengers. The forecast for 1965 was that 35 million would be flying international routes in increasingly crowded skies. Planes overtook the ocean-liner market much more quickly than anyone predicted.

At the age of 65, Trippe could have rested on his laurels; so could Allen at 66 after 20 years running Boeing. Instead, both men reached for a final prize: a wide-bodied jetliner twice as big as the 707, to carry 400 people—no, make it two and a half times as big, an utterly crazy notion given how many billions it had cost Boeing to develop the 707. Trippe primed Bill Allen by calling

The Boeing 747 *Clipper America,* ca. 1970. In 1981 it was rechristened *Clipper Juan T. Trippe*. At 150 feet, the economy section alone is longer than the Wright Brothers' first flight at Kitty Hawk.

him at Boeing to say, "This is a cup of tea that the Douglas fellows are interested in— they've dropped everything else over there." He had a bite that August when Trippe and Allen, with Betty Trippe and Mef Allen, went on a long-arranged fishing vacation in Alaska. They rented the elegant old classic of a yacht the *Wild Goose,* owned by John Wayne, and debated the relevant merits of Atlantic and Alaskan salmon. It was in this congenial atmosphere, Clive Irving records, that "an idea transmuted from a whim into an imperative almost without formal acknowledgement." Irving says it was put in terms of a dare: Trippe told Allen: If you build it, I'll buy it. And Allen responded: If you buy it, I'll build it.

With this exchange, the two men bet their companies on building the 747 (or the jumbo as this technological marvel came to be called by the British, who initially perceived it as an elephant of American grandiosity). It represented a breathtaking investment of $2 billion by Boeing and $550 million by Pan Am, the largest single under-

taking ever carried out by a commercial company. Trippe had already arranged the complex financing. The most significant intake of breath was in the White House, where President Lyndon Johnson and Defense Secretary Bob McNamara were alarmed that the deal would hazard the viability of Boeing, on which the military so depended. Trippe—never averse to dramatizing his diplomatic skills—let it out that he went to the White House with Allen and convinced Johnson. He convinced the Pan Am board to spend $18 million per 747 by taking Lindbergh into the meeting with him. "I don't think the plane would ever have been built," he wrote, "if Slim hadn't been there, adding the weight of his integrity, his insights and his prestige."

The odyssey of building the 747 is vividly explored in Clive Irving's book *Wide Body*. It was a cooperative triumph of the first order, in which Trippe's staff played a central role, working side by side with Boeing and Pratt and Whitney's engineers. Pratt and Whitney had to develop the revolu-

tionary fan-jet, the JT-9, delivering double the power of the J-75. Pan Am helped to bring space mission accuracy and reliability to commercial aviation with digital as well as hard sweep instrumentation on flight and engine instruments for safer flying. The result was not merely a flying machine but an advertisement: the most recognizable artifact of American technological achievement in the world.

As Boeing built the jumbo, Pan Am boomed. Trippe commissioned Walter Gropius to design the Pan American skyscraper in the middle of Park Avenue, financed by a British investment group. Touted as the largest corporate structure in the world in 1963 (and not universally popular with New Yorkers), its roof at 808 feet above Grand Central Station was the first pinnacle heliport for regular airline operation—a seven-minute ride to Kennedy airport. In 1966, Pan Am's profit was $83 million, 60 percent up on the year before. The man who had started offering beach rides in obsolete navy planes presided over a billion-dollar conglomerate that included "the world's most experienced airline," airports, his midtown skyscraper, hotels and much else. Pan Am, with 40,000 people employed, was flying a global network of 80,000 air miles linking the United States with 85 countries. On January 15, 1970, the first 747 delivered to Pan Am was christened *Clipper Young America* by Mrs. Richard Nixon, a week before a sensational inaugural scheduled flight to London that shrunk the world again. In less than one year, Pan Am introduced an entire fleet safely and efficiently, speeding 1.6 million passengers 4.5 billion passenger miles, and carrying almost 43,000 tons of mail, equal to 1,400 typical 707 freighter loads.

By then Trippe had already sprung another surprise on everyone—his retirement at the age of 68, announced in a monotone at the end of the annual meeting on May 7, 1968. The stunned president, the former pilot Harold Gray, moved to the microphone to praise Trippe's leadership. "As chairman of this meeting," said Trippe, "I rule your remarks out of order." Gray spoke the truth when he responded, "I seldom defy the boss." Trippe had run Pan Am as an imperial court, and he had groomed no successor.

The second phase of the jet age was the testing time for Pan Am—and Boeing. The inaugural flight was accompanied by the onset of a biting recession and plummeting stock prices, followed by the energy crisis that caused fuel costs to soar. Most of the airlines recovered. Pan Am never did. For Trippe it was horrible to watch his airline foundering. Frustrated by the inability to win domestic services, Pan Am was badly hurt when "open skies" deregulation provided the Carter Administration with the opportunity to certificate just about all hitherto domestic carriers with international routes. In the eighties, Trippe's successors made things worse with appalling managerial decisions, and then there was the terrorist outrage that crashed Pan Am flight 103 at Lockerbie in December 1988.

Trippe died in 1981, and the company he had taught to fly was sold off in pieces until nothing was left. His legacy endures in the 747. As the *Economist* noted, it was the first piece of transportation equipment that was not obsolete the day it went into ser-

TRIPPE'S TROPHIES

Of the 5 main eras of international air transport—infancy, to 1930; the flying boat era, 1930–1940; postwar transocean pistons, 1946–1958; first jet, 1958–1970; second jet age (jumbos) 1970–; Pan Am was not born for the first and was responsible for 3 of the other 4.

Among Pan American Airways' many "firsts," it was the first commercial airline to
• develop and operate four-engine flying boats (1931)
• develop and use long-range weather forecasting (1935)
• operate scheduled transpacific services (1935)
• operate international service with all-cargo aircraft (1942)
• install Ground Controlled Approach (GCA) in overseas operations (1946)
• provide tourist-class service outside the continental limits of the United States (1948)
• enter the Korean airlift (1950)
• open a scheduled round-the-world jet service (1959)
• make 100,000 transatlantic flights (1962)
• operate the Boeing 701-321C pure-jet freighters (1963)
• relay in-flight messages via Satellite Syncom III (1964)
• order the 747 Superjet (1966)
• make fully automatic approach and landing in scheduled service (1967)

vice. It came to the rescue of mass air travel at a time when the skies were dangerously congested; it effected a 3.5-to-1 increase in unit productivity while eliminating smoke and reducing noise pollution. It changed the economics of air travel and the nature of international tourism, making both feasible for ordinary Americans. Trippe's act of faith in aeronautics was the greatest ever made by an airline in technology. The 747 has been in frontline service for more than three decades and is still abreast of the market.

What drove him? It is too easy to say he did it for the money; he could have settled for much less and he was not a greedy man. It is obvious that he was competitive, but

not to the extent of many of our other innovators: He was competing with himself as often as with anyone else. It would be nice to say it was the democratic impulse to serve lunch to a Detroit mechanic on a jet to Acapulco; altruism should not be wholly discounted, but Trippe's heart never bled. A more compelling explanation may lie in the remarks of his intimate John Borger to Clive Irving. Borger confessed to finding Trippe as hard to read as did everyone else, but was impressed that no airplane was ever big enough for Trippe until he reached the 747 because he could always see the growth in the market that even plane makers could not trust themselves to believe. In short, he was propelled in the same way as Thomas Watson Jr. at IBM, able to see the potential of the machine but frustrated—seething with hope. Irving calls it the "fulfillment dream." Along the way, during the development of commercial aviation, Trippe saw it always being hijacked by people with different ideas—by the small-minded New Englanders at Colonial; by nations that protected flag carriers as necessary for the national ego and therefore were prepared to subsidize them; by elitists who regarded flight, like luxury liners, as a privilege that could be enjoyed only by the few; and more coarsely by the cartel operators who wanted to keep rigging prices because they thought that was where the higher margins would be.

Whatever Trippe's motivation, the democratization he effected as a visionary has been as real as that effected by Henry Ford, the mechanic. Fully 3.5 billion people, half the world's population, have been carried on a 747 since its inception. By 2004 the fleet had flown 35 billion statute miles, equal to 7,400 trips to the moon. It means that on any day, more than 900 Boeing 747s are in the air across the world, carrying close to 500,000 people, the population of a good-sized city. It means that 300,000 tons of cargo are in flight circumnavigating the globe with amazing ease.

It would never have happened—would not be happening today—without J. T. Trippe. And were he alive today, and in charge still of the world's only airline that was ever mega, he would have been first in line for the next generation, the 600-seater Airbus A380.

THE PEOPLE EXPRESS PIONEER

DONALD CALVIN BURR (1941–)

The revolutionary Donald Burr was an MBA graduate from Harvard who suffered "a great deal of pain" when he entered the business world. His experiences, he said, were too much at odds with the beliefs he had been taught as a child: "I had a mental model of people as all good, and when I went to work, I saw people who were greedy and suspicious, systems that were hierarchical and had poor value structure throughout."

His response in January 1980 was to quit as an executive (at one time the president) of Frank Lorenzo's Texas International Airlines, where tough direction was the order of the day. "Frank," he said, "is capable of any kind of behavior to win," to which Frank's answer was, "I'm not paid to be a candy ass."

With two other Texas executives, Burr had by April created a "humanistic" company to run a cut-rate airline, People Express.

His idea was to elevate every employee to the role of owner-manager relating laterally to other owner-managers on first-name terms. Nobody would be hired unless they bought 100 shares, available at a discount; there would be loans for anyone who did not have the cash. This idealistic company, starting up after deregulation in the first year of the Reagan presidency, excited investors, who oversubscribed the three million shares on offer, giving Burr $25 million. He snapped up 17 used but superbly maintained Boeing 737-100s from Lufthansa, and rented space cheaply in the abandoned North Terminal—a "ghost town" he called it—at Newark, New Jersey, an airport neglected by the majors. He had worked out a hub-and-spoke route system and on April 30, 1981, just a year since incorporation, three of the 737-100s began operating services to Buffalo, New York; Columbus, Ohio; and Norfolk, Virginia.

As well as a humanist, Burr was a minimalist. His concept was of the plane as a bus. You bought a ticket on board. There were no computerized reservations—in fact there were a lot of nos: no free baggage, no hot meal, no first class, no interline bookings. But People Express flew people at prices that made the majors gag. The simple standard fare was $35 on the first flights, no complications. He was able to keep prices down by maximizing space in his 737s. It cost People Express 5.28 cents to fly a passenger one mile against the industry average of 8.6 cents.

By the end of its first year People Express had carried 950,000 passengers. Within four years, it was the fifth largest airline in the United States, with 117 jetliners serving 107 North American cities— and London and Brussels. In 1985, it earned nearly a billion dollars carrying 12 million passengers.

As with so many innovators, Burr elaborated on the ideas of others. Kenneth G. Friedkin (1915– 1962) had made a success of the low-fare Pacific Southwest Airlines (PSA), operating within California from San Diego. Every one of the few hundred employees were on first-name terms, including Ken and Jean Friedkin. Burr's even more innovative approach to management was focused on eliminating bureaucracy, a ritual aspiration but in his case really achieved. No one, not even Burr himself, had a secretary or an assistant. Except for a few in critical areas of safety, nobody had a fixed position. Jobs were rotated: one week a desk job, another cleaning the plane. Pilots (called flight managers) would be rotated into administrative departments four days a month. Flight attendants (customer service managers) would switch to being ticket agents or recruiters. Pay was standardized.

Burr hoped to eliminate factionalism and discourage unionization. The salaries were low, but most of the staff entered into the spirit of the idea that it was their company. Burr set an example by paying himself a fraction of what other CEOs earned. In 1983, when People Express had profits of $10.2 million on sales of $292 million, Burr's salary was only $95,000.

Like Icarus, Burr may have flown too high too soon, encouraged by rapture in the business press. The radical expansion of the company, buying other airlines and adding routes, was at some cost to the family atmosphere and it certainly provoked the bigger airlines when he went head-to-head in their major markets. Robert Crandall at American Airlines led the counterattack. He invested in revenue/yield computer software programs that drew on historic data to show managers the permutations for pricing seats to fill a plane profitably. A number of rock-bottom fares could be offered to the budget-conscious while maintaining higher fares for the rest of the seats. The program was so sophisticated that American and then other airlines were able to operate profitably while offering discounts competitive with People Express but with the amenities that People Express could not afford.

DONALD BURR TO HIS PASSENGERS: Coffee stains on the flip-down trays means that we do our engine maintenance wrong. We have to make you think it's an important seat—because you're in it.

When People Express went bankrupt in 1986, those blessed with hindsight said they knew it could never work—but it did in the sense that it opened up air travel to new segments of the population and it galvanized the majors. It introduced a low-fare environment and underlined the critical value of information technology. The Smithsonian's airline authority, R. E. G. Davies, wrote, "No individual has entered the competitive arena of air transport in the United States and succeeded so quickly and so strikingly."

Burr was not easily consoled: "I will grieve about People Express all my life. It was one of those really rare, rare things. You get a shot once in your lifetime, and if you blow it, you never get over that."

GENERAL GEORGES DORIOT
Hundreds of bright ideas went begging until he started the first venture capital company

1899–1987

THE SCENE: French-born Georges Doriot is a dapper figure walking across Boston Common on the way to his factory at the same time every morning. Dark suit, Legion of Honor ribbon in his lapel, white shirt, black homburg, just the right amount of handkerchief displayed in his breast pocket. His calm countenance betrays nothing of his irritation at the sight of a woman—yet again!—tolerating her wretched poodle's assault on a flower bed. The complaint that something lamentable has happened to civic pride in America in the sixties is quietly vented on his secretary, Miss Dorothy Rowe, as he hands her a blue sheaf of three pages of notes. Georges Doriot is a man of fixed habits. Every night he is asleep at 11 in his elegant little house on Beacon Hill and at 2 a.m. he is awake worrying, writing notes on a blue pad without turning on the light so as not to disturb his adored wife, Edna. She is the recipient of hundreds of his love poems, but only Miss Rowe can decipher his nocturnal scribbles.

Doriot appears to be a cautious man; the sun is shining today but he invariably carries a neatly folded topcoat. It is natural that he should minimize risk in his personal life since his professional life is dedicated to risk-taking on a Herculean scale. The factory he runs in downtown Boston is, in fact, a risk factory. It is the headquarters of the world's first publicly owned venture capital investment company. Founded by him in 1947, Doriot's American Research and Development Corporation (ARD), was the first venture capital company to be listed on the New York Stock Exchange and the inspiration of hundreds of venture capital companies that have added untold billions to the American economy by seeding enterprises nobody else would touch—like Apple Computer, Federal Express, Lotus 1,2,3,

Compaq, Continental Cablevision, Genentech, Polaroid. Pretty much the entire biotech industry originated in "venture" dollars. Basically, before Doriot proved that systematic venture capital could work, innovators seeking money needed access to the few wealthy and adventurous families prepared to take the risks that frightened banking and investment institutions. Captain Eddie Rickenbacker went to the Rockefellers to start his Eastern Airlines, Juan Trippe to the Vanderbilts for his Pan American.

Doriot's headquarters is just a few square feet of office, sparsely but immaculately furnished with a wheat-yellow rug, a writing table holding several bins of blank paper, a tobacco box and a telephone, but here six days a week he is at the controls of a moving assembly line of ideas that will put thousands of men to work across the country. This Boston factory has its smoke, but it is aromatic. Doriot fills his pipe as often as he reaches for notepaper, which is more or less a constant motion since he fills every minute from 8:40 a.m. to 6:30 p.m. asking questions and scribbling; he likes representing hopes and expectations by drawing little charts. Scientists and technologists come here to unveil their brilliant or not so brilliant visions of making the world a better place by the manufacture and marketing of their innovations. They bring business plans for the dreams they hope he will finance, but Doriot's cool eye is focused on their character as much as their calculations. Men come first, schemes second. "A grade A man with a grade B idea," he says, "is better than a grade B man with a grade A idea." One day it's an ex-navy technician with a scheme for creating the first small computers for scientific and medical research; another day a Harvard chemist who foresees turning the deserts

green with a technique to make fresh water out of salt water; another morning two fresh-faced MIT-trained engineers in their twenties who claim to have invented a new type of computer-memory unit woven of copper wire; and the next, two physicists and an engineer who had started a company in a converted Cambridge parking garage to smash atoms; and such scientific experts may be followed in by a basement inventor who just knows that mothers by the million will rejoice in his safer, sturdier, cheaper pin for fastening diapers.

Disconcertingly, Doriot has a stopwatch on his desk. "Sometimes I use it to see how long it takes someone in a meeting to tell me the same thing three times." He looks for modesty and endurance, but has taken modest care to remind himself he can be wrong. Every time he looks up from note-taking, his eye catches a quote on his wall from Albert Schweitzer to the effect that it is almost impossible for one man to know what another is really like. "If I am right 50 percent of the time I judge men I think I'm damned good."

Doriot's visitors all fly their paper airplanes in the hope that Doriot will put an engine in there: not just money, but management advice for bumpy takeoffs, since Doriot is celebrated for the wisdom acquired and pithily dispensed as a professor of the Harvard Business School for 40 years (1926–1966) and his way of making things happen. When he joined the U.S. Army in the war, he found it irksome to wear a belt instead of his usual suspenders. He happened to be the deputy director of

The precise and dapper Doriot exchanged his French citizenship for American in 1940, saying that it was thanks to the United States that he had a wonderful wife, a good job and the ability to do something for them.

the Research and Development Division of the Quartermaster Corps, so he changed the rules: "I made suspenders items of Army issue so that no one could criticize me." His admiration for modesty did not inhibit his retention of wartime rank. He was "General Doriot" on his discharge in 1945, and ever after.

The general was a prudent investor, but only as prudent as befits a tightrope walker wheeling a barrow above Niagara Falls. In 1960, Doriot financed four ventures out of 166 examined by his small staff, in 1961 eight out of 186. Doriot and the scientists, engineers and financiers associated with him in making capital more accessible were not speculators in for the quick buck. They were excited by the rapid advances in technology following World War II, and while they were interested in making money in due course, they were moved as much by the patriotic thought that America would be held back in world competition if innovators were left to hit-or-miss social contacts or safety-first banking institutions. Doriot and his colleagues were democratizers, but not ideologues. They knew that traditional investment managers could not hope to identify the potential winners among the bewildering profusion of discoveries, and that anyway it was futile just to throw money at every Thomas Edison who came through the door. They also appreciated the difference between invention and innovation: Those who came up with the best ideas for new companies were not always ready to run them. As Doriot liked to put it, "There have been many fine scientists desperately trying to become poor businessmen."

Doriot and his associates saw themselves first and foremost as builders of businesses for the long term, equity holders, partner-managers, not debt collectors. Doriot was likely to turn up at a client plant, as he did at the High Voltage Engineering Corporation in Burlington, Massachusetts, in one breath berate the chairman's assistant for its untidiness and in the next sketch for Denis Robertson, the company's president, the benefits of building a plant in Europe to take advantage of the emerging Common Market. (They ultimately built it in the Netherlands to great profit.) Robertson recalls how Doriot insisted they kept their focus on technology. "At early board meetings I'd try and give an accurate account of the profit and loss. He would sort of look through me and ask what I really thought when I was shaving. He didn't want me to become a bookkeeper or accountant." High Voltage was a typical ARD venture that would never have got going if left to institutional lenders too impatient to wait the seven years it might take a technically oriented enterprise to begin repayments; as one engineer-industrialist phrased it, such investors treated a science-based company as "just another shoe store." Doriot scoffed: "Your sophisticated 'long term' stockholders make five points and sell out. But we have our hearts in our companies. We are really doctors of childhood diseases here. When bankers or brokers tell me I should sell an ailing company, I ask them, 'Would you sell a child running a temperature of 104?'"

Few at the time believed any of this was more than hot air. The renowned innovator Charles Kettering said ARD would go bust in five years. It didn't. Doriot's example became widely followed by his own Harvard students who went into venture capital and others who imitated them. But it was a struggle for the pathfinder. Behind the cool facade was a man in torment. It was not the arithmetic of risk so much as the algebra of government regulation. Just as ARD triumphed and investors and economists began appreciating the importance of venture capital for innovation, ARD itself fell afoul of others who did not: unimaginative regulators at the SEC and IRS. Only Doriot's personal papers reveal the bitterness of his tussle with yesterday's men.

GEORGES FREDERIC DORIOT was the son of a Swiss mother and a pioneering automotive engineer in Paris who helped design the first Peugeot. By all accounts, it was a pedagogic childhood in a strict Lutheran household. Denis Robertson observed, "I'm sure that when he came home in the afternoon, he was asked, 'What did you do today that was worthwhile?' And if it was worthwhile, he was told to get a good night's sleep so he could go out tomorrow and do it again." Doriot, always laconic about his early life in France, conceded that his father spanked him whenever he was not first in his lycée class. "I got from him a strong sense of duty, analysis and manual dexterity; from my mother I learned to be outgoing with people." When he was 11, he had a year in England at Lynton College, returning with an accentless English he forgot or discarded, speaking English with an engaging French accent for the rest of his life. After school in Courbevoie, he trained as a mechanic before taking a science and modern languages degree at the University of Paris, and then at 21 he did what so many Europeans did. He came to America and did not go home.

"Perhaps I stayed," he said, "because I have always been interested in the future." He got an MBA from Harvard, his studies paid for by his father, and at the age of 25 he acquired a taste of investment banking with a job in a small affiliate of Kuhn, Loeb and Company. Before he was 30 he had contrived a perfect mix of academia and business. Returning to Harvard Business School in 1926 as assistant dean and associate professor, he was invited to join the board of several companies, including a mining company, several technical companies and the Pressed Steel Car Company in Pittsburgh. Harvard appointed him to the chair of industrial management in 1929 and he kept the job for 37 years, at one stage shuffling between two days teaching in Boston and three days in Pennsylvania as president of McKeesport Tin Plate Corporation and chairman of National Can Corporation. (In a quote then off the record, he described how he got the presidency: "One day my predecessor got drunk at nine instead of eleven as usual.")

Perhaps it was as well he did not sleep much. His answer to bouts of insomnia was to play around with a palette knife and colors, which he refused to call "painting." Asked about a flower on his canvas, he would respond, "Have you been to Australia? I've found very few people go there, so I'm usually safe with Australian flowers." His house was crammed with books, but he deflected discussion of his obvious taste for history and sociology. "I have purchased more books that I have never read than anyone else I know. I love owning books.

But I just scan them. I just love to buy them."

His Harvard classes, ostensibly designed to instruct in the running of a company, were really an effort to imbue the students with his high ideals of how a businessman should behave. He was a Puritan. If you see a company president who is 38 and into yachts or racehorses, he advised, get out of his stock at once. One French word he did not know, said the head of one of his ARD companies, was camaraderie. He had no patience for kidding around, or washroom jokes, and he held a religious attitude to the value of time. He was a martinet. No sweaters in class. Be there two minutes before 2 or don't come at all. And if you are not coming, call my secretary—"You'd do it in business." Carry a notebook everywhere. Write in it for 10 or 15 minutes every day. Get into these habits, and it will liberate you for other things, more constructive things. But his long list of class rules on the first day was always introduced with a rider. "Never take anything I tell you as the truth. At best it's a conceptual scheme and you should improve on it." Students might groan about his disciplines, but they never forgot him. "I have a hard time relaxing," one remarked 10 years after graduation. "I learned from Professor Doriot how to spend each moment profitably and constructively and I can't break the habit." Philip Caldwell, a former chairman of Ford Motor Company, told *Forbes* magazine, "I can still hear him saying in his French-accented voice, 'Gentlemen, if you want to be a success in business, you must love your product.' He instilled that idea deeply in men."

In his first year as a full professor, he married Edna Blanche Allen, a Simmons college graduate and Harvard Business School research assistant, but decided there would be no children because they cost too much. "Edna and I decided we didn't have the financial means to raise children in the Depression. I never borrowed a cent in either my business or personal life." So, the 7,000 students he taught became his children. His one-year elective course was called Manufacturing but he would sharpen minds on any topic, pacing the aisles with his questions. What's worrying about the word "new"? Do boxcars have to be box-shaped? What's the most dangerous point in the life of a company? (Its first success.) Or try this: How do you pick a wife? Doriot said Harvard's discovery of the executive wife in 1955 was one of its great nonbusiness discoveries. He would meet with them regularly. "They are very bright and very important to their husbands. Some of them are brighter. You know a girl can absolutely make a man or break him. She has to know when to talk business with her husband and when he wants to forget it." Wives and girlfriends were the only visitors allowed in his classes "so they can get an idea what life with a future executive will be like." He paraded an army of aphorisms. "Always remember that someone, somewhere, is making a product that will make your product obsolete.... Always read the newspaper with the largest circulation because that's what your employees read." He set impossible standards and he was everlastingly proud of them. All his students were required to get out of the classroom and work with companies in Boston. At one end of his writing table at ARD, he would gesture to the pile of their technical reports. "Some of them earned $30,000 for that class work."

ONE OF DORIOT'S most popular courses was on start-ups. "Starting a company," he said, "is like getting married. Most of the problems are discovered after the honeymoon is over." He took his own first steps on the thin ice of venture capitalism in the New Deal era, when the big question was whether private enterprise was any longer capable of rescuing itself. In 1938 he served as adviser to a group of 20 Boston investors, among them Karl Compton, the president of MIT, and Ralph Flanders, a self-made Yankee mechanic from Vermont who became president of the Federal Reserve Bank of Boston

"I have never yet met an MBA who cannot analyze a company to the point where it is clearly not worth investing in."

(1944–46) and was destined for glory as the senator who had the courage to denounce Senator Joe McCarthy. Each bought 15 $100 shares in the launch of Doriot's Enterprise Associates. Unfortunately, Enterprise had only enough money to find and study projects—it had to "pass the hat" for capital. It got one project off the ground, a wobble plate internal combustion engine built by an MIT professor, but while the group was pondering its second investment, the Nazis invaded Holland. The morning after, most of the investors panicked and pulled out.

Doriot went to war. Six months before Pearl Harbor, when he was 42, he took leave from Harvard and gave up a new appointment as chairman of National Can to join the U.S. Army as a colonel in the War Department. He was given a big budget and teams of scientists. His friend army chief George Marshall told him that in every fight the United States had got mixed up in, there was always a shoe scandal of some sort and would he please make sure there was none in this war. He did. He made shoes that prevented frostbite and jungle foot and shoes that did not fall apart. He introduced new dehydrated foods, uniforms with innovative thermal qualities and less liable to tear. Then he looked into making a lightweight, nonmetallic body armor for combat, enlisting Dow Chemical, General Electric, Bakelite, Monsanto, Firestone, Westinghouse and Formica. The result, a fiberglass-plastic combination called Doron after him, was not in time for World War II, but the navy shipped 90,000 pieces to Marines in the Korean War and saved many lives.

Doriot's experiences at the Pentagon would greatly influence his venture firms. Creative people (Grade A men) had "self-stimulating glands." Told by private contractors that it would take a year or more to make a mask effective against liquid gas, he called in a young officer working for him. "I asked Henry which theater of war he would least like to have to fight in. He replied Burma and the East. I gave him the problem, told him find a solution in three weeks or I would ship him to Burma. I stimulated his glands and he came up with a solution." He concluded that the army paid too much attention to technology and not enough to the needs of individual soldiers, a conviction he took with him when he came back to Boston and picked up on discussions with the old Enterprise group. ARD was organized on June 6, 1946, but with the difference that this time the group set out to raise $5 million as a public company so that it would have enough funds to commit long term on its own accord.

The ARD board was a Who's Who of Boston's social, industrial and educational elite, with particularly close ties to MIT—its president, Compton, its treasurer, Horace S. Ford, Edward R. Gilliland, professor of chemical engineering, Jerome C. Hunsaker, head of MIT's department of mechanical and aeronautical engineering, and Frederick Blackall, who was also a director of the Federal Reserve of Boston. Doriot was chairman of the board of advisers. When Flanders, ARD's president, was appointed to an unexpired Senate term in December, Doriot took his position. The burden of raising the capital and finding the projects fell on Doriot and his two-person staff, Joseph W. Powell and Davis Dewey II. It was grueling. The stock market was falling. The initial public offering was on August 9, 1946, and by the end of December only 139,930 out of 200,000 shares at $25 had been issued. To protect investors, ARD limited public stockholders to those who could afford a minimum of $5,000 (a large amount of money at a time when only 1.3 percent of families was earning over $10,000). This limited the audience and all Doriot had to sell was faith in science and industry. The saving of ARD was a campaign Flanders had started for a small easement of the tax and security regulations that prohibited life insurance and nonprofit fiduciary trusts from making risky investments. Exemptions for 5 percent of the assets was agreed. Quickly, three universities (MIT, Penn and Rochester)

jumped aboard, as did John Hancock Mutual Life. By the end of 1947, insurers and educational institutions owned 49.6 percent of the shares.

By December 1946 three investments had been made: $150,000 to Circo Products of Cleveland, an existing company that had invented a new way of degreasing automobiles; $200,000 to the physicists Robert Van de Graaf and John Trump and the engineer Denis Robertson for High Voltage, and $150,000 to Tracelab, another new company staffed by MIT grads to sell radioactive material and make radiation detectors. Two more new companies were started in the next six months, for a total of five in the first year. They were all in the red. In 1948, ARD invested $50,000 for 75 percent of the stock of a company called Ionics, started by the Harvard chemist Dr. Walter Juda to make ion-exchange membranes that could make salt water potable. Millions of people across the globe could be saved from dying of thirst. But those who needed the water were too poor to pay for it and governments held back. The fiduciary institutions associated with ARD were resolute in support in these difficult early days for ARD. The "semiphilanthropic" character of ARD, as Patrick Liles put it, made losses tolerable in the short run, but it was touch and go whether ARD had a future. In the spring of 1949, no investment bank would underwrite an ARD offering to raise $4 million and it was left with 74 percent of its stock unsold.

Doriot vented his frustrations in a speech to the Harvard Business School Club of New York in January 1949. Castigating Wall Street for its inability to invest in new companies, he said: "Venture capital? Why get a speaker from Boston on the subject? A mental paralysis seems to have stricken (New York) financial circles—why? Those who made Wall Street useful and great should be ashamed of us. The trouble is Wall Street is useless or at least the rest of the nation thinks so. The brains are here,

but they are paralyzed—greedy—selfish—unconstructive."

The business press was more responsive than the street. *Business Week* gave ARD a cover story in 1949 and other favorable articles came out in *Barron's* and *Fortune*. And ARD started to fulfill its promises. By 1950 it was in the black and so were 10 of its new companies. Total sales for its 26 client companies were $20 million and 2,000 people were employed. The following year sales for client companies doubled to $40 million with 3,000 employed in companies that made among other things, high voltage accelerators, fire retardant paints, liquid fuel rockets for aircraft and guided missiles, ultrasonic generators, cancer therapy, hydraulic pumps, cargo palettes, pharmaceuticals, ceramics, optical scanning. Still, ARD shares continued to slump, down to $19 from the original $25. Doriot tried metaphor to make his point to Wall Street as if with a recalcitrant class of 1952. "Too many bankers and counselors have forgotten the history of the early years of our industrial giants of today. The first fifteen years of companies and of human beings are very much alike—hope, measles, failures, mumps, reorganizations, scarlet fever, executive troubles, whooping cough, etc. are parts of one's daily life. Hopes, disillusions, hard work, are all necessary, particularly during the first ten or fifteen years before a stable and healthy body or corporation can begin to exist."

In February 1953 Doriot organized an exhibition in Boston called "Products with a Future" that drew 4,000 people. If he was wowed by the technological wizardry of the companies he was financing, he reasoned, others would be too. *Business Week* wrote: "The spectacle made a hit with the public. Onlookers sampled shrimp, vein removed with special patented equipment, washed it down with a drink of fresh water drawn from an automatic de-salter that turns out 200 gallons an hour from salt water. They donned earphones to listen to the roar of a freight train produced by a two-dimensional sound recording, gaped at such items as rocket engines and a 'Chillow' to keep your face cool on summer nights."

All Doriot's faith was vindicated. After 1955, ARD reported consistent gains. By 1966, its net asset value had grown to $93 million from just under $3.4 million at the end of its first year. The most spectacular investment was $70,000 for 70 percent of Digital Equipment Corporation (DEC), worth $183 million to ARD in 1967. That one investment generated great wealth beyond the company in helping to make Route 128 outside Boston the technological powerhouse of the northeast and the hotbed of innovation. But the irony of Doriot's story is that he failed because he succeeded. On an arcane interpretation of the 1940 Investment Act, written before the existence of organized venture capital, SEC lawyers ruled that ARD officers could not have stock options in client companies. It was "against the public interest," though they never could explain why. It meant Doriot could not attract the best brains. The height of absurdity was when Ken Olsen, the founder of the glittering DEC, was not allowed to exercise options given him by his own company because he was also a director of ARD. The tension was constant between regulatory mechanisms set up to deal with the short term and Doriot's longer view. Insinuations of impropriety became commonplace. Time and again one government agency or another got a bee in its bonnet—often from misreading a newspaper article—ordered an investigation and then admitted it was in error. The regulators, unable to appreciate that venture capital investments take years to mature, also started second-guessing many of Doriot's decisions. "May I bring to your attention," Doriot acidly informed them, "that contrary to untutored public opinion, ten years is far too short a period in which to bring a new creation into really profitable maturity with some semblance of stability."

The fact that ARD had troubles with the SEC was reported at the time, but the scale and nature of the harassment is manifest only in Doriot's confidential notebooks, letters and journals at the French Library in Boston. In 1965 the agency suggested vaguely that ARD might be guilty of insider trading since it knew all about the operations of its client companies. Doriot, the Puritan, was so incensed he lost his supersmooth demeanor:

Sir, We try to help our new companies. Do you mind? Do you know about the problems, the vicissitudes,

the trouble connected with venture capital? We conduct this company on the highest ethical standard. I do not enjoy your very wide and general remarks. If you will give me specific cases and specific examples to which you object, then I can discuss the matter with you.

An SEC investigation came up with no further allegations and the matter was dropped. Still, interference continued. A November 1968 confidential memo records: "We had a long talk with the SEC chairman. He told us he understood our problems, but we never heard from him." What they got, on December 31, 1968, were two SEC investigators showing up unannounced for what they called a "surprise audit." Doriot's journal for January 6, 1969, notes: "Same two men appear unannounced at 1:50 p.m. Said they wanted to investigate our method of valuation. Showed them statement in our reports showing how valuations are arrived at according to SEC regulations. Less than 4% made by ARD directors. They had not read that! They said SEC was a big place. They disappeared during the afternoon."

Doriot kept hoping that ARD's very usefulness to the economy, to America itself, would enable it to navigate the maze of regulation. But the era in which ARD was born, the mid-forties, was far different from the suspicious, restrictive sixties. Doriot kept asking questions that deserved answers:

If the SEC states that ARD is not principally engaged in the venture capital business, will they please tell me what business we are in?

Why is it that officers of a so-called conglomerate can have options on their company stock and ARD officers cannot?

There were no answers. Things got worse when the Internal Revenue Service decided that ARD would lose its tax treatment as a regulated investment company if it did not sell off its stock in DEC by 1967. Someone writing the tax code in 1954 had decided that 10 years was the limit for an investment company to hold a significant portion of its assets in any one client company. It meant ARD would be cut off from future revenues of any company it

had seeded—revenues that Doriot had expected to use to seed more innovation— and it would no longer be in a position to offer its management expertise. The nonsense was that ARD stockholders were free to invest in such securities as GMAC, Allied Chemical, etc., but penalized if they kept it in the company they had started.

Doriot lobbied hard in Congress to get the 10-year rule eliminated. A number of congressmen took an interest, among them George H. W. Bush, whose future political career was based on his success with Zapata Oil, one of ARD's early investments. In the meantime, Doriot was forced to sell off its $183 million investment in DEC. Finally, legislation was passed in 1968 allowing ARD to hold stock in its other companies—but only until 1971. The SEC then sent word that any company receiving support from ARD could not give stock options to its employees. It was an out-of-the-blue ruling utterly without merit, and ARD would have been put out of business then and there had Doriot accepted it. No company would ever again look to ARD for money.

Ultimately there was little that could be done. Doriot ended ARD as an independent entity on January 12, 1972. To escape the war of attrition, he merged it with Textron. *Fortune* wrote of the merger: "Even if Digital Equipment is set aside, Doriot's record of investments in some 150 companies is impressive." It represented a return of nearly 15 percent.

But Doriot had done more than make money. He had blazed a path. Congress and the Carter Administration finally started passing legislation that Doriot had been urging for decades, freeing venture capital firms to invest more and to make more in capital gains. As William D. Bygrave and Jeffry A. Timmons conclude in their 1992 study of venture capital, ARD was the catalyst for a new structure for equity-based venture capitalism, more durable and less subject to short-term evaluation. But Professor Doriot might ask the class: What is the problem with this? Answer: All venture capital companies are organized as limited partnerships rather than as public companies so that the general public no longer reaps the financial fruits of some of venture capital's best investments.

There is work for another Pathfinder.

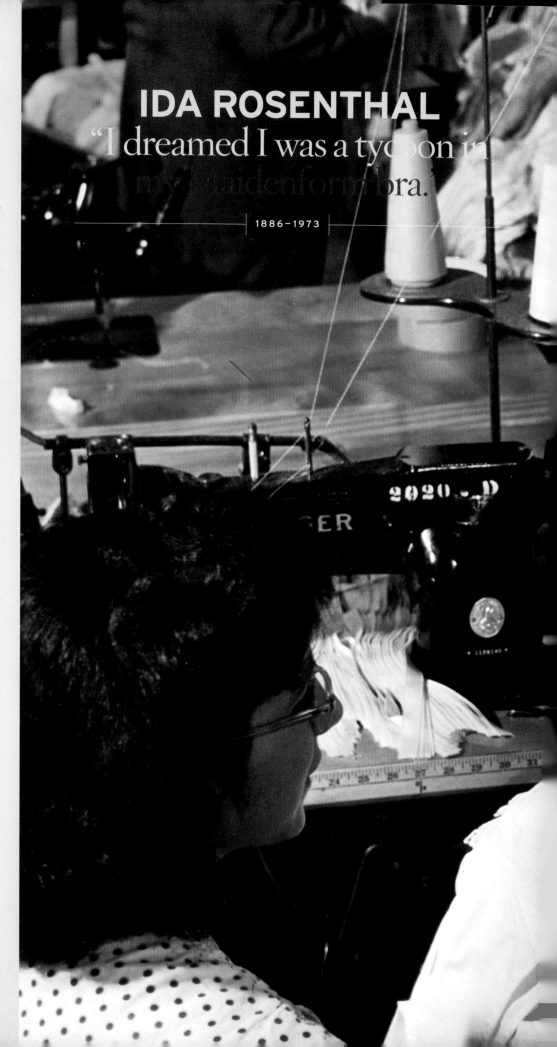

IDA ROSENTHAL
"I dreamed I was a tycoon in my Maidenform bra."

1886–1973

In 1920 Walt Disney was an out-of-work artist in Kansas. Today Disney is an international empire of movies, television, Web sites, theme parks, books, cruises and stores. Everybody in the world has heard of Disney. Hardly anybody has heard of Samuel Insull. He ran errands for Thomas Edison, then built a vast empire of light and power. Estelle Mentzer was a young mother in New York who persuaded passersby to let her put cream on their faces. As Estée Lauder, she founded the world's greatest family cosmetics company. Jean Nidetch was just a fat woman who wanted to be thin. Out of despair, she built an organization of hope for millions: Weight Watchers. Edwin Land set out to please his infant daughter and made social history with the Polaroid camera. Thomas Watson sold pianos off the back of a wagon, and the sequel was IBM.

All these innovators, and the others in this final section of part II, touched the lives of millions of people, and still do. Insull? Everyone who logs on to a computer, cooks dinner, irons a shirt—flicks any switch—enjoys his conviction that electricity should be cheap. That was a thought anybody could have, but the extrapolation of an idea into an industry, an enterprise into an empire, takes more than a brain wave. Just what it does take is the subject of these stories.

She could fret about two extra stitches that might make a bra uncomfortable but she had a bird's eye view of the whole business.

SUNDAY SCENE at New York's Coney Island beach in the '30s: In the sand, a man named William Rosenthal has fashioned a full-length nude woman with gently sculpted breasts. It is not a model of the woman on her knees in the sand next to him, his wife, Ida. She has a more voluptuous, fuller figure, and she is only four-foot-ten; putting it plainly, she is dumpy. The tall nude lying prone in the sand has a svelte hourglass body more like the Other Woman in their lives, an elegant clothes-horse and dancer by the name of Enid Bissett who is not with them on the beach outing. Yes, there is a triangular relationship here, but it is a happy one, and it is not about sex, it's about style. All three of these people are immigrants working in the rag trade—Ms. Bissett from a genteel England, the Rosenthals from revolutionary Russia. When they came together as innovators in the '20s, they were on the road to personal wealth and they were going to revolutionize the way women looked everywhere.

Ida's upbringing was a fusion of scholarship, commerce and subversion. She was born Ida Kaganovich on January 9, 1886, in the small town of Rakov near Minsk. Her father, Abraham, was a biblical scribe from a long line of Jewish intellectuals, wrapped up in his studies. The breadwinner was her mother, Sarah, one of the few literate women in the town. She ran a small grocery store with a sharp eye, and she was determined to give Ida and her sister, Ethel, the means of looking after themselves. Sarah apprenticed both to a local dressmaker. At 16, the nimble-fingered Ida found work as a seamstress in Warsaw while also attending high school classes to study mathematics and Russian. She returned a changed woman. Exposed to Warsaw's ferment of isms, she had chosen socialism because it seemed to promise peace, justice and rights for women, and on her return to Rakov she flung herself into the agitations leading to the 1905 attempt at revolution in Russia. The sight of this very tiny young woman standing up to make gutsy public speeches calling for the overthrow of the czar endeared her to a pale young revolutionary, William Rosenthal; it also speeded the local police chief round to Sarah Kaganovich's grocery to tell her

that one more speech like that and her daughter would be arrested. William was also in danger, about to be drafted into the Russian Army to fight the Japanese. He escaped to the United States and Ida followed soon afterward, promised shelter by an aunt and uncle in Hoboken, New Jersey. "I came after the man who was to be my husband," Ida said later. "I couldn't live without him." A granddaughter, Elizabeth Coleman, says another reason was her belief in women's rights, which were not

Ida and William Rosenthal, about six years after their 1906 marriage

accepted in Rakov's Orthodox Jewish community. "She was a real feminist, outspoken, really liberated," says Mrs. Coleman. "She did not function in an environment where people had to be very careful what they said." Many people in her home community were shocked, too, by her open liaison with William.

For millions of immigrants like Ida—known to her family and friends as Itel, says Mrs. Coleman—the next few years would have been 12-hour days in a factory, home at night to a grim tenement. Ida rebelled, all around. Her relatives had the notion that a penniless young woman of 18, fresh off the boat and knowing no English, needed guidance. Ida took that as bossiness, moved out and changed her last name to Cohen. She married William on June 10, 1906, when she was 20. Financially, her marriage was a rerun of her Rakov family. William was sick frequently—doctors thought it was tuber-

culosis—so she had to be the breadwinner. She would not consider a factory job: "I don't want to work for anyone else." Instead, she bought a Singer sewing machine on the installment plan and started making dresses in their house in Hoboken, working through an immediate pregnancy. She could not read English, or speak it much, but she looked at photographs of well-dressed women and she had flair. She priced her work as keenly as her mother had priced her bread and potatoes, and customers were charmed by her vivacity.

For the next 12 years, while William was in and out of sanitoriums, Ida built up her little business and raised two children—Lewis, born in 1907, and Beatrice, born in 1916. By 1912, the year she became an American citizen, she was employing six workers and earning from $6.50 to $7.50 for a dress; by 1919, she had 15 workers on two lower floors of a four-story house where the family lived.

And so the story of a fairly successful little business might have progressed. The first kink in that conventional road was the big snow of 1918, when William was marooned in New York and the Hoboken cop on the beat dropped by to tell Ida she had to clear the two sidewalks round her corner house. Ida recalled, "I couldn't ask a maid to do it, she would have quit. The cop said he was sorry, he'd help but he had too much to do." So Ida cleared two feet of snow, a big deal when you consider the banks were nearly half her own height. She was still speaking of the incident with irritation when interviewed by a *Fortune* magazine researcher 30 years later. "I wasn't meant for snow shoveling. I resolved right then that I wouldn't spend another winter in Hoboken." She moved her machines, materials and about a dozen of her employees to 141st and Broadway in Manhattan—and with her came her loyal clientele, who were now willing to pay $25 a dress.

The second kink in the road was a new customer, the director of five-year-old daughter Beatrice's nursery school. In 1921 the director happened to be wearing one of Rosenthal's $25 dresses when she looked round Ferle Heller's fashionable millinery shop at 36 West 57th Street between 5th and 6th Avenues, where women would pay

SAND SCULPTOR: For years, William Rosenthal sculpted the female figure in the sands of any beach he was on.

up to $300 a dress. There was an exclusive boutique in Heller's called Enid Frocks, run by Enid Bissett, who noticed the style and quality of the schoolwoman's dress. The enterprising Enid called on Ida's workshop. She was most impressed, according to Linda Jacobs Altman, by Ida's "deft interpretations of current styles." They made an arrangement for Ida to make some dresses for Enid Frocks, but the perceptive Enid soon came up with another proposition: They should team up as equal partners and dress the carriage trade from a new shop of their own at fashionable West 57th Street between 5th and 6th. *Why risk it?* was the response of friends and family. *Why not?* was the natural rebel's reply. Ida invested almost her entire life savings of $4,000.

Such are the curious ways of enterprise that what Enid and Joe Bissett did before meeting Ida proved important: They had

made a living as a vaudeville dance team. Stage friends were good and early customers for dresses costing from $125 to $300, but Enid and Ida didn't like the way their customers looked in their Enid frocks. The industrialist Bernard Baruch was largely to blame. When America entered World War I in 1917, women wore corsets; the corset was being undermined by the sinuous imperatives of the tango from 1911, followed by the American turkey trot and the bunny hug, but it was Baruch who was the catalyst for the transformation of the shape of women. He asked them to take off their corsets—in the national interest. He was chairman of the War Industries Board, and corset ribs took steel that was needed for weapons. Some 28,000 tons of metal were lethally diverted—enough to build an entire battleship.

The effect of Baruch's patriotic divestment was accelerated by politics. Women

had worked in the factories alongside men, and freedom and equality were equated with looking like a man. Women demanded the vote (and got it in 1920). They started to smoke in public. They cut their hair short. They took to the flat-chested flapper look, binding their breasts tightly with bandeaux, which resembled a bandage. Enid Bissett had the figure that suited the new "boyish form"; Ida didn't. But both of them were irritated that the dresses they made never fit perfectly since the bandeaux were not always wrapped with the same tightness. "It was a sad story," Ida lamented later. "Women wore those flat things like bandages. A towel with hooks in the back. And the companies used to advertise, 'Look like your brother.' Well, that's not possible. Why fight nature?"

It was Enid who took the first step. She experimented with cutting the bandeaux in half, putting a piece of elastic in the

CENTURIES OF CONSTRICTION

SEVERAL WOMEN lay claim to the invention of the modern brassiere. Mary Jacobs was the first to register a U.S. patent for a garment called a brassiere. From his research in the U.S. Patent Office, Hoag Levins awarded the palm to Marie Tucek, who patented a "breast supporter" in 1893. It had separate pockets for the breast, with over-the-shoulder straps fastening by the hook and eye. But Professor Jane Farrell-Beck's history credits Luman Chapman with getting the ball rolling in 1863 with a patented breast supporter, followed by Olivia Flynt in the 1880s, and Laura Lyon, who patented a sports bra in 1904. The French couturier Herminie Cadolle introduced a "corselete-gorge" in 1889 that offered breast support separate from the traditional tight-laced whalebone corset. She made ribbon-and-strap bras for the World War I spy Mata Hari.

But none of these inventions took off. The corset needed a stake through its heart and did not get one until Ida Rosenthal once and for all broke that garment's grip on a couple of centuries of women by establishing a mass market for uplift bras with standardized cups for women of all ages and shapes.

From classical Greece and Rome to the 20th century, with only brief changes of style, women were expected to bind the bust to the vanishing point. The stiff corset, combining a girdle and breast support, resisted all reformers.

French revolutionaries banned it as a symbol of aristocracy, since corsets laced from the back needed the assistance of a maidservant. Jean-Jacques Rousseau campaigned for a return to nature and simplicity. French and British doctors warned of the harmful effects of constriction and the "false and preposterous" attempts to attain a spider waist. The Pre-Raphaelite brotherhood of British artists in the mid-19th century told women to throw away their corsets and enjoy the freedom of flowing robes.

But nothing changed much. The fashion historian Beatrice Fontanel reports that corsetry reached "cruel and lunatic extremes" in the 19th century. Why, from the sixteenth to the late nineteenth century, did women never rebel? Corsets are said to have come back in the Napoleonic era (1804–15) because the idea took hold that they kept the vital organs in place. Fontanel suggests the corset survived because it was a badge of superiority. "Those wearing it were barred

from even the slightest useful exertion, thus reinforcing the prestige of the ruling class." There was also the problem that nobody was offering clothing to replace the corset.

The makers of corsets were prolific. There were nuptial corsets, wire satin corsets for balls, lightly boned morning corsets, stayless corsets for night wear, nursing corsets with drawbridge gussets, traveling corsets with tabs that could be let out for sleeping, riding corsets with elastic at the hips, corsets for singing, dancing and bathing at the seaside (boneless). In America, "Gibson Girls," immortalized by the artist Charles Dana Gibson, set the cause back by popularizing the S-curve. Their corsetry was touted as healthy, but with the bust thrust prominently forward and the buttocks pushed back, it just turned women into geese. The breasts were kept low and were not separately supported as in Rosenthal's uplift bras. By 1908 the most fashionably dressed women could not sit down.

Whatever the apocryphal bra-burners of the '60s thought they were about, the pioneers of the bra in the 20th century were the real liberators.

middle, and then fitting it into a dress so that two cups separated and supported the breasts but did not flatten them as did Caresse Crosby's 1913 brassiere.

This is where the sand sculptor comes into his own. In recovering his health, the quiet and gentle William had found satisfaction in obsessive sculpting: He was in the habit, when he went on vacation, of filling his suitcase with 30 pounds of clay. He was fitting dresses in the shop when he saw Enid's rudimentary brassiere, and the artist in him spoke out: "If you want to wear something like that, at least let me make you a nice one." The brassiere he produced was made of soft-knitted mesh with two pockets gathered in front but with room for curves; it was hardly the oomph of the sweater-girl era but it had distinct uplift. The partners started sewing William's cre-

ation into the dresses they made. They had no thought of making money from it. They just wanted women to look like women. Yet their priorities began to shift when clients started returning asking if they could have extra brassieres without having to buy a dress. It wasn't just the curves they liked—it was the freedom to move and breathe more easily.

The partners gave these early customers bras without charge, then started pricing them at a dollar apiece—and fairly quickly realized they might just possibly have a new business. The Rosenthals and Enid put up $4,500 as capital and Enid invented a brilliant name for the unboyish garments: Maidenform. They registered the title and in 1922 put a sign in the window. They made the bras in the attic of the shop and hired their first salesman.

How could they hope to build a business by word of mouth from one shop on 57th Street? They had no money for advertising. There was resistance in the trade. There was always the risk, given the vagaries of fashion, that they would expand to meet a demand that suddenly flamed out. The Bissetts' vaudeville days were critical in getting through these early years. They knew many actresses, and, according to William's brother, Moe, who worked with him, it was the fact that the stage people "were brave enough for uplift" that gave them their first clientele. Give my regards to Broadway. A Maidenform salesman, George Horn, recalls: "It was an extreme product, but it was accepted in the theatrical district. When I went out to the Bronx to show our bra, I would hear, 'It might sell on 42nd Street, but it will never sell here.' Strangely enough, it did sell. It sold very, very big. Several more salesmen were hired." Another salesman, Jack Zizmor, remembered: "When I showed a store a little bit of a bra, all hell would break loose. If it were the husband, he would call to his wife, 'Come over here and see what this crazy guy is trying to sell me.' That was the reaction we got in the beginning. They laughed and ridiculed us, and said, 'This is a fly-by-night thing. It will die out next week, next month, next year.'" And salesman Al Siegel: "Our bra was the reverse of fashion. They laughed at the bras, as they laughed at going to the moon." It is plausible that the reason the laughter evaporated as soon as it did was that Hollywood caught on to Broadway's embrace of uplift. After the success of busty Mae West in the '30s, Hollywood accepted and promoted other well-endowed women, such as Jane Russell, which no doubt influenced millions of ordinary women.

The business burst out of the dress shop attic. William enlisted relatives to come to the rescue; Rosenthals were everywhere. Two sisters sewed bras in their homes. His sister Masha Hammer gave up the kitchen of her house in Bayonne, New Jersey, for six or seven operators. Soon her whole apartment was flooded with women sewing bras. The partners were still making dresses but in 1925 talked about concentrating on bras alone. When Ida's brothers advised her to stick to the safe business of making dresses, she naturally staked everything on making bras full time. Ida and Enid rented a large manufacturing space in Bayonne and in 1926 had 40 sewing-machine operators working flat out. After the Wall Street crash of 1929, almost all the major dress houses went under, but Maidenform continued to expand, turning out half a million bras and more. They were making a product that didn't exist 20 years before and yet creating a demand that would help them surf through the Great Depression. In their sole bad year of 1932, the fighting zeal of Maidenform was represented in a National Recovery Act parade in Bayonne. The women workers marched with an enormous bra stretching from one side of Broadway to the other; the wind filled the cups and the spectators threw coins into them. By the end of the '30s, Maidenform

garments were sold in 95 percent of the country's department stores and all over the world.

The business, on Enid's retirement in 1930, had been divided, with William as president responsible for design and Ida as chairman for money, sales and strategy—but in reality involved in everything. She saw all the orders before they went to the factory so she could get the pulse of the business; by now there was competition. She was Tinker Bell, flitting from workshop to showroom to department store in little puffs of smoke—she smoked four packs a day until 1951, when she gave up overnight. She was always stylishly attired but refused to wear a long evening dress: "It makes me look like a mushroom." Entering an office, she would command the men to sit down so they did not tower over her. She was forthright and didactic. A "bra-zheer," she would say in her Russian accent, "is matter of engineering and psychology. A woman is very funny creature. You have to sell her right size and right type, but what she wants to hear is fashion." Ida told women they had been betrayed by American dress designers who caricatured French fashion. Paris minimized women's busts but didn't flatten them completely. "French women never got so unfeminine!" The universal testimony was to her charm: "She could romance a customer out of his shoes and socks. . . . In market week, she would float around the showrooms very gracefully. . . . She made it her business to say hello to every single person who came in."

For his part, William was never without a pencil stub, sketching new designs. Elizabeth Coleman says the family often felt he got short shrift in reports on the company. "Without his innovative designs, Ida would have had nothing to sell." His 1926 patent for a brassiere "to support the bust in a natural position" was just the beginning of the way he applied his art to commerce. He designed the first maternity bra, and the first scientifically designed nursing bra. His "Over-Ture" bra created uplift support using clever stitching instead of uncomfortable stays. He came up with brassiere sizes that evolved into the modern A, B, C and D cup bras, today's standard cup sizes. All the time he was insisting

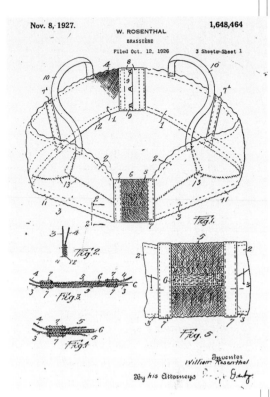

NOT SO SIMPLE: U.S. Patent N. 1,648,464 for Brassiere, November 8, 1927, suggests the intricacy of Maidenform designs. This bra was probably never manufactured, but parts of the design were incorporated in later bras.

on quality in a product made up of at least 20 different pieces, several no bigger than a quarter. "If Mr. Rosenthal didn't like a material," a staffer recalled, "he would say, 'It's too weak,' and then he would pick up a sample, hold it in his hands, and with his thumbs, break it apart. That ended the arguments."

Ida, too, could fret about two extra stitches that might make a bra uncomfortable, but her innovative gift was to have a bird's-eye view of the whole business—its future as well as its present intricacies. She measured every problem against her long-term goals, of which there were three: to produce a quality product; to make Maidenform a brand name all over the world; and to develop a corporate structure that would survive her. She succeeded on all three levels.

The two former revolutionaries from Russia knew all about the theories of the management innovator Frederick Taylor, and they acted. Picking up on the technique Masha Hammer had hit upon in her frantic

I dreamed I went shopping in my *maidenform* bra

I dreamed I was way out in my *maidenform* bra

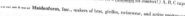

kitchen attempt to meet orders, they divided the making of a bra into a series of specific tasks, carried out time and motion studies, and invented new machinery for speed and testing. They were one of the first undergarment industries to use an assembly line and pay by piecework. Six months of negotiation, and lots of Ida charm, were required in 1937 to convince the International Ladies' Garment Workers Union that piecework would not harm workers or decrease jobs. It tripled output.

Ida was a masterful manager and saleswoman. In World War II she convinced the War Department that uplift bras would mean less fatigue for women in the WACs and WAVES and for nurses, so she got orders and priority shipments of scarce materials. One-third of Maidenform's production went to the war effort. She made pants, coats and shirts for the army, the only snag being the ribaldry provoked by the overeager stamping of MAIDENFORM BRA COMPANY on GI's shorts. The name was better kept in the public mind by the contract with the Army Signal Corps, who called up saying, "You people ought to know about designing something to hold a live, curved object. Can you make a carrier for a pigeon?" They could and did. The Maidenform carrier vest went into battle with paratroopers.

Despite all the diversions, Maidenform set a new production record in 1943 with the two millionth "Variation" bra. Through-out the war, Ida kept advertising Maidenform when other companies cut their budgets. She was vindicated when sales figures shot up once normal production was resumed. Maidenform had been one of the first soft goods businesses to go into advertising. The salespeople fought Ida; they wanted the money to buy floor space, but she had been insistent that advertising was vital to creating a brand name: To support their low-margin volumes, they needed to sell in the mass market. Widespread availability was crucial to this strategy; selling in a few exclusive stores would not do. Some 9,000 retailers had to be persuaded to carry a reasonable-size range of seven to eight Maidenform items, and they would only do that if people came in asking for the Maidenform brand.

Ida's gamble on advertising was an expensive dare that worked brilliantly. Ida had a conviction that Maidenform's future lay in repeat orders, so she continued old lines along with the new, unlike most other brassiere makers. In 1950 Maidenform was selling some items dating back to 1930, and old styles were nine-tenths of the business. But it was tricky—they had 15 items in production simultaneously, and so much money was needed to devise, ship and sell a new design that it took five years for Maidenform to make a profit on a new item. (Ida always waited a year before deciding whether a new line was a seller, which meant at least 300 sold in a week.) Then in 1950 they launched the bra that swept all before it. Dior's New Look of 1947 had emphasized the bosom, together with female curves, and the Chansonette was a response to the subsequent eagerness among women to accentuate the bust. Maidenform made more than a million Chansonettes in the first year and over 85 million by 1972 in 100 different combinations of size, cup, color and material.

Ida was selling more bras for more dollars than anyone in the United States, but she was also looking ahead. She had hoped that her son, Lewis, would succeed William; to her everlasting sorrow he died in 1930, and she could never bear to speak of it. But family remained a big part of the business. Ida brought her daughter, Beatrice, into the company as soon as she graduated from Barnard in 1938 and put her anonymously through all the jobs. And she recognized the potential of Beatrice's husband, Joseph Coleman, an unlikely piece of casting. Coleman was an ear-and-throat specialist who went to war as a captain in the medical corps. He was wooed by Ida and decided to give it a shot. He, too, was sat down at a sewing machine and worked his way up. Coleman pushed for taking on more salesmen, which meant cutting the territories of the existing staff. Ida was reluctant, believing it would harm the salesmen's livelihoods, and was just starting to give

I dreamed I played Cleopatra in my maidenform bra.

I dreamed I had a swinging time in my maidenform bra

chandising units in department stores. She returned to Russia as the only woman in the first American trade delegation, observing that the only bra available to Russian women was made of cloth, had no elastic and closed with buttons rather than hooks, with only four sizes, from small to "quite big"—a vast new market.

Ida never retired. In 1966 she was in Milwaukee persuading buyers to carry more Maidenform when she suffered a stroke that left her incapacitated until her death in 1973. In 1968, when Coleman died, Beatrice took over what her mother and father had begun. In 1980 she reported sales of $100 million for the family company.

The achievements of the Maidenform trio of Ida, William and Enid are manifold. They had the nerve to try and satisfy a perceived potential demand rather than a known market and had the management skills and stick-to-itiveness so often needed for innovation to succeed. Neither the original trio nor their successors were too proud to acknowledge, as armchair critics so often fail to allow, that every new product is a process. Ida told the *New York Post*, "We've made our bras more rounded, more pointed—oh, mama dear—like any new idea, the first airplane was not a jet." Ida, Enid, Masha and Beatrice proved that women executives could excel—at a time when most women in the workforce never thought of rising above the level of factory worker or secretary. They were democratizers. "Fashion," Ida told the *New York Times* in 1965, "is not for the few." The mass, high-quality-but-affordable output of Maidenform, coupled with advances in artificial fabrics, liberated millions of women, allowed them to find comfort and sensuality.

Maidenform was a very American innovation. The corset, so long associated with aristocratic pretensions, defined the human body as harmful and deficient, something man had to change and improve. William's 1926 patent for a brassiere to support the bust "in a natural position" was ultimately a Jeffersonian appeal to the natural order of things.

way when she and William went to see Arthur Miller's *Death of a Salesman*. She left the play distraught. "That finished it off," said her daughter, Beatrice, "but finally Joe Coleman convinced her." His idea worked, and Ida came to spend half her own time traveling to the markets.

It is curious, given Ida's vision and convictions, that she initially resisted Coleman again when he pushed the advertising campaign that was to take Maidenform to a whole new level. She was taken aback, as many people were in 1949, at the idea of photographing a young woman venturing into a public place wearing nothing above the waist except a Maidenform bra. The sauciness was complemented by a single line of copy linking a series of fantasy vignettes: I dreamed I did this and that in my Maidenform bra. It was conceived by a woman, Mary Fillius, at William Weintraub, and offered to Coleman only after it was turned down by a lingerie house that blushed and retreated, to its everlasting chagrin. The campaign was titillating, but it was also defiant. It was an assertion of a woman's right to do what she liked with her body, and it had direct appeal to women who felt frustrated going back to the routines of housewife, daughter and mother after they had tasted independence in the war.

The first ad in August 1949, featuring Allo-Ette in satin, was only one-third of a page, in black and white: "I dreamed I went shopping in my Maidenform bra." It was an immediate sensation. Coleman stuck with it in the furor. Top photographers such as Irving Penn and Richard Avedon, and designers such as Arnold Scaasi, Maximilian and Pierre Cardin, ensured that the series would maintain its wit and style. It entered the mainstream of American pop culture. (Parades invariably had someone doing a variation on the theme. Bing Crosby and Louis Armstrong spoofed the ads in the song "Dardanella" with the line, "She looks so dreamy in her Maidenform bra.") The campaign ran for 20 years and made Maidenform known round the world.

Fortune reported that, at 64, Ida Rosenthal was "as bright as a Christmas sparkler and as nicely rounded as a bagel." Maidenform had sales of $14 million, a tenth of all bra sales in America. Ten years later *Time* magazine estimated that 20 percent of all U.S. women—13 million—were wearing a Maidenform bra—an undergarment that shaped the contours of debutantes and matrons alike. The Rosenthals gave generously back to the community, believing philanthropy was as much a part of business as profits. After William died of a heart attack in 1958, Ida made Coleman president. She still worked nine to five, beginning every day by adding up the new orders to "see if the salesmen are working or playing golf." She pioneered self-service mer-

THE BLOW AGAINST THE CORSET

CARESSE CROSBY (1892–1970)

Late one night in 1913, as Ida Rosenthal crouched over her sewing machine in Hoboken, an elegant young lady in a rose-garland gown and matching roses in her hair danced with wild abandon at a high-society ball in Manhattan, captivating the blades and exciting the envy of her fellow debutantes. She had a secret. She was supple and alluring because she had abandoned the boxlike armor of whalebone and pink corsetry that from knee to armpit encased the other women under their silk-and-muslin glamour. In place of the corset, she had stretched two pocket handkerchiefs round her breasts, pulled taut by two pink ribbons stitched to the fabric and made fast at her back. She had invented the first modern brassiere to receive a patent.

Her name was Mary Phelps Jacobs, called Polly, and the inspiration struck her that rainy afternoon when she was getting ready for the ball with the help of her French maid, Marie. As they fussed with the headpiece of silk roses, the gown lay on the bed, and it reminded Polly of the last time she wore it, when the eyelet embroidery of her corset cover kept peeping embarrassingly through the roses around her bosom. Forty years later she recalled her thoughts with the bravura style that was to grace her life. "These [under]garments were forever having to be tucked or pinned and pushed out of sight once one was dressed for a party, and they were hellishly binding as well. If petting had been practiced in those days, it never could have gone very far, for even to get one's finger beneath the corset took a lot of wiggling. It was not a determination to ease this virgin state that motivated my invention, but rather a desire to move and sway and dance in comfort."

She announced to her startled maid, "I'm not going to wear that thing tonight. It spoils the entire effect."

"But, Mademoiselle cannot go without a *soutien-gorge*," Marie protested.

Polly had the maid fetch two pocket handkerchiefs, needle and thread, and pins. There and then, in front of the mirror, she pinned them together on the bias and knotted the ends around her waist. Marie stitched the ribbons to the two points so they could be tied to the back knot. Polly had her pull the ribbons tight, "the practice being to flatten down one's chest as much as possible so the truth that virgins had breasts should not be suspected."

She was 21, six years younger than the Russian immigrant across the river who, nine years later, was to make a fortune marketing the first uplift bra. Polly's bra was short and soft and divided the breasts naturally, but it did not offer uplift. She was a daring young woman. She loved the "nearly naked feeling" her bra gave her, without the bones, buttons and hooks of corsetry, but for all that, she noted, "In the glass I saw that I was flat and I was proper."

Her blow against the corset caught fire with her giddy circle. The young socialites came flocking to the dressing room afterward, so she gave them a peek and soon they were all wearing Polly's backless bra, as she called it. A lawyer encouraged her to seek patent protection. She thought so little of what she had done that she went off to Europe for a year with her parents and was amazed on return to find that the U.S. Patent Office, on November 3, 1914, had agreed that this was an original invention qualifying for protection. She rented a pair of sewing machines and secretly set up a little sweatshop with Marie as partner and two Italian girls doing the stitching. She went to see "Mr. Stern, Mr. Altman and Mr. Ginsberg," the department store titans of the city, and persuaded them to take a few dozen in glamorous packages. "It was such a revolutionary idea, the public might not accept it, they said. I felt rather like an anarchist."

Whether it was the conservatism of the women of the day, inertia in the stores, or lack of drive from Polly, she sold only a few hundred. She was more concerned with the arrangements for her marriage into the Peabody family, one of the most prominent in Massachusetts. When she bumped into the former Yale star quarterback Jack Field, who said he was making "jewel cases"—his jest for his job with Warner Brothers Corset Company in Bridgeport, Connecticut—she accepted his invitation to show Field's boss her samples. Warner straightaway offered her $1,500 for the sample and the patent. "To me this seemed not only adequate but magnificent. I signed on the line and went home in opulence." The best guess is that Warner made $15 million from her patent.

Ida Rosenthal would not have sold out so quickly. Polly's career in clothing was over, but not her innovating. She divorced Dick Peabody and married Harry Grew Crosby, a haunting figure of the '20s. He has been variously portrayed as the epitome of the Lost Generation, a tortured Gatsby in search of meaning in life and a rich brat in search of only his own pleasure; he was intermittently all of these, a spendthrift and also a disturbing writer. In Polly's vivid phrase, he was eclectic with rebellion. Two days after her September 9, 1922, marriage to Harry scandalized blue-blood Boston—she called it "the conquest of desire over obedience"—Mr. and Mrs. Crosby fled to Paris.

It was a time of fabled revelry. The woman who had gone corseted to balls in New York now dressed as an Incan princess for an art-student ball, riding bare-chested up the Champs-Elysee on a baby elephant "undraped and shameless." But she and Harry cultivated one of the most artistic networks of the 20th century. Anaïs Nin, a close friend, described her as a "pollen carrier who encouraged artistic and creative copulation in all its forms and expressions, who trailed behind her like a plume of peacocks a colorful and fabulous legend." In a house decorated with tiger skins and stuffed parrots, the couple entertained a catholic mix of friends: stars like Douglas Fairbanks and Mary Pickford, surrealist Salvador Dalí, and the expatriate literary glitterati—Edith Wharton, Ernest Hemingway, Malcolm Cowley, Archibald McLeish, Henry Miller, Hart Crane, T. S. Eliot and Ezra Pound. Harry gave Henri Cartier-Bresson his first camera. But they made something out of all these associations. Harry began to experiment with poetry and photography; when Polly, too, started writing, she said she didn't like the sound of Polly Crosby, so they came up with Caresse. In 1927 they started the Black Sun Press. Caresse—as we must now call her—came up with the idea of publishing paperback editions of young expatriate Americans and avant-garde European authors. This was an innovative venture. The pioneer of original paperback publishing, Sir Allen Lane in England, did not start his Penguin Press until 1935: The Random House Modern Library, begun in 1917, was all hardback.

Crosby Continental Editions successfully published Hemingway, Lawrence, Joyce, Faulkner, Antoine de Saint-Exupéry, Dorothy Parker, Raymond Radiguet and Robert McAlmon, and titles in French, German and Swedish for those countries at the equivalent of 25 to 50 cents. Caresse reasoned that paying $2 more for a hardback was a deterrent to reading. She came back to New York in 1932 to sell the idea to Dick Simon of Simon and Schuster and Bennett Cerf at Random House and to the Doubleday and Rinehart families. They all said the American public would not buy paperbacks. They were wrong and she was right, but she did not have the gift of patience. Disappointed by the timidity of American publishers, she discontinued the line in 1933.

Caresse's marriage ended with gunshots in December 1929. Harry, who had insisted they both have other lovers without concealment, was found lying dead on a bed with Josephine Noyes Bigelow at New York's Savoy Plaza Hotel. He had apparently shot her several hours before he shot himself in a suicide pact.

Caresse established a salon in a Virginian mansion built by Jefferson. She introduced surrealism and Salvador Dalí to America. She helped Miller edit *The Colossus of Maroussi*. In World War II, she opened an art gallery in Washington; billed as "the first modern-art gallery in this city," it exhibited

CHIC SURREAL: The purebred Hereford bull sprawled on the carpet was at Hampton Manor at the invitation of Salvador Dalí. The hostess, Mrs. Phelps Crosby, as she was known, is at the typewriter; Dalí is against the bookcase, writing, while his wife collates notes for him.

works by black artists. Just at the end of the war, she was off again on another adventure, launching a magazine called *Portfolio: An International Quarterly*, published in the United States, with alternating issues in Europe. It was a tour de force. She had contributions from Henry Miller, Gwendolyn Brooks, Kay Boyle, Henry Moore, Max Ernst and Man Ray. Picasso sent portraits. Albert Camus's appearance in *Portfolio* was his first in any English language magazine. Cartier-Bresson captured a decisive moment in its pages. Crosby published Tolstoy's last work, *The Law of Love and the Law of Violence*, never before published in English.

And then in 1948 she abruptly ended publication.

Caresse Crosby lacked the essential persistence of Ida Rosenthal, but she had insight into the nature of creative risk taking. Of her father, a passionate idealist always agog with wild business and social and real estate projects, she wrote: "I do not believe my mother ever understood the gossamer mantle in which the visionary must wrap himself for protection, a garment that can be as strong as faith or evanescent as a dream."

She wore her gossamer mantle with iridescent flair.

SAMUEL INSULL He was the sorcerer's apprentice who realized the magician's dream: to serve all America with cheap electricity

1859-1938

First Federal Trial, October–November 1934, Transcript

"Objection! Relevance?"
—U.S. attorney general Leslie Salter
"Overruled!"
—Judge James Wilkerson

The jury in the Chicago courtroom glared at the prosecuting attorney general. They were captivated by the story the frail old defendant on the witness stand had begun to tell. Even the judge seemed hooked. The prosecutor soon realized his attempts to cut off the testimony meant he risked losing the jury. Besides, he himself was growing intrigued by the defendant's unfolding story: how he had come to America as a young man; run Thomas Edison's businesses; ventured on his own to make electric power available cheaply to the masses; found himself enmeshed in one of the most notorious financial sensations of the day; and been kidnapped in Turkey while a fugitive from U.S. justice and brought forcibly back to America to stand trial. Perhaps it would be best if the objection were overruled.

The old man's story first took the jurors back to a New York dockside at nightfall on February 17, 1881.

The very proper Englishman soon after his arrival

Samuel Insull, a skinny shortsighted immigrant from England, pallid from eight days of seasickness, was 21 when he stepped nervously ashore from the *SS City of Chester.* Waiting for him on the pier was the only person he knew in the whole of America, Edward Johnson, Thomas Edison's chief engineer, who had himself just returned from setting up the first telephone exchange in London. Young Insull had been the first operator of that initial exchange, and Johnson had been impressed by his quick mind, his dawn-to-dusk organizing industry, his ability to write shorthand and his surprisingly encyclopedic knowledge of Edison. In the phrase of Robert Conot, young Sammy Insull had the demeanor of a shop clerk, but it was Johnson's hunch that he had the potential to bring some order to the life of the famous young genius, then hectically in the throes of developing an entire electrical network to exploit his newly invented incandescent bulb. Gaslight prevailed in streets and homes, steam in factories. Some public areas had begun to be illuminated by the electric arc lamps used in lighthouses, but the light was too blindingly bright for anything else.

Edison, at 34, and Insull had the same reaction at their first meeting at Edison's Fifth Avenue headquarters that February night: "My God, he's so young!" Insull, less than average height, did

Said Insull, "Here is an industry, which supplies convenience and comforts to the day laborer, which kings could not command half a century ago."

not make much of an impression on first sight. He was the very proper, stiff-white-collar son of struggling lower-middle-class parents in London, with a dreamy lay-preacher father in and out of work and a capable mother who shared her husband's lifelong conviction that drink was the devil's brew. Insull was too polite to admit it, but the bubbly Edison's Midwest torrent left few recognizable words in its wake, and for his part the hard-of-hearing inventor found Insull's cockney speech another of nature's challenging mysteries. Edison later recalled, "I thought I had made a hell of a mistake," then added, "but the next day I knew there was no mistake about it."

Insull showed an uncommon grasp of finance. At that first meeting, on the top floor of the brownstone at 65 Fifth Avenue, Edison fished a checkbook from his rumpled clothing, showed Insull the $78,000 on deposit in New York, and told him he needed as much as another $150,000. His American investors, led by JP Morgan, were balking at funding the three factories Edison needed to open to make dynamos, lamps and underground conductors for the world's first central power station on Pearl Street. How close to the $150,000 could he get by selling his European telephone stock? Where in Europe should Johnson go to get money? Insull, as secretary to Johnson, had studied Edison's European telephone operations and memorized every detail of their finances. By 4 or 5 a.m. he had gone through Edison's books, sorted the numbers arrayed in his mind and reported that by modifying a contract here and adjusting a clause there he could raise the full $150,000 without any need to sell stock. "Mr. Edison," he proudly related many years later, "got every cent in Europe I said he could." Insull dozed for a while as dawn broke, as Edison so often did, then Edison took him to look over a run-down building in lower Manhattan, where he fancied he might manufacture dynamos. It was at grimy 104–106 Goerck Street, swiftly and well acquired by Insull, that Edison, in the words of Edison biographer Matthew Josephson, planted the acorn out of which grew the vast oak of the General Electric Company.

Edison, with the classic Midwestern view of the English as effete snobs, was amazed by Insull. For the first time in his life he had met someone who worked harder than he did. Insull was at it 16 hours a day, apparently imperturbable on a dizzying variety of tasks. In England, he had been a dogged self-improver after family hardship forced him into work at 14. He taught himself accounting, keeping imaginary double-entry ledgers. He practiced shorthand by taking down sermons in Westminster Abbey and got glimpses of a more wicked world transcribing dictation late at night from Thomas G. Bowles, founder of England's irreverent satirical monthly *Vanity Fair*. He cultivated his memory by prodigious reading; he read while he shaved and he read while he walked from one appointment to the next. On cycling expeditions he made a point of memorizing the location of every road and business in the city and surrounding counties. But the robot had a heart. He lost himself in opera and literature. It was as an enthusiastic member of a literary society that he had formed his obsession with Edison. He read an article on Edison in *Scribner's Monthly*, followed every footnote to its source and delivered an uplifting lecture on the great inventor to his friends.

Edison called Insull "as tireless as the tides." It was Edison's habit, once his judgment was formed, to give unflinching trust to an associate. The awestruck Insull was startled to find himself not simply taking notes and running errands but in control of all the manifold business affairs of the prolific Edison while the inventor was off in his small laboratory at Fifth Avenue, or in a trench at Pearl Street reassuring the Irish laborers laying the underground cables. He gave Insull power of attorney and Insull did everything with utter devotion to his hero. He woke him up, wrote and signed his letters, made sure he ate, called his meetings, bought his clothes, coped with the stream of visitors, bankers and crazies, actresses and scientists, journalists and beggars. He negotiated real estate for central power stations in a dozen cities. He organized the credit and construction for more than a thousand "baby Edisons": isolated power plants in department stores, hotels, factories. He implemented Edison's dictum never to pay cash if you could buy on credit. (When he bought himself a $20 suit, he insisted on having a year to pay.) Whenever Edison wanted something—hundreds of thousands of dollars or his umbrella—he looked to the young Victorian enabler. In short, Insull was a devoted slave—but the quickest of studies, too. Like Edison, he knew what he didn't know. He made it his business to learn all there was to learn about the emerging electric power industry. For relaxation, he pondered a crucial factor in Edison's business plan, the numbers who might be induced to switch from gas to electricity.

Insull adapted quickly to his new society. The easygoing Americans around Edison teased him about his natty clothes and formal manner, neither of which combined well with his curled upper lip, pince-nez and bulging brown eyes that made people think he was sneering. Insull refused to give up his formal clothes—years later he went to jail wearing spats—but he shaved off his long sideburns and grew a mustache to conceal his upper lip. Out of respect for his mother, he declined to join anyone in a drink, but he camouflaged his churchly Englishness with a few Yankee profanities. He became an American citizen, and when he was 39 he married an American stage star, a tiny vivacious actress called Gladys Wallis, who succumbed to the passion in his letters more readily than she did in his embrace. In truth, Insull was made for America. The intoxication he freely imbibed was the rough equality of status, the expectation that small men from nowhere might do great things. In the hierarchy of English society, the Insulls were nobodies, their accents matters for disdain, their positions immutably fixed. (It was a reflection of Sammy Insull's insecurities that in midlife he would tell his only son, Samuel Insull Jr., he could go to any college except Harvard. He thought Harvard men snubbed him because of his lack of formal education.) At all times he yearned for respect. He became a model employer of thousands, but he would fire on the spot any worker who did not recognize him on his peregrinations and greet him with "Good morning, Mr. Insull."

In 1886, five years into their association, Edison entrusted Insull with establishing the Edison Machine Works in an old locomotive factory Insull had discovered way upstate in Schenectady. He told his protégé: "Do it big, Sammy. Make it either a big success or a big failure. Just do something. Make it go." He did, and he absorbed the advice in his bone marrow. "I knew little or nothing about the manufacturing business," he wrote, but over five years he grew the company from 200 to 6,000 workers, while making a return on investment of 30 percent. He was not an engineer, still less a scientist, but he had a genius for synthesis and he relished the exercise of authority in putting men and machines together. The historian James Tobin puts it neatly: "Inside this small nose-to-grindstone clerk there was a Napoleon." Inside his Schenectady factory was another acorn of General Electric and the behemoth's future headquarters.

Insull loved everything about his job in Schenectady except begging banks for capital to expand, especially going down to New York to pitch the JP Morgan group, where an oversensitive ear would catch echoes of the English class system. The experience, says his principal biographer Forrest McDonald, was "frantic, nerveracking and disgusting for him."

Anyone or anything that posed a threat to Edison instantly drew Insull's ire. In 1884, when he had perceived the Morgan chairman of Edison Electric Light Company to be all gas and no heat, Insull had secretly gathered proxy ballots to force him out and keep the direction of the company firmly under Edison. "There is no one more anxious after wealth than Samuel Insull," he wrote a friend about his coup, "but there are times when revenge is sweeter than money." But up in Schenectady Insull came to have a more shrewd appreciation of his master. "We never made a dollar," he reflected, "until we got the factory 180 miles away from Mr. Edison." He thus saw the point of the two steps that finally distanced Edison from operational control: first in 1889 the joint venture with Germany's Werner von Siemens, effected by the financial maverick Henry Villard, in which Edison's companies were absorbed into Edison General Electric, and then in 1891–92, Morgan's master stroke merging Edison General Electric with its competitor, the Thomson-Houston Company.

KINGPINS: The 50 most powerful men in the electrical industry, including Thomas Edison, threw Insull a farewell dinner at Delmonico's in 1892. There was a certain pity for a young man of 33 setting off for what they saw as the backwater of Chicago.

With that merger, Edison's name was dropped from the new colossus, the 50-million-dollar General Electric Company. Edison was sore about that and unhappy with Insull, but there was not much Insull could have done to protect Edison's wounded pride—Edison wanted the money anyway for other projects. Nevertheless, in light of Edison's feelings and his own ambitions, Insull decided to resign his senior vice presidency with General Electric, and with it the manufacture of electrical equipment, for the vastly different task of generating and distributing electrical power to the public. In March 1892 he chose to head for Chicago as president of a small generating company, Chicago Edison. It had a franchise to operate the Edison system, but it was a separate company, locally owned and run and just one of the windy city's 25 electric companies, with only 5,000 customers in a population of a million. The big-money men and electricity moguls in New York were rather condescending in their toasts to Insull at a farewell dinner at Delmonico's—"most of my intimate friends and intimate enemies," he called them in responding. In Chicago they told him he could never hope to have more than

25,000 customers. His company was capitalized at $885,000, more than 50 times smaller than GE. The salary of $12,000 he was accepting was only a third of that of the vice presidency at GE. Perhaps Insull was not so bright after all.

In fact, Insull, at 37, had bigger ideas than anyone appreciated. The Delmonico diners regarded it as an after-dinner witticism when he forecast that one day little Chicago Edison would exceed the worth of General Electric. What the sorcerer's apprentice had in mind, with Chicago as the showplace, was making bigger magic than the wizard. Edison's original conception was to deliver electricity to inner cities by a network of what were essentially local power stations. Insull longed to empower the whole population—metropolitan areas, suburbs and even rural areas. That was not the way the system was evolving in Chicago—or anywhere else—when he took over in July 1892. Electricity was too expensive, trapped in a straitjacket of technology, geography and finance, confining it to business districts, swell restaurants, department stores and the homes of the affluent. Even there, electric lights were turned on in the parlor only for guests, and

INSPIRATION: Newly arrived Insull walked the brilliant grounds of the Columbian Exposition in Chicago with one idea on his mind: This electricity must become a universal blessing.

when they had departed the household reverted to gas in the popular dual gas/electric chandeliers. Nobody was really thinking through just how fantastic the demand for electricity would be in everyday life because nobody could see a way to make it affordable.

In his demonic quest to break electricity's shackles of low expectations, Insull was not an instant Houdini. It took him seven years to conceive his far-reaching permutation of technology and marketing, 30 years to create the architecture of energy on which modern life is built, but he straight away came at the problem from the right end: mass production for a mass society. Large-scale generation of electricity spread the burden of the capital investment, reducing unit costs. In only 60 years, the city had expanded mightily, becoming the nation's key railroad junction, lumberyard and animal slaughterhouse for meat, thus making a vast Midwestern territory its empire. But electricity generation was split among 25 small central stations mostly

operated by "shoestring" utility companies and 500 establishments who made their own juice using isolated power plants sold to them by Insull in his days with Edison. Chicago Edison itself had three plants. The original one that Insull inherited at West Adams Street in the Loop, the heart of the business district, was so limited that it was overloaded just from lighting 50,000 bulbs in commercial businesses adventurous enough to light their offices electrically. It was, said one of the workers, "a Dante's Inferno . . . half-naked firemen, shoveling coal with feverish energy, made one feel as if an explosion might furnish a climax at any moment."

As a condition of accepting the presidency, Insull had insisted that the local directors float a $250,000 bond to finance a new power station. There turned out to be only one buyer for the bonds: Insull himself. (Marshall Field, the store developer, was so impressed by the vitality of the young immigrant applying for citizenship that he lent him the money.) Insull

appointed as his engineer a fellow Englishman, Fred Sargent, and the two of them sped the design and construction of a new power station in a disused railway yard at Harrison Street, along the Chicago River and thus accessible to coal barges. This was to be not any old power station but easily the biggest in the world, three times the capacity of West Adams Street. Throughout his entire career Insull was to follow this formula: Size equals power equals low unit cost. Refusing to leave the initiative to General Electric, he was forever badgering his old company for bigger and bigger equipment. Tomorrow was not just another day; it was today multiplied by ten.

A few months after Insull tied his destiny to Chicago, he walked into the future he intended to claim: a gleaming white city erected on the shoreline of Lake Michigan, its classical columns and portals "defined with celestial fire" by 93,000 incandescent lights. Twenty-two million Americans, 20 percent of the population, were drawn into the pavilions and courtyards of the World's

Columbian Exposition of 1893, one of the most stunning examples ever of something designed for fun blasting the way to a revolution in everyday life. Electricity was only part of the statement the city's leaders wanted to make about the vitality of their city—only 20 years before leveled by a famous fire—but it was a glimpse of the brilliant new world universal electricity might create that most excited the throngs. Chicago Edison had little to do with the expo; the current was supplied by Westinghouse Electric, the burgeoning new company set up in 1886 by George Westinghouse. But it was the effect that mattered. Insull was just one among the crowds entranced by the show: a fast elevated railway; a Ferris wheel taking 1,340 lights into the sky; 10,000 colored Edison bulbs in General Electric's 70-foot Tower of Light; a movable sidewalk; the swiftest of elevators; dazzling spotlights and fountains; speedboats; dream kitchens and workshops and the indispensable ladies' massaging corset: all powered by electricity.

It was all very exciting, but how was this wonderful genie to be safely and economically dispatched to streets and factories and millions of homes over areas much vaster than the 600-odd acres of the White City? Nobody had much of a clue. For Insull, the expo was both inspiring and frustrating. At Harrison Street, he was installing Edison's system of direct current (DC), which was limited in its range, so electrification of cities was by a patchwork of numerous small "backyard" power plants, all less efficient than a large plant at generating energy more cheaply than gas. Westinghouse, Edison's hated rival in electrical distribution, had lit the expo with the new technology of alternating current (AC). The superiority of AC was that it could be stepped up to thousands of volts, and such high-voltages could be transmitted very long distances cheaply and efficiently. The downside, dramatized by Edison in the "war of the currents," was safety; workmen repairing high-voltage cables were not infrequently electrocuted.

Hardly a blown fuse, however, interrupted Westinghouse's triumph in Chicago. Insull was determined to abandon his mentor's technology and embrace the hated rival's as soon as he could. Edison's spell

had been broken, he said later. Even at Schenectady, Insull had argued in favor of making dynamos to transmit AC. At the expo, he observed that Westinghouse used a new invention to convert his alternating current to the direct current necessary to drive the fair's elevated railway. Could he not do that in reverse, turn his DC into AC for long-distance transmission and then convert it back again to DC for home use? One of his young lieutenants, the MIT-trained Louis Ferguson, assured Insull that he could. Insull promptly ordered two of the new rotary converters, and Chicago Edison was the first American utility to install them in a commercial system. From August 1898, only five years after the exposition, Harrison Street was able to transmit electricity at 2,300 volts over a distance, and then convert it back at a substation to the safer, low-voltage DC fed into offices and homes on the Near South Side.

There was a more perplexing issue than technology: the economics of consumption. It had cost the exposition more than a million dollars to build the steam plant and dynamos. To generate electricity for a big city and then convey it along miles of cable was such an extraordinary expense that it was obvious to most people that when the Columbian lights were turned off, electricity would remain a luxury. Electricity was so little available in the rest of Chicago in 1893 that the entire city's consumption was only a third of the exposition's. It was a similar situation in other big cities. After its debut in New York under Edison, wiring went at a snail's pace and did not penetrate beyond the better-off districts.

Catering for the few, and charging what the traffic would bear, was how the power producers in America—and in London and Berlin—saw the future. It is Insull's singular contribution that he rebelled against this conception as both elitist and defeatist. He was a native-born Englishman but very American in aspiration. Fired anew by the exposition, he thought in the democratic spirit of illuminating every urban home, moving crowds swiftly through cities, empowering the farmer and his family in the twilight of the countryside. But how? Edison had once said, "We will make electric light so cheap that only the rich will

burn candles," but he had moved on to other excitements before getting anywhere close to doing that. The gas jet was an inferior and riskier light source, but it was much cheaper. Hotels, offices, stores and restaurants wishing to advertise their modernity by using electric light could generate their own by attaching dynamos to their steam engines. When Chicago's streetcar companies began retiring their horses in favor of electricity in 1892, they acquired their own generator. The more electricity such companies used, the more attractive it was for them to make their own juice, the less viable the central stations, and therefore the more expensive electricity was for everyone else.

Consolidation was Insull's first answer. Between 1893 and 1898 he bought up all the competing central stations in the Loop. It helped that before arriving in the city he had cunningly arranged with General Electric that it would give him exclusive right to the purchase of electrical equipment within the Loop. Still, the arithmetic was daunting. Electricity was a product unlike any other. It had to be manufactured, transported and consumed simultaneously. It could not be warehoused for future sale, like toasters and automobiles. What was not consumed, perished. Biographer McDonald calculates that Chicago Edison had invested about $1 million in Harrison Street to be able to generate a direct current total of 2,800 kilowatts. This translated at the bargain cost of 2 cents a kilowatt-hour, but the big new station operated at only about one-sixth capacity and the rest of the current was wasted. The true capital cost therefore was 12 cents, about the retail price of gas, and then electricity was priced still higher to cover operating costs. Shutting down part of the plant was no answer. Steam generators were mighty beasts, not built for tap-dancing to the variable rhythms of demand. They required hours to start up and then had to be kept running at full blast to cope with peak demand. Nothing could be worse than a sudden shutdown (as 50 million Americans would vouchsafe in 2003). The conventional wisdom of the industry was to grow only very slowly, if at all, since every new customer had to be supported by a risky increase in fixed costs. Insull adopted the opposite strategy: Grow as fast as possible, then find ways to sell every watt. Harold Platt calls it the "gospel of consumption." Insull was not the first in this. The philosophy of the railway builders was "Run the line and the settlers will come," but it took an extraordinary amount of ingenuity, as well as courage, to make the strategy work in electricity, where gas was a much more serious challenge than ever steamboat and coach were to rail.

Insull's first radical step was to cut prices below the estimated costs of production. When the newly built Great Northern Hotel said it would buy electricity but only, as Insull put it, at "a ridiculously low" price, his answer was to accept the price on condition that the hotel signed a long-term contract. His entry in his memoir is self-regarding but true: "The Great Northern Hotel contract was much criticized by managers of other central station companies in different parts of the country who had not the courage to cut so deeply. They were unwilling to take risks in trying to develop a real knowledge of the economic conditions governing the business." A number of big businesses signed up for the Great Northern deal.

But was Chicago Edison losing money on the extra cut-rate sales? Insull was finding out that to industry observers, he looked reckless, and in a sense he was, suspended in midflight by the gossamer thread of his faith in expansion. Given the imperative of keeping a plant at pretty well full stretch, what was the optimum rate needed to avoid bankruptcy while maintaining volume from a variety of customers making unpredictable demands? Insull had not figured it out when he took in the sea breezes in Brighton on an 1894 Christmas vacation in England. One lonely evening—he was as yet unmarried—he saw something he had never beheld in America: The shops were shut but every shop light was on. That would have been a mark of profligacy in most cities in Europe and America, so Insull tracked down the young head of the electric supply owned by the township. Arthur Wright was pleased to say the reason the shops could leave their lights on was a consequence of two inventions of his: first, a demand meter that measured not only the commonplace of total energy con-

sumption but also levels of demand through the day, and second, a rate structure in which he separated the cost of serving a customer into fixed and operating costs. His demand meter told him that even at the peak the shopkeepers in the antiques district imposed the tiniest of burdens on the utility's fixed costs; since it cost very little to supply their maximum demand, the rate charged to them could be low and still yield a profit to the utility.

Insull returned excitedly to Chicago with a demand meter and his head full of schemes to elaborate on Wright's concept. He did that brilliantly over the next five years, permutating rates and flow. He could do the numbers. He had no head for abstract mathematics, but quantitative arithmetic he ate for breakfast. He had seized on the deceptively simple secret of making electricity both profitable and a public service: Rate making was the key. Viability was based not on load—the total amount of electricity sold—but on what became called *the load factor.* The load factor was the percent of the system's capacity being sold at any given moment. The higher the load factor, the more the Harrison Street station maximized the use of its equipment and investment and the lower the unit cost of its electricity. He drew the load curve as a graph. It gave him a picture of peaks and valleys of demand for power: Chicago, asleep from midnight to 6 a.m. (using only 10,000 kilowatts); turning on the lights, rising and rushing to work on the streetcars at 8 to 9 a.m. (46,000 kilowatts); slackening off at lunchtime (36,000 kilowatts); going home and using appliances from 4 to 8 p.m. (46,000 kilowatts); switching off at midnight (18,000 kilowatts). On a dark and cold winter's day, demand might call for more than 90,000 kilowatts.

Insull's graphs highlighted how big an opportunity rested in those low-use hours. They clearly defined his management's task: Aggressively find customers whose demand cycles would fill the valleys. The more diverse the users were in their habits, and hence in the timing of their call on services, the more of them could be served by the same amount of capital investment. The higher *the diversity factor,* the greater the profit. He abandoned the flat-rate model of Edison and all the other electric

and gas suppliers, and from 1897 introduced two-tier billing. Householders who put little strain on the system found their bills reduced by 32 percent in 1898. He secured thousands of small users by offering to wire six lighting outlets free of charge in new or older houses. *Who would he not sign up?* one querulous industry insider asked. Insull's unhesitating response was that he would light a single customer's single 25-watt bulb if one was ever made that small. More and more residential areas signed up and used more and more power and paid less and less per unit. Private generating systems became uneconomic. Street railroads made their own power and sold no less than 47.4 percent of all electricity in the United States in 1902, the year before Insull changed the game. He astonished the Chicago companies by quoting rates far, far below their own costs. They grabbed the offer while it lasted: How on earth, they wondered, could he afford to virtually give it away? He could afford it because filling the valleys so dramatically drove down the unit cost of electricity—to public benefit and private profit. "Is it too much to predict," he asked at the turn of the century, "that in far less time than the succeeding twenty years electricity for all purposes will be within the reach of the smallest householder and the poorest citizen?"

Insull's success in selling electricity over wider and wider areas was so spectacular that within three years of his two-tier billing, his mighty Harrison Street plant, expanded to its limits, was running hot and so were all the other power stations he had bought up and brought online. He pressed his engineers for innovation. "The very best monument that any of you can erect," Insull told his managers, "is a first-class junk pile." He certainly regarded the reciprocating steam engines proudly installed in 1894 as junk. Their up-and-down pounding motion shook the building; doubling the size, as he wanted, would take up the entire area inside Chicago's downtown Loop and might well imperil the building. Through his European connections, Insull had heard that Sir Charles

BIGGER AND BIGGER: Insull's first 5,000-kilowatt turbine set in test, the largest ever at the time

Parsons, the British engineer, had made headway building smooth rotary turbine engines for racing boats. His staff told him large turbines were not feasible. He called on Charles Coffin, the president of General Electric. Same answer. (GE engineers, Insull liked to say, could prove anything impossible.) Insull bore down hard on Coffin for a 5,000-kilowatt generator, and GE finally took the risk of making one on condition that Insull took the risk of installation and start-up.

Seventeen months later, in October 1903, the world's largest turbine was started up in Chicago Edison's new Fisk Street station—and shut down at once. It shook so wildly that everyone was terrified. Visiting dignitaries were asked to step behind shelter before Fred Sargent, the engineer, tried again. He noticed Insull was still standing next to him. "Please take shelter," he told his boss. "This is a dangerous business." Insull replied, "Then why don't *you* leave?" Sargent pressed, "Look, Mr. Insull, this is my job. I have to stay here but you don't. Don't you understand, this damned thing might blow up."

General Electric Company, Schenectady, New York.

HOW LONG SHOULD A WIFE LIVE: This was one of Insull's campaigns for more electrical appliances in the home. He propagated the theme in his *Electric City Magazine* and advertisements along the lines of this General Electric photograph of a maid happily using an electric iron. He offered trials of free electric irons in exchange for old flatirons (2,000 in the pile at right).

Insull looked from Sargent to the turbine. "Well," he said, "if it blows up, I blow up with it anyway."

It did not explode, and, adjusted to run smoothly, Insull's large turbines, up to 12,000 kilowatts by 1905 and 35,000 by 1912, revolutionized the industry once again. They sharpened a downward spiral of prices that continued into the '60s, doubling the use of electricity every ten years for seven decades. How did Insull know that giant turbines could be built? He just did and was willing to bet his life on it, an inspired self-confidence characteristic of most great innovators. But the turbines were justified only by ever-expanding sales. Insull was the first utility operator in the United States, says the historian of electric power Harold Platt, to recognize that marketing strategy was more important than production technology. Insull was as ardent a missionary as he was a manager.

Insull's concept of load and diversity factors, fairly soon adapted by most American utilities, was the single most significant innovation in the single most important technological advance of the 20th century, the electrification of the continent. It justified Insull's belief in the democratization of electricity since more customers were in the end cheaper than fewer. But it also justified monopoly: Two power stations in the same marketplace would always provide more expensive electricity than one because they would both fail to maximize output. It is piquant that Insull's solution to the dilemma of the central power station was to marry his democratic social beliefs with the democratic bogey of monopoly. Here was the essential Insull paradox: He kept buying up competitors to achieve monopoly power, not to be able to charge higher prices but to be able to cut prices. He excoriated competition by individuals or municipalities as especially wasteful in electricity generation. "I know of no greater financial crime," he wrote, "than to spread along the streets of any city, whether it be Chicago or Springfield or Louisville or St. Louis, investment duplicated to afford exactly the same class of service. It is not possible for that to be right or economical."

He called what he sought "massing production," a phrase he coined before Ford popularized the concept of mass production. But Insull was not a typical robber baron of the Gilded Age—and nothing like the freebooting Enron-style traders of the 1990s, who exploited deregulation to buy up energy and charge what the market could be made to bear. He argued persistently for public regulation: "No monop-

METER MAIDS: They came to the door when you signed up with Insull's Minneapolis General Electric, 1918.

oly should be trusted to run itself." His campaign in favor of public regulation surfed the tide of the Progressive movement and found fulfillment in 1912 with the formation of the State Public Utility Commission of Illinois. It set a model for subsequent federal legislation protecting customers while assuring vertically integrated utilities a fair return on capital so that generating and transmitting facilities assuredly kept ahead of rising demand—a model that was abandoned in the deregulation experiments of the 1980s and '90s with very uneven consequences, resulting in lower rates in some states but also, in 2003, the most extensive blackout in the country's history.

Insull not only talked about the future, writes Platt, but also took practical steps to get there as soon as possible. Over the early decades of the 20th century, that meant wiring more and more customers. As the secretary of a literary society in his London youth, Insull had been able to persuade that emperor of persuaders, P. T. Barnum, to address his group on publicity technique. He infused his electricity campaigns with the spirit of Barnum. He sent more than a hundred salesmen on thousands of door-to-door calls carrying free electric irons for a six-month trial. He gave easy terms for wiring a house. He opened a chain of shops. Through advertising and a magazine, *Electric City,* he glamorized the life of an electric household whose sewing machines, ovens, heaters, fans, marshmallow toasters, vacuum cleaners and baby rockers, and later radios and refrigerators, would run on his cheap electric power. "How long should a wife live?" asked one of his advertisements. The home of the future would lay all the burdens on the shoulders of electrical machines so that mothers of the future would live to a good old age and keep their youth and beauty. His guide to salesmen extolled the psychology of envy: "In interviewing the lady of the house," they were told, "it is a good plan to mention what her neighbors are doing, and so play upon her social pride, insinuating in a delicate way that if they can afford it, she can. Explain how Miss so-and-so has now a lovely kettle for her afternoon tea, and declare, 'She could not live without one.'"

By 1907 his network of central and substations generated all the electricity in Chicago and beyond. In 1911 he merged 39 gas and electric companies to form the Public Service Company of Northern Illinois (PSCN), covering 6,000 square miles. He exchanged energy over high-tension transmission lines from Milwaukee in the North to the Mississippi in the West; Michigan and South Bend, Indiana, in the East and Southeast; and downstate Illinois in the South. It was one of the greatest pools of power in the world, all achieved without state or federal government planning and solely by the driving energy and vision of one man. And it was cheap. Within four years of starting Public Service, Insull had halved prices in the area and extended service to more than 100,000 new customers. His customers got their electricity at one-third the prices routinely paid by the consumers in New York, Boston, Milwaukee, Philadelphia and Baltimore. Nowhere else in the world was electricity this cheap. It was a major factor in building the muscle of Midwest industry in a country already expanding at a phenomenal rate. In 1916, the last summer of peace, Americans were enjoying an astonishing 13 percent increase in purchasing power over just two years. Within a few years of the end of World War I, the United States was producing more than the other six Great Powers taken together. Every development—radio, refrigeration and air-conditioning—spurred the growth of Insull's companies and ever-cheaper power. In 1933, for the first time, electric appliances in the home used more electricity than lighting: 50,000 Chicagoans bought refrigerators that year.

Insull carried his ideas of public service through supervised monopoly into transport. When all the city's elevated railways went broke in 1914, he took them over and employed the diversity factor to reduce and rationalize fares; he redecorated the stations, put in new coaches, enforced a universal transfer privilege, insisted on courtesy from employees and enjoyed good relations with the unions. Then he extended commuter lines into the suburbs to challenge steam.

Insull became a millionaire when he was 47, and his total annual income from dividends and salaries was $100,000. He bought a yacht—less as a nautical adventure than a social gesture, having in mind JP Morgan's dictum, "You can do business with anyone, but you can only sail a boat with a gentleman." He bred horses on a 4,000-acre farm at Libertyville, 38 miles outside the city, a pleasant place to relax with Gladys and seven-year-old Junior, but here again his obsessions intruded. Nobody more than a mile from town had electricity. The crusader for central power stations choked at installing a private plant; true to his faith that power should not be a privilege, he started wiring the nearby countryside to much ridicule from his peers. No bank would lend a cent for electricity for farmers. An attempt at full-scale wiring of rural areas, where most people still lived, had to await the New Deal, but as McDonald says, Insull was alone in taking the first steps to bring electricity to every village and farm in every corner of the land.

He had one failure in this period. He kept in touch with Edison all these years and eagerly followed his painstaking development of a storage battery for electric runabouts and light trucks. Edison wrote his protégé that it promised "to add many electric Pigs to your Big Electric Show." Insull tried for years to get Chicagoans enthused. He sank his own money into building garages where electric cars could be recharged. It was ideal business for a central station, recharging one car at profitable off-peak hours being equal to 20 residential customers. But Edison's friend, Henry Ford, won that one.

Around this time, "the Chief," as Insull was now called by everyone in the industry, lost interest in earning money. He found it more satisfying to give it away. He hated the idea of competitive giving, preferring to make anonymous gifts, maybe not more than $2,000 at a time to a widow, poor children, a proofreader gone blind, or books to workers to help them improve themselves the way he and Edison had done in their moldy digs. He was a soft touch for struggling actors, especially English indigents who had served king and country, and drunks who swore they would never touch another drop. He named things after Edison; one of the few places where the name Insull appeared was in a wing of the London Temperance Hospital in honor of his parents. He was passionate to make Chicago an even greater city. He was the leading patron of the arts, and his grandest gesture was to put $2 million into the Chicago Opera. He subsidized the splendid Chicago Civic Opera Building, still an ornament in Chicago. The teenager who, in London, had gone without supper to buy a seat in the Gods (the upper reaches of the theater) longed to make opera accessible to the people, and when the building was finished in the fall of 1919, the Chicago elite gasped: There were still box seats, but Insull had abolished the grander ones. Everyone, rich and poor, would have to sit, as McDonald put it, "in anonymous proletarian darkness just like people in the penny gallery." As befits someone who had seen his father struggle in his jobs, he was a generous employer, one of the best in the United States. He was deft in dealings with unions. He paid his workers more for their 46-hour week than was earned by workers at other companies who toiled for 60 to 70 hours every seven days, and provided his employees with free medical benefits, unemployment insurance, free night schooling and a stake in a real profit-sharing plan. He was way ahead of his time in hiring black workers and seeing that they had equality of status. He demanded much of everyone in return. He pushed immigrant workers to become American citizens on pain of losing pensions. High-ranking officers had to be active in community service or else. He valued loyalty but hated yes-men. He set traps by advancing patently absurd ideas and then publicly ridiculed the yessers. His home telephone number was freely available to anyone wanting to raise questions about their electricity or gas. He relished the popular esteem.

A man doing so much good inevitably attracted enemies. The New York banks resented the way he bypassed them, raising money by going to London and selling hundreds of thousands of bonds locally when very few Americans were stockholders of any kind. They were suspicious of Chicago standards; even before Prohibition spawned Al Capone, it was a notoriously corrupt and violent city. Chicago's elite Gold Coast, who knew better than to associate Insull with the mob, resented his pretensions. He would put tens of thousands of his own dollars into a hospital and call on Chicago's wealthy to chip in the rest. He shamed them into supporting African-American charities. From the moment

Insull had arrived, successive city and state politicians were peeved that they could not shake him down. Who did he think he was? Actually he had a practical as much as an ethical distaste for greasing palms. He had done it when working for Edison and came to the conclusion that no man was ever bought just once. In Chicago, he recognized, too, that any piece of legislation would come in the end at the price of the entire city council. He chose to make large and regular campaign donations to both

Still a bachelor in his midthirties in 1894

parties, believing these to be far more effective than payments just before important votes. Early on, he foiled a blackmail scheme by the city's "gray wolves," a bunch of corrupt Democrats. They formed a dummy corporation, the Commonwealth Electric Company (granting it a favorable license), with the idea of selling it to him for millions of dollars without having to raise a watt of energy. He called their bluff. When they tried to start in earnest, they found they could not buy dynamos from anyone: Insull had extended his early exclusive arrangement with General Electric to other major American manufacturers. For a mere $50,000 he bought Commonwealth Electric, along with its coveted 50-year franchise, and in 1907 merged it with Chicago Edison to create the giant Commonwealth Edison.

From 1912 there was a personal pall on his galloping success. Samuel Insull Jr. hovered three months at death's door, racked by scarlet fever. The illness strained a marriage not remarkable for its intimacy. Gladys nursed the boy back to health. She did everything she could to thwart her husband's desire to fashion the boy in his own image; she wanted Junior to be a writer, not a tycoon driving himself to the grave. She thrust a biography of Napoleon into her husband's hands, saying, "Sam, you should learn about that man and about what happened. If you don't that's what's going to happen to you." By the time the boy recovered, Gladys and Insull had grown apart. Gladys shut her bedroom door every night and never reopened it.

Insull was 53, miserable and lonely. Looking out on the city, as its lights went on at night, he found consolation in the idea that all this, and more, he could pass on to Junior, so, obsessively, he expanded his empire. He did it first by the device that was to become all the rage in the '20s, the holding company. It was easier for a big holding company to raise bank capital for expansion and it offered economies of scale in engineering and marketing. Insull also raised millions of dollars by offering low-cost bonds to the general public in return for shares—but no voting rights—in Middle West Utilities. He put his younger brother Martin in charge of Middle West. In 1912 the business controlled companies worth $90 million; by 1917 it was worth $400 million. Insull became a gas magnate, rescuing the People's Gas Light and Coke Company, a target of Progressive reformers for its corruption. He cleaned it up but got no thanks from the reformers for refusing rebates for the company's past misdeeds, which were not of his making.

He was inexhaustible. Even before Woodrow Wilson took America into war against Germany in 1917, Insull was rallying for Britain, discreetly at first in a city with so many Germans. He ran an underground railway to help men from all over the United States get into Canada for enlistment under British colors. On America's entry, President Wilson asked states to create defense councils, and Insull was invited to head one for Illinois. He commanded more than 380,000 workers in the war

effort. He deployed 3,000 soldiers in a mock battle for two million spectators and altogether helped to sell more than a million dollars in war bonds. When coal prices tripled 60 days into the war, he blandly told the owners he would seize their mines; they cut prices in half. He recruited and trained 20,000 city boys at the University of Illinois for work on the farms. He borrowed $1,250,000 for seed when a severe winter in 1917–18 threatened to destroy the corn

His bride of 1899, actress Gladys Wallis

crop. By the time the war ended he was a hero to many ordinary people.

When Insull in 1881 set foot in a New York without yet a skyscraper, the agrarian and frontier society of the 19th century—characterized by small towns serving a predominantly rural population dependent on human muscle and horse and steam power—was giving way to a dynamic, machine-based America of teeming metropolises and a whole clamoring continent of a market. By the '20s, when electric lighting had supplanted gas and kerosene and taken over America's shop floors with electric motors producing exactly the energy needed for each manufacturing step, Insull had done as much as anyone—any politician, any industrialist, even Edison—to propel the country forward at an accelerating pace.

The owlish, diffident young man who had disembarked in 1881 was now a stately, silver-haired and rather portly gentleman with a silver-topped cane and a panama hat, grown more formidable vaulting from one improbable rock face to another. Harold Ickes, a key figure in the Progressive Party who was to become the energy titan of the New Deal and World War II, sought a small favor from Insull in 1915. He wrote a friend: "He was at great pains to sell himself to me. I was struck by the man's forceful personality. There was no doubt, whether you agreed with him or not, that he was a real man, a force to be reckoned with if you crossed his path." Insull was borne up higher as the good times roared on from 1923. By 1927 every third home in America had a radio, and two-thirds had electricity. It was radio that persuaded the last holdouts to switch on. Gas might light a room, grill a steak, warm a kitchen, but it could not tell you what Babe Ruth or Charles Lindbergh was up to. In the onrushing bull markets of the '20s, Insull extended his empire, using the stock of one holding company or another. He got control of utilities in 14 eastern states. By 1929 he was supplying one-eighth of the entire nation's electricity and gas power, in 32 states, as much as the entire national power supply of any European country. British Prime Minister Stanley Baldwin invited him to come home and help set up the British grid. He declined so as to stay near Junior, now making his way in the company.

By now Insull was perceived as the most powerful businessman in America. Ickes wrote to a friend that Insull's political lawyer, Samuel Ettelson, "owns the city council, the state legislature, and the Illinois Commerce Commission. The governor of the state, the lieutenant governor of the state, the mayor of the city of Chicago, the president of the county board are all eager to carry out Insull's slightest wish." Ickes had the Progressives' conviction that anyone with money must be up to no good, compounded by a sense of personal slight. He had called Insull for a position on the Illinois Defense Council so often that Insull's aides referred to him as "that man Itches."

The overstretch that led to disaster was a raid by Cyrus S. Eaton, a buccaneering

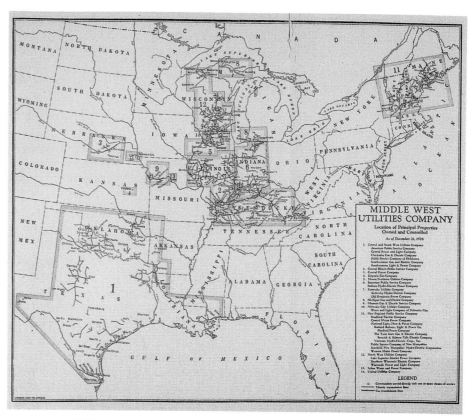

INSULL'S REACH: In outline are the 14 companies composing the Middle West Utilities Company as of December 31, 1926.

Cleveland capitalist who secretly began buying Middle West stock in 1927–28. Defensively, Insull established an investment trust in a pyramid through which he, his brother and his son could continue to run their 255 operating companies. It transformed his relationship to his operating companies, in the words of historian Thomas Hughes, from that of a manager to that of a proprietor. Crucially, it forced him into borrowing $20 million in something of a hurry. He could not organize permanent financing before inking the deal, so he was forced to borrow from the despised New York banks. They insisted on owning voting stock in the three major holding companies. It was a giddy time. In August 1929, Insull securities were appreciating at $7,000 a minute. His companies gained $500 million in two months. He was shocked by the amount of money he had become worth on paper. "My God, a hundred and fifty million dollars!" he exclaimed. "Do you know what? I'm going to buy me an ocean liner!" He did not do that; he set up another investment trust, more fuel for a market out of control.

On October 21, 1929, Insull spent an emotional evening with Edison, reiterating his first impression as a youth that he had met "one of the masterminds of the world." They had been reunited from time to time at conventions organized by Insull, where the living legend was acclaimed by thousands of delegates. This was special, a presidential gala to honor the 50th anniversary of the invention of the incandescent bulb. Edison was 81, ailing and now stone-deaf. Three days after the celebrations of Light's Golden Jubilee, the stock market boom collapsed. Before long the lights would go out for Insull.

The crash itself did not faze him. What was another financial panic to a Zeus who every day before breakfast flung shafts of lightning across a continent? With one hand, he got on with his business, opening an $80 million gas pipeline from Texas, spending $197 million on capital improvements in 1930. With the other he came to the rescue of the city he loved. He found $100 million to enable a stricken Chicago to pay for teachers, policemen and firefighters. In April, over the objections of

Progressive reformers, anti-utility on principle, voters passed a referendum allowing Insull to bail out and revitalize Chicago's transportation system for $500 million. He helped save countless smaller businesses and individuals, too. Demand for electricity kept growing through the Depression. Operating company earnings were higher in 1930 than 1929. By 1931 he seemed to have beaten the tide and the hysteria. President Hoover announced the Depression was over. Insull believed him because he believed in himself. He kept on expanding, but the recovery he expected did not come. He said later, "I was fooled—and so was the president." Since he had given a personal guarantee for many of the loans to his companies, Insull was now millions of dollars in debt himself.

A hostage to the borrowings to head off Eaton, natural in boom times but now clearly reckless, Insull was trapped. Every time the market dropped, the banks owned more voting stock in his empire. He ordered his top operator, Fred Scheel, to keep prices up by buying their own stock. Scheel favored the opposite tactic. He convinced Junior they should preemptively sell Insull stock short. Every time New York bears pushed the price lower, Scheel made money for Insull rather than lost it; if the bear squeeze went on long enough, he would reap billions and bankrupt every New York bank in the process. But when Insull found out, he was angry. "We can't do that," he exploded. "It would be immoral. We've got a responsibility to our stockholders. We can't let them down." As the historian of the '20s, Geoffrey Perrett writes, "With these words he delivered himself over to his enemies, hands bound, and with an apple in his mouth. No man ever had salvation so sweetly spiced with revenge handed to him on a platter."

In April 1932 Insull had to do what he hated: go back on bended knee to Wall Street to ask for more time on the $10 million due in June. New York banks owned only $20 million of holding companies worth $3 billion, but in the pyramid Insull had created it was a lever to rival Archimedes'. Owen Young, the chairman of General Electric and the New York Federal Reserve Bank, mediated a meeting with Insull and investors. It had barely begun when five men from Morgan banks entered the room and asked Insull to wait in the next room. An hour later Young came out and told Insull no further money or time would be forthcoming after June. "Does this mean receivership?" Insull asked. "It looks that way," said Young. "I'm sorry, Mr. Insull." Poignantly, Insull and Junior had just raised $10 million in relief funds for Chicago's poor.

Biographer McDonald, who was close to Samuel Insull Jr., believes he was the victim of a conspiracy by the House of Morgan. "All the weapons of the bear raid were at their disposal: short selling, tape advertising, wash and match sales, life or death power of the liquidation of brokerage accounts, and the deadliest weapon of all, the Wall Street rumor (Insull had committed suicide; Insull has been seen leaving a New York bank in tears; Insull is ill and his mind has snapped. . . .)."

For his own part, Insull wrote in his memoirs, "I am confident that if there had been any inclination on the part of the New York banks to make a substantial contribution to the fund established for the purpose of keeping the Middle West Utilities Company out of receivership that the Chicago banks would have been willing to do their share." Harold Platt's judgment is that the bankers were acting out of anxiety for their own investment rather than personal animosity: tension, yes, conspiracy, no. In the Insull archives, there is a note by Insull recognizing he was partly to blame for the tension by going to London for finance when Chicago bankers could not raise enough. "It was probably a mistake . . . but the London bankers understood my points of view better than they were understood in New York. The chances are that had I had paid more attention to dealing with New York bankers and less with London bankers, when the troubles came in 1932, I would probably have found the New York bankers more receptive. Their interest would have been with me instead of their occupying just the position of ordinary lender toward ordinary borrower."

In a single day, Insull was forced out of his 60 presidencies and directorships. On Monday, June 6, the still spruce "power wizard," as the newspapers called him, went out before a crowd of pressmen waiting for him outside his offices in the opera house building. "Well, gentlemen, here I am after forty years, a man without a job." In that, he was like millions of Americans, but the sensation of the collapse, the staggering amounts of money, the losses borne by half a million bondholders, the rumors from Wall Street . . . there was altogether too much blood in the water for Insull to be just a man without a job. The press was determined to make him a man on the run. And it was presidential election year, with Franklin Roosevelt challenging President Herbert Hoover. With the Democrats on the upswing, the Cook County prosecutor—Republican John Swanson—faced a tough election. On September 5, riding into Chicago on a train that had been built and operated by Insull, Swanson confided to his son-in-law: "You know Sam Insull is the greatest man I've ever known. No one has ever done more for Chicago and I know he has never taken a dishonest dollar . . . but Insull knows politics and he will understand. . . . I've got to do it."

What he had to do was to win headlines by announcing an investigation of Insull. Prompted by Harold Ickes, Franklin Roosevelt, in one of his demagogic less good moments, caricatured Insull as the capitalist evil, the "Ishmael or Insull whose hand is against every man's." It was grotesque (and FDR did not keep it up) if understandable in the hysteria of the time, but slack commentators resumed the vilification of Insull in the post-Enron era. Harold Platt's verdict must carry more weight: "Insull was no crook. He lost too much of his own money. He poured most of his own fortune into his own companies to shore them up. His prominence during the business boom of the twenties made him the perfect scapegoat for its collapse in the thirties."

Insull could see he was facing a lynch mob, trial by sensational newspapers. Gladys could not stand the strain of the press vendetta and hate mail. She opened a letter that read: "You can get ready to buy a cemetery lot as the gang will send you your crooked boy's head, you will pay as we have paid our good money that has been stolen by the dirty yellow Insull Jews." Insull, 73 now and a deeply depressed diabetic, fled with Gladys to Paris. Cut off

THE FUGITIVE: The 74-year-old Insull, returned to New York to face trial, is helped down a ladder from *SS Exilona* to the coast guard, May 8, 1934.

silver mustache and hair, took off his trademark pince-nez and navigated his way to a yacht chartered by London friends. He was not recognized and roamed the Mediterranean for two weeks, unsure where to land for fuel and food. Turkey seemed safe. It had no extradition treaty with the United States. But when he docked in Istanbul, he was kidnapped by Turkish agents at the request of the U.S. ambassador, imprisoned for several days after a mock Turkish trial and finally handed over under heavy guard to Burton Berry, a state department official.

Berry, escorting him to the United States aboard the *SS Exilona,* became attached to Insull. The old man tried to keep up appearances, turning up dapper on deck in a panama hat, but in fact he was close to suicide. Berry talked him out of it. He told him it would be taken as an admission of guilt. "Only when he spoke of his son," said Berry, "did he show any deep emotion." At the end of an excruciatingly long voyage, Insull whispered: "This mask which I have been wearing now for twenty-three days is getting very thin. It is with the greatest difficulty that I continue to wear it. But I am going to keep right on trying to keep a firm upper lip for my son in order to leave him a clear name. For myself there remains nothing in this life."

He was mobbed on arrival in New Jersey. "I made mistakes," he told the press, "but they were honest mistakes. You know only the charges of the prosecution. Not one word has been uttered in even a feeble defense of me." Then he rebuked the jostling photographers, "Keep quiet! There is plenty of time for taking pictures. This is my mug and I have a proprietary interest in it." In Chicago, Junior and defense counsel Floyd Thompson, a former Illinois Supreme Court judge, had arranged for bail of $100,000. At Cook County Jail, they found they had been double-crossed. Bail was now set at $200,000, with little time to raise the extra money. Insull, with an eye on gaining public sympathy, told Junior and Thompson not to bother. He spent the night in the pen with a murderer and hardened criminals.

The trial of the *United States v. Samuel Insull and Others* began on October 2, 1934, before Federal Judge James H. Wilkerson, a strong-minded judge who had sent Al

from all his income, his pension frozen by scared directors back home, he had less than $3,000. In his absence, he was indicted for selling securities at more than their worth. Insull managed to dodge the law for 19 months. The United States was dogged in its efforts to seize him, first in Paris, then Italy and finally in Greece. It canceled his passport and sent a special

prosecutor to Athens. Congress passed a special bill authorizing foreign countries to arrest Insull. Criminal charges and civil suits were brought against his son. The Greek judges dismissed the charges. The United States threatened the Greek government by claiming it would stop Greeks in America from sending money home, so one night the elderly fugitive darkened his

Capone to Alcatraz. The government's top prosecutor, Leslie Salter from New York, led for the prosecution, Thompson for the defense. The charge was that while Insull made enormous profits, he was a dishonest bookkeeper and so the "little people" lost their life savings.

The mood of the courtroom changed as the experts testified and were cross-examined. No, Insull had not taken a cent from the investment trusts. They were vehicles to ward off takeovers. His accounting methods were the same as the government's. An FBI investigator urged Salter to make much of Insull's income tax returns showing earnings averaging half a million in the five years before the collapse. When Thompson used the same tax returns to show that Insull had given away more money to charity than he had earned from all his various salaries, Salter turned to the agent and muttered loudly enough for the jury to hear, "You son of a bitch. Why didn't you tell me that was in there, too?"

On November 1, when Insull took the stand, business was almost stopped in downtown Chicago. Thompson was worried. In pretrial practice, Insull had seemed deeply tired and muddled. Once on the stand, though, he was charming and lucid. He spoke directly to the jury about his childhood in London, about his great idol Edison and how he had brought electricity to millions. Salter didn't realize what Thompson was up to, but when he figured it out his objections came too late. He tried stopping Insull's testimony almost apologetically the first time: "If the Court please, I do not like to interrupt, but while this is all very interesting, I just wonder if there isn't some proper limit so that we may hurry along and get down to the issues in this case. It seems to me it is taking quite a long while in the early stages."

After pages more of testimony Salter tried again: "For the life of me I cannot see the bearing which this testimony has on the issues in this case. I think we should hurry along somewhat to the issues in this case."

"What is the purpose of this testimony?" asked the judge.

"Perhaps counsel cannot see the purpose of it," Thompson replied calmly, "but the strongest test that I know of as to a man's actions in later years of his life is the character that is in him during the time that he is building his life, and his experience, and it is of course impossible for Mr. Insull to describe intelligently the course of his actions in the last two years of his life without giving the jury some knowledge of his actions during the first fifty years of his life." Judge Wilkerson was by this time curious himself. He allowed Insull to continue. Everyone wanted to hear about Edison; everyone wanted to know how Insull rose from rags to riches. He

INSULL ON TRIAL: The caption of March 4, 1935, said, "Seemed confident of acquittal on the day this picture was made."

explained to the jury, to the courtroom, to the whole world listening, the intricate economics of electricity. He described how he expanded his business and how he made electricity affordable to increasing numbers of Americans. The more Salter objected, the more he saw the jury was growing furious with his interruptions. Salter finally decided to stop objecting and listen as well. Even *he* became captivated . . . and confused. During a recess he approached Junior and said, "Say, you fellows were legitimate businessmen."

"That's what we've been trying to tell you," Junior replied.

The jury retired on Saturday, November 24, at 2:30 p.m. They reached their verdict in five minutes. They cleared the defendants of all the charges. The *Chicago Times* wrote the next day, "Insull and his fellow defendants—not guilty; the old order—guilty. That was the Insull defense, and the jury agreed with it."

Insull was a free man, but broken. The press continued to hound him unmercifully. He remained their whipping boy for all that had gone wrong in the Depression, the symbol of out-of-control capitalism. When Roosevelt and Congress created the Tennessee Valley Authority, officials were embarrassed at the prospect of copying the Insull systems they had subjected to so much obloquy. But they had no choice: His was the sensible approach. In fact, after the dust cleared from the scandal, the trial and the Depression, it turned out that about 40 percent of the stock of all American corporations was forfeited, whereas the stock of Insull's operating companies had fallen in value by less than 1 percent. None of his electrical or gas operating companies went bankrupt. Insull's creditors who held on to his stock were collectively $10 million better off at the end of the Depression. In the '60s Insull's companies were still supplying one-eighth of America's electricity and gas. Their prices were still among the lowest in the nation, and many were still run by old Insull employees.

Gladys refused to live in Chicago or London, and a bitter Insull moved to Paris to be with her. "I owe America nothing," he had told Berry. "She did only one thing for me. She gave me the opportunity. I did the rest and I repaid America many times for what she gave me."

On July 16, 1938, he dropped dead of a heart attack in a Paris metro station. For several hours, no one could identify him. There was nothing in his pockets except a silk handkerchief with the monogram "SI" and a few francs worth eight cents. Newspapers reported that he had died in poverty and made much of the riches-to-rags story. In fact, he had then about $10,000 to his name and often carried as much as $2,000 in his wallet. Someone must have pickpocketed the body. McDonald concludes, "And so in death, as in life, Samuel Insull was robbed, and nobody got the story straight."

PHILO T. FARNSWORTH The boy genius from nowhere who beat the world to the invention of modern television

1906–1971

FOLLOWING THE stock market crash of 1929, the San Francisco banker Jesse McCargar hurried over to a warehouse loft at 202 Green Street to turn off television— not a television set but the whole concept of the electronic transmission of sight and sound. Three years before, he and his associates at the Crocker Bank had invested $25,000 in a Mormon farm boy from Snake River Valley, Idaho, Philo T. Farnsworth, who had sold them on his conviction that a cathode-ray tube system of his invention would provide television broadcasting far superior to the existing mechanical scanning systems. When the well-dressed McCargar mounted the wooden stairs to the loft, Farnsworth, a pale, skinny and tousled 23-year-old in shirtsleeves, his hands stained with acids, stood in the doorway. George Everson, a business backer of Farnsworth from the inventor's teenage years in Salt Lake City, had tipped him off that the investors were on the warpath.

"Shut down today!" McCargar ordered Farnsworth. "Fire everyone!"

Nobody—not McCargar, not Everson, not the four or five lab workers—expected Farnsworth to react as he did. He was a mild, soft-spoken young man who deferred to authority, given to vehemence only when describing the principles of his invention. Now he was furiously defiant. He blocked McCargar's entrance. "No one," he shouted, "comes in here giving me orders as to what is to be done!"

He was on shaky ground. The bank owned the building and some 60 percent of the fledgling private company, Television Laboratories, Inc. McCargar was president, Farnsworth vice president of research, and Everson the treasurer. Month by exhausting month Farnsworth had struggled to convince the investors they would "soon" see a return on their money—somewhat

more than $60,000 by 1929. James Fagan, the executive vice president of the bank, was in the habit of squinting as if looking through a telescope and saying, "When are we going to see some dollars in Phil's gadget?" In May 1928 Farnsworth had responded by triumphantly calling them to Green Street to watch the camera he called an "Image Dissector" televise to another cathode-ray tube something he knew they understood: a dollar sign, in two dimensions—a considerable achievement. By accident in a trial run, the camera had caught smoke drifting from the cigarette of his brother-in-law, Cliff Gardner, and smoke blowing became part of the display as evidence of the potential to portray movement. Four months later, to keep up the bank's spirits, Farnsworth had gone public for the first time, showing the *San Francisco Chronicle* how he could televise a 30-second filmstrip of Mary Pickford combing her hair. It was the very first public demonstration of an all cathode-ray television—and a gamble that he would soon be protected by the patents he had applied for on January 7, 1927.

After hand-to-mouth years, Farnsworth longed to be rich, but according to his widow, Pem, interviewed by biographer Evan I. Schwartz when she was in her 90s, her husband was also inspired to persevere through all the disappointments by imagining the manifold benefits he might bequeath to the world. Prefiguring Ted Turner by eighty years, he believed that watching news as it happened would remove the risks of distortion in relying on middlemen to report and interpret. Beyond the entertainment value of televising movies, sports and concerts, Farnsworth was sure that television would become the world's greatest teaching tool, wiping out illiteracy and ushering in an era of world

peace: "If we were able to see people in other countries and learn about our differences, why would there be any misunderstandings? War would be a thing of the past."

These visions impressed the investors less than watching their dollars vanish into a machine they came to call "Jonah" for its absorptive capacities. They had been thrilled by the demonstration for the press in September, but it had not had the effect for which Farnsworth hoped: It had only made them keener to sell out to one of the big electrical companies, General Electric's RCA or Westinghouse or AT&T. These companies were all invested in television, but in a mechanical system based on a theory patented in 1884 by Paul Nipkow, a Russian working in Germany. (Light shining on an object would strike a spinning disk closely perforated from center to rim. Such light as passed through the holes would strike photoelectric cells, creating an electrical impulse. Wire would carry these impulses to a second spinning disk that would reverse the process, converting the electrical signals back into light, and the original image could then be projected on a screen.) The speed at which a perforated disk might revolve confined this kind of television to a flickering image comprised of 48 horizontal lines (compared with 500 lines in Farnsworth's design, 525 or 625 lines for standard TV today and 1,125 and 1,259 for high-definition television). Yet mechanical methods were the basis of the claims everyone in the '20s was excitedly making to "the first" practical demonstration of television broadcasting.

The Scotsman John Logic Baird in Britain had been the genuine first, transmitting outlines of blurred geometric shapes a few feet in a department store demonstration in London in 1925, followed

THE IMAGE: Farnsworth, at 27, got a chance to display his invention of television at the Franklin Institute in Philadelphia in August 1934. Joan Crawford appears on his cathode-ray tube.

by his televising of a talking face in January 1927. Two months later, the American inventor of the "Radiovisor," Charles Jenkins, had shown a whirling windmill; he applied for (and in 1928 won) the first federal license for a television station, W3XK. In April 1927 Herbert Ives at AT&T had rigged up two 15-inch Nipkow disks to convey to their Bell Laboratory in New York a small image of Herbert Hoover, the secretary of commerce (a central figure in allocating radio frequencies in the 1920s), talking on the telephone in Washington; AT&T was thinking of offering one-on-one radiotelephonic-television connections but came to think it was not worth the money. None of these "firsts" inhibited General Electric's chest beating in January 1928 when its renowned engineer, Ernst Alexanderson, whose alternator had made transocean radio possible, exhibited another tiny head smiling on a tiny screen. This was lauded by the *New York Times,* which should have known better, as the first indication of the potential for home television. As David and Marshall Fisher write in their definitive study of television technologies, "General Electric was playing to one of the standard rules of invention; bravado is just as important as achievement." By June 1928, when Farnsworth's progress was still secret, more than 20 television stations in America were broadcasting some form of rudimentary programming: a girl bouncing a ball, a statue revolving, a couple of blurry Lilliputian boxers—excitements that failed to sustain a flurry of sales of television sets and kits.

These broadcasters were all hurrying toward a dead end. Precociously brilliant, Farnsworth had seen as early as 1921, when he was a 14-year-old schoolboy, that it was futile to depend on a spinning disk when the vacuum tube was so much faster for transmitting current: Light could be switched off and on by a vacuum tube 10,000 times *a second*. McCargar's determination to close Farnsworth's electronic lab in 1929 was that of a banker with an eye to the wrong end of the telescope.

The doorway confrontation between him and Farnsworth might have ended in violence but for Everson, a stylish, dignified middle-aged man who had honed his diplomatic skills persuading companies in the West to invest in philanthropy. He stepped between the antagonists, and he suggested that Farnsworth's handful of staff, drawn to the scene by the clamor, might be willing to work without pay. They had already responded to earlier crises of confidence by accepting subsistence-level wages. Now they instantly agreed to no pay at all. They were bright young electrical engineers, graduates of MIT, Berkeley and Stanford, with impressive faith in Farnsworth. Everson reinforced his plea to McCargar for a reprieve by arguing that it would be hard for the investors to sell Television Laboratories without a television laboratory. McCargar grudgingly gave in, on the clear understanding that the bank would not put in another cent, and a buyer must urgently be sought. *Over my dead body* was Farnsworth's promise to himself. He was not about to become an employee when he was so close to making a practical reality of his vision.

Across the continent in New York, banking and business in 1929 were feeling just the same squeeze on liquidity, but a rival of Farnsworth's was enjoying a quite different response from his financial sponsor. Vladimir Kosma Zworykin, a Russian refugee scientist with bottle-thick glasses, had been invited in January 1929 to ascend the Woolworth Building to the well-guarded sanctum of the emperor of radio

manufacturing, the executive vice president of the Radio Corporation of America (RCA), David Sarnoff (1891–1971). Up to then, Zworykin had had a tough time. Working in Pittsburgh for Westinghouse, he had tried to convince the management there that electronic television was the way of the future. Harry Davis, the headman, was enamored of a mechanical television project headed by his golden boy, Frank Conrad, an engineer and ham operator. Conrad had, one Sunday, spontaneously broadcast music from his home and struck such a chord with other hams that Davis had erected a transmitter to serve the first radio broadcast station, KDKA, for Westinghouse, stealing a march on Sarnoff and everyone else. Having watched Zworykin transmit a flickering X by a mixed electronic and mechanical system, Davis had told Zworykin's boss, Samuel Kintner, "Put this guy to work on something useful." Davis was more immediately concerned with convincing the public that his new electric refrigerators were a good thing than with Zworykin's electric television. His skepticism was widely shared. At the Bell Telephone Laboratory, the chief of television development dismissed Zworykin's efforts: "The images are quite small and faint, and all the talk about the development promising the display of television to large audiences is quite wild." Would the interview with the mogul of RCA go just as badly?

Sarnoff, a paunchy 37, just two years older than Zworykin, was more receptive. Seated in his office, Sarnoff lit up one of his big fat cigars and invited him to hold forth. It was an intriguing encounter between two Russian-born immigrants from opposite social poles, surfacing after years in the American melting pot. The heavily accented Zworykin, scion of a wealthy family in sophisticated Saint Petersburg, had arrived as a refugee from the Bolsheviks at the age of 29, and sojourned at the plush Waldorf-Astoria. The smoothly spoken Sarnoff, from a poor family in a primitive village in Uzilan, near Minsk, had started his new life at the age of nine in a squalid rattling railway tenement in New York's Hell's Kitchen. Of all the millions of immigrants who have landed in America and found their way, none could

have assimilated faster, or with more determination, than David Sarnoff. Brought up in Russia to be a Talmudic scholar, he spoke only Yiddish on arrival but dedicated himself to the rapid mastery of spoken and written English. He sold newspapers, then sold himself, as an office boy, junior telegraphist and personal assistant to Guglielmo Marconi (facilitating the master's amorous trysts) and soon became chief inspector of the American Marconi Company. He was well placed when, in the cause of protecting American technology, the U.S. government sponsored a buyout in 1919 of American Marconi, a British subsidiary, so as to give General Electric a patent monopoly through the newly formed RCA. Sarnoff was named commercial manager the year Zworykin set foot in America.

In 1929 the very American Sarnoff had his eye on the future, and Zworykin was eager to explain his certainty that electronics would consign mechanical television to the museums. Sarnoff said nothing at first but beamed as Zworykin described how in 1910–11 he had assisted his professor at the Saint Petersburg Technical Institute, Boris Rosing, in transmitting an image through a Nipkow disk using a cold cathode-ray tube as a receiver. Rosing's best result, on May 9, 1911, was to transmit "four luminous bands." Zworykin said his own subsequent experiments had brought him close to making an effective receiver, but he needed funds and time to make an electronic camera with electrostatic focusing. Sarnoff of Uzilan gave a thumbs-up to Zworykin of Saint Petersburg. Without asking his board, Sarnoff there and then pledged the huge sum of $100,000 for two years of secret experiments, four times the initial capital behind Farnsworth. Zworykin was gratified by Sarnoff's adventurous spirit. Experience, said Zworykin, had conditioned him to fear that the American obsession with quick profits meant it was no longer possible to work on an idea in commercial research without camouflaging it.

Nine months after that meeting, when McCargar climbed the stairs in San Francisco to deliver his ultimatum to Farnsworth, Sarnoff's financial officers were likewise pressing him to end his research spending on television, a chimera by com-

THE TITANIC MYTH: Many writers have told how Sarnoff, a 21-year-old Marconi telegraphist, picked up the first news in the early hours of April 15, 1912, that the *Titanic* had hit an iceberg. He stayed heroically at his station on top of the Wanamaker store in Manhattan for three days and nights without sleep, "a horrified world hanging on his every word." This is a myth. In fact, the Wanamaker station was closed when the initial flash was received—at Cape Race. Sarnoff did not come on duty until the next day, but he knew the value of dramatizing his role in the press. As he rose up the corporate ladder at RCA (left), even *Fortune* magazine bought into it. Above, Sarnoff as assistant telegraph operator at the Marconi station at Siasconset, Nantucket Island, off Massachusetts, in 1908–9.

parison with his runaway success in manufacturing and licensing radios. RCA company profits were falling, his board was scared and pretty well every other company was slashing research spending, but Sarnoff kept his nerve. In the blackest years thereafter, as the Great Depression followed the crash and Sarnoff rose higher (RCA president from 1930), he never faltered in his support for Zworykin—he increased it. Eventually his faith would cost him almost $50 million before he made a dollar. Sarnoff had no scientific training, but he had developed a shrewd judgment of technological claims combined with a then-rare vision in business of what science might achieve. He had married high cultural aspirations with a sharp commercial sense in his very early advocacy of the mass production of a "Radio Music Box" at a time when radio was only Morse code and at best scratchy point-to-point speech (see Fessenden on page 219). His main idea for radio, probably first advanced in 1915 or 1916 and

fleshed out in 1920, had been to run it as a nonprofit service rather like today's NPR in content and structure, free of advertising and financed by advertising tax on radio sales. He had been ferocious in the acquisition of radio patents to give RCA its highly profitable manufacturing and licensing business and had followed up in 1926 by brilliantly outmaneuvering Walter Gifford—his former RCA boardroom colleague and new enemy—the chairman of AT&T.

Sarnoff had gambled that arbitration would uphold RCA's monopoly claim—and won. As a result he was able to take over AT&T's new chain of 13 radio stations named WEAF (for Wind, Earth, Air and Fire) as the basis for a nationwide radio broadcasting network, the National Broadcasting Corporation (owned 50 percent by RCA, 30 percent by General Electric and 20 percent by Westinghouse). Treating radio like its telephone service, AT&T had leased blocks of WEAF airtime to com-

merce, which outraged Sarnoff as much as Gifford's anti-Semitic whispering campaign against the "abrasive Jew" at RCA. But when Sarnoff took over Gifford's stations, he stuck with the scheme of selling time and never said another word about the virtues of the nonprofit, advertising-free broadcasting that he had espoused with such passion.

Farnsworth's little team had seen the shark's teeth. It was very different with Zworykin. His team, quickly established in a well staffed RCA lab in Camden, New Jersey, saw only the cheery countenance of a patron periodically moving among the experimenters in his shirtsleeves, making them feel they were doing the most important job in the world. Biographer Daniel Stashower is justified in writing: "It would be difficult to exaggerate the importance of Sarnoff's commitment, weighed against what would have seemed a distant and perhaps unattainable objective." But with that admirable willingness to commit, Sarnoff had combined an unsavory degree of cunning and a lethal mode of operating that boded ill for the boy from Idaho.

The legend about Philo Farnsworth is that he was inspired at the age of 14 with the idea of electronic line by line scanning with a cathode-ray tube when he surveyed the furrows of a potato field he had plowed for his father. No doubt Farnsworth believed this story himself (it is what he told his astounded high school science teacher Justin Tolman), but the idea was at least subliminally informed by the stacks of science and engineering magazines he found in an attic when his family lived with a relative for a time. He got up at 4 a.m. to read them before farm chores and riding horseback to school. He was struck by the contention in a letter from the Scottish inventor A. A. Campbell-Swinton in a 1908 issue of *Nature*. A Nipkow disk, Campbell-Swinton argued, would never spin smoothly or fast enough (or have enough apertures) to exploit the phenomenon known as visual persistence. The retina retains an image for about a tenth of second, and an illusion of movement can be created only if the next image is superimposed on the first within that time. This problem for television, wrote Campbell-Swinton, could most probably be solved by having two "kathode-ray" tubes, one to transmit an ultrafast beam and one to receive, "sweeping synchronously over the whole of the required surfaces within the one-tenth of a second necessary." By 1920 Campbell-Swinton had concluded that it was scarcely worth anybody's while to pursue this idea. He thought the problems were so intense that electronic television would never make sense financially. The world's big electrical firms were of the same mind, in light of the early achievements with spinning disks, but a poor, unknown farm boy decided to prove them all wrong.

Farnsworth was a polymath who might have done anything. He had the intellectual gifts to race ahead of his teachers at Rigby High School—he expounded Einstein's then-controversial theory of relativity to his junior classmates—and he was a gifted violinist, too. But for the early death of his father in 1924, visiting depression and hardship on his mother and the four younger children, Farnsworth would

THE BACKERS: Farnsworth at 22, happily sandwiched between the men who backed him as a teenager—Leslie Gorrell (left) and George Everson (right).

have flowered in a university. He joined the navy instead, hoping to study electronics on the cheap, then got an honorable discharge when he learned that anything he invented would belong to the government. By working as a janitor and securing a student loan, he was able to study for a year at Brigham Young University in Provo, Utah, before he had to abandon college entirely for full-time work.

It turned out to be a blessing. The next job, after sweeping the streets of Salt Lake City, was to stuff charitable appeals in envelopes for a community chest project run by two itinerant professional organizers, Californian buddies by the names of George Everson and Leslie Gorrell. When one of them asked Phil—as he now was called, having abandoned the "o" for the navy—whether he would go back to college, they were struck by a cataract of enthusiasm about his idea for television. "Farnsworth's personality seemed to change," wrote Everson. "His eyes, always pleasant, began burning with eagerness and conviction; his speech, which was usually halting, became fluent to the point of eloquence. . . ." Gorrell, who had taken some college courses in mechanical engineering, was impressed by the sketches Farnsworth made on any bit of paper he had. Everson became certain that he was in the presence of a genuine talent when Farnsworth fixed

the bearings on his Chandler roadster—a problem that had defeated a run of mechanics.

Neil Postman has observed that there were at the time no more than a handful of men on the planet who could have understood Farnsworth's electronic ideas. So the California pair were as adventurous as any forty-niners when they staked a claim on the 19-year-old would-be inventor they barely knew. "This is about as wild a gamble as I can imagine," said Everson. "I have about six thousand dollars in a special account in San Francisco. I'll put the money up. If I win, great. But if I lose it all, I won't squawk." If Farnsworth devoted himself full-time to television, he would have 50 percent of a partnership to be called Everson, Farnsworth and Gorrell for no investment and no liability for loss, and he would receive $200 a month for living expenses. Everson and Chandler expected to find backers in California and made it a condition that Farnsworth immediately install himself in a laboratory there. He had been enjoying himself for a change, letting his hair down, dancing the Charleston and hotting up his violin in ragtime jam sessions with 18-year-old Pem Gardner on piano and her brother Cliff on trombone. Three days after the community chest pact, on May 27, 1926, Phil and Pem got married and took the train west, still syncopating to

Al Jolson's "California, Here I Come" when they set up home—and laboratory—in a little studio flat at 1339 North New Hampshire Street in Hollywood.

It was, of course, a daunting task to transmit and reassemble thousands of elements of a moving television picture line by line at speeds beyond human comprehension. Farnsworth read up on chemistry and physics at Los Angeles Public Library, then set up experiments on his dining room table. Everson and Gorrell wound yards of copper wire to make magnetic coils. Cliff Gardner learned glassblowing, discovering an uncanny skill that enabled him to produce tubes of a shape and size the professionals said would implode—his biggest triumph was a cathode ray tube with a flat end for a screen. Farnsworth, into everything, found it hard to sleep. His backers found it hard to raise money. Every single bank in Los Angeles turned them down. There were many heartbreaking moments, too, in the three years from 1926, following the commitment by the Crocker Bank—an explosion that sprayed molten potassium in Gardner's eyes when he was attempting to purify it for a more efficient photoelectric surface for camera and screen; a fire; magnetic coils that failed the electrons; inadequate amplification from the Lee de Forest Audion vacuum tubes, then distortion when they used a number in series. When Everson and Gorrell dropped in on the dining room lab, their opening line came to be, "Hi, Phil! Got the damned thing working yet?"

Gorrell was not there on the morning of September 7, 1927, when Farnsworth tried for the twelfth time to transmit an image. "Put the slide in, Cliff," he called to Gardner in the next room. Gardner positioned the slide of a triangle in front of the camera. Farnsworth did not see a triangle on the small bluish square of light on the receiving tube, but there was a line—and when Gardner rotated the slide 90 degrees Farnsworth saw the line rotate 90 degrees. "That's it, folks," he said matter-of-factly. "We've done it. There you have electronic television." Pem gave him what she described as "a very big, unbusinesslike hug." The jubilant Everson and Farnsworth telegraphed Gorrell in Los Angeles: "The Damned Thing Works."

Farnsworth's rival, Vladimir Zworykin, with his receiver tube. When he saw Farnsworth's camera tube, he said, "This is a beautiful instrument. I wish I had invented it myself." Then he had it copied.

Dramatic progress had been made by the time McCargar gave his shutdown order and even more in the months after he was repulsed. In early 1930 Farnsworth sent a visual signal a mile by radio. In the lab, he showed a clip of Walt Disney's *Steamboat Willie* and archival film fragments of the sensational Jack Dempsey/Gene Tunney prizefight from September 1927. Green Street was honored by visits from Marconi, Lee de Forest and Ernest Lawrence, the emperor of atom smashers. Then Douglas Fairbanks Sr.

and his wife, Mary Pickford, came in order to be televised room to room, only to have their famous features mangled by a loose wire not detected until their disappointed, but gracious, departure. Extraspecial preparation was made to welcome the next important guest: Dr. Vladimir Zworykin.

Why did Farnsworth respond as warmly as he did to Zworykin's request to see the Green Street lab? He knew Zworykin was exploring electronic television at Westinghouse and he had been ever mindful of injunctions from his father and his science teacher to keep mum. Biographer Schwarz implies that Farnsworth did not know Zworykin was really working for Sarnoff, and Pem Farnsworth has written that it was "a few years" before they realized that Zworykin was on a spying mission at the behest of Sarnoff and RCA. This does not seem quite right. A month later, as Schwartz records, Green Street received another Sarnoff man, Albert Murray, who made no pretense that he was other than the head of advanced development; he even brought an RCA patent attorney with him. It is more probable that Farnsworth was aware of the Westinghouse-RCA connection in April, but he hoped he could sell Zworykin on the idea of Westinghouse taking a license on his patent, rather than buying it outright as his investors would have liked. He must have known it was a risk letting a competitor well informed on electronics see what he was up to, but he took comfort in his patents—and honor among men of science. Zworykin's application for patents registered in 1923 had gotten nowhere—he never submitted a model—and his challenges to Farnsworth's applications in the "interference" process seem to have fallen by the wayside. Farnsworth inferred that approval for his own applications was close, and indeed four months later, in August 1930, his confidence was vindicated by notice that he had been granted patent 1,773,980 for his camera and patent 1,773,981 for his receiver.

Zworykin was to put those four months to good use. When he arrived at Green Street for a three-day visit, and dinner with Phil and Pem, he was behind. He had invented a receiving tube he called a kinescope that was brighter than Farnsworth's Oscillite picture tube, but he was nowhere near solving the more difficult challenge of transmitting, as Farnsworth had done with an electron multiplier he invented to solve the problem of amplification from vacuum tubes. Farnsworth bubbled with good-natured excitement, describing every one of his achievements to the older man. He went so far as to tell Cliff Gardner to build an Image Dissector while Zworykin watched. When it was finished, Zworykin caressed the tube, saying, "This is a beautiful instrument. I wish I had invented it

myself." He then proceeded to simulate doing just that. He sent detailed instructions to the Westinghouse tube laboratory in Pittsburgh and had several copies of Farnsworth's tubes with him when he got back to the RCA's research lab at Camden.

Sarnoff backed Zworykin with more money and more men while McCargar and the other investors went round the country trying to find a buyer for the gold mine on Green Street: By April 1931 Farnsworth had all phases of the system working well. At RCA, the usual sycophants assured Sarnoff that RCA could do anything Farnsworth could, but Sarnoff's antennae were always attuned to patent law. He took it on himself to descend on Green Street in April 1931 to weigh how much of a nuisance this 24-year-old Farnsworth might be. Farnsworth was held up in New York on a possible sale of the company, so Sarnoff met with Everson. The immaculate big-city tycoon was cordial with the Green Street team, but he let everyone know that Zworykin's work made it possible to avoid the Farnsworth patents. As he said goodbye, Sarnoff told Everson, "There's nothing here we'll need." He was bluffing. Soon after, he bluffed some more with an offer of $100,000 for the entire company and the services of Farnsworth. Even McCargar and the nervous investors knew that Farnsworth was right in his instant rejection. From their viewpoint, it barely covered their investment; from Farnsworth's, it removed his candidacy for inclusion in the pantheon of Marconi, Edison and Morse. Biographer Stashower speculates that things might have developed differently if Sarnoff had met Farnsworth himself. Might he not have seen that the youngster had the "spark in the eye" Sarnoff looked for; might he have warmed to a youngster as determined as he had been? It is more likely that Sarnoff, surrounded by his millions and his minions at RCA, had become too used to issuing commands.

In June 1931, in great stealth, Farnsworth and the Crocker syndicate consummated a deal with the Philadelphia Storage Battery Company (Philco). It was the largest makers of radios, on which it paid a royalty to RCA, and it resented Sarnoff's obvious intentions to establish another monopoly. Very quickly, a secret deal was signed by which Philco employed Farnsworth and his

men for two years in return for a non-exclusive license to exploit his patents for the manufacture of television sets. A mile away from Philco, across the Delaware River, Zworykin worked on his systems. The secret that Philco was in television with Farnsworth did not survive the erection of a broadcasting tower on top of Philco's plant. Farnsworth had focused a camera on the University of Pennsylvania swimming pool. An RCA engineer picked up the signal, watched avidly, then phoned across the river to say, "Do you know some of those students are swimming naked?" Sarnoff, who was now proclaiming that RCA was leading the way in commercial electronic television, reached for his hatchet. (It was the competition he minded, not the nudes.) Schwartz credibly suggests that RCA blackmailed Philco, warning that it was risking its licensing arrangement with RCA to make radios. Philco did not renew its agreement with Farnsworth.

Farnsworth was now a stressed and sickly young man, stricken by the death of his 18-month-old son and strains in his marriage from overwork. But he kept innovating as far as his resources would permit. He developed a mobile version of his Image Dissector and installed it for ten days at the Franklin Institute in downtown Philadelphia, solving the problem of programming by letting the crowds wave at themselves. He put his camera on the roof and shot the moon, hailing it as "the first recorded use of television in astronomy." Farnsworth Television intermittently broadcast entertainment programming as W3XPF from the Philadelphia suburb of Wyndmoor, which hardly anybody saw because only Farnsworth and his associates had television sets.

All the time, Farnsworth was up against a not-so-quiet whispering campaign from RCA to stymie any more licensing deals. In exasperation, he filed suit against RCA. At stake was the control of television. Batteries of RCA lawyers opposed him, arguing the precedence of Zworykin's 1923 (unapproved) application. The electric moment in the trial before the patent examiners was when Farnsworth's attorney tracked down his high school teacher, Justin Tolman, who turned up in court with a worn piece of notebook paper retrieved from his attic. "This was made for me by

Philo in early 1922," said Tolman, unfolding the note from his breast pocket. It was the Image Dissector, predating Zworykin's patent application of 1923. In any event, the examiners declared on July 22, 1935, that, irrespective of the date of filing, Zworykin's application did not describe a credible electronic scanner. "Philo Taylor Farnsworth," said the ruling, "is awarded the priority of invention on his system of television."

Not for nothing had Sarnoff trained himself as a black-belt bureaucrat. He had his lawyers search for a pliant court where they might appeal, fighting for time, while enveloping Zworykin's work in secrecy. He was now in no hurry to press the Federal Communications Commission to approve national commercial television or to whet the public's appetite. Every year that went by was one more year off the 17-year life of Farnsworth's patent. Sarnoff refrained from staging another demonstration until July 1936, this time in the new Radio City headquarters he had brilliantly contrived in Rockefeller Plaza. The show, based on Zworykin's technology, did not impress E. B. White of *The New Yorker:* "President Roosevelt's face not only came and went, it came and went *underwater*."

Farnsworth did not have the resources to compete with Sarnoff's grandstanding, but he determined to stay ahead in the science. He filed no fewer than 22 patent applications the same year. He made a cross-licensing agreement with AT&T, giving him the right to send his signals over AT&T inner-city cables. He and Everson secured the services of the investment banking firm of Kuhn, Loeb and Company (one of the earliest backers of Edwin Land's Polaroid Corporation, page 384) for an initial public offering for Farnsworth Television and Radio. They recruited Edwin "Nick" Nicholas, RCA's licensing manager no less, and they made a bid for the Capehart Company in Fort Wayne, Indiana, makers of quality radios and phonographs. The intention was to continue radio production until FCC approval of standards for television transmission enabled them to make television sets. For more than 18 months—more time off Farnsworth's patent—the SEC maddeningly mired Farnsworth in red tape so that the stock

could not be floated until March 1939. He raised $3 million.

Sarnoff was furious that Nicholas had left him. Nicholas knew that if RCA went ahead with a service of electronic television broadcasting or the manufacture of sets it could be stopped and penalized for breaching Farnsworth's patent. But until RCA started a broadcasting service, Farnsworth, too, was prevented from full exploitation of his invention. It was a game of chicken, but Sarnoff, for his reputation and future profits, was determined to launch television with the RCA logo. He saw a dazzling opportunity in the 1939 world's fair. Ten days before President Roosevelt was due to open the fair, on the bright spring day of April 20, 1939, the portly 48-year-old Sarnoff walked to a podium to declare that NBC was launching the first regular electronic television schedules. "It is with a feeling of humbleness that I come to this moment to announcing the birth in this country of a new art so important in its implications that it is bound to affect all society. . . . Now, ladies and gentlemen, we add sight to sound." It was a surreal moment. Television cameras unveiled around the grounds showed television cameras unveiled around the grounds. The next morning, television cabinets set up for sale in the major stores showed television cabinets set up for sale. It was not a broadcasting system—just shortwave television over a short distance. And of course sight had been added to sound years before by Philo Farnsworth. There was no mention of his name, nor Zworykin's for that matter. It was Sarnoff's day. He was preempting the president, pre-empting the FCC, defying the patent examiners and erasing the inventor of television from history, and an amnesiac press lapped it up. It was a magnificent coup, it was inspired marketing, but it was bogus history.

On the fair's official opening day, Farnsworth felt sick when he passed a television in a store window and caught sight of Sarnoff at the podium with President Roosevelt. "Sarnoff was clearly taking credit for the invention," writes Schwartz, "in a way that Farnsworth knew he could never match, creating an impression that could never be erased. Sarnoff was doing this through the very power of television itself."

Sarnoff was masterful at traversing the high wire, but he knew he had to come down to the ground when the crowds had gone home. The following month RCA quietly opened negotiations with Nicholas to avoid suit for breach of patent. Sarnoff agreed to pay Farnsworth Television and Radio Corporation $1 million and a royalty on every set sold.

World War II changed everything. The manufacture of TV sets was banned and NBC's schedules put on ice. It cost Farnsworth years of royalty revenue. He sank into alcoholic depression. When he was a skeleton of 100 pounds, one doctor prescribed smoking to calm his nerves and another put him on addictive chloral hydrate. Sarnoff in 1944 became a brigadier general on Supreme Commander Eisenhower's staff in England, and thereafter anyone who forgot to call him "General" Sarnoff was in the deep freeze. Vanity may have impelled Sarnoff's spectacular rise, but his reputation in posterity was a hostage to his exaggerations and his craving for recognition. He was an important promoter of innovation. He deserved acclaim after the war for the way he pioneered a system of color TV compatible with black and white, defeating the noncompatible electromechanical system pushed by Bill Paley at CBS. But it was an honor too far when the Radio Television Manufacturers Association named him (at his suggestion) the Father of Television. (RCA staff were instructed to refer to Zworykin as the Inventor of Television. This was ironic, given that Zworykin was honored early in 1940 as an inventor on the level of Henry Ford, Willis Carrier, Edwin Land and Edwin Armstrong.)

With the love of his wife and family, Farnsworth recovered his health. His company made television sets in Fort Wayne. He disdained what he saw on them for the most part. (Zworykin would not let his children turn on a television.) But Farnsworth watched television with Pem on July 20, 1969, for a closer look at the moon than the one he had filmed in 1934. He saw the astronauts Neil Armstrong and Buzz Aldrin make their footprints, proud to note the pictures were coming from a miniature version of his Image Dissector. "Pem," he said, "this has made it all worthwhile."

WALT DISNEY How a wide-eyed boy from a Missouri farm beat the big bad wolves and branded an entertainment empire

1901–1966

Hi, folks! Here is a thought from my creator, Mr. Disney:

"You may not realize it when it happens, but a kick in the teeth may be the best thing in the world for you."

I know what he means. But Mr. Disney had more kicks than I did in his crazy cartoons, and he still made all that magic.

Mickey Mouse's quotation is accurate. Over 40 years of creative endeavor, Walt Disney had more than his fair share of kicks in the teeth, yet every setback catapulted him to a success he might not otherwise have enjoyed. Pure luck cannot explain why each frustrating diversion took him to the right higher road; it would not have been such a consistent feature of his life unless luck walked hand in hand with character—blind optimism matched with a vivid determination. "I function better when things are going badly," he reflected, "than when they're as smooth as whipped cream." But even that is only a partial explanation. Walt Disney was also a winner in the genetic lottery of siblings. In his steadfast older brother Roy O. Disney, he had a resourceful business manager loyally willing to stay in the shadows but ready to try and make the gambler in the footlights justify another throw of the dice. They had monumental rows but reconciled as only brothers can. Partnerships are often a key characteristic of innovation. This was one of the more enduring.

Their father, Elias Disney (1859–1941), was in on the building of the Union Pacific line through Colorado. He was an apprentice carpenter then and was everything else thereafter as he wandered between Florida and the Midwest as a mail carrier, lemon grower, fiddler, schoolteacher, contractor,

Cartoonist Walt, at 19, was eager to brand his name, practicing his signature over and over again.

cabinetmaker, farmer, newspaper distributor and manufacturer. He collected Populist opinions along the way and voted for the Socialist Eugene Debs. Elias was a disciplined man, a teetotaler and churchgoer, not very lucky in his enterprises except when, in a Chicago suburb, he almost single-handedly built his own wooden house and profitably sold two more, designed by his wife, Flora, who had taught school. Roy E., the son of Roy O., remembers her as a "dream grandmother, very warm with a great sense of fun."

Walter Elias Disney was born there on December 5, 1901, the fourth of five children, preceded by Herbert (1888), Raymond (1890) and Roy Oliver Disney (1893) and followed by Ruth Disney (1903).

Elias and Flora named Walt after the preacher who baptized him, the Reverend Walter Parr, the fundamentalist minister of their Congregational church (built by Elias). Bob Thomas, who penned authorized biographies of both Walt and Roy, writes that Elias "thought nothing of taking a switch to his son or the fat part of his belt." That was not at all unusual for the time, but Walt took the beatings badly. One can perhaps see here the germ of Walt's realization of the power of dramatic trans-

formation of the human form, of imagination and fairy tale as an escape from unpleasant reality. "Walt would bury his head in the bend of Roy's elbow," Thomas writes, "and ask if the man who beat him was really his father or just some mean old man who looked like him and wanted only to frighten or hurt him." Elias was clearly not a bundle of laughs—Roy E. describes his grandfather as straitlaced, puritanical, humorless—but stern Elias was a conscientious father: Concern for the children's welfare was the principal reason he moved the family out of the mean streets of Chicago in 1906 to a 40-acre farm in the small community of Marceline in Missouri. Elias and the three older sons did the grunt work of slopping the pigs, mending the fences, milking the cow. "Because Walt didn't have to work with animals the way the older boys did," Roy E. told the writer Tony Schwartz, "he became friends with the animals instead." Dairy farmers often give their cows a name, but Walt also conducted imaginary conversations in the cowshed, and he wandered off into the nearby forest, intrigued by the behavior of birds and rabbits, squirrels and foxes, chipmunks and raccoons, who would eventually show up in his animated movies. (He suffered terrible remorse when, after he was clawed by an owl, he instinctively knocked it to the ground and stamped it to death.)

It was a wrenching moment for Walt and Roy when the two eldest boys ran away, and Elias decided to sell up and move to Kansas City. Roy told biographer Thomas how he and Walt cried their hearts out when a little six-month-old colt they had tamed and broken in was sold off. "Later on that day, we were down in town and here was this farmer with his rig hitched up to the rigging rack . . . and that damn little colt saw us across the street, and he whinnied and whinnied and reared back on his tie-down. We went over and hugged him and cried over him and that was the last we ever saw of him."

In booming Kansas City, the restless Elias took up distribution of the *Kansas City Star* (whose bylines included Ernest Hemingway). Now at grammar school, Walt retreated from academics and spent his time sketching flowers and trees—often with faces instead of petals and arms in place of leaves. His artistic focus was still more intense when his father moved the family yet again, this time back to Chicago, where Walt entered night school and attended classes at the Illinois Institute of Art. A talent that might have become merely sentimental was toughened by the Red Cross. He was eight months' short of his 16th birthday on America's entry into World War I in April 1917, but when a recruiting drive for the American Ambulance Corps hit town a year later, he signed on, falsifying his birthday year with his mother's acquiescence. Stricken by influenza, he missed the first shipment to France and sailed a week after the November 11 armistice. He drove Red Cross supply trucks (like Hemingway) and chauffeured officers in Paris. He filled the long spare hours in between by drawing cartoons for soldiers to send home and for submission to magazines, which all rejected them. A husky five-foot-ten, he went through the rites of passage to adulthood; in an all-night poker game, he won the then-significant sum of $300. By the time he came home late in 1919, Walt was alive to the rewards of risk taking.

It is at this point one can discern the questing spirit that was to mark Walt's whole life—and the inner conviction that he would succeed. He disconcerted his father by spurning a $25-a-week job in his Chicago jelly factory (Elias's latest misadventure). Instead he went back to Kansas City, trying for a job at the *Kansas City Star,* first as an artist, then a copyboy, then a truck driver. He was turned down at every instance, but, unabashed, the tall, sleek-haired and smartly dressed Walt walked confidently with his portfolio into the Pesmen-Rubin Commercial Art Studio and was hired to prepare advertisements. The studio job did not last more than six weeks, but it was the scene of an auspicious

encounter: Walt Disney met Ubbe ("Ub" for short) Iwerks (1901–1971). A few months older than Walt, Iwerks (of Dutch extraction) was doing lettering work with a free-swinging brush and was happy to teach the teenage newcomer some tricks of the trade. According to his biographers, Leslie Iwerks (a granddaughter) and John Kenworthy in *The Hand Behind the Mouse*, Iwerks was amused that Walt did not follow the custom of practicing hand lettering by drawing the alphabet but instead kept designing his own name over and over again: "Walter

Iwerks joined his partner at the film ad company.

Both young artists were inseparable devotees of Charlie Chaplin and the early black-and-white Felix the Cat cartoons coming out of New York, and especially the films of Winsor McCay reacting in vaudeville to his finely drawn cartoon of an irascible dinosaur; *Gertie the Dinosaur* (1914) stimulated both Walt and Ub to think about ways of combining live action with animation. They haunted the Kansas City Public Library to pore over new instruc-

solution. He was a clever technician and a zanily brilliant draftsman: One of his fellow animators affectionately noted that "'Iwerks' is 'screwy' spelled backward." Walt, less capable as an artist, was the born storyteller and fizzed with ideas to sell to the local movie houses. The two friends moonlighted in a garage to make a series of one-minute animated jokes called Laugh-O-grams. Kansas City people had to wait a long time for streetcars—plagued by strikes and political corruption—so they were in the mood for the Disney-Iwerks

A CLOSE FAMILY:

(left) Baby Ruth always looked up to her big brother Walt, about two and four here, respectively. For the Greene biographers, Ruth's main childhood memory was "how very gritty he was." Walt would never admit pain.

(right) Walt and his mother, Flora. She held the family together.

(center) Roy (right) looked out for his kid brother, who was always tapping him for a quarter. This is Chicago 1917, just before Walt, too, joined the armed forces, lying about his age.

Disney, W. E. Disney, Walt Disney, Walter Elias Disney. Walter asked Ubbe which he preferred. 'Walt Disney,' Ub replied. Walt it was." Even at that early stage, Walt was into branding.

Pesmen-Rubin let Walt go after the Christmas advertising rush. He found work as a mailman, and a few weeks later Iwerks, too, was fired. Iwerks was astonished that Disney, with so little experience, was eager to set up in business; Iwerks-Disney Commercial Artists was formed in January 1920 and won a few commissions before Walt, with Ub's approval, took a well-paying job with the Kansas City Film Ad Company, leaving Ub to run their company. Acquiring business was not a talent of the shy and introverted Ub. Iwerks-Disney was bankrupt by March and

tion books on animation, learning most from Eadweard Muybridge's studies in motion photography (the source of Edison's inspiration for the movie camera, page 168). The animation practiced at the Kansas City Film Ad Company was crude, but they made the movements smoother, and Ub contrived a mechanical advance. The method was to cut human and animal figures out of paper, film them in one position, move them slightly and film them again to create the illusion of movement. This required the concentration of the two of them, one to crank the camera and the other to move the drawings. To make this easier, Ub rigged up a telegraph-key switch to activate the camera so that one of them could do everything while just sitting at the animation table. It was a classic Iwerks

cartoon of a young man at a streetcar stop metamorphosing into a long-bearded Methuselah and another of a young woman lost in flowers proliferating from the design on her stockings.

On May 23, 1922, the ambitious Walt left the advertising studio (and Iwerks) to set up as Laugh-O-gram Films, backed by $15,000 from Kansas City professionals. Walt had discovered his ability to charm and lead, but it was his concept of moving beyond single one- or two-minute gags to tell a proper story that attracted a dozen young people to work on a series of animated stories that included *Little Red Riding Hood, Jack and the Beanstalk,* and *The Four Musicians of Bremen.*

Walt did a distribution deal with a New York company called Pictorial Clubs, elated that they agreed to pay $1,800 for each of the first half-dozen cartoons. Iwerks joined him then, and the madcap crew of animators got busy on Walt's story lines, quickly shipping off eight or nine Laugh-O-grams and "Lafflets" and a three-reeler: *A Pirate for a Day.* But Pictorial had paid only a $100 deposit and never came through with the roughly $11,000 they owed; Pictorial and Laugh-O-Grams were both broke by the middle of 1923. So was Walt. Out of money for rent, he took to sleeping in the office and bathing at the Union Station showers. "It was probably the blackest time of my life," he later told an interviewer. "I really knew what hardship and hunger were." (His parents had followed their son Herbert to Portland in November 1921.)

Salvation appeared in the unlikely shape of a local dentist, who called just before the last curtain fell. He wanted to know if Walt could produce a film for a Kansas City dental institute on dental hygiene for children. Walt was invited to go over and sign a contract but instead explained his predicament: "I can't. I haven't got any shoes. I left them at the shoemaker and I don't have the dollar-fifty to pick them up." The dentist drove over to pick up Walt, paid to get the shoes and the two of them signed a contract for $500.

Tommy Tucker's Tooth was a temporary filling enabling Disney to pitch his next big idea, a winsome little Alice-in-Wonderland girl who would have adventures with animated characters (similar to *Who Framed Roger Rabbit* but decades ahead). M. J. (Margaret) Winkler in New York expressed interest in national distribution, but before Walt could finish the first adventure, *Alice's Wonderland,* the dental money ran out. Roy Disney was meanwhile on his back in a sanatorium in Sawtelle, California, a victim of tuberculosis acquired in World War I service in the navy. From his sickbed, Roy advised Walt to file for bankruptcy and try to find work in Hollywood. By the time he paid his one-way ticket from Kansas to Los Angeles in the summer of 1923, the skinny young Walt was down to his last $40—but his suitcase bulged with hope. "Tomorrow was always going to be the answer to all his problems," said Roy.

Hollywood in 1923 glistened with celebrity. Rudolph Valentino, Clara Bow,

William S. Hart, Tom Mix, Lillian Gish and Gloria Swanson were the silent stars; Cecil B. DeMille, D. W. Griffith, Erich von Stroheim were gods as directors; and Louis B. Mayer and Irving Thalberg were on the way to forming Metro-Goldwyn-Mayer. None of this stardust touched Disney. He talked his way onto the Universal lot brandishing a business card as the Kansas City representative of "Universal and Selznick Newsreels" and tried to sell himself as a director. The nearest he got to Tinseltown was as an extra hired for a cavalry charge, which was rained out, and when the sun shone the studio hired a new set of riders. Walt was forced back into cartooning: The idea that it was his life's dedication from day one is moonshine. Had he been given any job in Hollywood, he would have cheerfully abandoned animation in light of the intense competition from New York's cartoon factories. His father's brother, Robert, had given him board, and Walt set up orange crates in Uncle Bob's garage to sketch some more one-minute gags. He wrote a letter of elegant exaggeration to Margaret Winkler: "I am establishing a new studio in Los Angeles for the purpose of producing the

new and novel series of cartoons I have previously written you about. . . . I am taking with me a select number of my former staff and will in a very short time be producing at regular intervals."

He whistled in the dark very well. Near midnight on October 15, he walked down the row of beds on the screened porch of Roy's hospital to rouse him with a telegram from Winkler offering $1,500 for each of six "Alice Comedies" for national distribution. The amazing brother, dismissing his doctors' warnings, discharged himself from the hospital the next day. He invested the $285 he had saved from his navy disability pension and rounded up $2,500 from a mortgage on their parents' home in Portland and $500 from Uncle Robert. They rented a small storefront in Hollywood and stenciled a sign in the window: DISNEY BROTHERS STUDIO.

Roy E. reflects on this propitious partnership: "Without Walt, my father would probably have ended up as president of a bank in Kansas City. He was smarter than hell—as smart as Walt in his own way—but he had no ambitions for glory." When Walt, newly married to an inker at

It is 1926 and Walt has just opened his new studio on Hyperion Avenue. Little Margie Gay, who had replaced Virginia Davis in the Alice Comedies, brought along some of her friends, including Julius the Cat.

his studio named Lillian Bounds, came back from his honeymoon in July 1925, Roy O. sugested that he change the name of the company from Disney Brothers to Walt Disney Studios. It was a happy circumstance that Roy (also newly married to childhood sweetheart Edna Francis) felt no great need for center stage—and also that Walt was always ready to recognize his own limitations. Six episodes into the series, Winkler told him his ideas in the Alice series were wonderful, but his execution was not good enough. Not in the least put out, Walt set about seducing Ubbe Iwerks to abandon his well-paying job in Kansas and drive into the western sunrise. "Boy, you will never regret it," wrote Walt. "This is the place for you—a real country to work and play in—no kidding—don't change your mind—remember what ol' Horace Greeley said: 'Go West, young man.' . . . PS I wouldn't live in Kansas City now if you gave me the place. Yep. You bet. Hooray for Hollywood!" Iwerks brought not just his flair for animation but also his astounding productivity. He could finish as many as 700 rough drawings a day, filled in by others. Hugh Harman and Rudy Ising, animators from the Laugh-O-gram days, closed their own Arabian Nights Cartoon Studio in Kansas and trekked west to Disney's new location at 2719 Hyperion Avenue in the Silver Lake district, near Mack Sennett's comedy factory. There was no longer a need for Walt to animate. He recognized that Ubbe and the others were more talented, but he could not stop himself from snooping on the animators' desks when they had gone home for the night. It irritated them as much as his nagging perfectionism, combined with a contradictory insistence on speed.

The long-running Alice Comedies were a success, but Walt Disney Studios was still no more than a hand-to-mouth operation. So might it have remained but for Walt's reaction to the first of the Hollywood betrayals. Margaret Winkler had handed over her business to a new husband, Charlie Mintz, a former booking agent, and Mintz was a natural predator. The head of Universal Pictures, Carl Laemmle, had a rabbit in mind for a new cartoon series. Walt and Ub (as he now called himself) produced some sketches, Mintz came up with the name Oswald the Lucky Rabbit and Laemmle agreed to pay a $2,000 advance for the first film. He was not satisfied with the first effort. "Audiences like their characters young, trim and smart. This one is practically decrepit." Together Walt and Ub produced a sleeker "Ozzie." Sure enough, the second effort, *Trolley Troubles,* got rave reviews, but when the contract came up for renewal in February 1928, Mintz had prepared a nasty surprise for Walt, who was on his way to New York by rail from California: He had secretly invited Walt's entire staff of animators to join a proposed Charles Mintz Studios, and all except Ub and two apprentices had signed up with him. Walt had traveled to New York in good spirits with his wife, Lillian, expecting an increase in the fee and a share of profits. He was still only 26; he had grown a mustache and smoked a pipe to look older for contract negotiations. Mintz not only proposed to pay $400 less for each cartoon but demanded that he, Mintz, be a full partner with the Disney brothers on all future films. After all, said Mintz, he now owned Walt's animators. Delivering the coup de grâce, he pointed out that he and Universal owned the rights to Oswald. Walt had not bothered with the small print.

Mintz had Walt in a vise. Walking away from the deal would leave the Disney company with no characters, no contracts, no cash flowing in and virtually no animators. When Walt phoned Roy to confirm the wounding defections, Roy urged him to make the best settlement he could. Walt went back to Mintz's office with a different purpose in mind. "Here. You can have the little bastard!" he reportedly told Mintz, "He's all yours and good luck to you." His rejection of Mintz was the turning point in the history of the Walt Disney Studios. "Never again will I work for anyone else," Walt told Lillian. Taking the gamble of starting all over again was reckless, but cleaving to his independence became central to all of Disney's subsequent successes.

Walt did not wire Roy to tell him what he had done. He called on a few distributors and tried to dream up another character to replace the rabbit, but cartoon makers seemed to have emptied the menagerie. "About the only thing they hadn't featured," wrote Walt, "was the mouse."

Disney would often say later that the character he sketched on the long train ride back to California with Lillian was based on one particular mouse bold enough to amble across his desk in his old Kansas City studio: "He seemed to have a personality of his own." By the time the train reached the Midwest, Walt wrote, "I had dressed my dream mouse in a pair of red velvet pants with two pearly buttons." He was Mortimer Mouse for all of five minutes until Lillian said that sounded stuffy. "Why not Mickey Mouse?" Why not indeed.

Mickey's birth is described differently by the 2001 biographers of Ub Iwerks. In this version, before Walt went to New York he had suggested "in the spring of 1928," without specifically mentioning a mouse, that Ub start thinking up new character ideas. They say Ub rifled through stacks of magazines looking at animals while doodling circles on his drawing board. "An old publicity photo from the Alice days came to mind," they write. "In 1925 Hugh Harman had drawn some sketches of mice around a photograph of Walt. With minor alterations to the nose of Harman's mice, Ub took the basic design elements of Oswald and transformed his circles into a new character. The mouse was born." Ub, say his biographers, also created the female counterpart, named Minnie by Walt.

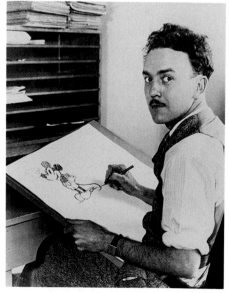

MICKEY TAKES OFF: Ub Iwerks was the fast, fast hand behind the mouse. For the first Mickey cartoon, based on Lindbergh's historic flight to Paris, Ub made an astounding 700 drawings a day.

The inspiration on the train has the gossamer of a fairy story but also a certain credibility. There seems no reason why Walt should ask Ub to look for other animals before he left for New York, because he went full of confidence that Oswald was their meal ticket. Whatever the case, there is no question that it was Ub who executed the animations of Mickey, while Walt wrote the stories. Mickey was remarkably similar to Ozzie, Mickey's flamboyantly circular ears being the main distinction from Ozzie's lugubrious lobes. Their first two Mickey Mouse cartoons, *Plane Crazy* (after the Lindbergh flight) and *The Gallopin' Gaucho* did not take off; the trial audiences laughed, but no distributor could be found. Walt was still willing to bet on himself even as he struggled to raise money for *Steamboat Willie,* a third Mickey Mouse adventure.

Steamboat Willie had already been finished as a silent short, based on a Buster Keaton comedy. Then Walt went to the movies and heard an electric Al Jolson singing in *The Jazz Singer,* the first real feature-length "talkie," which had been released late in 1927. Film executives were still divided about sound in 1927–28. "None of them are positive how it is all going to turn out," Walt wrote to Roy from New York, "but I have come to this definite conclusion: Sound effects and talking pictures are more than a mere novelty. They are here to stay and in time will develop into a wonderful thing." The critic Richard Schickel argues that Walt's distinction was to see sound as not just an addition to the movies but also a force that would fundamentally transform them. "He was the first moviemaker to resolve the aesthetically disruptive fight between sight and sound through the simple method of fusion, making them absolutely 'co-expressible,' with neither one dominant nor carrying more than a fair share of the film's weight."

Disney's immediate dilemma was less grand. It was how to postsynchronize music for *Steamboat Willie.* Silent film ran at 18 frames a second and sound film at 24 frames a second. A new apprentice animator, Wilfred Jackson, borrowed a metronome from his music-teacher mother with the idea of syncopating the tunes to the film frames. Walt tried it out. With Jackson playing the harmonica, Walt calculated on a score sheet how many frames should elapse to match every bar of the music. It was tricky work, but Walt, recalls John Hench, "was absolutely obsessed. . . . He had kind of a sense of destiny." A glimpse of his ear for sound and the intensity of effort can be seen in just one scene as written by Walt:

SCENE 2

Close-up of Mickey in cabin of wheelhouse, keeping time to last two measures of verse of "Steamboat Bill." With gesture he starts whistling the chorus in perfect time to music . . . his body keeping time with every other beat. At the end of every two measures he twirls wheel, which makes ratchet sound as it spins. He takes in breath at proper time according to music. When he finishes last measure, he reaches up and pulls on whistle cord above his head (use FIFE to imitate his whistle).

Before he took the marked-up print to New York to try and find a sound specialist, Walt assembled the animators' wives and girlfriends in the studio for a dry run. Roy projected Ub's animated Mickey onto a bedsheet across a doorway. Walt tried to breathe Mickey's squeals into the microphone at just the right moment and then voiced the squawk of a parrot announcing, "Man overboard!" Ub was on the washboard. Someone else was on percussion imitating the sound of a xylophone on a cow's teeth. One staffer played the theme tune on a harmonica; another imitated a musical piglet. It was chaos, and they had the time of their lives. The watchers were not impressed: "We had absolutely no idea what was going on," said Lillian. "And in any case it sounded terrible." According to Mildred Iwerks in *The Hand Behind the Mouse,* she was gossiping with the other wives in the lobby when Walt ran out and exclaimed, "You're here talking about babies and we're in there making history!"

But the wives were not alone. Walt waited in New York's film and recording offices—and waited. Nobody was interested in Walt's proposition. He wrote to Roy: "Personally, I am sick of this picture *Steamboat Willie.* I am very nervous and upset and I guess that has a lot to do with my attitude in the matter. This DAMN TOWN is enough to give anybody THE HEEBIE-JEEBIES. I sure wouldn't make a good traveling salesman. I can't mix with strangers and enjoy myself. . . . I have so much time to kill at night that I almost go nuts." All the same, he suggested that Ub go ahead with another musical cartoon, *The Barn Dance.*

It was in this vulnerable mood that Walt encountered Pat Powers, who owned the Cinephone sound system. Walt was warmed by his Irish charm and apparently intimate connections with the big names. "He is a dandy," wrote Walt, as wide-eyed as Pinocchio, telling Roy how this marvelous man Pat would protect them from the schemers in a city "just full of tricks that would fool a greenhorn." Powers got the distribution rights; Walt got access to the Cinephone system. The orchestra conductor nominated by Powers then proceeded to botch the first recording, insisting he did not need Walt's marked-up score, only to find his musicians could not keep time with the action. To pay for a second attempt, this time with the marked score, Walt had to sell his beloved Moon cabriolet roadster. Finding a distributor was another frustrating experience. The major distributors laughed, but they didn't buy. Finally, the ebullient Harry Reichenbach, who ran the Colony Theater in New York, persuaded Walt that a public screening was his best hope. "Those guys," said Reichenbach of the majors, "don't know what's good until the public tells them."

The public told them loud and clear on November 18, 1928: *Steamboat Willie* was a smash hit, and Mickey Mouse was on his way to iconic status. "He is not a little mouse," said Walt. "He only looks like one. He is Youth, the Great Unlicked, and

Uncontaminated." Scores of dissertations and psychoanalytic treatises have been written about his universal appeal, many of them analyzing his character and what it said about Walt himself. Of course, Walt provided Mickey's falsetto voice for many years, but the dual identity may be more significant than that. In an interview for this book, the modern-day Disney chief, Michael Eisner, said that the revelation for him was to listen to an audiotape of an unedited *Edgar Bergen/Charlie McCarthy Show*. "Normally Walt did Mickey's voice in a studio and did it over and over until he got it right. But this particular show was live and it became startlingly obvious that Walt and Mickey were one and the same person. He kept falling back into Walt when he was trying to be Mickey and back to Mickey when he was Walt, just as I later discovered that Kermit the Frog was part of Jim Henson and Charlie Brown was Charles Schultz. These great characters are not created by committee, nor are they fiction. They are real and alive and the alter egos of their creators."

This is an intriguing analysis, though there is admittedly not quite a perfect match. Mickey was a sunnier character than Walt; perhaps it would be truer to say that Walt projected onto Mickey the character he would want to be—always upbeat, resilient, trusting, clear about the difference between right and wrong, loyal in his relationships, "a nice fellow who never does anybody any harm." Walt had many of those virtues, if not as consistently on display as Mickey's. He was an egalitarian who hated pomp and sycophancy. He was remarkably without rancor against those who wronged him. He had not a malicious or jealous bone in his body: When the animator Norm Ferguson insisted on $300 a week, Walt told Fergie it was as much as he earned—and promptly paid it.

Critics have had a field day suggesting that the real Walt Disney was the antithesis of the Walt Disney Company image. "It was an old company joke," writes John Taylor, "that Walt Disney could not have been hired to work at Disneyland. Walt cursed vigorously, chain-smoked, liked a stiff drink at the end of the day, and wore a mustache. . . ." But the paradox is pushed too far when it drifts from personal habits into

values. Walt was not a cynic. He strenuously and sentimentally believed in the wholesome family virtues embodied in Disney entertainment. "All right, I'm corny," he said, "but I think there's just about a hundred and forty million people in this country that are just as corny as I am." He was not a prude, but he could not stand dirty jokes. He was a faithful husband and a loving father. It could be argued that he was too much in tune with Main Street America in the '20s. Most conservative businessmen then were not inclined to hire blacks or Jews, and he was no different. Was he anti-Semitic? Michael Eisner, who is of Jewish extraction, puts it in perspective. "I don't believe he was anti-Semitic, just anti–Hollywood mogul. He was just against being screwed, and it happened that a number of the early movie types who betrayed him were Jewish. It seems to me he despised them not for their faith but for what they did, and he was right to do so, right to conclude that he must at all times keep creative control and never hazard the name of Disney in partnerships."

Aubrey Menon, the Anglo-Indian novelist, wrote that Disney's public-relations people assured him Walt was affable, open and avuncular, but instead "I met a tall somber man who appeared to be under the lash of some private demon." Schickel considered him "a withdrawn and suspicious man." Given his early experiences it is surprising he was anything less than a full-blown paranoid by the '50s, which he wasn't. It is certainly true he could withdraw; an animator said you could put an arm round Roy's shoulder but not Walt's. Another remarked on how he would walk past without a greeting. But this is more the action of a man immersed in thought than a perverse loner. In the creative sessions on record he is thoroughly engaged and engaging, expecting to be treated as an equal but demanding the best and then some. His cutting remarks could reduce a staffer to tears, but biographer Thomas tells a story of how Ken Ferguson, proffering a light for a cigarette, burned off Walt's mustache and singed his nose and was invited to lunch the next day without any mention of the incident. Outside the studio he was convivial, even playful. The way to his heart was to talk about trains; he built a half-mile

model railway in the yard of his home in Holmby Hills and delighted in putting on his engineer's cap to take visitors for a ride. When he became hooked on polo, he induced half a dozen employees to join him, got a polo expert to sketch the maneuvers on a blackboard and erected a polo cage at the studio so the staff could vent frustrations by whacking the wooden ball. He was, at the same time, very protective of the privacy of his home life with Lilly and his two daughters (one adopted).

Confronted by so many varying perceptions of Disney at different stages of his life, we may find it easier to accept that in many ways Mickey's derring-do and cartoon struggles with adversity are a simple metaphor for Walt's whole career. On Walt's return from making the deal with Powers, Roy had asked, "Did you read the contract?" Walt had replied, "No, I didn't read it. What the hell, I want the equipment." His instinct that sound would transform his fortunes was correct, but his instinct for deal making was calamitous. Walt had committed to pay Powers $26,000 for the equipment, and he had trusted his new best friend to pay the distribution receipts promptly. He didn't. Instead, he went behind Walt's back to Ub Iwerks, secretly offering to start an Iwerks Studio. Powers had chosen his moment well. Iwerks's longtime loyalty to Walt had been stretched past the breaking point in trying to match Walt's ambitions for *The Skeleton Dance,* the first in the Silly Symphonies series suggested by Carl Stalling in which the music would inspire the characters, and not the characters the music. Ub's *The Skeleton Dance* was an inspired work, but Walt could not keep his hands off the animation timings, reworking them after Ub had gone home. "Don't you touch my drawings!" the mild-mannered Ub had shouted at Walt, and won, but Walt was a demon of pressure in telegrams from New York: "Listen, Ub. Show some of your old SPEED. Work like hell, BOY. It's our own BIG CHANCE to make a real killing. Forget a lot of the fancy curves and pretty-looking drawings and devote your time to the ACTION. You can do it—I know you can. . . . GIVE HER HELL. Don't tell me it can't be done. It has got to be done. . . . So quit acting nervous and fidgety. Forget everything. . . . get that DAMN picture back here in time."

Iwerks joined Powers. The varying fortunes thereafter of Ub and Walt suggest the constraints on pure creativity in building a business. Iwerks, the more talented artist and technician, made some fine films at the Iwerks Studio *(Flip the Frog, The Brave Tin Soldier)*. In 1934, to give depth of field in his cartoons, he hand built a horizontal multiplane camera mostly from bits of an old Chevrolet for just $350. But Ub had no Roy to manage a volatile staff and encourage the best: He had a harsh Pat Powers. And his genius drew him inward: He elevated technique over showmanship. His biographers write that he spent less and less time defining story line and characters whereas the Disney cartoons of the time proved that "strong plots and endearing characters, and not necessarily technical achievements on their own, were the key ingredients to box office success." Ub's studio closed after six years while Walt's roared on. Ub owned 20 percent of the Walt Disney Company. When he left he took it as a lump sum of $2,920. By the time he came back to Disney, welcome and promoted by the forgiving Walt, that stake would have been worth well in excess of $1 million. If Iwerks's heirs had inherited his original 20 percent, it would be worth around $5 billion today (assuming no inheritance taxes).

It was not plain sailing at Disney as the Great Depression rolled on. Mickey and the Silly Symphonies were acclaimed, and the brothers secured a new distribution contract with the legendary United Artists (Mary Pickford, Charlie Chaplin, Sam Goldwyn and Douglas Fairbanks), but animation was only marginally profitable at best, requiring 10,000 to 20,000 drawings for a single seven-minute film. Mickey Mouse Clubs, organized by movie houses, mushroomed to a million members, deals were made for merchandising and newspaper-strip cartoons, but Roy was still barely able to meet the payroll week to week for 187 — a staff of gag men, animators, inkers, painters, camera operators, sound men and musicians. The stress induced a nervous breakdown in Walt in 1931. He snapped at the animators at the slightest upset, lost track of the talk in the story conferences,

cried on the telephone, lay in bed staring sleepless at the ceiling. On doctor's advice, he took Lilly on a long trip, gradually relaxing in Havana and on a cruise through the Panama Canal and up the West Coast. Roy was pleased to see him back refreshed but not so pleased by the refueling of his ambitions. Walt wanted to splash out on the still-experimental Technicolor process, even restage a Silly Symphony he had already made, *Flowers and Trees*. Walt's attitude to Roy's anxieties was cavalier. "Why should money be so important? Maybe potatoes will become the medium of exchange and we can pay the boys in potatoes." Roy bowed before the vitality of Walt's insistent creativity. Had Walt instead bowed to Roy's intrinsic prudence, there would be no Disney today. Roy was tuned to the realities of accountancy in the Great Depression; without him, the company would almost certainly have gone broke. But in the '30s Walt was tuned to the heartbeat of a people aching for a little light and color. As Schickel wrote: "Cocky, and in his earliest incarnations sometimes cruelly mischievous but always an inventive problem solver, Mickey would become the symbol of the unconquerably chipper American spirit in the depths of the Depression."

Technicolor catapulted the Silly Symphonies to a new level. *Three Little Pigs* gave the nation a defiant hit song, "Who's Afraid of the Big Bad Wolf?" (Walt, always looking for something new, was very reluctant to do the two sequels. "You can't top pigs with pigs" became his mantra.)

In mid-1934 the Disney animators came from dinner at a café across Hyperion Avenue to find Walt waiting for them. Bob Thomas writes that he said, "C'mon in the soundstage. I've got something to tell you." On the bare stage, lit by a single bulb, Walt clutched his throat and fell to the floor in death throes. He had eaten a poisoned apple from the Wicked Witch. For two hours he literally acted out his ideas for

Snow White and the Seven Dwarfs. He twisted his mouth and arched his eyebrows and commanded an imaginary mirror. He collapsed his face as Grumpy, he squirmed and wriggled as Bashful, he dozed off as Sleepy. By all accounts it was a spellbinding performance and one that he would repeat in countless variations over the years. "Walt could have you in tears or rolling on the floor," said the longtime animator Frank Thomas. "He could do anything, even take an inanimate object like a tree or a stone and somehow make it come alive." Thomas says they had no idea *Snow White* would be as good as it turned out to be. "Walt just kept at us. He would say, 'Do you think we're missing something here?' or 'I'm not getting involved in this scene,' or 'I don't care about the characters here.' One of his greatest gifts was his ability to make you come up with things you didn't know you had in you, and that you'd have sworn you couldn't possibly do. When you did come up with an idea, he might reply, 'Yeah, that could work.' Or what you suggested would give him a new idea and he would build on it. He wouldn't praise you much, but if he saw you doing work he liked, he could call everybody else to look at it, and he would give you more to do." The animators and music composers learned to watch out for a single raised eyebrow or fingers drumming on the arm of his chair — signs they would have to go back to the drawing board. The notes of his storyboard conferences are detailed analyses and collaborative reworkings of the smallest gestures and sound.

Walt had prepared for a full-length feature film with an original initiative. Knowing that drawings of a quality beyond cartooning would be required, he paid for his artists to attend night classes at the Chouinard Art Institute in Los Angeles; he often drove them down there himself. Roy was appalled when Walt first came to him to say he wanted to spend $500,000 on *Snow White and the Seven Dwarfs*, their first full-length feature. Lilly was alarmed, too. But those three little pigs had worked their magic on young Joe Rosenberg, Disney's loan officer at Giannini's Bank of America (page 258). He was so impressed by their earnings that he provided a $1 million line of credit for *Snow White and the Seven*

THE SEVENTH DWARF: Nobody liked Walt's name for the seventh dwarf—Dopey, a sweet mute.
They said it sounded like a name for a drug addict. He found the word in a collected Shakespeare, so Dopey it was.

Dwarfs. Roy's misgivings on Walt's ability to stay in budget were justified. Walt was not profligate, but he was a perfectionist; his estimate was out by threefold. But Roy's estimate of the value of the picture was out rather more. Opening to sensational reviews on December 21, 1937, at the Carthay Circle Theater in Hollywood and several days later at the Radio City Music Hall in New York, it drew sellout crowds. In its first year of release, *Snow White and the Seven Dwarfs* earned $8.5 million, well over five times the cost, at a time when the price for a children's ticket was just ten cents.

Walt was celebrated on the cover of *Time* and congratulated with honorary degrees from Harvard and Yale; he received a special Academy Award in 1939. Even today its 83 minutes make *Snow White and the Seven Dwarfs* among the best of the classic animated films—and it is still earning.

The magic ingredient of *Snow White and the Seven Dwarfs* and the other timeless fairy-tale features is that they are endlessly recyclable. *Snow White and the Seven Dwarfs* produced so many millions in profit that Roy felt comfortable committing to a new studio, built on 50 acres adjacent to Griffith Park in Burbank. Through the '30s Pluto, Goofy and Donald Duck took their bows. Walt, meanwhile, launched still more ambitious projects. The day in 1938 that Roy E. first heard the story of the wooden puppet who wanted to be a real boy remains one of his most vivid childhood memories. At the age of nine or so, he was home sick with chicken pox when his uncle Walt came over on a Saturday evening and stopped upstairs to say hello. "He obviously decided to see how *Pinocchio* played with me, and for the next 40 minutes he acted out the whole story. It was completely mesmerizing. When I finally went to see the finished movie, I was actually disappointed. It was nowhere near as good as it was in Walt's telling." The animators wrought some brilliant touches in portraying Pinocchio as a little boy imprisoned in wood and in the personalities of Jiminy Cricket, the muscular puppet master Stromboli, the cunning Fox and the almost unbelievably cat-like Cat, but the movie's breakthrough was most of all in technique, with the multiplane camera that Walt had built at 20 times the cost of Ub's horizontal multiplane. It allowed the animators to give an illusion of depth right from the opening, when the camera sweeps across the rooftops of a Swiss village and zooms in on Geppetto's lighted workshop.

Walt originally conceived of *Fantasia,* his controversial extravaganza of animated characters responding to classical music, as a way of reviving the sagging fortunes of Mickey Mouse. Lazy Mickey would borrow the wizard's broom and command it to bring water from the well but would not be able to stop the cascade. Walt asked Leopold Stokowski to conduct the frenetic music of the French composer Paul Dukas, and then Disney and Stokowski got carried away like the mad broomstick, animating more and more classical music until we had Chinese mushrooms dancing to Tchaikovsky's "Nutcracker," devils cavorting to Moussorgsky's "Night on Bald Mountain," a ballet of hippos for Ponchielli's "Dance of the Hours" and much else involving, in the end, a production staff of 1,000.

Fantasia and Pinocchio, both completed in 1940 for costs in excess of $2 million, are classic examples of the prolific vitality of Walt's imagination and his readiness to adapt new technologies: The audio oscillators Disney used for Fantasia were the first product made by Bill Hewlett and David Packard in their Palo Alto garage. But for pressures from the bankers, Fantasia would have been shot in wide-screen with stereophonic sound. In the short term, however, the tepid box office put the studio back deeply in debt by the end of 1940.

that the only way to win was to "lick 'em with product."

"Oh no, Junior's got his hand in the cookie jar again," said Roy when he heard that Walt was planning to make three features (Alice in Wonderland, Peter Pan and Cinderella). Under the circumstances, Walt settled for one feature, Cinderella. But Roy—and everyone else—knew Walt had finally lost his mind when he came back from a vacation in Alaska in the summer of 1947 burbling on about the life cycles of Alaskan seals. RKO, which now distributed Disney films, told him with expert emphasis that there was no audience whatsoever for documentary nature films and no place

their business in the men's room of the Beverly Hills Hotel, Roy, with uncharacteristic boldness, told Walt he was going to set up their own distribution company, ending 30 years of dependence on others. They named it Buena Vista, and it got off to a dramatic start with the waltzing scorpions and other anthropomorphic creatures in The Living Desert. Produced for $500,000, it earned $5 million on its first release—and the profits no longer had to be shared.

The astonishing feature of Walt's career was how right he was on all the big imaginative decisions and how, despite his uncanny record, he had to keep on proving himself time and again. Michael Eis-

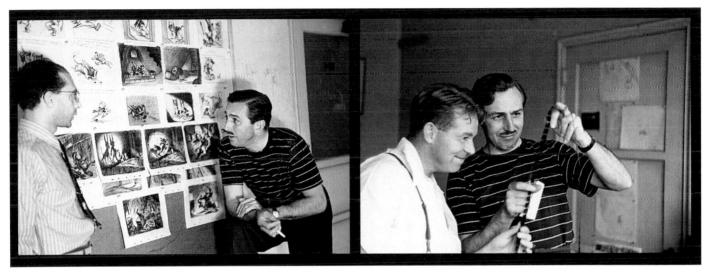

WALT AT WORK: (left) He controlled the story line and the gags through the storyboard. Sheets of sketches laid out like a comic strip could be reshuffled, written on and thrown away, and Walt did all of that. He was happy to credit this important invention to an animator, Webb Smith. (right) His animators devised a quicker way of scanning film; they photographed a trial sequence of drawings and projected them in small closet that came to be known as the "sweatbox."

The best solution, Roy told Walt, would be to make their first public offering of stock. Walt hated sharing ownership, but he agreed reluctantly—the obverse of the Walt-Roy creative dialogues—and in April 1940 Disney raised $3.5 million from its first shareholders. It could not forestall what was to be the company's darkest and most difficult decade: disappointment on the release of Bambi; a strike by the Screen Cartoonists' Guild that convinced Walt he was the victim of "communistic agitation"; and then the war, when large sections of his lot were commandeered by the U.S. Army. By 1945 Disney's debt to the Bank of America had climbed back to several million dollars. Walt went along with Roy in cutting staff by a third, while convinced

anywhere for a film that runs just half an hour, so naturally Walt went ahead and made Seal Island without RKO. Previewed at a movie house in Pasadena, the film received an overwhelming audience response and went on to win the 1949 Academy Award for Short Film: Live Action. Despite this vindication, RKO, which had been distributing Disney for nearly 17 years, jibbed yet again when later presented with the opportunity of placing a feature-length film. What on earth was he thinking—a 69-minute documentary, this time starring odious snakes and poisonous scorpions? But Roy was convinced that Walt was on to something. The eccentric Howard Hughes had taken over RKO. Having fallen in with Hughes's request that they conduct

ner, who, as chief executive from 1984, carried Walt's adventurous enthusiasms to a wholly new level, commented, "Looking back I am shocked by the number of times executives supposedly in the know almost snuffed out Walt's vision. I often remind myself of the latent power of genius when presented today with a dauntingly original and unusual project that I don't immediately understand."

The Bank of America's enterprising Amadeo Giannini was dead, and no bank was willing to fully finance Walt when he began talking about the $10 million—or so—he needed to build "an amusement park." Disgusted by the experiences of taking his daughters to fairs where the rides were tawdry, the employees hostile, and the grounds dirty, he envisaged something utterly different, but Roy was aghast at the idea of adding to their debt; $10,000 was the most the studio would advance. Walt found his own seed money for a feasibility study. He borrowed $100,000 on his life insurance, saying he could never convince the financiers "because dreams offer too little collateral." Walt's inspiration was to create a romantic miniature town, a showplace of nostalgia and happiness, a sealed-off rebuke to the sprawling urban blight outside. "I want the public to feel they're in another world. . . . Clocks will lose all meaning, for there is no present. There are only yesterday, tomorrow and the timeless land of fantasy." The inspiration for the shops along Main Street of his little town was an idealized version of places such as turn-of-the-century Marceline, Missouri. Everything on the street would be built to smaller scale to give it a child's perspective. At the head of the street his mind's eye saw a large castle, with four "themed" lands radiating from this central artery—Frontierland as homage to the Old West; Tomorrowland, a window on the future; Fantasyland, a world of make-believe; and Adventureland, a celebration of the exotic. All the rides would tell stories. And the place would be called Disneyland.

But how to find all the necessary millions—not 10, make it 15 or maybe 17? Despite his misgivings about the escalating estimates, Roy faithfully joined with Walt in approaching the fledgling television networks. All three had been hungering for a Disney-produced series—and Walt, after his early experiences, had always insisted on controlling television rights. The pitch was that whoever would invest in the Disneyland amusement park would be the preferred network for a Disney TV series. General David Sarnoff at NBC, then William Paley at CBS, strung them along for several months before backing out. On a call from Roy, Leonard Goldenson of ABC came right over to Roy's hotel. ABC had only 14 affiliates and needed programming more desperately than its rivals. The end of protracted negotiations was that ABC invested $500,000 in Disneyland and guaranteed loans up to $4.5 million. In return the network won a 35 percent stake in Disneyland and all of the profits from the concessions for ten years. Walt, on the other hand, agreed to provide a weekly one-hour Sunday night Disney show for ABC. The budget for the series was $5 million, but in addition Disney was given one minute of commercial time. (See Ruth Handler, page 388.)

The film industry was shocked that filmmakers would get into bed with television, the feared enemy. Walt was persuaded to host the shows himself, in avuncular mode, while Hollywood prayed for a flop. He was "scared to death" and worried about his husky smoker's voice, his Missouri twang and whether his diction was good enough for listening children. He opened on October 27, 1954, with a preview of *Disneyland* and soon stopped worrying. A Disney documentary about the filming of his *20,000 Leagues Under the Sea* drew an astounding 50 million viewers, one-half of all television households. They were still hooked when Walt presented his three Davy Crockett hour-long features. He rescued Fess Parker from bit parts in science fiction movies to play Davy Crockett and got one of his new studio composers, George Bruns, to compose "The Ballad of Davy Crockett," which was a number one song for 13 weeks. These inspirations were surpassed only by the incidental merchandising: Within weeks Walt had ten million Americans walking round in coonskin hats.

Disneyland, opening in Anaheim in July 1955, attracted a million visitors in seven weeks—double the number expected. Walt Disney Productions grew in five years from a gross income of $6 million to $27 million, and $70 million by the close of the decade. By the end of 1961, Walt and Roy had paid off the entire Bank of America debt. Once again Walt had blazed a trail—and firmly established a brand that could ride the roller coasters. "I'm not Walt Disney anymore," he said to Marty Sklar, a young employee who would eventually run Walt Disney Imagineering. "I do a lot of things the public doesn't want to know about. I smoke and drink and lose my temper. But Walt Disney is a thing, an image in the public mind. Disney is something they think of as a kind of entertainment, a family thing, and it's all wrapped up in the name Disney." He put the point forcefully when he recruited the talented Ken Anderson.

"You're new here," he told him, "and I want you to understand just one thing. What we're selling here is the name Walt Disney. If you can swallow that and always remember it, you'll be happy here. But if you've got any ideas about seeing the name Ken Anderson up there, it's best for you to leave right away."

Always provided they did not dilute the name of Disney, he was not grudging about credits. He always acknowledged that Webb Smith invented the fruitful idea of the storyboard to keep track of the match of image and narrative. But "Disney" had come to mean something special when it was associated with film. No other studio has ever achieved that sort of recognition before or since.

In the last year or two of his life, Walt showed undiminished creative genius after several years of lackluster feature films. He had for 20 years pursued P. (Pamela) L. Travers for the right to make a movie out of her story of a frosty middle-aged English nanny, *Mary Poppins.* She agreed in 1960. He threw himself into the project, using all his charm to keep the fastidious author happy as he turned her middle-aged nanny into the pretty and spunky character played by Julie Andrews (he had gone backstage at *Camelot* in New York to persuade her that her first film role should be as a flying nanny, with animated characters to boot). At the same time, when the British-born designer Peter Ellenshaw mentioned the English pub dance "Knees up, Mother Brown," Walt joined in the exuberant movements with Ellenshaw and the story editor Don DaGradi. Dick Van Dyke doing "Knees up" across the chimney tops of London was altogether memorable. Disney was continually adding visual comedy. ("How about having the vase fall off and the maid catches it with her toe?") According to biographers Katherine and Richard Greene, Walt shook his head when he first heard the Sherman brothers' song "Supercalifragilisticexpialidocious." "Something is wrong here. Why don't you try speeding it up a little bit?" *Mary Poppins,* Walt's last feature, entirely justified Richard Schickel's verdict that it had a cinematic excitement far in advance of most musicals, including *The Sound of Music,* "to which it is superior musically, directorially and theatrically, and

DISNEYLAND: It all started when he took his young daughters to amusement parks on Sundays in Los Angeles. "I sat on a bench eating peanuts and looking all round me, and I said to myself, Damnit, why can't there be a better place to take your children, where you can have fun together? Well, it took me about 15 years to develop the idea." Sharon Disney (Lund) is with him on one of his daily inspections. He asked for bigger and bigger trees and checked all the rides. At home he delighted in giving rides on a model railway in his garden.

even intellectually." It earned $44 million worldwide in its first year and encomiums from critics habituated to panning Disney's postwar features.

Walt refused to make a sequel, and he was not initially disposed to build a second theme park either, though Disneyland's popularity was so immense that fully a fourth of the nation's population had passed through its entrance by the release of *Mary Poppins.* He had by then persuaded himself that a second, more ambitious park would give him the opportunity to try out his ideas for a utopian future. Disneyland had invited sneers from some intellectuals (the word *Disneyfication* entered the language) who affected to pine for the strip malls and billboards of "real" urban life. Harvard's Margaret Crawford described it as the loss of real cities to rampant commercialism. "Disney's most profound innovation was to transform public space and the built environment into a commodity." Tom Vanderbilt, in an appreciation of the virtues of what Walt had tried to do—"a level of coherence unknown to most urban

planners"—wrote sardonically about the more extreme handwringers: "Over time I would learn to accept the received ideas of the school of Disney criticism, renouncing the site of my childhood fantasies as culturally bankrupt, homogeneous, soul destroying—it was a wonder I made it out alive." But there were many distinguished admirers (architectural critic Deyan Sudjic, futurist Ray Bradbury, novelist Aubrey Menen), perhaps most notably James Rouse, the city planner and developer, who told a Harvard planning conference, "The greatest piece of urban design in the United States today is Disneyland."

Disney, the showman turned educator, hoped to embody his most advanced ideas in the Experimental Prototype Community of Tomorrow (Epcot), dramatizing how social science and technology could relieve the stresses and sicknesses of modern society.

Before he could make yet another dream come true, he was dead at 65, killed by his lifetime of smoking. Roy fulfilled his brother's vision. Walt had endowed half his

estate to build the California Institute of the Arts (CalArts), and Roy saw it through. On October 1, 1971, he opened Walt Disney World Resort near Orlando with a musical joke Walt would have enjoyed. Roy had a marching band of a thousand parade down the fictional Main Street playing "Seventy-Six Trombones" from *The Music Man,* the musical comedy about an amiable charlatan who sells an illusion to River City, Iowa. Roy died three months later at the age of 78. The Walt Disney Company gradually sank, stultified by trying to top pigs with pigs—trying to guess what Walt would have done. *Fortune,* in a 1995 review, described it as by 1984 a "dilapidated company" (with only 1 percent of its income from movies), but then, said *Fortune,* it was "completely re-invented" by Michael Eisner and the late Frank Wells.

"I wanted something alive," said Walt of his theme parks, "something that could grow, something I could keep plussing with ideas. The park is that. Not only can I add things, but even the trees will keep growing." They have.

JEAN NIDETCH She found a new way to shed excess pounds — gorging on talk and more talk in Weight Watchers clubs

1923–

JEAN NIDETCH could never understand why she weighed so much. She never ate breakfast. Surely the midnight meat-loaf sandwiches did not contain so many calories; they were so unsatisfying that she was up at 3 a.m. for a light snack of pork chops and gelid baked beans. A couple of peanut-butter sandwiches on the way to the morning shop at the supermarket was just to keep the pangs of hunger at bay; she would not eat anything until lunchtime except pistachios from her capacious pockets, no more than a few handfuls.

But Jean Nidetch was undoubtedly fat. At the age of 38, she stood five-foot-seven and weighed 214 pounds. Her doctor said her problem was glandular, which was fine because it made it clear it wasn't her fault. After all, she had tried eating differently. She had lived off bananas and milk; then cottage cheese and peaches; then eggs and grapefruit; and she had tried oil capsules and weight-loss pills and hypnosis. She would lose 20 or 30 pounds and then immediately regain weight.

It had been this way all her life. At 16 or 17, growing up in Brooklyn as Jean Slutsky, she only dated boys as fat as she was. She had to quit the City College of New York for paid work when her cab-driver father died at the age of 42; her mother was only 36. Jean's job at the Internal Revenue Service was "coffee breaks all day long." On one of them, when she was 21, she went out midmorning for a Danish at a luncheonette, and there she met 27-year-old Marty Nidetch, fresh out of the army and so hungry he was pining even for army grub. They were a perfect match. Soon they were married, driving west in a 1942 Buick convertible to Tulsa, Oklahoma, which meant they swilled and guzzled their way across the country, testing the quality of the hot dogs, soft drinks and cotton candy at every car-

nival. They were back in New York in 1952, with a jolly baby. Marty, a hefty 265 pounds, drove a bus; Jean sold eggs door-to-door.

Jean gave birth to a second baby after they moved to Little Neck, Long Island, and was feeling good about herself the day that, in September 1961, a young woman acquaintance came up to her in the supermarket to say, "You look so *marvelous*. When are you due?"

"That really hurt," Nidetch wrote later. She walked home distraught, thinking, "I'm so fat I look nine months' pregnant." The same day she called the New York City Department of Health and made an appointment at the obesity clinic created by Dr. Norman Jolliffe, a heart specialist. On the day of her appointment, she did not tell anybody where she was going. She wore a loose coat to obscure her shape, put on high heels and fixed her hair nicely. She was too embarrassed to ask for the obesity clinic so she asked for the nutrition clinic. The woman behind the desk replied, "You want the obesity clinic. It's down the hall." In the room she eyed the other women waiting and ruminated, "How could they have let themselves go like that?"

A slim woman from the Department of Health entered the room at the very moment the thought of a nice ice-cream soda was subverting Jean's resolution. "Mrs. Nidetch," the woman commanded, "you will weigh one hundred forty-two pounds. You are five-feet-seven inches tall, you have a medium frame and this is what you will weigh." Nidetch protested, "That's utterly ridiculous." She said she would be happy if she dropped to 175 pounds, and she proposed substitutions in the official diet. She was promptly told that if she was not prepared to follow the program like everyone else, she could leave right now. Being terminated from a free public weight-loss pro-

gram was akin to failing at kindergarten, so she attempted the diet for several weeks. She lost two pounds a week. The director of the program expected much more and was not convinced by Jean's hypothesis that the weight must be due to the premenstrual retention of water.

In fact, Jean was lying. She could not confess to the nightly pillaging of two boxes of chocolate marshmallow cookies concealed in her bathroom laundry hamper. She felt so guilty, she called up six fat friends to talk about it. There was no shortage of fat jokes—"when I take off my girdle, it's like an explosion"—and none of the women said they noticed any difference in Jean. Nevertheless, she had lost 20 pounds over ten weeks, and the thought of the ordeal, minimally relieved by the cookies, gave emotional impetus to her announcement, "I'm only serving coffee this afternoon. No cake. No cookies. And you have to listen." She told them how she had lied to the nutritionist and was assured by everyone she had a perfect right to those cookies. They all admitted getting up in the middle of the night for light snacks of nuts and cookies.

At the end of the foodless afternoon, one woman asked if she could come back the following week. Nidetch said, "Sure. Let's make it a weekly thing. Let's just talk. This new diet is working for me. I know we can lose weight if we want to. We'll talk about it and maybe together we can make it." She gave her friends the diet sheet from the public clinic and insisted they show it to their doctors before they tried it.

The next Wednesday, the women returned—with three fat friends of theirs.

Within two months, 40 women were coming to Nidetch's meetings to confess to the éclairs in the oven, the cans of nuts hidden where the children wouldn't look. "We discovered something," says Nidetch. "We'd been kidding ourselves."

Nidetch stopped going to the clinic, but she ended sneaky snacks and started losing more than two pounds a week. The group pooled its money to buy a medical scale. Nidetch soon had hundreds of people wanting to come to her meetings. Men began coming, too. Nidetch did most of the talking, exhorting everyone not to cheat. People clamored for some sort of reward for sticking to their diets. "How about a corned beef sandwich?" suggested one dieter. Nidetch, sensing that the mental dam would break if the reward was food, called on a jeweler to make pins with a diamond chip, to be awarded for each ten pounds lost. The cost was $17; she asked everyone to contribute a quarter. It did not cover her outlay, but she was now a woman with a mission.

Calls came from all over New York City, and soon she was traveling to visit women who were too fat or too sick or too ashamed to come to her meetings. Everywhere people played her a record she had made herself: Their problems were "glandular" and they never ate lunch. Nidetch would insist that they eat. She explained, "The most difficult thing in the world is to teach fat people to eat—not to stop eating, but to start eating the right foods. You have to teach them to stay away from the wrong foods, to fill themselves up with healthy foods so they won't be hungry." She got into the habit of noting people's birthdays and reminding them beforehand to stay off cake. She was no longer lying about snacks. On October 30, 1962, at 4 p.m. the scales told her she was down to the 142 pounds demanded by the obesity clinic. Her husband, who had lamented the loss of his eating partner, finally got on the program after a bowling partner suggested only a slimmed-down Marty would have a remote chance of a strike.

Nidetch's extended speaking tours shed pounds for the groups she visited but left her in debt. The $1 everyone volunteered toward expenses did not cover the cost of so many pins awarded to women and jeweled tie clips to men. After a meeting, Felice (down 50 pounds) and Al (down 40 pounds) Lippert told Jean she had things the wrong way round: Everyone should be giving her presents. Al offered to help set up a business. Jean and Marty hesitated—they had so little money—but finally rented a movie theater in Little Neck for $75 a month. They incorporated Weight Watchers on May 15, 1963. The same day 200 people lined up down the block to get in. Nidetch divided them into groups and spoke to each for two hours at a time from a little platform built out of a Salvation Army table.

Nidetch had thought of asking a dollar admittance to cover costs; Al told her she should ask $2, as the movie house charged for films. The crowds kept coming to the chagrin of the owner of the pizza parlor downstairs, Romeo, who couldn't figure out why none of the hundreds lining up outside his shop would buy anything. (Nidetch later worked with him to create a Weight Watchers milk shake made of skim milk and strawberries or instant coffee or vanilla extract. He thrived on it.)

Nidetch developed simple rules for members: 1, You may not skip any meals; 2, No substitutions allowed; 3, No calorie counting; 4, No alcohol; 5, No appetite suppressants. A suggested breakfast menu was a poached egg on toast, a half grapefruit and a juice; for lunch, four ounces of salmon, lettuce and cucumber, lemon, toast and coffee; dinner, bean sprout soup, London broil, asparagus, brussels sprouts, fruit and a beverage. (Diet plans have been revised over the past four decades to take into account official recommendations.)

There was an insatiable appetite for the classes. Nidetch was thrilled by their egalitarian nature. Surgeons and litigators sat next to truck drivers and barbers. Nidetch, recruiting new group leaders, decided that only the formerly fat had the emotional knowledge to commune with the currently fat. Talking turned out to be as important as Dr. Jolliffe's original diet. Groups gasped when they heard the new slim Nidetch tell how she had once deceived her husband.

While he took a babysitter home, Jean ate an entire strawberry cake, and when he got back she said, "How do you like that? She ate the whole cake while we were out!" Nidetch saw being overweight as an emotional problem as much as a physical one. Confessing one's secret life of snacking turned out to be the key to leveling with oneself. The power of networks was also at work here, the bonding with other people overwhelmed by the urge to consume.

Within three years, Weight Watchers had done $160,000 of business. Lippert joined Nidetch as a business partner, and they opened Weight Watchers franchises around the country. In September 1968 their IPO was a sellout. By 1970 the company had grossed $8 million, there were two million members in 21,000 classes a month, and they started selling a successful line of prepared meals.

Nidetch herself no longer took classes. She was a missionary, speaking to enormous audiences, going on television. Her appearances became dieters' revival festivals. People wept. They blessed her. They kissed her. People who had been living as recluses emerged into the world svelte and newly confident. Nidetch, bleach blond and ever charismatic, stayed as the spokesperson until 1984, followed by the actress Lynn Redgrave and later Sarah Ferguson, the former Duchess of York.

In 1978 Weight Watchers was sold to H. J. Heinz Company and then in 1999 to the investment firm Artal Luxembourg. Today a million people a week in 30 countries attend Weight Watchers meetings. The company has revenues of a billion dollars and has more than 25 million graduates. There is clearly more than ever a need for the program. Obesity is a national epidemic that cuts many lives short: In 1970, 20 percent of Americans were overweight. According to the Centers for Disease Control, 63 percent are now overweight or obese—about 120 million people who just know they hardly eat a thing.

THOMAS WATSON
He was not an inventor or technically gifted, but his inspiration and drive turned a tiny company into the great icon of the information age, IBM

1874–1956

Y OU ARE 40, out of a job, a newlywed, your wife is expecting a baby, you don't own your home, you have no specialized qualifications, the only company you ever launched went bankrupt—and you have just been sentenced to a year in jail. This was Tom Watson in 1914—a highflier brought crashing to earth in Dayton, Ohio.

He was the son of a hard-luck lumberman from the wooded Finger Lakes region of New York State. At 16, with a brief immersion in a school of commerce, he kept books for a butcher and then made his way peddling pianos and sewing machines door-to-door. He thought big and fell hard. The idea of setting up a chain of butcher shops in Buffalo might have worked if his partner had not run off with their cash, bankrupting him. In 1897, at the age of 23, a tall, lean, very articulate man with high intelligence and a bubbling temper, Watson was a trainee salesman for the National Cash Register Company—known to many as "the Cash"—headquartered in Dayton. At 29, after six years as a star salesman, he was so admired by its perfectionist founder, the domineering John Patterson (see panel on page 359), that he was entrusted with a secret mission. He was asked to "resign" from NCR

EARLY DAYS:
The mustache was the trademark of the traveling salesman.

and set up on his own in New York with a million dollars of NCR money—not to sell cash registers but to stop competitors from selling them, especially dealers in secondhand machines. Watson executed his covert work well. Under the guise of a dummy company, he put out of business various traders in secondhand registers in New York, then Philadelphia and Chicago, by offering sky-high prices for secondhand machines and opening up adjacent showrooms. Mission accomplished, he returned to NCR and became Patterson's heir apparent.

It was partly for this guerrilla warfare, uncovered years later, that Watson, Patterson and 28 other NCR executives were in 1913 prosecuted and convicted for criminal conspiracy. They were the first businessmen to be sentenced to jail under the new anti-trust laws, and each was fined $5,000. Pending their appeal, due to be heard in 1915, they were allowed their freedom, on bail of $5,000. Patterson and the others were ready to negotiate a plea bargain settlement. Watson normally marched to Patterson's drum, but he refused a plea bargain, steadfastly maintaining he had done no wrong. He had risked much for Patterson, but in 1913 Patterson abruptly

GODFATHERS OF IBM: 1

HERMAN HOLLERITH (1860–1929)

THE MAN WHO gave IBM its technological backbone, Herman Hollerith, said the inspiration for his punch-card tabulator was a train ride. Hollerith, the son of German immigrants, had taught at the Massachusetts Institute of Technology since 1882, but on graduation from Columbia in 1880 he had initially worked at the U.S. Census Bureau and witnessed it struggle to count and analyze America's fast-growing population (it took nine years).

Hollerith first experimented with a machine in which pins selectively penetrated a thin roll of paper tape, activating an electric circuit, but on his train ride west, he was reminded of the superiority of stiff rectangular punch cards: The English inventor Charles Babbage (1791–1871) had thought of punch cards for automatic calculation. The train ticket Hollerith handed the conductor was also a form of identity card, called a punch photograph, which matched the presenter of the ticket with the purchaser. "The conductor punched out a description of the individual as 'light hair, dark eyes, large nose, etc.,'" said Hollerith, who then commented on his adaptation of this system to the census: "So you see, I only made a punch photograph of each person."

Hollerith's system had three parts. First, the unpunched cards. They were coded so that all the attributes of an individual (male, female, etc.) could be recorded as a pattern of holes. Second, a keyboard operator transferred the attributes noted in the census tally to the cards by punching a hole in the appropriate column. The clerk had an enlarged display to guide a penetrating stylus to the correct column of the card. One punch, one attribute, and then on to the next attribute until all had been registered. Third, the cards with their variable pattern of holes were placed in a tabulating machine resembling "an undersize upright piano." The tensile strength of the card was such that pins would only

penetrate where there were punched holes. The operator of a "pin box" penetrated cards where there were holes, sending electrical signals to a counter. At the end of this operation, cards were directed for further analysis to a sorting box with 26 compartments.

With Hollerith's system, it took only six weeks to count all 62,622,250 Americans in the 1890 census, instead of two years. But who were they? At the end of the tabulation and sorting, we also knew how many Caucasian (one punch) female (one punch) immigrants (one punch) living in New York (one punch) were married (one punch), etc., etc.

A sensational success, his system was swiftly adopted for censuses in Austria, Canada, Norway and Russia, and he kept on refining it. He was more an inventor than an innovator; he slipped up by not having a sales force like Patterson's. But two of Hollerith's key decisions became commercially crucial to IBM: his policy of leasing instead of selling machines and the separate sale of punch cards exclusively for use in Hollerith machines. In 1911 his company was absorbed in Charles Flint's Computer-Tabulating-Recording Company. It was performing poorly until 1914, when Thomas Watson came on board.

Hollerith tabulator in the census office, 1890

fired him, with a handshake of $50,000 and a fast Pierce-Arrow car, for the reason he always fired everyone who got close to the throne: Their wattage was exceeding his. It was time for Watson to fly solo, and he made it a life's ambition to excel Patterson's NCR. He was so ashamed of the conviction that he never told his son, Thomas Watson Jr., who learned of it in 1946 and then only from an associate of his father's. Still, the experience, in a sense, was the making of Watson. He knew the value of monopoly, but he learned the value of reputation.

Nevertheless, Watson's whole career after his eviction from NCR turned on the happy chance that Charles Ranlett Flint (1850–1934), the man who controlled the company that Watson would make into IBM, was himself an artful dodger of antitrust laws. Flint was known as the "king of trusts" for the nimble way he put together scores of companies to form monopolistic conglomerates in rubber, starch, bobbins, woolens and chewing gum. The little matter of running a dummy company was a bagatelle to Flint, who made much of his money as a genial merchant of death, selling guns, aircraft and warships—impartially to both sides, you understand. Flint was also an early participant in everything to do with speed—early aviator, early car racer, early steam yacht racer. He became a friend of Orville Wright and licensed Wright aircraft to the German kaiser, in time for the German aces to fly them in World War I against the British and Americans.

Two months after Patterson threw him out, Watson went to see Flint. Resplendent in his suite of offices at 25 Broad Street, bedecked with photographs of friends such as Theodore Roosevelt and Andrew Carnegie, the king of trusts was 64, a dapper little bewhiskered man in a vested pinstriped suit who was concerned about the slow pace of a conglomerate he had put together just three years before. It was a fusion of two trusts and a new company. In a trust called International Time, based in Endicott, New York, Flint had assembled a bunch of companies making punch-card time clocks for factories. The second trust, based in Watson's stamping ground of Dayton, was Computing Scale of America; it served food merchants with scales that computed prices based on weight. Having failed to get any traction with these two trusts, Flint had the glimmerings of a fine concept. Both trusts, he could say, dealt in business information. If he acquired Herman Hollerith's ailing Tabulating Machine Company, based in Washington, he could amalgamate it with the two trusts and have one big enterprise defined by its services in recording business information. But it was quite a stretch from meat scales to punch cards, and Flint had lumbered his new entity with an unmemorable title, the Computer-Tabulating-Recording Company (CTR). Three years after its launch in 1911, he badly needed someone to give it momentum and a sense of direction. Cue Watson.

One of Flint's largest shareholders, A. G. Ward, had spotted Watson as a comer. Flint liked Watson's dash and confidence; indeed, a man who drove a Pierce-Arrow and saw the virtues of monopoly was his kind of man. The other directors were less keen on having a potential jailbird take over. The compromise was to make Watson general manager, but not president, until he had been cleared by the courts—which he was ten months later when President Woodrow Wilson's new team at the Department of Justice dropped the case. For taking over CTR, Watson asked for and received what he called a "gentleman's salary" ($25,000), not a princely sum for reenergizing 1,200 demoralized people, but he negotiated 5 percent of worldwide profits and 1,220 shares, then worth around $36,000. With every spare dollar thereafter, he bought more shares.

The tabulation company in Watson's portfolio was a vestigial remnant of the company founded by the inventor of punch-card tabulation, Herman Hollerith (see panel, page 358), who had failed to maintain ascendancy against imitators and improvers. In his days at NCR, Watson had seen a Hollerith tabulator and sorter at Eastman Kodak in Rochester and had secretly rented one in the hope of untangling the bewildering variables behind the sales and revenues of district offices reporting to him as sales manager. When he did his show-and-tell of the results, the astonishment of the sales force at what he knew, not knowing how he knew it, had made a

GODFATHERS OF IBM: 2

JOHN PATTERSON (1844–1922)

THOMAS WATSON's mentor, and almost his nemesis, was the messianic John Patterson, patron saint (or devil) of American salesmen. It was Patterson who coined the slogan "Think," which later became IBM's mantra.

In Patterson's hometown of Dayton, Ohio, around 1879, a café-saloon keeper, James Ritty, patented the first mechanical cash register, inspired by the dexterity of his staff in stealing from him. Ritty's "Incorruptible Cashier" machine rang a bell with 35-year-old Patterson, who, having graduated from the Union Army, Dartmouth and the coal trade, had a store at Coalton, Ohio. He bought Ritty's patent and the entire company in 1884 for $6,500, renaming it National Cash Register.

Dayton considered Patterson a sucker. Ritty had sold very few machines. Patterson countered the prospect of humiliation by flinging himself into a sales campaign that the register paid for itself—and it did. It was a step toward more systematic accounting, and as such it was a business innovation of first importance.

Patterson was a capricious egomaniac, but he created America's first national sales force. He fought the image of the salesman as a shady, hard-drinking, lecherous "drummer" by training unsophisticated young men to look and behave like bankers. They were fired unless they turned up in dark suits, white shirts, and unostentatious ties. Top executives had to join him every morning in calisthenics and horseback riding. He burnished selling skills. As a practicing paranoid himself, he was expert in teaching NCR men how to play on the suspicions of traders that they were being robbed. He was ferocious with failures—one man arrived to find his desk and its contents incinerated on the lawn

outside the offices—but Patterson would also punish strivers who got too close to the throne. Not only would they be fired but their friends in the company, too.

At the same time, he was open-handed. He gave high achievers automobiles and paid vacations. NCR men on commission could make more than $30,000 a year (that's $300,000 if we use the multiplier of 10 from 1900 to 2000). As sales rose sharply—with annual sales of 25,000 registers by 1900, 100,000 by 1910—he hired still more salesmen. By 1902 he had 976, about a third of the entire NCR workforce, including manufacturing.

Patterson was the Elmer Gantry of salesmanship, conducting staff rallies round the country with the fervor of a revivalist tent crossed with the nitty-gritty of an auditors' seminar. Chalks and charts abounded. Quotas were set for exclusive territories. Quota-breakers were garlanded. Everyone chanted from the same bible, Patterson's pamphlet entitled "The NCR Primer: How I Sell a Cash Register." Slogans were hammered home. Visual aids were mandatory: "Remember, the optic nerve is twenty-two times stronger than the auditory nerve!"

Patterson emphasized sales and service, but he did not neglect technology altogether. In 1888 he set up a small "inventions department," one of the first formal research-and-development departments in the office-machine business.

By the time of his death in 1922, he had sold 22 million cash registers—and trained a new generation of businessmen. National Cash Register was a finishing school for corporate America: Graduation was getting fired. In 1984 it was estimated that one-sixth of the current CEO's had been NCR men. Tom Watson was the most significant.

How to spell *we* and *ox*—semaphore from the limbs of Watson and friend.

profound impression on Watson. Perhaps the single most important nanosecond in the historic genesis of IBM was the look on the faces of those salesmen. "I can see Walter Cool and Meyer Jacobs and Fred Hyde," Watson said years later, "because they were three men who just always knew everything about their territories. Finally, Mr. Hyde stood up, a great tall man, with great dignity and charm and everything else, and said, 'I take my hat off to you and your organization. I have always thought I was keeping track of my men, and you have told me a lot of things that I didn't know when I left home.'"

Watson's immediate if vague insight had been a vista of lots of unknown but knowable things: Technology somehow would make them known. At the Cash, he had been deeply impressed by one of his young friends, the engineer Charles Kettering, who had immediately made a working reality of Watson's suggestion that an electric cash register would be a winner. (In 1911 Kettering adapted the cash-register motor to a self-starter for Cadillac, replacing the laborious hand cranking.)

On his first day at work, May 4, 1914, Watson walked into a shabby little office at 50 Broad Street, 48 paces from Flint's fiefdom, intent on uniting the disparate squabbling divisions but certain that the whole company could move forward only on the synergy of research and sales. The emphasis on research was not appreciated by the two prima donnas inflicted on him. Hollerith himself was in Washington but a brooding presence as a grumpy consultant, ready to pick a fight if anyone, customer or no, suggested an improvement in his precious invention; in buying the company, Flint had conceded him the right to approve any design changes. While Hollerith nursed his ego, his original dominant position was eroded by the innovations of James Powers, who beat him for the 1910 census contract and made headway with insurance companies by adding a printer to the Powers machine. Watson's other roadblock was Congressman George Fairchild, who had become the largest shareholder and chairman; he was a prevaricator who maddeningly insisted on being informed of everything Watson wanted to do, and he favored International Time over CTR.

Watson's later associates would not have recognized their caustic, short-tempered boss as the sedulous step-by-step appeaser he had to be with Hollerith and Fairchild until the 1920s. Even so, without Flint's continual cheerful support, he would not have gotten very far. Flint endorsed Watson's wish to postpone dividends in 1914, over Fairchild's objections; Watson needed every dollar to fund the research lab urged on him by Hollerith's more adventurous deputy, Eugene Ford. Watson coined one of his better aphorisms when the Guaranty Corporation told him he could not have another $40,000 since the CTR already owed $4 million. Watson's winning answer was "Balance sheets reveal the past. This loan is for the future." He got his money and his research lab; to head it, he recruited the prodigious inventor and engineer James Wares Bryce (who ultimately secured 500 patents). Ford hired a number of engineers, including Clair Lake, who contributed a printing facility to Hollerith so that it at last became competitive with the Powers company.

These early years of Watson's at CTR, establishing an innovative thrust while coaxing and bullying a common spirit among the divided and suspicious staffs of the three divisions, have gone largely unappreciated in light of later dramas, but Kevin Maney has given us the flavor of his style. Here he is speaking to 13 executives: "We're going to have different cooperation in this business. There is going to be no more of this old woman's gossip, no more knocking, no more around-the-corner whisperings, no more backbiting. Everybody has got to put their cards on the table faceup who stays in this organization from now on, gentlemen." He was assiduous in trying to find talent within the company; he did not fire the old guard or swamp them with newcomers. But he was a raging demon for detail. He burst in on a sales meeting with the previous week's figures in his hand and rebuked them all: "To think that you men hold a Monday morning meeting here and you haven't got this. . . . I am ashamed of every member of the tabulating organization that attends this meeting." He lambasted them on the performance but misread the statistics, then later got madder still that nobody had been brave enough to correct him. It required bravery because Watson played emotional games, setting traps where a quick response to an apparently innocuous question released a guillotine.

Throughout these years, he was on a tear, working all hours at a desk overflowing with scattered papers, gulping down junk food (hot dogs, peanuts, doughnuts, lashings of coffee), chain-smoking cigars, setting impossible sales targets, incessantly enthusing exhausted groups: He fumed about yes-men, but only his

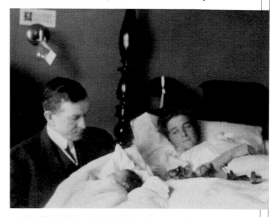

BABY TOM AND HIS DAD: The first of Jeannette (Kittredge) Watson's four children was raised to take over IBM.

gentle wife, Jeannette, dared to restrain him. On one occasion he stopped in mid-sentence before hundreds of employees and guests when she sent up a simple note: "Shut up!" The upside of Watson's habit of ambushing executives was that many of his surprises were pleasant: Those who caught his eye for performance might find they were off with their families on all-expenses-paid vacations. He was consistently warm-hearted, extending a helping hand to families in distress. He was singularly generous to the point of self-effacement in giving public recognition for work well done.

The business decisions Watson made were eminently sound but unsurprising and hardly original. The distinctive thing about him was how quickly he gave the fractious company he inherited an identity and sense of common purpose that was more and more focused on information and his own aspirations for "class." He was ultrasensitive to the impression he created personally. His son described to journalist Peter Petre how Watson Sr., an inveterate rail traveler (he was too scared to fly), always took care to clean the men's room wash-basin himself and always overtipped Pullman car conductors, porters, chauffeurs and waiters, not just because they deserved it but because they were among a "whole class of people who are in a position to poor-mouth you." He cultivated influential people, joined the right clubs, donated to conspicuous charities and broke all the records for name-dropping.

Almost by osmosis, since he gave no orders, Watson's executives came to dress soberly and deport themselves like him, shunning alcohol and treating women with very formal courtesy (Watson the-dragon by day was never happier than on the dance floor at night). He was an autocrat, but accessible: The rule was that the door to his unimposing office was open to anyone with a complaint or an idea or a personal problem. He wanted staff to feel they belonged—and he wanted to get rid of the fear of change, the arthritis of the grown business. He evolved responsibility on to small work groups with "managers" to help out rather than "foremen" who gave orders. He offered good pay to executives and salesmen, lifetime job security, opportunity at all levels, insurance, continuous

training and education programs for those who wanted to get on, creating a feeling of community and a sense of direction—all attributes that, in the judgment of management writer Peter Drucker, make him one of the great social innovators in American history.

Watson was not much concerned at first with the welfare of the factory staffs, who were a large bulk of the 3,000 employees in 1920, until he walked the floors of George F. Johnson shoe factories in Endicott and saw how Johnson's progressive management had produced a nonunion

force of 20,000 loyal and conscientious employees. As CRT gained momentum—$4 million profit on $14 million in sales by 1920, up from a sales revenue of $4.2 million in 1914—Watson gradually adopted many of Johnson's ideas. He based his stick-and-carrot leadership of the sales force on the lessons he had learned from Patterson at NCR, though for the most part he implemented them with irascible brio rather than random brutality. Watson's creation of the One Hundred Percent Club, made up of men who had reached their sales quota, was modeled on Patterson's

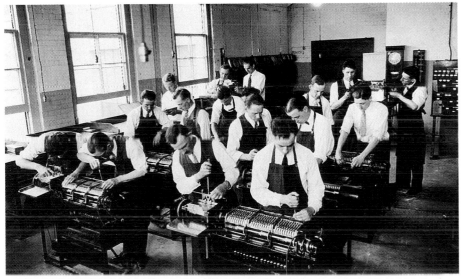

NCR DAYS: (above) Watson inspires his NCR salesmen in 1912. His own days there are numbered. (below) Learning to repair leased tabulators in 1924—the year he renamed his new company IBM.

THE REBEL: Thomas Watson Jr. (left) cannot bring himself to smile as the Watson family sets out on an ocean-liner vacation, 1929.

Hundred Point Club. In return for all this, he expected personal loyalty, and he got it. For all his searing style of leadership, he attracted affection, the more as he grew older. The cult of personality seems strange now, almost weird, but it worked in the '20s. President Calvin Coolidge was merely expressing the spirit of the era when he said, "The man who builds a factory builds a temple. And the man who works there worships there." Until the Great Depression, Americans believed that the utopia of humane capitalism would come not from political reform but inventive businessmen such Henry Ford and Thomas Edison. At sales rallies, Watson's staff was not embarrassed to belt out verses of the CTR song idolizing the father of their company:

> Mister Watson is the man we're
>   working for
> He's the leader of the CTR,
> He's the fairest, squarest man we
>   know;
> Sincere and true.
> He has shown us how to play the
>   game.
> And how to make the dough.

By the mid-1920s Watson's enemies had resigned or passed away, and he celebrated his ascension to full control by renaming CTR. Hereafter it was to be International Business Machines. IBM was still only a middle-size company. Remington Rand had three times IBM's sales revenue in 1928, having combined Remington Typewriter, Dalton Adding Machine, Rand Kardex, several office-supply companies and the Powers Accounting Machine Corporation, which had beaten Hollerith for the 1910 census contract. Watson was never tempted to diversify as Jim Rand Jr. hectically did; in fact, he came to resent the calls on his capital and energies from time clocks and scales. More and more the three things that mattered to Watson, to adapt the real estate mantra, were information, information, information. He rattled off many ideas that he expected his engineers to turn into reality. Give me a machine to automate banking transactions. Give me a machine to print railway tickets showing destination, time of boarding, etc. Give me faster machines. Give me machines immune to dampness.

These bright new horizons for information processing remained in Watson's

mind as the lights went out in 1929. The reaction of almost every other business to the Wall Street crash, and then the Depression, was to batten down the hatches in a total funk. Watson was supremely different. His response to stalled revenues was more innovation, more expansion, more research. He sold the scales division but bought the Electromatic Typewriter Company in Rochester, New York. His concentration on talent and technology was rewarded in 1932 when James Bryce presented the 405, a fabulous machine that could tabulate alphabetically (and not just digitally) at 150 cards a minute, print from card fields at 80 cards a minute and sort at the rate of 225 to 400 a minute.

But who was going to buy these machines when every month the depression deepened? Watson kept on manufacturing, enlarging the workforce from several hundred to more than 7,000, building up inventory at real risk to solvency. He was 58 and not slowing down, but personally, as well as professionally, he was on the rack. From his first month with CTR, he had invested everything he could in the company's stock—and lived high. "He had absolutely no impulse to hoard or even worry about money," his son Thomas told Petre. "He wanted to rise in the world, so he knew he had to spend." By 1932 IBM stock had lost more than 200 points and was still falling. He had bought the stock on margin, putting up only 10 percent of the price, so a collapse meant he had to make margin payments or sell at a huge loss. Another $3 fall in the stock would have left him bankrupt.

But the legislative child of the Depression was the New Deal, and the New Deal made IBM. All those machines in the warehouse were suddenly in demand—not just by the government but by every company in the country required to maintain the records of millions of workers and calculate how much of every employee's hourly and weekly earnings should be paid into the Social Security fund that began operating in 1935. None of Watson's competitors had the inventory to meet the demand, and they had not kept abreast of the technology. Watson won all the major contracts for processing information following the passage of the Social Security Act of 1935 and the Wage-Hours Act of 1937–38. To the conspiracy theorists of the '30s—of the few growth industries—the New Deal contracts were Watson's reward for supporting Franklin Roosevelt in his 1932 presidential campaign. He certainly had given money to the campaign, and he and his wife frequently took tea in Roosevelt's country home in Hyde Park and even slept a couple nights at the White House. Harry Hopkins, Roosevelt's closest aide, described Watson as "the only business friend that Roosevelt has," which was only a mild overstatement of the fact that throughout the '30s Roosevelt was anathema to the business community that was gathered most vociferously in the Du Pont family's American Liberty League, an organization of millionaires who denounced the New Deal as bolshevism, and FDR as a traitor to his class. Watson was too proud of the presidential connection to care about the frisson he might produce when he dipped into his pocket to display a handwritten note from "FDR." He told his son: "The average businessman's opinion on what is right and wrong for this country is almost always wrong."

Whatever the jealous suspicions about this relationship, it had nothing at all to do with Watson's New Deal triumph. Indeed, Roosevelt's Department of Justice shocked Watson by continuing an antitrust lawsuit against IBM begun during Hoover's presidency. The final Supreme Court ruling in 1935 forced IBM to open punch-card sales to competition. By then it was selling no fewer than three billion blank cards a year and had 85.7 percent of the nation's tabulating machines. Paradoxically, Watson's insistence on renting machines rather than selling them—the subject of another adverse antitrust ruling in 1953–56—was one of the reasons IBM was uniquely able to satisfy the requirements of the bureaucracy in 1935. A sold machine was out of sight, out of mind. A rented machine required IBM technicians to be on hand, repairing and updating, and inevitably they acquired an awareness of present—and future—needs. This liaison, developing into a partnership, was very much to the benefit of both sides. Social Security was a far more complicated task than the arithmetic and analysis of a digitized census. For Social

Security a file had to be created for every one of 30 million Americans, in alphabetical order, and organized in such a way that one file could be compared with another—a pension entitlement, say, compared with the record of date of birth (or disqualifying death). New Deal officials made no budget appropriation for machinery because they never thought the work could be done by punch cards. Only IBM was able to meet the government's imperative of automating the comparison of two sets of records. In two years, Social Security offices were running on more than 400 IBM accounting machines and 1,200 keypunches. Government offices sold their Burroughs, NCR and Remington Rand machines to lease IBM equipment. But the New Deal was more than a one-off coup. The ongoing contract made the government and IBM partners in research and development. IBM had entered a different universe. With sales doubling to $46 million in 1940, it was still only a relatively small firm, but the computer historians Martin Campbell-Kelly and William Aspray note that it was doing more business than any other office-machine firm in the world. The future seemed limitless.

IBM's success was Watson's. On the basis of a bonus of 5 percent of the worldwide profits, his earnings were $364,432 in 1934, inviting the label "Thousand-Dollar-a-Day-Man." He was the highest-paid man in America, earning more than the Hollywood moguls and the heads of the much bigger automobile companies, and was just $18 ahead of the iconic Will Rogers. Watson spread the wealth. He gave the gift of $1,000 life insurance policies to all 6,900 employees and announced a minimum pay increase of 37 percent. He supported more and more good causes—millions of dollars in his lifetime—and belonged to so many associations that he developed the longest entry in Who's Who. He acquired the site for a new Manhattan headquarters at 57th Street and Madison. For the first time—and he was now over 60—the press began to take seriously the predictions of an information society that had previously been dismissed as the delusions of a crank. He was called "an industrial giant" by the New York Times, the "most astute businessman in the world" by Time, the "master sales-

man" by Forbes and "a man of unusual vision" by Barron's. He lapped it all up, an appetite for flattery that was often to suborn his judgment.

In one area, however, Watson's visionary powers were circumscribed. Throughout the '30s, he was a defender of fascism, a subject that gives the most trouble to his biographers and commentators. Peter Drucker's elegant biographical essay skips mention of it, but it has to be noted that Thomas Watson, the man who backed FDR, the social innovator, the chairman of the Carnegie Endowment for International Peace, the chairman of the American section of the International Chamber of Commerce, the brilliant man so condescending about the conventional wisdom of the businessman, was a rube in the hands of Adolf Hitler. In 1937, when the odious nature of the regime was clearer with every month, he bent his neck in Berlin for the bestowal of Hitler's gift of the Merit Cross of the German Eagle. Two impulses seem to have been at work. A fervent desire for peace was one: "Peace through trade" was the slogan. Many people longed to believe that Hitler meant what he said. The other was simply business. IBM owned a tabulating subsidiary called Dehomag, set up by Hollerith before Hitler seized power, and Watson wanted to protect the asset.

Watson's conduct has to be seen in context. He had plenty of bad company, including Colonel Lindbergh, William Randolph Hearst, Henry Ford and John F. Kennedy's father, Joseph Kennedy, who was FDR's ambassador in London and was fired for defeatism. After the Kristallnacht murders and assaults on thousands of Jews in November 1938, Watson wrote a letter of remonstrance to Hitler and carried a copy in his pocket to show anyone who asked where he stood. What was required was a typical blast of Watsonian outrage; all the letter did was appeal to Hitler for kindliness to Jews on the grounds that public sentiment in America had changed in that direction. In June 1940 he belatedly followed up by returning the medal and was then reviled by the Nazis as a tool of the Jews. The hostage to history left by the Dehomag connection, elaborated in the researches by Edwin Black, was that IBM technology was the essential tool the Third

Reich used to identify, describe and track Jewish populations. The full extremity of those horrors was in the unimaginable future.

Watson made up for the misjudgment. Both patriotic and bitter at the way he had been humiliated by events, he immediately put all of IBM's manufacturing potential at the service of the United States. He pledged to limit profit margins to 1.5 percent and set aside that money for IBM's war dead and wounded and their families. As the war expanded, so did IBM. It was another bold gamble. The United States funded the building of factories and guaranteed the orders, but there was a risk that when the war ended demand would fall and IBM would be stranded, a bloated whale. It cannot have been Watson's original intention to build as many new factories as he did and to take on so many people that a workforce of 12,000 in 1940 became 22,000 by 1945, but he went with the momentum while driving engineers crazy for new ideas for war and peace. IBM factories made automatic rifles, gun directors, cannon, fire-control mechanisms and bombsights, but they most helped to win the war with punch cards. The military used IBM machines to churn through millions of computations. What materials do we need to build an airfield on Guam? Can the U.S. Army spare boots for the Red Army? Will these munitions last at different rates of fire? What is the net effect of strategic bombing? General Eisenhower's officers, trying to answer that question in London, asked for a Hollerith machine. Watson sent eight, with a staff to work them. The mathematical calculations for the atomic bomb were made at Los Alamos on IBM machines.

At the end of the war, Watson's risky expansion was vindicated by the boom times. The fallout from wartime technology in radar and combat radios was rich. Electronic circuits in vacuum tubes, the instrument of Edwin Armstrong's amplifying triumph in radio (page 216), were coming to be recognized as having astounding potential for calculation through off-and-on "flip-flop" circuits; electrons moved so fast that thousands of calculations might be made in a matter of seconds. Here, waiting to be developed, was the basic element of the logic circuits in a digital computer: A circuit that alternated between two electronic states could represent the binary digits of one and zero. In 1943 Watson had asked Bryce to find an outstanding professor of electronics for IBM. In the exigencies of war, none was available. Watson did not see vacuum tubes replacing punch-card machines, but he was forever restless if IBM was not on the frontiers—and seen to be there. He had thought even in the '30s how marvelous it would be if a calculating machine had a memory, if it could be

programmed to draw on its memory for a variety of tasks—if it could be a computer. Drucker pays Watson the compliment of saying he "saw and understood the computer fifteen years before the term was coined." And having thought of the ends, he had gone in search of the means. In 1938 Howard Aiken (1900–1973), a graduate student in physics at Harvard, had sketched an outline of a digital machine that would perform long sequences of calculations using pre-programmed instructions. Watson had taken him up on it, given him a crash course in the IBM lab at Endicott and assigned Bryce to oversee the detailed design and engineering with Lake and two other top engineer-inventors. Aiken went into the navy, but the IBM men continued work and shipped the finished machine to Harvard in February 1944—a sleek five tons of steel and glass that was 51-feet long and 8-feet high with 750,000 parts and hundreds of miles of wiring. It had cost IBM $500,000.

Watson was in a state of pleasurable expectancy when he took the train to Boston with Jeannette for the formal ceremony on a miserably wet day in August. By this time he was used to being garlanded wherever he went. It did not happen at Harvard. He was insulted. Not only did nobody greet him at the railway station but Aiken and Harvard had preempted the ceremony with a story in the newspapers, acclaiming Aiken's "invention" without mentioning the six years of independent work by the IBM men. An apoplectic Watson screamed at Aiken, "I'm just sick about the whole thing. You can't put IBM on as a postscript. I think about

IBM AT WAR: Under the camouflage is a mobile machine records unit, photographed by the U.S. Army Signal Corps.

IBM just as you Harvard fellows think about your university."

A history of IBM's early computers by Charles Bashe and others calls the Mark I a "technological dead end." Made without vacuum tubes, it was the high-water mark of the electromechanical tabulator technology introduced by Hollerith, but the snub galvanized Watson. He instructed his engineers to build a supercalculator—"the best, fastest, biggest" that would leave the Harvard machine in the dust—and insisted, "We're not doing enough in electronics." The result three years later was a machine with 12,500 tubes. It also had 21,400 clunky relays, but Watson's Selective Sequence Electronic Calculator (SSEC), nicknamed Poppa, fulfilled Hollywood's idea of a computer when he put it on display in the Madison Avenue offices early in 1948. Street crowds by the hundreds were in awe of flashing displays on three long walls filled with electrical consoles and panels, switches, meters and neon lights that blinked when the superbrain was thinking. It could be programmed by software, and it was 250 times faster than Aiken's Mark I. On average, according to Bashe and his IBM colleagues, it could perform 216 multiplications in one-fiftieth of a second, division in one-thirtieth of a second and addition or subtraction of 19-digit numbers in one–thirty-five hundredths of a second. Wallace Eckert, the astronomer head of the Watson Scientific Computing Laboratory at Columbia, gave it the task of computing the position of the moon for any given time in the past and future; his data was used in the first moon landing.

Watson dedicated Poppa on a nonprofit basis "to the use of science throughout the world." From a vacation in Florida he instructed his staff to get back to basics—to begin planning a "machine of the same type, with reduced capacity, to meet the requirements of the ordinary businesses we serve." The Mark I and Poppa were both far-too-expensive play. It was time for the next step, but a newcomer would have to take IBM into the fully electronic modern computer age—a man by the name of Thomas Watson Jr.

In 1956, a few weeks after handing over the company, Watson died at the age of 82. Perhaps it was well for peace on earth that he

CROWD PLEASER: Watson Sr.'s wall-to-wall Selective Sequence Electronic Calculator (SSEC) became everyone's idea of a computer when he put it on display in IBM's Madison Avenue showrooms in 1948. "Poppa," as it was called, had 12,500 tubes but 21,400 clunky relays.

was not around to hear so often attributed to him the statement that the commercial market in the United States would never require more than five or six computers. He never said anything of the kind. And the man who did, according to the computer historian Paul Ceruzzi, was Howard Aiken.

Thomas Watson Sr. left a remarkable legacy. The tiny CTR company he took over in 1914 had revenues of $1.3 million. When he finally retired after 42 years, the acorn was one of the tallest and most vigorous trees in the forest, with revenues of more than $700 million, and his personal fortune was around $100 million. If anyone had held on to 100 CTR shares from the day Watson had walked into the Broad Street office, the $2,750 would have been worth $2,164,000 plus $209,000 in dividends. But all these totals are insignificant to the betterment of daily life for millions of people afforded by the man who took those determined strides down the early miles of the information superhighway.

THOMAS WATSON JR. He was terrified of failure when he took over IBM from his father but he gave us systems to transform daily life through his faith in mainframe computers

1914–1993

THOMAS WATSON was a hard act to follow. Young Thomas Watson Jr. didn't even try. What was the point of trying to live up to the expectations of his charismatic father, who grew both more distant and more demanding as he climbed the social and business ladder and made IBM a world force? In the pleasant rural community of Short Hills, New Jersey, where Thomas Jr. grew up as the first of four children, he was known as Terrible Tommy for petty pilfering and piling up demerits at school. All the children had to be sent home the day he poured skunk juice into the ventilation system. His sweet-tempered mother administered the switching punishments at home because Watson Sr. got too angry—all the while never giving up hope that his son would mature enough to be the heir apparent at IBM. He paraded him at a sales conference at the age of 13 dressed as his body double. In the photograph with his father, Junior looks as if he would like to shrink beneath the derby hat and vanish into the folds of the long velvet-collared overcoat. He sobbed in his mother's arms, "I can't do it. I can't go to work at IBM." He suffered periodic depressions for six years.

Away in a boarding school near Princeton, and then in a fraternity at Brown, he was indulged with $300 a month pocket money, double the average American fam-

ALL BOYS TOGETHER: Father and son shared one enthusiasm at least: they were both big backers of the Boy Scouts. This is the launch of the 1955 campaign for funds at the Waldorf-Astoria hotel, with Watson Jr. in the second row (left) and Watson Sr. just behind in the third row. Mayor Robert F. Wagner is to his left. Watson Sr. administered the Scout oath as international Scout commissioner.

DUTY: Watson Jr. in 1937 at an IBM banquet. He would rather be at the 21 Club.

ily's earnings. He was Prince Hal, with demons of self-doubt, and soon escaped into the life of a rich F. Scott Fitzgerald playboy prankster, fast cars, pretty girls, partying all night, drinking in the speakeasies (Prohibition was in force until the end of 1933). He had been lucky to get into Brown on his father's influence, and he barely managed to pass. While at the IBM school for salesmen in Endicott, New York, he was still floundering with who he was and what he wanted to do with his life. He resented the cult worship of his father at work and his domination at home; he told the reporter Robert Lenzner, "I used to blow in his face at Thanksgiving dinner. I was trying to find a way of feeling even with him." He must have been insufferable.

His father's well-meaning efforts to advance him only enraged Watson more. After graduation from Endicott he was assigned the plum job of selling IBM machines in Manhattan's financial dis-

trict and felt demeaned by the successes arranged for him: His most spectacular was a sale on the first business day of 1940 that took care of his entire year's quota, instantly making him the company's number one salesman. "From then on," he told Peter Petre in his engaging memoir, "even though life outside of IBM seemed impossible to imagine, all I could think about was finding a way out."

Flying was the answer—his one youthful enthusiasm and the one area where he did not have to worry about being overshadowed by his father, for whom the very idea of flying was terrifying. Watson Jr. joined the National Guard as a private on the outbreak of World War II and won a pilot's commission in the Army Air Corps. To his dismay, he was posted to an irrelevant unit on the Pacific Coast and once again Dad had to come to the rescue, pulling strings to get him into a top military school. Here, at last, he buckled down and passed on merit while his marriage to Olive Field Cawley, a diminutive debutante, gave him the sweetness and love he needed. For the first time in his life, he was really trying. His persuasiveness in an official campaign to get pilots to learn blind flying on Link simulators won him an assignment as aide-de-camp to Follett Bradley, the head of the First Air Force, but then he almost blew it. He knew how to follow orders, to the letter, but he did not know how to give them. After several months in a besieged Moscow with Bradley, his flight crew made it clear they would rather be in combat than exposed to Watson's withering perfectionism. He had indulged in all the excesses he had so often deplored in his father. The realization changed him so much that his crew ended up volunteering again.

Several near-death experiences also gave the former playboy a different perspective on life. The closest call was when, flying into Russia as the copilot of a big B-24 bomber, he left the cockpit to check on the landing gear in the remote belly of the plane. His co-pilot, flustered by Russian fighters pressing him to land quickly, opened the wheel door prematurely. Watson's leg was trapped and he was spread-eagled a thousand feet over the Baku oil fields with the bomber making its approach to land. He found reserves of calm he never knew, first to hand signal a Russian navigator that he wanted to talk on the radio to the cockpit, then to explain his predicament and not panic as Bradley used a hacksaw to cut away at the metal holding his leg, freeing it only minutes before a touchdown that would have torn him apart. He volunteered for combat and frankly admits he was relieved when the top brass in the Army Air Force preferred to keep him as a personal pilot: One trip to evacuate wounded men in the Assam Valley "scared me enough to last the whole war." Everywhere he went in U.S. war zones, he saw his father's hand, the IBM punch-card units, the grease that kept the American war machine functioning.

Back from the war in 1945, Lieutenant Colonel Watson, as he had become, was pondering a job as a pilot for United Airlines when Bradley stopped him in his tracks with the remark that he expected Watson would go and run the IBM company. Watson wondered if he could do it. "When you put your mind to something," said Bradley, "I've never seen you fail." It was enough. Twenty-four hours later the more maturely ambitious Tom called his

RISKY DAYS: Pilot Watson found redemption in the hazards of World War II.

father to say he would like to join IBM after all, if IBM would still have him. "When I think of the difference in my general outlook now as against the 1937–1940 period," he wrote his father, "I am convinced I am now at least seventy-five percent better equipped to follow in your footsteps—as I intend to do." His father was overjoyed.

Watson Jr. was 32 and prematurely gray when he turned up for work on the first business day of 1946. Many were astonished at his metamorphosis. One of his attractive characteristics—one of the reasons he was to succeed—was his lack of pretension. He never pretended to know what he didn't, and looking back he never claimed he saw

the light of computer electronics from day one. Soon after turning up for work in the IBM regulation stiff paper collar, dark suit and quiet tie, he found himself perspiring heavily in a boiling hot room at the Moore School of Electrical Engineering at the University of Pennsylvania. The heat came from 18,000 vacuum tubes (and 70,000 resisters, 10,000 capacitators and 1,500 relays) occupying a room the size of a squash court in a wall-to-wall arrangement designed by John Presper Eckert, an engineer, and John Mauchly, a physicist. The brittle Eckert and the conceptual visionary Mauchly, both in their early 20s, were computer missionaries who in 1943 had undertaken a secret project for the Pentagon. Now they were proudly demonstrating the ability of ENIAC, as they called it—Electronic Numerical Integrator and Computer—to calculate the trajectory of an artillery shell through the air. ENIAC was a thousand times faster than the Harvard Mark I that Watson Sr. had subsidized and built; it could track a shell faster than the shell could fly.

In his memoir, Watson Jr. confessed: "I can't imagine that I didn't think, Good God, that's the future of the IBM company. But frankly I couldn't see this gigantic, costly, unreliable device as a piece of business equipment. I never stopped to think what would happen if the speed of electronic circuits could be harnessed for commercial purposes." He was there with his father's number two, Charley Kirk, who agreed it was unwieldy, expensive (at $450,000), unreliable and of no conceivable interest to IBM as a business machine. It did not give them pause that some eight or nine laboratories in the United States and Britain were on the same track as Mauchly and Eckert; call it a lack of vision, but these all seemed esoteric scientific exercises remote from the requirements of American business.

A few months following the Watson-Kirk visit, Eckert and Mauchly, denied tenure at Penn (which disliked their commercial ambitions), came to see Watson Sr. to sell themselves and their ideas for an improved computer. He met them with a number of IBM people, including a young salesman, Jim Birkenstock, who was to be an unexpectedly important figure in IBM's move to computers: He had just been catapulted to general sales manager—the youngest ever at the age of 33—on a sudden on-the-spot impulse of Watson Sr., who heard him telling salesmen to stop moaning about the unpopular downsizing of sales territories because that was better for customer service. Watson Sr. rejected Eckert and Mauchly's offer to buy them out, and they rejected his suggestion of starting a computer lab at IBM. Birkenstock, who was a lively 91 in 2003, recalls that when the pair of inventors left, Watson Sr. remarked, "I couldn't have Mauchly in the company anyway. Did you see his red socks?"

The U.S. Census Bureau and Prudential Insurance were more imaginative. They listened to Mauchly and Eckert's ideas for a new electronic digital computer and backed them. ENIAC could operate different programs only by the laborious process of disconnecting and reconnecting hundreds of patch cords to route electrical signals to different units of the machine. In effect, ENIAC had to be rebuilt for each new problem it was to solve, and that could take hours, even days. The slower Harvard Mark I could be swiftly reprogrammed by inserting new punch cards or paper tape. For their new effort, the whiz kids proposed a revolutionary computer that would enable a business such as an insurance company to dispense with millions of punch cards and store its own programs on magnetic tape, this time with far fewer vacuum tubes to burn out. The Smithsonian's Paul E. Ceruzzi observes that this was the crucial step leading to the establishment of "programming" (later "software") as something both separate from and as important as hardware design.

The distinct threat to IBM's business that Mauchly and Eckert called UNIVAC (Universal Automatic Computer) was to take them five years to build, at a cost of $1 million rather than the $300,000 they had charged the Census Bureau: As the computer historians Martin Campbell-Kelly and William Aspray note, "Like entrepreneurs before and after they wished away the financial problems in order to get on with their mission." While Mauchly and Eckert and their recruits fashioned a trapdoor for IBM by working all hours in two upper floors of a men's clothing store in Philadelphia, Watson Jr. was constantly being bugged by Birkenstock, who agreed with Mauchly and Eckert that punch cards were the past and present but not the future, an unthinkable heresy within any kind of psychic distance of Watson Sr. Punch cards were in so much demand that the plants were working three shifts. "IBM," Watson Sr. said in a speech, "is an institution that will live forever on the punch card." He was utterly sure he knew what his customers wanted; the problem, as described in Clayton Christensen's *The Innovator's Dilemma*, is that customers don't always know what they want. Someone has to figure out what they *need*.

The relationship of Birkenstock and Watson Jr. was critical to the future of IBM. Men of about the same age, both had difficult fathers, but otherwise they could not have come from more different backgrounds: one frittering his early years as a child of privilege while the other was working his way through the University of Iowa in the Great Depression years as a busboy, billboard painter, meat carver and cashier, graduating magna cum laude. Birkenstock had actually begun his working life at the age of nine as a golf caddy, became a master caddy to the greats and was a golf pro until bank collapses in Iowa cleared the courses of players. At IBM, Birkenstock was distinguished by a readiness not to conform to the party line, symbolized perhaps by a crenellated hairstyle at odds with the company's preference for the nondescript. As

THE BRAIN: The Watsons failed to see the potential of the all-electronic computer ENIAC invented by John Mauchly, a physicist, and John Presper Eckert, an engineer who was always worrying about his 18,000 vacuum tubes.

for Watson Jr., having been knocked into shape by the Army Air Force, he was very much the new boy on the block. Between Manhattan, Poughkeepsie and Endicott, there were 22,000 people in IBM—engineers, salesmen, factory workers—and nobody was going to tell the boss's son where the bodies were buried, so he relished straight talk from the forthright and ambitious Iowan, most of the time. "When Mr. Watson Jr. lost his temper with me," Birkenstock told us, "by the time I reached my own office a note of apology was already there."

Birkenstock was not an engineer, but he knew what the customers wanted while Watson Jr. found he had a natural understanding of technical matters. "Birkenstock," Watson Jr. wrote, "did more to put IBM into the computer business than any other man," which is an example of Watson Jr.'s generosity, since it is the accolade most often awarded to Watson Jr. himself—and justly.

Among the "fringe" people who enthused Birkenstock about electronics were Ralph Palmer and Stephen Dunwell, IBM engineers who had returned to the lab at Poughkeepsie after hush-hush wartime work for the navy and army, respectively, on codes and "other things." When Dunwell was asked what that meant, he replied, "I think they are forever classified." He felt "dismay and astonishment" after two years away to find vast numbers of vacuum tubes and relays but no sense of direction and also electrical engineers very much subordinate to Watson Sr.'s "inventors"—the mechanical engineers in the main laboratory on North Street in Endicott. Palmer, who like Birkenstock has enjoyed a long life, was able to form a small electronics group in the top floor of the Kenyon Mansion. (He is a man of unusual imagination: Since he can play pretty well every musical instrument, he formed Palmer's Band, which was Palmer recording himself on tape and intercutting so that it sounded as if 20 instrumentalists were going full blast.) He applied this original mind to experiments in electronic switching, counting and storage.

Palmer and his group were more than ready to follow up on the epiphany Watson Jr. experienced one day soon after his disenchantment with ENIAC. He was with his father and Charley Kirk when they dropped in on an office he had never seen before, labeled PATENT DEVELOPMENT. Inside, they saw a punch-card tabulating machine connected by a thick cable to a black container about the size of an upright piano. The tabulating machine was feeding the black box with 100 cards a minute, which was fast but was nothing compared to the speed by which the black box took the punch-card data and did multiplications in *thousandths of a second.* Watson looked at the 300 vacuum tubes inside the multiplier and told Petre: "That impressed me as though somebody had hit me on the head with a hammer because the multiplier looked like a relatively simple device. I left that room and said, 'It's fantastic, what that thing's doing. It's multiplying and coming out with totals and doing it all with tubes, Dad. We should put this thing on the market. Even if we only sell eight or ten, we'll be able to advertise the fact that we have the world's first commercial electronic calculator.'"

And that, Watson Jr. said, "is how IBM got into electronics."

Strictly speaking, the statement is misleading in its compression of history. Watson Sr. and the inventive Jim Bryce had given a young engineer called Halsey Dickinson permission to make a model—the IBM 603 Electronic Calculator—nearly two years before in the wake of the Harvard-Aiken debacle (pages 364–365). But Bryce had sold the idea to Watson Sr. specifically on the basis of its public relations value, with little expectation of going into production. The 603 had received no priority. In fact, biographer Kevin Maney says Dickinson started work on it in the basement of his home when electronics research at Poughkeepsie was abandoned during the war. The significance of the moment in 1946 was Watson Jr.'s recognition—"the most exciting moment of my business life"—that there was commercial potential in these vacuum tubes. All the 603 could do was multiply, but Watson's enthusiasm resulted in a production lot of 50 machines and a sales effort—a presentation at the National Business Show in New York in September, a full page advertisement in the *New York Times*—with the gratifying consequence that a hundred were rented out, instead of a handful. Even more importantly, Watson Jr. saw clearly that Palmer's mavericks in Poughkeepsie were stars. On their own initiative, they had already begun vastly enhancing the 603. Watson's "pet," the 604, delivered in the fall of 1948, was a major step in the transition from calculators to computers. It did not store its own programs, but it could execute 40, then 60 programs by reading a punch card, and Palmer's original concept of packaging circuits in separate pluggable units meant burned-out tubes could be easily replaced. It was, says the IBM historian Charles Bashe, "a fundamental contribution to the art of digital electronic equipment design." Renting at $550 a month, the 604 was a hit (eventually 5,600 were installed).

The notion that everybody at IBM was blind to electronics until Watson Jr. arrived misunderstands the nature of his achievement. In the country-house lab overlooking the Hudson in Poughkeepsie, there were a number of brilliant individuals dedicated to electronics years before Watson Jr. saw a vacuum tube; and his father did consider electronics as having an important place in the company. The difference was that Watson Jr. fairly quickly concluded that electronics *were* the company. He had no superior knowledge of either business machines or technology, but he was obeying the company dictum to listen and observe.

Watson Jr. more and more appreciated the way Birkenstock could "think down into the depth of things," so imagine his shock one day at the end of 1946 when he found that Birkenstock was no longer there. He had packed up his office and gone home full of "anger, disgust and disappointment" on being demoted from general sales manager on the grounds that he had too often opposed Watson Sr. He had been offered a position in the field, but he had resigned from the company and was too upset to come to the phone when Watson Jr. called. Watson Jr. went out on a limb, risking his father's displeasure, by meeting Birkenstock later in the week. He persuaded him to stay with IBM, reporting to him and not to Kirk or his father, with responsibility for a department called

Future Demands that reported on product planning.

Not long afterward, it looked as if the appointment was a blunder when Watson Jr., at home in Greenwich, watched a guest on a television program exult about a technological breakthrough that the man's company was about to achieve in document processing. Watson had Birkenstock on the carpet the next morning. Had he heard about this? What was he doing about it? Yes, said Birkenstock, he had heard and seen. But he had a story to tell. While Watson Jr. was away, the inventor of the device, one Chester Carlson, had been into the IBM offices seeking backing, and though it was a crude model, Birkenstock had been impressed enough to take it along immediately to Watson Sr., who asked him, "What has this to do with punch cards?" Birkenstock replied, "Nothing, but it will give us a new product for the office-machine market." Birkenstock recalls: "He responded, 'Now let me tell you something, young man. When my wife, Jeannette, tells me I'm the smartest man in the world, I respond, "No, Jeannette, I'm only smart in spots and I'm wise enough to stay on those spots." Birkenstock, you should know that the punch card is one of those spots and this Carlson invention isn't, so tell Mr. Carlson we're not interested in his invention.'" Thus did IBM kiss goodbye to what Joe Wilson, president of the Haloid Corporation, later turned into Xerox.

Watson Jr. tried to salvage a deal with the copy machine, but it was too late. It added fuel to his impatience with the way his father and Kirk were running the company. He hated the flurries of demotions and firings Kirk carried out, and he was appalled that 30 or 40 executives reported directly to his father; some of them were kept waiting in the outer office for days. In April 1947 he gave his father an ultimatum to choose Kirk or him. Watson Sr. sent them both on a European trip. A few days into it Kirk suffered a coronary and died. It did not ease things between father and son. "I'd always looked on Charley Kirk as a barrier between my father and me. It wasn't until after Kirk's death that I realized he had also been a buffer. Undiluted, T J Watson could be pretty hard to take." Both Watsons were all detonator and no fuse.

Scenes often ended in tears and remorse on both sides. It was not that they blew up over electronics; it was about the only thing they did not blow up over. It was that Watson Jr. was up against an entrenched hierarchy of marketing and financial men in New York and the seven senior engineers at Endicott beloved by his father as "the inventors." None of them understood electronics, and none of them took any notice of Watson Jr. Without his father's 100 percent backing, he was treading water. In September 1949 he was promoted to executive vice president with responsibility for manufacturing as well as sales, but his father was still liable to reach down the line and change a policy and was very resistant to expanding on borrowed money.

It was not good for Watson Jr.'s blood pressure that he kept picking up his own evidence about the veracity of Birkenstock as Cassandra. A vice president of Metropolitan Life Insurance invited him over to look at three floors of his building filled with punch cards. He had heard they could put all their records more economically on magnetic tape. Roy Larsen, the president of Time, Inc., told Watson Jr. the same thing. Punch-card systems were too slow to cope with increasing subscription lists.

It says something of the atmosphere at IBM at the time that even Watson Jr. felt it imprudent to repeat these observations to his father as the company was still doing a roaring business renting tabulators and selling punch cards. Watson contrived to set up a task force of 18 of the best systems experts to study magnetic tape, hoping they would break ice. They only made it more solid: Magnetic tape, they declared, had "no place" at IBM. The top salesmen said the same thing. Meanwhile, up in the Kenyon Mansion, Palmer, who started to experiment with magnetic tape for digital storage and high-speed input/output, had been stopped dead in his tracks by the head of engineering, who was concentrating all efforts on making giant punch cards to carry more information.

The block on Palmer was one of the last straws for Watson Jr. He still could not be sure that computers like UNIVAC would ever be dependable and economical enough for business, but the continued success of the 604, that nagging instinct for elec-

tronics and Birkenstock's "thinking in depth" provoked him to take his father head-on in 1949. "I criticized his organization in Endicott in the roughest possible way. I said to him, 'All you've got up there is a bunch of monkey-wrench engineers. The time for hacking machines out of metal is gone. Now you're getting into a field where you have to use oscilloscopes and understand the theory of electron streaming and scanning beams inside the tube. You've got to do theoretical things, you've got to do them with able people.'" Watson Sr. summoned the vice president for engineering, who told the old man what he wanted to hear, that they had the finest research organization in the world and not to worry. How the impasse would have been broken is unknowable, but broken it was. On his own initiative, the head of finance, Al Williams, did a study of comparative research investment between IBM, RCA and General Electric and found IBM nearly a third behind the others. The next day, Watson Sr. called in his top men. "I've been thinking about our efforts in research. I want you to go out there and build this up. Now Mr. Williams—Mr. Moneybags over there—may complain to you about the cost. But don't let him stop you."

No army commander ever moved more quickly through a gap in the enemy lines than Watson Jr. He caught the man he wanted to lead the expansion on the tennis court at the One Hundred Percent Club convention in Endicott. This was Wally McDowell, an MIT graduate and the Endicott lab manager. There and then, Watson Jr. gave him the instruction: "Come to New York and start hiring new engineers in quantity." McDowell asked, "What do you mean, in quantity? A few dozen, I could do that from up here." Watson's memorable answer: "No. I mean at least a few hundred, and perhaps a few thousand." (In the following six years IBM's 500 engineers were joined by 4,000 "double-domes" and "longhairs," as Endicott dubbed the doctors of electrical engineering, mathematics and physics.)

The tennis-court conversation was a turning point. The question then, in May 1950, was whether IBM had already lost the race to UNIVAC, said to be winning orders from insurance companies in anticipation of its completion—indeed whether the company was to survive at all in an electronic era. About 30 firms had entered the computer race by the early '50s. All were in the United States, except for the British Ferranti Mark I, based on a University of Manchester computer design that had established the practicability of a computer that could store programs. (Campbell-Kelly and Aspray write, "Unfortunately the enthusiasm for manufacturing computers in Britain was not matched with an enthusiasm for using them by its old-fashioned businesses, and by the end of the decade Britain's computer industry was fighting for survival.")

The Watsons were given a chance to catch up but let it slip through their fingers. Mauchly-Eckert lost part of their financing, and they came to New York to ask if IBM would buy their company. "When they came," wrote Watson Jr., "Mauchly slumped down on the couch and put his feet up on the coffee table—damned if he was going to show any respect for my father." The Watsons' answer was another no, but this time it was the law and not personalities that bothered them. UNIVAC was too prominent an American competitor, and Watson Sr. feared a bruising collision with antitrust law. Eckert and Mauchly's next stop was Florida and the yacht of James Rand Jr., the hard-charging president of Remington Rand. They came ashore as a wholly owned subsidiary of Rand, reporting to a black-belt bureaucrat, General Leslie R. Groves, chief builder of the Pentagon and wartime head of the Manhattan Project. It was not good news for IBM. Ten years later, Watson Jr. reflected on why they had fallen behind: "Unless management remains alert," he said in a lecture, "it can be stricken with complacency—one of the most insidious dangers we face in business. In most cases, it's hard to tell that you've even caught the disease until it is almost too late."

In June 1950 his father's patriotism enabled Watson Jr. to build dramatically on the momentum he had created. Only hours after the start of the Korean War in 1950, Watson Sr. called President Truman to say that an unlimited amount of IBM personnel and facilities would be available for any program directed and approved by the War Production Board in Washington.

Watson Jr. and Birkenstock were summoned to the old man's office, and if it had been anywhere other than sober IBM they would have emerged giving each other high fives. Birkenstock was authorized to meet U.S. defense needs by starting a new division, taking staff from wherever he liked. "When we came out of his father's office," Birkenstock told us, "Tom said to me, 'Here's your chance to go ahead with the large-scale computer development you've been wanting.'" Watson Jr. called on a brilliant academic mathematician, Cuthbert Hurd, who had joined IBM in March 1949 and was now director of a new Department of Applied Science. He sent Birkenstock and Hurd on a nationwide scouting mission talking to officials in the National Security Agency, generals, scientists and defense contractors—22 meetings in all about war games, weather forecasting, guided missiles, logistics, cryptanalysis.

Their return to New York months later was like a line of cavalry on the horizon for Palmer's beleaguered electronics team in Poughkeepsie, still held up in their ambitions to develop an electronic computer based on magnetic tape. IBM senior management had given the funds to the punch-card men.

Birkenstock and Hurd had an astounding proposal for Watson Jr. They urged him to skip a trial stage and immediately design and produce a prototype all-purpose mainframe computer that could be duplicated 25 to 30 times. Twenty-five machines! There were perhaps only 20 stored-program computers in the entire world. Hurd believed they could break away from the common one-of-a-kind machine and manufacture a computer that would serve all the varied uses he and Birkenstock reported. The prototype would cost $3 million, with production costing four or five times as much again. It was a daunting moment for Watson Jr., the apostle of innovation, when Birkenstock, Hurd and Palmer asked him

to commit on the basis of a bunch of barely comprehensible diagrams of black boxes connected by lines. He could not pretend to understand it well enough to explain to his father and justify embarking on a technical and financial gamble of the highest order—the highest in IBM's entire history: all the more so since he had been convinced by Birkenstock that they should fund it themselves to retain freedom rather than seek government subvention, the original idea. "Admittedly, this was a huge gamble," says Birkenstock now, "and a bold step never before attempted by IBM—or any other company. I played down the risk factor and emphasized our need to maintain our leadership in data processing."

Watson Jr. did not need them to tell him what the risks were. The IBM regular business was booming, but in 1950 the company owed the insurance companies and banks $85 million, they were going to need to borrow as much again and Watson Sr. hated debt. Watson Jr. told Birkenstock and Hurd to sound out military contractors and government departments—very discreetly—to see if they could rent the proposed computers for $8,000 a month. "When we returned with eighteen letters of intent," says Birkenstock, "Watson Jr. promptly gave his consent."

The role of leadership in a large corporation concerned with technical issues is especially hazardous. There are no formulas. There are times for innovation, times for consolidation: Coca-Cola will not forget its disastrous introduction of "New Coke." The mutineers, who are always in the woodwork, are not always right, and when they are wrong they are often catastrophically wrong. There are always contending interests, specialists who cancel each other out. The chief has to choose and not always on the basis of a mastery of the technicalities. Equivocation guarantees mediocrity. Not every chief can be a Thomas Edison or an Edwin Land, and he need not be: After all, George Eastman at Kodak did not have enough scientific training to understand the work of the chemists in his "invention" laboratory. Watson Jr., who had such trouble finding his destiny, concluded in his memoir, "You've got to feel what's going on in the world and then make the move yourself. It's purely

THINK! Thomas Watson Jr. retained his father's borrowed slogan, but shed the more formal trappings of office. He violated the dress code by turning up in a striped shirt.

visceral." Well, it is certainly partly visceral, but it was still altogether extraordinary that he summoned up the will while navigating the turbulence of his and his father's emotions. Perhaps he imbibed some faith in science as a pilot in the dark skies of wartime, committing his life to the validity of his instruments.

In any event, the innovators were cunning in conditioning Watson Sr. and the very resistant IBM establishment. They did not call their computer a computer; they called it a calculator to echo the earlier punch-card calculators. And they focused it on Watson Sr.'s desire to help Truman and the war effort. Thus it became the "Defense Calculator." Building it taxed all the brainpower Watson Jr. had recruited. Palmer's laboratory overflowed into an old pickle factory along the Hudson River. Watson Jr. encouraged himself by thinking of the Wright brothers, doggedly moving from problem to problem, any one of which could have grounded them for good. They were easing away from the relatively slow medium of punch cards that they understood very well to "something a hundred times faster we didn't understand. . . . We were trying to develop logic circuits, memory circuits, tape-handling devices, recording heads, card-to-tape transfer techniques and, in conjunction with other manufacturers, vacuum tubes and tapes themselves." By the time of the IBM annual stockholders' meeting in April 1952, when the machine was half built, Watson Sr. was a convert. He proudly announced the Defense Calculator as the IBM 701, boasting that it would occupy less than a fourth of the space of the hybrid Poppa machine he had put on show on Madison Avenue (page 365) and operate 25 times faster. It was the first step into a new era of data processing that initially supplemented and eventually replaced the punch-card systems. These old systems remained IBM's number one source of income until the 1960s, but had Watson Jr. not acted when he did, IBM would then have become a monument to obsolescence.

While Watson Jr. was driving IBM toward the new era, Mauchly and Eckert had not been slouches. Red LaMotte, one of the few old-timers who had ever dared to argue with Watson Sr., was head of sales in Washington, and he rushed to the airport when Watson Jr. was changing planes in 1950 to tell him that Remington Rand was on the point of delivering a UNIVAC to the Census Bureau and had orders for two more. In fact, the UNIVAC engineers did not run acceptance tests until March 1951, after sweltering in the 1950 summer in shorts and undershirts in the tremendous heat generated by 5,000 tubes. "I was terrified," Watson Jr. recalls. "I came back to New York in the late afternoon and called a meeting that stretched long into the night. There wasn't a single solitary soul in IBM who grasped a hundredth of the potential the computer had. I thought we were all on board the *Titanic*." His major push had two teams of engineers working three shifts round the clock. It did not spare him the anguish of November 1952, when he turned on CBS to watch Ed Murrow, Eric Sevareid and Walter Cronkite give the election news. They announced they would predict the result by calling on "that marvelous electronic brain, UNIVAC." It correctly forecast an Eisenhower landslide, such a surprising prediction that the Remington Rand engineers fiddled with the computer to suggest it was still a close race.

A month later, December 1952, the first model of the IBM 701 came off the line and was shown off in Madison Avenue. It was slower than the UNIVAC, and there were complaints. In response, Watson Jr. took what Birkenstock calls the "epic command decision" to retrofit all data processors with magnetic-core memory developed at MIT. Next, concerned about vacuum-tube failures, he ordered all 700 series machines to be retrofitted with transistorized circuits. By this daring crash program, IBM went a generation ahead of UNIVAC. For the 1956 election, the computer forecasting the result was an IBM computer. Remington Rand— soon to sell out to Sperry—got beaten because it was a conglomerate also busy selling electric shavers, farm machines and autopilots. Jim Rand never concentrated as both Watsons did, and IBM's specialty of after-sales service and training on the job was second to none.

Realizing the revolution in the works, *Time* magazine put Thomas Watson Jr. on the cover—something it had never done for his father. At 82, his father stepped down and was dead a month later. His son remembered with an ache how not long before he had ripped his arm away from a gesture by his father, shouting, "Goddamn you, old man! Can't you ever leave me alone?" Now Watson Jr., 42, felt very alone. "I didn't realize how much I still needed him emotionally. I remember standing in the corridor outside my office looking dumbly up the stairs that led to his. Fear of failure became the most powerful force in my life."

By a number of accounts, as IBM's chief, Watson Jr. operated on the theory that nice guys finish last, exhibiting rage and instilling fear in others. He promoted "scratchy, harsh" individuals and institutionalized conflict with a "contention" system whereby managers were encouraged to challenge one another. He told the One Hundred

Percent Club, "I just wish somebody would stick his head in my office and say, 'Tom, you're wrong!' I really would like to hear that. I don't want yes-men around me." But he did not go in for the maddening micromanagement his father had done; besides, there were something like 50,000 employees by now. With a minimum of disruption, Watson Jr. created a new, freer structure less censorious of personal habits. He caused a stir by showing up to work in a striped shirt, and he drank occasionally.

Early in the '50s, the air force had been in a panic that the United States was vulnerable to a surprise attack by Soviet bombers, a fear aggravated by the Soviet explosion of their first atomic bomb in September 1949 and their development of the hydrogen bomb from 1953–55. Nobody was

sure that what the air force wanted was at all possible: a network of huge digital computers to process input from ground-based radar, ships, early warning aircraft and ground observers so that if an attack was imminent, the human controllers could see the battle situation displayed on a screen. They called it SAGE, for Semiautomatic Ground Environment. (Watson Jr. was happy to have the business, but he was not a cold warrior. He was one of the few business leaders to speak out against the McCarthy witch hunts, and he and Birkenstock repelled CIA moves for covert agents to assume IBM cover.) During the war, many advances in military computation had been made by MIT's Servomechanisms Laboratory. A young engineer there, Jay Forester, was the brains behind Project Whirlwind, wrestling with stingy military procurement budgets from 1945 to develop digital computer technology instead of analog techniques, and with it real-time computing power. Watson set out to convince him that IBM was best equipped to work with MIT to adapt Whirlwind technology. IBM built a prototype—Forester was impressed by the adventurous spirit of the place and the fact that IBM then had three million vacuum tubes in operation—but still he hesitated. Watson Jr. clinched matters by saying he would build a factory on a handshake deal without waiting for the air force's formal commitment. He put thousands of engineers to work and built 48 computers, each having 49,000 tubes and a weight of 250 tons. The $8 billion system worked when it was finished in 1963, but it was made obsolete by the arrival of long-range missiles that were too fast to be monitored by airplanes.

If SAGE was a "costly fantasy," in Watson's words, he saw immense potential in the spin-off for serving the airlines, universities, banks, railroads, department stores, supermarkets, libraries. First out of SAGE came SABRE—another jaw-saving acronym for Semiautomatic Business Related Environment, but think of it as fixing your next flight. The airlines industry was up against the wall coping with the millions of people discovering air travel. The electromechanical systems were inadequate for the flow of bookings, cancellations, seat assignments, availability of seats, connecting flights and

THE IBM WAY: White-suited Soviet premier Nikita Khrushchev boasted of the Soviet lead in space on his tour of the United States in the fall of 1959 but was impressed by IBM's assembly lines in San Jose, California. Thomas Watson Jr. (just behind Krushchev) told him: "We believe good working conditions contribute to the dignity of our individual people."

all the other complications. The regular traveler had a one in twelve chance of finding that a reservation made the week before had vanished. American Airlines and IBM got together to solve the problem. Robert V. Head was in charge for IBM, having survived entry into the strait-laced company wearing blue shirts and bow ties. He recounts that American's objectives were "stringent." The system that would deal with 40,000 reservations daily was to be installed at more than 100 locations for use by 1,100 agents, supporting 83,000 phone calls a day, and information had to be retrievable within three seconds 90 percent of the time. Oh yes, and please keep a record of passengers' names and tastes.

Watson was told the technology did not exist to plot such a river of mercury. Solid state computers with core memory were needed, and a random-access disk storage unit was still in the laboratory stage. In 1957 he bet $40 million that technology would catch up under the impetus of a deadline. The company was in debt to an extent that would have horrified his father; it owed more money than any other American company, $300 million to Prudential alone, and it was going to need another $200 million. It raised the latter in the second largest stock sale in Wall Street history; in the same year, it entered the superleague of companies with a billion dollars a year revenue. The SABRE contract was built on two IBM 7090 mainframe computers with a storage capacity of 800 million characters, but it was much more than a conventional hardware deal. Watson had to gear IBM's traditional business of marketing hardware to developing software programs with on-site programmers and trainers. It took much longer and cost much more than he had hoped. SABRE was not fully operational until 1964, but then American recouped its share of the costs in a year: It was now able to handle ten million reservations a year and for the first time passengers could book seats without having to wait overnight for confirmation. For IBM, Campbell-Kelly and William Aspray note that an innovation for one airline became a competitive necessity for the others. Trippe's Pan American and Delta signed up. The industry was transformed. SABRE was a triumph of real-time systems long before the advent of the personal computer. It has survived conceptually unchanged into the 21st century.

By 1965 IBM had 80 percent of the market share of computers and UNIVAC less than 10 percent. With his father's old National Cash Register Company, among others, Watson Jr. provided the bar-code technology for Universal Product Scanning—for supermarket grocery checkouts. For business sales from 1954, he sold the IBM 702 as an "electronic data processing machine," Birkenstock's term. It was now vacuum tubes that were out of date. Palmer's people at Poughkeepsie were delighted to use transistors, with the punch card engineers resistant on grounds of unreliability. Watson Jr. told Petre: "I would go into the lab and say, 'Why not transistors?' hoping they would take a hint. But for months every design they sent down to New York was full of tubes. Finally, I issued a memorandum saying that after October 1 we will design no more machines using vacuum tubes. Each time I heard an engineer say that transistors were undependable I would pull a transistor radio out of my bag and challenge him to wear it out."

Watson could now have coasted, but as the boss in the mid-'60s with nobody to blame but himself, he launched an entirely new vision of computing. He envisaged a system called 360 to represent a full flexible circle of computer needs for every type of market, large or small, commercial or scientific. A company could start with a small computer and plug in others as its needs grew without having to buy new software, and all the computers would have printers and bigger hard-drive memories. The projected cost was $5 billion. *Fortune* magazine put up a headline with a lot of daunting zeros and called it the riskiest business venture of the age. It was compared to the Manhattan Project and the D-day landings. Robert Sobel comments: "On the face of it this was suicidal. The older machines would become obsolete, including thousands of machines on which they were collecting rent. The 707s were relatively young, but Watson decided to murder his darlings."

New factories were built, financed by another stock offering. Sixty thousand new workers were hired. Two thousand programmers wrote millions of lines of code. System/360 had a messy launch but it gave IBM dominance of the industry; RCA, Honeywell and General Electric dropped out, and by 1970 some 35,000 IBM computers were installed in a variety of government agencies, businesses and universities. By the time a heart attack forced Watson Jr. into retirement at the age of 57 in 1971, IBM was the world's largest computing company and owned two out of every three mainframe computers in the United States. Watson Sr. had taken a company with revenues of $1.3 million in 1914 to $700 million in 1956. His son fed those figures into an electronic multiplier. He grew sales more than ten times to $7.5 billion and the number of jobs from 72,500 to 270,000.

Jim Birkenstock, looking back on 60 years in 2003, summed it up in a sentence: "Thomas Watson Jr. was the savior of IBM." The company soared in the '80s, made record profits in 1990, but had become again the epitome of the complacency Watson Jr. detected in the late '40s, failing to lead in the era of the personal computer. When Lou Gerstner arrived in 1993 from RJ Reynolds and American Express, IBM was every commentator's favorite football. A few days into the job, he stepped into the car at his Greenwich home to go to work and found someone sitting in the backseat. It was a neighbor, Watson Jr. In the story of his successful makeover of IBM, *Who Says Elephants Can't Dance?*, Gerstner writes: "He said he was angry about what had happened to 'my company.' He said I needed to shake it up 'from top to bottom.'" He did, but Watson was not there to see it. He died on New Year's Eve at the age of 79.

Watson's 360 belonged to a world of mass—massive corporations with massive products built with massive amounts of capital. That world disappeared as technology started to become miniaturized. The belief that after World War II innovation took place only in large corporate research centers was wrong. Giant corporations, the principle feature on the American economic landscape, were still important, but they were no longer all-powerful. For a time, the elephants were the innovators and Tom Watson Jr. the master, but in America tiny start-ups and lone individuals who rebelled against the giants, who refused to be men in gray flannel suits, could still succeed.

A SAILOR'S GREAT IDEA—
THE INTERACTIVE MINICOMPUTER

KEN OLSEN (1926–)

KEN OLSEN had no idea what a computer was when he left the navy in 1946 at 20. But he knew the innards of a radio. A big hulking boy, Olsen had grown up in the Depression in a working-class area outside Bridgeport, Connecticut, where he and his more rebellious younger brother, Stan, built a one-tube radio out of pieces of garbage; they succeeded in interrupting the local station with their transmitter so Stan could sing his jingle, "Murphy's Meatballs."

The navy gave Olsen a year's education in electrical engineering and electronics in 1944–45—he was one of hundreds of thousands of technicians who would make the postwar economy hum. He ended up a seaman on a battle cruiser in the Pacific, too late for the shooting war and with ample time to marvel at the ship's radar system, which was made up of more vacuum tubes than he had ever seen before—all 150 of them. He was ever after grateful to the navy and the "uniquely wonderful" GI Bill. It enabled him to enter MIT as an undergraduate in February 1947 after a spell in a General Electric factory troubleshooting their FM radios.

MIT, Olsen recalled, felt there was not much future in electronics, so they diverted him into electrical engineering. In his four years with generators and magnets Olsen heard nothing of MIT's now-famous computers—so tight was cold-war security—but out of the blue he was offered a job in the computer lab. It recruited only the top 10 percent of the graduating year. "I didn't have quite the grades," he says, "but my love for electronics had caught the imagination of one of the professors."

Olsen was "awestruck" in 1950 when he was finally admitted to the huge room where Jay Forester was developing Whirlwind, the digital computer destined to be the seedbed for IBM's SAGE real-time air-defense system. On 18 very long racks, 11 feet high, there were no fewer than 10,000 vacuum tubes. "I had no concept of a computer. Entering the laboratory was a bit like going into a religious order as a neophyte." In the '50s, IBM's large mainframe computers were guarded by a priesthood of specially trained technicians who received batches of punch cards and returned the results later. Wesley Clark, an MIT programmer, observed that the very large IBM machine in the Institute's Computation Center was seen "not as a tool but a demigod."

Olsen was a methodical man. Hearing a preacher expound the qualifications for a wife—"when preachers used to preach practical things"—he carefully wrote down the requirements. Then he took a summer job in a ball-bearing factory in Sweden to pursue a Finnish girl he'd met. She matched up. He applied the same rigor of observation to his year at IBM's Poughkeepsie plant as liaison engineer on Whirlwind. He was appalled by the inefficiencies of that large organization—"It was," he recalled, "like going into a Communist state." Back at MIT in 1955 he told his supervisor, "Norm, I can beat those guys at their own game." As a result, he was given the management of TX-O, a project to build a research computer using the tiny, new, more powerful transistors instead of clunky vacuum tubes—provided he hired nobody and took no space! "I studied the rules carefully and found all the loopholes. We discovered that hallway was not space. So we moved my office into the hall and put walls round it. We discovered that part of the basement of the Lincoln Laboratory was just dirt. We talked people into pouring a concrete floor. When they discovered what we had done, they said never again."

The TX-O, designed by Clark with high-speed Philco transistors, was tough for Clark and his team to build. "If you combed your hair and touched one, you burned it out." So they designed new protective circuits. Olsen was determined that the new TX-O would be a reliable computer (something he had admired in Whirlwind), accessible by one person, inexpensive and low powered, but it also had to be compact, fast and exciting, with a monitor and a light pen—the equivalent of a mouse today. He added a loudspeaker and amplifier for music. The first night it ran, when everyone else had gone home, he hooked up the loudspeaker and stayed behind with his baby. "I went into the ladies' room and lay down on the sofa with the door open and fell asleep with my ear tuned to the sound so that I knew that it went all night long without a glitch, and that was a significant test."

Olsen published everything to alert the world to the promise of interactive, real-time computers that were small, rugged and inexpensive. "The commercial world just smiled at us and said we were academic. Just showing them what could be done was one of the reasons for going into business." The reaction forced a dramatic step for Olsen and his assistant, Harlan Anderson. Olsen was much influenced by his pastor, Harold Ockenga, a Boston evangelist who believed in technology as a way of spreading God's word. Olsen himself preached that science and Christianity were not in conflict—"It's obvious the main theme of both is the same, which is searching for truth, and that implies a certain humility." Knowledge was not "academic," and those who believed as much were, Olsen felt, going to be severely disadvantaged in the coming technological explosion.

In 1957 Olsen, with Anderson, left his decadelong shelter at MIT with two animating ideas—to manufacture interactive transistorized computers and do it in the MIT style of open debate. "We thought the world would be waiting with open arms for high-speed transistor computers," Olsen recalls, "but nobody cared. And it turns out it takes more than ideas. You've got to sell your idea." Olsen pitched to General Georges Doriot, president of the American Research and Development Corporation (page 302), spicing his presentation by playing a little Bach on his computer. Venture capital itself was a relatively untested idea then, but Doriot's board decided to gamble on Olsen, investing $70,000 for 77 percent of the start-up. Doriot worried that the word "computer" was still too new to call it the Digital Computer Corporation—so the Digital Equipment Corporation was born.

From August 1957, Olsen, with Anderson and Stan, ran DEC as a skinflint operation out of an old woolen mill in Maynard, Massachusetts. Office doors cost too much money, so they had none, not even for the bathrooms. Plastic bottle caps turned out to work fine as insulators for tiny pulse transformers. Aulikki Olsen, carried back from Scandinavia to become Mrs. Olsen, swept the floors.

At first it was slow selling interactivity. "Some people thought it was wrong. They almost spoke in ethical terms. Computers are serious, and you shouldn't treat them lightly. You shouldn't have fun with them." A break came when a federal department needed to analyze earth tremors. Congress had ruled that no more computers could be bought until all the computers in Washington were used 100 percent of the time, but Olsen got round that problem in 1959 by again shunning the taboo word. He called DEC's first model a Programmed Data Processor and made the sale. PDP-1 became the first commercial transistorized computer. It was attractively small—about the size of a refrigerator—but the key was that it allowed an individual to interact as we do today with the personal computer, without having to wait for a professional programmer. And it was relatively inexpensive—$125,000 to $150,000, compared with an IBM mainframe of between $1 and $3 million.

Remembering a piece of Doriot's advice that success too soon is fatal for companies, Olsen declined an order from NASA for 100 minicomputers; he was making two or three a month, and filling an order like that would have been a risky scramble. He stubbornly kept on the path of steady growth; he was as disciplined in his personal life, never smoking, drinking or cursing, and was a faithful attendee at a Boston prayer breakfast. Meanwhile, Olsen's emerging competition made what he called the classic mistake of looking at his current product and thinking they could beat it by offering modest improvements. "The thing they forgot is that we were working on new products." A week after the rivals unveiled their bulky mimics in 1964, Olsen brought out a small, light, greatly simplified computer at the bargain price of $18,000—plus a standardized Teletype printer he redesigned for continuous use. The latter was an important innovation, as printers before that had been prohibitively expensive.

This PDP-8, as Olsen's new model was known, was immediately popular. It ran chemical refineries and kept track of inventories; the navy put several on submarines; New York street crowds looking up at the neon news display in Times Square were seeing a PDP-8 in action. Caught up in the miniskirt fashion craze, journalists soon dubbed it the "minicomputer."

DEC went public in 1966, raising millions for itself and simultaneously validating the

OLSEN: "IBM was like a Communist state."

concept of venture capital. And Olsen's minicomputers with open architecture created an entirely new industry. Manufacturers found it easy to buy a PDP-8, attach new hardware, write software programs and resell the combination as a game, a typesetting system, a star-and-tide calculator and so on. Meanwhile, Olsen kept innovating. He pushed networking, standardizing technologies and communication protocols so that DEC computers could speak with one another, giving a number of people simultaneous access, something IBM machines could not do. By 1978 more than 40 percent of the minicomputers in the world were made by DEC, which had profits of $142 million and employed more than 100,000 people. IBM, which awoke to the minicomputer revolution in 1976, had only 2 percent of the market. Growing 30 percent a year for 19 straight years, DEC was regularly featured in business magazines as the most admired company in America.

The long glorious run was savagely interrupted in 1981. What should have been the next logical step for the company that first democratized computing eluded Olsen. When his engineers showed him early designs for a DEC personal computer, he asked: "Why should anyone need a computer of their own?" By the time he agreed they could go ahead, IBM was on the market in 1981 with its ground-breaking PC. Olsen and another engineer pried apart the IBM PC and laughed at how it had been slapped together. It took a year to get DEC's overengineered and undermarketed PCs into the market—three of them from $4,910 to $8,695—by which time IBM had set the standard. DEC had gone from computer democrat to computer aristocrat, and IBM had proved the maxim "second best wins."

Olsen slowly retreated from the PC market, attacked for poor marketing and staff defections—he could be very harsh—but by 1986 he had made a comeback. *Fortune* hailed the 60-year-old Olsen, then worth $260 million, as arguably the most successful entrepreneur in the history of American business. "In 29 years," wrote Peter Petre, "he has taken Digital Equipment Corp from nothing to $7.6 billion in annual revenues. DEC today is bigger, even adjusting for inflation, than Ford Motor Co. when death claimed Henry Ford, than U.S. Steel when Andrew Carnegie sold out, than Standard Oil when John D. Rockefeller stepped aside."

It was the high point. Speedy 32-bit microprocessors, offering the same power as minicomputers at far lower prices, proved another example of what Harvard's Clayton Christensen has called destructive innovation. Olsen fought back, but with an industry in recession he suffered big losses and was forced out in 1993 after 35 years at the helm. Five years later DEC was sold to Compaq for $9.6 billion.

Ken Olsen's contribution was immense. Many of the revolutionaries in personal computing cut their teeth on his systems—Gary Kildall for one (page 402). The process of democratizing the computer that continued through the PC revolution derived from Olsen's unique combination of faith in the glory of innovation and faith in the divine.

ESTÉE LAUDER Her talent for giving, as much as selling, founded the greatest family cosmetics company

1908–2004

H ow does a young woman mixing a few pots of homemade cream in a drab area of Queens, New York, during the Great Depression manage to found a glamorous international cosmetics business?

Josephine Esther Mentzer was the cosseted ninth child of immigrants from Hungary. They were a Jewish family in an Italian neighborhood, living over her father's hardware store. She was always at pains to say little about her origins; if the occasional mention of Vienna and European spas led others to infer aristocratic connections, all to the good in a business that thrived on exotic allure. She had advantages. She was lovely, with a gorgeous complexion; her marriage at 22 to a fine young man from Galicia, Joseph Lauter (with a "t" then) was providential; and her uncle, John Schutz, happened to be a struggling chemist with formulas for skin creams.

The Mentzers were hardly poor, but "Estelle"—transitioning to Estée—had a taste for real affluence, and her animating ambition was periodically inflamed by encounters with the rich. In 1985, the year she published a memoir, she still felt a twinge of the pain from 50 years before when she asked a smart woman in a beauty salon where she bought her fine blouse. Estelle: "She smiled. 'What difference could it possibly make?' she answered, looking straight into my eyes. 'You could never afford it.' I walked away, heart pounding, face burning."

In the moment of humiliation, Estelle resolved she would someday have whatever she wanted—"jewels, exquisite art, gracious homes, everything"—and so she did, because desire was served by an original talent for merchandising. She was devastatingly sincere about the products she peddled, and then developed, from the beauty repertoire of her uncle and her dainty mother. She just knew that if people tried them they would *glow*. It was this faith that made her so insistent in early forays to sell the jars of Uncle John's cream. Not to put too fine a point on it, she was pushy.

In chance encounters in elevators, on trains, in hotels and stores, trapped under the hair dryer, on the way to a Salvation Army meeting, wherever, no woman was safe from Estelle, cream jar in hand; those inclined to resist had their attention drawn to wrinkles they never knew they had. Her memoir is happily informed with the vanquishing of reluctant women who, thank you very much, did not at that precise moment want to have their faces treated. "Just give me five minutes," young Estelle would implore, and invariably they would succumb to the promise of instant rejuvenation from this charming elf.

The salon where Estelle had her hair washed and marcelled was the House of Ash Blondes on West 72nd Street in Manhattan. Its owner, Mrs. Florence Morris, yielded Estelle an important five minutes— or so: "First I applied some extrafine Cleansing Oil to Mrs. Morris's face," wrote Estée, "then gently removed it. Then, before she could change her mind, I patted on my Crème Pack. Her face began to glow almost instantly. . . . The original magic potion, my uncle's Super-Rich All-Purpose Cream, followed. After tissuing that off, I applied a light skin lotion. I brushed her face with the lightest and softest of face powders, which Uncle John and I had just developed, then on her cheeks and lips I used a bit of the new glow I had been testing."

Esther/Estelle transformed into Estée.

The glowing Mrs. Morris offered Estelle the beauty concession at her new salon at 39 East 60th Street. Every time a woman rose from the chair feeling good about herself, Estée, as she soon became, gave her a little free cream and asked if she would kindly tell another woman. The Tell-a-Woman campaign led to invitations from other salons in the city and hotels on Long Island, but it was slow. Friends and family, she writes, did not let a day go by without telling her she was wasting her time: "I cried more than I ate." She divorced Joseph in 1939, but after kicking up her heels as a girl-about-town in New York and Miami, she had the good sense to marry him again in 1942, and henceforth they were inseparable. He was a cool organizer and a gallant companion for Estée's developing social whirl (and social climbing). "We lead a very, very secluded life," said Joe. "We go out seven nights a week. Maximum."

The pivotal moment was in 1948 when Estée spoke at a charity lunch at the Waldorf-Astoria and gave away lipsticks in metallic sheathes at a time when plastic was commonly used. She had been in the habit of offering free samples, but this was a bigger gamble, carried out to impress the Saks Fifth Avenue cosmetics buyer, Bob Fiske, who had repeatedly declined to carry her cosmetics—and Saks was the recognition Estée craved. Right from her earliest days, with remarkable prescience, she had targeted the elite market and would not sell in drugstores. Fiske told biographer Lee Israel that at the end of the lunch, a line of women formed up across Park Avenue and across 50th Street into Saks asking for the Estée lipstick. Fiske gave Estée an order for $800.

To fulfill it, she and Joseph risked putting up six months' rent in advance to set up operations in a former restaurant on Central Park West, 1 West 64th street. On the old gas burners, they raced to complete the order in time, cooking and mixing the four creams and lotions, sterilizing and labeling the pale turquoise jars (a shade carefully chosen by Estée to complement bathroom decor). Their son and future longtime CEO, Leonard (born March 19, 1933), filled jars and rushed around on bicycle errands. (Another son, Ronald, arrived on February 26, 1944.)

Saks did well with Estée Lauder and she also sold into other prestige stores. At Neiman Marcus in Dallas, she went on the radio to woo women into the store. "I'm Estée Lauder, just in from Europe with the newest ideas for beauty, with a slogan: 'Start the New Year with a new face.'" When they had accumulated $50,000, Joe and Estée went to the BBD&O advertising agency (which handled the mighty Revlon account) to discuss how they might grow a quality business through advertising. They were laughed out of the office; $50,000, they were told, would barely buy a page in *Life,* then the hottest magazine. Estée hit upon the idea of spending their savings on a blockbuster gamble to give away a sample to end all samples—no less than three months' supply of a cream-based powder. Her conviction was that anyone who used it for three months would be so hooked on it that they would look for it again—and she was right.

Leonard recalls, "It is commonplace now at charity dinners to give out gifts, but my mother made a big thing of it. A band would play 'A Pretty Girl Is Like a Melody' and these statuesque models would come out wearing a sash saying ESTÉE LAUDER, and they would hand each person in the room a box of powder."

She still had to find a way of driving traffic into the stores. Credit cards had not arrived, but department stores had begun encouraging customers to open charge accounts. Their addresses were gold dust to Estée. She mailed a tasteful card inviting the customer to bring it to Saks for a free box of powder; later, it was a free lipstick, then a free compact. "Everyone in the trade said Estée Lauder would go broke giving so much away," recalls Leonard. Arriving to give out compacts at Bullocks Wiltshire, Los Angeles, he found 500 people in line. Estée extrapolated from the gift to gift-with-purchase schemes, offering a still more attractive freebie—but this time as a condition of buying something. The crowds came again, and now the Estée Lauder cosmetics company was growing at 40 to 50 percent a year.

In 1953 Estée had another original idea for venturing into fragrances. "Youth Dew" would be a very strong-scented bath oil with a perfume that would linger. (Lee Israel suggests that the formula was a gift from the king of fragrance manufacturing,

Arnold Louis van Ameringen, a friend made during her separation from Leonard.) Estée mailed blotters immersed in Youth Dew. The whole Lauder business increased tenfold. "The other concept that seems logical today but was totally illogical then," says Leonard, "was to target women to buy for themselves. They mostly got their fragrances as gifts; the advertising was for men to buy for women. My mother was the first to appeal directly to women, and we did it through the blotters and free samples, no advertising."

By 1960 Estée Lauder was an international corporation. Grossing around $6 million a year, it was still the new kid on the block against the mass marketing of Revlon by Charles Revson and the renown of Elizabeth Arden and Helena Rubenstein, but with their deaths it came to dominate the quality market. The gift-with-purchase scheme worked overseas as well as it had in America. When it was tried at Fortnum & Mason in London, they found—and made much of the fact—that one of the redeemed cards had been addressed to HM Queen Elizabeth II, Buckingham Palace. "We don't know if a lady-in-waiting had collected the gift," recalls Leonard, "or whether someone had filched it from a dustbin."

The innovations that carried Lauder higher and higher included the creation of new brands: Commonplace today but an innovation then were the cologne Aramis for men and—the most radical of all—the allergy tested, fragrance-free Clinique, itself a $140 million business by 1982, run by Carol Phillips.

Estée and Joe (who died in 1983) kept the business in the family. Estée promoted her "sweet boys" Leonard and Ronald, and their wives, Evelyn and Jo Carol respectively, were graceful assets. Leonard became president in 1972 (and his son William chief executive from July 2004), and under Leonard's graceful and innovative leadership Estée Lauder became the largest privately held cosmetics company in the world. When Estée died in 2004—aged 97 according to her family—the company she had founded had sales close to $5 billion. As Lee Israel notes, Estée's appointment of Leonard turned out to be one of the best justifications for nepotism in the history of business.

MALCOM McLEAN
He was a trucker who created our global marketplace with container shipping

1913–2001

How LONG does it take for a good idea to become an innovation? Malcom McLean was 24 when he loaded his truck with cotton bales in Fayetteville, North Carolina, during the Great Depression and drove to a pier in Hoboken, New Jersey, in time to catch a cargo ship. He had to sit around most of the day on the noisy dockside waiting his turn while worker ants muscled each crate and bundle off other trucks and into the slings that would lift them into the hold of the ship. On board the ship, with much yelling and arm waving, the stevedores unloaded each sling and saw its contents to the proper place in the hold.

It was a frustrating experience for McLean, not two years into the business of trucking and his income dependent on getting back to North Carolina for more loads. He recalled: "Suddenly the thought occurred to me. What a waste of time and money! Wouldn't it be great if my trailer could simply be lifted up and placed on the ship without its contents being touched?"

Yes, it would be great. It would be revolutionary. General nonbulk cargo had for centuries been shipped in the process he watched known as break-bulk shipping—boxes, bales and crates handled piece by piece. What McLean envisaged would have saved him a day, but it would have saved everyone else something like two weeks in

loading and unloading the ship: On average it was eight days to haul and distribute break-bulk shipments in the hold, plus another eight days at the other end to retrieve and distribute. Today, the concept that occurred to McLean is known as containerization, and it has done more than simply save a great deal of time and labor. It has created a thriving global marketplace. As the *Journal of Commerce* wrote, "Containerization's impact can be seen almost everywhere—the California lettuce in Paris restaurants, the imported beer sold in U.S. supermarkets, the Toyota plant in Kentucky that schedules its assembly line around just-in-time delivery of Japanese automobile parts." Moving cargoes to and from remote parts of the world is at minimal cost. Losses from pilferage and damage have virtually disappeared, and products are shipped in quantities that would have been prohibitively expensive when handled piece by piece.

McLean had his idea in 1937. The 24 year-old truck driver was 40 before he did anything about it. He was now a thrusting businessman who did sums faster in his head than colleagues with a calculator, chain-smoked Winstons and liked to utter snappy aphorisms: "Don't get up in the morning unless you can compete. Otherwise, you're going to waste the day." Assisted by his brother Jim and sister Clara, he had been busy building that one truck into the McLean Trucking Company, one of the biggest in America by the '50s with 1,700 trucks, 32 terminals across the country and annual revenue of $12 million. That was a long way from the rusting old pickup he had bought for $120 in 1934 to haul dirt and tobacco round Maxton, North Carolina. He had earned the money for that by pumping gas after high school; his father was a farmer and mail carrier, and McLean's

first earnings had been selling eggs on commission for his mother.

Competition was the spur that made McLean think again about his shipping idea. In the 1950s truckers and railroads were battling for the nation's shipping business. McLean went to Southern Railways with diagrams of how his trailers could be fitted on rail flatcars and then loaded on specially fitted ferries. Southern turned him down. "The whole thing sounded so obvious and natural," he said later. "I felt I had to try it. I kept saying to myself, 'What if someone else does it and I don't?'" He was not, in fact, the first with the idea. Even before his impatient day on the Hoboken dock, the Seatrain Lines Company had in 1929 carried railroad cars on specially converted ships between the United States and Cuba. In World War II, the U.S. military experimented with shipping small standardized containers, and after the war Andrew Jackson Higgins, who had made landing craft for the military, tried to interest investors in containers. Nobody had succeeded in making the innovation ubiquitous. Even the few familiar with the concept perceived too many difficulties, a foreign language to McLean. "He used to preach there were no complicated problems in business," said his longtime associate Paul Richardson. "He knew how to take a complicated problem and reduce it to simple form."

McLean had never been on a ship when he learned that oil tankers traveling to the Northeast from Houston, Texas, usually carried nothing above deck (and usually only ballast on the return). What about using the space to carry trailers? The thought was so obvious, it hurt; there must be a snag somewhere. McLean reckoned the best way to find out was to buy an oil tanker. "It's often the people who know all

INFINITE RICHES: Perhaps Edward Burtynsky's vivid photograph of the container ship *MSC RITA* in the Ceres Port, Halifax, Nova Scotia, was foreshadowed in the lines by the poet Christopher Marlowe (1564–1593):

Thus methinks should men of judgment frame
Their means of traffic from the vulgar trade,
And as their wealth increaseth, so enclose
Infinite riches in a little room.

about something that say it can't be done," he remarked. "I was totally ignorant, so I said, 'Why not give it a try?'" For an investment of $7 million in 1955, he came into possession of the oil tanker business of Pan Atlantic, a small shipping company that was a subsidiary of Waterman Steamship Corporation of Mobile, Alabama. He realized that Waterman controlled docking, shipbuilding and repair facilities that he would need, so he went along to see an account officer in the specialized industries division of Citibank in New York, one Walter Wriston, who would rise to be president. Wriston told *Business Week:* "We used to sit up all night trying to figure out how to put trailer bodies on ships and how to finance the acquisition of Waterman."

Wriston's loan enabled McLean to put down $42 million for Waterman and only $10,000 of his own money. The Interstate Commerce Commission warned McLean he could not be in both the shipping and the trucking business. He chose the unknown oceans over the well-traveled highways. "It was a very gutsy decision," said Richardson. "Not many people would have done that. A lot of people thought he was crazy." McLean brushed off the praise. "You know, Paul, what bothers me is that I never thought about it."

McLean worked first on enabling Pan Atlantic's two World War II T-2 tankers to carry trailers by installing a steel platform on the deck. His plan had been to stack trailers on top of one another. The wheels made the deck stack too high for safety, so he removed the wheelbase and then strengthened the trailers to cope with high winds and spray. The steel boxes that emerged from this process were the dimensions of a then trailer: 33 feet long, the legal maximum at the time, 8 feet wide and 8 feet high. They would stack easily on the ship, above or below deck, or on a railway flatcar and could be given a chassis for road travel. They were the first true shipping containers.

On April 26, 1956, nearly 20 years after his first insight, McLean's first container ship, the *Ideal X,* sailed from Shed 154 at Marsh Street in Port Newark. It was, as Oliver Allen wrote, a day that is universally recognized as the beginning of the container era. A reporter described the ship as

"an old bucket of bolts," but it carried 58 well-filled boxes. Their significance was better appreciated by another watcher, Freddie Fields, a top official at the International Longshoremen's Association who later went to jail on racketeering charges. A man walked up to him and asked, "What do you think of the ship?" Fields just shook his head and muttered, "I think they ought to sink the sonofabitch." McLean, who was then 43, did not reveal he was the owner of

MALCOM McLEAN: "I don't have much nostalgia for anything that loses money."

the ship they were watching. *Ideal X* was headed for a long journey down the East Coast of the United States into the Gulf of Mexico and on into Houston, and some predicted the containers would not survive the fierce gales.

They did. The bigger threat came from Fields's longshoremen, who nearly bankrupted McLean by refusing to work some of the first vessels. McLean pressed on, heartened by customer enthusiasm for the new speed and low rates. He expanded the service to Puerto Rico. His most urgent plan was to convert old ships to carry 226 containers and to fit the ships with their own gantry cranes to hook on to the corners of a container. He was in such a hurry to expand that his company—now called Sea-Land—suffered large losses. He went

round the New York banks for money to convert more ships, with the staunch support of Wriston and Citibank but some public skepticism. Richardson, later president of Sea-Land, remembers that on the morning of one important meeting with bankers the *New York Times* ran a story saying that containerization would never work. "It wasn't easy," said Richardson. "Almost all the rest of the industry was predicting we'd go bankrupt. We were borrowing money, building a highly capital-intensive business. We were working seven days a week. But Malcom was confident. He had total belief in his ideas."

McLean got the money for his conversions—and more. The customary fable of myopic bureaucrats holding back entrepreneurs did not play out. Soon after the sailing of the *Ideal X,* the Port of New York Authority administrators made a swift and farsighted decision to build the world's first container port on marshlands adjacent to Port Newark in Elizabeth, New Jersey, at a cost of $332 million. In the early '60s, such was the gathering success in the United States that McLean was willing to try shipping to foreign countries. Again, the general view was that he was inviting disaster since many foreign shippers could look to their governments for support. He planned his foray with care. The first thing was to build a container port in Europe. He chose Rotterdam and hired a young Dutchman, Frans Swarttouw (later head of Fokker Aircraft), to oversee the construction. Then he signed up 325 truckers in Europe and established a sales organization. Sea-Land's first container ship, the *SS Fairland,* left Port Elizabeth in April 1966, and its cargoes reached their final destinations four weeks faster than ever before. It was not a universally popular event. When McLean threw a party in Rotterdam, Dutch shipping executives who had been invited assailed his group with boos. Swarttouw burst into tears; McLean took it in stride. It was 29 years since he had first had the idea of container shipping, and its fulfillment gave him satisfaction enough. He never expected or sought the limelight: When the American Legion had voted to give him its Merchant Marine Award in 1959, he had forgotten to show up.

JIM SHERWOOD CATCHES A LINER

On the Saturday night sailings to New York from Le Havre in the early '60s, the SS *United States* and SS *America* invariably left without time to load boxes of valuable cargo from Switzerland—clocks, watches, cameras, aniline dyes and other high-value Swiss products. It was frustrating for Jim Sherwood, the young manager of the United States Lines for the French ports and Switzerland, that he saw the passengers and their luggage safely on board, but the varied commercial cargo could not be manhandled into the cargo hatches in time.

Sherwood was a Yale graduate who had arrived in commercial shipping after four years in Asia as a U.S. Navy office, "honing my bridge and running guns for the U.S. government," as he describes it. An admirer of McLean, he managed to get his hands on some marine containers and arranged for them to be delivered to the railway station in Basel, Switzerland. Customers filled the containers. They were lifted aboard a train leaving Basel at midnight and were alongside the passenger liner in Le Havre on Saturday evening, in good time for the quick and simple business of lifting them aboard.

Sherwood foresaw enormous demand for marine containers, so he left United States Lines and in 1985 started Sea Containers, Ltd., to own and lease containers to railways, ocean carriers and exporters. Today his company shares ownership with GE Capital of GE SeaCo, the world's largest lessor of marine containers, with a fleet of one million units having an original capital cost of $4 billion. But it is also a major player in luxury travel (with a half share in Orient-Express Hotels, Ltd.).

Sherwood, a jolly man with a taste for sharing the good things in life, bought the Hotel Cipriani in Venice in 1976, closely followed by restoration of the legendary Orient-Express train and the purchase of more than 40 properties, including the 21 Club in New York, all of which he put in Orient-Express Hotels, Ltd. He floated that company in 2000.

After Rotterdam, orders poured in from Europe. He turned his sights on Asia. His transpacific routes were a huge success, accelerated by the heating up of the war in Vietnam. Ports were so congested in South Vietnam that military logisticians grabbed McLean's service as a deliverance from chaos. Large shipping companies in America and then Europe followed McLean's course. The savings were so overwhelming that there was no alternative to what was being called the Container Revolution. Before containerization, transportation charges accounted for half the eventual cost of the good. Once McLean's innovation became widespread, shipping costs dropped to about 10 percent: In 1988 the shipping cost of a $200 VCR sent from Japan to the United States was down to about $2 a machine. Many unconverted shipping companies went broke, inviting a crusty response from McLean. "There are a lot of people in this business who don't know how to count." He became quite a bore, intoning, "You spell freight P-R-I-C-E," but he deserved his triumph. In 1968 Sea-Land, of which he owned 35 percent, netted $25 million from income of $227 million. In 1969 he sold his company to RJ Reynolds Industries for $500 million. Why did he sell out? "Greed, I guess." He pocketed $160 million and got a seat on RJR's board, but he hated it. "I am a builder and they are runners (managers). You cannot put a builder in with a bunch of runners. You just throw them out of kilter."

He was restless in supposed retirement. He invested in a machine to move hospital patients from beds to stretchers with comfort. In 1978, when he was worth $400 million, he ventured back into shipping with another dramatic idea to operate superlarge "econoships" able to hold no fewer than 2,240 containers 40 feet in length (in industry parlance, that is 4,480 "TEUs"—20-foot equivalent units). Like Pan Am's Juan Trippe, he believed in size and he borrowed $1.2 billion to achieve it. His concept was that his 12 energy-efficient leviathans, the largest vessels then afloat, would circle the globe at the equator while smaller ships serviced them, collecting containers and dropping them off at hubs. This is what is done today, but he was ahead of his time. His bet that rising oil prices, would give his fuel-efficient ships the edge was eroded by falling oil prices and then deregulation caused a flurry of rare cutting. The profit of $61.6 million on $959 million in 1984 was followed by a loss of $72 million on $1.2 billion. Bankruptcy followed in 1986. "I'm not making any excuses," he said. "We were the victims of a big capital program. We just guessed wrong." But as the *Baltimore Sun* wrote, "Today, every container shipping line has copied his move." New ships have 5,000 TEU capacities.

McLean enjoyed his later years before he died in relative obscurity on May 25, 2001, at the age of 87. His *New York Times* obituary was just 538 words and he is ignored in the reference biographies, but Walter Wriston was right when he said, "Malcom McLean is one of the few men who changed the world."

EDWIN LAND The willful optimist, brilliant scientist and enlightened manager behind Polaroid

1909–1991

ALMOST EVERYBODY faced with the dazzling intellect of Edwin Land felt in need of a polarizing filter. He allowed few visitors to the small, telephone-strewn office he called his "mole's hole." It was next to his lab at the corner of Main and Osborn Streets in Cambridge, Massachusetts, and here for almost 40 years he did much of the thinking that led to his 533 patents, a total putting him near Edison.

Land was a short good-looking man with jet-black hair, soft-spoken and charming but disconcertingly controlled and exacting in his expectations. His biographer and sometime colleague Victor McElheny, in an interview for this book, compared him to an explorer, like Amundsen at the South Pole. "Normally and exceptionally educated people meeting Land were alike astounded by the intensity of focus, the grasp of prodigious complexity. People who worked with him had to have a lot of self-confidence, good ego strength, and if they didn't they didn't survive."

From youth to old age, Land insisted on the impossible. Like the explorers, he drove his team to concentrate ferociously as they followed him into the unknown, forbidden ever to utter the word "problem." The striking thing about Land's instant camera, his best-known innovation, was that it was not a problem waiting to be solved, as Everest was a mountain waiting to be climbed. Nobody had even contemplated how nice it would be to have a photograph in hand a moment after it was taken. In a sense, making it happen was secondary to the very *idea* of a rapid self-developing camera, which says a great deal about Land's genius. His breakthrough was just to imagine the question: How, in the confines of a hand camera, do you develop the negative, rinse it, fix it, dry it, expose the positive, develop it, rinse it, fix it, wash it, dry it again and have the print ejected in 60 seconds? Land imagined all this as a single system.

The story of how the concept was perceived sounds apocryphal, but we have Land's word for it. On a short vacation seized from war work in December 1943, he strolled round Santa Fe with his wife and three-year-old daughter, Jennifer, taking pictures with a Rolleiflex camera. He recalled, "There was an inch of snow on the ground, the sun was bright and there was a marvelous smell from the pinewoods. I took a photograph of my daughter, and she wanted to know why she couldn't see the result immediately." Jennifer's beautifully naive query was the epitome of a Landian question, in the words of Polaroid's Peter Wensberg, in that it took nothing for granted and treated conventional wisdom as an oxymoron. Land wondered why he had never asked himself the question, and it must have been asked of thousands of other photographer dads. His first distinction was to appreciate the sublime nature of innocence; his second, instantly to pursue an answer. He says that by the end of a walk by himself later the same day, he had pretty well formulated a solution— "except for those details that took from 1943 to 1972." (The first of a series of cameras came out in 1947, climaxing in 1972 with the sophisticated SX-70.)

Land had been a precocious teenager, the only son in a happy family prospering in Bridgeport, Connecticut, on the father's trade in scrap metal. "Din," as his slightly older sister called him (her best shot at "Edwin"), slept with a copy of R.W. Wood's *Physical Optics* (2nd edition 1911) under his pillow and longed to find how he might make an intellectual contribution as beneficial to mankind as those of his heroes Michael Faraday and Thomas Edison. Edison was still a living exemplar when Land went to Harvard in 1926 at the age of 17. (Edison was then 79, George Eastman 72 and Henry Ford 63.) Land had a frustrating first semester at Harvard, but the compulsion to settle on an area of research that could be "tangibly embodied in a process or product" overcame him. He dropped out of Harvard and moved to Manhattan. There, walking along 5th Avenue—or Times Square, he kept changing the story—he noticed how little one could see crossing the road because of the blinding glare of headlights. Here was a danger that could yield to an idealistic inventor. He thought it would take him three months of work in a succession of Manhattan basements. It took three years.

Like so many innovators before him, Land visited the glorious main reading room of the New York Public Library almost daily. His obsession was Wood's obsession: everything ever published on the electromagnetic waves we call light. Without any intervention, light waves—vibrating electric charges—vibrate in a multitude of directions, and this we call *unpolarized* light. Roughly speaking half of the vibrations of unpolarized light are on a vertical plane and half on a horizontal plane. It is light vibrating on the horizontal plane that is the source of various kinds of glare. How could light be *polarized* so that its vibrations occurred on a single plane? He needed some kind of "gate" to sift the light. But what material—other than prohibitively expensive crystals— would absorb most of the disturbing horizontal components of light while transmitting the vertical polarized components?

EXPERIMENTER: (top right) Divided Land was a failed print in his experiments for the SX-70. Ever secretive, Land photographed himself to try out his various cameras.

NONGLARING LIGHT: Land's first great invention was a sheet of polarized plastic to get rid of glare. He hoped automobile makers would use it in the headlights of cars and in rearview visors. His aim was to have headlights bright enough to see obstacles like the man walking along the road but not bright enough to blind an oncoming driver. Biographer Victor McElheny writes that the idea of saving lives "gave him the moral energy to spend twenty years fighting for uses of polarizers beyond sunglasses and filters." The system was first exhibited in 1936 in Rockefeller Center, shortly before the formation of the Polaroid Corporation, but it was not taken up by the automobile industry.

Land was fascinated by an English doctor, William Bird Herapath, who in 1852 had discovered polarizers of light in ultrathin single crystals, formed by combining iodine with quinine salt, which he hoped would be useful in microscopes. Herapath's small crystals crumbled, so he tried larger crystals through frustrating decades of searching for a cheap polarizer. Land tried the large crystals too, and they crumbled. So he went to the other extreme, so often a pregnant practice in the history of invention. He ground the iodized crystals into submicroscopic fineness, smaller than the wavelength of light. But how could he rotate the crystals through various planes? He needed a 10,000 gauss electromagnet. The physics laboratory at Columbia University had one, but Land was not enrolled. One night he took the elevator to the sixth floor, found an open window and ventured out along a ledge to gain entry to the lab through another window. His clever bride-to-be, Helen Maislen, had the nerve to go with him, and once in the lab she aimed a bright light at the myriad of magnetized crystals in a glass cell. When Land examined the transmitted light with a Nicol prism, he could see it went from transparent to opaque as the prism was turned. Polarization! It was, Land later said, "the most exciting single event in my life."

To make a material cheap enough for large-scale production, the submicroscopic polarizing crystals had to be embedded in transparent plastic and magnetized to line up in the same direction. Land succeeded in doing this. He gave a polarized sheet to a fishing friend, who rushed back with a large trout, normally invisible in the sun's glare on the water. Land went back to Harvard in 1929, where his adventurous young physics teacher, George Wheelwright III, who had put aside a small fortune, proposed that they should be business partners. One semester short of graduation, Land left Harvard and, in a leased dairy barn in Wellesley Hills, developed a machine to make the sheets, now dubbed Polaroid. Later the partners opened a big basement workshop on Dartmouth Street in Boston's Back Bay and frantically filled a $10,000 order from Kodak for Polaroid camera filters laminated between two disks of optical glass.

Land discovered he had a flair for the theatrical. Seeking to interest the American Optical Company in making sunglasses with Polaroid, he invited its executives to a meeting in the Copley Plaza Hotel in Boston. He rented a room on the fourth floor, insisting on one facing west, and placed a bowl of goldfish on the window ledge. When three Optical officials entered the room against the glaring sunlight from the window, the boyish Land said, "I apologize for the glare. I imagine you can't even see the fish." Fish? What on earth was he talking about? Land then handed each of them a square of polarizer, and at once they saw six swimming goldfish. "This is what your new sunglasses will be made of," said Land cheekily. "It's called Polaroid."

It was a wonderful material, and orders poured in. Land polarizers were used in huge numbers of sunglasses and camera filters. In the '50s, millions watched 3-D movies through Polaroid's cardboard-and-plastic viewing glasses; from the 1990s his invention was used in the hundreds of millions of liquid crystal displays of pocket calculators and digital watches. But Land was thwarted in his original inspiration for making Polaroid and the original hope of his investors, who put up $375,000. For years he campaigned without success for the automobile industry to install polarized glass in headlights and thereby free night drivers "from the terror of blinding lights."

World War II fired Land to enlist science in the defeat of the enemies. Polaroid's 3-D Vectograph showed the American military the depth and contour of enemy defenses on the Pacific Islands and in Normandy for D-day. Polaroid employed 1,200 men and women making gun sights for tanks, photo-directed bombs and personal angle finders for warplanes. And General George Patton won his battles in Polaroid goggles. In the cold war, Land worked on a top secret mission to give the cameras in U-2 spy missions a higher resolution than the Soviets thought possible.

The end of the war business was the stimulus for Land to give an ultimatum to himself and everyone in the company to step up research on the vista opened by Jennifer's question in 1943. "If you sense a deep human need," he said, "then you go back to all the basic science." There were many

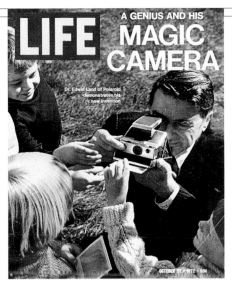

TAKE ME! TAKE ME!: In a media blitz to launch the SX-70 system, Land photographed children in a playground and was in turn photographed by Co Rentmeester. Land was happy to conceal his face behind the camera.

ingenuities in the systems he developed. The heart of his breakthrough camera was the pod, a small lead-foil container designed to crack under the high pressure of rollers moving the film and so release a tiny reservoir of chemicals to develop the image. Land called it a nice piece of plumbing. Kenneth Mees of Kodak, which supplied the negative film, was less inhibited. He held up a pod on an early demonstration and declared, "Gentlemen, this is the transcendent invention."

Land produced sepia-and-white photographs within a minute at a New York City conference of the Optical Society in February 1947. He knew his first Polaroid camera was not ready for commercial sale, but he staged the demonstration deliberately to terrify his people. As he had promised, the first Polaroid camera, made for Land by Samson United in Rochester, went on sale for $95 at Jordan Marsh in Boston the day after Thanksgiving in November 1948. All 56 units were sold within hours. He followed up relentlessly with lighter and faster cameras, pictures in true black and white in 1950, "Polacolor" in 1963 and the cheap Swinger portrait camera for around $20 in 1965. By 1956 he had sold one million Polaroid cameras, four million by 1962.

Land throughout was zealous and unpredictable; he would call technicians on a project at 5 a.m. every day, then not

speak to them for months. Nobody could keep up with his speed of mind. Pete Peterson (later U.S. secretary of commerce) was the young head of Bell & Howell in the early '60s and recalled being summoned to Land's cave, shown a Swinger camera and asked if he could make three million in a hurry. "Land was a very sensitive person," observed Peterson. "He guessed I had a doubt, and I told him that people as mechanically challenged as I was would not take good pictures because we would forget to focus the camera. Now, this was a minor thing, but it showed you how unbelievably quick he was. We were there for the day, and about two hours later he said, 'I've been thinking about that problem. And I've come up with a periscope.' And there and then he showed me the camera he had changed. He had figured out a way for the photographer to look down on the lens and get a reminder to set the distance."

Land and his teams struggled for years to realize his masterpiece in 1972, the SX-70, a light and elegant pocket-size folding single-lens reflex camera, featured on the covers of Time and Life. It sent an image through 13 layers of self-developing color film to deliver a dry color print in less than 60 seconds. Kodak, five times the size of Polaroid and Land's collaborator for so long, became concerned in the '60s at Land's escalating success—a half billion in sales, then a billion, then more—and brought out its own instant camera. In 1985 Land was triumphantly vindicated in a long breach of patent suit. Kodak, which had sold 16 million Instamatics, abandoned its instant business.

Land was a rare individual, a scientist able to grow a company based on his own successive inventions and the inventions he inspired. For nearly half a century he built shareholders' equity. In the 20 years from 1950 to 1970, he increased sales by nearly 100 times. He plowed the profits back and committed several hundred million dollars up front on faith in creativity. And while a demanding leader, he ran a civilized, enlightened company, encouraging education, diversity and philanthropy. As biographer McElheny writes, Land's lasting importance may be that he embodied, with unusual force, the true questing spirit of innovation.

RUTH HANDLER Millions of young girls round the world enjoy seeking identity through her creation, Barbie

1916–2002

RUTH HANDLER: Before the birth of Barbie, cowboys were a big thing.

THE NEWLYWED couple of Ruth Handler, aged 21, and Elliot, 22, stood in a disused Chinese laundry, a dingy little room in downtown Los Angeles, wondering if they could really afford to rent it for six months for $50. The year was 1936, a heartbreaking moment when the worst days of the Great Depression seemed to be returning. The Handlers were already $200 in debt to Sears, Roebuck for some cheap power tools they had bought on the installment plan so that Elliot could try his hand making coffee tables, trays and cigarette boxes. Since Ruth was working in Hollywood, a swift jump cut of 30 years is appropriate to 1966, when the company the young couple started is bringing in $100 million a year, and the Handlers are celebrated as the creators of an American icon—the Barbie doll. They helped mold the baby boomers into a generation of consumers luxuriating in newfound affluence after the decades of the Depression and war.

Another jump cut to the year 2000—a billion Barbies that the Handlers have cloned are out there in 45 different countries, and every second two more Barbies join them. The average American girl between three and eleven owns ten Barbies, the average British or Italian girl seven and French and Spanish girls five apiece, a fixation that variously intrigues and alarms sociologists and defeats censorious mothers. It is a business that takes in $2 billion a year, a sum that the *Economist* notes as being "a little ahead of Armani, just behind the *Wall Street Journal*."

How did the Handlers get from the Chinese laundry to Wall Street? It is, in many ways, a typically romantic story of American enterprise, the triumph of the children of steerage-class immigrants who had fled Russia to escape anti-Semitism and conscription in the czar's armies. But it is also a salutary tale of the things that can go wrong when innovators venture from their magic kingdom.

Ruth's father, Jacob Moskowicz, told the immigration authorities at Ellis Island in 1907 that he was a blacksmith, so they put him on a train to Denver for a job with the Union Pacific Railroad. Ruth was born in Denver, Colorado, on November 4, 1916, the tenth and final child of a frail, 40-year-old mother who never learned to read and could not cope with yet another child. Ruth was farmed out to an enterprising sister, Sarah, aged 20, who became her role model. Sarah and her husband had a drugstore in the prosperous '20s and at ten Ruth worked as a soda jerk, then as a helper in Sarah's luncheonette-liquor store. She planned to attend the University of Denver to be a lawyer like one of her brothers. On her 16th birthday, Sarah gave her a 1932 Ford coupe, and from it Ruth first saw Izzy on a street corner, as she drove stylishly round and round the blocks. She honked her horn at the slim, handsome young man, and a few weeks later they spent the night dancing at a B'nai B'rith charity where each dance cost a nickel. Izzy kept scrounging nickels from friends.

The family of Izzy (as he then was) Elliot Handler was less well-off. His father was a painter at the Denver Art Institute—up a ladder painting walls so that he could pay for Izzy to take art classes. Izzy was shy and introverted, but when Ruth took a summer job as a stenographer at Paramount in Los Angeles, Izzy pursued her, enrolled for art classes again and kept himself alive with an $18 a week job in a light-fixture company. They were married on June 26, 1938, and after a brief honeymoon driving across the desert in the wedding present of a new convertible, Ruth persuaded "Izzy" to become "Elliot" on the grounds that "Izzy" would only expose him to a lifetime of anti-Semitic sneers.

She was the dynamo and he was the dreamer. His musings about how he might make furniture for their sparse $37.50 a month apartment from the new acrylate plastic known as Lucite and another known as Plexiglas provoked her to say, "If you can make that stuff for us, you can make it to sell." In any free time between his art school and his job, Elliot poured Lucite into wooden molds, running between the shared garage and the kitchen oven—initiatives not appreciated by the occupant of the other half of the garage, who preferred his car without deposits of plastic and sawdust. The Handlers were invited to choose between having a workshop or a home. Their bold decision was not simply to rent the old laundry but also for Elliot to quit his job and school. "We knew we were taking a big, scary gamble," Ruth said years later, "but it felt right. I had such confidence in Elliot's talent."

Elliot was much too shy to try and sell his creations, so on her lunch break at Paramount, Ruth dragged an old suitcase full of samples to a chic store on Wilshire Boulevard. She was pretty and beguiling and determined. She got past the snooty saleswoman to the buyer, an old man with a thick European accent, and charmed him, winning an order worth $500. Then she heard that Douglas Aircraft was looking for a corporate Christmas gift. Elliot designed a clock with a Plexiglas face and a curved back, where a model DC-3 was poised for landing. Ruth got a deal for so many clocks that she had to borrow $1,500 from Sarah for the materials.

By 1943 Elliot's designs, perfectly catching the idiom of the late '30s, were bringing revenues of $2 million, with the help of Harold "Matt" Matson, Elliot's former foreman at the light company who had installed a proper baking oven in their new workplace. Ruth was frustrated to be absent much of the time being a mother. Barbara Handler was born on May 21, 1941, and Ken in March 1944. Ruth was back at work almost as soon as Ken arrived. "Domestic chores bored me silly," she wrote in her memoir. "I missed the fast-paced business world and the adrenaline rush that came with closing a tough sale and delivering a gigantic order on time." Little Barbara resented her mother's absences thereafter, and Ruth came to resent her protests that she wasn't like other mothers who stayed at home. "I cried myself to sleep on many, many nights. I wanted to please her, but whatever I did it wasn't good enough." In truth, though Ruth did not know it then, those few homebound years she disliked so much were more important than anything she did in the workplace on her return. Not that she was unproductive. Her first idea upon coming back was that picture frames would be in demand as more and more men and women were called into the armed services. She guessed right. She sold $6,000 worth and delivered them herself, driving a rented truck.

(left) The dynamo (Ruth) and the dreamer (Elliot) about the time of their marriage in 1938. (right) Busy Ruth had not much time for her baby, Barbara, who grew up to play a crucial role in the family fortunes.

In 1944, with Elliot facing the draft, they formed a company with Harold Matson, joining his surname with Elliot's to become Mattel Creations. "It never occurred to me that some part of 'Ruth' belonged in that name," she said, "since it was my idea to start the picture frames and I brought in that first big order. But this was 1944, and just as a woman got her identity in her personal life through her husband, should it not be so in business?" GI Elliot was lucky to be posted to Camp Robert, about 250 miles from Los Angeles. On his weekend passes, he came home and used up some scrap wood to make dollhouse furniture. Ruth packed a suitcase again but this time crossed the country by train to the annual toy fair in New York. She sold $100,000 worth of Elliot's miniatures.

It was clear that there was a great opportunity when the war ended—an almost toyless marketplace and young parents overjoyed to be reunited and employed. The trouble was that they were perpetually short of working capital. Ruth switched from a timid branch of the Union Bank to Giannini's Bank of America (page 258), which gave larger loans, but for bigger orders she still had to run around borrowing from family. It was a blow to their cash flow when Elliot's invention of a plastic Uke-a-Doodle toy ukulele, priced to sell at $1.49, was copied before the toy fair and the knockoff was priced 30 cents cheaper by Knickerbocker Plastics (whose spies had seen Elliot's doing well in the Ben Franklin stores). The price war with Knickerbocker, with Ruth and Elliot demonically saving every cent, was too traumatic for Harold Matson. He sold out. Ruth and Elliot were on their own as Mattel, but they won the price war, selling millions of ukuleles, and Elliot was on a roll. He designed a black miniature plastic piano, which produced notes from a piece of stamped metal held in place by a zinc bar. It was the hit of the toy fair in 1948, 600,000 selling at $3 against the inferior wooden miniatures at $5.

They lost $60,000 on the piano because Elliot had underestimated production and packing costs, but it proved a splendid investment in promotion. The name Mattel was now etched in the mind of the nation's toy buyers. It was this that brought to their door a movie-music composer, Ted

Duncan, who had been tinkering in his garage—where would we be without these garages?—with a toy music box. Swiss music boxes were exquisite clockwork devices costing up to $25. In Duncan's model, a hand crank moved a rubber band with knobs past 12 metal prongs mounted on a zinc block. Elliot loved the idea, believing its hand-crank feature would give children the feeling that they were actually playing the instrument. He refined it for mass production; his design skills ran to inventing assembly-line machinery that impressed observers as having the efficiency of the automobile manufacturers. Raising the money for the music-box venture was left to Ruth. She was resourceful and tough. (The joke among people awaiting her arrival for a meeting was, "It's time to face a moment of Ruth.") But the banks were still sticky, and once again Ruth had to turn to the family to borrow $20,000 each from two sisters.

The music box was a triumph. By 1952 they had sold 20 million and were impudently exporting them to Switzerland. They were making a few hundred thousand dollars but still not enough money to grow.

This is the dangerous moment for innovative companies. Ruth and Elliot were like the surfers on a big day off their California coast. They could stay with the little guys, enjoying themselves around the edges, or go out for the really big rollers—and know that if they missed they would get well and truly wiped out. Mattel's net worth at this point was about half a million dollars. The big wave rolling into them in 1955 was the suggestion by an ABC advertising man that they risk $500,000 on sponsoring 52 fifteen-minute episodes of a new show to be produced by Walt Disney, *The Mickey Mouse Club*. Television was still a young medium; nobody—not the agency, not Disney—could give any guarantees.

As it happened, another independent inventor had turned to Mattel, this one with a realistic toy machine gun modeled after a type used by paratroopers but emitting a comical burp rather than bullets. Ruth and Elliot decided to risk the company on Mickey and the Burp Gun. At the toy fair in March the gun did well with wholesalers, but when it reached the stores in the summer, four weeks before the tele-

vision show was due to air, the wholesalers started canceling orders in the wake of hundreds of returns from the first retail sales. Kids liked the gun, but they didn't know how to operate it well enough. They needed a demonstration—the kind they'd get in the television advertising when it ran. It all came down to Mickey Mouse. If the TV did not reverse the canceled orders, Mattel was well and truly dumped.

Not much happened on the first week of the show. Elliot and Ruth had to wait six agonizing weeks before it became apparent that they were riding the big roller. By the Monday after Thanksgiving, Mattel was inundated with wholesale orders and urgent pleas: "Cancel our cancellation. Send more Burp Guns." Every gun was sold. All that the factory had in the days before Christmas were two returns, both broken and both swiftly repaired—one for a sick boy in the hospital and one for President Eisenhower for his grandson (David Eisenhower grew up to be a historian and married Julie Nixon; Camp David is named after him). Mattel's TV advertisements tripled revenues—it was now the third-largest toy company. "We proved," said Ruth, "that both a toy and its brand name could be sold directly to the consumer, the child. In the past, parents bought toys by asking a salesperson for suggestions, and the toy or the manufacturer was rarely mentioned by name. By advertising every week we also created a year-round consumer demand and solved a major problem for us—80 percent of toys left to depend on the Christmas season." She resolved never to expose them again to the stress of waiting for information to move from salesclerk to retailer, to jobber's representative, to jobber, to factory representative, to manufacturing, so she hired her own "retail detail" representatives to visit stores across the country. "We reduced our time lag from six weeks to six minutes and that was crucial. Television speeded everything up." Mattel's marketing innovations were emulated by other toy companies, and in 1956 the *Saturday Evening Post* dubbed Elliot and Ruth—now aged 40 and 41—the whiz kids of the industry.

The exhausted whiz kids took the family on a European vacation in the summer of 1956—Barbara, now 15, and 12-year-old

Ken. It was not so much what they saw in the famous shop window in Lucerne, Switzerland, that mattered as what transpired when they went in. The window display was six 11-inch dolls of an alluring adult woman, a transmutation in flesh-toned pink plastic of a vamp named Lilli who started life as a cartoon in the German mass-market newspaper *Der Bild*. Lilli was sculpted by the German designer Max Weissbrodt as a plaything for men. Barbara wanted a doll in all six ski outfits on display. Ruth made a simple request of the saleswoman to buy one doll and six outfits. "She gave me a look. Only an American would ask such a stupid question." No, she was told, if you want to buy that outfit, you have to buy the doll with it. The doll and the outfit come together, don't you understand? The family was in Vienna the day afterward and there a store carried Lilli wearing a different ski outfit. Barbara sighed. "Oh, that's the prettiest. I wish I'd gotten that one." Ruth writes, "Something happened to me when she said that. My excitement began to build."

What happened was the inspiration that changing the adult clothes was the source of the fun. Indeed, the experience with Barbara in the shops in Lucerne and Vienna confirmed an insight that had lain more or less dormant since those frustrating times when she had to be at home looking after her children and not at work. She had observed that Barbara and her friends liked playing most of all with cardboard cutouts they could dress in different paper outfits. She had concluded they were role-playing: "It dawned on me that this was a basic much-needed play pattern that had never before been offered by the doll industry to little girls." Ruth had been brooding on making adult dolls for five years but had never been able to convince Elliot or the board of directors that any mother would buy her daughter a doll with breasts. "Children," she argued, "don't see it that way. They see breasts as normal. They observe

them all around them in adults. Breasts absolutely represent femininity to me."

Of course, she had a name for the doll she wanted to make — Barbie, after her daughter. But first she had to get her skeptical colleagues at Mattel to agree. Back in LA, the production people told her that they could never make an American version in that much detail at any kind of price that would sell. She took Lilli to Mattel's head of research and design, Jack Ryan, who

THE WHITE HOUSE
WASHINGTON

Dear Mr. Handler I Got the neat presents that you me and Gave my sisters, I Played Susies music Box and it was funny and I enjoy the Barpcun and I thank you very much. Love, David

A White House thank-you note from David Eisenhower, written long before he became a historian and married Julie Nixon.

had the right temperament for unconventional adventure: He was once married to Zsa Zsa Gabor and lived in a 36-room Hollywood castle. Ryan was about to go to Japan on a business trip, and Handler gave him a doll and said, "Jack, while you're over there, see if you can find someone who could make a doll of this approximate size. We'll sculpt our own face and body and design a line of clothes and accessories." In Japan, Ryan and another Mattel designer, Frank Nakamura, approached the Kokusai Bocki toy company, whose man in the United States, Tomio Tanabe, reported back to Japan that Mattel's bright, ultramodern 60,000-square-foot assembly line near the

Los Angeles airport was impressive, but what really overwhelmed him was Ruth. He later told her, "Your confidence and enthusiasm inspired me so much that I went back to Japan and got everybody to knock their brains out for Mattel."

Mattel went along reluctantly. The basic doll was one hurdle, the other was designing and making the tiny clothes she would wear that Ruth insisted would be sold separately. Ruth worked with a designer, Charlotte Johnson, two or three nights a week in Johnson's apartment to sketch 20 outfits that would see Barbie through various scenarios. Kokusai Boeki found nimble-fingered Japanese women who could sew the clothing. It took three years to design and test tiny bust darts, hems, zippers, snaps, buttons, and buttonholes. Ruth hired stylists to work full time on the stitched hair, and they came up with Barbie's ponytail. Translating a German sex doll into an all-American icon, Ruth diminished the come-hither look; she wanted a doll who was not too pretty, not too charismatic, someone with whom a young girl could identify. Barbie had to be everygirl.

When Ruth had the doll she wanted, she was sure the girls would respond, but she worried about the parents. She commissioned a Viennese-trained psychologist, Ernest Dichter, for $12,000 — a big sum then. He spent six months observing 23 fathers, 45 mothers and 357 children and reported that mothers were jealous of the doll and objected to it. He advised Ruth to use her television commercials as a means of reassuring mothers. Songs for the early Barbie tried to do that: "A little girl becomes a lovely lady. . . . Someday I'm going to be exactly like you." Still, the first Barbie was a little presumptuous when she made her debut before 16,000 wholesalers and buyers at the 1959 toy fair strutting her stuff in stiletto heels in a zebra-striped bathing suit and dark sunglasses.

Ruth had ordered a million dolls from Japan, expecting to sell 20,000 a week, and two million pieces of clothing. The response among buyers at the fair was so negative—"Ruth, little girls want baby dolls, they want to pretend to be mommies"—she called Japan to halve production, then returned to her hotel room and burst into tears. But it was the determined little girls who mattered, not the cynical buyers, and when they saw Barbie on Mattel's TV ads, they (and their mothers) stampeded the stores. Ruth reinstated her production, then doubled and tripled it. The dolls sold for only $3; the extra clothes at $1 to $3 were where the continuing excitement—and the profit—lay.

Barbie succeeded, in the view of the writer M. G. Lord, because she became an archetypal female figure upon whom girls projected their idealized selves: "I think she is the most potent icon of American culture in the late 20th century." In that regard the changes in Barbie have been significant reflections of the aspirations of each new generation. In the '50s, young girls (and their mothers) were content to dress Barbie in nice clothes, but she was a mom and homemaker. In the '60s, especially after Betty Friedan's call to rebel in *The Feminine Mystique*, Barbie became a careerist. Her 80 careers over the past 40 years reflect women's changing dreams: an astronaut in 1965, a young black woman at the height of the civil rights movement in 1968, a surgeon in 1976, a presidential candidate in 1992, an army heroine in the Gulf War, a rap musician in the '90s. She also keeps up with the fashions—a hippie headband in 1970—but she is not to be pinned down (or up) as a bimbo when she can turn herself into a TV news reporter, a paleontologist, a summit diplomat, a lifeguard, or a firefighter. As the *Economist* noted in its study of the Barbie phenomenon, "The secret of Barbie's eternal youth is reinvention."

Barbie had a brother, Ken (with a discreet undefined bulge). Toy-show buyers said to forget a boy doll. They were wrong again. Orders came in torrents. Ken and Barbie went surfing together. In 1966 they had siblings—Tutti and Todd—and Barbie started to talk in 1968: "Math class is hard. . . . Help me with my hair." Feminists fumed that this presented women as airheads, social critics lamented a stereotype of sexual expectation, mothers fretted about materialism. They were still fretting 40

years later, but as the *Economist* wrote, "Of all the forces against which resistance is futile, Barbie ranks right near the top."

Mattel went public in 1960, valued at $10 million. In 1966 it controlled 12 percent of the $2 billion toy market and had hit the Fortune 500. The Handlers had to find an answer to the problem of managing the spectacular growth. The stock price mattered to them. They hired a bright crop of MBAs as executive vice presidents. "Young tigers," Ruth called them. They certainly bit: The new executive vice president promptly told her she should not be called president of the company: "You're a woman, you're Jewish, and your style is all wrong. If you were to deal with the investment community, you wouldn't create the right impression." Ruth closed the doors to her office. Elliot found her crying. "I'm going to fire the son of a bitch," he said. It was Ruth who dissuaded him.

Over the next years, Ruth and Elliot basically lost control of their own company. They had been toy whiz kids. Now they were isolated among the financial whiz kids, who decentralized the company into divisions. Ruth's style of management had focused on marketing and product planning, responding to weekly reports. The tigers emphasized setting financial targets and sticking to them. Mattel fell into the Wall Street trap of ever-rising expectations. There was a limit to how many new toys Mattel could invent, so the young tigers went on an acquisition spree. They bought companies making pet products, playground equipment, audiotape equipment; they even bought parts of the Ringling Bros. circus. Most of them were mistakes.

Elliot spent much of his time devising new toys and traveling. Ruth was by then much occupied coping with breast cancer. After a mastectomy in 1970, she discovered how poor were the prostheses on offer and started making her own with a company called Nearly Me and a slogan: "The best man-made breast made by a woman." She

sold nearly one million prosthetic breasts, then sold the company in 1991 to Spenco Medical, a subsidiary of Kimberly-Clark. Ruth quit in 1975, and Elliot followed six months later. Price Waterhouse, imposed on Mattel by the SEC as special auditor, found fault with Mattel's previous accountant, Arthur Anderson and Company (shades of Enron). On February 17, 1978, five Mattel executives—including Ruth, but not Elliot—were indicted for various financial irregularities. One of the executives pleaded guilty, saying he knowingly claimed $10 million in sales that never took place.

Ruth vigorously protested her own innocence, but the strain of it all was telling on her. Counsel advised a plea of nolo contendere. She said she would accept that if she could also declare her innocence in court. The lawyers did not think it was possible, but on September 6, 1978, she was allowed to enter a plea of nolo contendere on ten counts and proclaim her innocence, to the fury of the prosecution. On December 11, Judge Robert Takasugi sentenced her to 41 years in prison—but it was a nominal breathtaker. He suspended the sentence on condition she paid $57,000 in reparations and performed 500 hours of community service under probation. She did so diligently. She started a foundation to help blue-collar probationers find work, and her probation was terminated a year and a half early.

Mattel struggled for some years after the Handlers quit. The only strong sellers were the toys Ruth and Elliot had invented. The company was rescued from near bankruptcy in the 1980s by another startling woman, Jill Elikann Barad (1951–), an actress from Queens, New York, who had played Miss Italian America in the 1974 movie *Crazy Joe* but who was also a brilliantly adventurous marketer. She put Barbie on a new trajectory with the ad campaign "We girls can do anything." In just over a decade, Barbie sales rose from $200 million to $1.9 billion by 1997. The Mattel of 2004 is a vigorous and diverse enterprise, but Barbie is one of the main reasons it is the world's number one company.

Ruth Handler died in April 2002 at the age of 85.

Barbie lives on.

Barbie kept pace with the aspirations of generations of young girls.

THE DIGITAL AGE

Life, liberty and the pursuit of happiness are vastly enhanced by the innovators in control of googols* of ones and twos

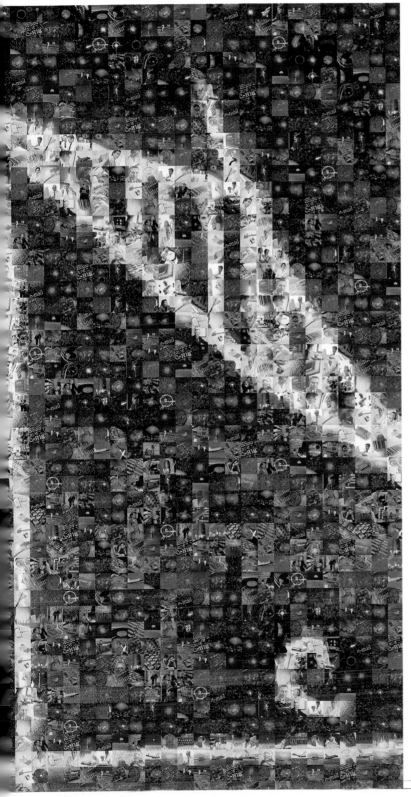

THE FATHER OF THE DIGITAL AGE was an unassuming American mathematician, Claude Shannon (1916–2001). Out of his concern for noise on the telephone lines came the conclusion that any kind of message—words, images, music—could be communicated by the binary digits one and zero, one representing an electrical switch turned on, and zero a switch turned off. He first noticed the similarity between Boolean algebra and telephone switching circuits while studying for a master's degree at MIT in 1940. Eight years later while at Bell Labs, he published his findings in the seminal paper "A Mathematical Theory of Communication."

Shannon, the son of a judge in the little town of Gaylord, Michigan, was a generous and joyful man. He liked to juggle beanbags and steel balls while riding a unicycle of his invention. His paper made him a star among scientists, but the vacuum-tube circuits of the day could not calculate his complex digital codes for practical application. We had to wait for the chip, advancing from the transistor and the semiconductor, before we could have CDs, satellite communications, cellular telephones, personal computers and the Web—and explore the inner workings of the human body in a new way.

The heroes of the digital age are the focus of this third part, and include a betrayed genius of computer code; the founder of biotech; visionaries who delivered on the promise of television; and impresarios of the Internet.

*A *googol?* See page 458

The lively Charles Shannon would no doubt have appreciated his digitized representation. Art catches up with technology in Robert Silvers' Image created for *They Made America*. Silvers invented his Photomosaics® at age 26 while a student at MIT's Media Lab and on graduation in 1996 founded Runaway Technology, Inc. The Photomosaic process compares the details in hundreds of individual images and arranges them to form a single mosaic, which when viewed from a distance resembles the portrait of Shannon.

The Electronic Elves

Bill Gates once said, "My first stop on any time-travel expedition would be Bell Labs in December 1947." On December 23, 1947, the voices of two Bell scientists boomed through the Murray Hill lab on the fringe of the Watchung Mountains in northern New Jersey, the main research center for AT&T. They were not shouting. **Walter Brattain (1902–1987)**, the experimenter, and **John Bardeen (1908–1991)**, the theorist, spoke in their normal voices. The sound waves were carried by ordinary wire to two minute gold wires, leading to a piece of processed germanium, a "brick" less than a sixteenth of an inch long, and a spring formed by an uncoiled office paper clip, and were therein amplified one hundred times.

This was the overture to the transistor era, and it was as dramatic as the opening of Wagner's *Ring*. It went unregarded by either media or academia for some time; **Robert Noyce (1927–1990)**, a young physics student at MIT in 1948, found little interest among his professors in "a novelty fabricated by a telephone company," though it had been ten years since the brilliant scientist **William Shockley (1910–1989)** had suggested that solid-state conductors were a promising field. "It has occurred to me," Shockley wrote in his lab notebook on December 29, 1939, "that an amplifier using semiconductors rather than a vacuum is in principle possible." His insight at that point was purely theoretical, but it would become the basis of all modern electronics.

Shockley, a driven and cantankerous man with glinting eyes and a smile that was at best enigmatic, was Brattain and Bardeen's boss at Bell Labs in 1947. He was envious but critical, too, thinking there was more to do. Shockley was born in 1910 to American parents living in London. He was a strange and clever child, given to tantrums so frequent his parents sent him to the Palo Alto Military Academy in California. After graduating from Caltech, he studied physics at MIT; in 1936 he started at Bell Labs, then located in lower Manhattan, where his scientific ability attracted attention, as did his taste for scaling the cafeteria's stone walls. Management wanted him to work on vacuum tubes, but apart from antisubmarine service in the navy, he spent his years studying that strange breed of matter called semiconductors. Conductors like metal easily transmit electricity because their electrons melt into a thin soup of negative charge. Insulators block current because their electrons are held in rigidly bound lattices. As their name indicates, semiconductors normally do not conduct electricity—in journalist David Kaplan's neat phrase this is a Roach Motel situation: Electrons can check in but they can't check out. However, when doped with tiny amounts of other elements, silicon and germanium are hospitable to electrons. They can let them go on their way as either positive or negative current, depending on the mixture. "N-type" semiconductors carry negative charge. "P-type" semiconductors are positive. The truly interesting thing is what happens when a P-type semiconductor is fitted next to an N-type. The semiconductor becomes a diode, or "rectifier," like John Fleming's original vacuum tube (see page 220); current can pass through it in only one direction, from the negative to the positive.

Bardeen and Brattain were perfect collaborators. The common assumption in transistor experiments had been that electricity traveled through all parts of a germanium block in the same way. Bardeen had a hunch electrons behaved differently on the surface of the metal. Brattain proved him right by ever so skillfully placing those two tiny gold wires on exactly the right spot on top of a P-N junction. But Shockley had disliked wires for more than aesthetic reasons. Attaching those two bits of gold wire by hand, two thousandths of an inch apart, would complicate mass production. Shockley worked alone without stop for the next five weeks to eliminate the wires. It was said he could see electrons, so powerful was his physical intuition. On January 23, 1948, he wrote down his ideas for a "junction transistor." Instead of a simple P-N junction with wires attached, Shockley's transistor was a sandwich of three semiconductors, an N-type, a thin layer of P-type and another N-type. Current flowing through the middle layer, or "gate," could control the flow of electricity in and out of the negative ends, blocking the current or amplifying it.

It was the beginning of a revolution. Semiconductors—unlike the hot, volatile vacuum tubes previously used as amplifiers—could be reduced to the scale of molecules, and in due course their fantastically fast on-off electrical signals would control watches and cell phones, steel mills and toys, streetlights and automobile instrumentation and, of course, the personal computer.

# of Silicon Valley

The main reason computers are useful is that they are extremely fast. They count as if they had only two fingers, one indicating "on," the other "off," requiring thousands of steps to add two plus two. With vacuum tubes, so many connections were needed to perform simple arithmetic that computers inevitably became complicated tangles of tubes and wires. But whereas the big hot vacuum tube could switch current off and on 10,000 times a second, transistors about 50 times smaller could switch on and off *billions of times a second*. Robert Noyce said, "After you become reconciled to the nanosecond, computer operations are conceptually fairly simple."

When the Bell Labs trio perfected the transistor, AT&T was once again feeling the hot breath of U.S. Department of Justice regulators eyeing AT&T's telephone monopoly. So 25 U.S. and ten foreign firms found themselves invited to a "Christmas" party in the spring of 1952: eight days of education, complete with transistor licenses, gift-wrapped at $25,000. In honor of Alexander Graham Bell, those who worked on transistors for hearing aids paid no fee.

THE RISE OF SHOCKLEY SEMICONDUCTOR

In 1956, Shockley went back to California. He was grieved that Bell Labs had not promoted him. "After all," he wrote to his second-wife-to-be, "it is obvious I am smarter, more energetic and understand people better than most of these other folks." His personality had grated on Bardeen (who left for academia) and Brattain—and they were not alone. Nonetheless, Shockley certainly had a nose for talent. When he set up his lab in an apricot storage shed at his new home in Mountain View, California, he hired a fine group of young scientists, all under 32, including Robert Noyce and **Gordon Moore (1929–)**.

Noyce radiated confidence. In a celebrated article in *Esquire,* Tom Wolfe wrote, "With his strong voice, his athlete's build, and the Gary Cooper manner, Bob Noyce projected what psychologists call the halo effect. People with the halo effect seem to know exactly what they're doing and, moreover, make you want to admire them for it." He was the son of a Congregationalist minister in Grinnell (population 7,000) in Iowa's corn belt, yet another who flourished in the atmosphere of middle America.

Congregationalists reject churchly hierarchies, and Noyce's father inspired in his son a democratic spirit that would pervade his future companies. Noyce conformed to type as a boy genius, developing powerful explosives, but he was also a star diver, sang like an angel, played the oboe and

acted. He went to nearby Grinnell College, where he was suspended for a semester for stealing a local farmer's pig for a luau. He confessed and was saved from jail by his physics teacher, Grant Gale. Noyce first became interested in semiconductors at Grinnell. Gale knew John Bardeen, who let him have two of the first transistors ever made to show his class.

Noyce got a Ph.D. in physics at MIT, then worked for Philco in Philadelphia, and by chance Shockley heard him give a paper on transistors. A month later Noyce got a call from Shockley asking him to apply for a job. "It was like picking up the phone and talking to God," Noyce remembered.

Gordon Moore was a modest, soft-spoken, laid-back 27-year-old Caltech-trained chemist with a dry wit. He was from the seaside village of Pescadero, 50 miles south of San Francisco, where his mother's family ran the only general store. When he got the call from Shockley, he was working on weapons propulsion research at Johns Hopkins and leapt at the chance to return to California. The others Shockley recruited were a dip in the melting pot of America: Julius Blank, a mechanical engineer from New York; Victor Grinich, an electrical engineer and navy man whose parents were immigrants from Croatia; Jean Hoerni, a theoretical chemist from Switzerland; Eugene Kleiner, a toolmaker and engineer who had fled the Nazis in Vienna in 1940; Jay Last, an optics specialist from Rochester; and Sheldon Roberts, a metallurgist at Dow Chemical.

All of Shockley's dozen or so hires were stellar, but one by one he made them sit down and stare at inkblots. This was a psychology test, along with an IQ test and lie detector examination. Shockley might have been able to see electrons, but he could not see people. His staff called the lab 391 Paranoid Place. He had an impulse to humiliate them—bawling them out for "not thinking," publicizing their salaries, getting them to rate one another. His mind, so dazzling intellectually, was clearly limited in emotional cognition, he was oblivious to the tension he was creating. Tom Wolfe has described the intensity of the lab: "Every day a dozen young Ph.D.s came to the shed at eight in the morning and began heating germanium and silicon, another common element, in kilns to temperatures ranging from 1,472 degrees to 2,552 degrees Fahrenheit. They wore white lab coats, goggles, and work gloves. When they opened the kiln doors weird streaks of orange and white light went across their faces, and they put in the germanium or the silicon, along with specks of aluminum, phosphorous, boron, and arsenic. . . . Then they lowered a small column into the goo so that crystals formed on the bottom of the column, and they pulled the crystal out and tried to get a grip on it with tweezers, and put it under microscopes and cut it with diamond cutters, among other things, the pale apricot light streaked over the goggles, the tweezers and diamond cutters flashed, the white coats flapped, the PhDs squinted through their microscopes, and Shockley moved between the tables conducting the arcane symphony."

The announcement on November 1, 1956, that Shockley had won the Nobel Prize with Brattain and Bardeen temporarily boosted morale. One of the few existing photos of Shockley Semiconductor shows Shockley and the group he had invited to a 7 a.m. champagne breakfast at Dinah's Shack in Palo Alto. But the bubbles soon burst. Shockley peremptorily announced they would now work only on diodes, not transistors. Mutiny erupted while Shockley was away on a vacation in Cape Cod in the summer of 1959: Seven (Moore, Blank, Hoerni, Kleiner, Last, Roberts and Grinich) asked Noyce to join them as their leader, the intention being to offer their services to another company.

All eight left on September 18. Shockley was so enraged he dubbed them the "traitorous eight" and never talked to them again. Fred Terman, president at Stanford, offered Shockley an endowed chair in the engineering department. Shockley taught aggressively and inflamed everyone around him. A few years later he ran for the U.S. Senate on a dysgenics platform, offering cash to people with IQs below 100 who agreed to be sterilized. He specifically urged African Americans to take the money. In the turmoil of the late '60s a permanent protest was encamped below his window at Stanford calling for him to be sterilized. The journalist T. R. Reid wrote that Shockley is perhaps the only person in history to see his own invention used to amplify calls for his death. Nevertheless, when the protestors' loudspeakers occasionally broke down, Shockley would go out and repair them. He died at 79, but his semen lives on in San Diego in a sperm bank for Nobel Laureates.

RISE OF THE INTEGRATED CIRCUIT

Eugene Kleiner's father, who owned a shoe factory, introduced the eight to the Wall Street investment firm of Hayden Stone. It dispatched a senior partner to see them, accompanied by a 31-year-old Harvard MBA named Arthur Rock. Rock was turned down by several dozen investors. The competition in transistors for portable radios and tape recorders was hot; ten years after Bardeen and Brattain's chorus, 30 million had been sold, and the price had come down to a few dollars compared with $50 or more initially. Finally, Rock managed to wring $1.5 million out of Sherman Fairchild, an inventor of aerial cameras and airplane brakes and head of Fairchild Camera. The eight had to invest $500 each. Fairchild had the option to buy them out within five years.

Fairchild Semiconductor opened in October 1957 in Palo Alto. It was the first venture capital investment in what would become Silicon Valley. Under Noyce's egalitarian atmosphere, people worked all hours, but there was almost no hierarchy, no reserved parking lots, no executive dining rooms and no private offices. The routine was everyone in by 8 a.m., brownies and whiskey at sales meetings. Fairchild had sales of $7 million the year Sherman Fairchild bought out the traitorous eight, but the company remained tiny. Texas Instruments had sales of $90 million. Other large companies, like GE, RCA, Philco, Westinghouse and Raytheon, had sales in the tens of millions of dollars. To survive, Fairchild Semiconductor had to offer something different.

The answer lay in a commonplace mineral that constitutes 90 percent of the earth's surface: silicon. With fear of a Soviet-U.S. missile gap growing—a false alarm as it turned out—the United States Air Force wanted rockets bristling with hundreds of thousands of transistors. The trouble was that germanium melted on the super-hot missiles as their warheads plunged back into the atmosphere. Only silicon could withstand the heat; silicon's high melting point is what kept the Santa Clara Valley from being known as Germanium Valley. But silicon was the devil to work with. It had to be baked to 1,000 degrees centigrade and then doped with traces of other elements: boron, which is slightly positive, turns silicon into a P-type semiconductor; phosphorous, which is slightly negative, turns it into an N-type. The process has been compared to smoking barbecued ribs, but the manufacturing required cleanliness higher than that in an operating room. Noyce said, "You've got to build in thousands of leads [connections] that are finer than a human hair, and every one has to be free of any defect. Well, how do you build a room that's free of dust?" Fairchild's Jean Hoerni found an answer to that. In 1958 he flattened a transistor into a thin layer and then oxidized the top to armor it against contaminants. Noyce said, "It's building a transistor inside a cocoon of silicon dioxide."

Noyce quickly patented the flattening technique which became known as the "planar process," over the objections of some who worried the devices were untested. He had an intuition that the protective planar arrangement might help to solve one of the technical dilemmas of the day. Electrical engineers were designing ever more complex circuits, adding to germanium or silicon wafers thousands of diodes, rectifiers and capacitors. Each of these components still had to be interconnected by hand-soldering thousands of bits of wire. Imagine peering through a microscope at something smaller than a particle of dust, manipulating diamond cutters to cut layers of the silicon apart, then taking tweezers to solder wires from the silicon to a multiplicity of components. The women who did this maddening work were extraordinarily nimble, but even the best had trouble soldering tens of thousands of microscopic connections, the "tyranny of numbers," as a Bell Lab scientist dubbed it.

Noyce was one of the two men who would solve this problem; the other was **Jack Kilby (1923–)**, of Texas Instruments.

It is piquant that Kilby wrestled with microscopic problems: He is a massive man, six foot six, with a massive appetite for knowledge. He reads three newspapers a day, numerous books and magazines and every one of the 60,000 patents granted annually by the U.S. Patent Office. The University of Illinois gave him only average grades in electrical engineering, a disappointment to his father, who ran an electrical company, and he failed to get into MIT. Cheerfully, he plunged into work at Centerlab, a Milwaukee company, developing ceramic-based, silk-screen circuits for consumer electronics. He earned 12 patents in ten years there as well as a master's. When he felt he had exhausted the research possibilities, he applied to Texas Instruments. He almost didn't get a job because he didn't have a Ph.D. He was 34 when he moved his family south, and he moved the earth when he got there.

Two months after his arrival, in July 1958, when everyone else took off for summer vacation, Kilby had not yet accrued enough service. Alone in his hot lab, he recalled circuit-printing techniques from Centerlab, and asked himself what stopped him from fabricating all the components in a single piece of semiconductor material, obviating messy soldering? Here's one why-not: A carbon resistor could be made for a cent; the same thing in silicon would cost $10. "It seemed foolish," said Kilby later, but he concluded it would cost less in the end to get rid of the soldering that produced so many defective transistors. And there was another potent thought: The size of a circuit could be shrunk even as it grew in complexity.

Kilby tested his concept with a circuit called a "phase-shift oscillator," which would turn direct current into alternating current. On September 12, he showed TI Chairman Mark Shepherd a crude device. It was just a sliver of germanium, with protruding wires, glued to a glass slide, but when Kilby pressed the switch, a sine curve undulated across the oscilloscope screen. It worked! Shepherd secretly committed TI to integrated circuits.

Noyce, unlike Kilby, had no eureka moment in January 1959: "I don't remember any time when a lightbulb went off and the whole thing was there. It was more like, every day, you would say, well, if I could do this, then maybe I could do that, and that would let me do this, and eventually you had the concept." He discussed each stage with Gordon Moore. Hoerni's planar process was the starting point. The silicon dioxide coating could hold the wires! But wait. Why wires? Lines of metal could simply be printed on top of the silicon coating, connecting different parts of the circuit without any soldering of wires. On January 23, 1959, Noyce wrote down in his lab notebook, "It would be desirable to make multiple devices on a single piece of silicon, in order to be able to make interconnections between devices as part of the manufacturing process, and thus reduce size, weight, etc., as well as cost per active element."

Noyce and Kilby were in ignorance of each other's breakthroughs when they applied for patents—Texas Instruments first on February 6, 1959, and Fairchild six months later. On April 26, 1961, the Patent Bureau spoke: Noyce was awarded patent number 2,981,877 for the integrated circuit. So began ten years and ten months of litigation. Kilby's integrated circuit had been first, but Noyce's had eliminated wires. The Supreme Court eventually sided with Noyce, but the legal outcome was moot. Years before the final ruling, the two pragmatic companies had worked out a licensing agreement to share royalties, earning each company hundreds of millions of dollars. Unlike the inventors of MRI, Noyce and Kilby respected each other, and both were satisfied with the title "coinventor."

There was, meanwhile, considerable skepticism in the industry. Noyce took no notice. He plunged into the new business. Despite TI's six-month advantage, Fairchild was first in the marketplace in April 1961, but it was the government that sustained both companies in those early days and created the industry. The air force's land-based Minuteman and the navy's submarine-borne Polaris missiles relied on thousands of integrated circuits. Each Apollo moon launch was stuffed with about a million Fairchild circuits. The fallout from defense was prolific. A hearing aid built by Zenith in 1964 was based on circuitry for a navy satellite.

Most integrated circuits remained pricey. To reach the mass market, the integrated circuit required one more innovation, and Robert Noyce provided it. He slashed prices—below the cost of production. Fairchild suffered losses at first but reaped enormous economies of scale. "That thinking permeated our industry," Moore added. "It's become the one driving factor that is really different about this business. Our standard solution to any problem has been to lower the price."

In 1970, when 300 million chips were sold, they were mostly bought by the civilian computer industry. Prices had already been tumbling. In 1963 a chip had cost on average $32; it fell to $18.50 in 1964, and $8.33 in 1965. Sales quadrupled every one of those years. In 1964 Gordon Moore made his famous lighthearted prediction that the number of transistors on a chip would double every year. It was more propaganda than prognostication. To his surprise his prediction turned out to be nearly accurate and became Moore's Law. Today the number doubles every 18 months to 2 years. Moore said, "At the time, I had no idea that anybody would expect us to keep doubling for ten more years. If you extrapolated out to 1975, that would mean we'd have 65,000 transistors on a single integrated circuit. It just seemed ridiculous."

Many high-tech companies in Silicon Valley can trace their genealogy to Fairchild and to Shockley Semiconductor. Shockley begat Fairchild. Fairchild begat dozens of other companies as engineers and scientists left in droves to form their own companies. By the late '60s Noyce and Moore were the only two of the founding traitorous eight still at Fairchild. Sherman Fairchild had died, and Noyce was passed over several times for CEO by a disputatious East Coast management, though he had managed the company well: By 1968 Fairchild had 32,000 employees and revenues of $130 million. It was time for him to make a break, and he urged Moore to join him. "The accidental entrepreneur like me," said Moore later, "has to fall into the opportunity or be pushed into it. . . . Then the entrepreneurial spirit eventually catches on." They dickered with names for their venture. Moore Noyce Electronics sounded like "more noise," so they settled on Integrated Electronics, which Noyce shortened to Intel.

THE RISE OF INTEL AND MEMORY CHIPS

Starting Intel was easy. Noyce just called up the laconic Arthur Rock, who had by now founded his own firm. Intel's business plan was only a page and a half. Such was the confidence in Noyce and Moore that within two afternoons every one of the 15 investors Rock called came in to back memory chips, a technology that did not exist. Noyce and Moore both contributed $250,000, about 10 percent of their net worth. Noyce made sure his old alma mater, Grinnell, was given a chance to invest—he had forgiven it for suspending him after the pig theft—and the investment went on to multiply the college's endowment many times.

Like Fairchild, Intel quickly became an intellectual powerhouse. "I don't think you could call it a relaxed atmosphere," Noyce told the *Harvard Business Review*. Noyce, with his hundred ideas a minute, put a premium on risk-taking. A technologist, he said, is "the kind of person who is comfortable with risk." Gordon Moore was the restrainer, a modest man with a sense of priorities in the swirl of business life: He would not come to the phone for friend or office while he was vacuuming his wife's multimillion-dollar home. He had little time for the hippies and yippies still prevalent in San Francisco, explaining, "We were the revolutionaries of the time."

As for making the memory chips, Moore called Intel's approach the "Goldilocks strategy." He described it to Robert Lenzner of *Forbes* magazine as choosing between an easy technology that could be quickly copied by rivals; a complicated one that might bankrupt them; or a moderately complicated one. They took the middle course. Intel's founders saw their opportunity in the speed of memory. Computers are made of two main parts—logic circuits, which do the actual calculations, and memory circuits, which store data and programs. Logic circuits used semiconductor chips, but in the '60s memory was stored in a far cruder way, in magnetic cores (devised most notably at MIT). **An Wang (1920–1990)**, a Chinese refugee who arrived in America in 1945, earned a Harvard Ph.D. in physics in 16 months and started one of the leading technology companies, storing stored memory in tiny disks of metal threaded on a mesh of wire. Core memory was two or three times cheaper than semiconductor memory, but Gordon Moore believed that the much faster semiconductors would get cheaper: "We were trying to change the technology." They did. Magnetic core memory became, well, a memory.

Four transistors were commonly needed to store one binary digit, or "bit"—a two-number combination of ones and zeroes. By 1967 Fairchild engineers had managed to squeeze 1,024 transistors on a single integrated circuit. A milestone was passed when Intel came out with a chip holding 1,024 bits in random-access memory, dubbed 1K of RAM. In the early 1970s Intel controlled nearly 100 percent of the memory chip market. Later, when it had a dozen competitors, Intel came out with a 16K RAM chip in 1975 and a 64K chip in 1980. "We reached a point where we could produce more complexity than we could use," said Moore.

THE RISE OF THE MICROPROCESSOR

Memory chips were profitable, but in the late 1960s, every firm wanted different circuits. It was a problem of excess like the tyranny of numbers in the '50s. If the trend continued, the number of circuits needed would proliferate beyond the number of circuit designers. It foreshadowed an economic as well as a technological crisis: Each new chip cost on average $50,000 to design. The more specially designed chips were made, the less companies could build economies of scale. What was needed was a Model T of integrated circuits, a one-size-fits-all model.

Marcian "Ted" Hoff (1937–) had been a teenage prodigy in Rochester. At 15, he had won a $400 scholarship in the Westinghouse science talent competition; then, while a sophomore at Rensselaer Polytechnic Institute in Troy, New York (where Theodore Judah studied, see page 126), he did summer work at the General Railway Signal Company and patented two circuits, one to protect against lightning and the other to detect trains by the audio frequencies transmitted along the track. Noyce recruited Hoff, who had earned his doctorate at Stanford, after he had a spell designing video games for Atari. This was the young man who gave us what Gordon Moore describes as "one of the most revolutionary products in the history of mankind." His Model T was the microprocessor—the third major revolution in microelectronics, after the transistor and the integrated circuit.

Hoff's microprocessor was born of the frenzy for portable electronic calculators in the late '60s. In 1969 a Japanese company, Busicom, asked Intel to help it design a new line of then-inexpensive $1,000 calculators. Hoff saw the 12-chip arrangement Busicom's engineers had sent him as needlessly complex. Just designing the thousands of overlapping circuits would require dozens of man-years. He heard the innovator's call: There must be a better way. Hoff's inspiration came from his desk, on which was one of Ken Olsen's minicomputers (see page 376). "I looked at the PDP-Eight, I looked at the Busicom plans," said Hoff, "and I wondered why the calculator should be so much more complex." Instead of building a specialized chip for arithmetic, his inspiration was to build a general, all-purpose circuit that could be programmed for other tasks, too.

Throughout the summer of 1970, Hoff designed a central processor unit on a chip. Coupled with two memory chips—one for data, the other for storing programs—Hoff's CPU was an entire computer in a tiny flake of silicon. Having devised the architecture, he had to wait six months while Intel found someone to put it in silicon—another immigrant in his early 20s. **Federico Faggin (1941–)** came fresh out of graduate school in Italy with a doctorate in solid state physics. Says Hoff: "He worked at just a furious pace, and in the space of about nine months designed the three major chips."

Intel has Texas Instruments to thank for what happened next. In 1971 it introduced a phenomenally cheap $150 calculator Jack Kilby had designed. Busicom immediately faced bankruptcy. It asked to renegotiate the price of the chips Hoff had designed. Hoff at once realized the magnitude of the opportunity, urging Noyce, "For God's sake, get us the right to sell these chips to other people." Hoff remembers outrage among the staff. "We've got diode salesmen out there trying to sell memories, and now you expect them to sell computers! You're crazy." Some saw the microprocessor as a dangerous departure from Intel's focus on memory chips, but Intel's new marketing manager, Ed Gelbach, realized that "general purpose" meant ubiquitous sales; the microprocessor, he said, could "insert intelligence into many products for the first time."

The 4004 chip was optimistically advertised in trade magazines for $200 as "a computer on a chip." Intel envisioned selling it for traffic lights, ovens, cars and watches. Oddly, the one use Noyce and Moore didn't see for the chip was home computers. When an engineer tried convincing Moore that Intel should build personal computers, Moore couldn't see the point. Wives could use them to store recipes, the engineer suggested. Moore immediately rejected the proposal, later saying, "I could just picture my own wife, Betty, at the stove with a computer beside it." Noyce envisioned microprocessors being used primarily for wristwatches.

Intel went public in October 1971, the same day as Playboy Enterprises. Both companies were valued about equally, but within a year Intel's value had doubled over Playboy's. As one Wall Street analyst said, "It's memories over mammaries." From the beginning, Intel fought for and achieved its goal of 10 percent for research and development and 10 percent in profit. Noyce and Moore were on their way to being billionaires. That same year Intel produced its first microprocessor, the 4004, which, with 2,300 transistors, was able to process four bits of information and could do 60,000 calculations a second. Like memory chips, microprocessors began obeying Moore's Law. By the year 2000 chips would contain tens of millions of transistors, each smaller than a bacterium, capable of performing billions of calculations per second. In 2001 alone the semiconductor industry sold $139 billion worth of microchips—about 60 million transistors for every person on the planet. Today's musical greeting cards pack more computing power than the gargantuan ENIAC (see page 368).

While the 4004 microprocessor was a crude device, Intel's 8080, an eight-bit microprocessor first sold in April 1974 for $360, transformed the industry. It was 20 times faster. Until then the computer industry thought microprocessors a joke. The 8080 was the first chip to approach the power of larger computers. (The $360 price was a jab at the IBM 360 mainframe.) Despite the long-run significance of the microprocessor, it remained a small part of Intel's business until the mid-1980s, when a radical decision by yet another immigrant saved the company.

Andris Grof (1936–), born in Budapest, survived both the Nazis and the Soviet crackdown on the 1956 Hungarian Revolution. His escape to Austria and then to the United States, where he became Andy Grove, was the subject of an engaging memoir, *Swimming Across,* in which he marvels that someone who a few years before was a refugee fleeing across plowed fields could become *Time* magazine's Man of the Year: "I've continued to be amazed by the fact that as I progressed through school and my career, no one has ever resented my success on account of my being an immigrant." He graduated at the top of his class in 1960 from New York's City College and had a Ph.D. in chemical engineering from Berkeley when Noyce and Moore appointed him Intel's director of operations in 1968. Grove had a psychedelic style of dress in the 1970s that Tom Wolfe described as "California groovy," but he was a very hard manager. It's debated whether the phrase Intel employees use to describe Grove's management style—"he'd fire his own mother"—is insult or compliment.

In mid-1985 Grove, alarmed by Japanese progress, went to see Gordon Moore, Intel's CEO and chairman, Noyce having stepped down. Grove remembers looking out of Moore's office window at the Ferris wheel turning in endless circles in the Great American amusement park. Grove looked at Moore gloomily and asked, "If we got kicked out and the board brought in a new CEO, what do you think he would do?" Moore answered right away, "He would get us out of memories." Grove writes, "I stared at him, numb, then said, 'Why shouldn't you and I walk out the door, come back and do it ourselves?'" They changed direction just before Japanese memory-chip makers flooded the market.

The question facing chip makers at the turn of the 21st century is how long they can innovate. Moore's Law is expected to hold until 2020. Then transistors will become so small they will reach the limits of atomic matter. (They are already considerably smaller than the shortest wavelength of visible light.) At that scale spooky and uncontrollable quantum effects will start taking place. Electrons will tunnel through solid matter. The transistor as designed by Shockley, Bardeen and Brattain will no longer work. Theoretical physicists are already talking about new types of computers using light, DNA or even the quantum states of individual atoms instead of electricity to store and process information. For now these technologies remain purely theoretical. Is this the end of progress, the end of Moore's Law? We'll know soon enough.

GARY KILDALL He saw the future and made it work. He was the true founder of the personal computer revolution and the father of PC software

1942–1994

Gary Kildall loved piloting his many aircraft, surfing his speedboats, roaring off on his motorcycles, riding the waves on his Jet Ski, racing his Lamborghini Countach S—at one time when he had more money than he knew what to do with he had the pick of 14 sports cars in his lakeside villa. But what Gary Kildall enjoyed most in his short life was sitting still for hours in a little office writing code for computers. "It's fun to sit at a terminal and let the code flow," he said. "It sounds strange, but it just comes out of my brain; once I'm started, I don't have to think about it." He would call colleagues in the middle of the night to tell them that a program had worked. "What a rush!" he'd shout. Author Robert Cringely's metaphor is apt: He wrote code as Mozart wrote concertos.

In the early '70s, he was utterly brilliant at programming—but that is an understatement of his crucial role in the personal computer revolution. He was the first person to realize that Intel's microprocessors could be used to build not just desk calculators, microwave ovens, traffic systems and digital watches but small personal computers with an unimaginable multiplicity of uses. Then, entirely out of his own head, without the backing of a research lab or anyone, he wrote the first language for a microcomputer operating system and the first floppy disk operating system before there was even a microcomputer, months before there was an Apple, years before IBM launched a personal computer. Kildall did it, moreover, in such a manner that programmers were no longer restricted by compatibility with the computer's hardware. In Kildall's system, anybody's application could run on anybody else's computer. It was the genesis of the

SWEETHEARTS:
Photo booth self-portraits of Gary Kildall and classmate Dorothy McEwen

whole third-party software industry. This alone would have been an astounding achievement.

Kildall stayed ten years ahead of his time and never stopped pushing the boundaries of technology up to his untimely death just as the Internet was beginning to take hold. He pushed for preemptive multitasking, window capabilities and menu-driven user interfaces. He laid down the basis for PC networking. He created the first computer interface for videodiscs to allow nonlinear playback and search capability, presaging today's interactive multimedia. He built the first consumer CD-ROM filing system and data structures for a PC. With all this inventiveness, the "Edison of computers" was also a dedicated teacher; as his son, Scott, noted, it was his devotion to creating tools to help the world, rather than

moneymaking, that led him to devote a great deal of time to a product called "Dr. Logo," an intuitive, nonabstract computer language program geared toward teaching kids to program, to use computers as learning tools, not merely game-playing machines. By the end of his life, he was working on wireless hardware connections. In all he did, he epitomized the openness of the early days of Silicon Valley, the zest for the next frontier, the conviction that the best technology would succeed in the marketplace on its own merits. He had the faith of the academic scientist that mankind advances less by the protection of knowledge than by its diffusion. Jacqui Morby, a venture capitalist, has an affectionate remembrance of his idealism from their first meeting. "He said to look for a red-bearded man in cowboy boots at San Joe airport, then he rolled up in a light plane and yelled from the cockpit for me to jump in for lunch at the Nut Tree. On a Nut Tree napkin he drew a visionary plan of an open industry in which the owner of the operating system would forswear going in for applications like word processing. He said that would create a dangerous monopoly and stifle innovation." Kidall was hardly a humorless missionary; he was unassuming, droll and generous. The bitterness that darkened the last decade of his life was similar to that inflicted on radio's Edwin Armstrong. Both men discovered that the sublime could come off a bad second to the mediocre, that misrepresentation and manipulation could prevail over truth and justice.

CODE KING: "It's fun to sit at a terminal and let the code flow. Once I'm started I don't have to think about it," writes Kildall, the endlessly innovative president of DRI, in 1982. Then he would put his feet up.

"The day Gary went flying" has entered legend as the explanation of how IBM came to market an inferior operating system from Microsoft's Bill Gates that became the foundation of Gates's fortunes: Kildall, so it is said, preferred a joyride to a meeting with IBM and was too prickly to sign IBM's standard confidentiality agreement. The story is propagated by Bill Gates and others, and swallowed whole by computer historians. It is false. IBM tricked Kildall. And in the end it was not Kildall who missed an opportunity but the rest of us. Had IBM backed Kildall's system, the majority of computer users would have had multitasking and windows a decade earlier. By adopting MS-DOS, which was based on QDOS, a slapdash clone of Kildall's system, IBM and Microsoft forced users to endure more than a decade of crashes with incalculable economic cost in lost data and lost opportunities.

At the end of his life, Kildall wrote an autobiography, "Computer Connections," which has never been published. It is incisive, unaffected, moving and funny, suffused by Kildall's romance with technology. It informs part of the narrative that follows and is the source of the Kildall quotations, but nothing may ever be enough to drive a stake through the heart of the appealing myth of how Kildall missed becoming the richest man in the world. In his manuscript, Kildall writes, "I think I'll make a cassette tape of the 'IBM Flying Story.' I'll carry a few copies in my jacket to give out on occasion. There's only one problem. I tell this story, and after I'm done, the same person says, 'yeah, but did you go flying and blow IBM off?'"

GARY KILDALL's precise seafaring father, Joseph, long dreamt of building a simple machine to take the tedium out of finding just where a ship was on the face of the earth. Having taken a sextant reading and checked a chronometer, a navigator still had tedious calculations to do based on tables from the *Nautical Almanac* to correlate the exact time and date. Joe, who taught navigation at the family nautical college, envisaged just punching the data into his machine of cams and gears and turning a crank for the answer. "It wasn't until the microcomputer was invented," writes Kildall, "that the 'crank' was truly feasible," but his father's idea stayed in his mind.

Gary was a poor performer at Seattle's Queen Anne High School. He applied his technological gifts to rebuilding old cars and boats and carrying out pranks. He managed to rewire neighborhood phone lines so as to eavesdrop on his sister's conversations with her boyfriend. He invented and patented a type of Morse code device. But his English grades at Queen Anne were so bad he had to stay back a year. It turned out to be a stroke of luck: When he squeezed his lanky frame into the desk for his repeat year he found himself sitting next to a beguiling and witty young woman, Dorothy McEwen. His focus on irregular verbs suffered—they talked so much they had to be moved to different corners of the classroom—but she became his bride a few years later. Dorothy remembers, "He was inventive. He was like a little kid in a candy shop."

After high school, Gary followed his father, who had followed *his* father, Harold, in becoming a teacher at the Kildall Nautical School. Teenage Gary taught navigation and trigonometry for several years alongside his father and Harold, who did not stop teaching until the week before he died at 92. The family tradition was strong, so Gary's father did his best to sabotage Gary's plans when, at the age of 21, he announced he was abandoning ship to go to college. His ambition ran afoul of not only his father's protests but the fact that his grades at high school hadn't been good enough to qualify for the University of Washington. He petitioned the University regents to take into account his teaching at the Kildall Nautical School, and "by entirely too close a margin," he was admitted in 1963, the year of his marriage to Dorothy. She supported him while he studied—and study hard he now did. "The Kildall Nautical School," he writes, "taught me processes that high school hadn't. Such as the ability to do mathematics of a sort and, most important, the mental tools to dissect and solve complicated problems, and to work from the beginning to the end in an organized fashion." He got nothing but top grades.

Kildall found himself in a pivotal moment in the history of computers. The 1960s were a time of transition between mechanical and digital computing. He studied both; of the mechanicals, he dryly remarked that after a lot of complicated button pushing, "sometimes the resulting number was correct."

Kildall's deepest passion was for an important piece of the computer software called a compiler. Compilers are translators. They take computer languages understandable by people and turn them into the famous binary digits—ones and zeros—called "bits" for short, that the computer understands:

0000111010101010000111111001010 0011010000001001, etc.

"They are sort of like natural language translators," writes Kildall, "who sit in a business conference and make English into Japanese. Compilers, when perfected, can be elegant to the point where you want to paste a printout on your wall, like artwork. OK, you have to be into writing compilers to get my meaning, but when your compiler works, you are very proud and want to show it off."

In 1966, the University of Washington bought a new Burroughs B5500, a computer powerful enough to run ALGOL, or Algorithmic Language—a series of procedures done by numbers. The computer follows algorithms to do mathematics much faster than people ever could. ALGOL was a precursor for today's PASCAL programming language. Kildall got himself a part-time job maintaining the Burroughs. He writes, "That old B5500 became my learning machine. I saw a ton of sunrises over that Computer Center." He became so gleeful having the computer to himself that at midnight he would put up a sign saying B5500 DOWN FOR MAINTENANCE. At 6 a.m. he would take down the sign after having played with the machine all night. "I learned from the architecture of the B5500 computer. In particular, I learned about data structures for organizing disk drive information."

His nocturnal exercises paid off. In 1967 he was accepted as one of 20 students in UW's first master's degree program in computer science. What the left hand of Providence bestowed, the right threatened to take away: He received a draft notice consigning him to the army and the Vietnam War. "Damn, all of a sudden visions of rice paddies flew through my head. I know you're not supposed to use connections, but quite frankly, I didn't want to get shot at. Dad connected me with one of his [navy] buddies, and I got a reprieve to finish my master's degree while I worked toward my commission as an officer." He spent two summers at the navy's Officer Candidate School in Newport, Rhode Island, in 1967 and 1968, became an ensign and, while waiting for assignment, taught data processing to sailors in Seattle. "It was a bummer. I was destined to become an officer on a destroyer tossing shells into the forests of Vietnam." Unbeknownst to Kildall, the president of the University of Washington, Dr. Charles Odegaard, had been impressed by Kildall's computer work in 1969 and arranged for him to have a decisive interview shortly before he was due to be posted (and he had graduated with honors). The navy captain he met with stared Kildall in the eye. "Mr. Kildall," he said, "you have a choice to make." He could become either an officer on a destroyer or an instructor in mathematics and computer science at the Naval Postgraduate School in Monterey, California. Kildall recalls: "This particular question made me understand the length of a microsecond. 'Well, sir,' I said, 'I would like dearly to serve my country in battle, but I think I shall take the second option, if you please.'" The captain warned him that if he taught at the Naval Postgraduate School, he would probably not reach the level of admiral. "I took a pensive stance for a moment and then told him that I would accept that risk."

He and Dorothy settled down to family life in the pleasantly sleepy town of Pacific Grove on the west edge of the Monterey peninsula. When his three-year tour of naval duty was up in 1972, Kildall kept a link with the school as an associate professor but returned to the University of Washington to pursue his Ph.D. His thesis topic was to optimize the translation of language into computer-readable form so as to reduce the amount of memory required. He called the project Global Flow Optimization. After several months, Kildall made a program that worked mathematically but he could not prove his process

was more efficient. He slept little, struggling vainly for an answer. "I just sat and sat and sat in my UW grad student office, resting my head in both hands until my eyes shut by themselves late [one] evening. Nothing. Then, in an instant, the proof came to me. I wasn't even paying attention to it. I awoke in an instant and wrote the entire proof of my central theorem, not finishing until sunrise. I guess that's why they put lightbulbs over cartoon characters. The discovery of this proof was one of the

KILDALL'S CHOICE: In a microsecond he lost the chance to be an admiral.

grandest experiences of my life, except, of course, for the time I visited Niagara Falls."

In 1972 a colleague showed him an ad in *Electronic Engineering Times* saying, "Intel Corporation offers a computer for $25." Actually, it was offering the four-bit computer chip, measuring approximately 0.8 to 0.3 inches, designed by Intel's young Ted Hoff for a Japanese desktop calculator but released for general sale at Hoff's urging. The cost was $25 only if you bought 10,000 of them; the price jumped to between $45 and $60 if you bought just one. But customers using the 4004 chip would first need to design a custom board-level or box-level system with memory, power suppliers, keyboard, display and cables. To help customers get started, Intel began selling various board-level "development systems" with enough memory to demonstrate chip operation and to run, test and design new programs. Still frustrated by the $3 million IBM mainframe, Kildall was intrigued. He had never heard of this "little chip company," but he sent for specifications for the first development system for the 4004. It was a little foot-square blue box called the SIM4-01, with

2,300 transistors on the chip and read-only memory (ROM), but the price was $1,000 plus $700 for a Teletype. He did not have enough money for both on his $20,000-a-year salary.

He got around this by faking the operation of the little 4004 on the big IBM 370. As he programmed the simulator, the limitations of the chip drove him crazy, but he saw the potential of escaping from the room-size IBM mainframes (and the refrigerator-size minicomputers from Ken Olsen's Digital Equipment Corporation). "This [4004] was a very primitive computer by anyone's standard, but it foretold the possibility of one's own personal computer that need not be shared by anyone else. It may be hard to believe, but this little processor started the whole damn industry. . . . There, in 1972, my dad's navigation 'crank' had arrived in the Intel 4004, but there appeared to be some major programming work to get the crank to actually work."

The 4004 had no trigonometric functions, so Kildall spent months programming the chip to find sines and cosines. After debugging the program on his simulator, Kildall knew he had something that might interest Intel. He called a friend there and offered to swap the 4004 simulator for a real chip, a $1000 SIM4-01. The Intel engineer was less interested in the simulator than in the trigonometric functions Kildall had written. They made the trade, and Kildall had his own 4004.

There was a long and tedious year's journey to make anything of a machine that could be fed data only four bits at a time and had no monitor. Kildall describes the process: shining a UV light through a quartz window for 30 minutes to erase 256 bytes of space on the EPROM (erasable programmable read-only memory) so there was room for his own little program; feeding paper tape into a Teletype and then line by line typing a program written in hexadecimal code, known as machine language; fixing the typing errors by going back to the beginning; running the corrected code to load each EPROM. "We pioneers had to do all this stuff two decades ago so you can enjoy your sweet little laptop while cruising placidly over Colorado at 37,000 feet. . . . For reference, an average JFK to

SFO flight takes about six hours. That's the time it takes to program twelve EPROMs of 256 bytes each, or a total of 3,072 bytes of memory."

A laptop today does the same in a fraction of a second.

Nonetheless, Kildall built a briefcase computer—"it may have been the first personal computer"—and took it around for demos, lugging with him the 60-pound Teletype. He inspired hundreds, one of them a young engineering graduate at the University of Washington, Tom Rolander, who later became important in his life. Intel, too, was impressed by Kildall's bubbling imagination and engaged him as a part-time consultant, initially to build a simulator for the new microprocessor the company was working on, which was to be more sophisticated than the 4004 and ten times faster. Software applications were a low priority then at Intel; the software "group" Kildall joined part-time was only two people, tucked away in a space the size of a small kitchen. Kildall devised a *Star Wars* kind of video game for his briefcase computer based on a 1972 idea by Intel engineer Stan Mazor, a codeveloper of the microprocessor, and the pair of them showed it off to one of the founders of Intel, Bob Noyce, a gentle, smiling presence who occasionally walked encouragingly through the little software corner in his white lab coat. Kildall writes, "Noyce peered at the LEDs blinking away on my 4004. He looked at Stan and me and said, bluntly, that the future is in digital watches, not in computer games." Intel had just bought Microma, one of the first digital watch companies, which was not long afterward beaten into the ground by a flood of Japanese digital watches. Intel thus passed up an opportunity to lead the video game industry. Kildall, in a judgment that would reverberate for him, too, writes of Noyce: "He, like all of us, made some decisions that are right, and some that could have made the future unfold in a different manner." What mattered to Kildall was that in building an industry in microprocessors, "Bob treated his people with dignity."

IN THE COCKPIT: Kildall writes: "Tom Rolander (right) was co-pilot in flying and in life. Many evenings we'd fly across the United States, watching the stars in pitch-black background. We'd talk about the beauty of nature, the inner feelings of relationships between us and our wives."

Intel was abuzz in 1973 with the triumph of the 8008 chip, which doubled the power of its first microprocessor, and Kildall was drawn to spend more and more time there. After his "eyeballs gave way," he would spend the night sleeping in his Volkswagen van in the parking lot. He became a trader in an electronic bazaar, swapping his software skills for Intel's development hardware. One morning, he knocked on the door of Hank Smith, the manager of the little software group, and told him he could make a compiler for the Intel 8008 microprocessor so that his customers would not need to go through the drag of low-level assembly language. Smith did not know what Kildall meant. Kildall showed how a compiler would enable an 8008 user to write the simple equation $x = y + z$ instead of several lines of low-level assembly language. The manager called a customer he was courting, put the phone down and with a big smile uttered three words of great significance for the development of the personal computer: "Go for it!"

The new program, which Kildall called PL/M, or Programming Language for Microcomputers, wrote microprocessor applications such as operating systems and utility programs, and Intel used it for decades afterward. Kildall's reward was Intel's small new computer system, the Intellec-8. It must have been the first commercial personal computer, Kildall notes, though no one thought of it as that. He

borrowed $1,700 to buy a printer and a video display. What irritated him was that he could not operate the Intellec independently of the expensive DEC PDP-10 minicomputer now installed in the navy's classroom at Monterey—unless he could contrive a way for the Intellec to store a great deal of data. As technology writer Al Fasoldt writes, without a disk operating system the computer is just too dumb to do anything useful.

Experiments with cassette tape did not work; then Memorex, just down the street from Intel, came up with an eight-inch floppy disk for mainframes. It held 250,000 characters, moved data at 10,000 characters a second (compared with ten characters a second with the Teletype paper reader) and in theory gave nearly instant access to any portion of the stored data without rewind or fast-forward. Wonderful—but the communication between Kildall's small computer and the disk drive needed a controller board to handle the complex electronics, and there was no such thing. "I sat and stared at that damned diskette drive for hours on end and played by turning the wheels by hand, trying to figure a way to make it fly. The absence of a controller for that floppy drive was the only thing between me and a self-hosted computer. It drove me nuts." The equipment sat in his office for a year, the software genius defeated by hardware. "I'd just look at it every once in a while. That didn't seem to work any better."

He went reluctantly back to his DEC minicomputer and built an operating system he called CP/M, or Control Program for Microcomputers, mimicking the name PL/M. (CP/M originally stood for control program/monitor.) He knew the program was sound, but he still could not get it to communicate with the disk. Desperate, he called his friend from the University of Washington, John Torode, who had a Ph.D. in electrical engineering. Torode worked on it for a few months and came up with a neat little microcontroller. Kildall held his breath: "We loaded my CP/M program from paper tape to the diskette, and 'booted' CP/M from the diskette, and up came the prompt:

*

"This may have been one of the most exciting days of my life, except, of course, when I visited Niagara Falls one day."

Kildall opened a file, stored it on the floppy and it appeared in the directory—commonplace stuff now but a dramatic achievement then, the world's first disk operating system for a microcomputer. Walking back to Kildall's home for a celebratory bottle of red wine, the programmer and the engineer told each other, "This is going to be a big thing." But where was the market? Ben Cooper, an entrepreneur from San Francisco, paid Kildall to write a program for an arcade astrology machine he was making: Put in a quarter, dial your birth date and out comes your future. Kildall built the software system in a small converted toolshed at the back of his home. When Ben mistakenly entered the command "del *.*" instead of "dir *.*" to get his files, he deleted all of the files on the diskette. And that is the origin of the prompt "Are you sure? (Y/N)."

Cooper finally got his machine installed on Fisherman's Wharf in San Francisco, and the entrepreneur and programmer sat on a bench one summer evening to see what would happen. A hand-in-hand couple put in a quarter, did not bother with the dial and walked off happily enough with someone else's horoscope. "Because of it," writes Kildall, "they are probably married with seven children to this day." But nobody wanted to buy the 200 machines Cooper had built.

Kildall's own happy marriage (with "two great kids, Scott and Kristin") hit a reef in 1974, but it was retrieved by Dorothy's willingness to help make a business out of the CP/M program. She had not had a formal college education, but she had worked in a phone company's customer service department and, as Kildall writes, often outsmarted the grads who came to him. Gary continued to teach at Monterey while Dorothy handled the early business, sending diskettes to customers responding to a $25 advertisement she and Gary had bought in the famous insider magazine *Dr. Dobbs' Journal of Computer Calisthenics and*

Orthodontia at the suggestion of its founding editor, Jim Warren. Demand for the diskettes was slow at first; the market was made up of early computer enthusiasts. "We started in a corner of the bedroom," Dorothy told us. "There was no long-term plan. We put no money into the operation. We didn't have much savings. We lived off Visa and MasterCard."

The first big break was a sale of a word-processing program in 1975 to Omron, which made cathode ray tubes (CRTs) for newspaper editing. It was the first company to build hardware using CP/M. Kildall and Torode split the $25,000.

Earlier in the year, in Albuquerque, New Mexico, Ed Roberts had come out with a mail-order kit for hobbyists for the first commercially successful personal computer, the Altair, which sold for $500. It had an Intel 8008 microprocessor inside with toggle switches on the front panel. It was notoriously difficult to use, with only 256 bytes of memory and no screen or keyboard.

A new company with wider ambitions to sell to the general public was formed in San Rafael, across the Golden Gate Bridge from Silicon Valley, calling itself IMSAI. It had promised delivery of a diskette operating system and had not even begun to figure it out when Glenn Ewing, a former naval student of Kildall's, engaged as a consultant, told IMSAI about CP/M. "Glenn came to my toolshed computer room in 1975," writes Kildall, "so we could 'adapt' CP/M to the IMSAI hardware. What this means is that I would rewrite the parts of CP/M that manage things like diskette controllers and CRTs (screens). Well, come on, I'd already done this so many times that the tips of my fingers were wearing thin, so I designed a general interface, which I called the BIOS (basic input/output system) that a good programmer could change on the spot for their hardware. This little BIOS arrangement was the secret to the success of CP/M."

Kildall had in essence created a digital pancake. The underside could be adapted to fit different hardware configurations. But the top part was truly revolutionary; it did not have to be rewritten. Kildall developed an instruction originally dubbed "Call 5" and later called "Int 21"; any application program could interface with his

operating system. This was a phenomenal advance. It liberated software from hardware. Any application could thereafter run on any computer.

According to Kildall, he and Ewing built the system on a lovely afternoon, sitting in the toolshed across from the house with its hummingbird feeders, a pastoral scene for a computer revolution, for that is what it portended. Kildall's friend and future partner, Tom Rolander, explains it well: "Think how horrible it was for the software vendors before that time. They would have to have different copies of their program configured to different pieces of hardware"—and there were scores of specialized pieces of hardware. Imagine a world where each model of car required a different kind of gasoline—that's what it was like for computer operators before Kildall's innovations. Kildall created the bedrock and subsoil out of which the PC software industry would grow. He licensed his system to IMSAI for $25,000 and felt rich.

Clearly, there was a business here, but Kildall found the transition from inventor to innovator wrenching.

He had fun with his classes at Monterey, where the graduates revved up on his enthusiasm and readiness to give everyone a chance. He led them through the steps to design a wristwatch computer that monitored a navy diver's nitrogen pressures at varying depths to avoid the "bends." His classroom, in the words of Michael Swaine, editor at large for *Dr. Dobbs' Journal,* was probably the world's first academic microcomputer lab. But it was time to move on.

"He just loved teaching," said Dorothy. "It was a hard decision for him to quit school full-time." But Dorothy encouraged the decision they made in 1976 to start a full-time mail-order business they called Intergalactic Digital Research—"intergalactic" only because someone else held claim to "Digital Research" for a couple of years. It happened—coincidentally says Kildall—that at this moment new management at Intel ended his consultancy. There is a story that he offered the whole system he was designing for $20,000 and that they missed a golden opportunity; the fact was, says Kildall, that Intel simply wanted to build their own operating system, and ultimately, that was a "godsend" for him. But it is clear from his memoir he was also disenchanted with Intel. Recounting how marketing manager Jim Lally racked up the price of the Intel 8080 microprocessor with paper tape and floppy disk drive to $12,000, he writes: "I was dumbfounded. This was a direct attempt to block advancement in our society's technology for the profit of Intel. It was a good lesson for me. I protested. I wasn't even listened to or even considered. But Jim Lally is now a very successful venture capitalist."

Nevertheless, Kildall's ethics proved shrewd marketing. He initially proposed selling his system for $29.95 a disk, i.e., giving it away. At Dorothy's insistence, he asked $70—which was still absurdly cheap. She remembers going down to the post office in 1976 hoping to find checks that would keep the company alive a little while longer, but by 1978 it was a roaring success, leaving other proprietary systems in its wake. CP/M made the Intel operating system look like a scam; in addition to its being cheap, Kildall's system was small, it was fast, and it would run on all Intel computers and competing Zilog Z80s. "No other software product had been priced our way before," Kildall writes. "OK, CP/M's price came up to $100 per copy with version 1.4, but no one seemed to care." That denomination was in itself another Kildall invention: The first digit was a "major" revision and the decimal point indicated a minor revision for update. "You charge the manufacturers and customers a "minor" fee to get the minor revision and then issue a "major" revision, like CP/M 2.0 and charge a major fee. That became the way microcomputer software was labeled and for that purpose only."

In 1978, when sales were $100,000 a month with a 57 percent profit margin, Gary and Dorothy moved into a spacious converted Victorian house in Pacific Grove overlooking the waves of Monterey Bay, where Gary worked under the cupola and Dorothy ran the business office on the ground floor, Dorothy abandoning the name Kildall for her maiden name, McEwen, to avoid the aroma of a mom-and-pop operation. "It was a very exciting time, and we were just very naive about everything, about starting a business, about the industry," Dorothy recalls. "We were

young. The grown-ups hadn't come yet." They gradually recruited a young staff, students, professors and friends and installed the programmers out of sight on the second level of the house. The atmosphere was zany; as Kildall put it, a lot of marriages, a bunch of babies. People came to work barefoot, in shorts, and in hippie dresses; anyone in a suit was a visitor. One candidate for interview with the boss found herself talking technology with a red-bearded Roman emperor in a toga. Tom Rolander, visiting Kildall after three years working as an engineer at Intel, remarked that as a pilot he recognized the model airplane on Kildall's desk. Within minutes, Kildall bundled him into a sports car for a fast drive to the airport and a flight in the real plane, a Cherokee 180. Two days later Rolander was at work in Pacific Grove writing the multitasking version of CP/M.

Rolander was with Kildall through all the emerging triumphs and crises. Kildall writes: "Tom and I had a knack about how we worked together. I would create new stuff, write programs, and he would clean things and make them products. Sometimes the products were good, and sometimes they weren't. But that's how this world works. You don't get a home run every time." Rolander, the son of a preacher, was described by one associate as Tom the Cannon. "What he meant is that you aim Tom in a particular direction and light the fuse. Tom really doesn't care what direction it is; he only wants to work 80 hours a week on an interesting software problem." He is today still a lean, focused man exuding fitness. Visitors to his office inclined to pick up his bicycle in a corridor find it impossible to move. Rolander loads it with heavy bricks to make sure he gets a proper workout. He might equally be called T-for-Thoroughness Tom. "Tom learned and practiced calligraphy," writes Kildall. "During our friendship, he wrote [out] *The Prophet* in calligraphy for me. I know it took him many, many hours to do this." The two men flew together, jogged the Asilomar Beach and confided in each other. Writes Kildall: "At the time he was my copilot in flying and in life." On one scary night flight, Rolander saved both of them. Kildall, distrusting his instruments, mistook a string of lights crossing Lake Ponchartrain outside

New Orleans for the horizon. They were half a second from crashing when Rolander, looking out of the right window, yelled the alarm. "With the airplane now in a bank," writes Kildall, "I went back on the gauges. Righting that Aerostar to 'instrument level' may have been the hardest thing I've ever done in my lifetime."

Kildall was not a daredevil pilot. He was fully instrument rated. But on the ground, he relished risky fun. For his 39th birthday in 1981, he was given a pair of roller skates, "the kind that look like tennis shoes mounted on a Formula One car." When the party ran out of champagne, he sped downhill on them to get some more, stumbling over small acorns to everyone's merriment. He liked the skates so much he rolled around on them in the corridors of the office. Alan Cooper, who made an accounting system using CP/M on an IMSAI computer, says Kildall got frustrated only when the company didn't function like a college. "Employees would come to him expecting him to solve business problems, marketing problems, personnel problems. He didn't know the answers; didn't really want to think about the problems. What he wanted to do was write code."

There was nothing wild and woolly about that. Flying more than 1,000 hours on business trips with Kildall, Rolander came to appreciate Kildall's very methodical approach to flying, whether for a brief bit of acrobatics in his Pitts biplane or for a journey across the country in a twin-engined Aerostar. "While my own personality would have prompted more spontaneous departures, Gary's would always be done after detailed weather briefings, fuel loading, and weight and balance calculations.

"Gary's programming was just as methodical. It always began with complete and detailed sketches of data structures on large sheets of paper. The coding never started until he had visualized and comprehended the overall design. From the preflight to landing, Gary was a consummate professional in his flying, paying attention to every detail and never getting flustered. He was always calm, confident and equally demanding of detail from his copilot. He would have me rehearse my ATC transmissions over and over so that I would sound like a professional. After all, we were

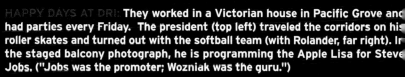

HAPPY DAYS AT DRI: **They worked in a Victorian house in Pacific Grove and had parties every Friday. The president (top left) traveled the corridors on his roller skates and turned out with the softball team (with Rolander, far right). In the staged balcony photograph, he is programming the Apple Lisa for Steve Jobs. ("Jobs was the promoter; Wozniak was the guru.")**

flying up at 25,000 feet, close to the big commercial jet traffic. Gary paid just as much attention to detail in his programming. Unlike other designers, who are often content to paint the broad picture and then let the more junior programmers fill in the details, Gary designed, implemented and debugged his products."

By 1980 Kildall had sold hundreds of thousands of copies of CP/M and had redesigned his system for the new hard drives. His was the standard operating system for most PCs. For the young couple, it was a heady time. Gerry Davis, who was then the Kildall attorney, remembers the bank calling to ask if DRI's profits were real. Davis said they were. "But they're making 85 percent profits. That's not possible." Davis assured the banker it was true. The Kildalls had a virtual monopoly. The natural question, then, is how Bill Gates got into the act.

Bill Gates was a 13-year-old hacker when Gary Kildall had already written his compiler and was pursuing his doctorate. Gates and Paul Allen famously simulated one of Ed Roberts's computers on the Harvard mainframe and installed on it a simple program invented at Dartmouth College by John Kemeny and Thomas Kurtz called "BASIC," meaning "Beginners' All-Purpose Symbolic Instruction Code." It was primitive, but it enabled hobbyists to write their own simple programs. Gates and Allen formed a company in 1975, originally called Microsoft, to sell this BASIC interpreter out of Albuquerque, not far from Roberts's factory, but two years into Microsoft, Gates wondered if Albuquerque was the right location for his little business.

Gates came to consult with Kildall, who drove him along the Central California coastline, and while commiserating about the speeding tickets they both routinely collected, they talked of merging their two companies. "We invited him to stay that night at our home. Dorothy fixed a nice roast chicken dinner," writes Kildall. But he adds, "For some reason I have always felt uneasy around Bill. I always kept my hand on my wallet, and the other on my program listings. I found his manner too abrasive and deterministic, although he mostly carried a smile through a discussion of any sort. Gates is more an opportunist

than a technical type. . . ." David Kaplan, the author of the engaging *The Silicon Boys,* says there seemed to be a gentleman's agreement that neither would get involved in the other's business. "DRI would stay away from languages, and Microsoft would leave operating systems alone."

Around this time, Kildall was sought out by Data General Corporation, located outside Boston, to write a whole new compiler for IBM's PL/I, "a dinosaur every bit as well done as Disney could have produced." He thought the project would take him nine months, but it ended up taking two years. It was by far the best compiler built for the Intel chip set, enabling a host of new applications, but it held him up making CP/M-86, a 16-bit version to run on Intel's 8086 chip—a delay that gave Bill Gates the opportunity of a lifetime.

Gates settled his enterprise near Seattle, Washington, of course. His breakthrough, in 1978, was Allen's design of a "Microsoft Softcard." This was an add-in board to the Wozniak-Jobs Apple IIe so that it would run CP/M and Microsoft Basic. The addition of CP/M gave Apple II users access to a large software base from the CP/M application suppliers. "I wanted a royalty," writes Kildall, "but Bill wanted a buyout and was stuck on that point. I sold him 10,000 copies for $2.50 each." Kildall adds with emphasis: "He signed agreements to protect the CP/M design under this license."

It was a wise precaution. Many people were pirating Kildall's design in the late '70s: Hundreds of "clones" had been made. Gerry Davis would issue warning letters, but Kildall found the most effective way to stop the rip-offs, instead of suing, was to drop in on the infringer and try a little shame. Roger Mellon bought an operating system from the Palo Alto Computer Store and was assured it was original. He was astounded when Kildall used the machine's built-in debugger to view Mellon's memory storage and embedded there was the message: "Copyright 1978, Digital Research." Mellon promptly signed up for a license. Kildall writes, "I put the copyright message in the object code for exactly that purpose, and you had to be a very sophisticated programmer to remove that message. Not only that, if it was removed, CP/M would not run because the operating

system checked to see if the message was there before starting, using an encryption scheme that worked quite well." (Kildall had learned the encryption techniques at the Naval Postgraduate School.) In the fall of 1979 Roger Billings was doing very well selling a computer system out of his company in Provo, Utah. Kildall and Rolander flew seven hours in single-engine Piper Archer, only to have Billings make them cool their heels in the waiting room. With nothing to do, Kildall played with a sample Billings computer in the waiting room. Using his debugger program, he quickly entered the innards of the computer operating system. There, again, was his copyright message. Kildall writes, "Roger became quite friendly all of a sudden."

Another participant in these little morality plays was Rod Brock, a neighbor of Bill Gates's in Redmond. Brock, who owned a small company called Seattle Computer Products (SCP), was impatient for the CP/M-86 Kildall was developing for the more powerful 8086 Intel chip. Brock's revenues were running down, so he hired Tim Paterson to fill the gap. Paterson did it by taking a ride on Kildall's system with a program he officially called "Seattle DOS," but which he also called QDOS, for Quick 'n' Dirty Operating System. Kildall writes: "Paterson's Seattle DOS was yet another one of the rip-offs of the CP/M design. The CP/M machine code was taken apart, using CP/M's own DDT [its debugger], to determine the internal workings of CP/M in order to make a clone of CP/M's operation." Paterson has denied using CP/M source code but admits making the two systems similar to help translate programs into QDOS. "Because of the completely different file-storage format, none of the internal workings has any corresponding relation to anything within CP/M," Paterson says. John Wharton, the former Intel engineer and computer specialist who became a friend of Kildall's, neatly sums up the ethics of that: "I can empathize somewhat with the bind in which SCP found itself: unable to sell its 8086 hardware for lack of software and unable to buy the software it wanted. But for Mr. Paterson to cite the unavailability of CP/M-86 as justification for appropriating the 'look and feel' of a competing operating system and its utilities seems to me analogous to telling a judge, 'I needed the car, Your Honor, and the plaintiff wouldn't sell me his, so I was forced to take it.'"

This would have all been a bagatelle, soon disposed of by lawsuit or shame, but for the curious behavior of IBM. Everybody in the computer world knew that Kildall had created CP/M—everybody, it seems, except the biggest beast in the mainframe jungle, in which personal computers had hitherto been almost invisible. In July 1980, IBM's top managers in Armonk, New York, set up a task force in Boca Raton, Florida, to report on the feasibility of mass-producing and mass-marketing a desktop computer. Philip "Don" Estridge was given just one year to get the secret project, code-named Project Chess, into the marketplace by buying components and an operating system with open architecture to facilitate add-ons—exactly as Kildall had designed. IBM chose an Intel processor. For the operating system they called not on Kildall and DRI in California, but Bill Gates in Seattle on the lazy assumption that he owned CP/M: Microsoft was then a tiny 40-person company selling a programming language that ran on CP/M. A whole IBM team of five headed by Jack Sams and Pat Harrington flew cross-country into Seattle on a Wednesday in August. Having ensured that Gates and his partner, Steve Ballmer, signed a tight confidentiality agreement, and a consulting agreement, they opened negotiations to buy a license for CP/M from Microsoft. Hello? Gates had to say it was not his system to license. According to Rolander, Gates phoned Kildall only to say that a "big client" was going to contact DRI and that Kildall should "treat them well." Then IBM phoned to schedule a meeting with DRI two days later.

This is where the myth begins. In his memoir, Kildall is quite specific (and Rolander confirms) that he arranged to meet the Project Chess team on a Friday afternoon. Knowing and explaining that he had a previously scheduled business trip on Friday morning (visiting an important CP/M distributor, Bill Godbout, at his factory in Oakland), he arranged an initial meeting between the visitors and Dorothy, who negotiated contracts; that very Friday morning, she signed an agreement with Hewlett-Packard.

The IBM team showed up as scheduled at 10 a.m. and the lawyer, well-known for his aggressive style, presented Dorothy with a ludicrously far-reaching nondisclosure agreement. According to Kildall, it stated, "All Ideas, Inventions, or other Information become the property of IBM." Anything IBM said would be confidential, whereas anything DRI said was not. Dorothy balked and gave the IBM team DRI's standard licensing agreement, which more than 1,000 manufacturers had already signed. There was a stalemate for a few hours. Dorothy would not sign IBM's broad agreement without knowing what IBM wanted. IBM would not reveal what it wanted until DRI signed. Dorothy sought the advice of Gerry Davis, who joined the meeting. He agreed with her that the undertaking asked for was too broad but thought it might be modified. He says, "Bill (Gates) signed that agreement because he had nothing to lose, because he didn't have an operating system."

Dorothy decided not to negotiate further until Kildall came back for the afternoon session. In the meantime, it appears, the IBM team fumed. There is an exponential arc to the revisionism that was to so amaze and dismay Kildall. In an interview with the *Times* of London in 1982, Gates said, "Gary was out flying when IBM came to meet, and that's why they didn't get the contract." *Accidental Empires,* a 1992 book by Robert Cringely and one of the seminal works on Silicon Valley, states that Kildall never bothered to show up at all and that IBM left DRI in annoyance without ever revealing what it wanted. Wrong, wrong, wrong, and this is a standard book on the origin of the PC. The Long Island newspaper *Newsday* wrote, "In a story often told, the starched-shirt IBM guys, after CP/M long hairs canceled an appointment, turned to an unknown company called Microsoft, headed by an unknown computer geek named Bill Gates." (On a smaller point of accuracy, Tom Rolander was quite bald by that time.) Maybe a source for the absent Kildall story is Jack Sams, who in 1992 told James Wallace and Jim Erikson that he was sure Kildall did not turn up for the meeting, "unless he was there pretending to be someone else." Alfred Chandler Jr., who does not doubt Kildall's

presence, writes in his 2001 book *Inventing the Electronic Century:* "Kildall was unwilling to sign the standard nondisclosure agreements on which IBM insisted. . . . If Kildall had been willing to accept the nondisclosure clause, and if Motorola's chip had been the first choice over Intel's commercially unpopular one, the underlying history of the personal computer during the critical decade of the 1980s might have remained much the same. But the industry's two most powerful players in the 1990s might not have been Intel and Microsoft." David Kaplan explains the credulity: "That's the Microsoft, and popular, version—and since winners tend to write history, it's the prevailing one."

In fact, when Kildall and Rolander arrived at Pacific Grove—on schedule—they met the IBM men in the early afternoon along with Dorothy and Davis. Once the nondisclosure agreement had been argued and agreed, with Kildall signing, IBM revealed its plans. Rolander demonstrated DRI's new MP/M-86, the brilliant new multitasking operating system that worked for Intel's 16-bit computers. (Gates long believed he persuaded IBM to adopt a 16-bit chip, but IBM had already decided on the 16-bit model according to Gates's biographers Stephen Manes and Paul Andrews.) In multitasking, Kildall was years ahead of anyone else. Rolander also discussed CP/M-86, the newest version of CP/M, which would be used to transition customers buying the new Intel chips. Ultimately, though, Kildall and Rolander wanted MP/M-86 to become the new standard. Kildall writes, "The new MP/M-86 was the operating system for the future, because it had built-in multitasking that supported the existing software base. It had built-in networking. Only today [1994] are we even considering these prospects. Clearly, the PC industry would be much more advanced if DRI had been allowed to introduce these products a decade ago."

Negotiations began over how much IBM would pay. According to Rolander, Kildall felt uncomfortable around the stiff, overdressed (by California standards) IBM men. They probably saw him as a hippie. DRI's earnings then were $5 million a year, mostly from CP/M. Kildall writes, "IBM offered to buy out CP/M-86 for its new PC for $250,000. You might be saying, 'Hey, Gary, sell the whole damn thing to IBM, then just wrap MP/M on top of that, say, hey?' That strategy may have worked, but our entire customer base wanted a smooth transition into 16-bit machines and we'd have lost them in a heartbeat. So I countered with an ongoing $10 per copy royalty for CP/M as was paid by all other manufacturers." Davis points out that DRI had contracts with "most favored nation" clauses, meaning that to sell CP/M to IBM for a flat fee might have caused DRI to be sued by its other customers. Kildall had to try to negotiate a contract similar to the ones signed by other customers.

IBM jibbed at this—it even insisted on renaming the Kildall system PC-DOS—but Kildall still believed they could strike a deal. As far as he knew, no one else had an operating system IBM could use. Kildall writes, "We broke from discussions, but nevertheless handshaking in general agreement on making a deal." Kildall and his family left that evening to vacation in the Caribbean. On the flight to Florida, they ran into the IBM team returning to Boca Raton. Kildall spent much of the flight discussing how he would adapt CP/M to IBM's needs. Dorothy describes the team as "friendly." "One of them kissed me on the cheek," she says. (Sams, the author of the "invisible Kildall" story, says she must have kissed one of his team going back to Boca Raton, but not him, because he went to Seattle, presumably to meet Gates again.)

The Kildalls were back in Monterey a week later. Kildall straightaway put in a call to the Boca Raton team—and another, and another. IBM had gone off the air; they'd gone back to Microsoft. Davis said DRI caught wind later that IBM was talking to Gates, but Kildall told him, "Bill's a friend of mine. He wouldn't cut my throat." But that is precisely what Bill did.

When IBM revisited Gates with news of the encounter with Kildall, Gates jumped in with the observation that Kildall had not yet finished designing CP/M for a 16-bit machine and that Microsoft could itself meet IBM's requirements. As soon as the IBM visitors went back to Boca Raton, Gates called Rod Brock and bought Tim Paterson's system for $75,000—an

initial $25,000 plus another $50,000 on closing the deal—without telling Brock why he wanted it so much (he later cited the nondisclosure agreement with IBM).

Gates was taking two gambles. The first was that Paterson's adaptation of Kildall's system risked a damaging legal suit: Gates never told IBM how close QDOS was to CP/M. The second was that IBM might pull out. They had done that before; back in 1974, IBM had made a $10,000 PC, the IBM 1500, which ran on Intel chips and failed to sell. "They seriously talked about canceling the project up until the last minute," says Gates, "and we had put so many of the company resources into this thing." But Gates was willing to bet everything. At the end of September, he and Ballmer flew to Boca Raton to present a proposal, much of it written by Kay Nishi, a Japanese employee, for using Paterson's version of Kildall's program—now renamed Microsoft DOS—and asking almost nothing in royalties. On the drive from the airport to the meeting, Gates panicked when he realized he had forgotten a tie. (They stopped at a department store on the way in.) Gates understood how to behave around IBM. His culture meshed with theirs far more than Kildall's did. Estridge, who died in a plane crash in 1985, told Gates over lunch that when IBM's new chief executive, John Opel, heard Microsoft might be involved with the PC, he enthused, "Oh, is that Mary Gates's boy's company?" Opel and Gates's mother served together on the board of the United Way. Gates signed an agreement with IBM in November 1980. He believes his mother's connection helped land him the contract for the new PC, which was code-named Acorn.

Kildall was relaxed about IBM's silence. He shared Silicon Valley's view of IBM as a dinosaur. "A lot of us in the microcomputer world in the early days," says Rolander, "saw IBM as all fluff and marketing, big, lumbering, slow, uninteresting, not clean, exciting, fast." In 1981 Kildall's CP/M ran on 90 percent of the roughly 500,000 or so Intel-chip-based personal computers in existence. (Apple was the exception, using chips made by MOS Technology and later Motorola.) Where else could IBM go? But about half a year later, Andy Johnson-Laird,

a friend of his who was a savvy consultant, showed Kildall a list of the PC-DOS API (application program interface) function calls for an IBM computer—the specifications for the software. These specifications had to be published so that programmers would know how to write new software for the upcoming IBM personal computer. Kildall was astonished to find how much of CP/M's proprietary list had been copied. He writes: "The first twenty-six function calls of the API in Gates's PC-DOS are identical to and taken directly from the CP/M proprietary documents [CP/M manuals]." He was irate, and with good cause. What Paterson essentially had done was rewrite the bottom part of the software—improving the way files were stored and adapting the program to a 16-bit machine—while copying most of the top part of Kildall's operating system (the Int 21 commands that allowed the operating system to interact with the application program). Even if QDOS and CP/M were 80 percent different, as Paterson has said, he took almost unaltered Kildall's Int-21 mechanism—the heart of his innovation. An independent examination of the two systems shows some blatant copies, some slight alterations. For instance, CP/M began each new line with:

A:

The DOS prompt was:

A>

Paterson copied Kildall's first 36 Int-21 functions into QDOS. He did rename Kildall's "Read Sequential" function "Sequential Read"; "write sequential" became "sequential write"; "Read Random" was called "Random Read." And so on.

In addition, CP/M's ED program was almost the same as PC-DOS's EDLIN editor program. Rolander says that "what Tim did was very clever. Ironically, an invention of Gary's was used against him." Gary's design was so good that he actually made it fairly simple for Paterson to "rip off" CP/M. Paterson in effect validated the significance of Gary's design. Applications were not tied to an implementation of an operating system (CP/M versus QDOS) or physical computer hardware (BIOS), but rather to a logical interface. Paterson's file system, Rolander acknowledges, was better for the larger disk, but he adds that mistakes were made in cloning Kildall's work. "QDOS didn't operate the way CP/M did just because of misunderstandings." John Wharton was an engineer at Intel when Gates visited Santa Clara to persuade Intel to abandon a joint development project with Kildall. "It was I who first informed Gates that the software he just bought was not, in fact, fully compatible with CP/M 2.2. At the time I had the distinct impression that, until then, he'd thought the entire operating system had been cloned." Wharton says that "the strong impression" he got of Microsoft programmers at the time was that they were "untrained, undisciplined, and content merely to replicate other people's ideas and that they did not seem to appreciate the importance of defining operating systems and user interfaces with an eye to the future. In the end, it was the latter vision, I feel, that set Gary Kildall so far apart from his peers." What Kildall saw, and what Paterson, Gates and IBM did not, is that CP/M-86 itself would soon be antiquated. The real problem was not that QDOS was similar to CP/M, but that it did not have the stable multitasking capabilities that Kildall was planning.

Kildall lost his traditional cool. This time he furiously got through to IBM. They immediately dispatched a manager and an attorney to Pacific Grove. "I showed the IBM attorney definitive evidence that PC-DOS was a clone of CP/M and immediately threatened a lawsuit for copyright infringement. The IBM attorney compared the API interface, and I can say clearly that he fairly blanched at the comparison and stated that he was not aware of the similarity. I told him that he should take note and become aware at the earliest opportunity, or else he should face a major lawsuit."

IBM knew it had to appease Kildall in some way since a lawsuit for "injunctive relief" could delay its entire secret venture, which was due to be launched in four months, in August 1981. They invited Kildall to fly to Boca Raton with Gerry Davis and there and then offered to market CP/M-86 alongside Microsoft's PC-DOS in similar packaging on the condition that Kildall would not sue IBM for infringement of CP/M copyrights. He accepted

that but did not agree to an undertaking to refrain from suing Microsoft. "We discussed pricing issues," writes Kildall, "but setting a price level, according to IBM, 'was a violation of antitrust.'" Pricing could not be set. Kildall thought he was getting exactly what he wanted. CP/M would not be changed to PC-DOS, and IBM accepted that it would pay DRI a standard royalty rate. Both PC-DOS, Microsoft's operating system, and CP/M would be released. Both operating systems would be sold in different boxes next to the new IBM. The marketplace would decide the victor. Gates was furious his old friend had been allowed back in the game, insisting that IBM had been "blackmailed into it."

IBM at least seemed to take DRI's involvement seriously. Kildall had never seen such strict security precautions. The prototype IBM loaned to DRI had to be chained to a desk in a locked room. No phone was allowed nearby. Any document printed out had to be shredded and burned. Several times, IBM technicians appeared on nearby roofs armed with special meters to detect if anyone was able to eavesdrop on the electromagnetic signals emitted by the new computer's keyboard.

In August 1981, IBM's PC finally came out. Rolander remembers driving with Kildall to the nearest store in the Bay Area, both of them brimming with excitement. They knew a knife had been plunged in their backs the moment they saw the labels on the software boxes: Microsoft's price advantage was a multiple of six. IBM asked $240 for CP/M-86 and only $40 for Microsoft's PC-DOS. Rolander says seeing the price difference was probably the biggest shock of his life. "It was just as if I were to reach across the table right now and give you a slap across the face, something completely off the wall. Looking at the price and knowing you had been completely screwed, that there was no intention whatsoever on their part to sell CP/M-86. No intention at all. There was such a trusting nature, especially in the academic world that was collegial. This was so big-business, aggressive, killer." He and Kildall felt so naive. They called IBM to demand the company reduce the price of CP/M, but no one called back. Davis says, "IBM clearly betrayed the impression they gave Gary and me."

Kildall writes: "The pricing difference set by IBM killed CP/M-86. I believe, to this day, the entire scenario was contrived by IBM to garner the existing standard at almost no cost. Fundamental to this conspiracy was the plan to obtain the waiver for their own PC-DOS produced by Microsoft." As psychiatrists like to say, even paranoids are persecuted. Kildall clearly was.

Gates continued the revisionism begun in the London *Times* in 1982. In an interview with *PC Magazine* in 1997, he said, "The IBM guys flew down there and they couldn't get the nondisclosure signed. Because IBM nondisclosures are pretty unreasonable. It's very one-sided. And we just went ahead and signed the thing. But they didn't. Subsequently, Digital Research woke up to the fact that this was a pretty important project and convinced IBM to also offer their product. But they priced it very high." There are, of course, two problems with these two sentences. The implication that DRI itself priced the retail product is untrue; and Kildall did sign agreements with IBM in both Pacific Grove and Boca Raton.

The obvious question is why Kildall did not sue Microsoft as he was free to do. Kildall flew to Seattle in August with his marketing vice president, John Katsaros, to confront Gates and Allen. He writes: "Allen was worried about a lawsuit and asked if DRI had ever sued anyone over copying CP/M. I said I hadn't. I was telling the truth. Paul is a gentle person, but he saw my chink and said that we were now engaged in OS-Wars."

The decision not to sue was a disastrous error. In the same year of 1981 venture capitalists invested in DRI—the attractive Jacqui Morby of TA Associates in Boston and the venture capital companies Hambrecht and Quist and Venrock Associates— and they helped DRI move into the big time with a bright new president, John Rowley, relieving Kildall of management. But the new board also dithered about suing as time ran out under the statute of limitations. At the time, no one had ever filed a lawsuit over the infringement of computer software copyright. The copyright law of 1976 was not amended until 1981, specifically to cover the look and feel of software. Gerry Davis himself won one of the first cases, putting a Bay Area infringer of CP/M out of business, but he had to advise the board and Kildall of the risks—the ignorance of judges at the time, the deep pockets of IBM. Was it a mistake to hold back? "Yeah," says Davis now, "what we should have done in retrospect was gone in and sued Microsoft very early on, even with the uncertainty of the law, because it would have stopped the development of a competitor. And if we had stopped them to begin with, they would never have gotten the foothold they have." Jacqui Morby agrees. "We could have won the first look and feel case and held up IBM." She recalls that the new board was not aware that during this period, while IBM and Gates kept very quiet, Microsoft's Steve Balmer was nonetheless continually on the telephone to DRI's project manager, Kathy Strutynski, asking for guidance on the internal engineering of the CPM operating system. "That was pure thievery." But aggression ran contrary to Kildall's character. Davis remembers him saying, "It's not nice to sue people, and we're going to succeed anyway." Everybody in the company was in denial for a couple of years, says Davis. "There was a lot of naïveté on the part of a lot of us, the board, me, and then the venture capital people."

The complacency at DRI is understandable. DRI seemed unstoppable. In 1981, CP/M was used worldwide in close to 200,000 installations with over 3,000 different hardware configurations. There were nearly 500 software products in the shops. The company doubled its space, moving from the Victorian house to offices on Central Avenue, and bought Gordon Eubanks's Compiler Systems, which produced CBASIC, a more commercial version of BASIC directly competitive with Microsoft. By the end of 1982, DRI employed more than 500 people and had operations in Europe and Asia. Revenues skyrocketed from $6 million in 1981 to $44.6 million in 1983. Everyone was confident because they knew that DRI's technology was superior, so it must surely prevail in the marketplace. Kildall and Rolander had moved beyond CP/M and MS-DOS, which was based on it, and they had a poor view of the IBM machine itself.

"That machine was a piece of crap," says Rolander, "compared to other machines. I would defy you to find anyone else who was around the industry 20 years ago who would have thought the IBM would be successful." When the IBM machine came out, Kildall and Rolander were working on an operating system that offered multitasking, multiprogramming and multiaccess. These are things computer users take for granted now, such as the ability to print a file while editing a spreadsheet, or cut and paste between spreadsheet and text. The IBM-Microsoft operating system, being single-tasking, yielded nothing like this. Only two months after the launch of the IBM machine, DRI began to ship MP/M-86, the multitasking 16-bit version of CP/M, and MP/M II, the 8-bit version. The key feature of MP/M-86 was that the

application program interface (API) was compatible with programs written for the eight-bit architecture, thus allowing programmers to easily adapt their applications.

But Kildall remained frustrated. The tensions reflected themselves in his personal life. He and Dorothy separated and then divorced after 18 good years together; she opened a lovely guest ranch in Carmel Valley. How utterly maddening it must have been. With Microsoft and IBM controlling the market, he could not push MP/M-86 with its multitasking capabilities. "I was competing with an operating system clone, MS-DOS, of my original design, and both operating systems were by this time completely out of date." Multitasking was thus delayed in America for a decade by the IBM-Microsoft hegemony. In Europe at least Kildall could push multitasking, which

TENSER DAYS AT DRI: John Rowley (left) came in as president, recruited by venture capitalists backing DRI, including Jacqui Morby, Venrock Associates, Hambrecht and Quist and Intel's Bob Noyce. "I call John a Boy Scout," writes Kildall (head of table). "He could talk so fast about products that your ears would flap."

is now of course the standard everywhere in the world. Digital Research had four European offices, two in England, not far from London, one in Paris and one in Munich. Foreign operations did better in many respects than DRI's American operations since IBM and Microsoft had much less market clout abroad. DRI's European operations kept the company afloat during the mid-80s. Paul Bailey, DRI's head of UK operations, beat out Microsoft for big accounts like Siemens and Nixdorf. DRI software was used to automate industry in Europe; Microsoft still could do only single-tasking. DRI's software allowed manufacturers to track multiple pieces of data.

While the salesmen fought the battle over operating systems, Kildall just could not stop inventing and innovating. Videodiscs were still new—they were the beginning of "multimedia"—and he and Rolander pushed the boundaries to fashion interactive videodisc hardware and software for the Commodore 64 computer. They labeled the system VidLink. Kildall astonished Grolier Publishing by storing Grolier's entire nine-million-word encyclopedia on a single videodisc. Grolier gave the go-ahead for Kildall to develop a commercial version of their *Academic American Encyclopedia* on videodisc. Ironically, the new management of DRI didn't take the

job, so Kildall and Rolander independently made the first encyclopedia videodisc in Kildall's garage. In 1984, Kildall set up a new company with Rolander called Activenture; the name later changed to KnowledgeSet. It was small, just like the early Digital Research, with Kildall and Rolander doing the engineering, and Kildall's new wife, Karen, doing the bookkeeping.

Kildall, ever prescient, set out in 1985 to build a CD-ROM version of the encyclopedia, called the *Grolier Electronic Encyclopedia.* Rolander remarks, "This was in June of '85. Here we are seventeen years later. At that point in time, we said, absolutely, every new computer will have a CD-ROM drive. You will not be able to buy a new computer without a CD-ROM drive. And it took at least ten years to get to the point where they were commonplace and twelve or thirteen before they were a standard device." Rolander's daughter, Kari, so astounded the teachers with her knowledge of Costa Rica from the CD-ROM searches she got an A+ on her paper.

Bill Gates, not realizing who KnowledgeSet really was, wrote the company a letter saying Microsoft might be interested in acquiring a CD-ROM firm. When he discovered that Kildall was the man behind it, he wrote him what Kildall describes as "a fine letter." It is not clear whether Kildall is paraphrasing, but he says it went like this: "Dear Gary, it has been a long time since we have been together. Next time you are in Seattle, maybe we can get together and go water-skiing, and talk about CD-ROMs." In the spring of 1985 Kildall visited Seattle to see his family and met Gates in a suite at the Olympic Four Seasons Hotel. The ever-generous Kildall writes that the meeting was pleasurable "and for some reason I opened up to Bill. I told him about the CD-ROM work that I was doing. We talked of standards. We talked for hours." Kildall mentioned his intention to hold a CD-ROM seminar at the Asilomar Conference Center, in Monterey, for publishers and was somewhat taken aback shortly afterward when Gates invited him to be the (unpaid) keynote speaker at a $1,000-a-head Microsoft CD-ROM conference. Only when he had given his speech did he hear from a Microsoft friend in the audience that Gates had come straight back to his office from the Four Seasons meeting to order a conference to preempt Kildall's. Kildall writes: "It was clever. It was divisive. It was manipulative. It is Bill Gates's nature. I must give him credit for being a very opportunistic person." Among other projects, KnowledgeSet worked closely with Sony to develop a Knowledge Retrieval System affording instantaneous searches of large databases. They made CD-ROMs containing the entire maintenance manual, complete with vector drawings for the Boeing 767.

By 1984 DRI was enabling PC users to link their computers through its Concurrent DOS program and Star Link software. You could buy one single IBM-compatible PC to serve as a hub to other PCs. Cables would connect all of the computers. Kildall writes, "The difference between multiple PCs and Concurrent with Star Link is that all workstations can share a common database. And that's what made it work. Take a typical VAR [value-added reseller] application such as a doctor's office. Patient records are stored at the hub and are made available to each authorized secretary for billing at a station or two. A nurse at the same time enters patient transactions, like drug dosage or the time spent with a doctor in the clinic. None of these tasks are difficult for even the most ancient IBM PC to perform. The issue was not speed, but simply sharing of common data in the office. If every workstation had a different PC, then the data for each patient could not be shared. And that, quite simply, just didn't work. But this need for common databases fostered PC networking."

Again, Kildall was a decade ahead with his Concurrent DOS.

By the middle of the decade, for all these innovations, for all the superior systems, for all the killer marketing efforts of John Rowley, DRI was losing its principal business against the muscle of IBM in alliance with Microsoft. The board fired Rowley and authorized Kildall to sell the company. Recognizing his responsibility to shareholders, he gritted his teeth and called Gates. Kildall flew his airplane to the San Francisco airport and met Gates in the United Red Carpet Room. "This is a very sticky situation," he writes. "Bill, although

once a good friend, had taken advantage of me at least twice. Bill appeared nearly on time at 2:00 in the afternoon. I learned what 'eating crow' means." No doubt fearing he might be taken advantage of again, Kildall gave Gates only public information and suggested $26 million would be a fair price. Gates replied that DRI was probably worth only $10 million. "We parted friends for some reason I don't understand today. However, this rejection by Bill was one of his big business mistakes."

Kildall made one deal with Atari's Jack Tramiel for its graphic display technology and another with Kay Nishi, the Japanese programmer and entrepreneur who had fallen out with Gates. Many people, like Nishi, wanted DRI to create an MS-DOS direct competitor. DRI was the only company that could legally parallel DOS, Kildall believed, because DRI had not forced suit; DOS was simply "a derived work of CP/M." Microsoft seemed vulnerable because it had not improved its operating system, had done nothing to support the new larger disk drives until Compaq moved in to do that and had failed to improve memory management for the larger applications programs (such as desktop publishing).

When DRI's first version of DR-DOS was released, Kildall must have loved the irony that the company he founded was now selling a clone of MS-DOS. The new single-tasking operating system was MS-DOS compatible and gave Microsoft a run for its money. On August 6, 1989, Bill Gates wrote in an e-mail to Steve Ballmer: "DOS being cloned has had a dramatic impact on our pricing for DOS. I wonder if we would have it around 30-40% higher if it wasn't cloned. I bet we would!" This was a loss of millions of dollars. Users started calling DRI's new operating system "Doctor Dos," not "Dee Are Dos," since it cured so many of the bugs found in MS-DOS. The August 1990 *Byte* magazine commented, "The latest incarnation of DR-DOS, Digital Research's MS-DOS clone, is an innovative and intriguing operating system that's thoughtfully designed. Version 5.0 is also packed with the extra features that Microsoft's own operating system should have (and might eventually have if the long-rumored MS-DOS 5.0 becomes a reality)."

Microsoft responded by announcing in May 1990 that within a few months it would issue a new release of MS-DOS, to be called MS-DOS 5.0, that would catch up on the DRI system. Industry experience indicates that it would have been near impossible for Microsoft to so soon develop and release a commercial version. Nonetheless, Microsoft repeated this vaporware announcement throughout the summer and into the fall of 1990. In fact, MS-DOS 5.0 was not released until June 1991, and when finally released, it did not offer the features Microsoft had promised.

On July 17, 1991, Ray Noorda, the founder of Novell, announced that his company was acquiring DRI—not for the $26 million Kildall had asked or the $10 million Gates had offered, but for $120 million. Using DR-DOS and its networking software, Novell became one of Microsoft's biggest rivals. Now Gates was up against a tougher opponent than Kildall. Noorda devoted himself to fighting Microsoft by acquiring a small start-up called Caldera, which employed the Linux system, and he used Caldera as a battering ram to sue Microsoft for monopolistic practices. He did not attempt to challenge the original cloning of CP/M, but focused exclusively on the "predatory" way Microsoft had cut DR-DOS sales by 91 percent. "This action," said Caldera's claim, "challenges illegal conduct by Microsoft calculated and intended to prevent and destroy competition in the computer software industry." Caldera alleged Microsoft would falsely announce new software that didn't exist, engage in exclusionary licensing, create false warning messages, criticize DR-DOS, use product tying, and threaten customers who used DR-DOS with retaliation. According to Judge Dee Benson, who oversaw the lawsuit, "On September 23, 1991, IBM officially endorsed DR-DOS 6.0, which was scheduled to be released to the public in September or October of the same year. Plaintiff alleges that in response to IBM's endorsement and in anticipation of an IBM/Novell alliance, Bill Gates publicly threatened retaliation against IBM should it choose DR-DOS. Caldera claims that as a result of the threatened retaliation and intense FUD [Fear, Uncertainty, Doubt campaign] concerning DR-DOS incompatibility with Windows, IBM withdrew its consideration of DR-DOS."

The lawsuit stretched on three and a half years. On January 10, 2000, just weeks before the case was to go to a jury, Caldera and Microsoft settled. The deal was secret, but Microsoft announced a one-time charge against earnings of three cents per share. Observers of the case quickly noted that since there were over five billion shares of Microsoft stock, that came to over $150 million. Many think the amount paid to Caldera was higher; the *Wall Street Journal* estimated it at $275 million, but some estimates go up to half a billion.

Kildall and his second wife, Karen, had moved to Austin, Texas, in 1991 after the sale to Novell. Again, Kildall was ahead of his time, provoked by technical conundrums encountered even by an undaunted computer wizard. His son, Scott, created a desktop publishing system using the Apple Macintosh, impressing Kildall enough to want to give it a try himself. He found setting up his own Macintosh "one of the worst (experiences) of my life, except for the day I visited Philadelphia." Then he wrestled with a Murata F-50 fax machine and found it "a switch-o-manic's nightmare," with 17 switches and such confusing instructions he ended up finding that his fax machine rang his personal phone day and night.

"OK," he writes, "so I am complaining about switches. How about proposing a solution to this stuff. I mean, plugging in a stereo these days seems to require a degree in electrical engineering. But there seems be something on the horizon that may help. It's called digital wireless." Kildall set up a company called Prometheus Light and Sound, working closely with Japanese company DDI, to exploit the fact that the $1 chips for cordless phones communicating at 32KB in a frequency range around 1.9 GHz, could also be used for stereos, VCRs, security systems, heating "and you name it, because for the local area you need no wires. . . . Buy a stereo at Macy's. Plug a unit into the wall and turn it on. No speaker connections. No CD player connections. No tuner connections. It just works. . . . It just works."

He predicted: "Switches, cables, wiring. We can't live with it in the future because of the complexity of the interconnections. Wireless will solve part of this. Some 'switch standards' will solve the rest." He might have made another fortune. But making money was never what drove him. He had a beautiful lakeside ranch in the West Lake Hills suburb of Austin, a mansion with a splendid sea view in Pebble Beach, California, and all his fast toys, but his second marriage was heading toward divorce. He got some satisfaction from charitable work for pediatric AIDS, but the continual anointment of Bill Gates as the founder of the PC revolution finally got to him. Rolander said, "The more the fortune and influence of Bill Gates grew, the more he became obsessed. Day and night, the film of that day played in his head. It wasn't a question of money. What really hurt him was the myth. Gary felt no one accorded any importance to what he had accomplished." Everywhere he went people would ask why he had "gone flying" the day IBM came. Cruelly, the University of Washington triggered an emotional decline. It invited Kildall—surely its most lustrous graduate—to attend the 25th anniversary celebration of UW's computer science program; he was mortified to hear that they had asked Bill Gates—"a generous donor"—to be the speaker that evening. When Kildall rang to question that, the chairman of the computer science department hung up on him. Kildall writes, "The UW Computer Science Department educated me so that I could produce compilers, like PL/M. Then, I made CP/M a success through millions of copies sold throughout the world, again using my knowledge gained through education at the UW. Gates takes my work and makes it his own through divisive measures, at best. He made his 'cash cow,' MS-DOS, from CP/M. So, Gates, representing wealth and being proud of the fact that he is a Harvard dropout, without requirement for an education, delivers a lecture at the twenty-fifth reunion of the computer science class. Well, it seems to me that he did have an education to get there. It happened to be mine, not his."

So Kildall ends his manuscript.

His health deteriorated. When he was afflicted with arrhythmia of the heart, his doctor banned him from flying. Kildall gave Rolander his pilot's helmet. It was a bittersweet moment. He had so loved flying. Now one of his last refuges was taken away from him.

During the summer of 1994, he returned to Monterey for a visit. Shortly before midnight on Friday July 8, 1994, he stumbled and hit his head inside the Franklin Street Bar and Grill in downtown Monterey. The place was packed, and he was found on the floor next to a video game. He went to the hospital twice over the weekend but was released. Doctors saw nothing wrong. Three days later, on July 11, he died of a cerebral hemorrhage. A blood clot had formed between his brain and skull.

He was 52. Three hundred people came to his memorial service at the Naval Postgraduate School. Bill Gates was not among them. Microsoft issued a statement that Kildall's passing was "a loss to the industry." Kildall's ashes were returned to Seattle to be buried not far from the lakefront where Gates was building his $60 million home.

Etched on Kildall's tombstone is a simple image: a floppy disk.

HOW STEVE JOBS PAVED THE WAY FOR BILL GATES

BILL GATES (1955–)   STEVE JOBS (1955–)

STEVE JOBS was not satisfied with having sold America on the personal computer in 1976; he has continued selling innovations for 30 years, rendering himself an icon in American culture and a torment to people working with him. "He would have made an excellent king of France," one Apple employee said. Jobs's contribution to the psychology of motivation is: "Nothing you say means anything to me. Why do you keep opening your mouth?"

On the other hand, the man with a low estimate of mankind has sublime taste. His real legacy may be a technological aesthetic. From the iPod to the iMac, from Pixar films to Apple IIe, everything he touches has style. He is not equipped to make an innovative contribution to technology in the sense of circuits and voltage, but his manufactures and his marketing represent an apogee of form and function. His ambition is to make "insanely great" products, and he flamboyantly succeeded in 2003 with his introduction of the iPod. The ability to download songs legally for 99 cents may be the biggest innovation in the way people listen to music since Sony introduced the Walkman. Jobs charmed music industry executives who had been terrified of online music ever since Napster started free downloading. He received the sincerest form of flattery as the heavyweights, Sony and Microsoft, copied his idea.

Steve Wozniak (1950–) designed the first Apple in Jobs's bedroom in Palo Alto and built it in Jobs's garage. It was a clunky wooden box priced at $666.66 (the mark of the beast, a "Woz" prank), and 175 were sold. Wozniak built an improved machine, the Apple II. Hewlett-Packard and Atari were not interested, so Jobs pitched his company to the venture capitalist Don Valentine, who asked the people who introduced them, "Why did you send me this renegade from the human race?" Valentine passed Jobs on to Armas Clifford Markkula Jr., an Intel engineer who had retired in his early 30s with

millions. Markkula invested and attracted others. Apple II was a runaway success. That lured IBM into computers, which made Bill Gates. Jobs's spellbinding salesmanship was not enough to offset Apple's high prices against Microsoft; Apple went from 20 percent of the computer market to 10 percent in two years. Jobs then failed with the Macintosh. It was wildly popular with students, but a financial disaster. In May 1985, after a showdown with Apple's president, John Sculley, Jobs was shorn of his power and quit the company in disgust.

During his years in the wilderness, Jobs invested in two new ventures that threatened to sap all his funds. He paid George Lucas $10 million for the Pixar film studios in 1986, and funded Pixar for the next decade. His persistence finally paid off when Pixar became a hit with animated movies like *Toy Story, Monsters Inc.* and *Finding Nemo.*

But his true ambition was to make a computer even better than the Apple. After quitting Apple, he founded NeXT, which made beautiful computers—a gleaming black cube on which the World Wide Web was invented—but again charged exorbitant prices. He might easily have lost all his money on the gamble, but in December 1996 Apple bought NeXT for $430 million and Jobs was back. One of his first steps was to swallow his pride and go to Seattle. Of course, he hadn't swallowed all of his pride. At Microsoft he said, "Bill, between us, we own one hundred percent of the desktop," acting as though they split the market down the middle when in fact Apple had about 3 percent of the market and Gates controlled the rest. Gates was stunned by his chutzpah. Still, his pitch worked. Gates agreed to invest $150 million in Apple. Jobs took to the stage to announce the news at a conference in Boston, while an enormous televised image of Gates appeared above on a giant screen, leading many to make Big Brother comparisons. People in the audience booed when Gates's smiling picture appeared, but Jobs told them to be grateful. Investors were. Apple stock rose 33 percent that day.

THE ODD COUPLE: Jobs and chairman Bill in November 1991

If Bill Gates had a dime for every time a Windows machine crashed . . . oh, wait a minute, he already does.

—JOKE GOING AROUND THE INTERNET

BILL GATES IS SO RICH he could buy Nigeria. He is so rich he could buy every major sports team and still be the richest man in America. He is so rich that all of his money laid end to end in dollar bills would stretch from here to the moon six times. He is so rich it wouldn't even be worth his while to spend three seconds picking up a $100 bill (he averages about $50 a second).

But is the richest man an anti-innovator? He has never invented any important, original software; hundreds of small innovative companies have died in Microsoft's bear hug; and Microsoft has never been first to market with any major innovation. The BASIC programming language Gates and Paul Allen first adopted for the personal computer in 1974 was invented ten years before by two Dartmouth professors. Microsoft developed Microsoft Word to compete with

WordPerfect; Microsoft Excel to compete with Lotus; Windows to compete with the Macintosh graphical user interface; DOS 6.0 to compete with Star Electronics' disk-compression software; Microsoft Net (1984) and LAN-man (1987) to compete with Novell's networking software; WinCE to compete with PalmPilot; Internet Explorer to compete with Netscape; MSN to compete with AOL; and the Xbox to compete with Sega's and Sony's home game systems. Hotmail, Microsoft's free e-mail Web site, was invented by Sabeer Bhatia, an Indian programmer, and the company plans to bundle a search engine with its new operating system in 2006 to compete with Google.

People magazine once dubbed him the "Edison" of software. He keeps a photo of Edison in his office, next to Einstein's and da Vinci's, but he is more like John D. Rockefeller. Microsoft's Windows operating system controls 90 percent of all personal computers; Standard Oil controlled 90 percent of the nation's oil refining capacity by 1879. Gates's originality was in being the first person to realize the importance of controlling the computer industry's standard operating system. In capitalizing on the perception, he was an imitator, not an inventor. DOS was based on Tim Paterson's QDOS, which derives from Gary Kildall's CP/M. Gates understood better than IBM the importance of owning its PC operating system. Within a decade Microsoft had overtaken IBM. From the beginning, Gates told *InfoWorld* magazine, "We were hoping a lot of other people would come along and do compatible machines." He figured out how to set up a tollbooth to computer technology, collecting half of every dollar generated by the PC industry.

What Gates lacks in invention, he makes up for in determination. No one works harder. He is so naturally competitive he buys jigsaw puzzles in sets of two, to race his wife to the finish. Gates routinely works 16-hour days. *BusinessWeek* visited Microsoft's offices and noted the piles of empty espresso cups and sleeping bags hanging from the backs of doors. An e-mail once warned employees, "If you find yourself relaxed and with your mind wandering, you are probably having a detrimental effect on the stock price."

Gates is a devastatingly effective businessman. His first venture capital backer, David Marquand, wrote after meeting him in 1980, "This guy knows more about his competitors' products than his competitors do." What's more remarkable than the fact that he was blindsided by the Internet is how quickly he adapted to it. "Chairman Bill Gates turned the software behemoth on a dime," wrote *Newsweek*. In three months the company created a 2,500-person Internet division and integrated its new Internet Explorer into the company's operating system. If at first Microsoft doesn't succeed—and it usually doesn't—it keeps throwing resources at the problem until eventually it solves it. "They're like the Chinese army," says Eric Schmidt, now Google's CEO. "They send wave after wave of soldiers at you, all of them expendable."

Microsoft has repeatedly used its monopoly to crush competitors. It threatened to cancel Compaq's Windows license if it installed Netscape Navigator on its computers. It bundled Internet Explorer on its operating system for free to undercut Netscape sales. It coded hidden bugs into Windows to cause rivals' software to crash. ("DOS ain't done 'til Lotus won't run" was one unofficial motto.) Two federal antitrust suits, and billions paid to settle lawsuits, have done little to slow it. In 1999, Judge Thomas Penfield Jackson found Microsoft to be a monopoly harmful to the American consumer. The incoming Bush Department of Justice let Microsoft off the hook, though the proposed split of the company would have been good news for shareholders. The European Union jumped in with a fine of $631 million, pocket change at Microsoft.

The man who has sold more CDs than Michael Jackson is about as highly esteemed. He is "evil," says Netscape's co-founder Jim Clark. "When you deal with Gates you feel raped," said Philippe Kahn, former head of the software-maker Borland. There are web pages that depict him as a Nazi, as Satan. Gates resents these taunts. After learning of the government's second legal action, he burst into tears, says author Gary Rivlin, asking, "Why does everybody hate me?" When Microsoft bought the rights to the *Funk & Wagnalls Encyclopedia* to create *Encarta*, it altered its leader's entry. No

longer was Gates "a tough competitor who seems to value winning in a competitive environment over money." Now he is "known for his personal and corporate contributions to charity and educational organizations." It's true. In 1998 alone Gates and his wife gave $100 million for a vaccination program for Third World children.

At Microsoft Gates inspires immense loyalty. *Fortune* magazine calls it the "Cult of Bill." Employees start walking, talking and dressing like him. His habit of spontaneously jumping up and down (it helps him think) led scores of early employees to buy trampolines. (*Time* magazine speculates he is autistic, others that he has Asperger's syndrome.) Company meetings have filled Seattle stadiums with thousands of employees thundering in unison, "Win-dows! Win-dows! Win-dows!"

Gates's colleagues think he is a genius, and are amused by his eccentricities. He had a habit of forgetting thousands of dollars, wallets, suits and vital contracts in hotel rooms. He can go to sleep anywhere, anytime, under his desk or under chairs in airports. His handlers sometimes lose him in such places. Gates has no small talk. Outside of conversations about his daughter, he is all business and technology. Walter Isaacson writes, "Broad discussions bore him. He shows little curiosity about other people, and he becomes disengaged when people use small talk to try to establish a personal rapport."

Gates so often orders cheeseburgers, fries and chocolate shakes from the local BurgerMaster that his secretary has the restaurant on her speed dialer. He also loves McDonald's. His friend Warren Buffett, a major McDonald's shareholder, sends him discount coupons—which Gates uses. He told *Playboy*: "In terms of fast food and deep understanding of the culture of fast food, I'm your man." And perhaps that is the key to understanding Gates's unfathomable success. He has created the software equivalent of fast food: standardized, ubiquitous and not always satisfying. He was only 20 when he coined his company's motto: "A computer on every desk and in every home, running Microsoft software." Bill Gates saw the future, and it was Microsoft.

HERBERT BOYER and ROBERT SWANSON
Over a casual beer in San Francisco, they put down $500 each to create the biotech industry

1936–	1947–1999

Marching through the streets of Berkeley, California, shouting anti-war and civil rights slogans of the sixties, Herbert Boyer did not look the sort who would be a caring father, given his tumbled mop of hair, his flamboyant moustache, his open-necked shirt, his faded jeans and his unbuttoned leather vest. But he was. He fretted that his first son might be on the small side for his age, so he took him to a pediatrician and accepted the doctor's suggestion that the boy should be hospitalized for 24 hours to have the levels of his growth hormones measured. Boyer and his wife, Marigrace, were relieved when the pediatrician reported that the tests suggested their son would grow to an acceptable size. "Then he told me," Boyer recalls, "how difficult it was to get human growth hormone to treat dwarfism. I didn't even know that kids were treated with human growth hormone."

Why should he have known? Boyer was by his own account a "small-town boy" who grew up in Derry, Pennsylvania, a coal mining community of 3,000 near Pittsburgh. His father worked as a miner and railway conductor—"about all you could do in western Pennsylvania in those days"—and Boyer went to Derry's small public high school, where one teacher taught four or five subjects. By the time he visited the pediatrician 20 years later, however, Boyer was not only a '60s rebel but a considerable scientist and one concerned, moreover, with molecular biology and the workings of compounds like hormones. He made up for his surprising lack of awareness about growth hormone therapy that day with the pediatrician. A few years later, in a brilliant piece of original genetic engineering, he and his associates grew a synthetic human growth hormone. Right afterward, he and his business partner, Robert Swanson,

scaled up to manufacture the hormone in such abundance, through their new company, Genentech, that treatments were no longer limited by the extreme scarcity of hormone drawn from cadavers. In 1985 the Food and Drug Administration (FDA) authorized Genentech—not an established pharmaceutical company—to manufacture and market the hormone. America's biotech industry, less than ten years old, was truly established as the latest and most dramatic harnessing of microbes. By then, another Genentech creation, human insulin, was already being manufactured under license by the pharmaceutical giant Eli Lilly. That insulin—the fruit of an intense race with Harvard scientists—was the first genetically engineered drug ever approved by the FDA.

Something positive clearly happened to Boyer in Derry. He was an eager guard for the Derry Borough High School football team (and, of course, a fan of the Steelers), coached by the high school teacher who taught chemistry, physics, math and biology. The perceptive coach convinced Boyer that although he was good at football, he should develop his talent for science. Francis Crick and James Watson helped, too. In 1953, as Boyer was about to attend St. Vincent's, a small Benedictine liberal arts college in nearby Latrobe, the famous Anglo-American duo shook the world with their double-helical model of the structure of deoxyribonucleic acid, known now to everyone as DNA, the substance of inheritance and metaphor: the genetic blueprint of life, the rope ladder hidden in every creature in the universe, the master molecule of existence, etc.

The reverberations of the discovery, offering a new, bewildering landscape for exploration, followed Boyer through his years studying bacterial genetics at the Uni-

versity of Pittsburgh, and for three more years at Yale (with time out for the civil rights movement). He was hooked. He named his Siamese cats Watson and Crick, and when he arrived at the University of California–San Francisco in 1966 as a 30-year-old assistant professor—with Marigrace, who was also a biologist—he really considered himself a molecular geneticist, a relatively new profession.

Boyer was one of the first scientists to recognize the commercial possibilities of genetic engineering. It was not a crowded field, despite the razzle-dazzle uproar that accompanied the 1968 publication of Watson's bestseller *The Double Helix,* "the most indiscreet memoir in the history of science" in the words of Watson biographer Victor McElheny. Boyer felt undersupported by the university. He had to go to another lab to purify enzymes. He had no centrifuge for a long time, not even an ice machine. Every day began with him carrying buckets of ice needed to preserve his fragile specimens. Things improved in 1968 when another scientist named Mike Bishop, also from Pennsylvania, joined the Department of Microbiology and gave Boyer the sort of intellectual camaraderie he sought. For a while Bishop and Boyer were the only two scientists working late into the night. "There weren't many others that had the same sort of crazy work habits we had," Boyer says in the oral history conducted by Sally Smith Hughes at the Bancroft Library. In 1969, the transformation at UCSF gathered pace when Bishop recruited a postdoctoral fellow called Harold Varmus (with whom he would share a Nobel Prize in 1989 for their discovery that normal cells contain genes capable of becoming cancer genes), and the university appointed William Rutter to fill the long vacant position of chairman of the

BOYER AND FRIENDS: Boyer converted the bacteria in his vial into a factory for mankind. A reluctant pause for photography in his lab at the University of Californa in December 1981.

Department of Biochemistry and Biophysics. Rutter saw the point of Boyer's work, and before very long, with more resources and researchers, UCSF would be the world's foremost center of DNA research.

Boyer had become fascinated by what might be done with the lab strains of the classic workhorse bacterium *Escherichia coli* (*E. Coli*). *E. coli* is a version of a microscopic single-celled bug that is less than one ten-thousandth of an inch long. If certain strains of the bug multiply too much, sickness and death can result. The up side is that overnight in a lab petri dish, *E. coli*'s facility to multiply explosively can yield quick answers to a scientific question posed only the previous afternoon.

The bacterium is a piece of protoplasm with a single chromosome holding something like 4,000 genes. Not many of these *E. coli* genes had been identified when Boyer began his work, but it was easier to get access to them than to the genes of eukaryotic cells, which are found in everything from flowers and mushrooms to amoebas and human beings. In a eukaryotic cell, the DNA is in a central nucleus coiled into the famous double helix: McElheny calls these "library stacks" of coded information, an appropriate image since the uncoiled DNA of a single human cell would stretch about six feet and there are hundreds of thousands of them. Instructions to make proteins—photocopies as it were—are carried by messenger RNA from the library stacks to the relatively distant manufacturing suburbs of the cell, the assembly lines of globular ribosomes, chains of different amino acids, usually hundreds of links long, that knit together to build more proteins. (Before all this was demonstrated, Watson's desk lamp at Harvard bore a scrap of lined paper with the slogan DNA MAKES RNA MAKES PROTEIN.) In the much simpler cells of bacteria like *E. coli*—so-called prokaryotes that lack a distinct nucleus—DNA rings float more accessibly in the cell cytoplasm.

Higher organisms, which do have cells with nuclei, or eukaryotes, pass on new helpful or unhelpful genes through sexual mating. Bacteria reproduce simply by dividing at a phenomenal rate, but they are not without the ability to transmit a gene—for instance, one that makes them resistant to a bug killer. One vehicle for this gene transfer is the plasmid, a tiny closed loop with a few genes floating in the bacterium that replicate independently of the bacterium's chromosomal DNA. Bacteria brushing against each other, or conjugating as the biologists say, may pick up a gene-bearing plasmid; this is the way previously vulnerable bacteria acquire immunity. Penicillin is not able to kill a bacterium that has had an affair with the *Staphylococcus aureus* plasmid that directs the production of an enzyme (penicillinase) to break down the attacking penicillin molecules. This copying facility is bad news for human hosts. The good news is that the defense mechanism of a bacterium, the drawbridge as it were, also offers access for invaders, for genetic engineers.

This was where the potential lay when Boyer immersed himself in molecular biology, encouraged by Marigrace. If the *E. coli* plasmids were so ready to receive and transmit new genetic information from their peers, perhaps they could be induced to combine with a gene from a higher organism. Such a recombination of DNA would offer amazing prospects of exploiting the normally malevolent capacity of a bacterium to make millions of copies of itself in a day. The hybrid plasmid—infiltrated with the appropriate gene—could in theory convert the bacterium into a pharmaceutical factory. But how could a plasmid be removed from bacteria so that it would be accessible to manipulation; how could its DNA be identified; how could one test the effect of inserting new DNA; would the host's ribosomes accept instructions from a higher organism infiltrator; and would the bacterium go on replicating with its new gene?

At Yale, Boyer had read an important paper published in 1962 in which the authors hypothesized that a certain enzyme—a chemical catalyst—had the ability to recognize and slice apart specific sequences of DNA. Boyer was inspired to study such enzymes, called restriction endonucleases, or restriction enzymes. One can think of endonucleases as molecular scissors, proteins able to cut up a DNA molecule into pieces. Boyer was specifically interested in enzymes that cut specific

parts of the DNA molecule. He says in the oral history, "There was developing at the time an identification of and knowledge about different types of nucleases in bacteria. But all the ones that had been described at that time were pretty much random in their cutting of DNA. They would just break down the DNA to small pieces, and it involved no specificity other than they recognized DNA per se. There was no recognition of unique sequences of DNA."

Boyer plunged into studying and purifying these novel enzymes. He identified one—from *E. coli* again. The *Eco*R1 was discriminating in its assault. It cut the DNA only when it found the nucleic acid sequence GAATTC. This was a significant step. We can think of genes as three-letter words made up of three of the four nucleic acid letters: A (adenine), T (thymine), G (guanine) and C (cytosine). DNA typically lines up in double strands, with A bonded to T and G to C. These four letters give us 64 possible three-letter words for the 20 basic amino acids that make up all living creatures. In DNAese CAT (cytosine-adenine-thymine) codes for the amino acid histidine. But though the vocabulary of life seems limited, these words can build up incredibly complex sentences, paragraphs and texts. If a single amino acid is altered within a long chain, the shape of the resulting protein completely changes. One mistaken letter in a gene can lead to all sorts of genetic diseases, even death. A single T in place of an A will change one of the 574 amino acids that compose hemoglobin and will thus cause sickle-cell anemia. A simple letter out of sync can have profound consequences. Moving just one letter at the end of each word can have a stupefying effect. Pity the poor protein that has to read this: Movin gjus ton elette ra tth en do feac hwor dca nhav ea stupefyin geffect.

By the summer of 1972, Boyer's next hope of advance depended on testing his perceptive enzyme on a receptive plasmid. As intriguing as Boyer and a few others found the possibility of cloning, he was working in the exurbs of biological science. He had little success trying to splice a plasmid called *lambda DV Ga*l with a gene coding for beta-galactosidase, an enzyme that helps bacteria break down milk sugar, or

lactose. Then he was introduced to Stanley Cohen by a colleague, Stan Falkow. A gentle, reticent intellectual, Cohen had considered becoming a rabbi before settling for medicine at the University of Pennsylvania. He had entered Stanford in 1968 as an assistant professor in the Department of Medicine, but he was spending more time on molecular research and had found a way to remove plasmids from bacteria and package them in test tubes. Cohen and Falkow urged Boyer to join them in Hawaii for a scientific summit on plasmids. These were of great interest to Japanese scientists because the overprescription of antibiotics by Japanese doctors had bred resistant strains of bacteria.

In Hawaii in November 1972, Cohen and Falkow described their work on a small plasmid called pSC (Cohen's initials) 101, which carried resistance to the antibiotic tetracycline. Boyer asked Falkow if he would collaborate on dissecting bacterial plasmids using Boyer's DNA-splicing enzyme. Falkow was not too interested but suggested Cohen might cooperate. There then took place one of the most productive scientific confabs of the 20th century. Over corned beef sandwiches late at night at a delicatessen across from Waikiki Beach, the ebullient Boyer and the quiet Cohen agreed to collaborate. Cohen's pSC101 was attractive to Boyer not only because it was accessible in a test tube, but also because its identifiable antibiotic gene presented the possibility of testing the efficacy of a genetic transfer: Any bacteria that picked up the gene would be immune to antibiotics. It would be a yes/no experiment. Boyer's enzyme was attractive to Cohen because it not only recognized the nucleic acid sequence GAATT, but it also left "sticky ends"—single strands of DNA not glued to each other. Instead of cleaving straight through both strands of the double helix, Boyer's *Eco*R1 sliced through the first strand and then left several molecules—nucleic acid letters—of the second strand dangling before cutting that too. With mortised sticky ends it would be much easier to reinsert this spliced fragment of DNA into other DNA molecules. However, mixing the right elements at the right temperature would be a tricky procedure. Enzymes, plasmids and DNA commingled in a volume of fluid about "the size of a human teardrop" (to adopt the description by science journalist Stephen Hall). A 1 percent success in slipping a plasmid into bacteria was a triumph then— acceptable because one cell would reproduce itself millions of times.

Boyer returned from Hawaii and visited a scientist at Cold Spring Harbor Laboratory (CSHL) on Long Island, which James Watson, having left Harvard, was improbably transforming into a major center of scientific research. At CSHL, Boyer learned a new, faster technique for studying DNA called gel electrophoresis, developed by Daniel Nathans at Johns Hopkins and refined by Joseph Sambrook and others at CSHL. Boyer watched his friend stain DNA fragments with a dye, then put them in thick acrylamide gel and force them to move with electric current. Different-sized DNA pieces have different mobilities in the gel and travel different distances, so the strands of DNA sorted themselves out by letter, A, T, C or G. In other words, he could quickly read sequences of DNA.

Boyer says in the oral history, "It was such a breakthrough in terms of how fast we could move. If we had to do this research with the techniques prior to that, it would have just taken forever. When I saw those gels back in Cold Spring Harbor, I was just so excited I couldn't wait to get home, get back to the laboratory."

One of Cohen's lab assistants who lived in San Francisco shuttled material between Cohen at Stanford and Boyer up in San Francisco. They chopped out the DNA inimical to tetracycline and added it to test tubes with nonresistant bacteria. The bacteria divided roughly every 20 minutes; by the end of a day the postdocs working with Boyer found they had many colonies of antibiotic-resistant bacteria. To test the process, they reversed it, this time asking the R1 enzyme to cut out the antibiotic-resistant genes. It did. Gel electrophoresis showed the bacteria restored to their original state: Boyer and Cohen had cloned a gene. The findings were published in the November 1973 *Proceedings of the National Academy of Sciences* (*PNAS*).

So far Cohen and Boyer had been working only with bacterial DNA. Now they wanted to experiment with more sophisti-

cated eukaryotic cells, the basic building blocks of higher life-forms. If eukaryotic DNA could be genetically engineered, it would mean that human DNA could eventually be used and all sorts of human proteins could be produced using this process. In the Bancroft oral history, Boyer says, "I think we were thinking about it at the time. Not in any grandiose ways, but sitting around the laboratory talking about 'Well now, if you can clone eukaryotic DNA, you can clone the gene for human insulin or human growth hormone or whatever you can think of, and should be able to make it in bacteria.' So I had these seeds of thought, fantasy more than anything."

The opportunity arrived when Boyer heard about the African clawed toad. John Morrow, a graduate student who had been working with geneticist Paul Berg at Stanford, told Boyer: "I have some amplified ribosomal DNA from *Xenopus laevis.*" (Boyer knew right away what he meant.)

Boyer and Cohen, with Morrow collaborating, took some of the toad's DNA, cut it into several pieces and recombined it with the DNA of an *E. coli* bacterial plasmid. They believed they had now combined eukaryotic and prokaryotic DNA—animals and bacteria had just been merged. But to prove they had really succeeded, the team then had to remove the DNA again. Boyer asks in the oral history, "The big question was How could we recover it?" The next steps were complicated. From the genetically engineered bacteria he cleaved out the eukaryotic toad DNA. Then he cut out the gene that made the bacteria tetracycline resistant. Next he mixed the two genes together in a test tube, using a polynucleotide ligase (from the Latin *ligare*, to bind), a needle-and-thread enzyme able to stitch two DNA strands together even if they have blunt ends. Once the two genes had been glued together, Boyer and his colleagues returned them to *E. coli* bacteria. The new bacteria had two new genes. One made them tetracycline resistant. The other was the gene from a toad. Since the genes were joined, those bacteria that resisted being killed by tetracycline in theory also carried the toad gene. Now Boyer removed the plasmids from these bacteria. He cut up the plasmid DNA with his enzyme scissors and compared the frag-

ments with the original toad DNA he had used. They were identical. They had added and removed eukaryotic DNA. And they found that 10 to 30 percent of the bacteria they studied carried the new genes.

The team did this with several different genes from eukaryotic cells, including rat DNA. They found that the genetically modified bacteria could now produce the DNA of the toad's ribosomal RNA gene. Boyer recalled: "So that was a little bit of icing on the cake that confirmed that we had cloned eukaryotic ribosomal DNA. It was a delicious moment. I can remember tears coming into my eyes, it was so nice."

Boyer and his team confirmed that the DNA replicated faithfully after several generations, a discovery with profound implications for biopharmaceuticals: Genetic engineering with human DNA could be contemplated. A front-page report by Victor McElheny, in the *New York Times* on May 20, 1974, focused on the practical therapeutic applications. The *Times* report was the trigger for Stanford University to begin a long, contentious and ethically complicated process of applying for a patent for the Cohen-Boyer method of cloning. A euphoric Boyer called Falkow, the man who had started it all by introducing Boyer and Cohen. Falkow was curious to know how Boyer and his team had been able to identify the small percentage of bacteria— about one in a million—that had the toad DNA. Falkow reported, "Herb just said he kissed every colony on the plate until one turned into a prince."

SWANSON'S COLD CALL

The prince in the practical world was Robert Swanson, who would effect a marvelous marriage of science and commerce before he died at the age of 52.

Swanson was born in Brooklyn, New York, but when he was an infant, his father, a crew leader of electrical maintenance at Eastern Airlines, moved to the airline's headquarters in Florida. His mother and father were keen that their only child should be the first in the family to graduate from college: He was allowed only one hour of television a week, something by Disney or *Wild Kingdom.*

He started at MIT in 1965, when Boyer was at Yale. He was a sociable and generous young man and fitted in well with the Sigma Chi fraternity; he credits his frat brothers with helping him pass his exams. "I learned not to be afraid of science or very complex problems. Probably the two most important things that came out of those early years were tackle things one at a time and manage your time." He ended his fourth year with all A's and a prize in chemistry. He stayed a fifth year to get a dual degree in chemistry and a master's degree at the Alfred P. Sloan School of Management. "The one area that was most interesting to me was how do you build teams of people to achieve things, climb mountains, whatever it is that needs to be done." His favorite course, on entrepreneurship, about the only one in the country then, was led by Dick Morse, who had helped start Minute Maid orange juice. Morse put him in touch with Phil Smith, the Citibank officer in charge of starting the bank's new venture capital group, who gave him $100 million to invest under guidance. "It was a great overview of others' mistakes and successes," Swanson recalled. "We actually made quite a bit of money for the bank." In 1973, he was chosen as one of the two people to set up Citibank's West Coast venture capital office, but he left Citibank after two years. His heart was in building something himself, and when Citibank fired 200 vice presidents in one day, he reckoned that entrepreneurship could hardly be riskier.

He was taken on in San Francisco by Eugene Kleiner of Kleiner Perkins, one of the most notable venture capital companies. (Thomas Perkins had been a student of Georges Doriot, see page 302.) At the end of the year, Kleiner and Perkins told him they'd like to confine the company to just the two of them, but he could have a desk and phone while he decided his next move. "This was a pretty scary period," said Swanson. "I looked at everything from joining Intel to a Stanford professor who had a way of concentrating radioactive waste."

He spent time in local university libraries reading technical monographs on the infant science of bioengineering. It was in the news because the Cohen-Boyer experiments had accelerated a concern among some scientists, spreading to the public, that hybrid molecules might run amok in the community: Michael

MAN IN SUIT: There were giggles when Robert Swanson turned up at Boyer's lab where ties and jackets were for graduation only, but he won respect for his business acumen. The blackboard drawings summarize how *E. coli* was subverted by its own plasmid to make new proteins.

Crichton's bestselling novel of 1969, *The Andromeda Strain*, was about viruses from outer space, but it haunted the popular imagination whenever genetic engineering was mentioned. Even James Watson urged a temporary end to experiments—he changed his stance later—and the initial call for the National Institutes of Health to intervene was from Stanford's Paul Berg, who had been the first to splice together two kinds of virus. A voluntary moratorium on recombination ensued. In February 1975, Berg and others organized a three-day international conference at Asilomar, on California's Monterey peninsula, to discuss biosafety guidelines so that the

moratorium might be lifted. Tempers ran high. To Boyer, the meeting was "a nightmare." "I thought it so emotionally charged that it was counterproductive," he says in the oral history. "There were a lot of accusations and shouting from the floor. I found it to be an absolutely disgusting week." James Watson, who had reversed himself and now objected to restraints on the research, called Asilomar "the worst week of my life." It was agreed that the moratorium would be succeeded by guidelines from the National Institutes of Health, but the debate was far from over when Swanson decided this was the area for him.

The first real biotechnology company, though not the first to do genetic engineering, was Cetus, founded by Don Glaser, the Nobel Prize–winning physicist who invented the bubble chamber to track subatomic particles. After winning the prize he became interested in microbial genetics and invented a machine that would scan colonies in petri dishes for unusual events, such as mutations or the production of antibiotics. He received some NIH funding and then used the machine to start Cetus, which contracted to pharmaceutical companies who used the machine to hunt for new antibiotics. Swanson went along to suggest they should pursue recombinant DNA technology. "No, we're not going to hire you," Glaser's associate Ron Cape told him. "We think this technology is going to be wonderful, but it's not going to happen for a long time. It wouldn't be fair to hire you and wait for it to happen." Swanson got the same answer from Syntex. "It's great, but it's not going to happen for a while."

Swanson, 28 and frustrated, was not going to wait. He started cold-calling names on scientific papers. He was brushed off. Venture capitalists, Hall writes, "were held in about as high esteem as ambulance chasers."

According to Boyer, Swanson called people alphabetically, which was why Boyer was one of the first people Swanson called. When Swanson came out with his patter that a company should be started to take the technology to the public, Boyer said, "Well, come by on Friday afternoon. I can spare you ten minutes."

On Friday January 17, 1976, Swanson dropped into Boyer's lab and epitomized the distance that existed between biology and business. One scientist remembered, "We were all standing out in the hallway laughing at this guy in the three-piece suit. We just didn't get people like that visiting us." But Swanson's ten-minute pitch was persuasive enough for the two men to walk out for a beer at Churchill's, a local hangout for molecular biologists. They wound up talking for a couple of hours and put down $500 each to start a pharmaceutical company to explore proteins that bacteria could be engineered to make. Boyer would have to borrow to do that. He told Swanson why

he was willing to go commercial: "I had the U.S. government funding my research for so many years in this area, letting me follow my nose in what I like to do. Commercial application would be an opportunity to give something back and see real benefits come of this research." (Indeed, in his school yearbook at Derry, Boyer had written that one of his aspirations was success in business, and eighteen months before Swanson's call he had been telling Stanford's patent lawyers human hormones were a promising area for commerce; in 1974, he was thinking of trying to make angiotensin II, a hormone that causes vasoconstriction.)

Hall uses a felicitous metaphor to describe the relationship between Swanson and Boyer: "Given the similarity and singularity of their interests, it was almost inevitable that Boyer and Swanson, like a pair of sticky ends, would ultimately seize onto each other."

Boyer rejected the title Boyer & Swanson, and Swanson rejected Swanson & Boyer and they both rejected Herbob. So they named their putative company Genentech, derived from Genetic Engineering Technology. Before Swanson went off in search of money, their first decision had to be what commercial proteins they might induce the bacteria to make. Swanson visited the science library and read a new book by Margaret Dayhoff on protein structure. "The obvious one that popped to the top of the list," he said, "was human insulin." The question then was whether synthetic insulin, if they could make it, would be competitive with insulin derived from animal pancreases. Swanson did the numbers and concluded they would at least be in the ballpark for winning some if not all of Eli Lilly's insulin requirements for its $400 million business.

While Boyer pondered the science, Swanson pondered his bank account. It was not at all a straight line from a beer at Churchill's to the Genentech boardroom. Neither man was sure just how committed he was. Buyer's remorse is the commonplace of transactions. Boyer kept teaching and Swanson kept running around looking for a job. For Swanson, Kleiner and Perkins made a difference—not by putting up money but by refusing to do so and thereby

forcing Swanson to contemplate his destiny. When he went straight back to them after his beer with Boyer and suggested they put him on salary while he formed Genentech, they gave him a flat No. Swanson, who would one day be one of the richest people in America, wound up on unemployment for the six months of 1976 after his meeting with Boyer. "I got $410 a month tax free. My half of an apartment in Pacific Heights was $250. My lease payment on the Datsun 240Z was $110, and the rest was peanut butter sandwiches and an occasional movie."

How did Swanson come to stop his job search and commit entirely to Genentech? He says in the oral history, "Finally I said to myself, Should I do this or not? The answer is—and I often recommend to other people to use this as a tool for making decisions because it incorporates all the logic as well as all the emotion—look at yourself as an old man or woman, say, eighty-five. Looking back over your past life, what would you want to have accomplished? And for me, the approach integrated everything, and it said, 'Look, I think this is important. If I don't do this, I'm not going to like myself so much for not having given it a shot.' So that was what made that decision."

Meanwhile, Boyer contemplated a decision that would make or break Genentech, though he did not know it at the time. If they got the money, he would first attempt to make a human insulin gene from scratch using chemicals he could buy from the shelf. It was an outlandish idea. The two-chain insulin molecule, smallest of proteins and the first to have its amino acid sequence worked out, had only 51 amino acids, but of course they had to be organized in the right sequence, and 51 x 51 x 51, etc., can yield 130,000 permutations. Other labs were going straight for the human gene, hoping to isolate it from messenger RNA. From this RNA, they would then copy a strand of complementary DNA or cDNA, which would be made of complementary nucleotides. In other words, creating an original document from a genetic photocopy. Purifying the RNA—getting a really reliable photocopy—would not be easy, but this was seen as the only practical way to isolate a gene. Boyer thought it trickier than others appreciated, but he had a political as well as a scientific misgiving: Natural DNA might fall afoul of genetic engineering regulations then being worked up by the National Institutes of Health, the source of major funding. They were likely to be very stringent for any work that involved a human genome, but a synthesized one would escape NIH regulation. Two other scientists had entered Boyer's life and were to make a major difference in this critical thought of going synthetic: Art Riggs and Keiichi Itakura of the City of Hope National Medical Center east of Los Angeles. Itakura, who had a Ph.D. from the University of Tokyo, had come to America because in Japanese science he said, "there's almost no freedom to do what you want to do." He believed he had a new way of creating synthetic DNA—assembling a gene letter by letter, nucleotide by nucleotide—ten times faster than the standard technique worked out in the 1960s. Soon after his meeting with Swanson, and with Swanson's agreement, Boyer called Riggs. "Hey, Art," he said, " I have this businessman friend of mine who thinks he can raise money to do insulin."

Riggs replied, "Oh, that's interesting."

Boyer: "Would you be willing to accept the contract to make the gene for insulin?"

"Well, there are a couple of things. Itakura and I are just in the middle of writing a grant to do somatostatin. Could I interest you in somatostatin?" asked Riggs.

Discovered only in 1973—years after Boyer's consultation with the pediatrician—somatostatin is a hormone secreted by the hypothalamus, located at the base of the brain, that serves to inhibit growth hormone. People without it suffer from the rare disorder gigantism. Making somatostatin, its being a small molecule of only 84 amino acids, could be a dry run for the trickier production of insulin. Boyer relayed Riggs's proposal to Swanson, who had to make his first big executive decision. He approved. They would finance the experiment, and Riggs and Itakura also would receive Genentech stock.

But Swanson still had to find the money.

Financial and scientific anxieties went hand in hand. Would a synthetic gene actually work inside a cell? Itakura had previously made a 21-base-pair piece of DNA that was part of the code for the lac gene, the one that cuts sugar in bacteria. Riggs and his team put it in *E. coli* bacteria, and Herb Heyneker, a young Dutch chemist working in Boyer's lab, devised an experiment that would turn bacterial colonies blue if the synthetic lac gene worked. Boyer said: "I can remember that we were in the lab late one night, around ten o'clock, and Herb Heyneker had done the experiment that morning. He brought plates in to look at. We knew there were clones if the clones turned blue. Herb said, 'It didn't work.' I said, 'Let's look a little bit more closely.' If you looked real closely, it was very obvious that they were turning blue. They were very faint, but there were quite a few blue colonies on the plate."

It was the first demonstration that manmade DNA could fool a real cell. It removed the last doubt Boyer had about going commercial. It induced Perkins, on April 7, 1976, to give Swanson $100,000 to start Genentech. Genentech gave Riggs and Itakura a contract for $300,000. They had made a just-in-case application to the National Institutes of Health for a grant of $401,426; they let it lie to see what would happen. The NIH said no. The Institute had no faith in DNA chemistry, thought the work had no practical purpose and judged it reckless to talk of making a gene in three years.

Two other groups of some of the best scientists in the country did not share that defeatism. The creation of Genentech was a starting bell in a race to make insulin. One rival group at Harvard was headed by the soft-spoken but very determined Walter (Wally) Gilbert, a theoretical physicist turned biologist; James Watson said his biggest achievement in science was "getting Wally into biology," and in 1980 Gilbert was to share a Nobel Prize for sequencing DNA. The second rival group was just five floors above Boyer at UCSF, headed by Howard Goodman and biochemistry department chair William Rutter. All three groups were encouraged by the interest of Eli Lilly, which feared that perhaps by 1990 or 2000 the supply of animal insulin would fall below the rising demand for treatment.

All three recognized that it would take four stages to produce a protein using recombinant DNA: (1) Isolate the message; (2) convert it into a gene; (3) clone it; and (4) express it as a protein. Boyer and Cohen had successfully done stage three. Still, no one had yet managed to put a mammalian gene into a bacterium and gotten it to express the protein coded by that piece of DNA. Most scientists thought doing this would take another ten years.

Unlike other researchers, Boyer was relying on synthetic DNA. That June his forebodings were vindicated. The NIH issued strict limits. Advanced work with human cDNA could be carried out only in so-called P3 high-security labs—the only higher grade, P4, was for germ-warfare experiments, and P4 labs could be used only under strict conditions. Boyer, working with synthetic DNA, was unimpeded; there was no mention of synthetic DNA, human or otherwise, because no one had thought it remotely likely. California legislated its own guidelines, but Cambridge, Massachusetts, proved more volatile. Through that summer of 1976, Frankenstein was being hatched in Harvard's labs in the imaginations of a populace whipped up by a demagogic mayor, Alfred Vellucci, and not discouraged by a number of scientists. Two packed and noisy sessions of the city council broadcast on television resulted in a six-month total ban until a special committee could report. The furor made rapid progress in recombinant DNA research almost impossible in Cambridge until after February 8, 1977, when the Cambridge City Council, after more extensive hearings, finally allowed recombinant DNA experiments to resume in the city.

Academic scientists dismissed the idea that corporate science would ever be able to clone a gene. Boyer's adventure, conducting industrial research in a publicly funded institution, divided the UCSF campus. Sally Smith Hughes writes: "The personal hostility directed at Boyer left enduring scars. He was an early target for the tensions that were to erupt throughout academia in coming years as universities and scientists at the forefront of molecular biology sought to capitalize on commercial opportunities."

Boyer and Swanson had committed Genentech to make somatostatin first, but

Alex Ulrich, a young German postdoc working upstairs in Howard Goodman's lab, was known to be making progress with rat insulin and Harvard would be a powerhouse once it got going. Swanson kept asking Riggs and Ikatura if they were sure they could make somatostatin; they kept saying yes, and he did his part splendidly, raising money for all of Boyer's work—$850,000 between December 1976 and February 1977 for staff and equipment. Kleiner and Perkins invested another $100,000. (It turned out to be a brilliant investment, returning 800 to 1 on their initial $200,000). A third round of Swanson funding raised the value of the company from $400,000 to about $3.3 million, and then to $11 million. But setbacks came again and again.

To make somatostatin, Itakura had to bond together 42 nucleotides into a single chain of DNA, attach their complementary nucleotides and then figure out how to mold the two chains into a double helix. He and Riggs had no luck cloning synthetic DNA in the summer of 1976. Mistakes in the genetic code frustrated them, but they radiated total confidence when Thomas Perkins summoned the scientists to a tense meeting with the business backers. Finally, in the spring of 1977, Itakura packed his best shot at synthetic DNA in dry ice and shipped it overnight to Boyer's lab. Boyer's assistants Herb Heyneker and John Shine cloned the DNA again and sequenced it. It was perfect. The trick now was to get a bacterium to accept the gene and produce a protein with it.

Simply putting a gene in a bacterium was not enough. It had to be placed in exactly the right position. If the bacteria started reading the gene in the wrong place, the entire message would be changed. Another problem was that the somatostatin molecule is so small that it would be immediately destroyed by the bacterium, which would not recognize the human protein. Boyer and Riggs invented a brilliant deception. Much research had already been done on the gene for beta-galactosidase (beta-gal), the lac gene previously mentioned. They decided they would tie the synthetic somatostatin gene to the well-understood and much larger beta-gal gene. The hope was that when the bacteria started reading

the beta-gal gene, it would just continue reading the somatostatin gene and not destroy the human protein. The problem, however, was that the resulting protein would be fused like the parent genes.

This was something that had stymied researchers for years, but Boyer and his associates devised a neat solution. Using a small disposable attachment they would tie the two genes together. The tie would be a short, three-nucleotide sequence, coded for the amino acid methionine, a chemical not present in somatostatin or beta-gal. In theory, that methionine link could later be detected and dissolved by cyanogen bromide. Imagine trying to produce the word *human* inside the word *bacteria*. The word *bacthumaneria* is unintelligible. But now imagine we have a special editing tool that allows us to dissolve the letter *p* shared by neither the word *bacteria* or *human*. First we create the word bac-p-human-p-teria. Then we dissolve the two *p*s, and the *human* floats free.

Boyer's postdocs glued Itakura's gene just "downstream" from the beta-gal inside a pBR322. "The person who was most concerned," said Riggs, "was the person that had the least detailed knowledge of the chemistry and recombinant DNA work that was going on, and had gambled the most. Had, in fact, staked his career on success. And that was Swanson." Swanson had to check himself into the hospital when the results came in on June 16, 1977: The bacteria had manufactured less than two molecules of somatostatin per cell. In the oral history, Swanson recalls, "Oh, God, everything that everybody thought might or should happen didn't happen." He was running out of money. Boyer too was crestfallen. His academic career had been tarnished by his association with Genentech. The critics seemed to have been right. One of his postdocs said, "Herb Boyer went through, and Genentech went through, a really down period." The only consolation was that the Harvard group was having not much luck either.

The Genentech team discussed the mystery of the missing somatostatin for more than a month (typical for an academic lab, but long for a struggling commercial enterprise). The most plausible explanation was that the bacteria recognized the protein as foreign and destroyed it. They

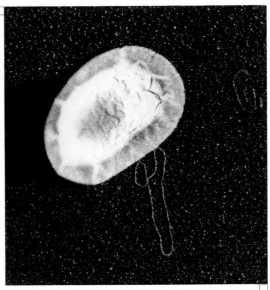

DNA on the march—magnified 100,000 times. The little "necklace" is recombinant DNA about to enter the bacterium.

conceived of camouflaging the gene still more. Boyer told Heyneker, "Let's bury it. Bury it. Bury it behind beta-gal." Other researchers at UCSF told Heyneker where he might better bury the somatostatin gene. Later on, these colleagues said they had given the Genentech team crucial information. Heyneker said he and Boyer already knew about this particular location, adding, "I was grateful for the information, but I did not consider it invaluable. I did consider it helpful." This was just one of many disputes Genentech would get itself into with academia—especially after the company became so successful. Clearly, Genentech did benefit significantly from its close ties to university labs.

Using the same methionine railroad-car technique, Boyer's team connected the somatostatin gene to the comparatively enormous beta-gal gene. It took two more months to redo the experiment. Swanson was in the lab every day. By August the lab had 11 colonies. The bacteria produced this strange new protein, a mix of somatostatin and beta-gal. The cells were sent down to City of Hope, where they were broken apart, the methionine dissolved, and the proteins collected and purified. Late on the afternoon of August 15, 1977 (just two months after Swanson's despairing hospital stay), Riggs and Itakura tested for somatostatin. There was plenty of it. They had expressed a human protein in bacteria, and they were the first in the world to do it. "Looks like we got it," said Riggs, shaking hands with Itakura; then he took his son to a baseball game.

THIS FIRST DEMONSTRATION that microbes could be put to work making foreign proteins was the dot on which the superstructure of the biotech industry was to rise. Swanson had the vision. Sitting in a little office in the Wells Fargo building in San Francisco with a rented desk, a telephone and a rented secretary, he envisaged Genentech as a fully integrated pharmaceutical company producing and selling a whole range of drugs, including human growth hormone, interferon, interleukins, tumor necrosis factor, animal vaccines and tissue plasminogen activator.

But Genentech then was nothing. It had financed the work in Boyer's lab and at City of Hope in southern California, but it had no lab of its own, no full-time scientists of its own, no salesmen, and Swanson was only too well aware that the Rutter-Goodman team at UCSF and Gilbert's Harvard group had an early start in making insulin. Despite the difficulties imposed by the NIH, either one of these groups might be the first to make insulin—which would destroy Swanson's dream of exciting the pharmaceutical industry enough to give them lift-off in big dollars.

In the summer of 1977, when Goodman cloned rat insulin, Swanson found the money for Keiichi Itakura's synthesizers at City of Hope to begin preparing fragments of synthetic insulin, work led by a young Italian organic chemist, Roberto Crea, whose progress was marked by his renderings of Neapolitan opera. At a celebration dinner for the somatostatin success, some of the group suggested it was crazy to think they could make insulin in anything less than two years. Crea piped up that they could do it in six months. "If you can make it in six months," said Swanson, "you go ahead."

Swanson went on a hiring spree. He and Boyer made a prize catch in March 1978 by attracting 26-year-old David Goeddel, mountain climber and ultracompetitive cloner. With him came a more established scientist named Dennis Kleid. Both of them were finally persuaded to come into commerce by Genentech's assurance that the corporation would allow them to pub-

lish research, a major difference from the big drug companies. They also relished the prospect of a race against the great Wally Gilbert. As soon as they arrived in March, they rushed into action with Itakura's group, working under individual ventilation hoods in the fume-filled lab. They decided to break down the job into smaller, more manageable tasks, making the insulin gene in separate A and B chains and then joining them together.

In June 1978, Genentech opened its own lab in the corner of a warehouse Swanson had leased in an industrial park in south San Francisco, near the airport. Boyer's lab at the university was an exciting place—small, crowded and bustling—and Genentech retained the same casual T-shirts-and-sneakers atmosphere. Guy wires held the ceiling and walls up. A polyethylene sheet was the roof over a room with two benches where Kleid and Goeddel worked fourteen-hour days when they were not doing the same thing at City of Hope. There was an office for Swanson, a vending machine for food, some equipment rooms and a small P3-level lab. A writer for *Esquire* found salespeople working side by side with technicians, and a sign announcing the visit of King Carl Gustav of Sweden next to the warning DO NOT PIPETTE BY MOUTH. Desks were cluttered with champagne bottles with dates of scientific discoveries scrawled on the sides. Goeddel wore a T-shirt: Clone or Die.

Because growing bacterial cultures and incubating experiments often took hours, Goeddel and Kleid would often go fishing behind the warehouse in the San Francisco Bay. They consulted Boyer on hard problems, but essentially the scientists had no boss, though they were haunted by Swanson's nervous vigilance. In the spring of 1978, when the synthesizers were roughly

halfway through their work, the news came through that the Harvard team had "made insulin." Their hearts sank—then rose again when it was learned they had made rat insulin, not human. The Harvard success essentially repeated Genentech's a year earlier with somatostatin, but the achievement was substantial. They were the first scientists to make a mammalian protein using cDNA.

On the West Coast, experiments continued 24 hours a day. Goeddel managed to splice the insulin A chain into the pBR322 plasmid but was having trouble with the B chain, which was longer—so long that it was itself being cloned in two pieces that would later be joined together. By mid-July 1978 the entire B chain was finished. It turned out the sequence was wrong; somewhere in the gene there was a mistake. Swanson, who had heard only that the B chain was finished but not that it was wrong, took Goeddel and Kleid to meet Perkins. The two scientists said nothing about the mistake and enjoyed dinner at the sumptuous Perkins home in Marin County, north of San Francisco. Perkins described all the companies he owned and the second Ferrari he had just bought. The scientists went straight from the opulence of the dinner to the bleakness of the lab and worked through two straight days to hunt for the error in the DNA and fix it, which they did. Now that the two chains were built and correctly sequenced, Genentech scientists stitched them to the back of a beta-galactosidase gene in their bacteria. They sent the bacteria down to City of Hope. It was up to the bacteria now. They refused to cooperate. One afternoon at the beginning of August, an impatient Tom Perkins drove up in his red Ferrari to find out what was causing the delay. Goeddel says, "He talked to Bob [Swanson] for a long time, and right after he left, Bob came in and in effect said to me, 'You're going down to the City of Hope. Don't come back until it's done.'"

For weeks and weeks Goeddel tried to get the two chains together, with poor results in the quality and quantity of insulin. To check his progress he was using a radioactive antibody that attached itself only to the correctly folded insulin molecule. He tried again during the wee hours of

August 23, and on the morning of the 24th he and Riggs stared anxiously at the radiation meter. They had their human insulin—20 nanograms of it. Goeddel told Hall: "I think we were actually jumping round a bit, even."

Swanson lost not a second contacting Eli Lilly. A contract was signed on August 25, but Swanson still had to cope with the scientific imperative that peer review of a paper for publication should come before any public announcement. To Swanson, it would be disastrous if they delayed long enough for one of their rivals to come out first. He issued what amounted to an ultimatum to the scientists by scheduling a press conference for September 6. Teams were drafted to write up the experiments; the paper was finished on September 4 and sent to the *Proceedings of the National Academy of Sciences* with the headline: "Expression in *Escherichia coli* of Chemically Synthesized Genes for Human Insulin." Because of the still-difficult American climate for recombinant DNA work, both of Genentech's rival groups were working in Europe. The Goodman-Rutter group was then decamped in a lab in France, and Gilbert was struggling with the gas masks and all the rigmarole of working in Britain's Porton Down biowarfare lab. (By the beginning of 1979 it became clear that they had been put through all this frustration for no good cause. Congress declined for the second year in a row to pass stricter rules, advised by scientists that many of the feared risks had not materialized.)

Making a big splash in the way Swanson did with his televised press conference was alien to the way the scientific community liked to work. But his publicity did what he intended it should do: It put Genentech on the map. The Ohio-based Lubrizol Corp. came forward with $10 million for Genentech to make interferon and other drugs. But the insulin story had some way to run. The difference between scientific research and biotech was that Lilly had to be assured the bacteria would march to the requirements of industrial production. Lilly, with experience in extracting insulin from animal organs dating back to the 1920s, presented Genentech with deadlines that even the best of the all-star groups doubted

could be met. Boyer says in the oral history, "We were so naive we never thought it couldn't be done. We knew that the pharmaceutical industry had a history of scaling-up products like antibiotics and amino acids, things of that sort. But this involves a whole different technology." Boyer says later in the oral history, "Naïveté was the extra added ingredient in biotechnology." But Boyer, ever focused, was a rallying figure. "We'll figure out how do it," he said, and so they did, just in time.

Early in 1980 Kleid took a precious package to Indianapolis, bacteria in which the insulin gene was tied to the tail of the small tryptophan gene, trpE. The bacteria, said Kleid, "really went bonkers" making what they thought was tryptophan but what was in fact a Trojan horse—tryptophan with a chain of insulin.

Two events of critical importance for the birth of the biotech industry followed. In the first, in June, the Supreme Court ruled in *Diamond v. Chakrabarty* that living organisms are patentable. The Yale historian Dan Kevles has argued persuasively that patents do not foster but actually penetrate industrial secrecy because they compel publication of the means and methods that lead to a patentable product. "Denying patents on life would throw corporate recombinant research deeper into the realm of trade secrets and away from public scrutiny, which would be unwise in the socially charged area of genetic engineering," explains Kevles. "Patents encouraged technological innovation." Genentech gradually changed the secretive ways of pharmaceutical companies, forcing them to allow more publication of basic research. The second major political event in 1980 was the Bayh-Dole Act. This followed up *Diamond v. Chakrabarty* by allowing universities to retain title to inventions made under federally funded research programs. That meant millions of dollars of income for basic science.

Genentech stock went on sale on October 14 of the same year. It was the first initial public offering (IPO) for a biotech company and one of the most spectacular IPOs in the history of Wall Street, not rivaled until the Internet boom of the mid '90s. Offering 1 million shares, Genentech raised $38.5 million in a few hours. Boyer

and Swanson each held shares worth $66 million.

That year, earnings for the entire company had amounted to only $80,000. But the outlook was bright. Lilly's sales of Genentech's insulin, called Humulin, went from $8 million in sales in 1983 to $90 to $100 million in 1986, $702 million by 1998, refuting predictions by other big chemical manufacturers, and even by those authoritative journals the *New Scientist* and *The Economist,* that the company would flop. Genentech's royalties jumped with Lilly's sales. It was still the only drug made by recombinant DNA in October 1985 when Genentech released its second drug, human growth hormone.

GENENTECH AFTER BOYER

Boyer lost little time retreating from direct scientific contact with Genentech, preferring to work in his lab at UCSF. He did not want to manage Genentech's scientists, or to overshadow newcomers. By normal standards, his name should have been on the insulin paper, but he insisted that the credit go to others. He was wounded by academic sniping, a mix of genuine concern at risking the openness of science to the imperatives of commercial secrecy and jealousy about the money. "There were people who were supportive," Boyer told Sally Hughes, "but over a period of time, I felt ostracized. The way the attacks went, I felt like I was just a criminal. I didn't think I was doing anything unethical or immoral. I never dreamed the financial rewards would amount to what they did." And again: "We had very idealistic and altruistic goals in terms of making products that would somehow benefit society. That was part of the original goal. Bob would always say, 'I'm not interested in hula hoops and tennis shoes.'"

Swanson proved a shrewd and inspiring young leader, "a short, chunky, chipmunk-cheeked thirty-six year old," in *Esquire*'s description in 1984. He hired the very best people he could find; he kept his promise to allow publication of research; he cultivated a common purpose in an open, unpretentious atmosphere, adopting the managerial philosophy of David Packard that made caring for staff a priority. *Business Week* said the innovative management style rivaled the achievements of the company's scientists. Swanson and Perkins invented what Perkins calls "brand-new financial securities" that were critical to the company's staying power. One was "junior stock," which helped to keep people like Dave Goeddel on board. The stock was sold to employees at a significant market discount but could not be traded until the company attained certain sales goals. Junior stock was widely copied by Silicon Valley start-ups. After the FDA approval of the human growth hormone, Swanson offered every employee—from bottle washers to vice presidents—options on 100 shares at that day's price. Genentech's incentives encouraged molecular biologists young and old to believe they could get rich quick, and many of them did: The day Genentech went public, a grad student at Caltech who had been paid in stock for some summer work suddenly found himself a millionaire.

Cynthia Robbins-Roth, who joined Genentech as "a typical postdoc" in 1980, has testified to the exciting sense of participation. In her book *From Alchemy to IPO,* she writes: "I still remember the in-house seminar . . . where the head of clinical development for Activase, Genentech's innovative heart attack drug, showed some of the first images of blocked blood vessels opening up after treatment. The message came through loud and clear—Genentech's research was keeping people alive, people who otherwise might have died. That was an incredible impetus." Jim Gower, head of marketing, recalled Swanson showing a party of Japanese through the Genentech halls and pausing to fix a leaky sink without dropping a beat in the conversation. Swanson's style appealed to the freewheeling scientists. Recruits were impressed that when they found themselves waiting at the company door while Swanson, who'd offered to drive them to lunch, hiked to the far end of the parking lot; he had no executive parking spot. Robbins-Roth said, "Many of us working at the bench late at night would look up and find Swanson striding in the lab door. He knew our names, knew what project we were working on, listened attentively to our descriptions of project status, and asked smart questions. That strong sense that the guy at the top knew and appreciated what we were doing kept the labs lit up far into the night."

Swanson had it all worked out why they should make Genentech an integrated pharmaceutical company. It was the only way they could make enough money to put prodigious sums into research. He plowed back 50 percent of revenue, a much higher percentage than the pharmaceutical industry average of 20 percent or less. It cost Genentech $200 million to develop Activase before they sold a single vial. With patent protection, they could achieve margins in the 80 to 85 percent range. "You compare that, say, to a ten percent royalty. If you are able to sell even ten millions of a product, you achieve a margin contribution of six million dollars, whereas a ten percent royalty on that would yield only one million. So over the long run that ability to capture greater value for your creativity in new drug development is going to be critical in terms of long-term survival."

Kleid developed an animal vaccine for foot-and-mouth disease. Goeddel, who became director of Genentech's molecular biology department at age 29 in 1980, led teams who cloned alpha interferon genes, beta interferon, gamma interferon, human growth hormone and other important genes. Genentech cloned blood factor VIII for the safe treatment of hemophilia. The company also teamed up with IDEC Pharmaceuticals, in San Diego, to make Rituxan, the first antibody ever approved for the treatment of a cancer, in this case non-Hodgkin's lymphoma.

Swanson was the first to identify and act on the fact that the basic science in molecular biology had progressed to the point where it could become a business. The big drug companies, growing out of the chemical industry on the East Coast and in the Midwest, had little need for extensive knowledge of biology to do what they did. In the phrase of Edward Penhoet, former CEO of the biotech company Chiron, they weren't mechanism based; they were phenomenologically based. As such, they were never going to lead the way in the complexities of genetic engineering as Boyer and Swanson did. The partners' achievements were liberating as well as therapeutic. They helped remove much of

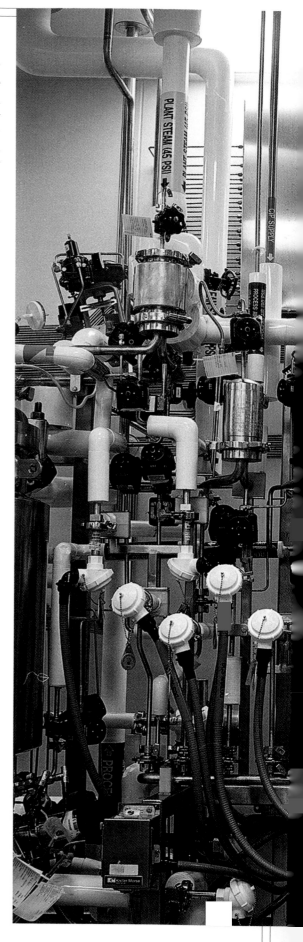

BIOTECH BREW: After the microscopic work comes the manufacturing. This is machinery for brewing biomedicines at Genentech's plant in 2003.

the stigma attached to commercial science. "One of the strengths of the United States," said Swanson, "and it hadn't been in the molecular biology field, has been the quick transfer of technology from academe to industry. We do it better than anywhere else in the world, and it's why we're so successful."

In 2001 Genentech broke ground in California on the largest biotechnology research facility in the world. There is a life-size metal sculpture in the quadrangle of two men sitting at a table, each with a beer, one in a suit leaning forward in his enthusiasm, the other in a denim vest and bell-bottom jeans leaning back, "skeptical but intrigued" in the words of a *Wall Street Journal* reporter. Robert Swanson did not live to see the representation of his meeting with Boyer. He stepped down as president and CEO in 1990, the year Roche Pharmaceuticals bought a controlling interest in Genentech for $2.1 billion. On December 6, 1999, he died at his home in Hillsborough, California, from glioblastoma, a type of brain cancer. At his bedside were his wife, Judy Church Swanson, their daughters; Katie, 16, and Erica, 11; and his mother, Arline Swanson.

It was a tragically early death for a man who had given so much to medicine. The biotechnology he and Boyer founded has saved and improved so many lives. There are now 1,500 biotech companies in the United States, and biotech is a 430-billion-dollar industry. The FDA has approved 120 biotech drugs. Genentech, a 25-billion-dollar company, accounts for 14 of these, including TNKase(r) (tenecteplase), a 90-second treatment for heart attacks, and Herceptin(r) (trastuzumab), a breakthrough treatment for certain forms of breast cancer that has been used to treat more than 25,000 women. More than 350 biotech drugs and vaccines are in human clinical trials aimed at treating, among other diseases, various cancers, Alzheimer's disease, heart disease, diabetes, multiple sclerosis, AIDS and obesity.

Swanson and Boyer opened a new chapter in the fight against human disease; we are still reading it.

sailing boat in a frostbite race late in 1961. His second marriage, of 25 years to Jane Smith, was troubled by his distractions (business and amorous); his third, to Jane Fonda in 1991, looked like a match made in heaven, but they separated in January 2000. "He needs someone to be there one hundred percent of the time," Fonda told the writer Ken Auletta. "He thinks that's love. It's not love. It's babysitting."

The turbulence associated with Turner gave him the reputation of a loner, a crazy loner, but he was perhaps the most consistently underrated businessman of his generation. Most of his innovative ideas were brilliant, executed with a big grin and a rebel yell. The gaffes that earned him the nickname Mouth of the South are a product of an essential innocence. He has no malice. He is not cynical. He is a romantic, a do-gooder, and proud of it. He says he loves people and it shows (particularly in unwisely loving every woman he meets). He is warm, generous, honest and a lot of fun. He is just incapable of filtering what he thinks at any particular moment. Turner's reputation has suffered from such episodes as sitting down at a business dinner with strangers and remarking of the host's wife, "Great tits!" He did not achieve universal popularity in England, when it was mourning the Fastnet disaster, by declaring that "everyone had a ball." Nor by telling millions of television watchers in Britain that Queen Elizabeth I's historic victory over Spain had been due to a similar freak of nature. "Storms like that happen. Look at the Spanish Armada. If it wasn't for a storm then, you Brits would all be speaking Español." His idea of consolation was to say that those who had gone up to that "great yacht race" in the sky wouldn't have to worry anymore about setting the storm trysail. But unfiltered Turner can sometimes strike a right note that could never emerge from corporate PR. Executives of Time Warner, who had taken control of Turner's company, were on pins in January 1999 when he was to address the American Chamber of Commerce in Berlin. They asked to have his script. He told them he didn't speak from scripts. So when he began, "You know you Germans had a bad century," there was panic. And then he went on: "You were on the wrong side in two wars. You were the losers. I know what that's like. When I bought the Atlanta Braves, we couldn't win either. You guys can turn it around. You can start making the right choices. If the Atlanta Braves could do it, then Germany can do it." The reception was rhapsodic.

The source of the competitiveness that has made Turner such a spectacular innovator is endlessly analyzed. To his biographer, Porter Bibb, he simply said, "I was interested in one thing and that was finding out what you could accomplish if you really tried." In our interview, he was disarmingly low-key. "When I was in my twenties, I used to tell people that I wanted to get to the top, and I wanted to get there in a hurry, not even knowing where the top was." The common explanation is that he is still trying to prove his worth to his disparaging father, Ed Turner, who shot himself in the bedroom of his South Carolina plantation home on March 5, 1963, and left his 24-year-old son to salvage the family company. Giving a talk once to an undergraduate audience at Georgetown University, Ted pulled out of his pocket a copy of a magazine called *Success* that had put him on the cover, looked up at the heavens and said, "Dad, are you satisfied now?" Hamlet had to deal with a ghost seeking vengeance; the ghost of Ed Turner seeks completion.

Ed Turner was a very bright, driven man whose father lost his Mississippi farm in the Great Depression and picked cotton as a sharecropper. Ed started with nothing in Cincinnati and built a billboard monopoly in five or six southern cities, headquartered in Savannah. He was a millionaire, but vulnerable to the fear of sharing his father's fate as he pursued an elusive image of success in the style of the Old South. Old money in Savannah looked down on Ed Turner as a parvenu. He smoked three packs of cigarettes a day and suffered from ulcers. He was volatile; as a father, he could be alternately domineering or indulgent with the son he loved. When stress led him to periodic bouts of heavy drinking and fairly persistent philandering, he was disgusted with himself—and took it out on Ted with abuse and beatings. He made Ted a gift of a Penguin class sailing dinghy, but every misdemeanor of his son's inflamed Ed's fear that Ted, too, would dissipate his talents.

On any score, Ted Turner has earned redemption from the haunting.

WHEN ED TURNER shot himself at the age of 53, his will left Turner Advertising to Ted. It was the largest billboard company operating in several cities in the Southeast. The snag was that in a last-minute funk, Ed had sold a plum acquisition in Atlanta to his best friend in the business, Robert Naegele of Minneapolis, disregarding his son's objections. Ed had hoped his son would go to Harvard, but after the Georgia Military Academy, Ted had gone to Brown, where to Ed's horror he wanted to major in classics, provoking a long caustic letter of rebuke about tangling with "those old bastards" Plato and Aristotle ("With whom would you communicate in Greek?"). Ted had tangled too much with women instead and been thrown out of Brown.

He had done well learning his father's business. He mixed easily with the poster crews, climbing the ladders, hammering the nails. He was a dashing six foot two, slim and gregarious, with small but brilliant blue eyes and an enjoyable way of escalating a smile into a guffaw. He sold space with gusto. After he became manager of the Macon branch office, he doubled sales in less than 24 months. Still, he could be a little wild, and Ed had not thought he could run the expanded company. Ted was sure he could. After Ed's death, his adviser and accountant, Irwin Mazo, warned Ted that he was ill-equipped to take over: "You don't have the line of credit your father had. You're only 24, and the banks don't know you. Besides, we've got estate tax problems, plus certain stipulations. . . . Even the things that were easy for your father wouldn't be available to you." Ted wasn't listening.

"Ted Turner must be out of his mind."

He flew to Palm Beach to ask Naegele to undo the deal. Naegele refused. From the airport, Ted called Atlanta and gave instructions to begin transferring all the Atlanta leases to the Macon office. Then he hired every member of the lease department in Atlanta from the company Naegele thought he owned. According to biographer Christian Williams, Ted also threatened to burn all the records unless Nacgele relented. He finally did. Ted Turner bought out com-

IN THE CNN NEWSROOM: Passion for CNN, but no political interference with the editors.

petitors and expanded the business nobody thought he could run. In 1968 he went into radio in Chattanooga, buying "the worst radio station in America" for $300,000, angry that he could have got it for less if he had jumped in sooner. He used empty billboards to advertise the station and did the same when he bought radio stations in Jacksonville and Charleston.

Jack Rice, an Atlanta coal merchant, owned an ultra-high-frequency (UHF) television station in Atlanta (Channel 17 WJRJ) that nobody watched. To receive its programs, such as they were, meant fiddling with one of those little antenna aerials on top of the TV. Only 5 percent of people in Greater Atlanta could summon up the energy to try. Why bother, when there were three affiliates of the national commercial networks? With its weak signal, Channel 17 was the least desirable of the five stations in Atlanta. It was losing $50,000 a month, its accumulated losses were more than $800,000 and its demise was expected any day. The fourth-place independent station, WATL, was owned by United Broadcasting, and it had lost $50 million trying to make UHF work. Unsurprisingly, the directors on Turner Communications, as the billboard radio company was now called, all objected in 1969 when Ted said he would like to buy Channel 17.

Irwin Mazo resigned. Turner got around that obstacle by swapping stock with Jack Rice in January 1970 so that he ended up owning the station for no cash. The snag was that the deal valued the television station at $2.5 million. He had bet the company on succeeding where everyone else had failed.

"Ted Turner must be out of his mind."

There was another UHF station in Charlotte, North Carolina, losing $30,000 a month with $3 million in liabilities. Turner wanted to buy that, too. When the board dug in its heels against vesting—Turner had only 48 percent of the stock—he borrowed $250,000 against his own signature and bought the Charlotte station on the courthouse steps as the sheriff went in to liquidate it. "It was a big bite for me at the time," he said in our interview, "when I realized billboards weren't going to do it for me. All they did was sell advertising space. Making money was always secondary to me. I was interested in the adventure and challenge of it all." Pause for reflection. "I enjoy being an adrenaline junkie."

Turner Communications had never posted a loss. It did at the end of 1970— $900,000.

Ted Turner knew how to sell advertising. Sure his programs were black-and-white, but that meant color advertisements

would stand out all the more. His best line, which gave him a dimple in his cheek to match the one in his chin, was to tell advertisers on his UHF channels that they would be reaching a superior class of customers, because only very bright people could figure out the antenna and once they had they would feel they owed it to themselves to keep watching. As for programming, he did that, too, counterprogramming movies against news and religion. He makes no claims to originality. "I was just following the formula that independent stations had used for success in markets like Chicago, WGN, or New York, WOR." But then he bought old movies outright instead of renting them as everybody else did. Paying top dollar made him a lot of friends who were eager to sell him more, so fairly soon he had a library he could run and run. He bought old black-and-white episodes of *I Love Lucy, Father Knows Best* and *Leave It to Beaver,* and packaged them as part of the family channel. He told people in Atlanta they were tired of violence and bad news. The number one station in Atlanta was an NBC affiliate, WSB. It dropped some network programs in favor of more profitable local shows, so Turner picked up five NBC sitcoms. Then United Broadcasting overnight shut down WATL, Channel 17's closest competition, a pure piece of luck that Turner celebrated as a reward to him from the good people of Atlanta. When he learned what sports programming might do, he used his charms on an old girlfriend, who got her husband to move his wrestling matches from an ABC affiliate.

All this was routine good business, but Turner hit on something that took him into a different league. The NBC affiliate, WSB, paid the Braves baseball team $200,000 to televise 20 games. Turner paid $600,000 on the condition that he could broadcast up to 60 games. Then he bought the team outright, and the Atlanta Hawks for basketball and the Flames for hockey. He had not just invested in sport. He had invested in guaranteed programming for his TV stations. The smart people in New York laughed when the redneck owner was photographed riding an ostrich into his first Braves game, but whose head was in the sand? In 1972 Turner's television stations turned in a profit of $1 million, rising fast. WTCG was the call name for his station (meaning Turner Communications Group), but for Turner it was Watch This Channel Grow. In two years, the station's value had grown from $2.5 million to more than $25 million.

Reading *Broadcasting* magazine one day in 1975, Turner came across an item that HBO was going to broadcast movies by new technology—satellite transmission. "I went straight up to New York to pick their brains," he said. He learned that if he could transmit a powerful signal from a big upturned parabolic disc to the satellite, his programs could be retransmitted to cable operators all over the nation if they could be induced to install receivers. He was quick into Rockefeller Plaza to ask RCA for a long-term lease on a transponder (it had 24 transponders on the satellite). RCA's executives were astonished that a small-time broadcaster would have the presumption to go on the satellite. Not a lot of people were lining up for transponders, and he was told yes, he could rent time on a transponder—but he would have to erect his own transmitting station at a cost of $750,000, and it was up to him to get approval from the Federal Communications Commission. Still unthwarted, he applied to the FCC, who said they would have to think about it; they had nothing in

their rule books to cover such a contingency. There would probably have to be hearings in Congress.

In his early sailboat races, Turner had observed the phenomenon that if you could build up an early lead, those behind, demoralized, fell even farther back. He had to maintain his momentum. Without waiting for the FCC, he started to build a transmitter and hired staff through an intermediate company. "It was impossibly complicated," recalls Turner's assistant Terry McGuirk. "We had committed millions of dollars not knowing what the FCC would do."

"Ted Turner must be out of his mind."

The networks and Hollywood went gunning for the pirate. They got a bill written that would kill satellite television. Turner gave evidence to the effect that the networks with their "bouncing-around-boobs" shows in prime time were corrupting this God-fearing country we all loved. His nonstop patter annoyed the House committee. He did better with the Senate, testifying as the aw-shucks homespun southern boy wanting fair play for the little man: "Boy Scouts never get covered, sir. How many times have they said the Eagle Scouts is on the increase? They never cover that. But let a heroin-soaked kid hold up a little store somewhere and shoot somebody, that makes the news." He beat the bill. On December 27, 1976, the FCC opened the door to satellite programming.

Turner had scored a liberating victory for the future of television in America. The UHF tower was junk. His sparkling 30-foot transmitter dish was the future. Channel 17, with its range of 45 miles, was superseded by the Superstation—his choice of name. But how would he attract cable viewers; how would he pay for it all? He had told Atlantans, "I bought the Braves because I'm tired of seeing them kicked around. I'm the little guy's hero." He took the same populist tack in a new campaign against the networks. On a long fast hike in the lowlands of South Carolina, he drummed the dogma into Christian Williams, a *Washington Post* reporter (who wrote an entertaining biography in 1981): "The networks are like the Mafia. The networks *are* the Mafia. Do you know they spent a quarter of

a million dollars in Washington to stop my Superstation showing movies and sports in people's houses? Well, their day is finished now. It's over. They've made themselves unbelievable profits, and what have they brought us? Mr. Whipple squeezing the toilet paper. The networks are run by a greedy bunch of jerks that have hoodwinked the American public, and now I'm riding in on a white horse."

Well, more like dapple gray. His programming was cheap and anodyne: reruns of *All in the Family,* baseball, cartoons, old movies. Oh yes, and news at 3 a.m. Turner had tried to get Showtime's Jeffrey Reiss to run the Superstation. To dramatize what good things he had in his film library, Turner had kicked off his shoes and jumped on the table in his stocking feet to reenact the Charge of the Light Brigade. Reiss had beaten a swift retreat. Even so, cable customers signed on for the Superstation at the rate of 250,000 a month, glad of any choice. Advertising was slower, but Turner had won the America's Cup in 1974—and was on his way to winning it again in 1977—and once Madison Avenue opened its doors to the sports hero, the Superstation supersalesman left everyone in his wake.

When he bought time on the transponder for the Superstation in 1975, Turner took an option on a second transponder without telling anyone what he intended to do with it. Reese Schonfeld, then working for a consortium of independent television stations called the Independent Television News Authority (ITNA), knew what he should do. Schonfeld lived for news. He had taken his law degree at Columbia while editing scripts for UP Movietone and had become more and more incensed at the way the network news divisions spent a fortune to overproduce 20 minutes of slick news while real breaking news, like the attempt to kill President Ford, was put on hold until the nightly 7 p.m. slot. He knew that Turner hated news. Once, on Channel 17, chairman Ted had walked into the studio, on air, and put a paper bag over the anchor's head.

Schonfeld was sure he was going to get his news channel with Time-Life. It had enjoyed a success downloading HBO movies from satellite, and Gerry Levin, Time-Life's head man, went into serious negotiation with ITNA. Levin and Schonfeld had spent a lot of time on the project in 1977, when early one morning Levin called Bob Weisberg, his point man on the news channel project. "I've just gotten out of the shower, and I get my best ideas in the shower," said Levin. "I've decided we're not going to do news." Weisberg's wife, Doris, summed it up best: "If I were Gerry Levin, I'd never take a shower again." (He did, on January 11, 2000—see page 439.)

Schonfeld began bugging Turner to join ITNA at every television convention. "Ordinarily, Ted would have a blonde on each arm, a couple of drinks in him and he would deliver a wonderful rant. 'I don't need news! It's all about rapes and murders. It makes people miserable!' he'd shout. We'd all laugh and move on. In those days nobody took Ted seriously."

So it was a big headline in Schonfeld's mind when, in September 1978, Turner called and said, "Reese, I want to do a 24-hour news network. Can it be done and will you do it with me?" A foolish consistency is the hobgoblin of small minds. Turner had changed his, but not as precipitately as it appeared. The common impression of Turner is that he is Jeb Stuart leading a wild cavalry charge. In business, he is more like Ulysses S. Grant, carefully surveying the terrain before making his move, and his photographic memory knows where the snipers are hidden. "Investment bankers will throw a 50-page contract at Turner," Porter Bibb said, "and he'll flip through it in 10, 20 seconds and say, 'I think you need to change paragraph 7 on page 13 and paragraph 22 on page 40. The rest of it is fine. Let's go.' And they'll think it is an act, but Turner knows exactly what he is asking for. Those were the key points in that agreement."

In our interview, Turner said the idea of news had been germinating in his mind as far back as 1975, when he took an option on a second transponder. He had always been impressed on visiting New York with the 24-hour radio news; and his encounters with regulators in Washington had intrigued him about the nature of political power. Bibb's view is that Turner was thinking about his own all-news service in 1977, a year before he called Schonfeld, but saying nothing—which was news in itself. Whatever the more plausible date, Turner understood very well that there was nowhere else he could go to maintain his satellite option, now a property much sought by others. He would rather do sports, but ESPN had preempted that. He would rather do movies, but HBO owned that. The broadcast networks had entertainment sewn up, so that left news. He was certainly a convert to news in 1978, with all the zeal of the born-again. Rick Kaplan, who was the CBS producer for Walter Cronkite's evening news (and who became president of MSNBC in 2004), remembers vividly how he and Cronkite were both bowled over by Turner's conviction when they talked with him about his plans for 24-hour news. "He wasn't just doing a number. It was like a golfer lining up a putt. He could just *see* it going in." Once he had Schonfeld on board, Turner went with Terry McGuirk to the Western Cable Convention with a detailed plan. He would spend $20 million of his own money—he was worth around $100 million by now—and once CNN was running, the cable operators would pay him 20¢ a month for each subscriber, 15¢ if they took the Superstation as well. Not a single operator signed on.

Schonfeld saw another dead end, but before long Turner's voice boomed over the phone line from Atlanta. "Hey, Reese! You want to do this thing? Okay, then get down here and let's do it!" There had been rumors that Scripps Howard or the *Washington Post* would pick up where Levin had left off, but it seems they were thinking on a smaller scale, more of a rip-and-read service. Turner's competitive juices flowed faster. As for the $50 million or so that might be needed for the start-up, he was sure the two biggest cable operators in the country, Time-Life and Teleprompter, would come in to share the risks. When he called to offer each one-third of CNN, first one and then the other declined. "I'd been conned," writes Schonfeld. "Ted didn't have the money to do CNN." Turner sold the Charlotte station for $23 million in a desperate attempt to fund his dream, but a $30 million line of credit with the First Chicago bank was canceled after a senior man went to a lunch where Turner gave a wild speech. The Turner Broadcasting chief financial officer was at his wit's end, but Turner told Schonfeld to go full speed ahead for a launch in under a year for a television operation never before attempted.

"Ted Turner must be out of his mind."

Ted Turner was not on his own at the helm in turbulent waters. Reese Schonfeld was at 1018 West Peachtree, a dingy area of Atlanta rather less redolent of the gracious Old South than its address implied. Inside what was once a bawdy house and rehab center, he had convened a gathering of the Seven Samurai of television: Burt Reinhardt, Bill McPhail, Ted Kavanau, Ed Turner (no relation), Stan Berk and reporter Daniel Schorr. Then, two months before showtime, he gave them a hundred young college kids who came to live rough, eager to become omnifunctional VJs—video journalists, news assistants and computer graphic artists. Turner, sleeping in his office above, would wake up and wander around in his dressing gown, electrifying the place. Awww-riight! Against the impossible deadline, Schonfeld ran the producers raw through the imperatives of staying live with the news. They tried out the state-of-the-art technology in a big open studio. Kavanau figured out a way to cope with computers that kept crashing and looked for on-air talent; he thought he might make something of one of the staffers with a little experience and gave her a chance exchanging badinage with him on the overnight between 3 and 4 a.m. This was Katie Couric, who stayed four years before taking another path that led her to the starring role on the *Today* show.

All this was put in jeopardy on December 6, 1979. The RCA satellite that was to be CNN's link to the world blew up. RCA adamantly insisted Turner could not have a place on the next satellite; they had sold all those transponders and they were going to draw lots to see who got a space. Draw lots for the life or death of CNN! Tell that to Ted Turner! Kavanau recalls Turner taking down a Confederate sword he kept on the wall. "He swung it round his head, shouting, 'We will not be stopped. No matter what it costs, we're going on!'" The ensuing meeting with RCA was enough to put Rockefeller Plaza into orbit. According to Reese Schonfeld, Turner for starters told Andy Inglis, who ran RCA Americom, he had better sell his RCA stock. "When I get done with you, it ain't gonna be worth a dollar a share." When Inglis resisted, Turner got up from the table, walked across the room and glowered down at him from his full height. "Sir," he yelled, "does your chairman know what you are doing? Does he realize where this is going to end? I'm a small company, and you guys may put me out of business, but for every drop of blood I shed, you will shed a barrel!" Inglis did not fancy being the meat in a corporate sandwich. The hush-hush resolution he proposed was that CNN would sue in Atlanta, and RCA would make it easy for them to win so it could tell its other aggrieved customers that it was bound by a legal ruling to give preference to Turner. In the meantime, CNN would be assigned a transponder on Satcom 1 for six months. On the way down in an elevator full of media ears, Turner bubbled over. "They're going to let us win, Reese! They're going to throw the case!" Nobody could shut him up, but nobody picked up on the indiscretion either, and CNN won the case.

The launch of CNN at 5:50 p.m. on Sunday, June 1, 1980, was a dazzling triumph of technology and journalism. There were more live feeds from home and overseas than ever before in television history. But there were only 1.75 million households signed up—half the number Ted had promised advertisers. The networks refused to admit CNN to the White House press pool and demanded an exorbitant rate for any pictures they supplied. CNN sued. When President Reagan was shot, the pool did not make the pictures available, so CNN defied them by taking pictures off ABC. For all his fiery talk, Ted Turner's good nature got the better of him. He met and liked Tom Wyman, the CBS number 2, and, according to Schonfeld, was "suckered" into a compromise. If CNN dropped its suit, it would be admitted to the overseas pool, where no union rules were in effect, and the network would not charge to the limit for pictures within the United States. Schonfeld was disappointed, not the first of his disagreements with Ted. "To this day my gorge rises," he wrote in 2001. "The son-of-a-bitch had sold me out."

Acid rain was falling early in 1982 when I visited Atlanta and all the networks in New York as a director of Goldcrest Films and Television charged with exploring the

TED AND FIDEL: Castro put up a dish to take pictures from CNN's satellite transmissions, then invited Turner to Cuba in 1981. Turner had no compunctions about going. He guessed he was propaganda-proof, and there was the chance of opening the door for a scoop interview; a year later, CNN's *Take Two* did broadcast a show live from Havana, the first time an American network had been live from Cuba since the revolution. The two men went duck hunting and shot the breeze about baseball and the global arena. Castro urged Turner to cover the world more thoroughly. During a long evening of cigars and wine, when Castro held forth at interminable length while everyone normally froze in their seats, Turner abruptly got up to say he was too tired to listen anymore and went to bed. When he got back, writes Schonfeld, he said, "Reese, Fidel ain't a communist. He's a dictator, just like me."

feasibility of a 24-hour cable news system for Europe. In "Black Rock," the CBS executive suites, and at ABC and NBC, too, I was advised to forget the whole idea. Twenty-four-hour news was a bust. Turner was in such trouble it would not be surprising if he took his father's .38 revolver and blew his brains out, too. He was a cracker redneck. CNN was the "Chicken Noodle Network"—an epithet ABC's Sam Donaldson ungallantly flung at CNN's experienced Karen Sugrue when she traveled on *Air Force One* with President Reagan but was barred from joining his onboard press conference. Not one of the network chiefs I talked with gave Turner a chance. He had just recently fired Schonfeld in a clash of egos (whose CNN was it?), and that was a genuine minus, but for the most part the criticisms were pure bile.

Television professionals everywhere were condescending about the apprentice who had, in their opinion, nothing but grief coming to him. It was the same story at the Edinburgh Festival later the same year, where I heard Turner give a rip-roaring address that covered every topic under the sun from a new angle. The fact that he had several blondes in tow made a bigger impact than his cogent defense of CNN. It took some years for the cruel jealousies to abate, for Turner to be recognized for his achievement. It took considerably less time to copy him. Bob Wright, the president and CEO of NBC, looks back and says: "He sees the obvious before most people do. And after he sees it, it becomes obvious to everyone."

Every year CNN proved the validity of its innovation. The professional naysayers had just known that it was journalistically and technically nonsensical to attempt 24-hour news on television. If it did become technically and journalistically possible, nobody would watch it. And if they did, there would be so few such news-obsessers and advertisers, the station would go broke. It was vanity television.

All the Cassandras were proved wrong, but CNN was more than a vindication of Turner's business sense. It was a vindication of his faith in the sense of the American people, in democracy really being a workable idea. Everyone who worked with him testifies to his respect for editorial integrity. He did not meddle with news angles. "He gave me ideas and tips," says Rick Kaplan, who became president in 1997, "but that was it. He left it to the editors to decide one way or the other." He was not into peddling any political ideology, or using CNN to advance extraneous commercial interests.

CNN had two effects that few anticipated. When CNN started picking up video from around the United States and around the world, the affiliates of the networks wanted those pictures, too: Who could ignore CNN's pictures of the *Challenger* disaster? Previously, all the images had to go through the network filter, and the network filtered no more than it needed for the compressed 20 minutes of nightly news. So the affiliates took the pictures off CNN, and the networks had to follow suit. Second, CNN opened America's eyes to the wider world. The networks had progressively abandoned foreign bureaus and consistent world reporting. CNN was alone in having the visual story of the Polish workers under Lech Walesa claiming their freedom, and far ahead in the extraordinary scenes of the student revolt in Peking's Tiananmen Square. On the networks, places like Afghanistan and Pakistan emerged from electronic oblivion only in a crisis when the networks relied on parachute reporting. It is cheaper than maintaining regular correspondents but inevitably less informed, less nuanced. Bill Paley, the architect of CBS television, had said that it would be a bad day when TV news became a profit center. The bad day came.

Ted Turner did not begin his career as an internationalist. He did not focus on detailed politics, but his political temperament was to the right in his youth. He was a born-again Christian who became an agnostic, devastated by God's failure to save his much younger sister from a premature death. His interest in the world outside Georgia was sparked by his yacht racing, beginning in 1966, when he sailed the Atlantic to Denmark for an international regatta and saw Russians "in little red jumpsuits." They looked human enough to him: His father, an ultraconservative Republican, had always drummed it into him that if the communists took over they were so demonic they'd kill everyone with more than $50. "For years," he told me, "I never carried more than $49 in my wallet." A shift of his political temperament from right to center was accompanied by alarm about how the world might stumble into war from stereotyping the other side, so by the time he started CNN he was a passionate believer in peoples speaking to peoples, a conviction that veered toward the sentimental, to the amused condescension of sophisticates. In my interview, he said, "Back in 1960 Khrushchev took off his shoe and hit the podium at the UN during the Cold War and I am absolutely certain as I sit here that if we didn't have a UN, we would have had a war and we'd all be dead now." Thirty years later, Turner made an important contribution to history, engaging the British producer Jeremy Isaacs to make a documentary series on the Cold War, 20 hour-long episodes notable for their insistent search for truth.

Turner fulfilled Marshall McLuhan's prediction of the global village. He was so intent on the concept he banned the word *foreign* because it was suggestive of alien, objectionable, hostile. Eason Jordan remembers a newscaster using the word and getting an immediate call from Turner. "Didn't I tell ya, don't use that word. It's *international,* darn it." It was pointed out that the newscaster could hardly avoid it when he was reporting remarks by a foreign minister. "Well, tell him to change his title!"

In 1991, when the United States was about to eject Saddam Hussein from Kuwait, the Bush White House and the Pentagon told Turner to take his reporters out. He kept them there. His new CNN president, Tom Johnson, told him they could not afford the cost of $1.8 million a day. Turner told him to spend whatever it took.

Ted Turner must be out of his mind.

The world watched the Gulf War of 1991 on CNN—President George Herbert Walker Bush and Saddam Hussein alike. "When the bombing started," Turner told me, "I was in Jane Fonda's bedroom—she was out working or exercising. I flipped to CBS and they had Dan Rather in the studio talking. I flipped to NBC and they had Tom Brokaw in the studio talking. I flipped to ABC, and they had Peter Jennings in the studio talking. I flipped back over to CNN and we had reports of the bombs going off live in Baghdad, and I was proud that this was the biggest scoop there has been in the history of television journalism." NBC's Tom Brokaw acknowledged as much, gallantly complimenting his competitor for pictures that had informed us all. CNN remained unrivaled in its world coverage well into the 21st century, with 28 overseas bureaus serving 165 million households.

In 1996 Turner, feeling that only the big would survive, accepted a bid from Gerry Levin to sell Turner Broadcasting to what was by then Time Warner, and joined its board. But in 2000, when Levin merged with the temporarily hot-stock America Online (see Mr. Levin's shower thoughts above), he sent a fax to Turner removing him from control of all of his creations—CNN, the TBS Superstation, Turner Classic Movies, the Cartoon Network, New Line Cinema studios and the sports teams. Turner threw himself into campaigns for the environment: He can identify all the wildlife on his vast estates. He pledged a billion dollars to the United Nations. "Horatio Nelson and Alexander the Great were my boyhood heroes, but I switched over about 30 years ago to Mahatma Gandhi and Martin Luther King because I changed from being a man of war to a man of peace." But he was no longer in control to guide his vessel from the Hurricane Roger [Ailes] blowing in from Fox News.

Ted Turner really did open up the world. He democratized information. For the first time in history, every world leader had access to the same information at the same time—and when they suppressed protest, they risked doing it in full view. Boris Yeltsin, who succeeded Mikhail Gorbachev as the leader of Russia, has testified that CNN's video of him climbing on a tank to defy the anti-Gorbachev plotters was the turning point in the failure of the coup. He believes it saved his life and with it the putative democracy in Russia.

The Cold War ended, but Ted Turner in 2004 still feared for a nuclear holocaust by accident or design: He sounded a very early alert to the menace of terrorism. Former senator Sam Nunn heads his activist organization Turner's Nuclear Threat Initiative. With this shadow more or less permanently on his mind, Turner has made arrangements for CNN to make its own final mark. On the day of his patriotic and sentimental launch of CNN, 300 guests stood in the sunshine under the Stars and Stripes and the flags of Georgia and the United Nations, and a marine band played the national anthem. Unseen by any of them, hours before the ceremony, Ted Turner had commissioned an all-services band to come along and play something special. He recorded the music but never broadcast it. The founder's instruction for the occasion when the secret tape may be taken out and played is simple: "We will stay on the air until the end of the world. We will cover the story and then we will sign off playing 'Nearer My God to Thee.'"

JOAN GANZ COONEY She led millions of us down where the air is sweet on Sesame Street

1929–

Bert doesn't get Ernie's joke, but Joan does. Sesame sleuths are invited to spot Jim Henson.

SUNDAY MORNINGS were Lloyd Morrisett's time to sleep in a little after a demanding week as a vice president of the Carnegie Corporation in New York. He was not thrilled on the Sunday in the early '60s when he was awakened early by some noise and found his three-year-old daughter, Sarah, sitting in front of the television—staring at a test pattern. She was waiting for a show to come on, any show.

Any show? They all seemed about the same. There was little original programming for children on the only two networks that mattered—NBC and CBS. There was *Captain Kangaroo* (CBS) and *Mister Rogers' Neighborhood* on public television, but the hours were mainly filled with cartoons. Racial and gender stereotypes were commonplace. Children were subjected to commercials twice as often as adults, one minute in every five, and the pitches were almost entirely for junk food. In brief, the kids were being shortchanged.

Morrisett, an arrow-slim man of 36 with piercing blue eyes, was imbued with the educational zeal of his father—a man who had begun his career teaching in a one-room country school in Oklahoma's dust bowl in the early years of the century and in 1941 advanced to a chair in education at the University of California at Los Angeles. Lloyd majored in philosophy at Oberlin College, in psychology at UCLA, took a doctorate in psychology at Yale and became an assistant professor of education at Berkeley, then abandoned that career path to do research on how people learned. He joined the staff at Carnegie at the age of 29, and later became vice president responsible for the foundation's studies in child development. Two conclusions became clear to him at Carnegie. First, the good news: The million dollars invested in experimental programs of preschool tuition proved that the test children did much better than their classmates in their early school years. Second, the bad news: All the programs had touched only a few thousand of the 12 million children aged five and under who had no form of preschooling. "There was a dissonance," says Morrisett, "between the goals we were trying to achieve and the mechanisms that we had available."

Sarah had pointed out a way to bridge the gap. If children like her were so addicted to TV, perhaps some of the viable preschool material from Carnegie's pilots could be infiltrated into television and reach millions of children.

Soon afterward, Morrisett was at a dinner party in Gramercy Park, Manhattan, that became a legend in television circles. It was given in Tim Cooney's apartment to celebrate his 36-year-old wife, Joan Ganz Cooney, who had just won an Emmy for one of her programs about the war on poverty. Joan was a producer at New York's WNET, Channel 13. She had started as a newspaper reporter at the *Arizona Republic* upon graduating from the University of Arizona with a degree in education. She switched from reporting to television in 1953, excited by the waves being made by NBC's innovative president Pat Weaver, a passionate believer in bringing culture to the common man. (Weaver was the first to broadcast opera; he was also the inventor of the *Today Show* and the father of the actress Sigourney Weaver.) Cooney imbibed much as a publicist for Weaver's shows until he was pushed out in 1955 by Robert Sarnoff, the son of David Sarnoff (page 336). Cooney took another publicity job at U.S. Steel on behalf of its CBS drama series, but she longed to move into production. Her slight stature, femininity and wit masked a discerning resolve. On the day the Educational Broadcasting Corporation won control of Channel 13, she was on the telephone to everyone she knew: "Get me to that general manager; I have to get a job there!" When the station said it needed producers more than publicists, she bluffed. "Sure, I can do that." She was willing to take a $3,000 pay cut to get the job.

The dinner party was a celebration of convergence. Cooney and Morrisett had met socially for several years, but surprisingly she did not know about his work in children's cognition, nor that this was a particular area of interest, too, for her forceful boss, Lewis Freedman, who was also at the dinner. Freedman was holding forth on the potential of television to educate when Morrisett raised the question, "Do you think television can be used to teach young children?" He recalls: "Joan said, 'I don't know, but I'd like to talk about it.' So I called Joan a few days later."

When Joan—and Freedman—met with Morrisett at Carnegie, Morrisett proposed that Carnegie finance a feasibility study with Joan as the lead investigator. "Well, Joan wouldn't be interested," said Freed-

man. Her chin went up, "Oh, yes, I would." She reflects, "I had done a show for Channel 13 called *A Chance at the Beginning,* on a Head Start–type project—before Head Start—and that was my passion." Still, there was the problem that her bosses at 13 had said no. They nominated another woman producer to Morrisett. Tim Cooney just happened to have lunch with Morrisett and enlightened him on Joan's predicament, wanting to work with Morrisett but be loyal to WNET. Morrisett lost no time making it clear to 13 that it was Joan he wanted to borrow.

For the next three months, Joan bounced about the United States in tiny airplanes. She watched with despair much of the kind of condescending programming satirized by Robin Williams in *Mrs. Doubtfire.* She spent weeks consulting 26 educators who had 260 opinions, but she came firmly to her advocacy of a series of daily, hour-long programs for three-, four- and five-year-olds that would be entertaining, but also teach solid, measurable skills in language and math. She was proud of the resulting paper "The Potential Uses of Television in Preschool Education." The president of Channel 13, Jack Kermeier, called her in, along with Freedman. He looked up

at her. "Who are you? I see this 'I' throughout the study. Why would anyone be interested in your opinion? I don't mean to put you down, but in fact you're not an expert in any of this. I just think it's crazy."

Freedman said Cooney could never hope to get any inner-city kids to watch public television.

Morrisett, however, was impressed by Cooney's imaginative proposal and invited her to join Carnegie. He could only offer a year for her to develop the project and help raise money, but she quit WNET. Morrisett was having his own problems getting the $1 million he wanted from Carnegie as a starting point for the $4 million they estimated they would need. "There was considerable staff resistance at Carnegie to doing anything in television," he recalls. "There was a whole variety of reasons—it didn't seem to fit with what Carnegie was doing; it cost too much money." With backing from the president, Alan Pifer, Morrisett finally got his million, and Morrisett and Cooney then went with high hopes to the Ford Foundation. They were repulsed. Morrisett called on CBS and Group W. They were cold, too. NBC's president held out a hope of some in-kind support, but no money.

"It was disappointing," said Morrisett in 2004, "but I felt we would get somewhere because the whole atmosphere in Washington was infused with the idealism of Johnson's Great Society and the imperative of reform. I doubt whether we could repeat today what we did then." Morrisett always kept a low profile, but he had a network of friends in other foundations, education and government, including Harold ("Doc") Howe, the U.S. commissioner of education, whom he had known when Howe was superintendent of schools in Scarsdale. Howe said he would listen: "We don't have the money to do this, I don't think, but come on down and we'll sit around the table with the department people, the research people, the preschool education people." They both vividly remember the meeting in Doc Howe's office. "Everyone asked questions, and everyone was negative," says Joan. "Then Doc went round the table, and *everyone* said, 'No, there isn't any money.' Then Doc said, in effect, 'The ayes have it.'" Howe came up with $4 million and called

McGeorge Bundy, the president of the Ford Foundation, who then reversed his staffer. Ford initially committed only $250,000 for planning seminars, but as Joan puts it, "we knew we were on magic time, and everything would keep rolling."

Cooney, now president of the Children's Television Workshop, with Morrisett as chairman, ended up with $8 million and 18 months to challenge conventional media wisdom. To create and then manage *Sesame Street,* a name they settled on even though none of them much cared for it, she recruited educators and a creative team fizzing with originality—Jon Stone, who suggested the street setting; Carroll Spinney, a kindly silver-haired actor who morphed into Big Bird and Oscar the Grouch; Joe Raposo, the musical wizard ("Bein' Green"); Chris Cerf, who came from Random House to turn television into revenue-earning books and write songs; and a strange, long-haired, heavily bearded man in a long overcoat and low leather cap who Cooney feared for a moment or two might be one of the notorious Weathermen student terrorists. This was Jim Henson, creator of the Muppet characters.

Five million children in America and untold millions more in 148 foreign countries watch *Sesame Street,* which entered its 35th year in 2004. Some 91 Emmys vouch for a show that has never been allowed to stagnate. More than a thousand academic studies have vindicated every hope of Morrisett and Cooney: Teenagers who watched when they were little have better grades in high school, read more books for pleasure and express less aggressive attitudes than teens who rarely watched. In 2004, no longer supported by the government, *Sesame Street* was paying for 70 percent of programming from licensing, publishing and grants. Who could have imagined how long and joyfully the show would run after November 10, 1969, when Big Bird blithely led children through Central Park to *Sesame Street* with the song:

Sunny day
Sweepin' the clouds away
On my way
to where the air is sweet
Can you tell me how to get
how to get to Sesame Street?

RAYMOND DAMADIAN Millions benefit from what he did first—take an image of the inside of the human body with an MRI scanner

1936–

COUNTLESS LIVES have been saved and enhanced by magnetic resonance imaging (MRI). The machines are such a familiar part of medical investigation it is at least curious that the doctor who first contrived a picture of people's insides by this technique was regarded by eminent scientists as a crackpot and a charlatan. He still is by some, though today 60 million patients a year worldwide benefit from scanning by MRI machines that now yield meticulously detailed images of soft tissue without any of the risks associated with X-rays.

Dr. Raymond Damadian was sorely tried. While he was struggling in the '70s to achieve one of the great medical breakthroughs of the 20th century, he was denied promotion at his university and his grant money was cut off. He persisted in his lonely course, drawing on the deepest emotional reserves and his faith in God, but John Gore, the director of the Institute of Imaging Science at Vanderbilt University, has suggested that the very obsessive character of Dr. Damadian may have clouded scientific judgment on his accomplishments. He upset a lot of people; like a number of the innovators in this book, he was relentless, egocentric, paranoid, abrasive, excitable and easily angered, qualities inflamed in 2003 by the decision of the Nobel Prize committee not to recognize him along with two men who invented and improved the fundamental basis for MRI for multiple purposes, Paul Lauterbur, a chemist at the University of Illinois at Urbana-Champaign, and Sir Peter Mansfield, a physicist at the University of Nottingham in England.

Nobel committee members admit that Damadian's personality worked against him. The years of hostile barbs he directed toward Lauterbur had an impact. According to the *Chronicle of Higher Education*, "Mr. Lauterbur let it be known that he would not accept the Nobel Prize alongside Dr. Damadian." Gary H. Glover, a professor of radiology at Stanford University, affirms: "Paul was adamant about not sharing the prize. He would rather give it up than share the prize with Dr. Damadian." Lauterbur—and others—had taken offense at Damadian's vociferous efforts to publicize his accomplishment. Academics ridiculed what they saw as showboating. Damadian's supporters point out that only someone with a personality like his would have had the animus to persist.

Lauterbur and Mansfield undoubtedly deserved their honors since the Nobel committee awarded its 2003 prize for discoveries "that led to the development of modern magnetic resonance imaging." Damadian's way of producing medical images was supplanted by the fundamentally different method of Lauterbur. Subsequent improvements took off from there. (MRI does not take photographs of the body. The signals created by nuclear magnetic resonance [NMR] do not produce any kind of graphic image that can be seen by the human eye. They are turned into images only through highly mathematical computer assessment and manipulation.) The technique today is immeasurably faster; there is also a new procedure called "functional MRI" that enables neurologists to study the brain at work in real time. Still, there can be no denying that Damadian's scientific contribution to all this was considerable—and he was in the classic mold of the innovator, determined to make and market a machine that accommodated a full-sized human being. Many chemists and physicists, Nobel Prize winners among them, failed for more than 30 years to see what he saw as a physician venturing into physics: that nuclear magnetic resonance had profound medical potential. "No one ever attempted a simple test-tube NMR measurement of cancer tissue," Damadian told us. Lauterbur and Mansfield, among others, were first turned on to the possibility of using NMR to make images by Damadian's 1971 *Science* paper. The images they made were far superior to his, but Damadian argues, "The Wright brothers had a very primitive technique for controlling flight. Yet that doesn't discredit the Wright brothers, because they did it first."

In the endless back-and-forth, detractors suggest the Wrights parallel is overdrawn because they worked out all the essential elements controlling flight and put them into practice, but in many ways the question of whether Damadian deserved to share the Nobel with Lauterbur and Mansfield is irrelevant to his innovation. Of the three scientists, Damadian was undoubtedly the innovator. As a scientist inventor, he was the first to discover a way of pinpointing an NMR signal; the first to create an MRI image of the human body; and the first to win an MRI patent. As an innovator, he was the first to start an MRI company, and he unveiled the world's first commercial MRI scanner. Moreover, he has remained an innovator into the 21st century. One problem with the standard lie-down MRI machines is that some people feel claustrophobic when shuttled into the cigar tube. Damadian's company, Fonar, in Melville, New York, is manufacturing a unique stand-up machine that has the additional advantage of scanning patients in a variety of MRI weight-bearing positions.

TELLTALE SIGNALS: Damadian, in 1971 at the National Magnet Laboratory at MIT, checks the oscilloscope for radio signals indicating a quantitative discrimination between tumor and normal tissue.

Even though the Nobel committee chose to overlook what he did in the '70s, his achievements continue to win acclaim—the National Medal of Technology from President Reagan in 1988, the Economist's Innovation Award in 2003 and the 2004 Franklin Institute Bower Award.

DAMADIAN WAS brought up in a working-class section of Queens, New York, the son of an Armenian father and a half-Armenian, half-French mother. His father, a photoengraver for the *New York World* (later *World Telegram*), had barely escaped with his life during the series of genocidal massacres of Armenians intermittently carried out by the Ottoman Empire from 1895 to 1918. Both parents played and loved the violin, and Raymond was so proficient that at the age of eight, he was accepted at the Juilliard School of Music while attending PS 101, an excellent public school. He dreamed of being a soloist.

The choice came when he was 15. At Forest Hills High School, where he excelled in math and science, he won a Ford Foundation scholarship that would allow him to enter a university before he finished high school—but it would preclude a professional music career. Juilliard's advice that he might make a career playing in an orchestra but was unlikely to be a soloist settled it for him. He went to the University of Wisconsin to major in math and minor in chemistry.

Damadian had been much moved when he was ten by the sufferings of his grandmother, who was dying of cancer, and the idea that he was destined to find a cure propelled him next to take a medical degree at the Albert Einstein College of Medicine, a newly created part of Yeshiva University. He paid his way by working over the summers as a tennis pro at the Dune Deck Hotel in Westhampton—he was that skilled in the sport—but he put his music to good use, too, serenading a young woman called Donna Terry for whole evenings of guitar and song. She found him "a lot of fun." They married in 1960 on his graduation from medical school, and she saw another side of him when he went on

to intern at the Downstate Medical Center in Brooklyn, assigned to Kings County Hospital. "I remember waking up one night because he was on his hands and knees banging the pillow," she told biographer Sonny Kleinfield. "In his dream, someone had had a cardiac arrest and he was trying to revive him. That's how intense he was about his work."

A medical mystery that stirred his fiercest concentration was how the kidney regulates the amount of sodium in the bloodstream. During a postdoctoral fellowship at Washington University in St. Louis, then at Harvard, and then at the School of Aerospace Medicine at Brooks Air Force Base in Texas after he was drafted, he pursued the search for the then-theoretical "sodium pump," which keeps sodium out of a cell and potassium inside. His fruitless hunt led him to read about Gilbert Ling, a controversial physiologist who argued, wrongly, that the sodium pump didn't exist. In innovation, however, even wrong ideas can be useful. Ling's correct corollary that the structure of water is different in healthy and cancerous cells led Damadian to the crucial next step. His random readings in chemistry, way outside the scope of his medical courses, got him thinking that it might be possible to distinguish cancer cells from normal cells by examining their chemical composition, instead of looking at their shapes through a microscope. Cancer cells were known to be unable to regulate their levels of potassium and sodium the way healthy cells do. The challenge then was finding the most effective way to measure these elements.

He returned to Downstate in 1967 as an assistant professor in the biophysical lab of the Department of Medicine and continued with potassium experiments, forcing bacteria to increase or decrease their intake of the chemical with the help of two students working on their Ph.D.s, Michael Goldsmith and Larry Minkoff. The bearded Goldsmith was a heavyset, sensitive man, competitive with the skinny and caustic Minkoff—an important physiological detail in the MRI story. In April 1969, Damadian went to a conference of the Federation of American Societies of Experimental Biology in Atlantic City. His encounter there

with a physician-physicist named Freeman Cope would be as significant for MRI as Herbert Boyer's meeting in Hawaii in 1972 with Samuel Cohen would be for biotechnology. Cope, a disciple of Ling's, had measured sodium in the brain and kidney using a spectrometer borrowed from Nuclear Magnetic Resonance Specialties Corporation, then a struggling 30-person company in New Kensington, just outside Pittsburgh. Damadian had first come into contact with NMR at Harvard, where he had taken a course in quantum mechanics taught by Edward Purcell, who had shared a 1952 Nobel Prize with Felix Bloch of Stanford. Purcell and Bloch had developed the 1938 discoveries of the Columbia physicist Isidor Isaac Rabi, who had measured the magnetic characteristics of atomic particles such as protons by subjecting them to an oscillating magnetic field.

Positively charged nuclei act like tiny magnets, lining up as a compass needle does in the direction of the external magnetic force. If millions of such nuclei are then bombarded with electromagnetic waves of just the right frequency, the nuclei flip over. This new coherent state lasts only a fraction of a second before the nuclei relax and right themselves again. As they do, they emit radio signals at the same frequency they received—they resonate. (Or in the words of a *New York Post* headline of 1939: "We Are All Radio Stations.") What makes magnetic resonance such a wondrous tool is that the relaxation time varies according to chemical composition so that NMR can determine the precise molecular makeup of chemical compounds.

Having succeeded in detecting sodium by magnetic spectrometer, Freeman Cope wanted to try detecting potassium, but it emits a much fainter signal. Damadian had an answer: He would get some bacteria from the Dead Sea with twenty times the normal level of potassium. Two weeks later, Damadian and Cope drove to NMR Specialties with a car filled with electronic and biological equipment—and the precious bacteria. The experiment worked; a potassium signal popped up on an oscilloscope the instant they put in the Dead Sea specimen.

Damadian said later, "It had a profound effect on me. I mean, wow! In a few sec-

onds we were taking a measurement that would usually take me weeks and sometimes months to do accurately. I was awed by something else. I observed with considerable excitement, 'Good heavens, Freeman, this machine is doing chemistry by wireless electronics.'"

A few days later, Damadian said to Freeman at breakfast, "If you could ever get this technology to provide chemistry of the human body the way it does for the chemist on a test-tube chloroform, you could spark an unprecedented revolution in medicine." Cope's response was that it might be possible in theory, but was not really practicable in terms of cost and time. The NMR machines in use were small, confined to holding something the size of a test tube or lab slide. Damadian understood Cope's skepticism. He later told Kleinfield, "You have to understand that we were talking about machines then that were wide enough to take something no bigger than a pencil. What I was talking about was like going from a paper glider that you tossed across the classroom to a 747."

The director of NMR Specialties, Paul Yajko, agreed Damadian could return to experiment on his own. He took a few days to master the machine—there was nobody to show him—then in June 1970 he put excised tumors of rats in the machine and compared the relaxation time with healthy tissue. He was stunned. The numbers were enormously different, so much so that he was sure he had made a mistake with the machine. He had not. The relaxation time of hydrogen nuclei of water in cancerous cells remained markedly distinct. He was giddy with excitement.

Damadian did not think of making images at this point nor did he mention imaging in the paper he published in the journal *Science* in March 1971. What he had in mind then was something more quantitative—using the NMR relaxation rates to build a cancer scanner. But that wasn't all. "Once I made the hit with cancer," he said, "it was immediately obvious to me that it would be good for all diseases—for heart disease, kidney disease, mental disease, the works." Under questioning by science writer Ed Edelson in the spring edition of the *Downstate Reporter,* Damadian envisaged distinguishing the different tissue by moving a magnet around so as to focus on one "sweet spot" at a time, a point-by-point scan. This was the first published mention of providing a means of localizing the information from a scan of a human body. He would later use this technique to create his first images.

Damadian was right in his prediction that he would generally be regarded as "a screaming lunatic" to seek funds for an NMR machine big enough to scan people. Years had passed since Rabi's initial discovery, but it still sounded very bizarre to suggest that the decaying signal of septillions of wobbling nuclei could turn human beings into radio transmitters. The New York City Health Research Council told him that the idea of full-body NMR scanners was "meaningless." The National Institutes of Health (NIH) rejected him, too. Damadian was so angry he wrote a letter to President Nixon, who had declared war on cancer, charging the NIH with "colossal stupidity" and begging the president to intervene. Nixon seems to have done so. In January 1972 Damadian was given funding for three years at $20,000 a year, and then another grant when that ran out. He bought a standard small NMR machine and continued his research into the relaxation rates of 28 different kinds of tissues, cancerous and normal, in human beings. He found, for example, that breast tumor tissue had a relaxation rate of 1.080 seconds, while normal breast tissue was only 0.367. He published his findings in the *Proceedings of the National Academy of Sciences*. He was now more determined than ever to build a full-body machine.

Although the medical establishment had no interest in Damadian's breakthrough, the chemists and physicists in the NMR community were intrigued. James Economou, a student working with Donald Hollis, an expert in NMR at Johns Hopkins University Hospital, suggested they should see if it was possible to detect cancer in humans, as Damadian had done in rats. "So I sat down and read Damadian's paper closely," Hollis told Kleinfield, "and except for suggesting that you could detect cancer, it didn't make much sense." He called Damadian's theory of structured water molecules "a totally harebrained theory. It's just madness." Damadian told us,

"Structuring of water within living cells and within tissue was universally rejected. Consequently, no NMR spectroscopists of the day, like Hollis, would have any reason to suspect the NMR water proton signal from cancer tissue would be different from healthy tissue." Economou and Hollis didn't have access to their own NMR equipment, so they visited none other than NMR Specialties in New Kensington, where Damadian had performed his experiments the year before. After two days of tests with their own live rats, they knew Damadian was on to something. Hollis wanted to study the question further, with many more types of cancer; he did not make the leap to try making a scanner that would accommodate full-sized human beings.

A few months later, in early September 1971, another Johns Hopkins researcher, Leon A. Saryan, a graduate student of Hollis's, brought more rats to NMR Specialties. His experiments were watched with great interest by Paul C. Lauterbur, a 42-year-old chemist who had just succeeded Yajko as president of the financially strapped company.

Born in Sidney, Ohio, about 40 miles from Dayton, the son of a mechanical engineer for the Peerless Bread Machine Company, Lauterbur grew up in rural America, enjoying outdoor pursuits of hunting, fishing and riding, but he was also a fiend for science. In high school in 1945 he gave lectures on the atomic bomb to a science club he had formed. He had good but not great grades, but took a degree in chemistry at Case Institute of Technology—later merged into Case Western Reserve University—and then pursued his Ph.D. while working as a research assistant at the Mellon Institute in Pittsburgh. Dow Corning funded his main scientific interest in silicon-based polymers, long chains of molecules with elastic properties like rubber. A lecture on NMR at the University of Pittsburgh made him realize what a useful tool it could be; then, on being drafted into the army, he was sent to the NMR lab in the Army Chemical Center in Edgewood, Maryland, to help research on chemical warfare. Coming out of the army, he started an NMR program at the State University of New York at Stony Brook, then took a year's sabbatical at Stanford University to do more

research on NMR. When he came to watch Saryan that day he was already pondering the principles of scanning anything. He was pressing on the frontiers of knowledge.

The signals from laboratory tissue were certainly interesting, but how would one be able to differentiate the source of the signals enough to make a chemical map? A sublime insight came to Lauterbur as he hunched up over a double-decker hamburger—a Big Boy—in a local diner called "Eat 'n' Run": a magnetic gradient. Think of a hot plate (perhaps subliminally he did). The temperature of the hot plate tends to increase as one moves toward the center. Thus there is a temperature gradient—the temperature at any given point depends on what part of the hot plate one chooses. In a normal NMR machine, the magnetic field is steady at all points, but Lauterbur reckoned that if he superimposed weak magnetic field gradients on the uniform magnetic field, the resonant frequency of the radio waves would be slightly different at every point along the object being scanned. A scanner would then be able to tune in to a different signal for every part of the object, telling a computer the exact spatial position of the source.

Mathematical analysis of the variations in the signal would permit the construction of an image in the sense of a picture. Combining shadows thrown up by gradients in different directions would be rather like combining different silhouettes of the same object seen through different windows of a house. Lauterbur did not work out any of the mathematics in his initial paper. Like Damadian's first paper, it described an insight whose ramifications had to be worked out over several years with the help of others skilled in math and physics. But an essential difference was this: While Damadian did note in 1971 that there were signal variations in normal tissue, his "principal concern," he told us, "was detecting cancer in the live human body." Lauterbur, by contrast, fastened on the signal variations from all tissues, normal as well as diseased. His central realization was that he could resolve the variations into a three-dimensional image.

Lauterbur wrote down his ideas in three pages of a notebook and asked Donald Vickers at NMR Specialties to sign his name and the date as a witness, a common procedure when a scientist thinks he might be able to patent his idea. Lauterbur was certainly aware of Damadian's initial diagnostic work. Bettyann Holtzmann Kevles says so in her book *Naked to the Bone,* and in that notebook entry of his for September 2, 1971, Lauterbur wrote: "For example, the distribution of mobile protons in tissues, and the differences in relaxation times that appear to be characteristic of malignant tumors (R. Damadian, *Science,* 171, (1971), 1151), should be measurable in an intact organism." Vickers signed the notebook the next day. Later, Kleinfield asked Lauterbur what he thought of the "remarkable combination of chance events" that led from Damadian to NMR Specialties to Hollis to Saryan to him. "Yes," said Lauterbur, "but life is full of things like that. If you turn left instead of right at the corner, you might not meet your wife."

Lauterbur insists the idea of building a scanner big enough for a person was his own and that he did not know of Damadian's intentions. On the other hand, the man who signed Lauterbur's notebook, Donald Vickers, says: "I think it was common knowledge at this time, at least throughout New Kensington, that Ray Damadian wanted to stuff people into NMR machines." Vickers later spelled out his understanding of the order of events in a letter to Damadian in 1987. According to Vickers, Damadian made four pivotal contributions that preceded Lauterbur's insight: (1) he was the first person to think of using NMR medically to differentiate healthy and cancerous tissue; (2) he was the first to prove this hypothesis in his experiments at NMR Specialties; (3) he was the first to think of building a full-body NMR scanner capable of diagnosing disease; (4) he was the first to invent a method for spatially localizing an NMR signal to make a point-by-point scan (which he used six years later to make the first MRI scan of a human body).

"At the time of Paul Yajko's departure, your work and intentions were common knowledge at NMR Specialties and were openly discussed," Vickers writes. "During the summer of 1971, and therefore after all of the above enumerated events, Paul Lauterbur, who succeeded Paul Yajko as president of NMR Specialties, discussed with me a plan for using an NMR spectrometer and field gradients for making pictures."

Is it possible that Lauterbur, president of a company abuzz with controversy over Damadian's idea for a scanner, decided to build a scanner without any awareness of Damadian's intent? Perhaps, but it hardly seems likely he was uninfluenced by the environment he worked in every day. In any event, Lauterbur was so excited by his own original idea for imaging (as distinct from medical diagnosis) that he almost immediately resigned from the soon-to-be bankrupt NMR Specialties and went back to Stony Brook to work on what he called "zeugmatography," derived from the Greek word *zeugma,* meaning "joining," the idea being that he was taking pictures by joining two magnetic fields. (This is reminiscent of Reginald Fessenden's choice of "heterodyne," formed from two Greek words to describe his own inventive superimposition, in his case two electromagnetic frequencies for the transmission of sound.) Lauterbur tried to set up a business. He broadcast his ideas at scientific meetings around the world while his team worked out the details, and he asked the Research Corporation, an organization supported by revenue from patents, to start a company. It declined on the grounds that it was unable "to identify a potential market of sufficient size to justify our undertaking the patenting and licensing of this invention." In the end, Lauterbur gave up being an entrepreneur.

Damadian filed for a patent on March 17, 1972. He had no more success with venture capital than Lauterbur. A Texas company, Hycel, sent a physical chemist and biologist to investigate and concluded it was not possible to build a full-body scanner. Damadian decided to go it alone, his competitive zeal reinforced by the accelerating suspicion that he was the victim of theft. He had brooded over the way his order for a superconductor magnet, placed with NMR Specialties, had been canceled by its president—Lauterbur. It was an innocuous enough intervention on Lauterbur's part; he knew the company could never fulfill it because it was going out of business. But Damadian was uncontainable

when Lauterbur published a report in the science journal *Nature* on March 16, 1973. He called it "Image Formation by Induced Local Interactions/Examples Employing Magnetic Resonance." Lauterbur's and Damadian's chosen titles emphasized their different foci of interest. Damadian wrote about "tumor detection" with NMR. Lauterbur described "image formation" with NMR. For the Nobel Prize committee in 2003 this divergence in emphasis would be crucial. Damadian was describing why NMR might be useful for medicine. Lauterbur was describing a specific technique for manipulating NMR. (Kleinfield says that Lauterbur's paper included the first NMR image ever made, but it seems this honor belongs to scientist Vsevolod Kudravcev, who had actually made a television image of an egg embryo in the 1950s before being told by his NIH supervisor to stop wasting his time.)

Nature had originally declined to publish Lauterbur's paper. It wanted him to spell out the "larger significance" of his method. He did so by pointing out its potential uses in science and technology as well as medicine. In revising the paper, he mentioned that the technique could be used to distinguish between normal and healthy tissue. He added: "A possible application of considerable interest at this time would be to the in vivo study of malignant tumors, which have been shown to give proton nuclear magnetic resonance signals with much longer water spin-lattice relaxation times than those in the corresponding normal tissues."

Who had done this research? Lauterbur had six citations in his five-paragraph paper. He mentioned several other articles that verified Damadian's initial research and cited Damadian's *Science* article—one of the other references cited Damadian's paper in its first paragraph—but Lauterbur himself forbore mentioning Damadian, despite having acknowledged Damadian's paper in his own notes signed by Donald Vickers. It was another decade before Lauterbur would acknowledge Damadian's work in public. In an entry for the *Encyclopedia of NMR,* he explained that he did not mention Damadian's experiments because he thought Damadian's "controls were inadequate and the publicity overdone."

Hollis, too, became more and more critical of Damadian, going so far as to publish numerous papers saying he saw little medical value in what became MRI. Hollis said of Damadian, "I don't know that he's what you'd call a scientist. He's a businessman or a public relations expert."

From the very beginning Damadian felt that he was excluded from an insiders' club of academics with Ph.D.s, a charge he repeated in many of his 2003 full-page ads in the nation's newspapers criticizing the Nobel Prize committee. His critics, on the other hand, accused him of making wild claims not justified by his research. His money ran out in November 1973, and he got nowhere with repeated appeals to the National Cancer Institute, a branch of NIH. The American Cancer Institute turned him down five times, saying his research was not high priority. Desperate for funding, he started a foundation, the Citizens' Campaign for New Approaches to Cancer. Damadian's father went cap in hand to the Armenian community and raised all of $65. Damadian wrote to Kirk Kerkorian, then head of MGM, as a fellow Armenian American. He wrote to other rich Armenians. "They didn't give me the time of day," he recalled. He spent a lot of time reading the Bible and praying. The man who came through was his brother-in-law, David Terry, a real estate and insurance agent who lived with his family in East Quogue. He blitzed the area with a campaign about "vested interests" holding back cancer research. Terry raised around $10,000 over the next few years. "Sometimes we'd go to a flea market and come back with fifty dollars. Little kids would toss in quarters. Pocket change from little kids that might have gone for bubble gum or Tootsie Rolls was paying the guys in the labs, making this thing happen." Pocket change paid to keep MRI alive.

Damadian was his own worst enemy in the struggle for funding. Kenneth Olson, director of the diagnostic branch of the National Cancer Institute, became a forceful advocate on the grants committee. "I

said that if Roentgen walked in that day saying he could see bones in the body with a machine they'd laugh him out of the room." But Damadian was so volatile, Olson told Kleinfield, he put people off. "Many times I had to hold the receiver away from my ear when he called at the NCI, and I had to calm him down before we could get on to any sort of productive conversation. In short, the man had a temper. As professionals, we should have seen

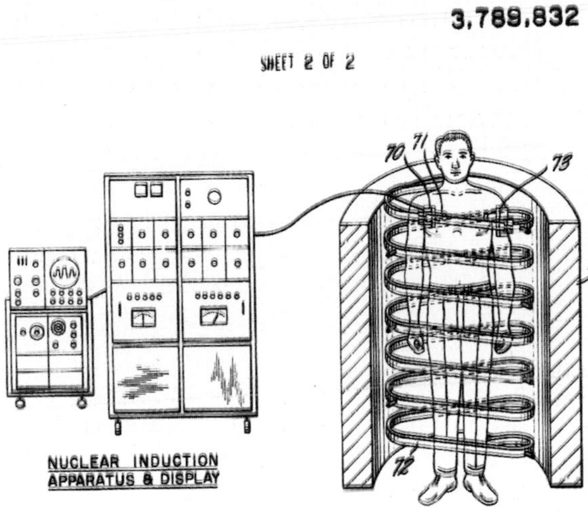

3,789,832

SHEET 2 OF 2

NUCLEAR INDUCTION
APPARATUS & DISPLAY

FIG. 2

PATENT NO. 3,789,832: Damadian's drawing for an "apparatus and method for detecting cancer in tissue"

through the kook in Damadian." Olson dismisses the idea that there was any kind of conspiracy to starve Damadian of money, but there was pettiness and jealousy. "Damadian was a physician, and he came in and discovered it. The physicists missed the boat. Nobody likes to miss the boat."

On February 7, 1974, Damadian was awarded Patent no. 3,789,832 for "Apparatus and method for detecting cancer in tissue." It was a pioneer patent, the only one in its class for the next seven years; by 1994 there would be 740 other patents in the grouping Damadian inaugurated. Olson says the NCI was not happy about scientists patenting their inventions; if they suspected he might use it in business, they were right. Thanks to Olson's efforts, the NCI did finally agree to fund three NMR researchers: Damadian at $100,000 dollars a year, Donald Hollis at $75,000 a year and Carlton Hazlewood of Baylor College

of Medicine at $50,000 a year. Damadian hired Joel Stutman to do the computer work, as well as two more graduate students, Michael Stanford and Jason Koutcher. Damadian recruited machinist Nean Hu, a Chinese refugee who had been imprisoned during the Cultural Revolution, when many scientists were ripped out of universities and their careers blasted.

But the opposition had not given up. Hollis, who had difficulty at Johns Hopkins repeating many of Damadian's experiments, remained particularly skeptical: The idea of scanning the full human body, he declared, was visionary nonsense. At a review with NIH officials in Bethesda, Maryland, Hollis said that using NMR to detect cancer was a dead end. Damadian interrupted him, shouting, "You're setting back science by ten years. If you don't believe in this, you ought to get out of the field." Hollis shouted back, "I like this field and I'll get out of it when I feel like it."

Damadian faced opposition at Downstate as well. "I was called a lunatic, a crackpot," he says. The pathology department refused to supply him tissue samples to scan. He scraped together cab fare to send graduate students with ice buckets to collect samples from other hospitals.

Lauterbur became the first researcher to image a living creature, a tiny clam his daughter had found on the beach, and the only animal that would fit in the four-millimeter hole of Lauterbur's magnet. He progressed to nuts, green peppers and a mouse, which he showed Damadian. "I think he showed it to me to impress me," Damadian says. "I was impressed." Damadian raced to catch up. In February 1976 he put an anaesthetized mouse, dubbed Pioneer Mouse, in his own small magnet. As he increased the current to the tiny coil around the rodent's belly, Damadian began to smell something. "I cooked him. It was just a screwup on my part." Pioneer Mouse Number 2 survived his scan a month later, and the blurry image Damadian took wound up on the December 24, 1976, issue of *Science*. The editors thought it looked like a Christmas ornament. Damadian then put a monkey into a larger machine, but it woke up from its anesthesia in the middle of a long scan and crawled out. Minkoff searched for hours before finding the monkey sitting atop the machine and staring down at him with folded arms.

Animal testing was but an intermediate step. The grail everyone in the field sought was a scan of a living person. Only that, Damadian felt, would convince the medical community of MRI's importance. He also hoped to reap the lion's share of the glory. It would be no simple feat. He would require a magnet larger and more powerful than any used before in NMR. It would also require a lot more money.

Unfortunately for Damadian, Olson left NIH, and the forces inimical to Damadian and NMR itself saw to it that the grant was not renewed for 1976. It meant Damadian had no money to buy a magnet powerful enough. In April 1976 it finally dawned on him that he would have to give up or attempt to build one on his own on the spot—a major undertaking. For a start, the ceiling in the lab was too low. Damadian, Goldsmith and Minkoff exhausted themselves attacking the ceiling's foot-thick granite with sledgehammers and chisels before eking out money for a construction crew. The magnet would be made by sending a current through coils of superconducting wire cooled by liquid helium to near absolute zero, minus 273.15 degrees centigrade. The ends of the wire would then be fused together, and without any electrical resistance, the current would flow around forever, creating a powerful magnetic field. The slightest flaw in the dozens of miles of wire or a poorly wrapped coil would ruin the image and eventually weaken the magnet. Handling the helium was especially tricky; it is volatile and could easily evaporate into the atmosphere. Too little helium risked the superconducting coils melting within seconds.

Damadian enrolled in a course at the RCA Institute of Electronics and got advice and software from the physics department at nearby Brookhaven National Laboratory. He needed nearly 150,000 feet of superconducting wire—nearly 30 miles—at an unaffordable dollar a foot, but he was in luck. Westinghouse had abandoned its Bal-timore superconductor business, and the physicist in charge had exactly 150,000 feet he would let go for ten cents a foot. The next day Minkoff and Goldsmith were on their way to pick it up in a rented U-Haul. Next, all three of them crawled around the Brooklyn Navy Yard, where battleships used to be assembled, in the hope of finding metal they could use for spooling the wire. Damadian spotted some junked metal bookcases he thought would work in the basement of Downstate Medical Center, so they hauled them up to his lab. For sixteen hours a day, six days a week, Goldsmith wound the wire, dizzy from watching the rotating bookcase spools they had constructed. A Downstate professor rebuked him: "Why don't you stop what you're doing, because it's an embarrassment to the university, before it's too late. The guy is a failure." Even at his home base, Damadian was not taken seriously. He was that weirdo working with giant magnets and vats of liquid helium, activities apparently remote from medicine. In normal NMR scans for chemicals, test tubes were spun around inside the magnet hundreds of times a minute to make the chemicals more homogenous. The one-liner making the rounds in NMR circles nationally was "Well, how fast, Dr. Damadian, are you going to spin the patient?" The joke was ubiquitous. Countless times after giving a speech, someone in the audience would raise his hand and ask, "Dr. Damadian, would you mind telling the audience how fast you plan to spin the patient?" It produced a predictable guffaw every time.

While Goldsmith made the magnetic coils, Damadian and Minkoff spent a full year on the delicate welding of insulated vessels to contain the helium. These "dewars" were metal doughnuts ten feet tall, six feet wide and eighteen inches deep, and a single pinhole to the outside would ruin them; in the process, they found and repaired dozens of leaks.

Damadian was worried that he would be beaten by Lauterbur or by Peter Mansfield at the University of Nottingham. Lauterbur and Mansfield were both known to be working on rendering images from NMR data, and both had ordered magnets from specialist companies. To Lauterbur's consternation, the magnet with a 24-inch-diameter bore hole he had ordered from Walker Scientific, a Massachusetts company, arrived with a bore hole of only 16 inches, too small for a human body. Mansfield was still awaiting his magnet through 1977. Unknown to Damadian, there was a third challenger; the British firm EMI, pioneer of CT scanners, had funded a secret project, code-named Neptune, for a full-body scanner.

Emotions were running high in the long exhausting hours at Downstate. Goldsmith urged Damadian, "Let's just do an arm and let's publish that." Damadian would settle for nothing less than a body scan. He was utterly ruthless with everyone's lives, including his own. He had no money left to pay the assistants. He was encouraged to fly to Plains, Georgia, to meet with president-elect Jimmy Carter. He did; nothing came of it. The white knight turned out to be a wealthy Tennesseean named Bill Akers, a retired civil engineer who had made a fortune from the sale of his asphalt company. Akers heard of Damadian's research through a scientist he met while raising money for Vanderbilt's engineering school. He traveled to Downstate and found Damadian. No big foundations responded to Akers's appeals, so he wrote a personal $10,000 check, then got his brother and two friends to give Damadian $10,000 each. The $40,000 kept the assistants and paid for liquid helium, which evaporated at a rate of $2,000 a week, despite the focus on fixing leaks.

The final hurdle was making a radio coil that would wrap around a patient to pick up the signals from the nuclei. Damadian did not have the mathematical training to match an antenna with the right set of capacitors to detect radio waves, and information from NMR Specialties proved inappropriate for the scale of their operations. Damadian pushed Goldsmith to the limit, telling him he would just have to arrange capacitors by trial and error. Goldsmith, as massive as his taste for understatement, commented: "A lot of fiddling was involved." It was not until June 1977 that he got a radio coil to work at 14 inches in diameter. It was made of garbage bin cardboard and copper foil tape. He couldn't get anything larger to work. It would just have to do.

Considering everything he and his team had had to go through to build the machine, Damadian dubbed it the "Indomitable." It was a primitive monster. The subject, strapped in the cardboard radio coil, would have to sit on a narrow wooden plank running through the center of the giant magnet. No one knew what effect the intense magnetic field and bombardment by radio waves would have on a human being. In theory, nothing would happen, but no one really knew. Damadian felt it was his duty to take the risk. Downstate regulations required that Damadian ask for permission before experimenting on himself. He foresaw months of ponderous deliberation and told nobody. He also reckoned that the obstructive department chairman (Alfred Bollet) would not really care if he dropped dead.

On the evening of May 10, Damadian, still nervous about having himself locked in the embrace of the giant magnet, scanned a dead turkey instead. The 26-minute scan suggested the magnet's 500 gauss strength—less than he had hoped to build—was enough to scan a human being. The next day, May 11, 1977, Damadian asked several doctors from the hospital to come into the lab with a defibrillator and an EKG. He wrote in his notebook, "30 minutes in the field if I can stand it, 15 if I can't." Damadian took off his shirt. The doctors monitored his heartbeat and blood pressure and took an EKG of his brain. A picture shows him looking fatalistically into the camera.

The tight coil was squeezed down around Damadian's chest. He had to take shallow breaths; too deep an inhalation and he could have burst the coil. The machine was turned on at 8:55 p.m. There was no signal. Goldsmith and Minkoff found a broken wire, repaired it, and seven minutes later Damadian was back inside the mag-

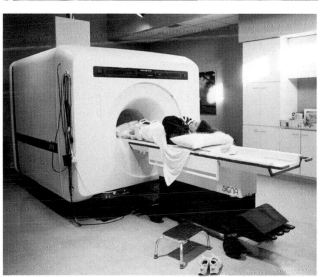

GUINEA PIG: The reluctant Lawrence Minkoff finally submitted himself to Damadian's primitive machine for the first-ever scan of a human body on July 2, 1977. Above: Figure skater Nancy Kerrigan undergoes a scan of her knee after it was injured by a club-wielding assailant.

net. There was still no signal. Over the next few hours, they positioned Damadian in all sorts of ways, but nothing showed on the oscilloscope. He noted: "Virtually no adverse reaction to the time of the field. No headaches. No eye pain. Some flushing around the ears, No dizziness. Not even the usual confusion from the magnet

headache." But history was not going to be made that night.

Damadian was sure the answer to the puzzle would come to him during sleep. It didn't. The next day they reassembled to brood. No one could figure it out until Goldsmith, no stick insect himself, told Damadian, "Maybe you're too fat." In truth, his belly was not small, and since the body conducts electricity, it was possible that the resistance and energy absorption of Damadian's mass was overloading the antenna. They would need a skinnier volunteer. All eyes turned to Minkoff. No one had to say it out loud. Minkoff, then 26, recoiled. Everyone wooed Minkoff in the name of science, posterity, fame, fortune. None of it worked. Goldsmith was less subtle: "Larry, get in the damned machine already." Minkoff told them where to get off.

On July 2, 1977, a month and a half after the failure of the original test, Minkoff told Damadian he would take the risk. At midnight Minkoff removed his shirt and slid apprehensively into the cardboard radio coil. Even on him it was snug. With Minkoff squeezed into the magnet, Damadian said to Goldsmith, "Well, Mike, you've finally got Larry where you've wanted him all these years." The laughter broke the tension; then they were all focused on the oscilloscope. Suddenly, it picked up a signal. Minkoff's body was doing its work, emitting radio signals. Minkoff thought the scanning would last a few minutes. Damadian had other ideas. He had a graduate student at each end of the plank and kept telling them, "Move him one inch to the right, and again. . . ." They recorded the data for a couple of minutes, then moved Minkoff another inch, scanning a line point by point across the width of his chest. It was still not enough for Damadian. They kept moving Minkoff backward and forward, scanning more lines, point by point. Throughout all this, a chilled Minkoff had to keep his arms raised above his head; otherwise they would lose the signal in the coil. Every time Minkoff started to droop his arms, Goldsmith enjoyed yelling at him. After an hour, Damadian hung a piece of rope in front of the machine so Minkoff could rest his arms on it. At two hours, Damadian called for a break. Minkoff had an ice cream before Indomitable claimed him again. At 3 a.m., Minkoff was so cold he insisted on putting his shirt back on. The graduate students were also tired from pushing the plank around for hours. Yet Damadian would not stop. Only at 4:45 a.m., July 3, after almost five hours, did he relent. He had moved Minkoff to 106 different positions, taking twenty or thirty readings in each position. Goldsmith sketched out the data on graph paper. The image he drew, based on the readings, was the first of a chest cavity of a live man. Goldsmith wrote in his notebook:

FANTASTIC SUCCESS!
4:45 a.m. First human image.
Complete in amazing detail
Showing heart
Lungs
Vertebrae
Musculature
Image taken at Minkoff nipple level.

Damadian lit a cigar and pulled out a bottle of white wine. The dawning day was July 3, the day fixed for a large Independence Day picnic party at the home of his wife's family in East Quogue. The celebrants were regaled with the Minkoff interior, the drawing Goldsmith had made from the radio data. "I'd never seen him quite so happy," said Damadian's wife, Donna.

As soon as the revelers had departed, Damadian got down to writing a paper for the journal *Physiological Chemistry and Physics*. He was in a hurry to publish; his fear of submitting his results to a better-known journal was that someone would preempt him, especially since both Lauterbur and Mansfield were better funded. On July 20, 1977, using a public relations firm, Damadian summoned 20 reporters to the lab. The press release said, in part, "Dr. Damadian finally achieved a major medical breakthrough in cancer detection WITHOUT NECESSITATING X-RAY, NOR MAJOR SURGERY. Here, at this press conference, attendees will witness the application of this technique to the first live human being. The human subject will be introduced into this giant magnet for the scanning of his anatomy and the formation of the world's first image of living human organs on the video screen." It was an assertion too far, a description of all the things Damadian hoped his scanner would one day achieve rather than a statement of what he had accomplished. Reporters were quickly annoyed when they found they were not actually going to see a live scan but the old image of Minkoff's chest done two weeks before. A local TV news reporter that night said, "Now at today's press conference Damadian seemed to be making some very extravagant claims." The reporter quoted several experts in the field who said, anonymously, "If true, terrific. Needs more investigation"; "cannot detect malignant tumors"; "claims are premature and inflated"; "doesn't publish his details." The *New York Times* ran the headline "New York researcher asserts Nuclear Magnetic Technology can detect cancer, but doubts are raised." The story began,

A New York City medical researcher announced yesterday at a news conference that he had developed 'a new technique for the nonsurgical detection of cancer anywhere in the human body.' However, in an interview later in the day, he retracted the contention that he had already used the technique on a cancer patient. And other cancer experts expressed skepticism that the technique had reached the stage where it could be used in diagnosing cancer.

The story then described how the National Cancer Institute had withdrawn its support of Damadian, quoting an NCI spokesman, Larry Blaser: "We don't look on nuclear magnetic resonance as a promising area of diagnosis."

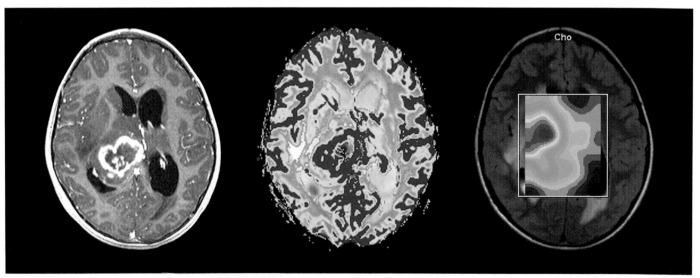

IN DEADLY COLOR: Different types of MRIs can help diagnose disease. The first image is a standard MRI of a brain tumor. MR perfusion imaging in the second picture shows increased blood flow. Damadian's dream of analyzing the body's chemical composition through NMR comes true in the third image—MR spectroscopy shows increased metabolism and reveals the tumor to be malignant.

Many researchers who looked at Damadian's image were puzzled by his seemingly extravagant claims. "The first MRI images were far cruder than those produced by existing CAT-scan technology of the day," Damadian told us. Many saw little need to supplement an imaging technology that looked so promising. Damadian was both blessed and cursed by his ability to see MRI's potential. It spurred him on but caused those who lacked his vision to think him mad. The grudging *Times* report gave Damadian apoplexy.

There was a trickle of congratulations. Albert Szent-Gyorgyi, the Nobel Laureate in biochemistry who discovered vitamin C, wrote, "Delighted with your wonderful achievement." Edward Purcell wrote: "I congratulate you and shall keep the picture as a perpetual reminder of how little one can foresee the fruitful applications of any new physics." Competitive researchers in NMR from the University of Nottingham wrote generously: "Your progress is impressive. There are probably a few people suffering from envy now. This is an important achievement and a valuable step forward in applying NMR imaging to medicine." Peter Mansfield, who would share the 2003 Nobel Prize for MRI with Lauterbur, wrote, "Let me congratulate you on a very nice piece of work." Later, he remarked, "The effort Damadian put in was monumental. You have to admire the guy. He set his mind to it and he pulled it off." (In April 1978, Sir Peter became the second person to make a human scan.)

These encomia were rare. Lauterbur himself told Kleinfield he didn't think much of the image: "The technique itself was an obvious dead end. It was slow and produced low resolution and poorly defined images. . . ." Damadian's technique was indeed much inferior to Lauterbur's—as Lauterbur's was to be somewhat inferior to Mansfield's, as both were to Ernst's, which was improved by researchers at the University of Aberdeen. But Damadian did it first, with no help and no funding. He had to publicize himself à la Edison because no one else would. No one else took him seriously. Donald Hollis said of the first image: "It was nothing but a publicity stunt. It's not good enough to even be sure it's

human. I don't recognize it as a picture. I claim he did not take a picture of the human body." Any medical resident, however, can identify the image as a human chest cavity. Minkoff tested this. So did we. Hollis was plain wrong. Not long afterward, he gave up science to raise chickens and sheep.

Damadian was fully aware his first image was just a start and that he would face setbacks. In late July 1977, several days after the disastrous press conference, he tried and failed to image an actual cancer patient, a friend of his from his Forest Hills Church. Meanwhile, he had no means of repaying the $40,000 owed his department, and chairman Alfred Bollet had no patience. On November 15, 1977, he cut off Damadian's telephone. Famous scientists ringing in from around the world were told Damadian's exchange was no longer working. Damadian had tenure, but he heard rumors his locks were going to be changed. Emergency money arrived just in time to prevent that, but the damage was done. The *New York Times* article had wrecked his chances of the promotion to a full professorship that would have given him more staff and money.

Just when his work and life at Downstate had become intolerable, Damadian overcame the teething troubles. In January 1978, the successful scan time of Minkoff's chest dropped to 35 minutes. On February 5, 1978, Damadian scanned his first cancer patient and was able to demonstrate the difference between healthy and cancerous tissue. Given the hostility of Bollet and the still lackluster prospects for funding, Damadian decided that making scanners and selling them was the only way he could pay for his continuing research. A friend who had business dealings with General Electric suggested introducing him to some GE people. "But they'll steal my idea," Damadian fretted. "No, don't worry about that," his friend said. An executive from GE medical systems visited Downstate, told Damadian he was interested but did not follow up. Soon Damadian heard GE was going to develop its own NMR scanner. The lawsuits that followed would not be resolved for another 20 years. In 1997, the United States Supreme Court

ruled against the giant General Electric in Damadian's favor for $128.7 million (about ten times Fonar's annual revenues then of $13 million).

Damadian read a book on how to start a company for $50 or less, mailing a form to the author along with $50. He called his company Field Focusing Nuclear Magnetic Resonance (Fonar) and mortgaged his house to invest $50,000. Damadian had 51 percent of the stock, Minkoff and Goldsmith 7 percent each and Stutman, 3.5 percent. In March 1980, he launched the commercial NMR scanner, then placed machines in hospitals in Mexico, Italy and Japan. Although the machine was still awaiting FDA approval, in June 1981 Fonar had its IPO. Stocks, offered at $5 a share, were quickly bought (making the NMR pioneers wealthy men). Within a year of Damadian entering the market, major corporations invested heavily in MRI—as it came to be called, dropping the "nuclear" because it worried patients fearful of being exposed to something radioactive. By 1983 machines were being made by Technicare (the subsidiary of Johnson & Johnson), General Electric, Bruker Medical Instruments, CGR Medical Corporation, Diasonics, Elscint Limited, Intermagnetics General Corporation, Oxford Research Systems, Philips Medical Systems, Picker International, Siemens Medical Systems, and Toshiba Medical Systems.

Although small compared to the other players in the field, Fonar has continued to innovate. In 1983 it created a mobile MRI scanner that could be carried on a large truck. Its researchers invented a way of taking images using oblique planes through the body rather than only at right angles. Its stand-up scanner is increasingly popular with hospitals and patients.

The loss of a share in the Nobel Prize pains Damadian deeply. "I've been stricken from history," he says. "My life's work has been stricken." Posterity is likely to have a kinder judgment. Damadian's machine is now in the Smithsonian—not far from a similarly crude first effort that for a long time received no recognition in the Smithsonian, the Wright brothers' *Flyer.* Both resonate with hope.

RUSSELL SIMMONS

He created a cultural movement in music, fashion, movies, comedy, poetry, television—and social action. He's the marketing maestro of hip-hop

1957–

THE MAN who would become the Reverend Joey Simmons was on the stage at New York's Madison Square Garden. At his command, 20,000 people in the audience, caught in spotlights, took off their unlaced sneakers and shook them in the air, hooting and hollering. Almost everyone was in sneakers—Adidas sneakers—because that was the regular dress code for the hip-hop crowds and they were there to rock it out for an explosive, hot hip-hop number called "My Adidas," which was heading for the top of the charts in the summer of 1986.

The whole event had been orchestrated by hip-hop master marketer Russell Simmons, Joey's elder brother by seven years, for whom sneakers were a serious issue:

"I've never had dirty sneakers in my life," he wrote in his 2001 memoir with Nelson George. "Not when I was a teenager and not now." The future Rev. Simmons had recorded "My Adidas" with his partners, Darryl McDaniels and Jason Mizell, though at the time the reverend was better known as Run for his ability to run off at the mouth on any subject; McDaniels was better

Pretty much everything Russell Simmons touched turned to gold. He is a vegan ("I won't eat anything that runs away from me") and keen on the practice of yoga: "One of our practices is to cause the least amount of harm possible to all sentient beings and the planet." He lives on a $14 million estate in suburban New Jersey.

known as DMC, Mizell as Jam Master Jay. Their group, Run-DMC, celebrated for the percussive, in-your-face dynamic of its call-and-respond rhythms, was the most generative act in hip-hop's history. They were three black kids out of Hollis, Queens; along with their shell-toe Adidas, they all wore the ghetto uniform of leather suits with black velour fedoras and gold neck chains. The Adidas executives who had flown in from Germany for the event were falling over themselves to spend one million dollars and more to sponsor Run-DMC national tours. This was the first time the apparel company had ventured beyond athletic sponsorship, and the decision to do so was a portent. In the words of the British critic and author Alex Ogg, the "future synthesis of music, fashion and business that made hip-hop the dominant cultural trend of the nineties was prefigured by the deal."

The supreme synthesizer was Russell Simmons, better known as Rush, and the Adidas deal was negotiated with the "precision of a bank heist" by his associate Lyor Cohen, a lanky Israeli whom Simmons describes as "the glue that held the management together." Joey Simmons had a long run as emcee with Run-DMC, but at the height of his success, he had a revelation while sitting in his Jacuzzi eating pancakes, "smoking a joint and waiting for a ho to call." He writes, "The syrup was falling in the tub, ashes were falling in the pancakes. I thought I had all the riches but I was really poor. It was then that the Lord came into my life and raised me up, and I started to feel better and all those worldly things didn't mean anything to me anymore." He was ordained by Zoe Ministries Community Congregation in New York in 1994.

Darryl McDaniels, after rehab for alcoholism, was ordained as a deacon.

Jam Master Jay, however, was not raised up. He was gunned down in his Queens recording studio in 2002, apparently over an unresolved debt.

Rush, the prolifically enterprising, affable and garrulously profane manager of the group, has gone on and on, spinning golden webs in hip-hop enterprises and preaching his own brand of social conscience. He is the innovator who discovered hip hop when nobody outside inner-city New York had heard of it. He is often compared to Motown's Berry Gordy (see page 470), which does justice to neither man. Gordy had a unique ear for talent; his musical legacy is larger and has endured for 40 years. Simmons's distinction is less in picking stars—though he jump-started the careers of LL Cool J and "the Fresh Prince" Will Smith—than in his very early perception of what hip-hop signified, and his perspicacity as the impresario who pushed it in its pure form in music, and pushed it some more in comedy and movies, then pushed it some more in fashion and was still pushing it some more in 2004 in getting people to register to vote. Bill Adler, an entrepreneur in music, calls him "the Moses of hip-hop" for leading the hip-hop generation to the promised land of American success and prosperity. "The only big crossover artist at the time who was black was Michael Jackson, and he was clearly not the black guy next door; he's a brother from another planet," says Adler. "Russell's genius was to broker the artists as they were. They didn't have to be white. They didn't have to have white mannerisms. They didn't have to speak standard English. Russell's point was that they would get over because they were authentic, true to themselves."

Adler credits Rush with an influence beyond music. What he achieved, in Adler's phrase, is the browning of America. "In the seventies, if you were a white kid, you loved Aerosmith and Led Zeppelin, and if you were a black kid, you were listening to George Clinton and the Delfonics. There was a vast and deplorable gulf in the culture, and what hip-hop has done is to reunite the races on a shared affection for this music and this culture. I think this is a much more integrated society at ground level because of Russell Simmons."

RUSSELL SIMMONS had a great start in life—and almost threw it away. His father was a history professor at Pace University, his mother a Howard University graduate in sociology and the family lived in a pleasant part of Queens. The white population was on the move out, but Rush attended an integrated school that he says had a very important effect. "It taught me to see people for what they were. I never let race affect my judgment of a person's charac-ter." The family said prayers every day and went to church on Sunday. But hidden in the bushes of the front garden there were little bags of marijuana Russell would sell for $5 apiece to indulge his passion for fly clothing—three-stripe Pro-Ked sneakers, $32 silk-and-wool pants. "At the time I thought having the right sneakers and pants meant I was a future Rolls-Royce nigga." Inevitably, the drug pushing led to running with a gang, in his case the Seven Immortals, a Queens branch of the more violent Bronx founding chapter. "For my father," Rush writes in his memoir, "everything I was into . . . meant I was heading straight to prison." He knows he was lucky to come through. When he was robbed once, one of his gang put a .45 in his hand, and he took a wild shot at the fleeing robber. "In my heart I knew missing Red was the best thing I ever did," he said in an interview. "I only shot a gun one time. I shot over his head. I told everybody I tried to kill him. He was murdered a week later, he and his brother. The sad thing is how many of my Hollis crew peers were killed by the drug lifestyle. Some got hit in the head by thieves one time too many. Some got shot. Some died of AIDS in jail."

Around the time he headed off to City College (located in Harlem) in 1975, the ghettos of South Bronx and Harlem were undergoing a sea change. Coming from the parks and community centers of these written-off, depressed, burned-out neighborhoods was a song unlike any heard before. Led by Afrika Bambaataa's Zulu Nation, the hip-hop culture slowly began to replace the gang warfare of knives and guns with battles of sound systems, words and dance moves. At first the music sounded like the disco and funk hits being played in the downtown nightclubs, but the drumbeats and the chants were like nothing else.

Rush describes the epiphany he had one night hanging out in Charles' Gallery on 125th Street when Eddie Cheeba started rhyming and the crowd of black and Hispanic college kids chanted along, moving harder to the beat because the emcee said so. "It wasn't a sophisticated rhyme by current standards, but seeing Eddie spitting them flames on the mike made me feel I'd just witnessed the invention of the wheel.

A lot of people were mad at him, he says, for managing the white hip-hop group the Beastie Boys, but to him they were honest. They were doing their version of black music. "They weren't rip-off artists . . . they were ultimately accepted by black people because they were good."

Just like that I saw how I could turn my life into another, better way. All the street entrepreneurship I'd learned I decided to put into promoting music."

He stopped selling drugs. Instead, he promoted club parties, sticking "Rush Productions" flyers on lamp posts, talking up a storm about artists who could only be heard in person: The recording industry had no interest in the rappers who inspired Rush. "I just looked up to people like the great DJ Hollywood, Melle Mel, Lovebug Starski, Kurtis Blow and Grandmaster Flash (*"introducing the disco dream on the mean machine, a Darth Vader on the slide fader. No one cuts it straighter, than New York's number one creator"*). One afternoon in Queens, *Billboard* magazine's black music writer Robert "Rocky" Ford spotted Rush's brother Joey hanging posters along Jamaica Avenue. He hooked up with Simmons, who says he learned marketing from Ford. "He taught me to have integrity, build the artists, build the truth. A lot of the disco records out there were fake. They put together a group and that was the end of it."

The first artist Rush managed was Kurtis Blow, the name he gave to Curtis Walker, a student at City College. He sold him in Queens as "a smooth Harlem native with

a booming low tenor voice and real charisma," but he could not get him on radio or vinyl. The shock that propelled Rush into even more determined efforts came in October 1979. At the Armory in Queens, a DJ played a new record, "Rapper's Delight," from a New Jersey group, the Sugarhill Gang; it went on to sell 14 million copies. "We were just stunned," writes Rush. "Someone had taken our rhymes, our attitude, our culture and made a record. And not one of us in the community had anything to do with it. I remember after the record played, DJ Starski got on the mike and said, 'Y'all know we started this shit. Don't worry. We're gonna go to the moon.'"

Four credits short of his sociology degree, Rush upset his father by pulling out of school to concentrate on hip-hop. He and Ford cut a record with Kurtis Blow spoofing "The Night Before Christmas" (*"he was roly, he was poly, and I said, 'Holy molie,' you've got a lot of whiskas on your chinny, chin chin"*). The clubs loved "Christmas Rappin'" but the buyers in the record industry had all concluded that the big sales for "Rapper's Delight" were a fluke. Rush and Ford jimmied open the door by going into retailers and placing fake orders for "Christmas Rappin'," saying it was a Polygram recording. Dealers ordered from Polygram, which persuaded the company buyers to have second thoughts about hip-hop. Kurtis Blow became the first rapper on a major label. The song was a hit. The door was open.

In 1983, Rush got together with his brother and McDaniels to form Run-DMC. They wanted to call themselves the Sure Shot Two; he insisted on the odd name to establish that a new genre was taking wing. Mizell came onboard as the duo's DJ. "Of all the stylistic innovations to shape modern rap," writes Ogg, "the most indelible contribution came from Run-DMC." The group members were bright, well-adjusted suburban kids, he notes, "but their brooding and confrontational records spoke of an alternative reality that fellow suburbanites were happy to take as the gospel of the street." Rush says, "The band and I weren't concerned with reaching blacks or whites but with making new sounds for people who wanted to listen to

them." By 1984, Simmons thought Rush Management, Inc., ruled the scene; he managed many acts, including Blow, Whodini, Stetsasonic and Jekyll and Hyde. Then he heard a record that rocked him back with its raw quality, "It's Yours" by T La Rock and Jazzy J. He was even more astounded when he met the producer—Rick Rubin, a white, long-haired punk rocker from a well-to-do Jewish family in Lido Beach, Long Island. Rubin had organized the hit out of his scruffy dorm room at New York University, the headquarters of a label he had called Def Jam, *def* being the vernacular for definitive (and also hip-hop slang for "cool"), and *jam* from the freewheeling breakouts in classic jazz.

Simmons's strength has always been his readiness to recognize and then associate with people whose talents complemented his own. It was Rubin who would see the promise in 16-year-old James Todd Smith (LL Cool J) when Rush didn't. The two of them, for an investment of $4,000 each, started Def Jam Records in the fall of 1984. Over the four years of their association, they developed stars like Public Enemy, a group with a conscience and antidrug message; Slick Rick; and the first successful white hip-hop act, the Beastie Boys. Run-DMC, then on the Profile label, could earn $150,000 a night on tour. Rush talked hip-hop onto MTV. He pointed out that the channel had no black artists to speak of beyond Michael Jackson, and a suburban audience that had been isolated from the urban movement. He and Rubin made a

One of the future Beastie Boys found a demo tape made by 16-year-old LL (or Ladies Love) Cool J, left, in the dorm room of Rick Rubin, and Rubin sold his talent to Simmons.

RUSH COMMUNICATIONS HEADQUARTERS: A focus on the important sneaker

video featuring Run-DMC and the already well-known rock band Aerosmith. It bowled over MTV's Ted Demme. "It was fantastic. They took hip-hop into the video millennium."

In 1985, Rush and Rubin signed a major distribution deal with CBS for $600,000. Several coownership deals followed—Rubin departed in 1988—then in 1999, the label was sold to Universal for $120 million. In 2002, Def Jam—under Lyor Cohen—was the second largest label in the industry, with sales of more than $700 million. Nelson George commented: "Def Jam has become an institution and Simmons a role model for a generation of hip-hop entrepreneurs."

After a few false starts in Hollywood, Simmons again found smashing success with *Russell Simmons' Def Comedy Jam,* the HBO comedy series developed, he likes to point out, by two black men (Simmons and veteran television producer Stan Lathan) and two white men (agents Brad Grey and Bernie Brillstein). Originally slated as a miniseries, it aired for seven years, providing HBO with its highest rated weekly show until *The Sopranos.*

So much for entertainment. Rush Communications has recognized no boundaries, venturing into every area of popular cul-ture. In 1992, having enjoyed chasing fashion models, Rush married one of the most beautiful, Kimora Lee, and started a clothing line with hip-hop inflections based on a flagship boutique he opened in Soho. He called it Phat Farm. It was a worrying struggle. Year after year he lost money he could not afford to lose. He refused to sell into big-city department stores that intended to place Phat Fashions in "ethnic (aka 'nigger clothing') sections." In the sixth year, with $10 million invested, he turned a profit, and by then Phat Farm had bred eight divisions, including Baby Phat, a line for women and babies run by Kimora Lee, and Phat Farm Footwear, run by Run, i.e., the Reverend Joey Simmons. Phat Farm clothing was finally integrated in stores with the likes of Polo, Hilfiger, Klein and Nautica. He sold Phat Fashions for $140 million in January 2004 to Kellwood, one of the largest and staidest of manufacturers. He remained CEO.

According to *Business Week,* in 2003 one-quarter of all discretionary spending in the United States was influenced by hip-hop. "Coke, Pepsi, Heineken, Courvoisier, McDonald's, Motorola, Gap, CoverGirl—even milk: They all use hip-hop to sell themselves." Simmons has become the go-to guy for corporations like these trying to reach the young urban and suburban market.

The reason it is hard to get ahold of him, though he seems to have a cell phone permanently stitched to both ears, is that while he is doing all this, he is also thinking about the *Def Poetry Jam* enterprise he started in 2001, doing for "slam" poets what he did for comedians; or his DefCon 3 carbonated beverage in 7-Eleven stores; or his Rush Philanthropic Arts Foundation, which distributes $350,000 a year to stimulate arts education for disadvantaged youth. *Def Poetry Jam* on HBO won a Peabody Award in 2002. He is proud of that. "If you had asked kids in a class five years ago how many of y'all write poetry, nobody would have raised a hand. Now sixty to eighty percent of kids in inner-city schools write poetry."

Bill Stephney, president of StepSun Records and cofounder of Public Enemy, describes what happens when he drops in on classrooms of young black kids. "If you ask them what they want to do, and I do it rather frequently, they say, 'Well, I want to run my own business.' The notion of controlling your destiny, as much as you possibly can, basically did not exist within the community until Russell came along."

PIERRE OMIDYAR His unworldly ideals
led to the electronic bazaar known as eBay

1936–

IF YOU HAVE ever bought or sold anything on the eBay auction Web site—and many millions of people have—you may marvel at the business acumen that has made its founder a billionaire many times over. Pierre Omidyar is certainly a high-flying moneymaker, but that is the least important thing about him and his creation. More than anything else, eBay is a manifestation of trust, a celebration of community, an affirmation of equality. Had it been otherwise, it would not have become, and remain, a wildfire success. As such it is a vindication of all the ideals you would associate with a starry-eyed Silicon Valley computer nerd in ponytail, beard and glasses, i.e., Omidyar as a young man.

Omidyar was born in Paris but was brought up in Maryland after his Iranian-born father took up a medical residency at Johns Hopkins. At school, Omidyar habitually sneaked out of gym class to fiddle with the Science Department's Radio Shack TRS-80, so the school got him to program the printing of the library's catalog cards. He majored in computer science at Tufts and Berkeley, then designed a program enabling Mac users to draw pictures and took it to Claris, an Apple spin-off that Donna Dubinsky (later of PalmPilot fame) had helped to start.

Omidyar left Claris when Apple reabsorbed the company. Like Dubinsky, he got into pen-based computers, but with less success. The Ink Development Corporation he and some friends from Claris started in 1991 was dead in 18 months. What survived were some on-line software tools they had developed, and they diverted these into an electronic retailing site, e-Shop, that sold goods over proprietary networks. Omidyar was more interested in how the free Internet might develop, so he left in 1994 (though he kept his stock, which made him a millionaire in 1996).

Living in Silicon Valley in the heyday of the Internet bubble, Omidyar was uneasy

at the disparity of wealth IPOs were creating. The biographer of e-Bay, Adam Cohen, writes: "Financial markets were supposed to be free and open, but everywhere he looked he saw well-connected insiders profiting from information and access that were denied to ordinary people." An auction seemed to Omidyar the perfect antithesis and the Internet the perfect vehicle for its expression: Everybody who took part had an equal chance at the same time and could see what everybody else was doing. With a few clicks of a mouse, seemingly distant sellers and buyers would be able to interact equitably. But Omidyar dreamed of a Web site that was more than a marketplace. "The first commercial efforts," he says, "were from larger companies that were saying, 'Gee, we can use the Internet to sell stuff to people.' Clearly, if you're coming from a democratic, libertarian point of view, having corporations just cram more products down people's throats doesn't seem like a lot of fun. I really wanted to give the individual the power to be a producer as well. It was letting the users take responsibility for building the community—even building the Web site."

That sounds like a recipe for chaos, but seminal studies by Columbia University's Douglas Watts and Cornell University's Steven Strogatz have suggested that "small world" networks are more resilient than

top-down hierarchies. There is a name for the spontaneous rise of structure out of random connections: "emergence." It can be seen throughout nature, in civilizations, in the bustling stasis of cities, in the neuronal networks of the brain. So Omidyar was in tune with a growing body of thinking, though he did not know it when he mused about a digital democracy.

He founded a Web consulting company he called Echo Bay Technology Group— "it just sounded cool"—but for his personal Web site he had to have a different name than Echo Bay Technology, which was already taken, so he called it eBay.com. It was a mishmash. One subsite expressed his preoccupation with the killer Ebola virus, another was his fiancée's promotion of a small biotech company and what he called "Auction Web" became a third element, all run out of a spare bedroom as a hobby, a break from his day job in programming. He circulated Auction Web invitations on the Usenet groups: "The most fun buying and selling on the Web." Omidyar's first inkling that he might be on to something came after the laser pointer he had bought for $30 stopped emitting the red dot he had used to make his cat pounce. On Auction Web, Omidyar typed "Broken Laser Pointer" and asked for bids starting at $1. The first week there were no bidders. The second week he had an offer of $3, then $4. At the end of two weeks, he let it go for $14, amazed that people were willing to pay good money for stuff others consider junk.

It took several days for a few visitors to come to the site after it officially opened on September 12, 1995, with an utterly unpredictable list of items from sellers: a Superman lunch box for $22; autographed Marky Mark underwear for $400; an autographed Michael Jackson poster for $400; an '89 Toyota Tercel for $3,200; a health club membership in Chicago for $400.

Omidyar had no evidence for basing a Web site on the assumption that total strangers would commit to innumerable acts of trust. It was a pure act of faith. He says, "My mother always taught me to treat other people the way I want to be treated."

But he did invent one device that both checked on occasional bad guys and reinforced the idealism of an electronic social contract. Very early on he wrote the code for a "Feedback Forum" inviting users to rate other buyers and sellers. More than 99 percent of these evaluations proved positive. Anyone getting more than a few bad reviews was and is barred from the site—kicked off the island, as it were. "Most people are honest and trustworthy," he says. "They were grateful for the chance to give praise where it was due. It's become a kind of virtuous circle that's encouraged good behavior."

Auction Web generated so much traffic, even in the early months, that the service hosting the site demanded $250 a month. "That's when I said, you know, this is kind of a fun hobby, but $250 a month is a serious lot of money." He started charging a percentage on all sales: 5 percent of the final price for anything below $25, and 2.5 percent for anything above. In March 1996, envelopes arrived stuffed with cash. In April he netted $2,500, in June, $10,000. He could not cope. "I had a vague idea of what I needed to do as an entrepreneur, but I knew I wasn't going to be able to put together a business plan." He approached Jeff Skoll, a Canadian with a Stanford MBA who thought it was an absurd idea to believe people who had never seen each other or the goods were engaging in so much trading on trust. He was convinced only when he saw the cash piling up in Omidyar's spare bedroom. He joined in August 1996. They moved the transactions to Skoll's house, then to an office in Campbell, California. Skoll argued that the Web site should be devoted exclusively to Auction Web, i.e., they should lose the Ebola page. Omidyar parted reluctantly with his viruses. Omidyar and Skoll were patient and deliberate, flying below the radar, careful not to awaken Web tigers like Yahoo! and America Online; the paradoxical assignment they gave the public relations officer they appointed, Mary Lou Song, a young Korean-American graduate from Stanford, was to keep the company from getting noticed.

In the mid-1990s, there were other auction services on the Internet, but they were not interactive like Auction Web (sellers and buyers on these other sites had to keep in touch by streams of private e-mail). Nor were they as welcoming. The major threat was Onsale, started by Silicon Valley entrepreneur Jerry Kaplan. In 1996 it handled $30 million in sales, but its top-down model was the opposite of Auction Web's. What was for sale was dictated by Onsale's management, not by what Internet users around the world wanted to offer.

Auction Web grew by word of mouth, fingers on keyboards. Omidyar started a chat room, eBay Café, whose name was voted on by auction webbers. In the month of January 1997 alone eBay hosted 200,000 auctions. Omidyar and Skoll kept wondering when it would taper off. The toy billionaire Ty Warner made that a ridiculous concern when he discontinued a line of older characters in his Beanie Baby series. Collectors went wild to find them. Instead of having to search one another out somehow, they conglomerated on Auction Web—in April 1997 more than 2,500 Beanie Babies changed hands at prices three or four times the $5 original price. Other collectors flocked to this stock market in esoterica—numismatists, philatelists, bibliophiles, lovers of Chocolate the Moose. Finding a special type of hunting decoy made only by a carver in Michigan persuaded a partner of Bruce Dunlevie's Benchmark Capital to advocate investing: He had been looking for such a decoy for five years. Benchmark Capital put up $5 million for 21.5 percent of the company.

Omidyar and Skoll were now able to upgrade the technology. Auction Web became eBay after a survey of user preferences. EBay took on a company lawyer who was shocked that Omidyar had patented nothing, not the Feedback Forum, not the bidding algorithms. The libertarian Omidyar was against patents; he believed software should be shared. Competitors just copied. In October 1997, Auction Universe appeared, backed by the Times Mirror Group, and Kaplan's Onsale transformed itself into an intermediary, like eBay, instead of selling merchandise direct. Onsale Exchange, as it was now called, sent "bots," little computer programs that act as spies, to probe eBay for users' e-mail addresses. Then it spammed the eBay customers, hoping to persuade them to try Onsale. EBay was upset, then relieved. The spamming alienated many eBay users and left them more loyal than ever. As Song put it, "EBay was part of the equation, but they were really loyal to each other."

EBay consolidated its lead and advanced further in February 1998 when it wooed Meg Whitman, a tall blonde with a Harvard MBA and rich experience in marketing and management. She became CEO, and Omidyar moved up to chairman. A few days later, eBay had its six millionth listing; its gross margins were 85 percent and it was growing 30 percent a month. In September 1998, a public stock offering for about 9 percent of the company was gobbled up despite a disruptive announcement by Yahoo! that it was going to start a free auction service.

The stock market value of the little company started in a spare bedroom was $2 billion. The entire staff had been offered stock options, and two-thirds emerged multimillionaires. The following summer *Fortune* pegged Omidyar's wealth at $10 billion, Skoll's at $4.8 billion and Whitman's at $1 billion. They gave 100,000 shares for philanthropic work through the eBay Foundation. One of the foundation's earliest efforts supplied computers to poor native craftsmen in Guatemala so they could cut out the middlemen and sell direct over the Internet at American prices. A small act perhaps, but symbolic of the wider significance of eBay; in the words of Omidyar, the "tremendous" financial empowerment you can bring people by giving them access to an efficient market. EBay itself is a powerful economic engine. The *New York Times* calculated that in 2002, more than 75,000 people were making their living exclusively by selling over eBay. In 2004 the community of McCracken, Kansas, would have given away an empty school for $1. Auctioned on eBay, it fetched $49,500—from two eBay members who intended to make the community of McCracken a distribution center for their eBay business.

What is Omidyar up to now? Check him out at http://pierre.typepad.com.

LARRY PAGE and SERGEY BRIN
It's omnivorous, it's democratic, it's fast, it's fun, it's indispensable, it's Google™

1972–	1973–

ITEM: Terry Chilton, a 52-year-old home builder in Plattsburgh, New York, was surfing the Internet one day when he felt pain in his chest and shooting down his arms. He took a Tums, but the pain didn't go away. As he later wrote to the search engine Google, "I started to get anxious. I went to AltaVista to look up 'heart attack symptoms,' and a bunch of pictures came up as I waited for the site to download. I was worried, so I switched to Google and quickly got the American Heart Association site. They had a diagram of what the symptoms were that sounded like me—I got myself to the hospital quick." An emergency triple bypass operation saved his life.

ITEM: In September 2002, the Chinese government blocked Google from the mainland. There was no official announcement, but the outcry was so great that normal Google service was allowed to resume ten days later.

IT IS AWKWARD writing a sentence that is out of date before the last keystroke. So take this one as an approximation of the unimaginable: In 2004 Google was answering 200 million queries a day from its store of 4.5 billion Web text pages and 800 million picture pages—and it is growing exponentially to keep pace with the doubling of the number of pages on the Web every few months. Google's 10,000 computers search Web pages in about 100 languages, including Latin, pig Latin, Urdu and Klingon (an extraterrestrial language from *Star Trek*), and it offers automatic translation of Web pages in other languages. Users misspelling items are politely asked, "Did you mean . . . ?" Apparently, there are 600 ways people misspell "Britney Spears," the pop princess; they all get answered. Between 2002 and 2005 more content, wise and wacky, will appear on the Internet than was published in the previous 40,000 years of human history, and Google is the principal means of finding any particular grain of sand in the sandstorm of information. (Google runs 34.9 percent of all Web searches in the United States, Yahoo! 27.7 percent. Worldwide, 43 percent of all searches are through Google.)

The name Google comes from "googol," the number 1 followed by 100 zeros. A googol is so large that there is not a googol of anything in the universe, not a googol of dust particles, stars or atoms. The word has been around since 1938, when it was introduced into mathematics by Columbia University's Edward Kasner (1878–1955)—invented, he said, by his nine-year-old nephew, Milton Sirotta—but the new kid on the block, Google, has transcended the etymology. It is more than a corporate name. It is a cultural phenomenon. Like Xerox and Federal Express, it is a verb as well as a noun—we Fedex packages and Xerox documents—but Google's range runs from the routine to the exotic. Blind dates routinely google each other before meeting. Businesspeople google the competition before presentations. Bird-watchers find the names of exotic species they see in nature and google for the characteristics. Teenager Isabel Evans googles for science homework on tectonic plates "and pictures of hot guys." Chefs Google-cook, finding interesting recipes just by googling a list of ingredients. Some fancy hotels google guests to see if they can learn their preferences to make their stay more enjoyable. People who google themselves are said to go "ego surfing." When ABC aired *Who Wants to Be a Millionaire,* Google got tens of thousands of extra hits from viewers wanting to find the answers. As the *Washington Post* says, "Google is the first [search engine] to become a utility, a basic piece of societal infrastructure like the power grid, sewer lines and the Internet itself." The rival Yahoo!, which gets half a billion page views per day, tried to verbify in its ads: "Do you Yahoo?" That was fun, but *yahooing* sounded too loutish to appeal.

I googled some of our innovators:

Henry Ford 3,160,000 entries in 0.14 seconds (8,070 images in 0.10 seconds)

Walt Disney 2,370,000 in 0.11 seconds (36,500 images in 0.07 seconds)

The Wright brothers 1,330,000 in 0.10 seconds (6,120 images in 0.16 seconds)

Thomas Edison 730,000 entries in 0.12 seconds (4,750 images in 0.08 seconds)

Too much! But Google is almost as fast in multiword sifting. Here is a declension from Thomas Edison:

"Thomas Edison phonograph" yields 26,100 entries in 0.24 seconds.
"Thomas Edison phonograph music" yields 16,700 entries in 0.17 seconds.
"Thomas Edison phonograph most popular music" yields 4,420 entries in 0.15 seconds.
"Thomas Edison phonograph most popular music available now" yields 1,900 entries in 0.14 seconds.

Of course, this is an indication of information available, true or false. The fact that Henry Ford has more entries does not tell us he made a greater contribution to

SOLID BUBBLES: There was every promise in 2004 that Google would not be like one of the Internet bubble companies of the '90s. Millions thought so, clamoring for the stock offering that made instant billionaires of founders Sergey Brin (left) and Larry Page. To appease photographer William Mercer McLeod, they tried out a Jacuzzi in the house where they set up their first headquarters. For further research on bubbles, see bubbles.org (alas, 2,400,000 entries).

civilization than Thomas Edison, the Wright brothers or Walt Disney, only that this is the level of attention each has received. Both Steve Jobs (3,890,000) and Bill Gates (3,330,000) have even more references, while some of the innovators judged of great significance in this book scale lower. The life-saving microbiologist Herbert Boyer has 71,300 entries; Britney Spears 6,090,000—double the number of entries for Henry Ford. Gary Kildall, so important in the genesis of computer operations, has only 6,370 references of any kind; Samuel Insull, who gave us the cheap electricity on which Google and everything else depends, has only 3,140 entries. When

this book is absorbed by Google, these numbers will change! But at least as important to every user as volume and speed is quality, the precision of the answer to the search, the exclusion of "junk results" and intrusive advertising disguised as editorial. How Google beats back the infiltrators of the citadel of knowledge is a central part of its story, though the battle will rage on and on as Web sites multiply.

The two Stanford University dropouts who created the most mathematically sophisticated search engine on the Web, Larry Page and Sergey Brin, were barely into their 30s in 2004. Page, the deviser of Google's secret search techniques, is the

son of a Michigan State University computer science professor, Carl Victor Page, an expert in artificial intelligence who died in 1996 at age 58. "We had computers really early," Larry says. "The first computer we owned as a family was in 1978, the Exidy Sorcerer." He grew up in Lansing, Michigan, always playing with electronic gadgets. He rides around his corporate headquarters with his cofounder on a Segway and owns a tzero, a fast and expensive electric car. He does his best thinking, he says, "Rollerblading; in Hawaii; when I'm hanging out and chatting with friends. Anywhere but the office." Page went to college at the University of Michigan, where

That garage again. Actually, they didn't start here because it was too cold, but why spoil a good myth?

he earned a bachelor of science degree in computer engineering and graduated with honors. He served as president of the university's Eta Kappa Nu Honor Society and built an Inkjet printer out of LEGO blocks. After college he worked as a software developer for Advanced Management Systems in Washington, D.C., and for CogniTek in Evanston, Illinois. "I really love technology. I like to see every possible product," he says. He scours the Web for wacky scientific and technological news. He owns almost every single type of PalmPilot. His office has three computer screens, a browser on one, a schedule on another and e-mail on a third. He can drag things across the different screens. He also has a projector connected to his computer so he can show things on a big screen.

Sergey Brin was born in Moscow but came to the United States at the age of six, when his family fled Soviet anti-Semitism. His father taught math at the University of Maryland at College Park. Brin, too, was weaned on computers. He got his first, a Commodore 64, when he was nine. As a freshman at the University of Maryland, he tutored his professors on how to write computer programs. He graduated with honors in mathematics and computer science and has published more than a dozen scientific papers.

Brin and Page first met in the spring of 1995 in San Francisco during a welcome weekend for applicants to Stanford's computer science Ph.D. program. Brin, then 23 and already a Stanford graduate student, was assigned to show Page, 24, around. The two argued for the rest of the weekend about almost every topic. Brin liked Page more as a result. The following fall Page arrived to start at Stanford. Google's personality is reflected in its creators'. Both are playful and driven. They affect to have fun tossing barbs at each other. As *Technology Review* wrote in 2000, "Talk with Page and the Russian-born Brin, both 27, and one-liners are more likely to roll off their tongues than algorithms." Their division of duties follows their personalities. Brin says, "Larry focuses a little more on the operations side—computers and things like that. I focus on research and marketing. But for the most part we're interchangeable. It's a great convenience for us; it gives us more flexibility."

Both are highly idealistic. When in 2004, they announced that they were opening Google to public subscription, the five-page letter of intent was a warm California zephyr down the cold canyons of Wall Street. The document, written in the first person by Larry Page and inspired by Warren Buffet, promised to see fair play between large investors and the small fry, who usually gets squeezed out, by conducting the sale through an auction. They were emphatic that the Google triumvirate would maintain a long-term view and keep innovating. "In our opinion, outside pressures too often tempt companies to sacrifice long-term opportunities to meet quarterly market expectations. We won't 'smooth' quarterly or annual results. If earnings figures are lumpy when they reach headquarters, they will be lumpy when they reach you." All this only enhanced Google's popularity with the general public while irritating Wall Street; the *New York Times* was moved to raise half an eyebrow: "Let's not get too cute." But there was no doubt of the sincerity of the founders. It was what they had been saying for a long time. Page had told National Public Radio, "We always kind of figured that if we did a good job of providing the right information for everybody in the world, all the time, that would be an important thing to do." He added, "I'm always amazed that there aren't more people who work on technology and science because I think that's the easiest way to change the world." They would be immensely rich after the sale, but Google was born out of their intellectual curiosity, not a desire to start another search engine or the ambition to make a killing associated with the dot-com era. They made a major contribution to understanding the Web in a joint paper in 1998 called "The Anatomy of a Large-Scale Hypertextual Web Search Engine," which became the tenth most accessed paper of all time in scientific bibliographical searches. They noted that "despite the importance of large-scale search engines on the Web very little academic research has been done on them."

That was true, and Page and Sergey deserve their renown, but others must be acknowledged first in the genesis of the Web and Web searching. Tim Berners-Lee devised three key tools, and Marc

Andreessen and Jim Clark cofounded the Web browser, Netscape, which did so much to popularize the Web. Jerry Yang, a Taiwanese native raised in San Jose, California, and David Filo, from Moss Bluff, Louisiana, two other Stanford Ph.D. dropouts, founded Yahoo! in 1994 as a high-quality human indexing Web page. Netscape got people onto the Web. Yahoo! helped them find what was on it. Then there are two brilliant mathematicians with particular relevance to Google, one the joyous Claude Shannon, the other a Russian probability theorist with a passion for poetry.

In his Theory of Communication, Shannon described ways to compress information by removing repetition. "Cn u rd ths sntnce?" is more efficiently transmitted than a sentence using the full words. Google developed Shannon's idea. "Google," writes Page, "stores all of the actual documents it crawls [downloads] in compressed form. One of our main goals in designing Google was to set up an environment where other researchers can come in quickly, process large chunks of the web, and produce interesting results that would have been very difficult to produce otherwise."

One of the major mathematical tools Shannon used to describe the transmission of information in bits was invented at the turn of the century by St. Petersburg's Andrei Markov (1856–1922). So-called Markov chains are sequences of random variables in which the future variable is determined by the present variable without reference to the way the present variable was created.

Markov was the inspiration for Page when in 1995, purely for academic interest, he set out to map the proliferations of Web sites and links. How did they all link together? Of course, even then, with Web sites growing from a few hundred pages in 1994 to tens of millions of pages a few years later, it was too big for a drawing the size of Yankee Stadium (the frontispiece to this book is more art than science). "There were too many Web pages, too many blocks, too many arrows," said Craig Silverstein, another Stanford dropout who was in at the start and in 2004 was director of technology. "So Larry tried to simplify the Web

in some way mathematically and came up with this idea of analyzing the links by using Markov chains, mathematical structures that can have lots of equations and variables. Markov's work is used in all sorts of mathematical modeling. Larry applied it to the Web, which is new. No one had thought of this before. People had analyzed links on the Web in all sorts of different ways before, the most obvious being to count the number of words on a page, a technique very easy to use but not very successful."

The eureka moment came when Page and Brin realized that the mapping exercise had produced a mathematical ranking of connections that could be regarded as a high-precision tool for searching the Web. "Intuitively," said Larry, pages that are well cited from many places around the Web are worth looking at. Also, pages that have perhaps only one citation from something like the Yahoo! home page are also generally worth looking at. If a page is not high quality, it is quite likely that Yahoo!'s home page would not link to it."

They called this system PageRanking—after Larry, not the page. It is not just a count of the links on any Web site. One link will be considered more important than another if other high-ranking documents link to it. And those documents, in turn, are affected by the PageRanking of other documents linked to them. The PageRank of a document is thus always determined recursively. It is a dynamic system, based in the end on the linking structure of the whole Web. In their explanatory paper, Page and Brin wrote: "We have created maps containing as many as 518 million of these hyperlinks, a significant sample of the total. These maps allowed rapid calculation of a Web page's PageRank, an objective measure of its citation importance that corresponds well with people's subjective idea of importance."

In their review of systems in 1995 and 1996, Brin and Page acknowledged that human-maintained lists like Yahoo! covered popular topics effectively—"but they are subjective, expensive to build and maintain, slow to improve, and cannot cover esoteric topics." They were more critical of the automated class of search engines, such as AltaVista, Lycos, HotBot, Infoseek

and Ask Jeeves, all of which employed the same technique to find information, searching for repetitions of a key word in the query. This invited manipulation: To get their Web pages ranked highly, Web site creators simply had to repeat searched-for words multiple times. One major search engine geared to returning the document most closely approximate to the query (a standard information retrieval system) responded to "Bill Clinton" with a page containing only "Bill Clinton sucks" and a picture. "Some argue that on the Web, users should specify more accurately what they want and add more words to their query," wrote Page and Brin. "We disagree vehemently with this position. If a user issues a query like 'Bill Clinton,' they should get reasonable results since there is an enormous amount of high-quality information available on this topic." A reliable search engine, Brin and Page believed, should take a user to the White House first. They were also quickly convinced that the ability to return ever-greater numbers of search results did not help Web searchers if there was no way of sorting through these results by relevance. "The number of documents in the indices has been increasing by many orders of magnitude, but the user's ability to look at documents has not. People are still only willing to look at the first few tens of results. The bigger the Web gets you care less and less about recall—how many documents get returned—and more and more about whether they're the right document."

Page's Markov chain equation thus tries to account for several things: the number of pages linking to a certain Web page; the number of pages leading out of that page; and the chance that someone surfing the Web will not always jump directly from one page to another linked to it but may begin on a completely new Web page. All of this information is boiled down to the PageRank formula, simply a mathematical equation of a kind of Web consensus:

The PageRank of page A, or $PR(A)$ for short, $= (1-d) + d \left(PR(T_1)/C(T_1) + \ldots + PR(T_n)/C(T_n) \right)$

To spell it out, $PR(A)$ is the PageRank of page A, $PR(T)$ is the PageRank of page T_1, which links to page A. $C(T)$ is the number of outward-bound links on page T_1. Finally

d is a damping factor, which can be set between 0 and 1. The exact formula is kept secret by Google, although it is known that the company assumes the damping factor *d* is 0.85. Page says, "We look at what the Web thinks." What do other Web pages say? It's a pretty good measure that it's the right page, because the whole world thinks it's the right page." It's important to note one caveat: Sometimes the whole world is wrong, or at least popular opinion misinformed. What Google has a harder time doing—and what no search engine does well at this point—is finding information prized by an elite few. Google is not an editor. Much as people tend to trust the information they find, it can often be wrong. As research for this book has found, many Web sites simply cut and paste information from other Web sites without verifying whether or not the statements are true. The ease with which information spreads on the Internet applies to false information as well as true. As Peter Lyman, a UC Berkeley professor, told the *Washington Post,* "There's been a culture war between librarians and computer scientists. Google won."

In 1996 Brin and Page put Markov to work, launching their first search engine. They called it BackRub because it analyzed the "back links" that point to a Web site. By 1997 word of mouth made it well known around the campus at Stanford and then on the Internet. Their first data center was in Larry's dorm room, which quickly filled up with computers and fans to keep them cool. During the year, BackRub yielded to www.google.Stanford.edu. Craig Silverstein says, "One of the many luck factors of our success is that we changed the name before it became popular. Here again there's innovation of a sort. Larry had criteria that he wanted the name to have. Again from his human-computer interaction perspective, he wanted the name to be short so that people could type it easily, and easy to pronounce so people could tell their friends about it." Says Page: "It was fun, it was short, it was reasonably easy to spell."

Larry and Sergey saved money with their tinkering skills. At their local Fry's electronic store, they tore apart cheap PCs, threw out all extraneous chips and cards and turned their Frankenstein machines into efficient data servers specially crafted to carry Google's Web traffic. Spreading the cost on three credit cards, they invested $15,000 in computer disks—a terabyte, or one million megabytes, of memory.

Next, they rang up their cheap contraptions alongside other people's million-dollar electronic beauties in a seven-feet-by-eight-feet room they rented in a data center in Santa Clara, California. A proper test of their formula meant exploring as much of the Web as inhumanly possible: They needed more and more computers. Page remembers, "At Stanford we'd stand on the loading dock in the computer science building and try to snag computers as they came in. We would see who got twenty computers and ask them if they could spare one." Brin elaborated, "We'd go down to the loading dock, sort of borrow the ones that were sitting there, before, you know, the people who had really got them on their grants, like, got around to it." In other words they would filch other people's computers until the other people started to complain. Expressing himself in the language of computers, Brin added, "Eventually, that became kind of unscalable." (*Scalable* meaning having the capacity to scale up for the impending torrents of information.)

With the help of Stanford's Office of Technology Licensing, Page and Brin approached Silicon Valley's Excite, Infoseek, Yahoo! and AltaVista. They hoped to license the system in return for money to continue research, without abandoning their Ph.D. studies. "We spent a lot of time talking to people like George Bell, CEO of Excite, and Steven Kirsch, founder of Infoseek, trying to convince them our technology would give them an edge," Page says. "But no one really engaged with us. The crunch came when the CEO of a major search company said that search didn't matter to him that much. He told us, 'Our search is eighty-five percent as good as the next engine.' That astonished me." That company is now out of business. They talked to their friend from Stanford, Yahoo! founder David Filo, who told them, "When it's fully developed and scalable, let's talk then." According to the Stanford Office of Technology Licensing (OTL), there was a little more interest than Page and Brin now let on. Says the OTL, "Companies were interested. Several responded to OTL, and more than one pursued in-depth negotiations for the technology. One company even bid a significant amount of money for Google, but none of the offers equaled Google's potential and no companies looked to be its ideal developers."

In the end starting a company seemed to be the only solution if they wanted to keep pace with the growth of the Web. More computers! More computers!

But it was not so simple. Larry and Sergey remained diffident. Starting their own company was daunting. Brin says, "We were quite lazy about it. Stanford was pretty comfy. Being a graduate student there, you didn't really get paid well, but you got to spend time with a lot of interesting people." Page says, "We wanted to get our Ph.D.s, graduate and stay researchers." They were preempted one morning on the porch of the house of their professor David Cheriton, a cofounder of Granite Systems in Palo Alto. They gave a swift demo to Andy Bechtolsheim, a founder of Sun Microsystems and Stanford alumnus who left in a hurry for another appointment. But as he left, he said, "Instead of us discussing all the details, why don't I just write you a check?" It was made out to Google, Inc., and was for $100,000.

Google the company did not yet exist, and Brin and Page had to incorporate before they could deposit the $100,000. It stayed in Page's desk drawer for two weeks, a hotter and hotter challenge. Finally, they took the plunge, scrambling to put together their legal papers and raise more money from family, friends and Stanford professors and students. There was enthusiasm. With additional backing from both Sequoia Capital and Kleiner Perkins, two of Silicon Valley's most prestigious venture capital firms, Page and Brin raised $1 million before the company was even incorporated. That happened on September 7, 1998. Page became president and Brin CEO. Craig Silverstein told us, "We didn't even cheer to say Google's done, we're launching it now. We had Google running with the Stanford address. One day we changed the name to Google.com, but it was just a name change. We had thirty or forty people and we had salespeople, who like to celebrate. So we brought some champagne into Google and some food from McDonald's. Champagne and hamburgers, that was the celebration. Everyone worked after dinner."

The official story is that Google was started in the garage of Susan Wojcicki, a friend in Menlo Park, California. Actually, says Craig Silverstein, they didn't use the garage because it was soon winter and it was not heated. "Luckily we had access to some rooms inside as well." Brin told NPR, "The house came with a hot tub, washer, dryer, shower and refrigerator." Page and Brin installed a cache of candy and snacks, eight more phone lines, a cable modem, and a DSL line. The company's first employee became Craig Silverstein, and they made three other hires. The hot tub lay unused until a German reporter came to do a story on the company, and Sergey and Larry posed in the pool. Both now all of 25, they told their landlady, Susan, "A year from now people will know about our search engine." Susan, who became Google's director of product management, thought, "If you're so important, why are you renting my garage?"

In February 1999, they moved their offices to University Avenue, Palo Alto's main drag, with 30 employees. Desks were old doors resting atop sawhorses. The company now had 500 servers and kept them in a data center. Red Hat, a Linux software company that had been enticed by Google's own use of the Linux operating system, became Google's first customer. Soon AOL and Netscape incorporated Google's technology. In June 1999, Google raised another $25 million from the VC companies on Sand Hill Road. John Doerr, a Kleiner Perkins partner, joined Mike Moritz of Sequoia Capital and Ram Shriram, CEO of Junglee, at the Ping-Pong table that was the centerpiece of Google's corporate boardroom. Michael Moritz, the Sequoia partner in charge of the Google deal, told *Fortune,* "It was very unfashionable to invest in a company like Google back then. But we thought Larry and Sergey were taking a fresh look." Sequoia chipped in $10 million. To help market Google and to test it, Brin gave free customized search engines to universities that year. This move helped spread its reputation for fast, efficient search among early adopters of the innovation.

The staff, as well as the data, was heavily compressed in the University Avenue office. Nobody could stand up without asking somebody else to move. In October, when the former student project was accomplishing 3.5 million searches a day, they moved offices again to Mountain View, California, about ten minutes from Stanford, where Google was born. They now had 39 employees and were growing the company fast while winning multiple awards for Web search, design and innovation. By 2000 the company had indexed 1 billion Web pages out of an estimated 2.6 billion then on the Web, making it the largest search engine on the Web. In June, Yahoo! chose Google for its default Web searcher when Google was handling 7 percent of all searches, an arrangement that lasted until February 2004. Yahoo! then had a panoply of new search-engine companies

at its disposal, and it dropped Google as its default search engine and became its official rival. Google has opened 21 offices around the world, all computers running on Linux servers. The staff is only about 1,300, including scores of Ph.Ds. Brin told NPR, "When you think about it, as to what impact those thousand people have on the world, it's actually an amazing effect per person."

In 2000 Google's board began pressuring Larry and Sergey to bring in an experienced executive to run the day-to-day operations. Eric Schmidt, the former chief technology officer of Sun and former chairman and CEO of the networking company Novell, became CEO of Google. Schmidt had earned his Ph.D. at Berkeley, where some of his academic work on computer compilers was inspired by the late Gary Kildall's dissertation, says Gary's partner, Tom Rolander. Schmidt seemed ideally

poised to lead Google, having a serious computing background but also years of experience in the corporate world. Schmidt told *Forbes,* "My job was to impose a little order. I made it clear that I wasn't coming in to get rid of the founders." Still, his vision for the company caused tensions. He told *Fortune Small Business* that Page and Brin "disagreed with everything I believe." They wanted Google to remain a small company. They didn't understand why earnings had to keep growing. "I couldn't decide if this was refreshing insight or naïveté on their part," Schmidt says. These days, he says, the relationship couldn't be better: "Larry and Sergey have become my best friends. We have lunch every Saturday at Quiznos [a sub chain]. Larry Rollerblades in these itty-bitty shorts. He still looks like a college kid. And Sergey comes in from a diving lesson. These are special times. . . . We seldom disagree. If we do, we put it to a vote, and whoever gets two votes wins. If it's a particularly egregious disagreement and it's important enough, I'll override them and they'll be mad at me for a while."

Google's official Web site proclaims, "You can make money without doing evil." Google donates free Web search software to schools, universities and other educational institutions. In Germany, it blocks access to Holocaust denial sites, as is required by law, but it does not interfere with its search results even when the word *Jew* brings odious results to the top of the list. Google's policy is to warn viewers that the results are "very disturbing" but explain that it cannot interfere without destroying the computer-generated objectivity of its whole search mechanism. It is a hard explanation to accept entirely since Google does not block some hard-core porn, but Schmidt says, "We've taken a blood oath on the issue."

So how does Google make money without doing evil? It runs services for company Web sites, indexing them for swifter everyday access. Early buyers were the *Washington Post,* Cisco Systems and Virgin.net, and Google also does searches for AOL and Amazon. It has set its face against intrusive pop-up and banner advertising. In a 1998 article, the partners wrote that Web advertising was "insidious," and Page has been overheard to say that other search engines are "bastardizing" search through commercial exploitation. Google's solution, borrowed from its rival Overture, is Adswords, short text pitches set aside from the main text and all related to the subject of the search.

The response rate, five times better than the rate of traditional banner advertising, took Google into profit in 2001. Brin celebrated by splurging on furniture for his Palo Alto bachelor apartment. "Now I have a nightstand! That will make a big improvement in my quality of life." In 2003 they had a net profit of $106.5 million, a profit margin of 11 percent compared with Yahoo!'s 15 percent, Amazon's 1 percent, and eBay's 21 percent. But Google was just hitting its stride.

Silverstein believes that a key factor in the success of Google has been its restraint. Brin designed the clean, colorful main page and logo of Google with advertising constrained—so much more user-friendly than Yahoo!'s crowded, commercially oriented main page (living up to Gulliver's Yahoos?). "I would never underestimate the significance of a concern for the user," says Silverstein. "It's exactly in these factors of the spare Web site design, the care we take in introducing the features and not making the site harder to navigate as a result, the lack of graphics that would make it slow to download. There's a whole slew of issues that we think about all the time."

At Google's headquarters in Mountain View, a screen hangs next to the receptionist's desk displaying real-time search queries on Google from around the world (with the sex-related searches expurgated). Like the main page, the offices have a playful atmosphere. Visitors are greeted by a friendly, jelly-bean-eating assistant and entertained by a monitor that continuously scrolls a selection of in-process Google searches. Full-time masseuses are on hand, along with yoga classes, free Ben & Jerry's bars, saunas and two free organic meals a day prepared by the former chef of the Grateful Dead. (When the subject of free employee lunches first came up, Page addressed the problem by coming up with a plan to eliminate world hunger before working out how to give his employees free food.) According to the *Seattle Times,* "They said the free, healthful meals only came about after calculating reduced health-care costs and the time saved from driving off-site to eat. There's even a Web cam trained on the lunch line so that employees can avoid a long wait." There's pool, shuffleboard and Ping-Pong, as well as arcade games and a gym. Roller-hockey games echo through the corridors. Lava lamps, hammocks and beanbag chairs adorn the corridors, where friendly dogs are allowed to roam. (Brin loves lava lamps: "I like to think they're very beautiful—you know, they're only forty bucks each. For four thousand dollars you'd have a hundred Lava lamps.") Once a year the staff goes off to Lake Tahoe. Twice a week sections of the parking lot are roped off for roller-hockey games.

The company also pays very well. But most important to the bottom line, Brin and Page feel, are strong corporate values. They are proud that AdSense and Google News have emerged from the 20 percent of time Google employees are encouraged to spend being creative and interactive. They try to structure the working environment so that creative people feel comfortable and have to interact. Page says, "For example, since most communication is informal, we have deliberately limited the seating in the cafeteria so you have to sit wherever you can. It encourages workers to meet random individuals and not be so isolated in their jobs." It also encourages extremely long working days. Twelve hours is considered standard. Openness is carried to the extent of telling employees every week how Google is performing financially; they are given confidential information, a strong sign of trust in a company contemplating going public. Page says, "Our staffers hear the same presentations we give to the board, and none of that information has ever leaked out. I believe if you give people a lot of responsibility and respect, then the company gets that back."

Essentially that same attitude permeates the PageRankings. "Democracy on the Web works," is the proclamation on Google's official site. It is a good note to close the last chapter of a book telling of the triumph of innovators over two centuries, men and women nurtured in freedom and enhancing it with their brave creativity. The search continues.

Ten Lessons

WHAT CAN BE LEARNED FROM HISTORY'S INNOVATORS

When a true genius appears in the world, you may know him by this sign, that the dunces are all in confederacy against him.

— JONATHAN SWIFT

1 MAKE NO ASSUMPTIONS. Edwin Armstrong was constantly trying to do things in radio circuitry he was told by the experts were impossible. He kept reminding them of one of the sayings of Henry Wheeler Shaw (1818–85), the Lanesboro, Massachusetts, lecturer and humorist known as Josh Billings: "It ain't ignorance that causes all the trouble in this world. It's the things people know that ain't so." Time and again in this study we find that breakthroughs come from the discarding of assumptions. Ignorance that ignites curiosity is a better starting point than half-knowledge. "No man becomes a fool," said the immigrant electrical engineer Charles Proteus Steinmetz, "until he stops asking questions."

2 FIRST ISN'T ALWAYS BEST. John Fitch was the first to operate a steamboat service, not the commonly presumed Robert Fulton. But Fulton learned from Fitch's mistakes and triumphed on a different river at a more propitious time. Similarly with Henry Ford, who was late to automobile production, Charles Goodyear with vulcanized rubber and Isaac Singer and his sewing machine. Mark Gumz at Olympus has the point: "Understanding comes from failure; success comes from understanding failure and acting upon this knowledge." The path beaten to somebody's door will reveal the potholes.

3 IT'S OK TO STEAL. The corollary to item 2. More innovations come from borrowing and combination than simple invention. Henry Ford said, "I invented nothing new. I simply assembled into a car the discoveries of other men behind whom were centuries of work." The imaginative association of ideas previously considered separate is a hallmark of innovation. Jean Nidetch did not invent the diet she used for Weight Watchers. Nolan Bushnell did not invent the first home video game. Lewis Tappan, a model of moral probity, did not start the first credit agency. Ruth Handler stole the Barbie doll from a German sex doll named Lilli. "Good artists borrow, great artists steal," said Picasso, who may have lifted the quote from someplace else.

4 DIFFIDENCE WON'T DO. An idea may work only when pushed to its limits. Samuel Insull and Juan Trippe went to extremes in the size of the engines they demanded respectively for power generation and aircraft. Halfway measures would not have yielded cheap power and air fares. There is a parallel in some of the sciences.

Nineteenth-century experimenters trying to make cheap polarizers from crystals found that small crystals crumbled, so they moved to larger crystals and still failed. In the 1920s the youthful Edwin Land succeeded by going entirely in the other direction to crystals of microscopic size. Rubber, that ubiquitous substance, was unusable for decades because on hot days it turned to a sticky, smelly goo. Charles Goodyear accidentally exposed a treated sample to prolonged heat in his wife's oven, and vulcanization became a commercial prospect.

5 NOTHING WORKS THE FIRST TIME. An impatient society and media expect instant results. Wall Street's fetish with quarterly earnings makes it worse. Georges Doriot, the founder of venture capitalism, observed that too many bankers and counselors had forgotten the history of our industrial giants. "The first fifteen years of companies and of human beings are very much alike — hope, measles, failures, mumps, reorganizations, scarlet fever, executive troubles, whooping cough, etc., are part of one's daily life." His question for bankers and brokers who told him he should sell an ailing company was "Would you sell a child running a temperature of 104?" *USA Today,* Amazon.com and CNN ran fevers for years but came good because the basic ideas were sound. Innovation is often cut off too soon because backers fail to appreciate that it takes time to work out the wrinkles. See the Wright brothers.

6 NEW IDEAS DISTURB. There are a hundred Cassandras for every change. Raymond Damadian was a "screaming lunatic" for thinking nuclear magnetic resonance might be used for medicine. Theodore Judah was "crazy Judah" for advocating a railway line over the High Sierra; Edwin Drake "crazy" for believing he could drill for oil. Giannini was a hothead (and worse) for thinking banking should be for the masses. The flat-earth dogmatists can never remember their predictions when success is achieved — but they serve a purpose. Something like 90 percent of new ventures do fail, so the odds are with the naysayers; they are only going to be disproved once in ten times. Ted Turner put it this way to me: "If you've got an innovative idea, and the majority does not pooh-pooh it, then the odds are you must not have a very good idea. When people thought I was loony, it did not bother me at all. In fact, I considered that I must really be onto something."

7 CROSS-POLLINATION WORKS. Leo Baekeland borrowed from photographic chemistry to invent plastics. Raymond Smith borrowed from carnival techniques to make gambling a major

American industry. Elisha Otis built his safety elevators using springs he had learned about making carriages. Wilbur Wright figured out how to turn an airplane by thinking about the way he turned a bicycle. Boundary transgressors are able to mobilize knowledge more flexibly and selectively.

8 SUCCESS IS RISKY. Georges Doriot was fond of saying that the most dangerous moment in the life of a company was when it had succeeded. His experience told him that this was when the company would customarily stop innovating, whereas only continual improvement and innovation could protect it from imitators. Ken Olsen both taught and received the lesson. He kept innovating with his minicomputers so that when the imitative competition emerged, he was ready to leapfrog with a wholly new, less expensive design. Then he was blindsided by the personal computer. Similarly, CNN allowed itself to be overtaken by Fox News. And Edison got stuck on direct current and was overtaken by George Westinghouse. Large corporations are especially vulnerable because they are not natural risk takers. Gordon Moore, leaving Fairchild to start up Intel, thought of large corporations as oil tankers, very hard to turn on a dime. Bureaucracies breed elaborate defenses. Innovators are by definition barrier-breaking troublemakers. Every company needs them and few tolerate them.

9 WHEN ONE PLUS ONE EQUALS THREE. We see many of our innovators flourishing in partnerships. But the psychology has to be right. Two people with the same abilities are less likely to make a good mix. Pete Peterson, the chairman of Sony United States who was a secretary of commerce and chairman of Bell Howell as a young man, observed: "If you have two people who have identical abilities they usually get into trouble because they have turf battles. So I'm a great believer in partnerships that are complementary, what I call 'interlocking neuroses.'" The illustrative partnerships are Wozniak and Jobs at Apple, Ida and William Rosenthal at Maidenform, Marc Andreesen and Jim Clark at Netscape, Singer and Clark, Colt and Root, Ford and Couzens, the Wright brothers and Sergey Brin and Larry Page at Google.

10 PLUGGING INTO NETWORKS. Isolated innovators may produce wonders—see Philo Farnsworth's invention of television—but they are more likely to succeed in a knowledge network, whether connected by geography or electronics. One of the reasons Leo Baekeland, inventor of plastics, settled in America was that the universities in his native Belgium disdained commerce. Time and again in this survey we see how proximity to knowledge centers assists business innovators: The biotech industry sprang out of San Francisco University, electronic advances from connections to Stanford University.

ONE GOOD IDEA DESERVES ANOTHER

Connections between innovators are ubiquitous—one good innovation deserves another. LEO BAEKELAND corresponded with EDISON, the WRIGHT BROTHERS, FORD, WILLIS WHITNEY of General Electric, the DUPONTS and BELL. That the engineering potential of Bakelite might be greater than the chemical was ELMER SPERRY's contribution. SAMUEL COLT was friends with SAMUEL MORSE. GIANNINI's Bank of America took a risk by investing in WALT DISNEY's early movies, including *Fantasia* in 1940. Disney hired the young BILL HEWLETT and DAVID PACKARD to build some of its first electronic equipment for *Fantasia*. THOMAS WATSON SR. was probably the first passenger in a car with an electric self-starter built by his friend CHARLES KETTERING, who needed Leo Baekeland's new plastic for insulation. HENRY FORD used Bakelite for his fenders. Ford's assembly lines were inspired by the beef and pork disassembly lines in the Chicago factories of the meatpacking innovators ARMOUR and SWIFT. ARNOLD BECKMAN backed WILLIAM SHOCKLEY who hired ROBERT NOYCE and GORDON MOORE whose circuits were used by NOLAN BUSHNELL, the founder of Atari, who hired STEVE JOBS. And TED HUFF, creator of the chip that made personal computers possible, worked for Intel, which called in GARY KILDALL, the most important innovator in computer operating systems, who was betrayed by IBM.

Innovators Gallery

ONE HUNDRED ONE WHO ALSO MADE A DIFFERENCE

Mary Kay Ash (1915–2001): Door-to-door cosmetics queen who loved pink and turned a $5,000 investment into a billion-dollar corporation.

Frank Ball (1857–1943): Started a canning business after buying John L. Mason's patent for Mason jars and slowly changed America's food habits by preventing spoilage; the Ball Corporation became a billion-dollar company.

Burton Baskin (1913–1967) and Irvine Robbins (1917–): Chocolate, vanilla, and strawberry were practically the only flavors of ice cream until Baskin and Robbins offered 31 varieties at their national chain, one for each day of the month.

Arnold O. Beckman (1900–): Invented pH meter and founded Beckman Instruments in 1935. Beckman's investment in William Shockley's transistor company helped turn Silicon Valley into a hub of innovation.

Edward Bernays (1891–1995): Self-described father of public relations and nephew of Sigmund Freud who adapted his uncle's science to manipulating public opinion.

William Bernbach (1911–1982): Adman who revolutionized his industry in the 1960s by harnessing public mistrust of consumerism to promote consumerism.

Clarence Birdseye (1886–1956): Arctic naturalist who noticed that fresh-caught fish froze instantly on ice; perfected a method for flash-freezing food to preserve taste, founding Birdseye Seafoods.

Henry Wollman Bloch (1922–): Founded H&R Block and democratized the nation's tax preparation. He pioneered electronic filing with the IRS. (H&R Block now accounts for half of all electronic filings.) Also owned CompuServe.

Michael Bloomberg (1942–): Michael Bloomberg's $4 billion fortune matches the budget deficit of the city he led to revival after being elected mayor in the wake of 9/11. He roiled bureaucracies to good effect the same way he revolutionized the financial world. He created the Bloomberg electronic business services in 1981 after he was sacked from Salomon Brothers.

Marvin Bower (1903–2003): In the depths of the Depression he was a young lawyer who noted that committees in charge of bankrupt companies ignored basic issues of management. "No one asked why these companies had failed," Bower said. "I saw the need for a professional firm that could give independent managing advice." He joined James McKinsey, a Chicago accountant, took over his New York office and transformed what was then known as industrial engineering into the fledgling field of management consulting. A graduate of Harvard's law and business schools, Bower modeled McKinsey & Co. after a law firm. Other consulting companies hired semiretired industrial executives. Bower hired young but brilliant recent graduates. He was known for his blunt advice. "The problem with this company is you," Bower

JEFF BEZOS (1964–): Amazon.com has defied the nayers and helped democratize commerce as e-tailing changes retailing. In 2004, 39 million shopped at Amazon for books and just about everything else. Bezos gives a booming laugh that has been compared to "a jackass gargling with bumblebees" when publishers complain about negative reviews. "Some of my proudest moments are when customers tell me that we talked them out of buying something." Bezos first heard of the Internet in a Princeton astrophysics class in 1985. He made millions as a Wall Street banker before launching Amazon from Seattle in 1994. He churned out a business plan on his laptop, choosing books because they were an easy search and there were millions of them, more than even the biggest stores could carry. Word of mouth quickly popularized the site, and frenzied investors drove up the stock until billion-dollar losses led critics to call it Amazon.bomb. Other Internet companies fell, but Amazon finally reported profits in 2003. "If I had a nickel for every time a potential investor told me it wouldn't work . . ." said Bezos when he had 100 billion of them.

once told an obstructive CEO of a floundering company. He lost the client but made it up to his partners. Between 1950 and 1967, as McKinsey's managing director, Bower doubled annual revenues. More important was the reputation for excellence he cultivated.

Bill Bowerman (1909–1999) and Phil Knight (1938–): Track coach and runner who cofounded Nike and shod the world in sneakers.

Warren Buffett (1930–): Warren Buffett did what is theoretically impossible: He outperformed the stock market. No powerful computers were used, no crack batteries of Ph.D.s. The plainspoken sage of Omaha became one of the richest men in the country through simple principles of investing. Biographer Roger Lowenstein writes: "Buffett could explain it like a general-store clerk discussing the weather." The secret to finance is that deep down it's all about ordinary companies. At a time of roller-coaster securities prices, Buffett's innovation was not to follow the fads.

DONNA DUBINSKY (1955–) and JEFF HAWKINS (1957–): Dubinsky and Hawkins organized America: No one got handheld computers right until they started Palm in 1992. Hawkins, the technical genius, figured out how to make handwriting-recognition software work when he realized humans were smarter than computers. If computers can't learn to read well, people at least can learn to write differently. Dubinsky, an articulate Harvard MBA, got the hard job—figuring out how to raise money for a product financiers considered dead. Apple had already made a $200 million mistake. The first PalmPilots were pared down compared to the Newton, but they were elegant and they worked. Buyers clamored for more. When 3Com acquired Palm, Dubinsky and Hawkins again set off on their own. HandSpring, their new venture, surpassed Palm in innovation, but fierce competition from Microsoft forced the inevitable. The two smaller companies merged in late 2003.

Nolan Bushnell (1943–): Nolan Bushnell learned about games as a carnival barker one summer. As an engineering student at the University of Utah, the 6-foot-4 bearded Mormon was captivated by a program on the department's mainframe called Spacewar. After graduation the microcomputer revolution led him to Silicon Valley, where he mixed his knowledge of electronics and showmanship to create Pong, the first commercial video game. "I had to come up with a game people already knew," Bushnell said, "something so simple that any drunk in any bar could play." The game appealed equally to the sober. The morning after he installed Pong at Andy Cap's, a Sunnyvale, California, bar, there was a line out the door. The game broke that night—too many quarters clogged the coin slot. Bushnell pooled $500 with a colleague and founded Atari, for "checkmate" in the Japanese game Go. It was an instant success, ushering in the video-game era.

Asa Candler (1851–1929): Atlanta businessman who bought the recipe for Coca-Cola from pharmacist John Pemberton and made it a national beverage through innovative advertising and marketing.

Willis Carrier (1876–1950): Invented air conditioning, making possible large-scale development of the southern U.S.

Steve Case (1958–): "AOL E Morte" ran an Italian headline from the early '90s, just one of many predictions of imminent demise, but America Online thrived under Steve Case, luring millions of Americans onto the Internet with its ubiquitous free start-up discs. AOL billed itself as family oriented; its dirty little secret was the immense popularity of adult chat rooms. Case's purchase of Time Warner was his undoing and a financial debacle for Time Warner when the Internet bubble collapsed.

James E. Casey (1888–1983): Failed Nevada gold prospector who started a messenger service with six employees and two bicycles and built it into United Parcel Service.

Liz Claiborne (1929–): Designer of a working women's clothing line who kept prices low and fashions sensible as women entered the workforce in increasing numbers.

Josephine Cochrane (1839–1913): Invented the dishwasher to save women from kitchen toil and formed a company that became a key part of Whirlpool.

John Deere (1804–1886): Blacksmith who speeded up the seeding of crops. Before 1840 it took two days to turn the soil of an acre with a yoke of oxen and a wooden plow. Many had tried to improve the plow and failed to make much impact on the heavy, sticky soil covered with prairie grass. After moving west from Vermont to Grand Detour, Illinois, Deere found the answer in 1837 in a broken steel saw in a sawmill. He chiseled off the teeth, heated and shaped it and made it the cutting edge of a wrought-iron moldboard that proved to be self-scouring. The Deere plow cut so sweetly through the soil, it was called the singing plow. He sold 700 plows in 1848 and 2,136 in 1849. John Deere and Company, incorporated in 1868, became one of the world's largest makers of farm equipment.

Michael Dell (1965–): Brought mass customization to personal computers through direct selling and just-in-time manufacturing.

Donald Douglas (1892–1981): Founded a company that produced the first modern airliner, the DC-3.

Charles Dow (1851–1902): Journalist who invented the Dow Jones Industrial Average, which opened at 40.94 in 1896 to track 12 industrial stocks; also founded with Edward Jones the Dow Jones financial news service in 1881 and the *Wall Street Journal* in 1889.

Charles Drew (1904–1950): This black doctor invented the blood bank and then bled to death after an auto accident when a white southern hospital would not admit him.

John Dryden (1839–1911): In 1873 John Dryden democratized life insurance in America by selling it to the poor for the first time. Inspired by the Prudential Company in Britain, he broke down payments into small weekly installments. Business boomed, but with millions of pennies flowing in regularly, the firm's clerks and accountants became overwhelmed with paperwork. Dryden saved the American company and revolutionized the industry by investing heavily in typewriters and mechanical tabulating machines. In 1891 Dryden became the first commercial customer of Herman Hollerith, whose company became the backbone of IBM. In 1875 Prudential insured 2 percent of the U.S. population; by the end of the century 17 percent. Prudential played a leading role in the evolution of information technology.

James Duke (1856–1925): Monopolized the American cigarette market after perfecting a cigarette rolling machine and getting smokers to buy packaged cigarettes rather than rolling their own.

Pierre du Pont (1870–1954): Transformed DuPont from family munitions company into a modern chemical corporation that later created plastics, cellophane, Kevlar and nylon; also helped found General Motors.

Richard Fairbank (1950–) and Nigel Morris (1959–): In the mid-'80s, Fairbank and Morris were management consultants who knew nothing about credit cards. After studying the industry for a single day, they concluded that one-size-fits-all interest rates of 19.8 percent were unacceptable. They founded Capital One and revolutionized credit cards. Instead of single, standardized cards, they offered thousands of varieties, all with different interest rates and incentives. Cutting-edge computers analyzed how consumers bought, borrowed and repaid debt. "Scientific tests on a massive scale," as Fairbank put it, allowed them to mass customize. They also innovated balance transfers, introductory teaser rates, mileage awards and multicolored cards. Credit card debt exploded as ever greater numbers of Americans bought on credit. Many new Capital One products haven't succeeded, but, as Fairbank puts it, "Failure is information too."

Cyrus Field (1819–1892): Laid the first transatlantic cable and helped start the New York City subway system.

Henry Flagler (1830–1913): Rockefeller associate who turned Florida from swamp into a vibrant economy with hotels and railroads.

Paul Galvin (1895–1959) and Robert W. Galvin (1922–): Father who founded Motorola and son who transformed it from a radio to a cellular phone company.

Wilbert Gore (1912–1986): DuPont engineer who invented Gore-Tex and founded an innovative company that makes everything from waterproof clothing to dental floss.

Wilson Greatbatch (1919–): Invented cardiac pacemaker and founded Greatbatch Technologies.

Charles Martin Hall (1863–1914): Aluminum was once so expensive Napoleon III commissioned a set of cutlery from the metal that cost more than gold; Charles Hall, an Oberlin College student, invented a cheap way of extracting the element and cofounded Alcoa, the Aluminum Corporation of America.

Joyce Hall (1891–1982): Founded Hallmark, the world's leading greeting card company, and helped create many American holidays.

Hugh Hefner (1926–): Magazine publisher who got porn out from behind counters and onto newsstand display racks. Hefner didn't invent selling sex, but he was the first to turn it into a corporation. The $3 billion porn industry stems in large part from *Playboy's* success and includes some of the largest corporations.

William Harley (1880–1943) and Arthur Davidson (1881–1950): Inventors of the Harley-Davidson motorcycle and creators of an American icon.

Howard Head (1914–1991): Aerospace engineer who created high-tech skis and tennis rackets, popularizing two sports.

William Randolph Hearst (1863–1951): "The Chief" was a party-loving Harvard dropout at 23, but when his father gave him control of the *San Francisco Examiner,* he made it a sensational success by narrative journalism and campaigns. He misused his power often (incidentally whipping up hysteria to provoke the Spanish-American War), but his energy and innovations gave him control of the world's largest publishing empire by 1930, with 28 newspapers, the Cosmopolitan Picture Studio, radio stations and magazines. He forged a synergistic link between film and print, getting his magazine short-story writers to give him movie rights. His excesses were dramatized by Orson Welles in *Citizen Kane,* but the company remained powerful in the 21st century.

Milton Snavely Hershey (1884–1946): Candy maker who built the largest chocolate factory in the world, and a Pennsylvania town to service it; proved milk chocolate could be mass-produced.

William R. Hewlett (1913–2001) and David Packard (1912–1996): Stanford graduates who started Hewlett-Packard in the much storied Silicon Valley garage and created one of the most enlightened companies in America.

Howard Hughes (1905–1963): The millionaire flyer, movie mogul and enigma was an innovative aircraft constructor. He made TWA a great airline and created its flagship, the famous Lockheed Constellation, the forerunner of pressurized 300-mph airliners. Described as "truly phenomenal" by the Smithsonian's R. E. G. Davies, Hughes was a pilot in Lindbergh's class; by the time he

---

MICHAEL MILKEN (1946–): Some 400 million cell phone users owe Milken a call—to thank him for the very industry he virtually created. He found the money to do that as he had found the money to reenergize American home building in the '70s, install a network of fiber-optic lines for cable television and the Internet, set off an explosion in more efficient alternatives to public health care facilities and generally liberate the creative energies of thousands of people who were frustrated by the old-boy network of finance—Ted Turner and CNN among them. Milken was a youth from Los Angeles who got a job collecting garbage in a New York investment bank in 1969 and catapulted the bank from obscurity to the front rank by his brilliantly researched mobilization of high-yield junk bonds. He raised the capital to start new firms whose only collateral was a cache of unbankable innovative ideas, and to restructure inert ones that had been given up for dead by Wall Street. He was a democratizer of capital. Milken went to jail for 22 months. Does a jailbird deserve to be in a roll of honor? The dissection of the case by Daniel Fischel, a professor of law at the University of Chicago (*Payback*, HarperBusiness, 1995), suggested he was the scapegoat for the savings and loan fiasco. Milken was not guilty of insider trading. He was punished for minor technical violations that had never been prosecuted before and have never been prosecuted since. Jesse Kornbluth (in *Highly Confident*, William Morrow, 1992) writes, "It is difficult to comprehend the evil in his genius when the court that sentenced him finds that his economic damage is less than $320,000." Milken devotes his life to philanthropy—not as a penance but a mission begun in the early '70s when his wife's mother was diagnosed with breast cancer. The Milken Family Foundation, established in 1982, works innovatively with 1,000 organizations around the world.

was 42 he had flown faster, further and with larger aircraft than anyone else in the world. In movies, he made *Hell's Angels, Scarface* and *The Outlaw,* the latter probably attracting more ink than anything else in a brilliant career that ended in mental illness.

Wayne Huizenga (1937–): Garbage truck driver who made Blockbuster Entertainment a videocassette giant.

Ken Iverson (1925–2002): President of Nucor who pioneered the use of mini-mills, rescuing the moribund American steel industry and showing that small computerized factories could produce steel cheaper and more efficiently than industry behemoths.

John H. Johnson (1918–): Publisher who recognized the growing black middle class with *Ebony, Jet* and *EM* magazines.

Robert Wood Johnson (1845–1910): Until the 1880s surgeons killed patients as often as they saved them, typically operating in bloodsplattered street clothes and dressing wounds in dirty cotton. Robert Wood Johnson, a Brooklyn pharmacist, heard a lecture on the theory of airborne germs and became convinced sterilization could solve the problem. He developed antiseptic dressings sealed in germ-resistant packets and promoted the theory of airborne germs. He moved his company, Johnson and Johnson, to a former wallpaper factory in New Jersey and in 1891 founded a bacteriological lab that invented a steam-and-pressure method of sterilizing cotton gauze. Doctors and hospitals began ordering in bulk. James Johnson took over after his brother's death and oversaw the invention of Band-Aids.

BERRY GORDY (1929–): Millions of white Americans listen to black music because of Berry Gordy. Motown Records, which he founded in 1959, was the first large black-owned music company in the U.S. and launched the careers of Michael Jackson, Lionel Ritchie, Diana Ross, Stevie Wonder, Marvin Gaye and Smoky Robinson. Gordy dropped out of high school in the 11th grade to become a professional boxer but realized he loved music more. When his Detroit record store failed, he took up songwriting. He also began managing musicians, publishing music and distributing his artists' records. His hits lit up the charts. To appeal to larger audiences, he applied the mass-production techniques of nearby auto factories to music: standardized songwriting, an in-house rhythm section, quality control. The Motown sound he created was songlike and soulful with powerful rhythms. Gordy moved the company to Los Angeles in 1971 and in 1988, with falling sales, sold out to MCA for $61 million.

Henry Kaiser (1882–1967): The U.S. Navy gave Henry Kaiser 18 minutes to make his presentation on ship building and promptly turned him down. A lucky introduction to FDR won him a contract to build the first 50 Liberty ships. Kaiser devised a method for churning out whole ships every few days with preassembled, mass-produced sections. The *John Fitch* was finished in 24 hours. Kaiser spurred the entire western economy and after battling the American Medical Association founded the country's largest health maintenance organization.

Herb Kelleher (1931–): Cofounder of Southwest Airlines who pioneered no-frills, low-fare travel.

Charles Kettering (1876–1958): Ford built cars for the common man, but Kettering's electric self-starter got women driving by obviating the difficult and dangerous hand crank; he founded Delco and headed General Motors' research.

Margaret Knight (1838–1914): Prolific inventor who devised a machine to mass-produce paper bags, now ubiquitous in grocery stores, and founded the Eastern Paper Bag Company in 1870.

Henry Kravis (1944–): Bought and sold scores of companies by pioneering the leveraged buyout at Kohlberg, Kravis, Roberts.

Ray Kroc (1902–1984): Started McDonald's restaurant franchise.

Brian Lamb (1941–): Cable innovator and cerebral on-air host who took advantage of new satellite technology in 1979 to create C–Span, the free, unfiltered public-affairs channel, and a second network in June 1986. Lamb's contribution to the reading of history is incalculable.

William Lear (1902–1978): Founded Learjet, invented the eight-track cassette and automatic pilot and helped start Motorola.

Stan Lee (1922–): Superpublisher of Marvel Comics who invented Spider-Man.

William Levitt (1907–1994): Ex-Seabee who came home from war to do for housing what Henry Ford had done for cars; he built middle-class houses more cheaply and quickly than ever before by subdividing the building process, first at Levittown, Long Island. Creator of the American suburb.

Marcus Loew (1870–1927): Fur trader who partnered with Adolph Zukor in 1905 to open a chain of movie theaters and later bought Metro Pictures and Samuel Goldwyn's and Louis B. Mayer's studios to form MGM.

George Lucas (1944–): Director and producer who changed the movie industry with special effects, most notably with *Star Wars*.

Henry Luce (1898–1967): Launched *Time* magazine in 1923 with Briton Hadden. Hadden died a few years later, but Luce lived to reshape American journalism, later founding *Fortune, Life* and *Sports Illustrated*.

Rowland H. Macy (1822–1877): Retired whaler who turned his dry-goods shop into one of the largest department stores by banning haggling and pioneering the price tag—much like John Wanamaker in Philadelphia.

Hiram Maxim (1869–1936): Prolific inventor whose big success was the first machine gun. "If you really want to make your fortune," a friend told him, "invent something that will help these fool Europeans kill each other more quickly." He sold his weapon to all sides on the eve of WWI.

Bill McGowan (1927–1992): His MCI fought AT&T and broke "Ma Bell's" long-distance monopoly, slashing rates.

William L. McKnight (1887–1978): Minnesota Mining and Manufacturing was facing collapse until it hired him at 20 as a bookkeeper in 1907. McKnight turned 3M around, rising through management and making the company one of the world's most systematically innovative. Under McKnight, 3M created hundreds of new products: waterproof sandpaper, masking tape, Scotch Tape, Scotchguard. He instituted rules to foster innovation: 30 percent of revenue must come from products created in the last four years; employees can spend 15 percent of their time on personal projects without telling management; the company may grant $50,000 for new ideas that have already been rejected. Post-it notes, Thinsulate and advanced fiber optics are some fruit of McKnight's system. "If you put fences around people you get sheep," he said. "Give people the room they need."

Frank McNamara (1917–1957): Invented the first general-purpose credit card, legend has it, after discovering in a New York restaurant he had left his wallet in another suit. In 1950 he helped found Diners Club.

George Merck IV (1894–1957): Merck took over the family drug company at age 32 and quickly transformed it into a scientific powerhouse. He fostered research and development by allowing scientists to publish their work. Other pharmaceutical companies kept findings secret. Top academics flocked to Merck & Co., where they extracted vitamin B_{12}, commercialized Selman Waksman's streptomycin, and mass-produced cortisone.

Charles Merrill (1885–1956): Founded Merrill Lynch and brought Main Street to Wall Street.

Joy Morton (1855–1934): Salt clumped together in rainy weather until Morton added an anticaking agent in 1911. "When it rains, it pours."

Condé Nast (1873–1942): Built first magazine chain in America, by featuring the rich for those who wanted to be like them. A more diverse and flourishing empire in the 21st century.

Al Neuharth (1924–): Gannett CEO creator of *USA Today,* the first truly national paper.

Paul Newman (1925–): Actor who gave neighbors bottles of homemade salad dressing for Christmas founded Newman's Own in 1982. The company came to donate 100 percent of its profits and had then raised $150 million for charity by 2004. Newman's motto: "Shameless exploitation for the common good."

Lucien L. Nunn (1853–1925): Engineer who contracted with George Westinghouse and Nikola Tesla to set up the first high-voltage alternating-current power plants in Colorado and Utah.

William Paley (1901–1990): Bought a failing chain of radio stations in 1929 and turned Columbia Broadcasting System (CBS) into the programming leader of the new television age.

Mary Pickford (1893–1979): "America's Sweetheart," actress and producer who founded United Artists with Charlie Chaplin, D. W. Griffith and husband Douglas Fairbanks.

Charles William Post (1854–1914): Businessman who suffered a nervous breakdown but felt better after stealing the idea for breakfast cereal from Dr. John Kellogg's and William Kellogg's sanitarium; started selling Grape-Nuts and launched the first nationwide ad campaign.

Philo Remington (1816–1889): Inventor who took over family arms business, perfected the breech-loading rifle and entered the typewriter business with a machine that had a shift key for capital and lowercase letters. Mark Twain used a Remington: *Life on the Mississippi* was the first typed novel to be published.

Leonard Riggio (1941–): His innovation was first one of scale. The young man who could not afford to attend college full-time and began his career as a clerk in a university bookstore became the biggest bookseller in the United States, with more than 1,000 sites enriching community life. With financing from innovator Michael Milken, he moved from mall stores to create the Barnes and Noble superstores, offering 100,000 titles or more, Starbucks coffee, snacks, magazine racks and comfortable chairs. Borders had led the way with the concept of a bookstore as an urban commons, but Riggio pushed it harder with many merchandising innovations. His second innovation is for a bookseller to dip a toe into publishing—backlist classics out of copyright. (Small independent booksellers have found it hard to compete, but the best survive by personal service, "hand-selling" books they love.)

John D. Rockefeller (1839–1937): Oil tycoon who became one of the richest and most powerful men in history. His innovation was to realize that although thousands of prospectors were digging oil wells, true riches lay at the point of refinery. Standard Oil, the massive vertically integrated company Rockefeller created, controlled 95 percent of U.S. refineries by 1878 but was split apart by the Supreme Court in 1911. Its progeny still dominate the industry.

Julius Schmid (1865–1939): *Fortune* dubbed Julius Schmid "the grand old man" of condoms in 1938 but didn't mention he was a convicted criminal. He had been arrested for peddling contraceptives in 1890 by Anthony Comstock's vice police. The half-paralyzed German-Jewish immigrant had arrived in New York in 1882 at the age of 17. Penniless and hobbling around on crutches, Schmid finally landed a job at a sausage casing factory where he found a lucrative, albeit illegal, use for the animal skins. He raised funds for what would become one of the largest condom companies in the world and wasn't deterred by jail: Ramses and Sheik were two of the most popular brands he invented. His greatest innovation was to manufacture safe vulcanized-rubber condoms. By the time condoms became legal in 1918—as long as they were used to prevent disease, not conception, a judge ruled—Schmid was already selling condoms to Allied troops, but not to American soldiers, who were ordered to practice "moral prophylaxis" in France. Almost 10 percent of G.I.s came down with venereal disease. By WWII the government learned its lesson, making Schmid the official condom supplier to the U.S. armed forces.

Howard Schultz (1953–): Marketer who was inspired by Italians and turned a Seattle coffee bean shop, Starbucks, into a nationwide hangout for the well caffeinated.

Charles Schwab (1937–): In the mid-1970s Charles Schwab democratized the brokerage firm by slashing fees to about a quarter of those charged on Wall Street. His three core tenets were no advice, no soliciting trades and no commissions. Customers were expected to know what they wanted. "The old brokerage companies, the old banks had gross inefficiencies built into their systems," said Schwab. His San Francisco firm was quick to embrace new technology and started one of the first online brokerages. Small-time investors flocked to Schwab, making it one of the largest brokerages in the United States.

Leonard Shoen (1916–1999): Founder of U-Haul who invented the self-moving industry but got embroiled in lawsuits and accusations of murder after his sons seized the company. His death in a Las Vegas car crash may have been a suicide.

Alfred P. Sloan Jr. (1875–1966): When Alfred Sloan took over William Durant's General Motors it was a sprawling mess with plummeting stock. With Pierre du Pont, Sloan revitalized the company through cooperative decentralization. Separate divisions sold different cars to different market segments, but all were involved in the overall bottom line. Sloan pioneered the production of new yearly models to boost demand. By the late 1920s his strategy of building "a car for every purse and purpose" displaced Ford—whose Model Ts came in any color "as long as it's black."

Albert Spalding (1850–1915): Red Sox player who built a sporting-goods chain and mythologized baseball.

George Squier (1865–1934): U.S. Army general with a Ph.D. who invented Muzak in 1928 by devising a way to pipe music over telephone wires and into elevators.

Julian Stein (1896–1981): Ophthalmologist founder of MCA, Music Corporation of America, one of the most powerful entertainment companies in the United States.

FRED SMITH (1944–): Fred Smith was an undergraduate at Yale in the mid-1960s when he wrote an economics paper that changed the world. In an automated society, he said, critical parts had to be delivered faster and more reliably than the Post Office guaranteed. The myth is that Smith got a C; he doesn't remember his grade. He considered going to business school after graduation but wound up being sent to Vietnam, where he led a rifle marine platoon and flew over 200 missions as a pilot. He was 27 when he mustered out and decided to put his delivery-system idea into action. The key would be a hub-and-spoke network (the airline industry later adopted Smith's system). Starting small was not possible. Smith raised $90 million, coolly telling each nervous investor he was the last holdout. Federal Express started on April 17, 1973, with a 25-city network. It lost $29 million in its first 26 months—at one point Smith met his payroll by winning $29,000 at a Las Vegas blackjack table—but finally turned the corner. Smith said he owes his success to lessons learned in the marines. Motivating employees is vital. "You're not delivering sand and gravel," he tells his ubiquitous couriers. "You're delivering someone's pacemaker, chemotherapy treatment for cancer drugs, the part that keeps the F-18s flying, or the legal brief that decides the case." Fed Ex entered the 21st century as a $20 billion corporation, operating in 211 countries and employing 215,000 people who make sure our parcels get there absolutely, positively overnight.

Gertrude Tenderich: Founded the Tampax company in 1936 and started mass-producing tampons from her home after buying the patent from their inventor, Dr. Earle Haas (1888–1981).

Earl S. Tupper (1908–1983): Invented Tupperware and founded a company to make it, but attracted large sales only after a single mother, Brownie Wise, started selling his products at home parties. Tupper quickly hired her as vice president.

Craig Venter (1947–): Founder and former president of Celera Genomics who used computers to achieve a dead heat with the U.S. government in decoding the human genome.

Sam Walton (1918–1992): Relentless discount retailer who started Wal-Mart in 1962 (the year Sebastian Kresge started KMart) and created one of the most powerful corporations in America by investing heavily in innovation: a computer database second only to the Pentagon's, handheld scanners, satellite communications and a ceaseless circulation of supply trucks between stores and warehouses. His business vision transformed America, driving out many smaller stores and paving the way for other "category-killer" superstores like Arthur Blank and Bernard Marcus's Home Depot and Tom Stemberg's Staples.

Sidney Weinberg (1891–1969) and Walter Sachs (1884–1980): Helped transform Goldman, Sachs into a modern investment bank.

Jack Welch (1935–): What may be called Welch's Law says that when the rate of change in an institution becomes slower than the rate of change outside, the end is in sight. For 20 years he acted on that mantra to give the behemoth the flexibility of a ballet dancer, diversifying from manufacturing into services in engineering, capital, computer networks, electronic business, medical systems and television: He acquired NBC, and CNBC and MSNBC were spin-offs. He "went nuts" about focusing outward on the consumer, pursuing the Six Sigma strategy of insisting on 99.99966 percent perfection (against a general industrial target of 97 percent revenues), as concerned about a slow call on repairing a faulty washing machine as the speed of global-positioning systems. His continuous innovations were notable for elevating ideas and intellect over hierarchy and tradition; finding leaders who lived the values of the company was more important to him than finding those who made the numbers. But the numbers were sensational. When he retired from GE in 2001 after 20 years, he had multiplied earnings tenfold and built it to a company with a market value of $450 billion. The son of a housewife and a railway conductor who sold newspapers and shoes in Salem, Massachusetts, Welch was long rated as America's most admired CEO.

George Westinghouse (1846–1914): Inventor of train air brakes who championed AC power in a pitched battle with Edison's DC system and presided over an electrical conglomerate.

Joseph Wilson (1909–1971): Xerox CEO who risked millions over years developing Chester Carlson's photocopying technology.

Kemmons Wilson (1913–2003): Bad hotels ruined a family vacation. He made his Holiday Inn chain clean and friendly, named it after a Bing Crosby movie and created the modern hotel.

Oprah Winfrey (1954–): Billionaire producer, actress and television host who has revolutionized the talk show business and the publishing world. She was crowned Miss Fire Prevention at age 17 in Nashville, Tennessee, and was soon hired as the city's first black TV news anchor. She moved to Chicago in 1984 to start her own show.

Evelyn Wood (1909–1995): "Mother of speed-reading" who founded Reading Dynamics in 1959 and was able to read 15,000 words a minute, slowed only by having to turn pages.

Frank Zamboni (1901–1988): Invented ice resurfacing machine ubiquitous at skating rinks.

SELECTED BIBLIOGRAPHY

GENERAL

Alderman, John. *Sonic Boom: Napster, MP3, and the New Pioneers of Music*. New York: Perseus, 2001.

Ambrose, Stephen and Douglas Brinkley. *Witness to America*. New York: HarperCollins, 1999.

American Social History Project: *Who Built America? Working People & the Nation's Economy, Politics, Culture & Society*. New York: Pantheon Books, 1992.

Anderson, Robert, et al., eds. *Innovation Systems in a Global Context: The North American Experience*. Montreal, Quebec: McGill-Queen's University Press, 1998.

Andrist, Ralph K., ed. *The American Heritage History of the Making of the Nation*. New York: American Heritage/Bonanza Books, 1968.

Asbell, Bernard. *The Pill: A Biography of the Drug That Changed the World*. New York: Random House, 1995.

Ashton, T. S. *The Industrial Revolution, 1760–1830*. 1948. New York: Oxford University Press, 1996.

Bailyn, Bernard. *To Begin the World Anew: The Genius and Ambiguities of the American Founders*. New York: Knopf, 2003.

Beatty, Jack, ed. *Colossus: How the Corporation Changed America*. New York: Broadway Books, 2001.

Bell, Daniel. *The Coming of Post-Industrial Society*. New York: Basic Books, 1999.

Bernstein, Peter L. *Against the Gods: The Remarkable Story of Risk*. New York: John Wiley & Sons, 1997.

——. *Capital Ideas: The Improbable Origins of Modern Wall Street*. New York: Free Press, 1993.

Bhidé, Amar V. *The Origin and Evolution of New Business*. Oxford University Press, 2000.

Biggs, Lindy. *The Rational Factory: Architecture, Technology, and Work in America's Age of Mass Production*. Baltimore, MD: Johns Hopkins University Press, 2003.

Boesen, Victor. *They Said It Couldn't Be Done: The Incredible Story of Bill Lear*. New York: Doubleday, 1971.

Boorstin, Daniel J. *The Americans: The Democratic Experience*. New York: Vintage Books, 1973.

——. *The Americans: The National Experience*. New York: History Book Club, 2002.

Bourne, Russell. *Invention in America*. Golden, CO: Fulcrum Publishing, 1996.

Bower, Marvin. *Perspectives on McKinsey*. Privately published, 1979.

——. Unpublished memoirs, McKinsey & Co.

——. *The Will to Lead: Running a Business with a Network of Leaders*. Boston: Harvard Business School Press, 1997.

——. *The Will to Manage: Corporate Success Through Programmed Management*. New York: McGraw-Hill, 1966.

Boyer, Paul S., et al. *The Enduring Vision: A History of the American People*. Vols. 1 & 2. Lexington, MA: D. C. Heath, 1996.

——. *The Oxford Companion to United States History*. Oxford: Oxford University Press, 2001.

Branscomb, Lewis, and James J. Keller, eds. *Investing in Innovation: Creating a Research and Innovation Policy That Works*. Cambridge, MA: MIT Press, 1999.

Brogan, Hugh. *The Penguin History of the United States*. New York: Penguin, 1990.

——. *Telephone: The First Hundred Years*. New York: Harper & Row, 1976.

Brooks, John, ed. *The Autobiography of American Business: The Story Told by Those Who Made It*. New York: Doubleday, 1974.

Brown, David F. *Inventing Modern America*. Cambridge, MA: MIT Press, 2002.

Bruce, Robert V. *The Launching of American Science*. New York: Knopf, 1997.

Buderi, Robert. *Engines of Tomorrow: How the World's Best Companies Are Using Their Research Labs to Win the Future*. New York: Simon and Schuster, 2000.

Bunch, Bryan, and Alexander Hellemans. *The Timetables of Technology*. New York: Simon and Schuster, 1993.

Burlingame, Roger. *Backgrounds of Power: The Human Story of Mass Production*. New York/London: Scribner's, 1949.

——. *March of the Iron Men*. New York: Scribner's, 1938.

Burrough, Bryan, and John Helyar. *Barbarians at the Gate: The Fall of RJR Nabisco*. New York: Harper & Row, 1990.

Burt, Ronald S. *Structural Holes: The Social Structure of Competition*. Cambridge, MA: Belknap Press, 1995.

Bush, Vannevar. *Science Is Not Enough*. New York: William Morrow, 1965.

Calder, Ritchie. *The Evolution of the Machine*. New York: American Heritage Publishing, 1968.

Carnes, Mark, ed. *Invisible Giants*. New York: Oxford University Press, 2002.

Castells, Manuel. *The Rise of the Network Society: The Information Age: Economy, Society, and Culture*. Oxford: Blackwell, 1996.

Chandler, Alfred D., Jr. *The Visible Hand: The Managerial Revolution in American Business*. Cambridge, MA: Harvard University Press, 1977.

Chernow, Ron. *Alexander Hamilton*. New York: Penguin, 2004.

———. *Titan: The Life of John D. Rockefeller, Sr.* New York: Random House, 1998.

Christensen, Clayton. *The Innovator's Dilemma*. New York: HarperBusiness, 2003.

Cohen, Wesley M., and Daniel A. Levinthal. *Absorptive Capacity: A New Perspective on Learning and Innovation*. Ithaca, NY: Cornell University Press, 1990.

Collins, Gail. *America's Women: 400 Years of Dolls, Drudges, Helpmates and Heroines*. New York: HarperCollins, 2003.

Collins, Jim, and Jerry I. Porras. *Built to Last: Successful Habits of Visionary Companies*. New York: HarperBusiness, 2002.

Cooke, Alistair. *Alistair Cooke's America*. New York: Knopf, 1974.

Cowan, Ruth Schwartz. *More Work for Mother: The Ironies of Household Technology from the Open Hearth to the Microwave*. New York: Basic Books, 1985.

——. *The Social History of American Technology*. Oxford University Press, 1997.

Cox, Archibald. *The Court and the Constitution*. Boston: Houghton Mifflin, 1987.

Crichton, Judy. *America 1900: The Turning Point*. New York: Henry Holt, 1998.

Davie, Emily. *Profile of America*. New York: Thomas Y. Crowell, 1954.

Diamond, Jared. *Guns, Germs, and Steel: The Fate of Human Societies*. New York: W. W. Norton, 1997.

Dickens, Charles. *American Notes*. Introduction by Christopher Hitchens. New York: Random House, 1995.

Dorsey, Gary. *Silicon Sky: How One Small Start-Up Went Over the Top to Beat the Big Boys into Satellite Heaven*. New York: Perseus Books, 1999.

Drucker, Peter. *Innovation and Entrepreneurship: Practice and Principles*. New York: Harper & Row, 1984.

Evans, Harold. *The American Century*. New York: Alfred A. Knopf, 2000.

Faulkner, Harold Underwood. *American Economic History*. New York: Harper and Brothers, 1954.

Fischel, Daniel. *Payback: The Conspiracy to Destroy Michael Milken and his Financial Revolution*. New York: HarperBusiness, 1995.

Fischer, Claude S. *America Calling: A Social History of the Telephone to 1940*. Berkeley, CA: University of California Press, 1994.

Flatow, Ira. *They All Laughed . . . From Light Bulbs to Lasers: The Fascinating Stories Behind the Great Inventions That Have Changed Our Lives*. New York: HarperCollins, 1992.

Foster, Richard. *Innovation: The Attacker's Advantage*. New York: Summit Books, 1996.

Foster, Richard, and Sarah Kaplan. *Creative Destruction: Why Companies That Are Built to Last Underperform the Market—and How to Successfully Transform Them*. Currency, 2001.

Frank, Thomas. *The Conquest of Cool: Business, Culture, Counterculture, and the Rise of Hip Consumerism*. Chicago: University of Chicago Press, 1997.

Freeman, Chris, and Luc Soete. *The Economics of Industrial Innovation*. Cambridge, MA: MIT Press, 1999.

Gabor, Andrea. *The Capitalist Philosophers: The Geniuses of Modern Business—Their Lives, Times, and Ideas*. New York: Times Books, 2000.

Gladwell, Malcolm. *The Tipping Point: How Little Things Can Make a Big Difference*. New York: Little, Brown, 2002.

Gordon, John Steele. *The Business of America*. New York: Walker & Company, 2001.

Gordy, Berry. *To Be Loved: The Music, the Magic, the Memories of Motown*. New York: Warner Books, 1994.

Green, Constance. *Holyoke, Massachusetts: A Case History of the Industrial Revolution in America*. New Haven, CT: Yale University Press, 1939.

Hafner, Katie, and Matthew Lyon. *Where Wizards Stay Up Late: The Origins of the Internet*. New York: Touchstone, 1996.

Hall, Kermit L., et al. *The Oxford Companion to American Law*. Oxford: Oxford University Press, 2002.

Hallett, Anthony, and Diane Hallet. *Entrepreneur Magazine Encyclopedia of Entrepreneurs*. New York: John Wiley and Sons, 1997.

Hamilton, Neil A. *American Business Leaders: From Colonial Times to the Present*. 2 Volumes. ABC-CLIO, 1999.

Hann, Judith. *How Science Works*. Pleasantville, NY: Reader's Digest, 1991.

Harvard Business Review: Interviews with CEOs. Cambridge, MA: Harvard Business School Press, 1995.

Hessen, Robert. *Steel Titan: The Life of Charles M. Schwab*. Pittsburgh: University of Pittsburgh Press, 1975.

Hindle, Brooke. *Emulation and Invention*. New York: New York University Press, 1981.

Hindle, Brooke, and Steven Lubar. *Engines of Change*. Washington, DC: Smithsonian, 1986.

Hounshell, David A. *From the American System to Mass Production, 1800–1932: The Development of Manufacturing Technology in the United States*. Baltimore and London: Johns Hopkins University Press, 1984.

——. *Science and Corporate Strategy: Du Pont R&D, 1902–1980*. Cambridge: Cambridge University Press, 1988.

Hughes, Thomas, P. *American Genesis*. New York: Viking Penguin, 1989

——. *Elmer Sperry: Inventor and Engineer*. Baltimore, MD: Johns Hopkins University Press, 1993.

——. *Rescuing Prometheus*. New York: Pantheon Books, 1998.

Hunt, Alfred. *The Management Consultant*. New York: Ronald Press Company, 1977.

Hunter, Louis C. *A History of Industrial Power in the United States, 1780–1930*. 3 vols. Charlottesville, VA: University Press of Virginia, 1979–1991.

Hyman, Paula E., and Deborah Dash Moore, eds. *Jewish Women in America: An Historical Encyclopedia*. 2 vols. New York: Routledge, 1997.

Inkster, Ian. *Science and Technology in History: An Approach to Industrial Development*. Piscataway, NJ: Rutgers University Press, 1991.

Isaacson, Walter. *Benjamin Franklin: An American Life*. New York: Simon and Schuster, 2003.

Israel, Paul. *From Machine Shop to Industrial Laboratory: Telegraphy and the Changing Context of American Invention, 1830–1920*. Baltimore: Johns Hopkins University Press, 1992.

James, Henry. *The American Scene*. New York: Penguin Books, 1994.

James, Marquis. *Merchant Adventurer: The Story of W. R. Grace*. Wilmington, DE: Scholarly Resources, Inc., 1993.

Jennings, Walter Wilson. *20 Giants of American Business*. New York: Exposition Press, 1953.

Josephson, Matthew. *The Robber Barons*. New York: Harcourt, Brace, 1962.

Kador, John. *Charles Schwab: How One Company Beat Wall Street and Reinvented the Brokerage Industry*. New York: John Wiley & Sons, 2002.

Kaempffert, Waldemar, ed. *A Popular History of American Invention*. 2 vols. New York: A. L. Burt Co. and Charles Scribner's Sons, 1924.

Kahn, E. J., Jr. *The Problem Solvers: A History of Arthur D. Little, Inc*. Boston: Little, Brown, 1986.

Kanter, Rosabeth Moss. *The Change Masters: Innovation & Entrepreneurship in the American Corporation*. New York: Touchstone, 1983.

Katzenbach, Jon R. *Peak Performance: Aligning the Hearts and Minds of Your Employees*. Cambridge, MA: Harvard Business School Press, 2000.

Klein, Maury. *The Change Makers: From Carnegie to Gates, How the Great Entrepreneurs Transformed Ideas into Industries*. New York: Times Books, 2003.

Kleiner, Art. *The Age of Heretics: Heroes, Outlaws, and the Forerunners of Corporate Change*. Currency, 1996.

Koehn, Nancy F. *Brand New: How Entrepreneurs Earned Consumers' Trust from Wedgwood to Dell*. Cambridge, MA: Harvard Business School Press, 2001.

Kornbluth, Jesse. *Highly Confident: The True Story of the Crime and Punishment of Michael Milken*. New York: William Morro, 1992.

Krugman, Paul. *The Great Unraveling: Losing Our Way in the New Century*. New York: W. W. Norton, 2003.

Kuhn, Thomas S. *The Structure of Scientific Revolutions*. Chicago, IL: University of Chicago Press, 1962.

Lagasse, Paul, gen. ed. *The Columbia Encyclopedia*. 6th ed. New York: Columbia University Press, 2000.

Latour, Bruno. *The Pasteurization of France*. Cambridge, MA: Harvard University Press, 1988.

Lienhard, John H. *The Engines of Our Ingenuity*. Oxford: Oxford University Press, 2000. Also: www.egr.uh.edu/me/faculty/lienhard

Livesay, Harold C., *American Made: Men Who Shaped the American Economy*. Boston: Little, Brown, 1979.

Macaulay, David, with Neil Ardley. *The New Way Things Work*. Boston: Houghton Mifflin, 1998.

Maier, Pauline, Merritt Roe Smith, Alexander Keyssar, and Daniel J. Kevles. *Inventing America: A History of the United States*. 2 vols. New York: W. W. Norton, 2003.

Mark, J. Paul. *The Empire Builders: Inside The Harvard Business School*. New York: William Morrow and Co., 1987.

Martin, Roger. *The Responsibility Virus: How Control Freaks, Shrinking Violets—and the Rest of Us—Can Harness the Power of True Partnership*. New York: Basic Books, 2002.

Marx, Leo. *The Machine in the Garden: Technology and the Pastoral Ideal in America*. Oxford: Oxford University Press, 1964.

Mayr, Otto, and Robert C. Post. *Yankee Enterprise: The Rise of the American System of Manufactures*. Washington, DC: Smithsonian Institution Press, 1981.

McCraw, Thomas K. *Creating Modern Capitalism: How Entrepreneurs, Companies, and Countries Triumphed in Three Industrial Revolutions*. Cambridge, MA: Harvard University Press, 1995.

McKibben, Gordon. *Cutting Edge: Gillette's Journey to Global Leadership*. Boston: Harvard Business School Press, 1998.

Meikle, Jeffrey L. *Twentieth Century Limited: Industrial Design in America, 1925–1939*. Philadelphia: Temple University Press, 1998.

Micklethwait, John, and Adrian Wooldridge. *The Company: A Short History of a Revolutionary Idea*. New York: Random House, 2003.

Miller, William, ed. *Men in Business: Essays in Entrepreneurship*. Cambridge, MA: Harvard University Press, 1952.

Misa, Thomas J. *A Nation of Steel: The Making of Modern America, 1865–1925*. Baltimore, MD: Johns Hopkins University Press, 1998.

Morgenson, Gretchen, ed. *Forbes Great Minds of Business*. New York: John Wiley and Sons, 1998.

Mulgan, Geoff. *Connexity: How to Live in a Connected World*. Cambridge, MA: Harvard Business School Press, 1997.

Negroponte, Nicholas. *Being Digital*. New York: Alfred A. Knopf, 1995.

Newman, Paul, and A. E. Hotchner. *Shameless Exploitation in Pursuit of the Common Good*. New York: Doubleday, 2003.

Noble, David. *America by Design: Science, Technology, and the Rise of Corporate Capitalism*. New York: Knopf, 1977.

Nye, David. *American Technology Sublime*. Cambridge, MA: MIT Press, 1994.

——. *Consuming Power: A Social History of American Energies*. Cambridge, MA: MIT Press, 1999.

Oxford Analytica. *America in Perspective: Major Trends in the United States Through the 1990s*. Boston: Houghton Mifflin, 1986.

Perkins, Edwin J. *Wall Street to Main Street: Charles Merrill and Middle-Class Investors*. Cambridge: Cambridge University Press, 1999.

Peters, Thomas J. *The Circle of Innovation*. New York: Knopf, 1997.

Peters, Thomas J., and Robert H. Waterman Jr. *In Search of Excellence: Lessons from America's Best-Run Companies*. New York: Warner Books, 1982.

Petzinger, Thomas, Jr. *The New Pioneers: The Men and Women Who Are Transforming the Workplace and Marketplace*. New York: Simon and Schuster, 1999.

Pine, B. Joseph, II. *Mass Customization: The New Frontier in Business Competition*. Cambridge, MA: Harvard Business School Press, 1993.

Preston, Richard. *American Steel*. New York: Avon, 2000.

Reader's Digest. *How in the World? A Fascinating Journey Through the World of Human Ingenuity*. Pleasantville, NY: Reader's Digest, 1990.

Rhodes, Richard, ed. *Visions of Technology*. New York: Touchstone, 2000.

Riordan, Michael, and Lillian Hoddeson. *Crystal Fire: The Birth of the Information Age*. New York: W. W. Norton, 1997.

Rosenberg, Nathan. *Inside the Black Box: Technology and Economics*. Cambridge: Cambridge University Press, 1994.

Sandler, Martin W. *American Image*. Chicago: Contemporary Books, 1999.

Scherer, F. M. *New Perspectives on Economic Growth and Technological Innovation*. Washington, DC: Brookings Institution Press, 1999.

Scott, Shane. "Cultural Influences on National Rates of Innovation." *Journal of Business Venturing* 8 (1993): 59–73

Sellers, Charles. *The Market Revolution: Jacksonian America, 1815–1846*. New York: Oxford University Press, 1991.

Sinclair, Upton, *The Jungle*. New York: Penguin Books, 1985.

Smith, Merritt Roe, and Leo Marx. *Does Technology Drive History? The Dilemma of Technology Determinism*. Cambridge, MA: MIT Press, 1998.

Sobel, Robert. *The Entrepreneurs: Explorations Within the American Business Tradition*. New York: Weybright and Talley, 1974.

Sobel, Robert, and David B. Sicilia. *The Entrepreneurs: An American Adventure*. Boston: Houghton Mifflin, 1986.

The Statistical History of the United States. New York: Basic Books, 1976.

Stewart, James B. *Den of Thieves*. New York: Touchstone, 1991.

Strasser, Susan. *Satisfaction Guaranteed: The Making of the American Mass Market*. Washington, DC: Smithsonian Institution Press, 1989.

Strouse, Jean. *Morgan: American Financier*. New York: Random House, 1999.

Tedlow, Richard S. *New and Improved: The Story of Mass Marketing in America*. New York: Basic Books, 1990.

Thompson, Holland. *The Age of Invention: A Chronicle of Mechanical Conquest*. IndyPublish.com, 2001.

Thorndike, Joseph J., ed. *Great Stories of American Businessmen*. New York: American Heritage, 1972.

Tone, Andrea. *Devices & Desires: A History of Contraceptives in America*. New York: Hill and Wang, 2002.

Tye, Larry. *The Father of Spin: Edward L. Bernays & the Birth of Public Relations*. New York: Crown Publishers, 1988.

Usher, Abbott Payson. *A History of Mechanical Inventions*. New York: Dover Publications, 1988.

Utterback, James M. *Mastering the Dynamics of Innovation*. Cambridge, MA: Harvard Business School Press, 1994.

Veblen, Thorstein. *The Theory of the Leisure Class*. New York: Penguin Books, 1979.

Wang, Charles B. *Techno Vision*. New York: McGraw-Hill, 1994.

Way Things Work, The: An Illustrated Encyclopedia of Technology. New York: Simon and Schuster, 1967.

Welch, Jack, with John Al Byrne. *Jack: Straight from the Gut*. New York: Warner Books, 2001.

Wawro, Thaddeus. *Radicals and Visionaries: Entrepreneurs Who Revolutionized the 20th Century*. Irvine, CA: Enterprise Press, 2000.

Weber, Robert J., and David N. Perkins. *Inventive Minds: Creativity in Technology*. New York/London: Oxford University Press, 1992.

Williams, Trevor T. *A History of Invention*. New York: Facts on File, 2000.

Wilson, Mitchell. *American Science and Invention*. New York: Simon and Schuster, 1954.

Wolf, Michael J. *The Entertainment Economy*. New York: Times Books, 1999.

Wolff, Michael. *Burn Rate: How I Survived the Gold Rush Years on the Internet*. New York: Simon and Schuster, 1998.

Wood, Gordon S. *The Radicalism of the American Revolution*. New York: Vintage Books, 1993.

Wright, Michael, and Mukul Patel, eds. *How Things Work Today*. New York: Crown, 2000.

PART 1
STEAMBOATS

Albion, Robert Greenhalgh, with Jennie Barnes Pope. *The Rise of New York Port: 1815–1860*. New York: Charles Scribner's Sons, 1938.

Barnes, Joseph. *Remarks on Mr. John Fitch's Reply to Mr. Rumsey*. Philadelphia: 1788 pamphlet.

Baxter, Maurice. *Daniel Webster & the Supreme Court*. Amherst, MA: University of Massachusetts Press, 1966.

———. *The Steamship Monopoly*. New York: Alfred A. Knopf, 1972.

Boyd, Thomas. *Poor John Fitch: Inventor of the Steamboat*. New York: Putnam, 1935.

Burton, Anthony. *Richard Trevithick: The Man and His Machine*. London: Aurum Press, 2000.

Dangerfield, George. *Chancellor Robert R. Livingston, 1746–1813*. New York: Harcourt Brace and Company, 1960.

Dickinson, Henry Winram. *Robert Fulton, Engineer and Artist*. London: John Lane, 1913.

Fitch, John. *Autobiography*. Philadelphia: The American Philosophical Society, 1976.

———. *The Original Steamboat Supported or a Reply to Mr. James Rumsey's Pamphlet*. Philadelphia: 1788. Reprinted in *Documentary History of the State of New York*, Albany, 1849.

Flexner, James Thomas. *Steamboats Come True*. New York: Fordham University Press, 1993.

Fulton, Robert. Letter to Oliver Evans. 10 Jan. 1812. Manuscript Department. The New-York Historical Society, New York, NY.

———. Letter to Robert Livingston. 20 Nov. 1807. Manuscript Department. The New-York Historical Society, New York, NY.

———. Notes. Manuscript Department. The New-York Historical Society, New York, NY.

Havighurst, Walter. *Voices on the River: The Story of the Mississippi Waterways*. Minneapolis: University of Minnesota Press, 2003.

Hodge, James. *Richard Trevithick*. Buckinghamshire, UK: Shire Publications Ltd., 1995.

Hunter, Louis C. *Steamboats on the Western Rivers*. Dover Publications, Inc., 1977.

McCall, Edith. *Conquering the Rivers*. Baton Rouge, LA, and London: Louisiana State University Press, 1984.

Morgan, John Smith. *Robert Fulton*. New York: Mason/Charter, 1977.

Morrison, John H. *History of American Steam Navigation*. New York: Stephen Daye Press, 1958.

Philip, Cynthia Owen. *Robert Fulton*. New York: Franklin Watts, 1985.

Sale, Kirkpatrick. *The Fire of His Genius: Robert Fulton and the American Dream*. New York: Simon and Schuster, 2001.

Shepherdstown, WV, Public Library Web site. James Rumsey page, http://www.lib.shepherdstown.wv.us/sin/rumsey/html.

Smith, Jean Edward. *John Marshall*. New York: Henry Holt, 1996.

Sutcliffe, Alice Crary. *Robert Fulton and the Clermont*. New York: The Century Company, 1909.

———. "Robert Fulton, the Man." Ms. New-York Historical Society, New York, NY.

Turnbull, Archibald Douglas. *John Stevens: An American Record*. New York: Century Co., 1928.

Turner, Ella M. *James Rumsey: Pioneer in Steam Navigation*. Publisher unknown. Scottsdale, PA, 1930.

Twain, Mark. *Life on the Mississippi*. New York: The Modern Library, 1994.

Warren, Charles. *The Supreme Court in United States History, Volume I, 1789–1835*. Boston: Little, Brown, 1932.

Westcott, Thompson. *The Life of John Fitch: The Inventor of the Steamboat*. Philadelphia: J. B. Lippincott, 1857.

Works Project Administration. *A Maritime History of New York*. New York: Doubleday, Doran, 1941.

OLIVER EVANS

Bathe, Greville, and Dorothy Bathe. *Oliver Evans: A Chronicle of Early American Engineering*. New York: Arno Press, 1972.

Boker, George Henry. *Oliver Evans*. Boston: Twayne Publishers, 1984.

Evans, Oliver. *The Abortion of the Young Steam Engineer's Guide*. Philadelphia, PA: Fry and Kammerer, 1805.

———. *The Young Mill-Wright and Miller's Guide*. Philadelphia: first edition 1795; 13th edition, Lea and Blanchard, 1850.

Ferguson, Eugene S. *Oliver Evans: Inventive Genius of the American Industrial Revolution*. Wilmington, DE: Hagley Museum & Library, 1980.

Hazen, Theodore R. Pond Lily Mill Restorations Web site, http://www.angelfire.com/journal/pondlilymill/index.html.

Mortensen, C. W. *Oliver Evans, A Genius from the First State*. Newark, DE: 1984.

Pursell, C. W., Jr. *Early Stationary Steam Engines in America*. Washington, DC: Smithsonian Institution Press, 1969.

Thurston, Robert Henry. *A History of the Growth of the Steam Engine*. New York: Appleton and Co., 1884.

———. *A Manual of the Steam Engine*. 2 vols. New York: John Wiley & Sons, 1891.

ELI WHITNEY

Battison, Edwin A. "Eli Whitney and the Milling Machine." *Smithsonian Journal of History* 1, no. 2 (Summer 1966): 9–34.

Cooper, Carolyn. "Myth, Rumor, and History: The Yankee Whittling Boy as Hero and Villain." *Technology and Culture*, 44.1, 2003, 82–96.

Eli Whitney papers at Yale University Library, New Haven, CT.

Fuller, Claude E. *The Whitney Firearms*. Huntington, WV: Standard Publications Inc., 1946.

Green, Constance. *Eli Whitney and the Birth of American Technology*. 1956.

Jefferson, Thomas. *Writings*. New York: Library of America, 1984.

McCullough, David. *John Adams*. New York: Simon and Schuster, 2001.

Mirsky, Jeannette, and Allan Nevins. *The World of Eli Whitney*. New York: Macmillan, 1952.

Roe Smith, Merritt. *Harpers Ferry Armory and the New Technology: The Challenge of Change*. Ithaca, NY, and London: Cornell University Press, 1977.

Woodbury, Robert S. "The Legend of Eli Whitney and Interchangeable Parts." *Technology and Culture* 1 (Summer 1960): 235–253.

SAMUEL SLATER AND FRANCIS CABOT LOWELL

Conrad, James L., Jr. "Drive that branch: Samuel Slater, the power loom, and the writing of America's textile history." In *Technology & American History: A Historical Anthology from Technology & Culture*, edited by Stephen H. Cutcliffe and Terry S. Reynolds. Chicago: University of Chicago Press, 1997.

Dalzell, Robert F., Jr. *Enterprising Elite: The Boston Associates and the World They Made*. New York: W. W. Norton, 1987.

Gibb, George S. *The Saco-Lowell Shops: Textile Machinery Building in New England*. Cambridge, MA: Harvard University Press, 1950.

Jeremy, David. *Transatlantic Industrial Revolution: The Diffusion of Textile Technology Between Britain and America, 1790–1830s*. Cambridge, MA: Harvard University Press, 1981.

Leavitt, Sarah, ed. *Slater Mill*. Dover, NH: Arcadia Publishing, 1997.

Rivard, Paul E. *A New Order of Things: How the Textile Industry Transformed New England*. University Press of New England, 2002.

Tucker, Barbara. *Samuel Slater and the Origins of the American Textile Industry, 1790–1860*. Ithaca, NY: Cornell University Press, 1984.

Ware, Caroline F. *The Early New England Cotton Manufacture in the United States*. New York: Russell & Russell, 1966.

White, George Savage. *Memoir of Samuel Slater: The Father of American Manufactures*. New York: A. M. Kelley, 1967.

SAM COLT

Edwards, William B. *The Story of Colt's Revolver: The Biography of Col. Samuel Colt*. Harrisburg, PA: The Stackpole Company, 1957.

Garavaglia, Louis, and Charles Worman. *Firearms of the American West: 1803–1865*. Albuquerque, NM: University of New Mexico Press, 1984.

Grant, Ellsworth S. *The Colt Legacy: The Colt Armory in Hartford, 1855–1980*. Lincoln, RI: Andrew Mowbray Publishers, 1982.

———. "Samuel Colt." West Hartford, CT: Connecticut's Heritage Gateway, www.ctheritage.org.

Hosley, William. *Colt: The Making of an American Legend*. Amherst, MA: University of Massachusetts Press, 1996.

Keating, Bern. *The Flamboyant Mr. Colt and His Deadly Six-Shooter*. New York: Doubleday, 1978.

Markham, George. *Guns of the Wild West: Firearms of the American Frontier, 1849–1917*. London: Arms & Armor Press, 1991.

Rohan, Jack. *Yankee Arms Maker: The Incredible Career of Samuel Colt*. New York: Harper and Brothers, 1935.

Simon, Kenneth A. "Colt: Legend and Legacy." Transcript, 1997, http://www.simonpure.com/colt.html.

Wilson, R. L., and Sid Latham. *The Colt Heritage: The Official History of Colt Firearms from 1836 to the Present*. New York: Simon and Schuster, 1979.

———. *The Peacemakers*. New York: Random House, 1992.

SAMUEL FINLEY BREESE MORSE

Coe, Lewis. *The Telegraph: A History of Morse's Invention and Its Predecessors in the United States*. Jefferson, NC: McFarland and Company, 1993.

Cookson, Gillian. *The Cable that Changed the World: The Story of the Atlantic Telegraph*. Gloucester, UK: Tempus, 2003.

Coulson, Thomas. *Joseph Henry: His Life and Work*. Princeton, NJ: Princeton University Press, 1950.

Cox, Samuel Sullivan. *Memorial Eulogies Delivered in the House of Representatives of the United States*. Washington, DC: Government Printing Office, 1883.

Mabee, Carleton. *The American Leonardo: A Life of Samuel F. B. Morse*. New York: Alfred A. Knopf, 1943.

Morse, Samuel F. B. A Register of His Papers in the Library of Congress, Manuscript Division, 2001, http://memory.loc.gov/ammem/sfbmhtml/sfbmhome.html.

Prime, Samuel Irenaeus. *Samuel F. B. Morse, LL. D., Inventor of the Electromagnetic Recording Telegraph*. New York: Appleton & Company, 1875.

Reid, James D. *The Telegraph in America: Its Founders, Promoters, and Noted Men*. New York: Derby Brothers, 1879.

Taylor, William B. *An Historical Sketch of Henry's Contribution to the Electromagnetic Telegraph with an Account of the Origin and Development of Prof. Morse's Invention*. From the Smithsonian Report for 1878. Washington, DC: Government Printing Office, 1879.

Trefoil, James, ed. *Encyclopedia of Science and Technology*. New York: Routledge, 2001.

CYRUS MCCORMICK

Casson, Herbert N. *Cyrus Hall McCormick: His Life and Times*. A. C. McClurg, 1909.

Hutchinson, William T. *Cyrus Hall McCormick: Harvest, 1856–1884*. New York: D. Appleton-Century Company, 1935.

——. *Cyrus Hall McCormick: Seed-Time, 1809–1856*. New York: The Century Co., 1930.

McCormick, Cyrus. *Century of the Reaper*. Houghton Mifflin, 1931.

——. Letter to the Editor (with diagram). *Mechanics' Magazine and Register of Inventions and Improvements*. 4 (Oct. 11, 1834).

Shenandoah Agricultural Research and Extension Center Web site. www.vaes.vt.edu/steeles/mccormick/mccormick.html.

ISAAC MERRITT SINGER

Abbot, Charles Greeley. *Great Inventions*. Vol. 12. Washington, DC: Smithsonian Scientific Series, 1932.

Bishop, J. Leander. *A History of American Manufactures from 1608–1860*. Vol. 2. New York: Augustus M. Kelley, 1966.

Bissell, Don. *The First Conglomerate: 145 Years of the Singer Sewing Machine Company*. Brunswick, ME: Audenreed Press, 1999.

Brandon, Ruth. *Singer and the Sewing Machine: A Capitalist Romance*.

Philadelphia and New York: J. B. Lippincott Company, 1977.

Cooper, Grace Rogers. *The Sewing Machine: Its Invention and Development*. 2nd edition, revised and expanded. Washington, DC: Smithsonian Institution, 1976.

Ewers, William, H. W. Baylor, with H. H. Kenaga. *Sincere's Sewing Machine Service Book*. 3rd edition. Phoenix, Arizona: Sincere Press, 1971.

Godfrey, Frank P. *An International History of the Sewing Machine*. London: Robert Hale, 1982.

"Hunt v. Howe." *Federal Cases*. Book 12. St. Paul: West Publishing Co., 1895.

Iles, George. *Leading American Innovators*. New York: Henry Holt and Co., 1912.

Jack, Andrew B. "The Channels of Distribution for an Innovation: The Sewing-Machine Industry in America, 1860–1865." In *Explorations in Entrepreneurial History*, Vol. 9, No. 1, 113–141. Cambridge, MA: Harvard University, 1956.

Lewton, Frederick L. *The Servant in the House: A Brief History of the Sewing Machine*. Smithsonian Publication 3056. Washington, DC: Government Printing Office, 1930. 24 pages, 8 plates. Reprinted from the *Smithsonian Annual Report* for 1929. Found at http://www.sil.si.edu/digitalcollections/hst/lewton/.

Lyon, Peter. "Isaac Singer and His Wonderful Sewing Machine." *American Heritage* 9 (October 1958).

O'Brien, Walter. "Sewing Machines." *Textile American* (March 1931).

Parton, James. "History of the Sewing-Machine." *Atlantic Monthly* 19, issue 115 (May 1867): 527–544. Found at http://cdl.library.cornell.edu/moa.

"Portraits of the People No. 514: Mr. Isaac M. Singer, the Inventor." *The Atlas* 1 (March 20, 1853).

Singer Company Records. Hagley Museum and Library, Wilmington, DE.

"Who Invented Sewing Machines?" *The Galaxy* 4 (1867): 471–82.

CHARLES GOODYEAR

Beard, Mayall, et al. *Tallis's History and Description of the Crystal Palace, and the Exhibition of the World's Industry in 1851*. Vols. I, II. London: John Tallis and Co., 1851.

Dutton, S. W. S. *A Discourse, Commemorative of the Life of Charles Goodyear, the Inventor*. New Haven: Thomas J. Stafford, 1860.

Goodyear, Charles. *Gum-Elastic and Its Varieties*. New Haven, CT: Original privately published, 1853, 1855. Reprinted by Maclaren & Sons, London, 1937.

Hancock, Thomas. *Personal Narrative of the Origin and Progress of Caoutchouc or India-Rubber Manufacture in En-

gland*. London: Longman, Brown, Green, Longmans, & Roberts, 1857.

Hayward, Nathaniel. *Petition of Nathaniel Hayward for an Extension of his Invention*. Norwich, CT: Bulletin Job Office, 1864.

——. *Some Account of Nathaniel Hayward's Experiments with India Rubber*. Norwich, CT: Bulletin Job Office, 1865.

Huke, D.W. *Introduction to Natural and Synthetic Rubbers*. London: Hutchinson & Co., 1961.

Korman, Richard. *The Goodyear Story*. San Francisco: Encounter Books, 2002.

Reports by the Juries Vol. III (Crystal Palace Exhibition). London: Spicer W. Clowes and Sons, 1851.

Slack, Charles. *Noble Obsession: Charles Goodyear, Thomas Hancock, and the Race to Unlock the Greatest Industrial Secret of the Nineteenth Century*. New York: Hyperion, 2002.

Stempel, Guido H. *American Chemists and Chemical Engineers*. Wyndham D. Miles, ed. Washington, DC: American Chemical Society, 1976.

Vollmert, Bruno. *Polymer Chemistry*. New York: Springer-Verlag, 1973.

Webster, Daniel. *Speech of the Hon. Daniel Webster, in the Great India Rubber Suit, Heard at Trenton, New-Jersey, in March, 1852, in the Circuit Court of the United States, before the Hon. Robert C. Grier, and Philomon Dickerson, Judges of that Court*. New York: Arthur & Burnet, 1852.

Wolf, Ralph F. *India Rubber Man*. Caldwell, OH: Caxton Printers, 1940.

ALBERT AUGUSTUS POPE

Crouch, Tom D. "How the Bicycle Took Wing," http://polaris.umuc.edu-fbetz/references/Crouch.html.

Epperson, Bruce. "Failed Colossus: Strategic Errors at the Pope Manufacturing Company, 1878–1900." *Technology and Culture* 41.2 (2000): 300–320.

Goddard, Stephen B., *Colonel Albert Pope and His American Dream Machines*. Jefferson, NC, and London: McFarland and Co., 2000.

EDWIN DRAKE

Asbury, Herbert. *The Golden Flood: An Informal History of America's First Oil Field*. New York: Knopf, 1942.

Dolson, Hildegarde. *The Great Oildorado: The Gaudy and Turbulent Years of the First Oil Rush; Pennsylvania, 1859–1880*. New York: Random House, 1959.

"Edwin L. Drake." *American Petroleum Institute* (1959): 18–19.

Pees, Samuel T. "Oil History," www.oil-history.com.

Williamson, Harold F., and Arnold R. Daum. *The American Petroleum Industry: The Age of Illumination, 1859–1899*.

Evanston, IL: Northwestern University Press, 1959.

Yergin, Daniel. *The Prize: The Epic Quest for Oil, Money and Power*. New York: Simon and Schuster, 1991.

LEVI STRAUSS

Cray, Ed. *Levi's*. Boston: Houghton Mifflin Company, 1978.

Emerson, Gloria. "Jeans Resist Any Change in 108 Years." *New York Times*, July 10, 1958.

Fleming, Alice. "Everybody's Keen on Jeans." *Pictorial Living* 7 (Feb. 20, 1966).

Henry, Sondra, and Emily Taitz. *Everyone Wears His Name: A Biography of Levi Strauss*. Minneapolis: Dillon Press, 1990.

James, Richard D. "Levi Strauss Builds Its Success on a 'Fad' That Has Not Faded." *Wall Street Journal*, Feb. 7, 1977.

Kendall, Elaine. "Men's Fashions, Too, Reflected the Times." *New York Times*, Aug. 2, 1964.

Koshetz, Herbert. "A Look That Refuses to Quit." *New York Times*, Jan. 9, 1977.

Mackendrick, Russ. "Numismatics." *New York Times*, Feb. 18, 1979.

Van Steenwyk, Elizabeth. *Levi Strauss: The Blue Jeans Man*. New York: Walker and Company, 1988.

www.levistrauss.com

www.levi.com

ELISHA OTIS

Cooper, David A. "History of the Escalator." *Lift Report* (Feb. 2000).

Gavois, Jean. *Going Up: An Informal History of the Elevator from the Pyramids to the Present*. Otis Elevator Company, 1983.

Goodwin, Jason. *Otis: Giving Rise to the Modern City*. Chicago: Ivan R. Dee, 2001.

Jackson, Donald Dale. "Elevating Thoughts from Elisha Otis and Fellow Uplifters." *Smithsonian Magazine* (Nov. 1989).

Latvala, Eino K. *Evolution of Elevator Technology*. Cambridge, MA: The Winthrop Group, 1991.

Otis, Charles. Pamphlet commemorating the centennial of his father's birth. Otis Elevator Co., 1911.

Otis Elevator Archives. Otis Elevator Corporate Headquarters, Farmington, CT.

Petersen, Leroy A. *Elisha Graves Otis, 1811–1861, and His Influence Upon Vertical Transportation*. New York: Newcomen Society of England, 1945.

Vogel, Robert M. "Elevator Systems of the Eiffel Tower, 1889." *United States National Museum Bulletin 228*. Washington, DC: Smithsonian Institution, 1961.

Worthington, William, Jr. "Early Risers." *American Heritage of Invention and Technology* (Winter 1989).

LEWIS TAPPAN

Amistad Research Center. Tulane University, New Orleans, LA.

Blackman, Travis B. *A Chronicle of Credit Reporting: The Story of Dun & Bradstreet*. Murray Hill, NJ: Dun & Bradstreet, 1987.

Madison, James H. "The Evolution of Commercial Credit Reporting Agencies in Nineteenth-Century America." *Business History Review* 48 (Summer 1974): 164–186.

Norris, James D. *R. G. Dun and Company, 1841–1900: The Development of Credit Reporting in the Nineteenth Century*. Westport, CT: Greenwood Publishing, 1978.

Olegario, Rowena. "Credit and Business Culture: The American Experience in the Nineteenth Century." Thesis. Harvard University, 1998.

R. G. Dun ledgers. Harvard Business School, Cambridge, MA.

Vose, Edward N. *Seventy-Five Years of the Mercantile Agency: R. G. Dun & Co., 1841–1916*. New York: R. G. Dun & Co., 1916.

Wyatt-Brown, Bertram. "God and Dun & Bradstreet, 1841–1851." *Business Review of History* 40 (Winter 1966): 432–450.

———. *Lewis Tappan and the Evangelical War Against Slavery*. Baton Rouge: Louisiana State University Press, 1997.

RAILROADS

Ambrose, Stephen. *Nothing Like It in the World*. New York: Simon and Schuster, 2000.

Ames, Charles Edgar. *Pioneering the Union Pacific: A Reappraisal of the Builders of the Railroad*. Des Moines, IA: Meredith Corporation, 1969.

Bain, David Haward. *Empire Express: Building the First Transcontinental Railroad*. New York: Viking, 1999.

Cochrane, Thomas C. *Railroad Leaders, 1845–1890*. New York: Russell & Russell, 1966.

Dodge, Grenville M. *How We Built the Union Pacific Railway*. New York: Microprint Corp., 1966.

Galloway, John Debo. *The First Transcontinental Railroad: Central Pacific, Union Pacific*. New York: Dorset Press, 1989.

Hill, Forest Garrett. *Roads, Rails & Waterways*. Norman, OK: University of Oklahoma Press, 1957.

Holland, Rupert Sargent. *Historic Railroads*. Philadelphia: Macrae Smith, 1927.

Howard, Robert West. *The Great Iron Trail*. New York: G. P. Putnam & Sons, 1962.

Jones, Helen Hinckley. *Rails from the West: A Biography of Theodore D. Judah*. San Marino, CA: Golden West Books, 1969.

Judah, Theodore. *A Practical Plan for Building the Pacific Railroad*. San Francisco: Museum of the City of San Francisco, 1857.

Lavender, David. *The Great Persuader: the Biography of Collis P. Huntington*. Boulder, CO: University Press of Colorado, 1998.

Lewis, Oscar. *The Big Four*. New York: Alfred Knopf, 1938.

Martin, Albo. *Railroads Triumphant: The Growth, Rejection and Rebirth of a Vital American Force*. New York: Oxford, 1992.

Mountfield, David. *The Railway Barons*. New York: W. W. Norton, 1979.

Perkins, J. R. *Trails, Rails and War: The Life of General G. M. Dodge*. Indianapolis, IN: Bobbs Merill, 1929.

Riegel, Robert Edgar. *The Story of the Western Railroads*. Lincoln, NE, and London: University of Nebraska Press, 1926.

Theodore Judah Papers. Bancroft Library, Berkeley, CA.

Weisberger, Bernard A., et al. *The Age of Steel and Steam*. New York: Time/Life Books, 1964.

Wheeler, Keith. *The Railroaders*. New York: Time/Life Books, 1973.

White, John. *A History of the American Locomotive: Its Development, 1830–1880*. Baltimore: Johns Hopkins University, 1968.

Williams, John Hoyt. *A Great and Shining Road: The Epic Story of the Transcontinental Railroad*. New York: Crown, 1988.

PART 2
THOMAS ALVA EDISON

Baldwin, Neil. *Edison: Inventing the Century*. New York: Hyperion, 1999.

Cheney, Margaret. *Tesla: Man Out of Time*. New York: Touchstone, 2001.

Clark, Ronald W. *Edison: The Man of Who Made the Future*. New York: G. P. Putnam and Sons, 1977.

Conot, Robert. *A Streak of Luck*. New York: Seaview Books, 1979.

Ford, Henry, with Samuel Crowther. *Edison As I Know Him*. New York: Cosmopolitan Book Corporation, 1930.

Friedel, Robert, and Paul Israel, with Bernard S. Finn. *Edison's Electric Light: Biography of an Invention*. Piscataway, NJ: Rutgers University Press, 1986.

Hughes, Thomas P. "Edison's Method." In *Technology at the Turning Point*, edited by William B. Pickett. San Francisco: San Francisco Press, 1977.

Israel, Paul. *Edison: A Life of Invention*. New York: John Wiley & Sons, 1998.

Jehl, Francis. *Menlo Park Reminiscences*. Vol. 1. New York: Dover Publications, 1990.

Jonnes, Jill. *Empires of Light: Edison, Tesla, Westinghouse, and the Race to Electrify the World*. New York: Random House, 2003.

Josephson, Matthew. *Edison: A Biography*. New York: McGraw-Hill, 1959.

Kline, Ronald R. *Steinmetz: Engineer and Socialist*. Baltimore: Johns Hopkins University Press, 1992.

McAuliffe, Kathleen. "The Undiscovered World of Thomas Edison." *Atlantic Monthly* 276, no. 6 (December 1996): 80–93.

Millard, Andre. *Edison and the Business of Innovation*. Baltimore, MD: Johns Hopkins University Press, 1990.

Nye, David E. *Electrifying America: Social Meanings of a New Technology, 1880–1940*. Cambridge, MA: MIT Press, 1992.

Tate, Alfred O. *Edison's Open Door: The Life Story of Thomas A. Edison, A Great Individualist*. New York: E. P. Dutton & Co., 1938.

Thomas Edison papers, documentary editing project by Rutgers, the State University of New Jersey, the National Park Service, the Smithsonian Institution and the New Jersey Historical Commission.

Vanderbilt, Byron M. *Thomas Edison, Chemist*. Washington, DC: American Chemical Society, 1971.

LEO HENDRIK BAEKELAND

Amato, Ivan. *Stuff: The Materials the World Is Made Of*. New York: HarperCollins, 1997.

Bijker, Wiebe E. *Of Bicycles, Bakelites, and Bulbs: Toward a Theory of Sociotechnical Change*. Cambridge, MA: MIT Press, 1997.

Kaufmann, Carl. Unpublished master's thesis.

Leo H. Baekeland papers. The Archives Center, National Museum of American History, Smithsonian Institution, Washington, DC.

Mees, C. E. Kenneth. "Leo Hendrik Baekeland and Photographic Printing." Society of Chemical Industry meeting. Royal Institution, London. 2 June 1994.

Meikle, Jeffrey L. *American Plastic: A Cultural History*. Piscataway, NJ: Rutgers University Press, 1997.

Roll, Peter B. "Leo Hendrik Baekeland." Ms. *Fortune* archives.

WILBUR AND ORVILLE WRIGHT

Crouch, Tom D. *The Bishop's Boys: A Life of Wilbur and Orville Wright*. New York: W. W. Norton, 1989.

Freedman, Russell. *The Wright Brothers: How They Invented the Airplane*. New York: Scholastic, 1991.

Howard, Fred. *Wilbur and Orville: A Biography of the Wright Brothers*. New York: Knopf, 1987.

Jakab, Peter L. *Visions of a Flying Machine: The Wright Brothers and the Process of Invention*. Washington, DC: Smithsonian Institution Press, 1997.

Kelly, Fred C., ed. *Miracle at Kitty Hawk: The Letters of Wilbur and Orville Wright*. New York: Farrar, Straus and Young, 1951.

———. *The Wright Brothers: A Biography*. Mineola, NY: Dover, 1989.

Mitchell, Charles, and Kirk W. House. *Glenn Curtiss: Aviation Pioneer*. Charleston, SC: Arcadia, 2001.

"Re-Living the Wright Way." National Aeronautics and Space Administration (NASA) Online, http://wright.nasa.gov/.

Shulman, Seth. *Unlocking the Sky: Glenn Curtiss and the Race to Invent the Airplane*. New York: Perennial, 2003.

Tobin, James. *To Conquer the Air: The Wright Brothers and the Great Race for Flight*. New York: Free Press, 2003.

Walsh, John Evangelist. *One Day at Kitty Hawk: The Untold Story of the Wright Brothers*. New York: Thomas Y. Crowell Company, 1975.

Wright, Orville. *How We Invented the Airplane: An Illustrated History*, edited by Fred C. Kelly. New York: Dover Publications, 1988.

GARRETT AUGUSTUS MORGAN

Aaseng, Nathan. *Black Inventors*. New York: Facts on File, 1997.

Brodie, James Michael. *Created Equal: The Lives and Ideas of Black American Innovators*. New York: William Morrow, 1992.

Garrett A. Morgan Papers. MSS 3534 Microfilm Edition. Western Reserve Historical Society, Cleveland, OH.

Morgan, Karen (granddaughter). "The Life of Garrett A. Morgan Sr.: Husband, Father, Inventor, Life Saver." Outline for report, November 26, 1963.

Wilson, David. "Garrett Morgan: The Hidden Man." *Blacks in Ohio: Seven Portraits*, edited by John A. McCluskey. Cleveland, OH: New Day Press, 1976.

EDWIN HOWARD ARMSTRONG

Aitken, Hugh G. J. *The Continuous Wave: Technology and American Radio, 1900–1932*. Princeton, NJ: Princeton University Press, 1985.

Douglas, Susan J. *Inventing American Broadcasting 1899–1922*. Baltimore, MD: Johns Hopkins University Press, 1997.

Edwin H. Armstrong Collection. Box 1, catalogued correspondence. Columbia University, New York.

Erickson, Don. *Armstrong's Fight for FM Broadcasting*. Tuscaloosa, AL: University of Alabama Press, 1973.

Kiver, Milton. *F-M Simplified*. New York: Van Nostrand Company, 1947.

Lewis, Tom. *Empire of the Air: The Men Who Made Radio*. New York: HarperCollins, 1991.

Lessing, Lawrence. *Man of High Fidelity.* Philadelphia and New York: J. B. Lippincott, 1956.

Maclaurin, W. Rupert. *Invention and Innovation in the Radio Industry.* New York: Macmillan, 1949.

Radio Club of America, The. *The Legacies of Edwin Howard Armstrong.* Red Bank, NJ: Radio Club of America, 1990.

Rider, John, and Seymour Uslan. *FM Transmission and Reception.* New York: John F. Rider Publisher, 1948.

Terman, Frederick. *Electronic and Radio Engineering.* New York: McGraw-Hill, 1955.

HENRY FORD

Banham, Russ. *The Ford Century: Ford Motor Company and the Innovations that Shaped the World.* Thousand Oaks, CA: Artisan Sales, 2002.

Batchelor, Ray. *Henry Ford: Mass Production, Modernism and Design.* Manchester: Manchester University Press, 1995.

Beatty, Jack. *Colossus: How the Corporation Changed America.* New York: Broadway, 2001.

Brinkley, Douglas. *Wheels for the World: Henry Ford, His Company, and a Century of Progress, 1903–2003.* New York: Viking, 2003.

Burlingame, Roger. *Henry Ford.* New York: The New American Library, 1954.

Caute, David. *Great American Families,* edited by Magnus Linklater and Francis Wyndham, London: Times Books, 1977.

Greenleaf, William. *Monopoly on Wheels: Henry Ford and the Selden Automobile Patent.* Detroit, MI: Wayne State University Press, 1961.

Flink, James. *The Automobile Age.* Cambridge, MA: MIT Press, 1990.

Ford, Henry, with Samuel Crowther. *My Life and Work.* Garden City, NY: Garden City Publishing Company, 1927.

Lacey, Robert. *Ford: The Men and the Machine.* Boston: Little, Brown, 1986.

McShane, Clay. *Down the Asphalt Path: The Automobile and the American City.* New York: Columbia University Press, 1994.

Nevins, Allan, and Frank Ernest Hill. *Ford.* 3 vols. New York: Charles Scribner's Sons, 1954–1962.

Scharchburg, Richard P. *Carriages Without Horses: J. Frank Duryea and the Birth of the American Automobile Industry.* Warrendale, PA: The Society of Automotive Engineers, 1993.

Sward, Keith T. *The Legend of Henry Ford.* New York: Rinehart, 1948.

Tedlow, Richard S. *Giants of Enterprise: Seven Business Innovators and the Empires They Built.* New York: HarperBusiness, 2001.

Veenswijk, Virginia Kays. *The Coudert Brothers: A Legacy in Law.* New York: Dutton, 1994.

GEORGE EASTMAN

Ackerman, Carl W. *George Eastman: Founder of Kodak and the Photography Business.* Introduction by Edwin R. Seligman. Boston: Houghton Mifflin, 1930.

Brayer, Elizabeth. *George Eastman: A Biography.* Baltimore, MD, and London: Johns Hopkins University Press, 1996.

Collins, Douglas. *The Story of Kodak.* New York: Henry N. Abrams, 1990.

Ford, Colin. "Pioneers brought into focus." *Times Higher Education Supplement* 23 (August 2, 1996).

George Eastman House, www.eastmanhouse.org.

Roberts, Pam. *Photo Historica: Landmarks in Photography.* New York: Artisan, 2000.

Szarkowski, John. *Photography Until Now.* New York: Museum of Modern Art, 1989.

"The Wizard of Photography." *The American Experience,* http://www.pbs.org/wgbh/amex/eastman/filmmore/index.html.

SARAH BREEDLOVE WALKER

Bundles, A'Lelia. Lecture on Madam Walker at New York Institute of Technology, March 11, 2002.

———. *On Her Own Ground: The Life and Times of Madam C. J. Walker.* New York: Washington Square Press, 2001.

Gates, Henry Louis, Jr. "Madam's Crusade." *Time,* Dec. 7, 1998.

"Harlem Renaissance." *Encarta Africana,* Feb. 19, 2004, http://www.africana.com/research/encarta/tt_387.asp.

Hellman, Peter. "Unpacking Harlem History." *New York Times,* May 8, 2003.

Lowry, Beverly. *Her Dream of Dreams: The Rise and Triumph of Madam C. J. Walker.* New York: Knopf, 2003.

Thomas, Paulette. "Madam Walker's Post-Bellum." *Wall Street Journal,* June 14, 1999.

Two Dollars and a Dream, directed by Stanley Nelson. Filmmakers Library, 1989.

Walker, Madam C. J. "Instructions to Agents Before 1919." Madam C. J. Walker Papers, Indiana Historical Society. Series 6: Walker Manufacturing Company Records, Box 7, Folder 5.

Wells, Ida B. *Crusade for Justice,* edited by Alfreda M. Duster. Chicago and London: University of Chicago Press, 1970.

AMADEO PETER GIANNINI

Beckett, Henry. "Big Boss of the World's Biggest Bank System." *New York Post,* June 25, 1947.

Bernays, Edward L. "Men at the Top: A Bernays'-eye View." *Fortune* (Oct. 1965).

Bonadio, Felice A. *A. P. Giannini: Banker of America.* Berkeley, CA: University of California Press, 1994.

Forbes, B. C. "Forbes Relates Story of Giannini's Rise. *Forbes* (Sept. 10, 1928).

Hector, Gary. *Breaking the Bank: The Decline of BankAmerica.* Boston: Little, Brown, 1989.

Hollie, Pamela G. "Coast Bank Going Interstate." *New York Times,* Apr. 9, 1981.

James, Marquis, and Bessie R. James. *Biography of a Bank: The Story of Bank of America NT & SA, 1904–1953.* San Francisco, CA: BankAmerica Corporation, 1954.

Nash, Gerald D. *A. P. Giannini and the Bank of America.* University of Oklahoma Press, 1992.

Ostrow, Al. "Californian Who Made His Bank the Largest of Its Kind in World." *St. Louis Post-Dispatch,* May 18, 1947.

Pollack, Andrew. "BankAmerica Cut Is First in Over 50 Years." *New York Times,* Aug. 6, 1989.

Porter, Sylvia F. "Eccles' Ouster." *New York Post,* Feb. 3, 1948.

Time/Life Archives.

MARTHA MATILDA HARPER

Auch, Herm. "Born in Rochester: The Harper shampoo chair." *Democrat & Chronicle* (Rochester, NY), Nov. 30, 1999.

"Martha M. Harper, Pioneer Beautician." (obituary) *New York Times,* Aug. 5, 1950: 15.

Martha Matilda Harper papers. Rochester Museum & Science Center Collection, Rochester, NY.

Parker, Sally. "Martha Matilda Harper and the American Dream." *Rochester Review* (Fall 2000): 11–15.

Plitt, Jane R. *Martha Matilda Harper and the American Dream.* Syracuse, NY: Syracuse University Press, 2000.

Robinson, Diana. "The Top Ten Business Concepts We Can Emulate that Were Embodied by Martha Matilda Harper in 1888." *Agora Business Center,* Mar. 17, 2003, http://agora-business-center.com/businessconcepts.htm.

RAYMOND "PAPPY" INGRAM SMITH

Armstrong, Bryn. Notes for *Time* magazine story. April 30, 1953. Time/Life Archives.

Ostrander, Gilman M. *Nevada: The Great Rotten Borough, 1859–1964.* New York: Alfred A. Knopf, 1966.

Smith Harold S., Sr., with John Wesley Noble. *I Want to Quit Winners.* Englewood Cliffs, NJ: Prentice-Hall, 1961.

Turner, Wallace. "Sale of Harolds Club Is Climax of $16,000,000 Reno Bonanza." *New York Times,* July 17, 1972.

Vogle, Ed. "Page from Past: Nostalgia is not powerful enough to allow old clubs such as Harolds to survive into the next millennium." *Las Vegas Review-Herald,* Dec. 13, 1999.

JUAN TERRY TRIPPE

Banning, Gene, ed. *Airlines of Pan American Since 1927.* McLean, VA: Paladwr Press, 2001.

Becker, William H. "The Airline Industry," *Encyclopedia of American Business History and Biography.* London: Bruccoli Clark Layman, 1992.

Bender, Marylin, and Selig Altschul. *The Chosen Instrument: Pan Am, Juan Trippe, the Rise and Fall of an American Entrepreneur.* New York: Simon and Schuster, 1982.

Conrad, Barnaby III. *Pan Am: An Aviation Legend.* Emeryville, CA: Woodford Press, 1999.

Daley, Robert. *An American Saga: Juan Trippe and His Pan Am Empire.* New York: Random House, 1980.

Davies, R. E. G. *Airlines of Latin America Since 1919.* McLean, VA: Paladwr Press, 1997.

———. *Airlines of the United States Since 1914.* Washington, DC: Smithsonian Institution Press, 1972, 1984.

———. *Rebels and Reformers of the Airways.* Washington, DC: Smithsonian Institution Press, 1987.

———. *TWA, An Airline and Its Aircraft.* Illustrated by Mike Machat. Wiltshire, UK: Airlife Publishing Ltd., 2001.

Irving, Clive. *Wide Body: The Triumph of the 747.* New York: William Morrow, 1993.

Kissel, Gary. *Poor Sailor's Airline: The Story of Kenny Friedkins' Pacific Southwest Airlines.* McLean, VA: Paladwr Press, 2002.

Leary, William A., ed. *The Airline Industry.* New York: Facts on File, 1992.

———. *From Airships to Airbus: The History of Civil and Commercial Aviation: Infrastructure and Environment.* Washington, DC: Smithsonian Institution Press, 1995.

Trippe, Betty Stettinius, R. E. G. Davies, ed. *Pan Am's First Lady: The Diary of Betty Stettinius Trippe.* McLean, VA: Paladwr Press, 1996.

Van der Linden, F. Robert. *Airlines and Airmail: The Post Office and the Birth of the Commercial Aviation Industry.* Lexington, KY: University Press of Kentucky, 2002.

GENERAL GEORGES DORIOT

Bohr, Peter. "Georges Frederic Doriot." *Fortune* (May 26, 1979).

———. Notes for "Georges Frederic Doriot." Interview with Georges Doriot, 1979. Time/Life Archives.

Bygrave, William D., and Jeffry A. Timmons. *Venture Capital at the Cross-*

roads. Boston: Harvard Business School Press, 1992.

Bylinsky, Gene. "General Doriot's Dream Factory." *Fortune* (August 1967).

———. *The Innovation Millionaires: How They Succeed.* New York: Charles Scribner's Sons, 1976.

Georges Doriot Papers. French Library, Boston, MA.

Georges F. Doriot Papers. Manuscript Division, Library of Congress, Washington, DC.

Liles, Patrick. *Sustaining the Venture Capital Firm.* Cambridge, MA: Management Analysis Center, 1977.

———. *The Use of Outside Help in Starting High-Potential Ventures.* Unpublished dissertation, June 1970. Harvard Business School.

Petre, Peter. "America's Most Successful Entrepreneur." *Fortune* (Oct. 29, 1986).

Robertson, Wyndham. Notes for "General Doriot's Dream Factory." Taken from interviews with Walter Juda, William Congleton, Harlan Anderson, Denis Robertson, Stewart Cowan and Georges Doriot, 1967. Time/Life Archives.

Southwick, Karen. *The Kingmakers: Venture Capital and the Money Behind the Net.* New York: John Wiley, 2001.

IDA ROSENTHAL

Altman, Linda Jacobs. *Women Inventors.* New York: Facts on File, 1997.

Burstyn, Joan N., ed. *Past and Promise: Lives of New Jersey Women.* The Women's Project of New Jersey. Metuchen, NJ: Scarecrow Press, 1990. (Ida Rosenthal section by Grace A. Aqualina and Margaret Dooley Nitka.)

Cook, Joan. "A Maidenform Dream Come True." *New York Times,* Dec. 9, 1965.

Ettorre, Barbara. "The Maidenform Woman Returns." *New York Times,* June 1, 1980.

Ewing, Elizabeth. *Dress and Undress: A History of Women's Underwear.* New York: Drama Book Specialists, 1978.

Fontanel, Beatrice. *Support and Seduction: The History of Corsets and Bras.* Translated from French by Willard Wood. New York: Harry N. Abrams, 1997.

Magill, Frank N., ed. *Great Lives from History: American Women Series,* Vol. 4. Pasadena, CA: Salem Press, 1995.

Maidenform Collection, 1922–1997. Archives Center, National Museum of American History, Smithsonian Institution, Washington, DC.

"Maidenform's Mrs. R." *Fortune* (July 1950): 75–76, 130, 132.

Michaelson, Judy. "Our Town's Leading Business Women, Part 6: Ida Rosenthal." *New York Post,* Sept. 6, 1964.

"Profile of a Company: Maidenform, 1922–1987." Brochure published by

Maidenform in honor of its 65th anniversary.

Sacco, Joe. "Dreams for Sale: How the one for Maidenform came true." *Advertising Age,* Sept. 12, 1977.

Time/Life Archives

Vare, Ethlie Ann, and Greg Ptacek. *Mothers of Invention: From the Bra to the Bomb; Forgotten Women & Their Unforgettable Ideas.* New York: William Morrow, 1987.

SAMUEL INSULL

Berry, Burton Y. *Conversations* (1934). Privately published by Samuel Insull, Jr., 1962.

Granovetter, Mark, and Patrick McGuire. "The Making of an Industry: Electricity in the United States." In *The Laws of the Markets,* edited by Michel Callon. Oxford: Blackwell Publishers, 1998.

Hughes, Thomas P. *Networks of Power: Electrification in Western Society, 1880–1930.* Baltimore, MD: Johns Hopkins University Press, 1983.

Insull, Samuel. *Central-Station Electric Service: Its Commercial Development and Economic Significance as Set Forth in the Public Addresses (1897–1914) of Samuel Insull,* edited by William Eugene Keily. Privately published, 1915.

———. *Public Control and Private Operation of Public Service Industries and Municipal Ownership.* The Other Side Publishing Co., 1899.

———. *Public Utilities in Modern Life: Selected Speeches (1914–1923).* Privately published, 1924.

———. Unpublished memoirs, 1934.

McDonald, Forrest. *Insull.* Chicago: University of Chicago Press, 1962.

Perrett, Geoffrey. *America in the Twenties: A History.* New York: Simon and Schuster, 1982.

Platt, Harold. *The Electric City: Energy and the Growth of the Chicago Area, 1880–1930.* University of Chicago Press, 1991.

Tobin, James. *Great Projects: The Epic Story of the Building of America.* New York: The Free Press, 2001.

Watkins, T. H. *Righteous Pilgrim: The Life and Times of Harold L. Ickes, 1874–1952.* New York: Henry Holt, 1990.

PHILO T. FARNSWORTH AND DAVID SARNOFF

Bilby, Kenneth. *The General: David Sarnoff and the Rise of the Communications Industry.* New York: Harper and Row, 1986.

Carsey, Marcy, and Tom Werner. "Father of Broadcasting: David Sarnoff." *Time* (Dec. 7, 1998): 48–50.

Dreher, Carl. *Sarnoff, an American Success.* New York: Quadrangle/The New York Times Book Company, 1977.

Early Reports on Radio. Vol. 1, 1914–1924. David Sarnoff Library, Princeton, NJ.

Fisher, David E., and Marshall Jon Fisher. *Tube: The Invention of Television.* Washington, DC: Counterpoint, 1996.

Gould, Jack. "CBS Jettisons Monopoly on Color Video Production." *New York Times,* Mar. 26, 1953.

———. "Sarnoff: Mr. Do-It of Broadcasting." *New York Times,* Dec. 13, 1971.

Sarnoff, David. "Enroute: Sealing Expedition off Labrador Coast, February 12, 1911–April 28, 1911." David Sarnoff Library, Princeton, NJ.

———. "Turn the Cold War Tide in America's Favor." *Life* (June 6, 1960): 108, 110, 117–118.

Schwartz, Evan I. *The Last Lone Inventor: A Tale of Genius, Deceit, and the Birth of Television.* New York: Harper Collins, 2002.

———. "Televisionary." *Wired* (Apr. 2002): 68–74.

Stashower, Daniel. *The Boy Genius and the Mogul: the Untold Story of Television.* New York: Broadway Books, 2002.

WALT DISNEY

Eisner, Michael, with Tony Schwartz. *Work in Progress.* New York: Random House, 1998.

Heide, Robert, and John Gilman. *Mickey Mouse: The Evolution, the Legend, the Phenomenon.* New York: Hyperion, 2001.

Iwerks, Leslie, and John Kenworthy. *The Hand Behind the Mouse: An Intimate Biography of Ub Iwerks.* New York: Disney, 2001.

Schickel, Richard. *The Disney Version: The Life, Times, Art and Commerce of Walt Disney.* New York: Avon, 1968.

Smith, Dave, ed. *The Quotable Disney.* New York: Disney Corp., 2001.

Taylor, John. *Storming the Magic Kingdom.* New York: Knopf, 1987.

Thomas, Bob. *Building a Company: Roy O. Disney and the Creation of an Entertainment Empire.* New York: Hyperion, 1998.

———. *Walt Disney: An American Original.* New York: Hyperion, 1994.

JEAN NIDETCH

Navarro, Rafael Garcia. "Real Life: Weigh to Go." *Observer,* August 10, 2003.

Nidetch, Jean, as told to Joan Rattner Heilman. *The Story of Weight Watchers.* New York: W/W Twentyfirst Corp., 1970.

www.weightwatchers.com

http://www3.weightwatchers.com/international/aus/about/corporate_history.html

http://www.naafa.org/press_room/history_obesity.html

THOMAS WATSON AND THOMAS WATSON JR.

Bashe, Charles J., Lyle R. Johnson, John H. Palmer, and Emerson W. Pugh. *IBM's Early Computers.* Cambridge, MA: MIT Press, 1986.

Black, Edwin. *IBM and the Holocaust: The Strategic Alliance Between Nazi Germany and America's Most Powerful Corporation.* New York: Crown Publishers, 2001.

Campbell-Kelly, Martin, and William Aspray. *Computer: A History of the Information Machine.* New York: Basic Books, 1996.

DeLamarter, Richard Thomas. *Big Blue: IBM's Use and Abuse of Power.* New York: Dodd, Mead, and Co., 1986.

Gerstner, Louis V. *Who Says Elephants Can't Dance? Inside IBM's Historic Turnaround.* New York: HarperBusiness, 2002.

Maney, Kevin. *The Maverick and His Machine: Thomas Watson, Sr. and the Making of IBM.* New York: John Wiley, 2003.

Sobel, Robert. *IBM: Colossus in Transition.* New York: Times Books, 1981.

Tedlow, Richard S. *The Watson Dynasty: The Fiery Reign and Troubled Legacy of IBM's Founding Father and Son.* New York: HarperBusiness, 2003.

Watson, Thomas J., Jr., and Peter Petre. *Father, Son, & Co.: My Life at IBM and Beyond.* New York: Bantam Books, 1990.

Zygmont, Jeffrey. *Microchip: An Idea, Its Genesis, and the Revolution It Created.* Reading, MA: Perseus Publishing, 2003.

KEN OLSEN

Computer History Museum Web site, www.computerhistory.org/events/hall_of_fellows/olsen/.

Olsen, Ken. Interview with David Allison at Digital Equipment Corporation. Division of Information Technology & Society, National Museum of American History, Washington, DC. 28–29 Sept. 1988. Found at http://americanhistory.si.edu/csr/comphist/olsen.html.

Petre, Peter. "America's Most Successful Entrepreneur." *Fortune* (Oct. 27, 1986).

Rifkin, Glenn, and George Harrar. *The Ultimate Entrepreneur: The Story of Ken Olsen and Digital Equipment Corporation.* Chicago and New York: Contemporary Books, 1988.

Waldrop, M. Mitchell. *The Dream Machine: J. C. R. Licklider and the Revolution that Made Computing Personal.* New York: Penguin Books, 2001.

ESTÉE LAUDER

Israel, Lee. *Estée Lauder: Behind the Magic.* New York: MacMillan, 1985.

Koehn, Nancy F. *Brand New: How Entrepreneurs Earned Consumers' Trust from Wedgwood to Dell*. Boston, MA: Harvard Business School Press, 2001.

Lauder, Estée. *Estée: A Success Story*. New York: Ballantine, 1986. Rpt. of *Estée: An Intimate Memoir*. New York: Random House, 1985.

Lauder, Leonard. Personal interview with Harold Evans, Nov. 19, 2003. Estée Lauder offices, New York, NY.

Tannen, Mary. "When Charlie Met Estée." *New York Times Magazine*, Feb. 24, 2002.

Warren, Catherine. "Estée and Joe Lauder: In this 1983 interview, the Lauders talk about the secret of success—in business and marriage." *Women's Wear Daily* (Sept. 13, 1999): 60+.

MALCOM MCLEAN

Allen, Oliver E. "The Man Who Put Boxes on Ships." *Audacity* (Spring 1994).

Bangsberg, P. T. "Boxes That Have Transformed Shipping." *Business Times*, June 30, 2002.

Beck, Bill. "Intermodal Transportation Grows on a Novel Idea." *Area Development Online* (August 2001): http://www.area-development.com/past/0801/features/intermodal.html.

Sherwood, James. Interview and correspondence with Harold Evans, April 2004.

www.seacontainers.com

EDWIN LAND

Bello, Francis. "An Astonishing New Theory of Color." *Fortune* (May 1959): 144+.

——. "The Magic That Made Polaroid." *Fortune* (April 1959): 124+.

Chakravarty, Subrata. "As I See It: An Interview with Dr. Edwin Land." *Forbes* (June 1, 1975): 48–50.

Cordtz, Dan. "How Polaroid Bet Its Future on the SX-70." *Fortune* (Jan. 1974): 83+.

"Dr. Edwin H. Land." *Biographical Memoirs of Fellows of the Royal Society* 40 (1994), 195–219, http://www.rowland.org/land/land.html.

Garwin, Richard L. "Edwin H. Land: Science and Public Policy." Light and Life: A Symposium in Honor of Edwin Land. American Academy of Arts and Sciences, Cambridge, MA, Nov. 9, 1991, http://www.fas.org/rlg/land.htm.

Karwatka, Dennis. "Edwin Land: Inventor of the Sheet Polarizer and Instant Photography." *Tech Directions* (Apr. 2000): 16.

Kostelanetz, Richard. "A Wide-Angle View and Close-Up Portrait of Edwin Land and His Polaroid Cameras." *Lithopinion* (Spring 1974): 48–57.

Land, Edwin H. Farewell address as Polaroid CEO. Annual Shareholders' Meeting, Polaroid Corp., Symphony Hall, Boston, MA, Apr. 23–24, 1980.

Lenzner, Robert. "Edwin Land on Polaroid—and the Kodak Patent Suit." *Boston Globe*, Oct. 18, 1976.

——. "Land: The Man Behind the Camera." *Boston Globe*, Oct. 17, 1976.

Lyons, Vincent. "Precocity of Edwin H. Land, Now only 30 Years Old, Develops Polarized Light for Man Practical Uses." *New York World Telegram*, Mar. 1, 1940.

McElheny, Victor K. *Insisting on the Impossible: The Life of Edwin Land, Inventor of Instant Photography*. Reading, MA: Perseus Books, 1998.

——. Personal interview with Peter Wohlsen, June 17, 2002.

Peterson, Pete. Personal interview with Harold Evans, Oct. 23, 2003.

Pringle, Peter. "Mr. Polaroid's Portrait." *Sunday Times*, May 2, 1976.

Reed, Douglas. "'Crazy' ideas stoke creative invention." *Raleigh-Durham Business Journal*, Aug. 30, 2002.

Reinhold, Robert. "Land Achieves His Dream with New Polaroid SX-70." *New York Times*, Oct. 30, 1972.

Rowell, Galen. "Polarizing Visions." *Outdoor Photographer* (June 2000).

Siekman, Philip. "Kodak and Polaroid: An End to Peaceful Coexistence." *Fortune* (Nov. 1980): 82+.

Wensberg, Peter C. *Land's Polaroid: A Company and the Man Who Invented It*. Boston: Houghton Mifflin Company, 1987.

RUTH HANDLER

Barbie Nation: An Unauthorized Tour. By Susan Stern. New Day Films. El Rio Productions, 1998.

Handler, Elliot. "The Impossible Really Is Possible: The Story of Mattel." Address to the Newcomen Society, New York, Mar. 14, 1968.

Handler, Ruth, with Jacqueline Shannon. *Dream Doll: The Ruth Handler Story*. Stamford, CT: Longmeadow Press, 1994.

Lord, M. G. "En Garde, Princess!" *Salon* (Oct. 27, 2000).

——. *Forever Barbie: The Unauthorized Biography of a Real Doll*. New York: William Morrow, 1994.

Rand, Erica. *Barbie's Queer Accessories*. Durham, NC: Duke University Press, 1995.

Ruth Handler papers. Harvard University, Radcliffe Institute for Advanced Study, Schlesinger Library, Cambridge, MA.

PART 3
ELECTRONIC REVOLUTION

Berners-Lee, Tim, with Mark Fischetti. *Weaving the Web: The Original Design and Ultimate Destiny of the World Wide Web by Its Inventor*. San Francisco: Harper San Francisco, 1999.

Borsook, Paulina. *Cyberselfish: A Critical Romp Through the Terribly Libertarian Culture of High Tech*. New York: Public Affairs, 2000

Campbell-Kelly, Martin. *Computer: A History of the Information Machine*. New York: Basic Books, 1997.

Cerruzi, Paul. *A History of Modern Computing*. 2nd ed. Cambridge, MA: MIT Press, 2003.

Chandler, Alfred D., Jr. *Inventing the Electronic Century*. New York: Free Press, 2001.

Clark, Jim, with Owen Edwards. *Netscape Time: The Making of the Billion-Dollar Start-Up that Took on Microsoft*. New York: St. Martin's Press, 1999.

Dell, Michael. *Direct from Dell*. New York: HarperBusiness, 1998.

Grove, Andrew S. *Only the Paranoid Survive: How to Exploit the Crisis Points That Challenge Every Company*. New York: Doubleday, 1999.

——. *Swimming Across: A Memoir*. New York: Warner, 2001.

Hanson, Dirk. *The New Alchemists: Silicon Valley and the Microelectronics Revolution*. Boston: Little, Brown, 1980.

Herz, J. C. *Joystick Nation: How Videogames Ate Our Quarters, Won Our Hearts, and Rewired Our Minds*. New York: Little, Brown, 1997.

——. *Surfing the Internet*. New York: Little, Brown, 1995.

Hoddeson, Lillian, and Vicki Daitch. *True Genius: The Life and Science of John Bardeen*. Washington, DC: National Academy Press, 2002.

Jackson, Tim. *Inside Intel: Andy Grove and the Rise of the World's Most Powerful Chip Company*. New York: Dutton, 1997.

Kaplan, David A. *The Silicon Boys and Their Valley of Dreams*. New York: William Morrow, 1999.

Lewis, Michael. *The New New Thing: A Silicon Valley Story*. New York: W. W. Norton, 2000.

Microsoft Corp. *Inside Out: Microsoft in Our Own Words*. New York: Warner Books, 2000.

Packard, David. *The HP Way: How Bill Hewlett and I Built Our Company*. New York: HarperBusiness, 1995.

Pearson, Jamie Parker, with Ken Olsen. *Digital at Work: Snapshots of the First Thirty-Five Years*. Burlington, MA: Digital Press, 1992.

Queisser, Hans. *The Conquest of the Microchip: Science and Business in the Silicon Age*. Translated from the German by Diane Crawford-Burkhardt. Cambridge, MA: Harvard University Press, 1988.

Reid, T. R. *The Chip: How Two Americans Invented the Microchip and Launched a Revolution*. New York: Simon and Schuster, 1984.

Rogers, Everett M., and Judith K. Larsen. *Silicon Valley Fever: Growth of High-Technology Culture*. New York: Basic Books, 1984.

Shurkin, Joel L. *Engines of the Mind: The Evolution of the Computer from Mainframes to Microprocessors*. New York: W. W. Norton, 1996.

Swisher, Kara. *AOL.com: How Steve Case Beat Bill Gates, Nailed the Netheads, and Made Millions in the War for the Web*. New York: Times Books, 1998.

GARY KILDALL

Cringely, Robert X. *Accidental Empires: How the Boys of Silicon Valley Make Their Millions, Battle Foreign Competition, and Still Can't Get a Date*. Boston: Addison-Wesley, 1992.

Deutschman, Alan. *The Second Coming of Steve Jobs*. New York: Broadway Books, 2000.

Freiberger, Paul, and Michael Swaine. *Fire in the Valley: The Making of the Personal Computer*. 2nd ed. New York: McGraw Hill, 2000.

Gates, Bill. *Business @ The Speed of Thought: Succeeding in a Digital Economy*. New York: Warner Books, 1999.

——. *The Road Ahead*. New York: Viking, 1995.

Gilroy, John. "As an Old System Fades Away, MS Is Hidden." *Washington Post*, June 14, 2000.

——. "Windows: How to Reinstall It and What's Under the Hood." *Washington Post*, June 29, 2001.

Kildall, Gary. *Computer Connections*. Unpublished memoirs.

——. "CP/M: A Family of 8-bit and 16-bit Operating Systems." *Byte* (June 1981).

Leven, Antov. "History of MS-DOS," 1996, http://www.maxframe.com/HISZMSD.

Manes, Stephen, and Andrews, Paul. *Gates: How Microsoft's Mogul Reinvented an Industry—and Made Himself the Richest Man in America*. New York: Doubleday, 1993.

Pollack, Andrew. "Debate Grows Over the Role an Operating System Plays." *New York Times*, July 20, 1998.

Rivlin, Gary. *The Plot to Get Bill Gates: An Irreverent Investigation of the World's Richest Man . . . and the People Who Hate Him*. New York: Times Books, 1999.

Rohm, Wendy Goldman. *The Microsoft File: The Secret Case Against Bill Gates*. New York: Random House, 1998.

Sigismund, Charles G. *Champions of Silicon Valley: Visionary Thinking from Today's Technology Pioneers*. New York: John Wiley and Sons, 2000.

Swaine, Michael. "Gary Kildall and Collegial Entrepreneurship." *Dr. Dobb's Special Report* (Spring 1997).

Wallace, James, *Overdrive: Bill Gates and the Race to Control Cyberspace*. New York: John Wiley and Sons, 1997.

Wallace, James, and Jim Erickson. *Hard Drive: Bill Gates and the Making of the*

Microsoft Empire. New York: John Wiley and Sons, 1992.

HERBERT BOYER AND ROBERT SWANSON

Boyer, Herbert. Interview with Sally Smith Hughes, 2001. Bancroft Library, University of California, Berkeley, CA.

Campbell, Neil A., and Jane B. Reece. *Biology*. 6th ed. Boston: Benjamin Cummings, 2002.

Hall, Stephen. *Invisible Frontiers: The Race to Synthesize a Human Gene*. Boston: Atlantic Monthly Press, 1987.

Hughes, Sally Smith. *Making Dollars out of DNA*. Gainesville, FL: History of Science Society, 2001.

Krimsky, Sheldon. *Biotechnics & Society: The Rise of Industrial Genetics*. Westport, CT: Praeger, 1991.

Ptashne, Mark, and Alexander Gann. *Genes & Signals*. Woodbury, NY: Cold Spring Harbor Press, 2002.

Robbins-Roth, Cynthia, *From Alchemy to IPO: The Business of Biotechnology*. New York: Perseus Publishing, 2000.

Swanson, Robert. Interview with Sally Smith Hughes, 2001. Bancroft Library, University of California, Berkeley, CA.

Thackery, Arnold, ed. *Private Science: Biotechnology and the Rise of the Molecular Sciences*. University of Pennsylvania Press, 1998.

TED TURNER

Auletta, Ken. *Backstory: Inside the Business of News*. New York: Penguin, 2003.
———. Interview with Linda Garmon, Feb. 17, 2004, New York, NY.
———. *Three Blind Mice: How the TV Networks Lost Their Way*. New York: Random House, 1991.

Bibb, Porter. Interview with Linda Garmon, Feb. 5, 2004, New York, NY.
———. *Ted Turner: It Ain't as Easy as It Looks*. Boulder, CO: Johnson Books, 1997.

Goldberg, Robert. Interview with Linda Garmon, Feb. 6, 2004, New York, NY.

Kavanau, Ted. Interview with Linda Garmon, Feb. 6, 2004, New York, NY.

Kenyon, Sandy. Interview with Linda Garmon, Feb. 5 2004, New York, NY.

Lowe, Janet. *Ted Turner Speaks: Insights from the World's Greatest Maverick*. New York: John Wiley and Sons, 1999.

Mayer, Martin. *Making News*. Cambridge, MA: Harvard Business School Press, 1993.

Miklaszewski, Jim. Interview with Linda Garmon, Mar. 29, 2004, New York, NY.

O'Gorman, Pat. Interview with Linda Garmon, Feb. 6, 2004, New York, NY.

Schonfeld, Reese. Interview with Linda Garmon, Feb. 5, 2004, New York, NY.
———. *Me and Ted Against the World: The Unauthorized Story of the Founding of CNN*. New York: HarperCollins, 2001.

Shaw, Bernard. Interview with Linda Garmon, Mar. 29, 2004, New York, NY.

Stefoff, Rebecca. *Ted Turner: Television's Triumphant Tiger*. Ada, OK: Garrett Educational Corp., 1992.

Turner, Ted. Interview with Harold Evans, Apr. 22, 2004, New York, NY.
———. Interview with Linda Garmon, Mar. 9, 2004, Atlanta, GA.
———. Telephone interview with Harold Evans, Apr. 7, 2004.

Williams, Christian. *Lead, Follow or Get Out of the Way: The Story of Ted Turner*. New York: Times Books, 1981.

Williams, Mary Alice. Interview with Linda Garmon, Feb. 6, 2004, New York, NY.

RAYMOND DAMADIAN

Heitzman, E. Robert. *The Mediastinum: Radiologic Correlations with Anatomy and Pathology*. St. Louis: C. V. Mosby, 1977.

Kevles, Bettyann Holtzman. *Naked to the Bone: Medical Imaging in the Twentieth Century*. Piscataway, NJ: Rutgers University Press, 1997.

Kleinfeld, Sonny, *A Machine Called Indomitable*. New York: Times Books, 1985.

Mattson, James, and Merrill Simon. *The Pioneers of NMR and Magnetic Resonance in Medicine: The Story of MRI*. Ramat-Gan, Israel: Bar-Ilan University Press, 1996.

McMurry, John. *Organic Chemistry*. Pacific Grove, CA: Brooks/Cole, 1988.

Olendorf, William, and William Olendorf Jr. *Basics of Magnetic Resonance Imaging*. Boston: Martinus Nijhoff Publishing, 1988.

Rinck, Peter A., ed. *Magnetic Resonance in Medicine: The Basic Textbook of the European Magnetic Resonance Forum*. Oxford: Blackwell Scientific Publications, 1993.

Young, Stuart W. *Magnetic Resonance Imaging: Basic Principles*. New York: Raven Press, 1984.

RUSSELL SIMMONS

Adler, Bill. Interview with Linda Garmon, Feb. 3, 2004, New York, NY.

Cohen, Lyor. Interview with Linda Garmon, Feb. 3, 2004, New York, NY.

Fricke, Jim, and Charlie Ahearn. *Yes Yes Y'all: The Experience Music Project Oral History of Hip Hop's First Decade*. Oxford, UK: Perseus Press, 2002.

George, Nelson. Interview with Linda Garmon, Feb. 3, 2004, New York, NY.

Lathan, Stan. Interview with Linda Garmon, Feb. 3, 2004, New York, NY.

Light, Alan. Interview with Linda Garmon, Feb. 2, 2004, New York, NY.

McDaniels, Darryl, with Bruce Haring. *King of Rock: Respect, Responsibility, and My Life with Run-DMC*. New York: Thomas Dunne Books, 2001.

Ogg, Alexander. *The Men Behind Def Jam: The Radical Rise of Russell Simmons and Rick Rubin*. London: Omnibus Press, 2002.

Simmons, Russell. Interview with Linda Garmon, Feb. 17, 2004, New York, NY.

Simmons, Russell, and Nelson George. *Life and Def: Sex, Drugs, Money, and God*. New York: Crown Publishers, 2001.

Stephney, Bill. Interview with Linda Garmon, Feb. 2, 2004, New York, NY.

PIERRE OMIDYAR

Cohen, Adam. *The Perfect Store: Inside eBay*. New York: Little, Brown, 2002.

Hardy, Quentin. "The Radical Philanthropist." *Forbes* (May 1, 2000).

Hof, Robert D. "The People's Company." *Business Week* (Dec. 3, 2001).

Johnson, Steven. *Emergence: The Connected Lives of Ants, Brains, Cities, and Software*. New York: Scribner, 2001.

Kelly, James. "That Man in the Cardboard Box." *Time* (Dec. 27, 1999).

SERGEY BRIN AND LARRY PAGE

Brin, Sergey, and Larry Page. "The Anatomy of a Large-Scale Hypertextual Web Search Engine." Computer Science Department, Stanford University, Stanford, CA, http://www-db.Stanford.edu/-backrub/google.html.

Craven, Phil. "Google's PageRank Explained, and How to Make the Most of It," www.webworkshop.net/pagerank.html.

Haveliwala, Taher H., and Sepandar D. Kamvar. "The Second Eigenvalue of the Google Matrix." Computer Science Department, Stanford University, Stanford, CA.

Russell, Stuart J., and Norvig, Peter. *Artificial Intelligence: A Modern Approach*. Upper Saddle River, NJ: Prentice Hall, 1995.

Silverstein, Craig. Interviewed by David Lefer, Feb. 2004, CA.

Shannon, Claude E. "A Mathematical Theory of Communication." *The Bell System Technical Journal* (July–October 1948).

ADDITIONAL RESOURCES

Audacity magazine
Business Week
Columbia Encyclopedia
Economist
Encyclopaedia Britannica
Garraty, John A., and Mark C. Carnes, gen. eds. *American National Biography*. 24 vols. New York/Oxford: Oxford University Press, 1999.
http://inventors.about.com
http://web.mit.edu/invent
Journal of Commerce
Malone, Dumas, ed. *Dictionary of American Biography*. Vols. 1–10. New York: Scribner's, 1946.
The New Yorker
Strategy & Business
Technology Review, MIT's magazine of innovation
U.S. News & World Report
www.invent.org

ILLUSTRATION CREDITS

History would be ill served without the dedicated keepers of our visual past. Space does not allow mention of all the hundreds of archivists, historians and librarians who helped my research. And, because *They Made America* covers 200 years, I went from digging in "dusty" archives to visiting contemporary photographers and galleries for images evocative of innovation. Thanks to all the custodians of our pictorial heritage and to the photographers whose work defines an era.

I could not have succeeded in getting pictures from repository to printed page without the talents of special people. Peter Wohlsen, a former student of mine at Cooper Union, became my teacher in all things digital. A fine artist in his own right, he was truly a jack-of-all-trades, helping with picture and historical research and organizing the thousands of images we collected for review by Harold Evans. Marilyn Doof made certain that every production detail, from paper to print quality, was at the highest level. Her devotion to excellence is sincerely appreciated.

I was fortunate to have the support of the wonderful people in the Time Inc. Picture Collection: Kathi Doak, director, and Joan Shweky and Cornelis Verwaal, researchers. As a photographic historian, I knew the pictures taken for *Fortune* magazine since 1930 were the pinnacle of excellence in areas related to industry, technology and business, and I am grateful to John Huey for enabling me to research the *Fortune* archives and to the current picture editor of *Fortune,* Michele F. McNally, who guided me toward many of the gems. Rona Tuccillo, first while she was at Timepix and then later at Getty Images, was always patient and supportive.

My friend the collector Stephen White generously made his unique photographic collection available. Between Stephen and Walt Burton, we had access to the finest personal collections of Wright brothers photographs. Charles Schwartz, a major photography collector, allowed us to copy rare 19th-century photographs from volumes in his collection.

No magazine cares more about the confluence of history and photography than *American Heritage.* I was fortunate to be given access to their vast picture archive by the editor in chief, Richard Snow. His assistant, Jeannette Baik, has the patience of a saint. The specialists at the Smithsonian Institution's National Museum of American History, whether they are in the departments of Maritime, Automotive, Electrical, Computer, Photographic History or the Archives Center, helped me uncover treasure in their areas. Frank Goodyear of the National Portrait Gallery, whose ancestor is featured in the book, greatly facilitated my research at his museum.

Visual research takes all forms, including a special performance of the life of Theodore Judah by the accomplished actor and playwright Stu Richel. A more traditional route is visiting collections, and I acknowledge the assistance of these archival guides: Chris Hunter, Schenectady Museum; Nasrin Rohani, Polaroid Archives; Paul Israel, Edison Paper Projects and Ed Wirth, Edison National Historic Site; Craig S. Likness and his staff, University of Miami Archives and Special Collections; Cynthia Read-Miller and her colleagues, Henry Ford Museum and Greenfield Village; Alex Magoun, David Sarnoff Library; Jeanette Wagner, Sally Susman and Melissa Bedolis Cattanach, Estée Lauder Company; Bill Hosley, Antiquarian & Landmarks Society, Hartford, Connecticut; Paul Lasewicz and Dawn Hugh, IBM Archives; Dr. Michele Aldrich, Otis Elevator Archives; Robin Sndyer, Genentech; Jane Knowles and her colleagues, Schlesinger Library; Eileen Flanagan, Norman Currie and Donna Daily of Corbis.

I thank the following for their expert advice: Deborah Willis-Kennedy, Carol Squiers, Ellen Yeske, Janet Testa, Sandy Taylor, Peter Wilson, Barnaby Conrad III, Robin Snyder, Catherine Brawer, Elizabeth Coleman, Steven Masket, Larry Wilson, Gary Kurtz, Bill Adler, Tom Rolander and Drs. Harvey Blanch, Mervyn Turner, Federico Capasso and Lorimer Miller. In addition, Victor McElheny and Daniel Kevles and his wife, Bettyann Holtzmann Kevles, answered many questions pertaining to my photographic research. Sarah Frank and Jonathan Marder, friends and colleagues, always provide generous support.

Gail Buckland, 2004

ABBREVIATIONS

AH: American Heritage

BB: Brown Brothers

CHS: The Connecticut Historical Society, Hartford, Connecticut

B-C : Bettmann-CORBIS

ENHS: U.S. Department of the Interior. National Park Service. Edison National Historic Site

GEH: International Museum of Photography and Film at George Eastman House

T/L/Getty: Time/Life Pictures/Getty Images

HF/GV: From the Collections of Henry Ford Museum & Greenfield Village

LC: Library of Congress

Loyola: Loyola University of Chicago Archives: Samuel Insull Papers

N-YHS: New-York Historical Society

NMAH/Archives: The National Museum of American History, Archives Center, Smithsonian Institution

NMAH: Smithsonian Institution National Museum of American History, Smithsonian Institution

U of M: Archives and Special Collections Department, Otto G. Richter Library, University of Miami, Coral Gables, Florida

WA: Wadsworth Atheneum, Hartford. Bequest of Elizabeth Hart Jarvis Colt

WSU: Special Collections & Archives, Wright State University

The name of the photographer or artist, when known, is listed before the name of the collection. The authors have made every attempt to contact the artists, photographers and copyright holders. Illustrations are copyright and/or courtesy the following individuals and collections:

Endpapers: United States Patent Office: http://www.uspto.gov/

Frontmatter: 2–3 HF/GV #0.393; 4 CAIDA (Cooperative Association for Internet Data Analysis, www.caida.org), Young Hyun "Walrus" Internet Map, 2002, © 2002 The Regents of the University of California. Courtesy University of California, San Diego.; 8–9 Greg Miller.

Part I: 16–17 William Walker/ Philip and Tina DeNormandie; 20 (top left) Gustavus Hesselius/Rare Book and Special Collections Division, Library of Congress, (top right) AH, (middle) Rare Book and Special Collections Division, Library of Congress, (bottom) NMAH #31631; 21 Harvard Map Collection; 22 AH; 24 The New Jersey Historical Society; 25 The Mariners' Museum, Newport News, VA.; 27 Independence National Historical Park; 28 (top) Jean Antoine Houdon, N-YHS #2062, (2nd from top) Sharples, N-YHS #76018D, (3rd from top) Benjamin West, Fenimore Art Museum, Cooperstown, New York. Photograph by Richard Walker, (bottom) Robert Fulton, The Pierpont Morgan Library/Art Resource, NY. The Pierpont Morgan Library, New York, NY, U.S.A.; 30 (top) CHS; (bottom) Robert Fulton, N-YHS #36657; 31 (top) N-YHS #76017D, (bottom) Williams, N-YHS portrait file 052, box 86; Robert Fulton, N-YHS #1924.7; 33 Edward Lamson Henry, Print Collection, The New York Public Library, Astor, Lenox and Tilden Foundation/Art Resource, NY. The New York Public Library, New York, NY, USA; 34 Culver Pictures; 35 Library Special Collections, Stevens

Institute of Technology, Hoboken, NJ; 36, 37, 38, 40 (both), Hagley Museum and Library; 42 Lloyd Hawthorne, The R.W. Norton Art Gallery, Shreveport, LA.; 44 (right) "Explosion of the Washington, 1816" from James T. Lloyd's *Lloyd's Steamboat Directory, 1856;* 45 LC –Z62-38180; 47 AH; 48–49 LC-USZ62-56499 (b/w negative); 50 (left) James Frothingham (attr.) American, 1786–1814 Catherine Littlefield Greene Miller ca. 1800–1814 oil on panel 32¾ x 25¾ in. Telfair Museum of Art, Savannah, Georgia Museum purchase, 1947; (right) American Textile History Museum; 51 The Georgia Historical Society; 52 S. B. Dayton, The Georgia Historical Society; 53 (both) The New Haven Colony Historical Society; 55 The Eli Whitney Museum; 56, 57 (all) Museum of American Textile History; 58–59 Arthur F. Tait, *The Pursuit,* 1855 oil on canvas, 30 x 44⅜ in., Milwaukee Art Museum, Gift of Edward S. Tallmadge as a Memorial to the Men who Loyally and Selflessly Gave Their Lives for Our Country in World War II, M1971.24a; 60, 61 WA; 62 courtesy Bill Hosley; 63 (left) John N. McWilliams Collection; 63 (top), 64 (all) Museum of Connecticut History; 67 CHS; 68 Denver Public Library, Western History Collection, X-22154; 69 Seaver Center for Western History Research, Los Angeles County Museum of Natural History; 70 LC/dag #1278; 71 NMAH #29,651; 73 Chicago Historical Society; 75 (left) A. J. Russell, Huntington Library, 75 (right), 76 SI/Archives, Western Union Collection; 77 SI/Archives, Warshaw Collection; 78 Chicago Historical Society; 79 Charles Loring Elliott, National Portrait Gallery, Smithsonian Institution/Art Resource, NY. National Portrait Gallery, Smithsonian Institution, Washington, D.C.; 81 from *The Illustrated Exhibitor . . . Sketches, by Pen and Pencil, of the Principal Objects in the Great Exhibition of the Industry of all Nations,* 1851, courtesy Charles Schwartz; 82 American Antiquarian/AH; 84 LC-USW3-011493 E, 85 Edward Harrison May, National Portrait Gallery, Smithsonian Institution/Art Resource, NY. National Portrait Gallery, Smithsonian Institution, Washington, D.C.; 86 from (Greene Co.) *History of Greene County,* 67 Clara Goldberg Schiffer Collection, The Radcliffe University for Advanced Study, Schlesinger Library, Harvard University; 88 BB; 89 Albert Sands Southworth (American, 1811–1894) and Josiah Johnson Hawes (American, 1808–1901) Elias Howe, ca. 1850, daguerreotype, 14.0 x 10.8 cm (5 ½ x 4 ¼ in.) The Metropolitan Museum of Art, Gift of I. N. Phelps Stokes, Edward S. Hawes, Alice Mary Hawes, Marion Augusta Hawes, 1937. Photograph © 1996 The Metropolitan Museum of Art; 91 from Brandon, Ruth. *Singer and the Sewing Machine: A Capitalist Romance.*

Philadelphia and New York: J.B. Lippincott Company, 1977; 92 AH; 93 Albert Sands Southworth and Josiah Johnson Hawes, Society for the Preservation of New England Antiquities, Boston/AH; 95 Courtesy Charles Schwartz; 96, from Goodyear, Charles. *Gum-Elastic and Its Varieties.* New Haven, CT, 1853, 1855. Reprinted London, 1937; 97 AH; 98 *Fortune,* September, 1930, p. 98; Russell Lee, LC-USF-34-T01-35601; 101 CHS; 102 John Mather/Drake Well Museum; 103 John Mather/Michael G. Wilson and the Wilson Centre for Photography; 104 John Mather/Drake Well Museum; 105 Baker Library, Harvard Business School, Harvard University; 106 BB; 108 AH; 109 Andre deDienes, courtesy Shirley T. deDienes/Art Brokers of America, Beverly Hills, CA.; 111 George H. Johnson/courtesy Matthew R. Isenberg; 113 J. E. Stimson/from the J. E. Stimson Collection, Wyoming State Archives; 114 Otis Historical Archive No. 1578, United Technologies; 115 Gavois Collection No. 7257, Otis Elevator Archives, United Technologies; 116 Otis Historical Archive No. 388, United Technologies; 117, Otis Historical Archive No. 771, United Technologies; 118 J. C. Buttre, N-YHS #76016D; 119, 120 C-B; 122 LC-USZ62-2277; 123 Landauer Collection PR 031, N-YHS #76049D; 125 (top) American Antiquarian Society; 125 (bottom left) CHS, 125 (bottom center and right) Beinecke Rare Book and Manuscript Library, Yale University; 127 The Huntington Library; 129 (top left) The Huntington Library, (bottom left, center, right) California History Section, California State Library; 130 de Grummond Children's Literature Collection, the University of Southern Mississippi; 133 The Mariners' Museum, Newport News, VA.; 134, 135 California History Section, California State Library.

Part II: 136–137 Diego M. Rivera/Gift of Edsel B, Ford, photograph © 2001 The Detroit Institute of Art; 140–141, 142–143, 144–145 Andrew J. Russell/The Andrew J. Russell Collection, the Oakland Museum of California; 145–147 J. E. Stimson/from the J. E. Stimson Collection, Wyoming State Archives; 150–151 Hammer Collection, NMAH/Archives; 153 HF/GV #G2665; 155 Edison Paper Project; 156, 159, 160 ENHS; 162 Museum of the City of New York; 163 (top) ENHS; 163 (bottom) Schenectady Museum; 164 ENHS; 166 Consolidated Edison of New York; 167 (both) ENHS; 168 (left) LC, (right) IEEE History Center, Rutgers University; 170 National Archives RG 306 PS A Box 7; 172 Baekeland Collection, NMAH/Archives; 173 BB; 174, 175, 176 Baekeland Collection, NMAH/Archives; 177 (left) BB, (right) HF/GV #188.28273; 178 *Fortune,* April 1939;

179 Baekeland Collection, NMAH/Archives; 180–181 Association des Amis de Marey et des Musées de Beaune; 182 (top) LC-DIG-ppprs-005366; (bottom) HF/GV #1623/8; 185 (top) National Air and Space Museum, SI #2003-30820; 185 (bottom) Culver Pictures; 186 (top and bottom) LC; 187 http://www.centennialofflight; 190, 191 LC; 192 (both), 193, 194 (left) WSU; 195 (both), 196, 200 LC; 206, 207 (top) Stephen White, Collection II, (bottom) LC; 208 Walt Burton Collections, LLC; 209 WSU; 210 N-YHS; 210–211, 211 WSU; 212, 213, 214 The Western Reserve Historical Society; 216 Columbia University Archives-Columbiana Library; 217 IEEE History Center, Rutgers University; 219 B-C; 221 LC-Z62-65479; 219 B-C, 221 LC-Z62-65479; 223 Paul Thompson, ca. 1915/AH and New York Public Library Picture Collection; 226 IEEE History Center, Rutgers University; 227, 228, 231 Columbia University Archives-Columbiana Library; 232 Harris & Ewing/Stock Montage; 234–235 HF/GV 188.4778; 236 (top) HF/GV #JIT, 236 (left and center) #K561, K562, (right) #K563; 237 (l-r) #K563, 564, 565, 566, 567; 238 Detroit Public Library, National Automotive History Collection; 239 HF/GV #0.574; 241 #833.21089; 242 HF/GV #189.11074; 243 Norman Rockwell/ "The Revolution that Started in a Shed at Night," 1952, HF/GV #0.4686; 247 Russell Aikins/ T/L Getty; 248 HF/GV #P.B.24582; 249 Detroit Public Library, National Automotive History Collection; 250–253 (all) GEH; 254–257 (all) Indiana Historical Society, Madam C. J. Walker Collection; 258 LC-USZ62-58337; 259 Culver Pictures; 260 (top) Culver Pictures, (bottom) B-C; 261 (top) Culver Pictures, (bottom) B-C; 262-263 B-C; 264 J. R. Eyerman/T/L Getty; 265 Jon Brenneis; 266, 268, 271 J. R. Eyerman/T/L Getty; 275 Jon Brenneis; 276–277 (all) courtesy Jane R. Plitt, Harper biographer; 278 Charles Bennett/T/L Getty; 278–279 J. R. Eyerman/T/L Getty; 280 Hans Oswald Wild/T/L Getty; 282–283 Jon Brenneis; 284–285, 286 (top) UofM; 286 (bottom) National Air and Space Museum/SI #79-10995; 287 B-C; 288 Historical Society of Southern Florida; 289 San Francisco Airport Museums, gift of Barnaby Conrad III; 290 UofM; 291 Collection Barnaby Conrad III; 292 Museum of Flight/Patrick Kam Photographer; 295 San Diego Aerospace Museum; 296 Pan American World Airlines Collection of The Museum of Flight; 297 Collection of Barnaby Conrad III; 298 UofM; 299 © 1999 The Boeing Company, Boeing Historical Archives; 301 Kim Steele; 303 Edward Burks; 308–309 T/L Getty; 310 Catherine Brawer; 311 Maidenform Collection, NMAH/Archives; 313 Catherine Brawer; 314–315 Maidenform Collection, NMAH/Archives; 317 Eric

Schaal, courtesy Miriam Schaal; 318 Loyola; 318–319 Schenectady Museum; 321 Loyola; 322 Chicago Historical Society; 325 Schenectady Museum; 326 (left) LC, (top) LC-USZ62-67635, (bottom) Schenectady Museum; 327 National Archives #86-G-11D-1; 329 (both), 330 Loyola; 332 Chicago Historical Society; 333 B-C; 335, 337 (left) B-C, (right) David Sarnoff Library; 338 Special Collections Department, J. Willard Marriot Library, University of Utah; 339 David Sarnoff Library; 342 © Disney Enterprises, Inc.; 343 Alfred Eisenstaedt/T/L Getty. Used by permission from Disney Enterprises, Inc.; 344–353 (all) © Disney Enterprises, Inc.; 354–355 (all) LC; 356 IBM Archives, 356–357 Lisa Larsen/T/L Getty; 358–368 (all) IBM Archives; 369 Hagley Museum and Library; 373 Alfred Eisenstaedt (detail) T/L Getty; 374 IBM Archives; 377 Fred Burell; 378 Estée Lauder Companies; 380 Edward Burtynsky "Container Ports #17," Ceres Port, Halifax 2001; 382 T/L/ Getty; 383 courtesy Jim Sherwood; 385–387 (all) Polaroid Corporate Archives, Polaroid Corporation; 388–391 (all) The Radcliffe University for Advanced Study, Schlesinger Library, Harvard University; 393 David Levinthal.

Part III: 394–395 Photomosaic-TM by Robert Silvers/www.photomosaic.com based on photograph of Claude Shannon from MIT Museum/photograph digitally altered and reversed; 402 courtesy Dorothy McEwen and Kristin Kildall; 403 Rick Browne, Wishing Well Productions; 405 courtesy Dorothy McEwen and Kristin Kildall; 406 courtesy Tom Rolander; 409 (top left) Tom O'Neal/TGO Photography, (middle left) Mickey Pfleger, (bottom left, top right, bottom right) courtesy Dorothy McEwen and Kristin Kildall; 414–415 courtesy Pattie McCracken; 417 courtesy Tom Rolander; 419 George Lange/CORBIS Outline; 421 Steve Northrup; 424 Genentech; 427 Huntington Potter and David Dressler; 430–431 Timothy Archibald; 432 George Silk/T/L Getty; 433, 435, 438 Turner Broadcasting System; 440 Robert Fuhring/Sesame Workshop; 443 Bob West; 447, 449 (top) Dr. Raymond Damadian, Fonar Corporation; 449 (bottom) Neal Preston/CORBIS; 450 courtesy Janice Ford-Benner and Dr. Meng Law, NYU Department of Radiology; 452 Glen E. Friedman; 454 (top) Peter Anderson/ S.I.N., (bottom) Oggi Ogburn; 455 C. Taylor Crothers; 456 Matt Stroshane/Bloomberg/Landov; 459, 460 William Mercer McLeod; 463 (left) Opsware, Inc., (right) Donna Coveney/MIT; 467 Rex Rystedt; 468 Eric Millette Photography; 469 Steve Smith; 470 West Grand Media, LLC; 472 Slick Lawson, courtesy Clearance Quest.

BIOGRAPHIES

SIR HAROLD EVANS is the author of the acclaimed *The American Century*, a study of the great people, movements and events from 1889–1989 (with 900 illustrations collected for the book by the photographic historian Gail Buckland). He was the president and publisher of Random House Trade Group until 1997. Previously, he was editor of the *Sunday Times of London* from 1967 to 1981 and then editor of the *Times* until he resigned in 1982. His account of those editing years, *Good Times, Bad Times,* was a number 1 bestseller in Britain.

Evans first came to the United States in 1956 as a Harkness Fellow at the University of Chicago and Stanford, following graduation with honors from England's Durham University. His thesis on the Suez crisis of 1956–7 was awarded a master of arts degree (and the university later honored him with a doctorate in civil law). He traveled widely in 40 states from 1956 to 1958 and reported for the *Guardian* and the *Manchester Evening News;* some of his experiences were later reflected in *The American Century*. On his return to Britain in 1958, Evans was for five years editor of the *Northern Echo,* where his work led to the establishment of the first national program for the detection of cervical cancer and the pardon of a man wrongfully executed. His years as editor of the *Sunday Times* and the *Times* were marked by numerous prizes for the newspapers and their editor. He was awarded the European Gold Medal of the Institute of Journalists for his investigations and campaigns for thalidomide children, and his successful appeal to the European Court of Human Rights against the suppression of articles by the House of Lords. For his editorship of the *Times,* he was named Editor of the Year by Granada Television's *What the Papers Say*. The Royal Photographic Society presented him with the Hood Medal for his work in photojournalism.

On his return to the United States in 1984, Evans taught for a term as a visiting professor at Duke University. He subsequently served as editor in chief of the Atlantic Monthly Press and editorial director of *U.S. News & World Report* until 1985, when he joined Conde Nast Publications as the founding editor of *Condé Nast Traveler* magazine. After his seven years at Random House, he returned to *U.S. News & World Report* as editorial director and vice chairman also of the *New York Daily News,* the *Atlantic Monthly* and *Fast Company.* In 2000 he took up full-time writing while remaining a contributing editor at *U.S. News* and a consultant editor to *The Week* magazine. He wrote the monograph *War Stories: Reporting in the Time of Conflict* for the Newseum's exhibition of that name, which he created.

In 2000 Evans received the Gold Award for Lifetime Achievement from the U.K. Press Awards Committee and was named one of 50 "world press heroes" by the International Press Institute, Vienna. The following year, in a professional poll, he was named the greatest all-time British newspaper editor. He was knighted in Queen Elizabeth's New Year Honors list in 2004 for services to journalism.

Sir Harold lives in New York with his wife, Tina Brown, and their two children.

BOOKS BY HAROLD EVANS: *The American Century; The Index Lecture: View From Ground Zero; War Stories; Good Times, Bad Times; The Freedom of the Press* (with Katharine Graham and Lord Windlesham); *Pictures on a Page; Essential English; Newspaper Design; Text Typography; Newspaper Headlines; The Active Newsroom; We Learned to Ski* (with Brian Jackman and Mark Ottoway); *Suffer the Children: The Story of Thalidomide* (with the *Sunday Times* Insight Team); *Eyewitness*

GAIL BUCKLAND is professor of the history of photography at the Cooper Union, New York City. She has taught at Columbia College, Chicago; Pratt Institute, Brooklyn; and Sarah Lawrence College, Bronxville, New York, where she held the Nobel Chair in Art and Cultural History in 1991. She is the former curator of the Royal Photographic Society of Great Britain and has organized numerous exhibitions on both sides of the Atlantic, including From Today Painting Is Dead at the Victoria and Albert Museum, Fox Talbot and the Invention of Photography at the Pierpont Morgan Library, Visions of Liberty at the New-York Historical Society and Shots in the Dark: True Crime Pictures at the Chelsea Art Museum. In 2002 she worked with Al and Tipper Gore on their *The Spirit of Family*. She is the author of nine books on photography and history, including a twelve-year collaboration with Harold Evans and Kevin Baker on *The American Century*.

She lives in Westfield, New Jersey, and Warwick, New York, and is the mother of Alaina and Kevin.

BOOKS BY GAIL BUCKLAND: *The Spirit of Family* (by Al and Tipper Gore); *Shots in the Dark: True Crime Pictures; The American Century* (by Harold Evans); *The White House in Miniature; The Golden Summer: The Edwardian Photographs of Horace W. Nicholls; Travelers in Ancient Lands* (with Louis Vaczek); *Cecil Beaton War Photographs; First Photographs; Fox Talbot and the Invention of Photography; The Magic Image* (with Cecil Beaton); *Reality Recorded*

DAVID LEFER studied astrophysics and English at Harvard, literature at Oxford and the University of Paris and journalism at Columbia, where he got his master's degree. He was a New York City auxiliary police officer for three years and has worked at *Newsday,* the *Pittsburgh Post-Gazette,* the PBS talk show *The Digital Age,* the *China News* in Taiwan and the *New York Daily News,* where as an investigative reporter he spent three days on the streets to write about the homeless.